Barry C. Arnold • Ashis SenGupta • Gratiela Boca
Editors

# Likelihood Ratio Tests and Related Topics in Multivariate Analysis

## Selected Contributions of Carlos A. Coelho

Springer

*Editors*
Barry C. Arnold
Department of Statistics
University of California, Riverside
Riverside, CA, USA

Ashis SenGupta
Applied Statistics Unit
Indian Statistical Institute
Kolkata, West Bengal, India

Gratiela Boca
Department of Economic and Physical Sciences
Technical University of Cluj Napoca
Baia Mare, Romania

ISSN 1431-1968 ISSN 2628-8966 (electronic)
Contributions to Statistics
ISBN 978-3-031-68007-6 ISBN 978-3-031-68008-3 (eBook)
https://doi.org/10.1007/978-3-031-68008-3



This Springer imprint is published by the registered company Springer Nature Switzerland AG
The registered company address is: Gewerbestrasse 11, 6330 Cham, Switzerland

# Contributions to Statistics

The series **Contributions to Statistics** features edited and conference volumes in theoretical and applied statistics. Composed of refereed selected contributions, the volumes present the latest developments in all the exciting areas of contemporary statistical research.

# Preface

Carlos Coelho is currently a Full Professor at the Mathematics Department of NOVA College of Science and Technology of NOVA University of Lisbon. This volume shows a significant part of his research activity, namely his contributions to the area of likelihood ratio tests (LRTs) in Multivariate Analysis. The chapters in the book follow their chronological order of publication, which also shows a bit of the research path followed by Carlos. Most of Carlos' research activity focused on developing LRTs for elaborate covariance structures and for tests of the equality of mean vectors, when the covariance matrices have elaborate structures, obtaining then the distributions of the associated statistics in a closed finite form or obtaining sharp approximations for their distributions when it is not possible to obtain the exact distribution in closed form. His work also focused in obtaining manageable forms for the distributions of statistics associated with existing LRTs, used in Multivariate Analysis. This volume will be useful for advanced students as well as researchers and practitioners who deal with likelihood ratio tests in Multivariate Analysis.

Riverside, CA, USA
Kolkata, India
Baia Mare, Romania
June 2024

Barry C. Arnold
Ashis SenGupta
Gratiela Boca

# Contents

# Biography

**Barry Arnold, Ashis SenGupta, and Gratiela Boca**

Carlos Manuel Agra Coelho (Carlos A. Coelho as an author, or just Carlos Coelho or Carlos for friends) is currently a Full Professor at the Mathematics Department of NOVA College of Science and Technology of NOVA University of Lisbon. He was born on November 22, 1958, in what was then a small town on the shore of a quite large bay on the south banks of the Tagus river called Barreiro, where a flourishing chemical industry was thriving. His father, Joaquim Borrego

B. Arnold (✉)
Department of Statistics, University of California, Riverside, Riverside, CA, USA
e-mail: barry.arnold@ucr.edu

A. SenGupta
Applied Statistics Unit, Indian Statistical Institute, Kolkata, West Bengal, India
e-mail: amsseng@gmail.com

G. Boca
Department of Economic and Physical Sciences, Technical University of Cluj Napoca, Baia Mare, Romania
e-mail: Gratiela.BOCA@econ.utcluj.ro



B. C. Arnold et al. (eds.), *Likelihood Ratio Tests and Related Topics in Multivariate Analysis*, Contributions to Statistics,
https://doi.org/10.1007/978-3-031-68008-3_1

Coelho, and his mother, Maria Luiza Brás Gonçalves Agra Coelho, were chemical analysts at CUF (Companhia União Fabril), the largest chemical company installed in that small town. Since he was quite young Carlos Coelho liked both animals and nature, as well as all that might be related with mathematics. This ended up showing in his high school grades, making it difficult for him to choose between the areas of biology and mathematics when entering the University. He first enrolled in the Mathematics undergraduate program at the then Classical University of Lisbon, but eventually ended up attending the Agronomy/Forestry program at the Agronomy College of the then other University of Lisbon, the Lisbon University of Technology (since 2012 these two Universities coalesced to form the University of Lisbon). The two first years were common between the undergraduate degrees of Agronomy and Forestry, both with a curricula of 5 years plus a professional/technical internship. Carlos Coelho ended up choosing the Forestry program, mainly because it had a Dendrometry course where Multiple Regression was taught. In the meantime, during the two first years a quite strong formation in Analysis and Algebra was granted through the two annual courses in Mathematics and Calculus, for the dismay of most of his colleagues and the great pleasure of Carlos. He ended up following an internship at the Agronomy College itself, which lead him into the use of Multiple Regression and Analysis of Variance techniques and half-accidentally into starting teaching Analysis and Calculus courses at the Mathematics Department of this same College in the year 1983, given the very good grades he had earned in his Mathematics undergraduate courses. He had never done much in Forestry and at that time he decided to enter the enticing world of Mathematics and Statistics, which was actually, since the beginning, his first love in terms of science. After trying to enter a Master's level course in Statistics, which he was barred to do because of being an Engineer, he decided to take by himself an examination on Teaching and Scientific Capability, which would allow him to pursue the Teaching Career without a Master's level course. He had to write a Research Work/Thesis, which he did on "Factorial Methods of Data Analysis", where, using a mostly algebraic-geometric approach, the Generalized Canonical Analysis was the base method used as an umbrella method for a host of other multivariate linear methods. By then he had his first contact with the book "Multivariate Analysis" from Professor Anant M. Kshirsagar, which definitely influenced his intention of pursuing the inferential aspects of these methods and models and to try to get to the U.S.A. for a Ph.D. in Statistics. He was lucky enough to be able to be a Fulbrighter at the University of Michigan in Ann Arbor, Michigan, U.S.A., having also received grants from FLAD/LADF (Fundação Luso-Americana para o Desenvolvimento/Luso-American Development Foundation) and the INVOTAN Standing Committee. He started at the Statistics Department and then moved to the Biostatistics Department, where he ended up being a Ph.D. student of Professor Anant M. Kshirsagar. His experience at the University of Michigan was a most enriching one in all ways and he will never forget the Professors he had, the colleagues he had, and the friends he made, many of them just people not connected to the University but living in Ann Arbor. A summary of his experience and his views of this period may be read in "My experience as a Fulbrighter – an appraisal of the Fulbright

Program" in Programa Fulbright – Volume Comemorativo, edited by Professor Laura Pires (pages 41-50), which was also published online as part of Volume 18 of the Gaudium Sciendi online Journal (http://www2.ucp.pt/resources/Documentos/SCUCP/GaudiumSciendi/GaudiumSciendi_N18/08CarlosCoelho.pdf).

After coming back from the U.S.A., Carlos Coelho pursued his career at the Agronomy College of the then Lisbon University of Technology, as an Assistant Professor, at first mainly dealing with applications of Statistics in the quite large variety of research areas at this College. He was Vice-President of the Mathematics Department and Head of the Research Unit in Applied Mathematics of the Agronomy College, during a time this Unit had a classification of 'Very Good' at the national level. By that time, being quite tired of working a lot in several types of applications and getting from the researchers in the application areas little recognition in terms of publications, he decided to put to a better use the knowledge and background acquired at the University of Michigan, which enabled him to develop a career in Multivariate Analysis and Multivariate Statistics, embracing more formal approaches. These more formal approaches lead him to develop the exact distribution of the Wilks Lambda and the generalized Wilks Lambda statistic in the form of a Generalized Integer Gamma distribution (Coelho, C.A. (1998): The Generalized Integer Gamma distribution – a basis for distributions in Multivariate Statistics. Journal of Multivariate Analysis, 64, 86-102 (https://doi.org/10.1006/jmva.1997.1710)), when at most one of the sets has an odd number of variables, and also to use it as the exact distribution of a quite large number of other likelihood ratio statistics used in Multivariate Analysis (Coelho, C.A., Arnold, B.C. (2019): *Finite Form Representations for Meijer G and Fox H Functions – Applied to Multivariate Likelihood Ratio Tests using Mathematica®, Maxima and R*, Lecture Notes in Statistics, Springer, Cham, Switzerland, 515+xviii pp. (https://doi.org/10.1007/978-3-030-28790-0)). He also developed what he called 'near-exact' distributions, which are asymptotic distributions based on a factorization of the characteristic function of the logarithm of the likelihood ratio statistics in such a way that 'most' of the distribution is left untouched and only a 'smaller' part is then approximated, so that the 'near-exact' distribution obtained is manageable. This concept was developed in the paper "Coelho, C.A. (2004). The Generalized Near-Integer Gamma distribution – a basis for "near-exact" approximations to the distributions of statistics which are the product of an odd number of particular independent Beta random variables. J. Multivariate Anal., 89, 2, 191-218. (https://doi.org/10.1016/j.jmva.2003.12.001)", and it was applied to a wide variety of likelihood ratio statistics used in Multivariate Analysis.

Two of the most frequent co-authors of Carlos have been Professor Barry Arnold, who besides being a co-author is also a good friend, and Carlos' former Ph.D. student Filipe Marques, who is at present an Associate Professor of Statistics at the Mathematics Department of NOVA School of Science and Technology of NOVA University of Lisbon. And when referring to Barry Arnold, it is also not possible to forget another good friend whose first name is Barry, that is Barry Nussbaum, who was ASA's 112th President in 2017 and with whom Carlos crossed paths at a couple of conferences, always sharing some good moments (see Figs. 1 and 5). Also Luís

**Fig. 1** Carlos Coelho with Barry Arnold at the 2000 National Congress of the Portuguese Society for Statistics (left) and in 2016 at the 10-th Probability and Statistics Day at the Mathematics and Statistics Department at UMBC – University of Maryland Baltimore County, together with Barry Nussbaum (right), and with Barry Nussbaum in 2015 at the 9-th Probability and Statistics Day at the Mathematics and Statistics Department at UMBC (bottom).

Grilo and Rui Alberto, two of Carlos former Ph.D. students have been co-authors of Carlos in a number of papers (see Fig. 5).

One other long lasting friendship is with Ashis SenGupta who Carlos Coelho had the chance to visit in Calcutta in the year 2000, when Ashis SenGupta was involved in the organization of a Conference in honor of the 80th birthday of Professor C. R. Rao. Ashis and Carlos then met in a number of other Conferences, some of them organized in Turkey (see Fig. 2).

It was at the Conference organized by Ashis SenGupta in Calcutta that Carlos met Professor C. R. Rao and that was where a very interesting friendship started. Professor C. R. Rao attended the session where Carlos was presenting and where he introduced the idea of a 'near-exact' distribution, and Professor C. R. Rao ended up being one the persons responsible for Carlos pursuing his work on these distributions with the incentive he gave him.

In September 2006, when Carlos had already moved to the Mathematics Department of NOVA School of Science and Technology of NOVA University of Lisbon, he was the Speaker at the ceremony of nomination of Professor C. R. Rao as an Honoris Causa Doctor of NOVA University of Lisbon. This was one of the high moments of what was already by then a quite long lasting relation of esteem and appreciation for one of the highest exponents in the area of Statistics and his wife Mrs. Bhargavi Rao (see Fig. 3).

**Fig. 2** upper left (from left to right): Birdal Şenoğlu, Carlos Coelho, Ashis SenGupta, Ivette Gomes, Ozlem Turksen, Dimitrios Konstantinides (December 2017 at Ankara University, for the Conference celebrating the 25th anniversary of the Turkish Statistical Association); upper right (from left to right): Talha Arslan, Ashis SenGupta, Aysun Ikiz, Gratiela Boca, Carlos Coelho (at a Conference in Van, Türkiye, in 2019); bottom (from left to right): Kadir Emir, Talha Arslan, Birdal Şenoğlu, Carlos Coelho (July 2017, by the Tagus river – Lisbon)

**Fig. 3** Carlos Coelho as the Speaker at the Honoris Causa ceremony for Professor C. R. Rao (left: Carlos Coelho, C.R. Rao, João T. Mexia, from left to right) and Carlos Coelho with Professor C. R. Rao and his wife, Mrs. Bhargavi Rao, at the SCRA 2006 Conference (right)

**Fig. 4** left: Carlos Coelho with Bikas Sinha (left) and Bimal Sinha (right) in 2013, at the 7-th Probability and Statistics Day at UMBC, right: Carlos Coelho after receiving from the hands of Bimal Sinha a plate commemorating his participation as an invited speaker at the 9th Probability and Statistics Day at UMBC in 2015 (Abdulkadir Hussein, Carlos Coelho, Bimal Sinha, from left to right)

Mentioning Turkey, there is no way we can forget Kadir Emir, Talha Arslan, and Birdal Senoglu, Ph.D. advisor of Talha Arslan. Birdal Senoglu and Talha Arslan were two of the main organizers of the 2017 Conference which took place at the University of Ankara to commemorate the 25th Anniversary of the Turkish Statistical Association, where Carlos was one of the invited speakers. Talha Arslan later also organized a significant Conference in Van, a very interesting city in eastern Turkey (see Fig. 2).

In fact, Carlos' scientific activity lead him to participate in a quite large number of Conferences, and it was during many of the Conferences he attended, quite often as an invited speaker, that he made many of his close friends, among which there were some of the most well-known names in the area of Statistics.

Some of these conferences were the Probability and Statistics Day Conferences organized at the Mathematics and Statistics Department of the University of Maryland Baltimore County by Bimal Sinha and his colleagues, one other long standing friendship that could not be overlooked, actually with both Bimal and Bikas Sinha, the twin Sinha brothers (see Fig. 4).

Bimal Sinha was also the main Ph.D. advisor of Ricardo Moura, one of Carlos Coelho's Ph.D. students, who benefitted from a Fulbright grant for one year to pursue his Ph.D. studies at the Mathematics and Statistics Department at UMBC and who actively participated and also gave a helping hand in organizing the 9-th Probability and Statistics Day (PSDay) in 2015 at UMBC. In Fig. 5 there is a picture with Ricardo, Peter McCullagh and Guido Knapp at the 9-th PSDay. It was with Bimal Sinha and Ricardo that Carlos Coelho entered the enticing area of Data Protection and Statistical Disclosure Control, where they published a series of papers.

Carlos also attended a few of the IWMS (International Workshop on Matrices and Statistics) Conferences, during which he had the chance to meet Theodore W. Anderson and a few more friends like George Styan and Michael Perlman (see Fig. 6).

**Fig. 5** left: Luís Grilo, Carlos Coelho and Filipe Marques at the Castelo do Bode dam; right: from left to right, Peter McCullagh, Carlos Coelho, Ricardo Moura and Guido Knapp at the 9-th Probability and Statistics Day at UMBC in 2015

**Fig. 6** Carlos Coelho, George Styan, Theodore W. Anderson and Michael Perlman (from left to right) in 2008 at lunch time on a boat on the Castelo de Bode dam, at the time of the 2008 IWMS Conference held in Tomar, at the Polytechnic Institute, to commemorate the 90th birthday of T. W. Anderson

Not less important for Carlos is the friendship of a number of other colleagues, who also played a very important role in his life, as it is the case of Viktor Witkovský, from the Institute of Measurement Science of the Slovak Academy of Sciences in Bratislava, Ben Kedem, from the University of Maryland, U.S.A., and Gratiela Boca from the University of Cluj in Baia Mare, Romania (see Fig. 7).

Another group of friends, some of them related with the organization of the IWMS series of conferences, is the group that has been organizing several Conferences on Linear and Mixed models and Linear Algebra, most of them in Bedlewo, Poland, for a number of years. Among others, we may note the friendship with Roman Zmyslony and Augustyn Markiewicz (see Fig. 8).

As already mentioned, Carlos Coelho's main area of work has been in Multivariate Analysis in developing manageable closed form expressions for the probability density and cumulative distribution functions of many likelihood ratio statistics used in Multivariate Analysis, and also the development of the so-called near-exact distributions. His main interests have been in likelihood ratio tests for elaborate covariance structures and also for tests of equality of mean vectors, when the

**Fig. 7** Carlos Coelho with Viktor Witkovský at the Devín castle in Slovakia (upper left), with Ben Kedem at the University of Maryland campus, USA (lower left) and with Gratiela Boca in Sintra, Portugal (right)

**Fig. 8** Carlos Coelho with Augustyn Markiewicz at the 2007 IWMS (left) and with Roman Zmyslony in 2010, in the gardens of Rogalin palace, Poland (right)

**Fig. 9** Carlos Coelho with Bells, a wonderful former wild female dog adopted by a friend in Algarve (upper left), with a cheetah, after a Conference in South Africa (upper right), with Van cats, after a Conference in Van (lower left) and with a falcon, after a Conference in Poland (lower right)

covariance matrices have specific structures. Lately he has been interested in some other related areas as for example the use of random sample sizes in likelihood ratio tests in Multivariate Analysis, the use of Multivariate Analysis techniques in Data Protection and Statistical Disclosure Control problems and even more recently in the development of likelihood ratio tests that may be used in high-dimensional settings in Multivariate Analysis, namely for the fit of the MANOVA and the multivariate linear models, including the test for nested multivariate linear models.

Some years ago he also made an incursion into computer science and natural language studies, an area in which he published a few papers with a number of colleagues from the Computer Science Department from NOVA School of Science and Technology of NOVA University.

Carlos' already mentioned passion for animals still endures (see Fig. 9). Bells was a wonderful former wild female dog that was adopted by a friend from Algarve. She was a special companion of many extraordinary walks in the hills, valleys and small rivers of Algarve, a much less known, but not less fantastic facet of the southernmost province of Portugal.

# Introduction

**Barry Arnold, Ashis SenGupta, and Gratiela Boca**

The book shows a significant part of Carlos' research activity, namely his contributions to the area of LRTs (likelihood ratio tests) in Multivariate Analysis. The chapters in the book follow their chronological order of publication, which also shows a bit of the research path followed by Carlos. Most of Carlos' research activity focused on obtaining manageable forms for the distributions of statistics associated with existing LRTs, used in Multivariate Analysis, and also in developing LRTs for elaborate covariance structures and for tests of equality of mean vectors, when the covariance matrices have elaborate structures, obtaining then the distributions of the associated statistics in a closed finite form or obtaining sharp approximations for their distributions when it is not possible to obtain the exact distribution in closed form.

"The Generalized Integer Gamma Distribution – A Basis for Distributions in Multivariate Statistics", is the Chapter that reproduces the paper in which Carlos established the GIG (Generalized Integer Gamma) distribution as a basis for representing the exact distributions of several LRT statistics used in Multivariate Analysis (Coelho, 1998). The GIG distribution is defined as the distribution of the sum of a set of independent r.v.'s (random variables) with Gamma distributions, all

B. Arnold (✉)
Department of Statistics, University of California, Riverside, Riverside, CA, USA
e-mail: barry.arnold@ucr.edu

A. SenGupta
Applied Statistics Unit, Indian Statistical Institute, Kharagpur, West Bengal, India
e-mail: amsseng@gmail.com

G. Boca
Department of Economic and Physical Sciences, Technical University of Cluj Napoca, Baia Mare, Romania
e-mail: Gratiela.BOCA@econ.utcluj.ro

 
B. C. Arnold et al. (eds.), *Likelihood Ratio Tests and Related Topics in Multivariate Analysis*, Contributions to Statistics,
https://doi.org/10.1007/978-3-031-68008-3_2

with integer shape parameters, and different rate parameters, the important point being that the expressions for both its p.d.f. (probability density function) and its c.d.f. (cumulative distribution function) have closed finite forms, which do not involve any unsolved integrals nor any series, thus enabling precise computation of quantiles and p-values. Using the notation introduced by Carlos A. Coelho, we say that a r.v. $X$ has a Gamma distribution with a shape parameters $r$ and a rate parameter $\lambda$ if its p.d.f. can be written as

$$f_X(x) = \frac{\lambda^r}{\Gamma(r)}\, e^{-\lambda x} x^{r-1}\,, \qquad (r, \lambda > 0;\ x > 0)\,,$$

and we will denote this fact by $X \sim \Gamma(r, \lambda)$. Then, if we take

$$Z = \sum_{j=1}^{p} X_j\,,$$

with $X_j \sim \Gamma(r_j, \lambda_j)$, where $r_j \in \mathbb{N}$ and all $\lambda_j$ are different, $Z$ will have a GIG distribution with p.d.f. and c.d.f. respectively given by

$$f_Z(z) = K \sum_{j=1}^{p} P_j(z)\, e^{-\lambda_j z}\,, \quad (z > 0)$$

and

$$F_Z(z) = 1 - K \sum_{j=1}^{p} P_j^*(z)\, e^{-\lambda_j z}\,, \quad (z > 0)$$

where

$$K = \prod_{j=1}^{p} \lambda_j^{r_j}\,, \quad P_j(z) = \sum_{k=1}^{r_j} c_{j,k}\, z^{k-1}$$

and

$$P_j^*(z) = \sum_{k=1}^{r_j} c_{j,k} (k-1)! \sum_{i=0}^{k-1} \frac{z^i}{i!\,\lambda_j^{k-i}}\,,$$

with

$$c_{j,r_j} = \frac{1}{(r_j - 1)!} \prod_{i=1 \, i \neq j}^{p} (\lambda_i - \lambda_j)^{-r_i}\,, \quad j = 1, \ldots, p\,,$$

and, for $k = 1, \ldots, r_j - 1$ and $j = 1, \ldots, p$,

$$c_{j,r_j-k} = \frac{1}{k} \sum_{i=1}^{k} \frac{r_j - k + i - 1)!}{(r_j - k - 1)!} R(i, j, p) c_{j,r_j-(k-i)} ,$$

where

$$R(i, j, p) = \sum_{\substack{k=1 \\ k \neq j}}^{p} r_k (\lambda_j - \lambda_k)^{-i} \quad (i = 1, \ldots, r_j - 1) .$$

Of course, given the unavoidable slight complexity of the distribution, reflected in the expressions for some functions of its parameters that appear in the expressions of the p.d.f. and the c.d.f., the computation of these two functions usually calls for the use of adequate software or computational languages, which, mainly because of the $c_{j,k}$ functions of the parameters need to be able to use extended precision, since sometimes the double precision, commonly used by software packages, is not enough.

Anyway most software packages and computer languages used nowadays have this capability, as it is for example the case of Mathematica® and MAXIMA, or it may be easily implemented through the use of an adequate package as it is the case of R. Indeed modules for the computation of the p.d.f. and the c.d.f., and as such of p-values and quantiles, of the GIG distribution are available in these three software packages or computer languages (see Coelho and Arnold (2019) and the repository https://github.com/meijerfoxfiniteforms).

But it still remains to fully explain why the GIG distribution is thought to be such a useful tool. It is a known fact, for quite some time now, that most of the LRT statistics used in Multivariate analysis have, under the corresponding null hypotheses, a distribution that is the same as that of a product of independent Beta distributed r.v.'s (Tukey and Wilks, 1946; Wilks, 1960; Wilks, 1963; Khatri, 1965a,b; Schatzoff, 1966; Olkin and Press, 1969; Gupta, 1971; Pillai and Jouris, 1971; Mathai and Rathie, 1971; Kshirsagar, 1972, Sec. 8.2; Rao, 1973, Chap. 8; Mathai, 1973; Mathai and Saxena, 1973, Sec. 5.2; Bhargava, 1975; Bhargava and Khatri, 1981; Tang and Gupta, 1984; Anderson, 2003, Sects. 8.4, 8.5, 9.3; Muirhead, 2005, Sects. 10.5, 11.2; Liang, 2015). But this distribution was seen as unavoidably leading to p.d.f.'s and c.d.f.'s that have expressions that involved infinite series or special functions (Nagarsenker and Das, 1975; Springer and Thompson, 1966a; Carter and Springer, 1977; Springer, 1979; Nandi, 1980; Bhargava and Khatri, 1981; Nagarsenker and Suniaga, 1983; Tang and Gupta, 1984; Pederzoli, 1985; see the Chapter entitled "On the Distribution of the Product of Independent Beta Random Variables – Applications" for a more detailed discussion on this subject). There were however a few authors that spotted that in some situations it was possible to obtain finite form representations for the p.d.f. and c.d.f. of these distributions, but they were not able to obtain general representations that would cover all cases neither general representations for all the functions of the parameters involved (Wald and Brookner, 1941; Schatzoff, 1966; Tang and Gupta, 1986).

The development of the GIG distribution solves this problem in two ways. It gives a rather simple basis for the representation of the p.d.f.'s and c.d.f.'s of such LRT statistics, for the cases where these representations may have finite forms, and at the same time it helps in identifying the situations in which this happens. In the original paper where the GIG distribution was established (Coelho, 1998), it was applied to, or used as, the exact distribution of the logarithm of the generalized Wilks $\Lambda$ statistic (Wilks, 1935), that is the LRT statistic used to test the independence of several groups of normally distributed random variables, in the situation where at most one of this groups has an odd number of variables. By setting up this application, the author clearly was willing to show that similar applications might be possible for a host of other LRT statistics that exhibit somewhat similar structures. Indeed there are a quite large number of LRT statistics used in Multivariate Analysis whose logarithms have p.d.f.'s and c.d.f.'s that may be expressed in finite forms by using the GIG distribution, many of which are depicted in the recent book by Coelho and Arnold (2019). See also Coelho (2022) on the existence of several LRT statistics used in Multivariate Analysis that have a similar structure to that of a Wilks $\Lambda$ statistic.

Later on, Arnold et al. (2013), see the Chapter entitled "The Distribution of the Product of Powers of Independent Uniform Random Variables – A Simple but Useful Tool to Address and Better Understand the Structure of Some Distributions" in this book, developed a distribution which they called the EGIG (Exponentiated Generalized Integer Gamma) distribution, that is, as its name indicates, the distribution of a random variable whose logarithm, or rather, negative logarithm, has a GIG distribution, and whose distribution was linked to that of a product of powers of Uniform(0,1) random variables. This distribution is used as the exact distribution of several LRT statistics used in Multivariate Analysis (see several of the Chapters in this book and also Coelho and Arnold (2019)).

But it may be that, at least for the reader who is less familiar with the endeavors related with the study of the distributions of the LRT statistics used in Multivariate Analysis, the link of the GIG distribution and the distribution of the logarithm, or more precisely, the negative logarithm of LRT statistics used in Multivariate Analysis may not be too clear. We stated that the distribution of many of such LRT statistics is the same as that of the product of a number of independent Beta distributed r.v.'s, while the GIG distribution is the distribution of a sum of independent Gamma distributed r.v.'s, all with integer shape parameters. But these are quite different distributions, so what is the link?

Let $X$ be a r.v. with a $Beta(a, b)$ distribution. Then the $h$-th moment of $X$ is given by

$$E(X^h) = \frac{\Gamma(a+h)}{\Gamma(a)} \frac{\Gamma(a+b)}{\Gamma(a+b+h)},$$

expression that is valid for $h > -a$, showing that it is defined for $h$ in a neighborhood of zero. This is a most important fact, since this means that we have the m.g.f. (moment generating function) and c.f. (characteristic function) of $Y = -\log X$

readily available, respectively given by

$$M_Y(t) = E\left(e^{tY}\right) = E\left(e^{-t\log X}\right) = E(X^{-t}) = \frac{\Gamma(a-t)\,\Gamma(a+b)}{\Gamma(a)\,\Gamma(a+b-t)}\,, \quad t < a\,, \tag{1}$$

and, for $\mathrm{i} = \sqrt{-1}$,

$$\Phi_Y(t) = E\left(e^{\mathrm{i}tY}\right) = E\left(e^{-\mathrm{i}t\log X}\right) = E\left(X^{-\mathrm{i}t}\right) = \frac{\Gamma(a-\mathrm{i}t)\,\Gamma(a+b)}{\Gamma(a)\,\Gamma(a+b-\mathrm{i}t)}\,, \quad t \in \mathbb{R}\,. \tag{2}$$

Although for all the r.v.'s Carlos uses in his papers and book chapters the m.g.f.'s do exist, he always preferred to use the c.f.'s, due to the fact that simpler and more direct inversion theorems and formulas exist for the inversion of c.f.'s than for m.g.f.'s. If we look carefully at expressions (1) and (2) we can see that at least in cases where $b$ is an integer we can easily express, for example, the c.f. in (2) as a product of c.f.'s by using the fact that for the Gamma function we have the well-known result that

$$\Gamma(r+1) = r\,\Gamma(r)\,,$$

which being applied successively $n$ times yields for positive integer $n$ and any real or complex $a$, such that $a+n$ is not a negative integer,

$$\frac{\Gamma(a+n)}{\Gamma(a)} = \prod_{\ell=0}^{n-1}(a+\ell)\,. \tag{3}$$

Using this result we may then write the c.f. in (2) as

$$\Phi_Y(t) = \frac{\Gamma(a+b)}{\Gamma(a)}\,\frac{\Gamma(a-\mathrm{i}t)}{\Gamma(a+b-\mathrm{i}t)} = \prod_{\ell=0}^{b-1}(a+\ell)\,(a+\ell-\mathrm{i}t)^{-1}\,,$$

where the factors in the product are now c.f.'s of Exponential r.v.'s with parameters $a+\ell$ $(\ell = 0,\ldots,b-1)$, showing that in this case, where $b$ in an integer, the distribution of $Y$ is also that of a sum of $b$ independent Exponential r.v.'s with parameters $a+\ell$ $(\ell = 0,\ldots,b-1)$. If instead of having just a single Bet r.v. $X$ we have a product of, say $p$, of these r.v.'s, that is, say of

$$\Lambda = \prod_{j=1}^{p} X_j \tag{4}$$

where $X_j \sim Beta(a_j, b_j)$, with $b_j \in \mathbb{N}$, then if we take $W = -\log\Lambda$, the c.f. of $W$ becomes that of a double sum of independent Exponential r.v.'s, where, given the structure of the parameters $a_j$, commonly related with the sample size, subtracted from some index which runs from a lower bound to a given upper bound related with the number of variables in the study, some of the Exponential r.v.'s will exhibit the same parameters. Then, given the well-known fact that the sum of independent Exponential r.v.'s with the same parameters is a Gamma distribution with that same rate parameter and a shape parameter equal to the number of Exponential r.v.'s being added, the c.f. of $W$ becomes then that of a sum of independent Gamma r.v.'s with integer shape parameters, and the exact distribution of $W$ is thus a GIG distribution and that of $\Lambda$, by simple transformation, an EGIG distribution.

The case of integer second parameters $b_j$ in the Beta r.v.'s $X_j$ is not the only one yielding a GIG distribution as the exact distribution of $W = -\log \Lambda$ for $\Lambda$ in (4). There are many cases where $b_j$ being rational, the exact distribution of $W$ still is a GIG distribution. Many of these situations are depicted in Chapter 3 of Coelho and Arnold (2019), and an extension of Theorem 3.2 in this reference is brought to the reader in Coelho (2021), reproduced as the Chapter entitled "Testing Equality of Mean Vectors with Block-Circular and Block Compound-Symmetric Covariance Matrices" in this book.

The reader may refer to the Chapters "Near-Exact Distributions for Certain Likelihood Ratio Test Statistics", "A General Near-Exact Distribution Theory for the Most Common Likelihood Ratio Test Statistics Used in Multivariate Analysis" and "The Distribution of the Product of Powers of Independent Uniform Random Variables – A Simple but Useful Tool to Address and Better Understand the Structure of Some Distributions" through "The Likelihood Ratio Test of Equality of Mean Vectors with a Doubly Exchangeable Covariance Matrix" in the present book, and also to Coelho (2022), and to Chapter 5 of Coelho and Arnold (2019) for a more thorough study and elaboration on a large number of cases where the technique just presented is used to identify the distribution of several LRT statistics used in Multivariate Analysis.

Furthermore, as shown in Coelho and Arnold (2019, Chaps. 2–4), the p.d.f. and c.d.f. of the GIG distribution may be used to obtain closed form expressions for instances of the Meijer G (Meijer, 1936a,b, 1941, 1946; Erdélyi et al., 1953, Sect. 5.3; Luke, 1969, Sect. 5.2; Luke, 1975, Sect. 5.3; Askey and Daalhaus, 2010; Mathai and Saxena, 1973, Sect. 1.1; Mathai, 1993, Sect. 2.1, 3.11; Mathai and Haubold, 2008, Sect. 1.8, 1.9; Prudnikov et al., 1990, Sect. 8.2, 8.3) and Fox H functions (Fox, 1961; Braaksma, 1963; Springer, 1979, Sect. 6.2; Mathai et al., 2009, Sect. 1.2).

The GIG distribution is referred to in a number of books in Multivariate Analysis and Probability (Bilodeau and Brenner, 1999; Timm, 2002; Paolella, 2007), and it also finds applications in a host of other areas, ranging from wireless networks and telecommunications (Karagiannidis et al., 2006; Suraweera et al., 2008; Surobhi, 2009; Kaiser and Zheng, 2010, Sect. 6.6; Parruca and Gross, 2014; Wang and Liang, 2013; Liu et al., 2015; Ibrahim et al., 2016; Liu et al., 2017; Müller, 2018; Bedoui et al., 2019; Psomas and Krikidis, 2016, 2018; Ghofrani-Jahromi et al., 2020; Ruby et al., 2022), to wireless power transfer (Tegos et al., 2022), to transportation (Müller et al., 2015; Müller and Safiulin, 2016), reliability (Mohri and Takeshita, 2022) and finance and risk theory (Constantinescu et al., 2019) and even to bacterial cell division (Ghusinga et al., 2016, supplementary material).

The Chapter "Addendum to the Paper "The Generalized Integer Gamma Distribution – A Basis for Distributions in Multivariate Statistics"" is a reprint of a paper that is an addenda to the paper in the previous Chapter. In this addenda it is shown that the distribution of

$$\prod_{j=1}^{p} Y_j \quad \text{where} \quad Y_j \sim Beta\left(\ell - \frac{b}{2} - \frac{j}{2}, \frac{b}{2}\right) \tag{5}$$

form a set of $p$ independent r.v.'s, and that of

$$\prod_{j=1}^{b} Y_j^* \quad \text{where} \quad Y_j^* \sim Beta\left(\ell - \frac{p}{2} - \frac{j}{2}, \frac{p}{2}\right) \tag{6}$$

form a set of $b$ independent r.v.'s, are the same, that is, in (5) and (6) $p$ and $b$ are interchangeable, with no requirements on their magnitude, and the distribution of such products has a closed finite form representation as long as at most one of them is even, while in Coelho (1998) it was said that they had to be both even. This result, which may seem to be of lesser importance, is indeed of key importance in handling the distribution of many LRT statistics, given the fact that quite often their distributions may arise, at least at an initial stage, in just one of the forms in either (5) or (6).

The Chapter "The Generalized Near-Integer Gamma Distribution: A Basis for 'Near-Exact' Approximations to the Distribution of Statistics Which Are the Product of an Odd Number of Independent Beta Random Variables" is a reprint of a paper where Carlos developed what he called "near-exact" distributions (Coelho, 2004) and where he developed a distribution he called a GNIG (Generalized Near-Integer Gamma) distribution, which may be seen as a companion distribution to the GIG distribution. It is the distribution of the sum of a GIG distributed r.v. with an independent Gamma r.v., this one with a non-integer shape parameter. This distribution was developed with the aim of being used, either by itself or in mixtures of several of these distributions, in the so-called "near-exact" distributions. These are asymptotic distributions developed under a different concept, which involves keeping "most" of the original c.f. of the logarithm of the LRT statistic intact and approximating only a "smaller" part of that c.f. with an asymptotic result with very good convergence properties (towards the exact part it replaces), in such a way that the resulting c.f. yields a known and manageable distribution from which it will be easy to compute quantiles and p-values. The concept is most easily applied to situations where the c.f. of $W = -\log \Lambda$ (where $\Lambda$ denotes the LRT statistic) has a product structure or a series structure, which is the case of almost all the LRT statistics used in Multivariate Analysis.

Let us suppose that the c.f. of $W = -\log \Lambda$ may, after some witted handling, be written as

$$\Phi_W(t) = \Phi_1(t)\,\Phi_2(t)$$

where we have to assure that both $\Phi_1(t)$ and $\Phi_2(t)$ are legitimate c.f.'s, with

$$\frac{1}{2\pi}\int_{-\infty}^{+\infty}\left|\frac{\Phi_W(t)-\Phi_1(t)}{t}\right|dt << \frac{1}{2\pi}\int_{-\infty}^{+\infty}\left|\frac{\Phi_W(t)-\Phi_2(t)}{t}\right|dt\,, \tag{7}$$

while at the same time $\Phi_1(t)$ corresponds to a known and manageable distribution (which will commonly be a GIG distribution), and $\Phi_2(t)$ corresponds to a non-manageable distribution (which will commonly be that of the distribution of a sum of logarithms of Beta r.v.'s, but which we are not able to express as a GIG distribution). Then our aim is to leave $\Phi_1(t)$ unchanged and approximate $\Phi_2(t)$ by an asymptotic result with good convergence properties (towards $\Phi_2(t)$). The relation in (7) means that $\Phi_1(t)$ accounts for the "major" part of the distribution of $W$, while $\Phi_2(t)$ accounts for a "smaller" part. In (7), the expressions on each side of the inequality are really very tight upper bounds on the difference between the corresponding c.d.f.'s, that is, for example

$$\frac{1}{2\pi}\int_{-\infty}^{+\infty}\left|\frac{\Phi_W(t)-\Phi_1(t)}{t}\right|dt \geq \max_{w>0}|F_W(w)-F_1(w)|$$

where $F_W(w)$ is the c.d.f. of the exact distribution of $W$, corresponding to $\Phi_W(t)$, and $F_1(t)$ is the c.d.f. corresponding to $\Phi_1(t)$.

Let us call $\Phi_2^*(t)$ the c.f. used as an asymptotic replacement for $\Phi_2(t)$. It is possible to show (see for example Chapters entitled "The Generalized Near-Integer Gamma Distribution: A Basis for 'Near-Exact' Approximations to the Distribution of Statistics Which Are the Product of an Odd Number of Independent Beta Random Variables", "Near-Exact Distributions for Certain Likelihood Ratio Test Statistics" and "A General Near-Exact Distribution Theory for the Most Common Likelihood Ratio Test Statistics Used in Multivariate Analysis" in this book) that an adequate form for $\Phi_2^*(t)$ is either the c.f. of a Gamma distribution or the c.f. of a mixture of Gamma distributions, ideally with $\Phi_2^*(t)$ equating the first few derivatives of $\Phi_2(t)$ at $t = 0$, which will make the near-exact distribution to match that many moments of the exact distribution. Carlos calls

$$\Phi_1(t)\,\Phi_2^*(t)$$

the near-exact c.f. of $W$, since most of it is indeed equal to the exact c.f. of $W$, and the corresponding distribution, a near-exact distribution of $W$, noting that we may obtain different near-exact distributions for $W$ for different $\Phi_2^*(t)$ c.f.'s. Then, by a classical transformation of variables we easily obtain the corresponding near-exact p.d.f.'s and c.d.f.'s of $\Lambda$.

Most commonly, when dealing with LRT statistics in Multivariate Analysis, $\Phi_1(t)$ appears as the c.f. of a GIG distribution, and $\Phi_2^*(t)$ as the c.f. of a Gamma distribution, or of mixture of Gamma distributions, with non-integer shape parameters, thus yielding either a GNIG or a mixture of GNIG distributions as the near-exact distribution for these statistics.

Near-exact distributions have been successfully developed for a quite large number of LRT statistics with applications in Multivariate Analysis (Coelho et al., 2006, 2010, 2015, 2016, 2022; Alberto and Coelho, 2007; Grilo and Coelho, 2007, 2010a,b, 2012, 2013, 2017; Marques and Coelho, 2008, 2010, 2011, 2012a,b,c, 2013a,b,c, 2015, 2016, 2018, 2020; Coelho and Marques, 2009, 2010, 2011, 2012, 2013a,b; Marques et al., 2011, 2015; Coelho and Arnold, 2014; Coelho, 2017, 2018, 2022; Coelho and Roy, 2017, 2020; Correia et al., 2018; Coelho and Singull, 2020; Liang et al., 2022; Serim et al., 2022), besides those in Chapters entitled "The Advantage of Decomposing Elaborate Hypotheses on Covariance Matrices into Conditionally Independent Hypotheses in Building Near-Exact Distributions for the Test Statistics" through "A General Near-Exact Distribution Theory for the Most Common Likelihood Ratio Test Statistics Used in Multivariate Analysis" and "The Exact and Near-Exact Distributions of the Main Likelihood Ratio Test Statistics Used in the Complex Multivariate Normal Setting" through "The Likelihood Ratio Test of Equality of Mean Vectors with a Doubly Exchangeable Covariance Matrix" of the present book.

Near-exact distributions exhibit several advantages over the classical asymptotic distributions. They yield very sharp approximations even for very small sample sizes, with this approximation improving for increasing numbers of variables involved, as opposed to classical asymptotic distributions, which worsen their performances for larger numbers of variables involved. In general, near-exact distributions are asymptotic for all parameters involved in the distributions, that is, classically they are asymptotic for increasing sample sizes and also for increasing numbers of variables and populations involved in the tests.

The Chapter "The Wrapped Gamma Distribution and Wrapped Sums and Linear Combinations of Independent Gamma and Laplace Distributions" is the reprint of a paper which is actually Carlos' only paper on circular distributions. In this paper he develops the circular or wrapped Gamma distribution and a number of other related distributions, as the wrapped versions of the generalized Laplace and GIG distributions, as well as the wrapped distribution of the difference of two independent r.v.'s with independent GIG distributions. At the time these distributions were developed, they were a novelty, and the interest in this paper resides also in the fact that the circular distributions depicted in the paper are able to yield quite symmetric p.d.f.'s as well as quite asymmetrical ones, with the mixtures being able to exhibit multimodal p.d.f.'s. Although these distributions have found applications in several areas, as for example in health and medicine (Fukaya et al., 2017), we feel that clearly not all the potential of these distributions has already been used, and this is one further reason to include this paper in the book.

One other paper that we may say falls a little off the mainstream of the book, but that we thought should be included in it, is the paper reproduced in the Chapter entitled "On the Distribution of the Product and Ratio of Independent Generalized Gamma-Ratio Random Variables", which deals with the distribution of the product of independent Gamma-ratio or $F$ distributed r.v.'s, and which at first look may seem to fall a bit off the mainstream of the book. The point is that there is a quite well-known relation that exists between an F distribution, or the distribution of the ratio of two independent Gamma distributed r.v.'s, and the distribution of a Beta r.v.. Let us suppose that the r.v. $X$ has an $F_{m,n}$ distribution. Then we may say that

$$X = \frac{X_1/m}{X_2/n} = \frac{Y_1}{Y_2}$$

where, for our notation of the Gamma distribution, set at the beginning of this Foreword, $X_1 \sim \chi^2_m \equiv \Gamma\left(\frac{m}{2}, \frac{1}{2}\right)$ and $X_2 \sim \chi^2_n \equiv \Gamma\left(\frac{n}{2}, \frac{1}{2}\right)$, or equivalently, where $Y_1 = \frac{X_1}{m} \sim \Gamma\left(\frac{m}{2}, \frac{m}{2}\right)$ and $Y_2 = \frac{X_2}{n} \sim \Gamma\left(\frac{n}{2}, \frac{n}{2}\right)$, with $X_1$ and $X_2$, and as such also $Y_1$ and $Y_2$ two independent r.v.'s. Let us now take the r.v.

$$Z = \frac{Y_1}{Y_1 + Y_2} = \frac{X_1/m}{\frac{X_1}{m} + \frac{X_2}{n}}.$$

The r.v. $Z$ has a $Beta\left(\frac{m}{2}, \frac{n}{2}\right)$ distribution. Thus, in many situations we may interchangeably use either a Beta distributed r.v. or an $F$, or a Gamma-ratio, r.v. in many inference procedures. This is what happens in many univariate linear models, namely in simple or multiple Regression models, where to make inferences about the fit of a model with $p$ explanatory variables we commonly use an $F$ statistic with an $F_{p,n-p-1}$ distribution, but where we could alternatively use the so-called $R^2$, the coefficient of determination, which is the square of the multiple correlation coefficient between the response variable and the set of explanatory variables, which, for a sample of size $n$ and a set of $p$ explanatory variables, has a $Beta\left(\frac{p}{2}, \frac{n-p-1}{2}\right)$ distribution. It happens that the two statistics are indeed interchangeable, it being quite easy to express any one of them as a function of the other, and thus one can exchangeably use the quantiles or p-values of any one of them to make inference about the fit of the model.

Although it is not quite the case that Carlos advocates the use of a product of $F$ r.v.'s , or anything similar, to handle the inference procedures usually handled with LRT's in Multivariate Analysis, since this would in general lead to rather intricate approaches, we still thought it to be useful to include the paper republished in this Chapter, given the fact that statistics that have the distribution of products of independent $F$ or Gamma-ratio distributed r.v.'s find useful applications in inferential procedures related with many linear univariate and multivariate models. This is the case for i) tests in random effects models with balanced cross-nesting, ii) tests in meta-analysis, iii) the test of independence of a set of variables, and iv) the test for equality of two generalized variances. One other reason to include this chapter is that this distribution has found applications in many different areas, from wireless communications and wireless networks (Akhtar et al., 2020; ElSawy et al., 2013; Ai and Cheffena, 2017; Koç et al., 2016; Praveenraj and Krishnan, 2019; Wang et al., 2015; Taranetz, 2015; Tanbourgi, 2015; Chen et al., 2013; Gupta et al., 2014; Wang, 2014; Brendel et al., 2014; Jaramillo-Ramírez, 2015; Jaramillo-Ramírez et al., 2015) and telecommunications technology (Sharma and Garg, 2014) to inferential models for prediction (Martin and Lingham, 2016), econometry (Defever and Riaño, 2022) and models for fire claim data (Nadarajah and Bakar, 2015), and also in the setup of clinical trials (Phadnis et al., 2017; Phadnis and Mayo, 2020), and in the genotype-phenotype map (Daalman, 2020), a conceptual model in genetic architecture that models the interdependency of genotype and phenotype. All this besides the fact that in this Chapter the distribution of the difference of two independent GIG distributed r.v.'s is obtained, a distribution that would latter find an useful application in the distribution of linear combinations of chi-square r.v.'s, depicted in the Chapter entitled "On the Distribution of Linear Combinations of Chi-Square Random Variables".

The Chapter entitled "The Advantage of Decomposing Elaborate Hypotheses on Covariance Matrices into Conditionally Independent Hypotheses in Building Near-Exact Distributions for the Test Statistics", is the reprint of a paper (Coelho and Marques, 2009) where it is shown that we can easily build near-exact distributions for LRT statistics for elaborate tests, or rather, for tests of elaborate hypotheses, provided that we can split the original null hypothesis into a sequence of "nested conditionally independent" sub-hypotheses, since then we will be able to use the factorization of the c.f.'s of the logarithms of the LRT statistics associated with these sub-hypotheses to quite easily build near-exact distributions for the LRT statistic of the overall null hypothesis. This act of profiting from the work previously done on the LRT statistics for the sub-hypotheses is something that is not possible to be done for common asymptotic distributions. Let us present a general example to better illustrate the technique and also to illustrate what we mean by "nested conditionally independent" hypotheses (although for a more thorough development we direct the reader to this Chapter of the book).

Let us suppose that we can write the overall null hypothesis $H_0$ as

$$H_0 \equiv H_{0c|ab} \circ H_{0b|a} \circ H_{0a} \,, \tag{8}$$

where '$\equiv$' means 'equivalent to' and 'o' means 'after', or 'composed with', and $H_{0b|a}$ means $H_{0b}$ assuming $H_{0a}$, and $H_{0c|ab}$ means $H_{0c}$ assuming $H_{0a}$ and $H_{0b}$ (see Coelho and Singull (2020); Marques and Coelho (2012a,c); Coelho and Marques (2013b); Correia et al. (2018) for examples). Let us suppose that $\Lambda_a$, $\Lambda_b$ and $\Lambda_c$ are the LRT statistics used to test $H_{0a}$, $H_{0b|a}$ and $H_{0c|ab}$, respectively, and let $W_a = -\log \Lambda_a$, $W_b = -\log \Lambda_b$ and $W_c = -\log \Lambda_c$. Then

$$\Lambda = \Lambda_a \, \Lambda_b \, \Lambda_c$$

is the LRT statistic to test $H_0$ in (8), and if the null sub-hypotheses $H_{0a}$, $H_{0b|a}$ and $H_{0c|ab}$ were correctly nested, $\Lambda_a$, $\Lambda_b$ and $\Lambda_c$ are independently distributed, and we will say that the null hypotheses $H_{0a}$, $H_{0b|a}$ and $H_{0c|ab}$ are "nested conditionally independent" hypotheses. The correct nesting of the these null sub-hypotheses has to assure that

$$\Omega_{H_0} \equiv \Omega_{H_{0c|ab}} \subset \Omega_{H_{1c|ab}} \equiv \Omega_{H_{0b|a}} \subset \Omega_{H_{1b|a}} \equiv \Omega_{H_{0a}} \subset \Omega_{H_{1a}} \equiv \Omega_{H_1}\ (\equiv \Omega)$$

where $\Omega$ indexed by some hypothesis denotes the parameter space for that sub-hypothesis, and where given that we are dealing with LRTs, the parameter space of the null hypothesis has to be a subspace of the parameter space of the corresponding alternative hypothesis (see Anderson (2003, Lemma 10.3.1) and Coelho and Singull (2020, Appendix 2)). Then, for $W = -\log \Lambda$ we have its c.f. $\Phi_W(t)$ given by

$$\Phi_W(t) = \Phi_{W_a}(t)\,\Phi_{W_b}(t)\,\Phi_{W_c}(t)\,,$$

where $\Phi_{W_i}(t)$ are the c.f.'s of $W_i$ $(i = a, b, c)$. Let us suppose that we have for each $\Phi_{W_i}(t)$ $(i = a, b, c)$ a factorization

$$\Phi_{W_i}(t) = \Phi^{(1)}_{W_i}(t)\,\Phi^{(2)}_{W_i}(t)\,, \quad i = a, b, c$$

where $\Phi^{(1)}_{W_i}(t)$ correspond to known manageable distributions and also to the 'major' part of each $\Phi_{W_i}(t)$, in the spirit of expression (7), and $\Phi^{(2)}_{W_i}(t)$ correspond to non-manageable distributions, and at the same time to 'smaller' parts of each $\Phi_{W_i}(t)$, also in the spirit of (7). Then, a sharp near-exact distribution for $W$ may be obtained by keeping the $\Phi^{(1)}_{W_i}(t)$ $(i = a, b, c)$ c.f.'s unchanged, and replacing the product

$$\Phi^{(2)}_{W_a}(t)\,\Phi^{(2)}_{W_b}(t)\,\Phi^{(2)}_{W_c}(t)$$

by a c.f. $\Phi^*(t)$, which approximates the above product asymptotically for increasing sample sizes, taking then

$$\Phi^*_W(t) = \Phi^{(1)}_{W_a}(t)\,\Phi^{(1)}_{W_b}(t)\,\Phi^{(1)}_{W_c}(t)\,\Phi^*(t)$$

as a near-exact c.f. of $W$, which should correspond to a known manageable distribution, from which we can easily compute quantiles and p-values. In this way we can use the work done in the factorization of the c.f.'s of $W_a$, $W_b$ and $W_c$ in building a near-exact distribution for $W$ in a much faster and easier way than if we would try to handle directly the whole c.f. of $W$. This technique was used in a number of papers to obtain sharp and manageable near-exact distributions for several LRT statistics used in Multivariate Analysis as it is the case of the LRT statistics in the above mentioned references and also those in the Chapters "Testing the Hypothesis of a Double Exchangeable Covariance Matrix", "Testing Equality of Mean Vectors with Block-Circular and Block Compound-Symmetric Covariance Matrices" and "The Likelihood Ratio Test of Equality of Mean Vectors with a Doubly Exchangeable Covariance Matrix".

The following two Chapters of the book are somewhat similar, since they both intend to show that although in the landmark books in Multivariate Analysis (Kshirsagar, 1972; Anderson, 2003; Muirhead, 2005) the LRTs to test i) independence of sets of variates, ii) equality of mean vectors, iii) sphericity of the covariance matrix, and iv) equality of covariance matrices, are usually addressed separately, it is possible to devise a common setup for the development of near-exact distributions for the LRT statistics for these four classical tests. The two papers were developed in a intertwining timing, with the paper in the Chapter entitled "Near-Exact Distributions for Certain Likelihood Ratio Test Statistics" (Coelho et al., 2010), published in 2010, being really written after the Technical Report that served as basis for the paper in the Chapter "A General Near-Exact Distribution Theory for the Most Common Likelihood Ratio Test Statistics Used in Multivariate Analysis", published in 2011, to which Coelho et al. (2010) indeed makes a reference to. The base technique used in Coelho et al. (2010) is the one that has been used more recently in developing near-exact distributions for other LRT statistics. It really allows for some extra flexibility in building these near-exact distributions, allowing for a quite easy way to obtain the asymptotic approximation that has to be developed for the part of the c.f.'s that corresponds to a non-manageable distribution, enabling the user to obtain near-exact distributions that equate almost any number of the first exact moments of the logarithm of the LRT statistics for which the near-exact distributions are developed. Furthermore, the results in these two chapters, in terms of the decomposition of the c.f.'s of the logarithms of the LRT statistics addressed, were later used, and still continue to be used, together with the technique described above and detailed in the Chapter entitled "The Advantage of Decomposing Elaborate Hypotheses on Covariance Matrices into Conditionally Independent Hypotheses in Building Near-Exact Distributions for the Test Statistics", to build near-exact distributions for several LRT statistics for tests of elaborate covariance structures and for tests on the equality of mean vectors when the covariance matrices have such elaborate structures, as it is the case of the developments in the Chapters entitled "Testing the Hypothesis of a Double Exchangeable Covariance Matrix", "Testing Equality of Mean Vectors with Block-Circular and Block Compound-Symmetric Covariance Matrices" and "The Likelihood Ratio Test of Equality of Mean Vectors with a Doubly Exchangeable Covariance Matrix".

In the Chapter entitled "The Distribution of the Product of Powers of Independent Uniform Random Variables – A Simple but Useful Tool to Address and Better Understand the Structure of Some Distributions", the distribution of say $V = e^{-W}$, where $W$ has a GIG distribution is established as the EGIG (Exponentiated Generalized Integer Gamma) distribution, and is shown to be the exact distribution of many LRT statistics used in Multivariate Analysis. Although this distribution may be obtained by direct transformation of the GIG distribution of $W$, the interest of this Chapter relies on the derivation of this distribution as the distribution of products of powers of Uniform(0,1) r.v.'s, allowing for a wider interpretation of its relations with other distributions and its possible applications. On the other hand, it is shown how some particular distributions of products of powers of Uniform(0,1) r.v.'s may appear as the exact distributions of a wide range of LRT statistics used in Multivariate Analysis. The EGIG distribution, together with the GIG distribution, form the basis for the results in the quite recent book by Coelho and Arnold (2019) and the results in this Chapter found a variety of applications, in areas from image encryption (Bhnassy et al., 2019), to Mathematical Finance, more precisely, in $n$-player games of optimal stopping (Tan, 2019; Nutz et al., 2020) and to population dynamics (Brilhante et al., 2012).

In the Chapter "The Exact and Near-Exact Distributions of the Main Likelihood Ratio Test Statistics Used in the Complex Multivariate Normal Setting" the distributions for the more commonly used LRT statistics involving Normal complex random vectors are obtained in very manageable forms. While some of these distributions are obtained on finite closed forms for all cases, other are sharply approximated with highly manageable distributions. The results may find applications in many areas, and it was also interesting to have in a single volume the results concerning the exact and near-exact distributions for the four most common LRTs in Multivariate Analysis both for real and complex distributions for the underlying data. As noted in the chapter, the complex multivariate normal distribution has applications in a wide range of areas, from crystallography to spectral analysis in time series and to studies on satellite images, signal processing, communications and information theory. Multivariate complex random variables occupy an important place in Statistics, as it may be seen from the bibliography collected by Subramanian and Subramanian (1973), Krishnaiah (1976) and Saxena (1978). These bibliographies were collected in the 1970's, but if we note that such importance was already attained a few decades ago, we will not be surprised that complex random variables are recently being applied in more and more areas as for example in biomedical signal processing and communications (Eriksson et al., 2009), in geophysical data analysis (Menke, 2018), and namely Normal complex random variables, in spatio-temporal processes (Rao and Terdik, 2012), in physics, when studying the scattering from dispersed particles (Piazza and Degiorgio, 2005), in optics (Goodman, 2000; Lahiri, 2016, namely Chap. 7). We feel that not all the potential in this Chapter has already been used, and as such we feel it will be useful to include it in this collection. Besides all this, we feel that the derivations in the appendices in the supplementary material of the chapter not only have interest for the developments in the chapter itself, as well as they are of high instructional utility not only for an audience at large but more specifically for graduate and later undergraduate students in the area of Statistics and Multivariate Analysis.

The Chapter "Testing the Hypothesis of a Double Exchangeable Covariance Matrix" addresses a LRT for a quite elaborate, and at the same time also a quite general covariance structure, that includes as particular cases several covariance structures of interest, as for example, the block compound symmetric structure, quite widely used in longitudinal and repeated measures analysis models (Fitzmaurice et al., 2011, Chap. 3), and the double compound symmetry (Coelho and Singull, 2020), the block-matrix spherical (Marques and Coelho, 2011) and the spherical covariance structures. The double exchangeable covariance structure is an example of a structure that has the property of modeling the covariance matrix using fewer parameters than the positive-definite covariance structure, which may be an important feature in situations where obtaining large samples may be a problem. Since the use of such structures uses fewer degrees of freedom to model the covariance matrices, it allows for more degrees of freedom to be used for the error term of the tests, and thus allowing for the use of smaller samples in order to be able to carry out the tests. This chapter also has the aim to illustrate the application of the concepts and techniques developed in "The Generalized Near-Integer Gamma Distribution: A Basis for 'Near-Exact' Approximations to the Distribution of Statistics Which Are the Product of an Odd Number of Independent Beta Random Variables" and "The Advantage of Decomposing Elaborate Hypotheses on Covariance Matrices into Conditionally Independent Hypotheses in Building Near-Exact Distributions for the Test Statistics", and the use of some of the results in Chapters "Near-Exact Distributions for Certain Likelihood Ratio Test Statistics" and "A General Near-Exact Distribution Theory for the Most Common Likelihood Ratio Test Statistics Used in Multivariate Analysis". Also, the test developed in this Chapter should be used before the test in the Chapter "The Likelihood Ratio Test of Equality of Mean Vectors with a Doubly Exchangeable Covariance Matrix" is implemented, since it is the test for the covariance structure assumed for the covariance matrices in this latter Chapter.

In the Chapter "On the Distribution of Linear Combinations of Chi-Square Random Variables" the author addresses the distribution of linear combinations of chi-square random variables, a distribution that, as it is highlighted in the chapter itself, has many connections with other important distributions in Statistics, as is the case of the distribution of quadratic forms in Normal r.v.'s. Mathai and Provost (1992, Sect. 4.2i) state that "Since the exact density of a general linear combination of chi-square variates can be obtained from Theorem 4.2i.1, this theorem gives a representation of the density of indefinite quadratic forms in central normal variables.", and when referring to the importance of the distribution of quadratic forms in Normal variables, the same authors (Mathai and Provost, 1992, Chap. 7) remark that "The test statistics employed in many tests and estimation procedures related to normal distributions and Gaussian processes are expressible as quadratic forms in normal variables. In some other cases these statistics are asymptotically equivalent to such quadratic forms". Moschopoulos and Canada (1984) state that the distribution of a linear combination of chi-squares "is important in a variety of problems in statistical inference and applied probability", and the importance of this distribution is corroborated by the fact that Johnson et al. (1994, Chap. 18, Sect. 8) dedicate to it a full section in their book, where its link with the distribution of quadratic forms is also evoked, while the same authors, in their second volume (Johnson et al., 1995, Chap. 22, Sect. 9) give an example of the application of this distribution to methods of inference related with extreme value distributions. Rao (1973, Chap. 6, sect. 6d.2), referring to the distribution of the statistic to test independence in a contingency table, when cells are pooled, mentions that Chernoff and Lehmann (1954) "show it to be the distribution of a linear combination of $\chi^2$ variables", to state then that "Since it is difficult to use such a distribution the following approach is suggested. ...", showing well the difficulties that may arise when trying to use this distribution. In this Chapter of the book we want to show that in some cases it is possible through a quite straightforward use of the GIG distribution in the Chapter "The Generalized Integer Gamma Distribution – A Basis for Distributions in Multivariate Statistics", the GNIG distribution in the Chapter "The Generalized Near-Integer Gamma Distribution: A Basis for 'Near-Exact' Approximations to the Distribution of Statistics Which Are the Product of an Odd Number of Independent Beta Random Variables", and the distribution of the difference of two independent GIG distributed r.v.'s developed in the Chapter entitled "On the Distribution of the Product and Ratio of Independent Generalized Gamma-Ratio Random Variables", to obtain closed form representations for the p.d.f. and c.d.f. of a number of cases of linear combinations of chi-squared r.v.s, even with real (that is, with both positive and negative) coefficients, while for the other cases, not covered by these finite form representations, it is possible to obtain very sharp and manageable approximations. Based on these results, the distributions of the statistic used in the Behrens-Fisher problem and that of statistics used to test for ratios of quadratic forms are then also addressed. We believe that, as it is the case with other papers reproduced in this book, not all the usefulness of the results in this Chapter has already been used by ourselves or by other researchers.

The Chapter "On the Distribution of the Product of Independent Beta Random Variables – Applications" addresses the distribution of the product of independent Beta r.v.'s, a distribution that, as already remarked above, is directly related with the distribution of many of the LRT statistics used in Multivariate Analysis, such as the Wilks Lambda statistic (Kshirsagar, 1972, Sect. 8.2) and the LRT statistics to test: compound symmetry (Tukey and Wilks, 1946), circularity (Olkin and Press, 1969), the general linear hypothesis and the equality of mean vectors (MANOVA), independence of sets of variables, sphericity, equality of covariance matrices (Tang and Gupta, 1984; Anderson, 2003, Sects. 8.4, 9.6), and equality of generalized variances (SenGupta, 1987), as well as the ones in the Chapters entitled "The Exact and Near-Exact Distributions of the Main Likelihood Ratio Test Statistics Used in the Complex Multivariate Normal Setting", "Testing the Hypothesis of a Double Exchangeable Covariance Matrix", "Testing Equality of Mean Vectors with Block-Circular and Block Compound-Symmetric Covariance Matrices" and "The Likelihood Ratio Test of Equality of Mean Vectors with a Doubly Exchangeable Covariance Matrix". But this distribution also finds applications in other areas as in "the computation of Bayesian confidence limits for the product of an arbitrary number of binomial parameters" (Springer and Thompson, 1966b) and in studies in reliability theory of series and parallel systems when using a Bayesian approach (Tang et al., 1997) and in reliability studies in for one-shot systems (Guo et al., 2011). The results obtained may be applied in an omnibus way, as it is illustrated with its application to the distribution of a number of LRT statistics, with these omnibus approximations performing not much worse than the dedicated approximations previously developed for the same LRT statistics in the Chapters "The Generalized Near-Integer Gamma Distribution: A Basis for 'Near-Exact' Approximations to the Distribution of Statistics Which Are the Product of an Odd Number of Independent Beta Random Variables", "Near-Exact Distributions for

Certain Likelihood Ratio Test Statistics" and "A General Near-Exact Distribution Theory for the Most Common Likelihood Ratio Test Statistics Used in Multivariate Analysis", and even with some gains in some very particular situations. The results developed in this Chapter may thus find applications in many areas and they still represent an improvement over existing results in the area.

In the two last Chapters, "Testing Equality of Mean Vectors with Block-Circular and Block Compound-Symmetric Covariance Matrices" and "The Likelihood Ratio Test of Equality of Mean Vectors with a Doubly Exchangeable Covariance Matrix", are presented LRTs for the tests of equality of mean vectors, when the distributions have respectively a block exchangeable and a block circular structure. These LRTs are new in the literature and they represent generalizations of the LRTs previously published in Coelho (2017) and Coelho (2018). It is believed that the LRTs in these four publications may find applications in situations of high-dimensionality, that is, situations where the number of observations is smaller than the number of variables in the study, being thus the only LRTs known to date that may be applied both in the low-dimensional and in the high-dimensional cases. This is an area where the author started to work recently and where he expects to be able to develop further generalizations of these tests, with possible applications in the high-dimensionality case, even when extremely small samples are used.

The editors want to call the reader's attention to the fact that in some of the Chapters, that is, for some of the papers in the book, there are links to files with modules used to implement the distributions developed in those papers. It happened that all these were on Google pages, which with the reformulation carried out by Google disappeared. These files are now harbored in the GitHub repository https://github.com/Carlos-Coelho/LRTs-and-Related-Topics-in-Multivariate-Analysis. The best way to reach a file is by going to this repository and then choosing the folder with the Chapter number to which corresponds the paper in question and then select the required file. For example, for the file `Mathematica_Modules_for_Exact_distr_stats_Sec_2.nb`, mentioned in the Chapter entitled "On the Distribution of Linear Combinations of Chi-Square Random Variables" (the 12-th paper in the book), one has, after using the link above, to click on the folder `Chapter-12` and then look for the file in that folder. In case one wants to have a more direct access to the file one can use a link similar to the above one to which you add /tree/main/Chapter-12, or even a link similar to this latter one to which you add /`<filename>`, in case one wants to access directly the file.

## References

Ai, Y., Cheffena, M.: Performance Analysis of Hybrid-ARQ over Full-Duplex Relaying Network Subject to Loop Interference under Nakagami-m Fading Channels. 2017 IEEE 85th Vehicular Technology Conference, Sydney, NSW, Australia, pp. 1–5 (2017)

Akhtar, M.W., Hassan, S.A., Saleem, S., Jung, H.: STBC-Aided Cooperative NOMA With Timing Offsets, Imperfect Successive Interference Cancellation, and Imperfect Channel State Information. IEEE Transactions on Vehicular Technology, 69, 11712–11727 (2020)

Alberto, R. P., Coelho, C. A. (2007). Study of the quality of several asymptotic and near-exact approximations based on moments for the distribution of the Wilks Lambda statistic. Journal of Statistical Planning and Inference, 137, 5, 1612–1626.

Anderson, T. W.: An Introduction to Multivariate Statistical Analysis, 3rd ed. J. Wiley & Sons, New York (2003)

Arnold, B.C., Coelho, C.A., Marques, F.J.: The distribution of the product of powers of independent uniform random variables – a simple but useful tool to address and better understand the structure of some distributions. J. Multivariate. Anal. 113, 19–36 (2013)

Askey, R.A., Daalhaus, A.B.O.: Generalized Hypergeometric Functions and Meijer G-function. In: Olver, W.J., Lozier, D.W., Boisvert, R.F., Clark, C.W. (eds.) NIST Handbook of Mathematical Functions, pp. 403–418. National Institute of Standards and Technology, U.S. Dept. of Commerce and Cambridge University Press, New York (2010)

Bedoui, A., Et-tolba, M., Nouasria, H.: An MMSE Integrated Equalization for HARQ Chase Combining in OQAM-FBMC systems. 2019 15th International Wireless Communications & Mobile Computing Conference, Tangier, Morocco, pp. 811–816 (2019)

Bhargava, R.P.: Some one-sample hypothesis testing problems when there is a monotone sample from a multivariate Normal population. Ann. Inst. Stat. Math. 27, 327–339 (1975)

Bhargava, R.P., Khatri, C.G.: The distribution of independent Beta random variables with application to multivariate analysis. Ann. Inst. Statist. Math. 33, 287–296 (1981)

Bhnassy, M.A., Hagras, E.A.A., El-Badawy, E.-S.A., Mokhtar, M.A., Aly, M.H.: Image encryption and watermarking combined dynamic chaotic hopping pattern with double random phase encoding DRPE. Optical and Quantum Electronics, 51, 246, 18pp. (2019)

Bilodeau, M., Brenner, D.: Theory of Multivariate Statistics. Springer Texts in Statistics. Springer, New York, NY (1999)

Braaksma, B.L.J.: Asymptotic expansions and analytic continuations for a class of Barnes-integrals. Compos. Math. 15, 239–341 (1963)

Brendel, J., Riess, S., Schroeter, S., Weigel, R., Fischer, G.: Signal detection and cooperative sensing with sensor nodes with limited dynamic. 1st International Workshop on Cognitive Cellular Systems, Germany, pp. 1–5 (2014)

Brilhante, M.F., Gomes, M.I., Pestana, D.: Extensions of Verhulst Model in Population Dynamics and Extremes. Proceedings, 5th Chaotic Modeling and Simulation International Conference, 12–15 June 2012, Athens Greece, 115–122 (2012)

Carter, B. D., Springer, M. D.: The distribution of products, quotients and powers of independent H-function variates. SIAM J. Appl. Math. 33, 542–558 (1977)

Chen, Y., Mucchi, L., Wang, R.: Angular spectrum and second order statistics of interference in wireless networks. 2013 IEEE International Conference on Communication, Networks and Satellite, Yogyakarta, Indonesia, pp. 41–45 (2013)

Chernoff, H., Lehmann, E.L. (1954). The use of maximum likelihood estimates in $\chi^2$ tests for goodness of fit. Ann. Math. Statist., 25, 579–586.

Coelho, C.A. (1998). The Generalized Integer Gamma Distribution – a Basis for Distributions in Multivariate Statistics. J. Multivariate Anal., 64, 86–102.

Coelho, C.A. (2004). The generalized near-integer Gamma distribution: a basis for 'near-exact' approximations to the distribution of statistics which are the product of an odd number of independent Beta random variables. *J. Multivariate Anal.*, 89, 191–218.

Coelho, C.A. (2017). The Likelihood Ratio Test for Equality of Mean Vectors with Compound Symmetric Covariance Matrices, in Computational Science and Its Applications, Gervasi, O., Murgante, B., Misra, S., Borruso, G., Torre, C. M., Rocha, A. M. A. C., Taniar, D., Apduhan, B. O., Stankova, E., Cuzzocrea, A. (eds.), Lecture Notes in Computer Science 10408, Vol. V, Springer, pp. 20–32

Coelho, C.A. (2018). Likelihood ratio tests for equality of mean vectors with circular covariance matrices, in Oliveira, T. A., Kitsos, C., Oliveira, A., Grilo, L. M. (eds.), "Recent Studies on Risk Analysis and Statistical Modeling", Contributions to Statistics, Springer, pp. 255–269.

Coelho, C.A. (2021). Testing equality of mean vectors with block-circular and block compound-symmetric covariance matrices, in 'Multivariate, Multilinear and Mixed Linear Models',

Filipiak, K., Markiewicz, A., von Rosen, D. (eds.), Springer series in Contributions to Statistics, 157–201.

Coelho, C.A. (2022). Likelihood ratio tests for elaborate covariance structures and for MANOVA models with elaborate covariance structures – a review, Journal of the Indian Institute of Science, 102, 1219–1257.

Coelho, C.A., Arnold, B.C. (2014). On the exact and near-exact distributions of the product of generalized Gamma random variables and the generalized variance, Communications in Statistics – Theory and Methods, 43, 2007–2033.

Coelho, C.A., Arnold, B.C. (2019). Finite Forms Representations for Meijer G and Fox H Functions – Applied to Multivariate Likelihood Ratio Tests Using Mathematica®, MAXIMA and R. Lecture Notes in Statistics, Springer.

Coelho, C. A., Marques, F. J. (2009) The advantage of decomposing elaborate hypotheses on covariance matrices into conditionally independent hypotheses in building near-exact distributions for the test statistics. Linear Algebra and its Applications, 430, 2592–2606.

Coelho, C. A., Marques, F. J. (2010) Near-exact distributions for the independence and sphericity likelihood ratio test statistics. Journal of Multivariate Analysis, 101, 583–593.

Coelho, C. A., Marques, F. J. (2011) On the exact, asymptotic and near-exact distributions for the likelihood ratio statistics to test equality of several Exponential distributions. AIP Conference Proceedings, 1389, 1471–1474.

Coelho, C. A., Marques, F. J. (2012). Near-exact distributions for the likelihood ratio test statistic to test equality of several variance-covariance matrices in elliptically contoured distributions. Computational Statistics, 27, 627–659.

Coelho, C. A., Marques, F. J. (2013a). Near-exact distributions for the block equicorrelation and equivariance likelihood ratio test statistic, AIP Conference Proceedings, 1557, 429–433.

Coelho, C. A., Marques, F. J. (2013b). The multi-sample block-scalar sphericity test: exact and near-exact distributions for its likelihood ratio test statistic, Communications in Statistics – Theory and Methods, 42, 1153–1175.

Coelho, C.A., Roy, A. (2017). Testing the hypothesis of a block compound symmetric covariance matrix for elliptically contoured distributions, TEST, 26, 308–330.

Coelho, C. A., Roy, A. (2020). Testing the hypothesis of a doubly exchangeable covariance matrix. Metrika, 83, 45–68.

Coelho, C.A., Singull, M.: Testing for Double Complete Symmetry, In: Holgersson, T., Singull, M. (eds.) Recent Developments in Multivariate and Random Matrix Analysis – Festschrift in Honour of Dietrich von Rosen, pp. 17–39 (2020)

Coelho, C. A., Alberto, R. P. and Grilo, L. M. (2006). A mixture of Generalized Integer Gamma distributions as the exact distribution of the product of an odd number of independent Beta random variables: applications. Journal of Interdisciplinary Mathematics, 9, 2, 229–248.

Coelho, C. A., Arnold, B. C., Marques, F. J. (2010) Near-exact distributions for certain likelihood ratio test statistics. Journal of Statistical Theory and Practice, 4, 4, 711–725 (invited paper for the special memorial issue in honor of H. C. Gupta, guest-edited by C. R. Rao).

Coelho, C. A., Marques, F. J., Arnold, B. C. (2015). The exact and near-exact distributions of the main likelihood ratio test statistics used in the complex multivariate normal setting, Test, 24, 386–416 + supplementary material, 14pp.

Coelho, C. A., Marques, F. J., Oliveira, S. (2016). Near-exact distributions for likelihood ratio statistics used in the simultaneous test of conditions on mean vectors and patterns of covariance matrices, Mathematical Problems in Engineering, Volume 2016, Article ID 8975902, 25 pages.

Coelho, C. A., Marques, F. J., Jorge, N., Nunes, C. (2022). On the distribution of the likelihood ratio test of independence for random sample size – a computational approach, Journal of Computational and Applied Mathematics, 404, Paper No. 113394, 20 pp.

Constantinescu, C.D., Ramirez, J.M., Zhu, W.R.: An application of fractional differential equations to risk theory. Finance Stoch 23, 1001–1024 (2019)

Correia, B. R., Coelho, C. A., Marques, F. J. (2018). Likelihood ratio test for the hyper-block matrix sphericity covariance structure – characterization of the exact distribution and development of near-exact distributions for the test statistic, REVSTAT, 16, 365–403.

Daalman, W.K.-G.: Coupling of genotype-phenotype maps to noise-driven adaptation, showcased in yeast polarity, Doctoral Thesis, Delft University of Technology, Delft, Netherlands (2020)
Defever, F., Riaño, A.: Firm-destination heterogeneity and the distribution of export intensity. Economics Letters 219, 110810 (2022)
ElSawy, H., Hossain, E., Camorlinga, S.: Traffic offloading techniques in two-tier femtocell networks. 2013 IEEE International Conference on Communications, Budapest, Hungary, pp. 6086–6090 (2013)
Erdélyi, A., Magnus, W., Oberhettinger, F., Tricomi, F.G.: Higher Transcendental Functions (The Bateman Manuscript Project), vol. I. McGraw-Hill, New York (1953)
Eriksson, J., Ollila, E., Koivunen, V.: Statistics for complex random variables revisited. 2009 IEEE International Conference on Acoustics, Speech and Signal Processing, Taipei, Taiwan, 3565–3568 (2009)
Fitzmaurice, G.M., Laird, N.M., Ware, J.H. (2011). Applied Longitudinal Analysis, 2nd ed. Wiley, New York
Fox, C.: The G and H functions as symmetrical kernels. Trans. Amer. Math. Soc. 98, 395–429 (1961)
Fukaya, K., Kawamori, A., Osada, Y., Kitazawac, M., Ishiguro, M.: The forecasting of menstruation based on a state-space modeling of basal body temperature time series. Stat. Med. 36, 3361–3379 (2017)
Ghofrani-Jahromi, S., Zolghadrasli, A., Neinavaie, M.: Performance analysis of multihop decode-and-forward relay networks with diversity in Nakagami fading channels. AEU - International Journal of Electronics and Communications, 113, 152973 (2020)
Ghusinga, K.R., Vargas-Garcia, C.A., Singh, A.: A mechanistic stochastic framework for regulating bacterial cell division. Sci. Rep. 6, 30229 (2016), supplementary material.
Goodman, J.W.: Statistical Optics. Wiley Classics Library Edition, Wiley, New York (2000)
Grilo, L. M., Coelho, C. A. (2007). Development and study of two near-exact approximations to the distribution of the product of an odd number of independent Beta random variables. Journal of Statistical Planning and Inference, 137, 5, 1560–1575.
Grilo, L. M., Coelho, C. A. (2010a) The exact and near-exact distributions for the Wilks Lambda statistic used in the test of independence of two sets of variables. American Journal of Mathematical and Management Sciences, 30, 1–2, 111–145.
Grilo, L. M., Coelho, C. A. (2010b) Near-exact distributions for the generalized Wilks Lambda statistic. Discussiones Mathematicae - Probability and Statistics, 30, 53–86.
Grilo, L. M., Coelho, C. A. (2012). A family of near-exact distributions based on truncations of the exact distribution for the generalized Wilks Lambda statistic. Communications in Statistics - Theory and Methods, 41, 2321–2341.
Grilo, L. M., Coelho, C. A. (2013). Near-exact distributions for the likelihood ratio statistic used to test the reality of a covariance matrix, AIP Conference Proceedings, 1558, 797–800.
Grilo, L.M., Coelho, C.A. (2017). The exact and near-exact distributions for the statistic used to test the reality of covariance matrix in a complex normal distribution, in Applied and Computational Matrix Analysis, Bebiano, N. (ed.), Springer Proceedings in Mathematics & Statistics 192, Springer, pp. 295–315
Guo, H., Jin, T., Mettas, A.: Designing reliability demonstration tests for one-shot systems under zero component faiures. IEEE Transactions on Reliability 60, 286–294 (2011)
Gupta, A. K.: Distribution of Wilks' likelihood-ratio criterion in the complex case. Ann. Inst. Statist. Math. 23, 77–87 (1971)
Gupta, A., Mathur, S., Singhal, N., Sharma, P.K., Garg, P.: Outage analysis of full duplex relaying with two decode-and-forward relays. 2014 Annual IEEE India Conference, Pune, India, pp. 1–4 (2014)
Ibrahim, A.M., Ercetin, O., ElBatt, T.: Stability Analysis of Slotted Aloha With Opportunistic RF Energy Harvesting. IEEE Journal on Selected Areas in Communications, vol. 34, no. 5, pp. 1477–1490 (2016)
Jaramillo-Ramírez, D.: Interference mitigation techniques for 4G networks. Thèse de doctorat en Télécommunications. Ecole Doctorale Sciences et Technologies de l'Information, des Télécommunications et des Systèmes, Orsay, Essonne (2015)

Jaramillo-Ramírez, D., Kountouris, M., Hardouin, E.: Coordinated Multi-Point Transmission With Imperfect CSI and Other-Cell Interference. IEEE Transactions on Wireless Communications, vol. 14, no. 4, pp. 1882–1896 (2015)

Johnson, N.L., Kotz, S., Balakrishnan, N. (1994). Continuous Univariate Distributions, Vol. 1, 2nd ed. Wiley, New York

Johnson, N.L., Kotz, S., Balakrishnan, N. (1995). Continuous Univariate Distributions, Vol. 2, 2nd ed. Wiley, New York

Kaiser, T., Zheng, F.: Ultra Wideband Systems with MIMO, Wiley (2010)

Karagiannidis, G.K., Sagias, N.C., Tsiftsis, T.A.: Closed-form statistics for the sum of squared Nakagami-m variates and its applications. IEEE Transactions on Communications, 54, 1353–1359 (2006)

Khatri, C. G.: Classical Statistical Analysis Based on a Certain Multivariate Complex Gaussian Distribution. Ann. Math. Statist. 36, 98–114 (1965a)

Khatri, C. G.: A test for reality of a covariance matrix in a certain complex Gaussian distribution. Ann. Math. Statist. 36, 115–119. (1965b)

Koç, A., Altunbas, I., Yongacoglu, A.: Outage Probability of Two-Way Full-Duplex AF Relay Systems over Nakagami-m Fading Channels. 2016 IEEE 84th Vehicular Technology Conference, Montreal, QC, Canada, pp. 1–5 (2016)

Krishnaiah, P.R.: Some recent developments on Complex Multivariate distributions. J- Multivariate Anal. 6, 1–30 (1976)

Kshirsagar, A.: Multivariate Analysis. Marcel Dekker, New York (1972)

Lahiri, A.: Basic Optics - Principles and Concepts. Elsevier, Amsterdam (2016)

Liang, Y.: Contributions to Estimation and Testing Block Covariance Structures in Multivariate Normal Models. Ph.D. Thesis, Stockholm University, Stockholm, Sweden (2015)

Liang, Y., Coelho, C. A., von Rosen, T. (2022). Hypothesis Testing in Multivariate Normal Models with Block Circular Covariance Structures, Biometrical Journal, 64, 557–576.

Liu, F., Riihijärvi, J., Petrova, M.: Robust data rate estimation with stochastic SINR modeling in multi-interference OFDMA networks. 12th Annual IEEE International Conference on Sensing, Communication, and Networking, Seattle, WA, USA, 211–219 (2015)

Liu, Y., Qin, Z., Elkashlan, M., Gao, Y., Hanzo, L.: Enhancing the Physical Layer Security of Non-Orthogonal Multiple Access in Large-Scale Networks. IEEE Transactions on Wireless Communications, 16, 1656–1672 (2017)

Luke, Y.L.: The Special Functions and Their Approximations, vol. I. Academic Press, New York (1969)

Luke, Y.L.: Mathematical Functions and Their Approximations. Academic Press, New York (1975)

Marques, F. J., Coelho, C. A. (2008) Near-exact distributions for the sphericity likelihood ratio test statistic. Journal of Statistical Planning and Inference, 138, 726–741.

Marques, F. J., Coelho, C. A. (2010) The exact and near-exact distributions of the likelihood ratio statistic for the block sphericity test. AIP Conference Proceedings, 1281, 1237–1240.

Marques, F. J., Coelho, C. A. (2011) The multi-sample block-matrix sphericity test. AIP Conference Proceedings, 1389, 1479–1482.

Marques, F. J., Coelho, C. A. (2012a). The block sphericity test – exact and near-exact distributions for the likelihood ratio statistic. Mathematical Methods in the Applied Sciences, 35, 373–383.

Marques, F. J., Coelho, C. A. (2012b). The multi-sample independence test. AIP Conference Proceedings, 1479, 1129–1132.

Marques, F. J., Coelho, C. A. (2012c). Near-exact distributions for the likelihood ratio test statistic of the multi-sample block-matrix sphericity test. Applied Mathematics and Computation, 219, 2861–2874.

Marques, F.J., Coelho, C.A. (2013a). The multi-sample block-scalar sphericity test under the complex multivariate normal case, AIP Conference Proceedings, 1557, 420–423.

Marques, F.J., Coelho, C.A., (2013b). The multisample block-diagonal equicorrelation and equivariance test, AIP Conference Proceedings, 1558, 793–796.

Marques, F.J., Coelho, C.A. (2013c). Obtaining the exact and near-exact distributions of the likelihood ratio statistic to test circular symmetry through the use of characteristic functions, Computational Statistics, 28, 2091–2115.

Marques, F. J., Coelho, C. A. (2015). Near-exact Distributions for the Likelihood Ratio Test Statistic for Testing Multi-Sample Independence – The Real and Complex Cases, Journal of Statistical Theory and Practice, 9, 1, 37–58.
Marques, F. J., Coelho, C. A. (2016). Near-exact Distributions for Positive Linear Combinations of Independent Non-central Gamma Random Variables, AIP Conference Proceedings, 1738, 190005-1–190005-4.
Marques, F. J., Coelho, C. A. (2018). The simultaneous test of equality and circularity of several covariance matrices, Journal of Statistical Theory and Practice, 12, 861–885
Marques, F. J., Coelho, C. A. (2020). Testing simultaneously different covariance block diagonal structures – the multi-sample case, Journal of Applied Statistics, 47, 2765–2784.
Marques, F. J., Coelho, C. A., Arnold, B. C. (2011) A general near-exact distribution theory for the most common likelihood ratio test statistics used in Multivariate Analysis. TEST, 20, 1, 180–203.
Marques, F. J., Coelho, C. A., de Carvalho, M. (2015). On the distribution of linear combinations of independent Gumbel random variables, Statistics and Computing, 25, 3, 683–701 +supplementary material, 5pp.
Martin, R., Lingham, R.T.: Prior-Free Probabilistic Prediction of Future Observations, Technometrics, 58:2, 225–235 (2016)
Mathai, A.M.: A review of different techniques used for deriving the exact distributions of multivariate test criteria. Sankhyā, Ser. A 35, 39–60 (1973)
Mathai, A.M.: A Handbook of Generalized Special Functions for Statistical and Physical Sciences. Oxford University Press, New York (1993)
Mathai, A.M., Haubold, H.J.: Special Functions for Applied Scientists. Springer, New York (2008)
Mathai, A.M., Provost, S. B.: Quadratic Forms in Random Variables – Theory and Applications. Marcel Dekker, New York (1992)
Mathai, A.M., Rathie, P.N.: The exact distribution of Wilks' criterion. Ann. Math. Stat., 42, 1010–1019 (1971)
Mathai, A.M., Saxena, R.K.: Generalized Hypergeometric Functions with Applications in Statistics and Physical Sciences. Lecture Notes in Mathematics, vol. 348. Springer, New York (1973)
Mathai, A.M., Saxena, R.K., Haubold, H.J.: The H-function – Theory and Applications. Springer, New York (2009)
Meijer, C.S.: Über Whittakersche bzw. Besselsche Funktionen und deren Produkte. Nieuw Archief voor Wiskunde 18(2), 10–29 (1936a)
Meijer, C.S.: Neue Integraldarstellungen aus der Theorie der Whittakerschen und Hankelschen Funktionen. Math. Annalen 112(2), 469–489 (1936b)
Meijer, C.S.: Multiplikationstheoreme für die Funktion $G_{p,q}^{m,n}(z)$. Proc. Koninklijk Nederlandse Akademie van Weteenschappen 44, 1062–1070 (1941)
Meijer, C.S.: On the G-function I–VIII. Proc. Koninklijk Nederlandse Akademie van Weteenschappen 49, 227–237, 344–356, 457–469, 632–641, 765–772, 936–943, 1063–1072, 1165–1175 (1946)
Menke, W.: Geophysical Data Analysis – Discrete Inverse Theory, 4th ed., Academic Press, Elsevier, Amsterdam (2018)
Mohri, H., Takeshita, J.: Catastrophic Failure and Cumulative Damage Models Involving Two Types of Extended Exponential Distributions. International Journal of Reliability, Quality and Safety Engineering, 29, 2141004 (2022)
Moschopoulos, P.G., Canada, W.B. (1984). The distribution function of a linear combination of chi-squares. Comput. Math. Appl., 10, 383–386.
Muirhead, R. J.: Aspects of Multivariate Statistical Theory, 2nd ed.. J. Wiley & Sons, New York (2005)
Müller, M.K. (2018). System level investigations for mobile and indoor users in future cellular networks. Ph.D. Thesis, Faculty of Electronics and Informatics, Technical University of Wien
Müller, M., Safiulin, I.: Advanced System Level Applications. In: Rupp, M., Schwarz, S., Taranetz, M. (eds.) The Vienna LTE-Advanced Simulators, 335–354 (2016)

Müller, M.K., Taranetz, M., Rupp, M.: Performance of Remote Unit Collaboration Schemes in High Speed Train Scenarios. 2015 IEEE 82nd Vehicular Technology Conference, Boston, MA, USA, pp. 1–5 (2015)

Nadarajah, S., Bakar, S.A.A.: New Folded Models for the Log-Transformed Norwegian Fire Claim Data, Communications in Statistics – Theory and Methods, 44:20, 4408–4440 (2015)

Nagarsenker, B. N., Das, M. M.: Exact distribution of sphericity criterion in the complex case and its percentage points. Comm. Statist. 4, 363–374 (1975)

Nagarsenker, B. N., Suniaga, J.: Distributions of a class of statistics useful in multivariate analysis. J. Amer. Statist. Assoc. 78, 472–475 (1983)

Nandi, S. B.: The exact distributions of certain likelihood ratio criteria. Sankhya, Ser. B 42, 187–196 (1980)

Nutz, M., San Martin, J., Tan, X.: Convergence to the mean field game limit: a case study. Annals of Applied Probability 30, 259–286 (2020)

Olkin, I., Press, S. J.: Testing and estimation for a circular stationary model. Ann. Math. Statist. 40, 1358–1373 (1969)

Paolella, M. S.: Intermediate Probability - A Computational Approach. J. Wiley & Sons, Chichester, England (2007)

Parruca, D., Gross, J.: On the interference as noise approximation in OFDMA/LTE networks. 2014 IEEE International Conference on Communications, Sydney, NSW, Australia, pp. 2513–2519 (2014)

Pederzoli, G.: Some properties of beta functions and the distribution for the product of independent beta random variables. Trabajos de Estadistica y de Investigacion Operativa 36, 122–128 (1985)

Phadnis, M.A., Wetmore, J.G., Mayo, M.S.: A clinical trial design using the concept of proportional time using the generalized gamma ratio distribution. Statistics in Medicine, 36, 4121–4140 (2017)

Phadnis, M.A., Mayo, M.S.: Group sequential design for time-to-event data using the concept of proportional time. Statistical Methods in Medical Research, 29, 1867–1890 (2020)

Piazza, R., Degiorgio, V.: Scattering, Rayleigh. In: Encyclopedia of Condensed Matter Physics, Bassani, F., Liedl, G.L., Wyder, P. (eds.), Vol. 5, 234–242, Academic Press, Elsevier, Amsterdam (2005)

Pillai, K. C. S., Jouris, G. M.: Distribution of Wilks' likelihood-ratio criterion in the complex case. Ann. Math. Statist. 42, 517–525 (1971)

Praveenraj, S., Krishnan, K.V.: The Outage Performance of Dual-Hop Full Duplex AF Relay Network with MRC Technique over Nakagami-m Fading Channel. 2019 International Conference on Vision Towards Emerging Trends in Communication and Networking, Vellore, India, pp. 1–5 (2019)

Prudnikov, A.P., Brychkov, Y.A., Marichev, O.J.: Integrals and Series, Vol. 3: More Special Functions. Gordon and Breach, Newark (1990)

Psomas, C., Krikidis, I.: Adaptive variable-length feedback using wireless power transfer for opportunistic beamforming. 2016 IEEE International Conference on Communications, Kuala Lumpur, Malaysia, pp. 1–6 (2016)

Psomas, C., Krikidis, I.: A Wireless Powered Feedback Protocol for Opportunistic Beamforming Using Rectenna Arrays. IEEE Transactions on Green Communications and Networking, 2, 100–113 (2018)

Rao, C.R.: Linear Statistical Inference and Its Applications, 2nd ed. J. Wiley & Sons, New Tork (1973)

Rao, T.S., Terdik, G.: Statistical Analysis of Spatio-Temporal Models and Their Applications. In: Time Series Analysis: Methods and Applications, Rao, T.S., Rao, S.S., Rao, C.R. (eds.), Handbook of Statistics, Vol. 30, 521–540, North Holland, Elsevier, Amsterdam (2012)

Ruby, R., Pham, Q.-V., Wu, K., Heidari, A.A., Chen, H., ElHalawany, B.M.: Enhancing Secrecy Performance of Cooperative NOMA-Based IoT Networks via Multiantenna-Aided Artificial Noise. IEEE Internet of Things Journal, 9, 5108–5127 (2022)

Saxena, A.K.: Complex multivariate statistical analysis: an anntated bibliography. Int. Stat. Review 46, 209–214 (1978)

Schatzoff, M.: Exact distributions of Wilks's likelihood ratio criterion. Biometrika, 53, 347–358 (1966)
SenGupta, A.: Tests for standardized generalized variances of multivariate Normal populations of possibly different dimensions. J. Multivariate Anal. 23, 209–219 (1987)
Serim, H., Coelho, C. A., Park, J. (2022). An Exact and Near-Exact Distribution Approach to the Behrens–Fisher Problem, Mathematics, 10, 2953 (18 pp.)
Sharma, P.K., Garg, P.: Performance analysis of full duplex decode-and-forward cooperative relaying over Nakagami-m fading channels. Trans. Emerging Tel. Tech., 25, 905–913 (2014)
Springer, M.D.: The Algebra of Random Variables. J. Wiley & Sons, New York (1979)
Springer, M.D., Thompson, W.E.: The distribution of products of independent random variables. SIAM J. Appl. Math. 14, 511–526 (1966a)
Springer, M.D., Thompson, W.E.: Bayesian confidence limits for the product of $N$ binomial parameters. Biometrika 53, 611–613 (1966b)
Subramanian, K., Subramanian, K.: Multivariate Analysis: A Selected and Abstracted Bibliography 1957–1972. Marcel Dekker Inc., New York (1973)
Suraweera, H.A., Smith, P.J., Surobhi, N.A.: Exact Outage Probability of Cooperative Diversity with Opportunistic Spectrum Access. 2008 IEEE International Conference on Communications Workshops, Beijing, China, pp. 79–84 (2008)
Surobhi, N.A.: Outage Performance of Cooperative Cognitive Relay Networks. MSc Thesis, Victoria University, Melbourne, Australia (2009)
Tan, X.: Optimal Transport and Equilibrium Problems in Mathematical Finance, Ph.D. Thesis, Columbia University, New York, USA (2019)
Tanbourgi, R.: Diversity Combining under Interference Correlation in Wireless Networks. Doktor-Ingenieurs Thesis, Karlsruher Institut für Technologie, Karlsruher, Germany (2015)
Tang, J., Gupta, A.K.: On the distribution of the product of independent Beta random variables. Stat. Prob. Letters 2, 165–168 (1984)
Tang, J., Gupta, A. K.: Exact distribution of certain general test statistics in multivariate analysis. Aust. J. Stat. 28, 107–114 (1986)
Tang, J., Tang, K, Moskowitz, H.: Exact Bayesian estimation of system reliability from component test data. Naval Research Logistics 44, 127–146 (1997)
Taranetz, M. (2015) On the Product and Ratio of Pareto and Maxwell Random Variables Ph.D. Thesis, On the Product and Ratio of Pareto and Maxwell Random Variables
Tegos, S.A., Karagiannidis, G.K., Diamantoulakis, P.D., Chatzidiamantis, N.D.: New Results for Pearson Type III Family of Distributions and Application in Wireless Power Transfer, IEEE Internet of Things Journal, 9, 24038–24050 (2022)
Timm, N.H.: Applied Multivariate Analysis. Springer Texts in Statistics. Springer, New York, NY. (2002)
Tukey, J.W., Wilks, S.S.: Approximation of the distribution of the product of beta variables by a single beta variable. Ann. Math. Statist. 17 318–324 (1946)
Wald, A., Brookner, R.J.: On the distribution of Wilks' statistic for testing the independence of several groups of variates. Ann. Math. Statist. 12, 137–152 (1941)
Wang, K.: Performance Analysis for Cooperative Wireless Communications. Ph.D. Thesis, University of Warwick, Warwick, UK (2014)
Wang, X., Liang, Q.: On the ergodic capacity of cooperative-diversity networks with decode-and-forward relaying over Nakagami-m fading channels. 2013 IEEE Global Communications Conference (GLOBECOM), Atlanta, GA, USA, pp. 3832–3837 (2013)
Wang, K., Wang, T., Chen, Y., Alouini, M.-S.: Statistics of $\alpha - \mu$ Random Variables and Their Applications in Wireless Multihop Relaying and Multiple Scattering Channels, IEEE Transactions on Vehicular Technology, 64, 2754–2759 (2015)
Wilks, S.S.: On the independence of $k$ sets of normally distributed statistical variables. Econometrica 3, 309–326 (1935)
Wilks, S.S.: Multidimensional statistical scatter. In: Contributions to Probability and Statistics: Essays in Honor of Harold Hotelling. Olkin, I., Ghurye, S.G., Hoeffding, W., Madow, W.G., Mann, H.B., Stanford University Press, Stanford, California, pp. 486–503 (1960)
Wilks, S.S.: Multivariate Statistical Outliers. Sankhyā, Ser. A 25, 407–426 (1963)

# The Generalized Integer Gamma Distribution – A Basis for Distributions in Multivariate Statistics

**Carlos A. Coelho**

**Abstract** In this paper the distribution of the sum of independent $\Gamma$ variables with different scale parameters is obtained by direct integration, without involving a series expansion. One of its particular cases is the distribution of the product of some particular independent Beta variables. Both distributions are obtained in a concise manageable form readily applicable to research work in the area of multivariate statistics distributions. The exact distribution of the generalized Wilks' $\Lambda$ statistic is then obtained as a direct application of the results.

**Keywords** Independent Gamma variables · Different scale parameters · Integer shape parameters · Independent Beta variables · Wilks' Lambda · Likelihood ratio statistic

**AMS 1991 Subject Classification Numbers** 62E15, 62H10

---

Reprinted from Journal of Multivariate Analysis, with permission from Elsevier.
First published as Coelho, C. A.: The Generalized Integer Gamma Distribution – a Basis for Distributions in Multivariate Statistics. Journal of Multivariate Analysis, 64, 86–102 (1998). (https://doi.org/10.1006/jmva.1997.1710)

---

C. A. Coelho (✉)
Universidade Técnica de Lisboa, Lisbon, Portugal

Current Address: Mathematics Department, NOVA School of Science and Technology, NOVA University of Lisbon, Caparica, Portugal
e-mail: cmac@fct.unl.pt

 
B. C. Arnold et al. (eds.), *Likelihood Ratio Tests and Related Topics in Multivariate Analysis*, Contributions to Statistics,
https://doi.org/10.1007/978-3-031-68008-3_3

## 1 Introduction

The distribution of the sum of independent Gamma random variables with different scale parameters and the distribution of the product of particular independent Beta random variables has been studied by a few authors. All the results have been obtained under the form of series expansions. Indeed, Kabe (1962) by inverting the characteristic function of a linear combination with all positive coefficients of independent Gamma variates, obtained its distribution in terms of a hypergeometric series. Tretter and Walster (1975) and Nandi (1977) expressed the distribution of the product of particular independent Beta variates as a mixture of incomplete Beta distributions, involving a series representation. Gupta and Richards (1979) by expanding the exponential term and using term by term integration presented the distribution of the sum of independent Gamma variables with different parameters also as a series expansion.

Let

$$Y_i \sim \Gamma\left(r_i, \lambda_i\right), \quad r_i, \lambda_i > 0$$

be a short notation for the fact that the probability density function (p.d.f.) of $Y_i$ is a Gamma distribution with shape parameter $r_i$ and scale parameter $\lambda_i$, this is

$$f_{Y_i}\left(y_i\right) = \frac{\lambda_i^{r_i}}{\Gamma\left(r_i\right)} y_i^{r_i - 1} e^{-\lambda_i y_i}, \quad y_i > 0, \quad r_i, \lambda_i > 0\,. \tag{1}$$

Now let $Y_1, \ldots, Y_n$ be independent random variables having distributions given by (1), with $\lambda_1 = \cdots = \lambda_n = \lambda$, and let $Y = Y_1 + \cdots + Y_n$. Then it is a well known fact that

$$Y \sim \Gamma(r, \lambda)$$

with $r = r_1 + \cdots + r_n$.

In this paper we will obtain the distribution of $Y = Y_1 + \cdots + Y_n$ under the situation where all the $\lambda_i$'s are different, without involving a series expansion, as long as all the $r_i'$ s are integer. We will call such distribution a Generalized Integer (GI) Gamma distribution and after some simplifications it is presented in a concise and easily manageable form. Not only is this distribution a generalization of the Gamma distribution, but also one of its particular cases is the distribution of the logarithm of the product of an even number of particular independent Beta random variables. Therefore such particular GI Gamma distribution may be used to obtain the distributions of a number of multivariate statistics.

## 2 A Preliminary Result

A key result for the developments ahead is the following.

*Result 1* For integer $p$ and $q$, and for $z > 0$

$$
\begin{aligned}
&\int_0^z (z-x)^{q-1} x^{p-1} e^{-kx}\, dx \\
&\qquad = (-1)^q (q-1)! \left( \sum_{j=0}^{p-1} \frac{(p-1)!}{j!} a_{p-j,q} z^j k^{j-q-p+1} \right) e^{-kz} \\
&\qquad\qquad + (p-1)! \sum_{j=0}^{q-1} (-1)^{q-1-j} \frac{(q-1)!}{j!} a_{q-j,p} z^j k^{j-q-p+1} \\
&\qquad = (p-1)!(q-1)! \left\{ (-1)^q \left( \sum_{j=1}^{p} \frac{a_{p-j+1,q}}{(j-1)!} z^{j-1} k^{j-q-p} \right) e^{-kz} \right. \\
&\qquad\qquad \left. + (-1)^p \left( \sum_{j=1}^{q} \frac{a_{q-j+1,p}}{(j-1)!} z^{j-1} (-k)^{j-p-q} \right) \right\} \quad (\text{for } k \neq 0) \\
&\qquad \left( = \frac{z^{p+q-1}}{a_{p,q}(p+q-1)} \quad \text{for } k = 0 \right)
\end{aligned}
$$

where

$$
\begin{aligned}
a_{j,p} &= \sum_{i_p=1}^{j} \sum_{i_{p-1}=1}^{i_p} \cdots \sum_{i_2=1}^{i_3} \sum_{i_1=1}^{i_2} 1 = \sum_{i=1}^{j} a_{i_i,p-1} = a_{p,j} \\
&= \binom{j+p-2}{p-1} = \binom{j+p-2}{j-1}, \qquad j, p \geqslant 1
\end{aligned} \tag{2}
$$

with

$$
a_{i,1} = a_{1,i} = 1
$$

for all $i \geq 1$. □

The above result is obtained on integrating successively by parts.

## 3 Two Useful Distributions

Using Result 1 we may easily obtain the distribution of the sum of independent Gamma random variables with different scale parameters and integer shape parameters.

We have then the following theorem.

**Theorem 1** *Let*

$$Y_i \sim \Gamma(r_i, \lambda_i) \quad i = 1, 2$$

*be two independent Gamma random variables.*

*Then, if $r_i$ $(i = 1, 2)$ are integer, the p.d.f. of $Z = Y_1 + Y_2$ is, for $z > 0$*

$$f_Z(z) = \left(\prod_{i=1}^{2} \lambda_i^{r_i}\right) \sum_{i=1}^{2} (-1)^{S-r_i} \left(\sum_{j=1}^{r_i} \frac{a_{r_i-j+1,S-r_i}}{(j-1)!} z^{j-1} (2\lambda_i - L)^{j-S}\right) e^{-\lambda_i z}$$
$$(\text{for } \lambda_1 \neq \lambda_2)$$
$$\left(= \frac{\lambda^S}{\Gamma(S)} z^{S-1} e^{-\lambda z} \ \text{ for } \lambda_1 = \lambda_2 = \lambda\right), \tag{3}$$

*where*

$$S = r_1 + r_2, \quad L = \lambda_1 + \lambda_2$$

*and the $a_{r_i-j+1,s-r_i}$ are defined as in (2).*

*Proof* From (1), the distribution of $Z = Y_1 + Y_2$, for $\lambda_1 \neq \lambda_2$, taking into account the independence of $Y_1$ and $Y_2$, and Result 1, is

$$f_Z(z) = \int_0^z \frac{\lambda_1^{r_1} \lambda_2^{r_2}}{\Gamma(r_1)\Gamma(r_2)} y_1^{r_1-1} e^{-\lambda_1 y_1} (z - y_1)^{r_2-1} e^{-\lambda_2(z-y_1)} dy_1$$

$$= \frac{\lambda_1^{r_1} \lambda_2^{r_2} e^{-\lambda_2 z}}{(r_1-1)!\,(r_2-1)!} \int_0^z (z - y_1)^{r_2-1} y_1^{r_1-1} e^{-(\lambda_1-\lambda_2)y_1} dy_1$$

$$= \lambda_1^{r_1} \lambda_2^{r_2} \left\{ (-1)^{r_2} \left(\sum_{j=1}^{r_1} \frac{a_{r_1-j+1,r_2}}{(j-1)!} z^{j-1} (\lambda_1 - \lambda_2)^{j-r_1-r_2}\right) e^{-\lambda_1 z} \right.$$

$$\left. + (-1)^{r_1} \left(\sum_{j=1}^{r_2} \frac{a_{r_2-j+1,r_1}}{(j-1)!} z^{j-1} (\lambda_2 - \lambda_1)^{j-r_2-r_1}\right) e^{-\lambda_2 z} \right\}. \quad ∎$$

**Corollary 1** *Let*

$$Y_i \sim \Gamma(r_i, \lambda_i), \quad i = 1, \ldots, g \geqslant 2,$$

*be g independent Gamma random variables.*

*Then, if $r_i$ $(i = 1, \ldots, g)$ are integer, the p.d.f. of $Z = Y_1 + \cdots + Y_g$ is, for $z > 0$,*

$$\begin{aligned} f_Z(z) = K_i^g \sum_{i=1}^{g} \Bigg( S_i \sum_{j_1=1}^{r_i} a_{r_i-j_1+1, r_1^{*i}} \left(\lambda_i - \lambda_1^{*i}\right)^{j_1-r_i-r_1^{*i}} \\ \sum_{j_2=1}^{j_1} a_{j_1-j_2+1, r_2^{*}} \left(\lambda_i - \lambda_2^{*i}\right)^{j_2-j_1-r_2^{*i}} \\ \cdots \sum_{j_{g-1}=1}^{j_{g-2}} \frac{a_{j_{g-2}-j_{g-1}+1, r_{g-1}^{*i}}}{(j_{g-1}-1)!} \left(\lambda_i - \lambda_{g-1}^{*i}\right)^{j_{g-1}-j_{g-2}-r_{g-1}^{*i}} z^{j_{g-1}-1} \Bigg) e^{-\lambda_i z} \\ \left(\text{for } \lambda_i \neq \lambda_{i'}, i, i' \in \{1, \ldots, g\}, i \neq i'\right) \\ \left( = \frac{\lambda^S}{\Gamma(S)} z^{S-1} e^{-\lambda z} \quad \textit{for } \lambda_1 = \cdots = \lambda_g = \lambda \right) \end{aligned} \tag{4}$$

*where $f_Z(z)$ is $f_Z\left(z; \lambda_1, \ldots, \lambda_g; r_1, \ldots, r_g\right)$, the coefficients $a_{j,r^*}$ are given by (2),*

$$K_i^g = \prod_{i=1}^{g} \lambda_i^{r_i}, \quad S_i = (-1)^{S-r_i}, \quad S = \sum_{i=1}^{g} r_i \tag{5}$$

*$\lambda_j^{*i}$ is the jth element of the set $\{\lambda_1, \ldots, \lambda_g\} \setminus \{\lambda_i\}$ and similarly $r_j^{*i}$ is the jth element of the set $\{r_1, \ldots, r_g\} \setminus \{r_i\}$, where "\" denotes set difference. Equivalently $\lambda_j^{*i}$ and $r_j^{*i}$ may be defined as*

$$\lambda_j^{*i} = \begin{cases} \lambda_j & i > j \\ \lambda_{j+1} & i \leqslant j \end{cases} \quad r_j^{*i} = \begin{cases} r_j & i > j \\ r_{j+1} & i \leqslant j \end{cases} \quad i = 1, \ldots, g; \quad j = 1, \ldots, g-1 .$$

*Proof* Theorem 1 enables us to obtain the distribution of $Z_1 = Y_1 + Y_2$. Then applying Result 1 to the joint distribution of $Z_1$ and $Y_3$ we get the distribution of $Z_2 = Z_1 + Y_3 = Y_1 + Y_2 + Y_3$, and so on. For $\lambda_1 = \cdots = \lambda_g = \lambda$ the result is known. ▮

Distributions (3) and (4) may be seen as generalizations of the common Gamma distribution. Indeed, if $\lambda_1 = \lambda_2$ in (3) or $\lambda_1 = \cdots = \lambda_g$ in (4) then we have a common Gamma distribution and if $r_1 = r_2 = 1$ in (3) or $r_1 = \cdots = r_g = 1$ in (4) we have the sum of independent exponentials. So we may call the distributions (3) and (4) as Generalized Integer (GI) Gamma distributions.

As a matter of fact further simple changes in the summation order yield distribution (4) under the form

$$f_Z(z) = K_i^g \sum_{i=1}^{g} P_i(z) e^{-\lambda_i z} \quad (z > 0) \tag{6}$$

where $P_i(z)$ is a polynomial of degree $r_i - 1$ in $z$ which may be written as

$$P_i(z) = \sum_{k=1}^{r_i} c_{i,k}(g, \underline{r}, \underline{\lambda}) z^{k-1} \tag{7}$$

where

$$c_{i,k}(g, \underline{r}, \underline{\lambda}) = S_i \sum_{j_1=k}^{r_i} \sum_{j_2=k}^{j_1} \cdots \sum_{j_{g-2}=k}^{j_{g-3}} \alpha_{j_1} \beta_{j_2} \cdots \gamma_{j_{g-2}} \delta_k \tag{8}$$

with

$$\begin{aligned} \underline{r} &= \left(r_1, \ldots, r_g\right)' \\ \underline{\lambda} &= \left(\lambda_1, \ldots, \lambda_g\right)' \end{aligned} \tag{9}$$

and

$$\begin{aligned} \alpha_{j_1} &= a_{r_i - j_1 + 1, r_1^{*i}} \left(\lambda_i - \lambda_1^{*i}\right)^{j_1 - r_i - r_1^{*i}} \\ \beta_{j_2} &= a_{j_1 - j_2 + 1, r_2^{*i}} \left(\lambda_i - \lambda_2^{*i}\right)^{j_2 - j_1 - r_2^{*i}} \\ \gamma_{j_{g-2}} &= a_{j_{g-3} - j_{g-2} + 1, r_{g-2}^{*i}} \left(\lambda_i - \lambda_{g-2}^{*i}\right)^{j_{g-2} - j_{g-3} - r_{g-2}^{*i}} \\ \delta_k &= \frac{a_{j_{z-2} - k + 1, r_{g-1}^{*i}}}{(k-1)!} \left(\lambda_i - \lambda_{g-1}^{*i}\right)^{k - j_{g-2} - r_{g-1}^{*i}} . \end{aligned} \tag{10}$$

Expression (6) is easily obtained from (4) by simply interchanging the summations. We may note that, using the notations in (10) and $k$ for $j_{g-1}$, distribution (4) may be written as

$$
\begin{aligned}
f_Z(z) &= K_i^g \sum_{i=1}^{g}\left( S_i \sum_{j_1=1}^{r_i} \alpha_{j_1} \sum_{j_2=1}^{j_1} \beta_{j_2} \cdots \sum_{j_{g-2}=1}^{j_{g-3}} \gamma_{j_{g-2}} \sum_{k=1}^{j_{g-2}} \delta_k z^{k-1} \right) e^{-\lambda_i z} \\
&= K_i^g \sum_{i=1}^{g}\left( S_i \sum_{j_1=1}^{r_i} \sum_{j_2=1}^{j_1} \cdots \sum_{j_{g-2}=1}^{j_{g-3}} \sum_{k=1}^{j_{g-2}} \alpha_{j_1}\beta_{j_2}\cdots\gamma_{j_{g-2}}\delta_k z^{k-1} \right) e^{-\lambda_i z} \\
&= K_i^g \sum_{i=1}^{g}\left\{ \sum_{k=1}^{r_i} \underbrace{\left( S_i \sum_{j_1=k}^{r_i} \sum_{j_2=k}^{j_1} \cdots \sum_{j_{g-2}=k}^{j_{g-3}} \alpha_{j_1}\beta_{j_2}\cdots\gamma_{j_{g-2}}\delta_k \right)}_{c_{i,k}} z^{k-1} \right\} e^{-\lambda_i z}.
\end{aligned}
$$

Expression (8) is quite easy to compute but may become a bit long even for moderately large values of $r_i$, given the nesting of the summations. However, a little algebra will allow us to get a much faster and easier way to compute the coefficients $c_{i,k}(g, \underline{r}, \underline{\lambda})$ $(k = 1, \ldots, r_i)$, so that in many cases the coefficients may even be computed by hand.

From (8) we can see that, for a given $i$, the easiest coefficient to compute is the one associated with the highest degree in the polynomial, which may be given by

$$
c_{i,r_i}(g, \underline{r}, \underline{\lambda}) = \frac{1}{(r_i - 1)!} \prod_{\substack{j=1 \\ j \neq i}}^{g} \left(\lambda_j - \lambda_i\right)^{-r_j} . \tag{11}
$$

Then, for $k = 1, \ldots, r_i - 1$, we have

$$
c_{i,r_i-k}(g, \underline{r}, \underline{\lambda}) = \frac{1}{k} \sum_{j=1}^{k} \frac{(r_i - k + j - 1)!}{(r_i - k - 1)!} R(j-1, i, g, \underline{r}, \underline{\lambda})\, c_{i,r_i-(k-j)}(g, \underline{r}, \underline{\lambda}), \tag{12}
$$

where

$$
R(n, j, g, \underline{r}, \underline{\lambda}) = \sum_{\substack{i=1 \\ i \neq j}}^{g} r_i \left(\lambda_j - \lambda_i\right)^{-n-1}, \quad (n = 0, \ldots, r_i - 1). \tag{13}
$$

Using the above notation it is then very easy to get the cumulative distribution function (c.d.f.) of $Z$. For non-negative integer $j$ and real $\lambda$, we obtain

$$\int_0^w z^j e^{-\lambda z} dz = \frac{j!}{\lambda^{j+1}} \left\{ 1 - \left( \sum_{i=0}^{j} \frac{\lambda^i w^i}{i!} \right) e^{-\lambda w} \right\}, \quad w > 0 \tag{14}$$

on integrating by parts. Then, from (6), (7) and (14), the c.d.f. of $Z$ is

$$F_Z(z) = K_i^g \sum_{i=1}^{g} P_i^*(z) \quad (z > 0) \tag{15}$$

where $P_i^*(z)$ is a polynomial of degree $r_i - 1$ in $z$, with

$$P_i^*(z) = \sum_{k=1}^{r_i} c_{i,k}(g, \underline{r}, \underline{\lambda}) \frac{(k-1)!}{\lambda_i^k} \left\{ 1 - \left( \sum_{j=0}^{k-1} \frac{\lambda_i^j z^j}{j!} \right) e^{-\lambda_i z} \right\}. \tag{16}$$

Expressions (15) and (16) show that $F_Z(z)$ may be seen as a generalization of the Incomplete Gamma function, as one would expect.

Indeed, the expression for the c.d.f. of $Z$ may be further simplified since from (15) and (16)

$$F_Z(z) = K_i^g \sum_{i=1}^{g} \sum_{k=1}^{r_i} c_{i,k}(g, \underline{r}, \underline{\lambda}) \frac{(k-1)!}{\lambda_i^k} - K_i^g \sum_{i=1}^{g} e^{-\lambda_i z} \sum_{k=1}^{r_i} c_{i,k}(g, \underline{r}, \underline{\lambda})(k-1)! \sum_{j=0}^{k-1} \frac{z^j}{j! \, \lambda_i^{k-j}}$$

where, after some algebraic manipulation we may show that

$$\sum_{i=1}^{g} \sum_{k=1}^{r_i} c_{i,k}(g, \underline{r}, \underline{\lambda}) \frac{(k-1)!}{\lambda_i^k} = \prod_{i=1}^{g} \lambda_i^{-r_i} = \left(K_i^g\right)^{-1}.$$

Thus $F_z(z)$ may be further written as

$$\begin{aligned} F_Z(z) &= 1 - K_i^g \sum_{i=1}^{g} e^{-\lambda_i z} \sum_{k=1}^{r_i} c_{i,k}(g, \underline{r}, \underline{\lambda})(k-1)! \sum_{j=0}^{k-1} \frac{1}{j!} \frac{z^j}{\lambda_i^{k-j}} \\ &= 1 - K_i^g \sum_{i=1}^{g} P_i^{**}(z) e^{-\lambda_i z} \quad (z > 0) \end{aligned} \tag{17}$$

where

$$P_i^{**}(z) = \sum_{k=1}^{r_i} c_{i,k}(g, \underline{r}, \underline{\lambda})(k-1)! \sum_{j=0}^{k-1} \frac{1}{j!} \frac{z^j}{\lambda_i^{k-j}} \tag{18}$$

is a polynomial of degree $r_i - 1$ in $z$.

Now $F_Z(z)$ is clearly shown to be a generalization of the Incomplete Gamma Function. In some cases, as we will see in Theorem 2, the distribution of $U = e^{-z}$ may be also sought. From the p.d.f. of $Z$ we can readily obtain the p.d.f. of $U$, through the transformation of variable $Z = -\log(U)$ since the Jacobian of the transformation is $1/u$, then

$$f_U(u) = K_i^g \sum_{i=1}^{g} P_i(-\log u) u^{\lambda_i - 1} \qquad (0 < u < 1) \tag{19}$$

where $P_i(-\log u)$, still given by (7), is now a polynomial of degree $r_i - 1$ in $(-\log u)$ with coefficients $c_{i,k}$.

Then the c.d.f. of $U$ is given by

$$F_U(u) = K_i^g \sum_{i=1}^{g} P_i^{**}(-\log u) u^{\lambda_i} \qquad (0 < u < 1) \tag{20}$$

where $P_i^{**}(-\log u)$, a polynomial of degree $r_i - 1$ in $(-\log u)$, is given by (18).

The above expression for $F_U(u)$ may be obtained either from (19) and (18), using, for non-negative integer $j$ and real $\lambda$,

$$\int_0^w (-\log l)^j l^\lambda dl = w^{\lambda+1} j! \sum_{i=0}^{j} \frac{1}{i!} \frac{(-\log w)^i}{(\lambda+1)^{j-i+1}}, \qquad 0 < w < 1$$

which is easily obtained on integrating by parts or, equivalently, through the change of variable $z = -\log t$

$$\begin{aligned} F_U(u) &= \int_0^u f_U(t)dt = \int_0^u f_{-\log U}(-\log t)\frac{1}{t}dt = \int_u^0 -f_{-\log U}(-\log t)\frac{1}{t}dt \\ &= \int_{-\log u}^{+\infty} f_Z(z)dz = \int_0^{+\infty} f_Z(z)dz - \int_0^{-\log u} f_Z(z)dz \\ &= 1 - F_Z(-\log u). \end{aligned} \tag{21}$$

Expression (21) clearly shows that it is indeed equivalent to use $Z = -\log U$ or $U = e^{-Z}$ to carry out tests. For example, using (21) above, we see that the

$\alpha$-percentile of $U$ is equal to the exponential of the symmetrical of the $(1-\alpha)$-percentile of $Z$, and conversely, the $\alpha$-percentile of $Z$ is equal to the symmetrical of the logarithm of the $(1-\alpha)$-percentile of $U$.

A particular and important GI Gamma distribution arises in the following situation.

**Theorem 2** *Let $p = 2m$ be an even integer, with $m \geqslant 1$, and*

$$Y_j \sim B\left(a_j, \frac{b}{2}\right) \qquad j = 1, \ldots, p$$

*be independent random variables with Beta distributions, where $b$ is a positive integer and $a_j = k - j/2$ $(j = 1, \ldots, p)$, with $k > p/2$. Further let*

$$W' = \prod_{j=1}^{p} Y_j$$

*and*

$$W = -\log W' = -\sum_{j=1}^{p} \log Y_j\,.$$

*Then $W$ is the sum of $p+b-2$ independent Gamma distributions with parameters $\lambda_j = k + (j-p-1)/2$ $(j = 1, \ldots, p+b-2)$ and integer parameters $r_j$ given by*

$$r_j = \begin{cases} h_j & j = 1, 2 \\ r_{j-2} + h_j & j = 3, \ldots, p+b-2 \end{cases} \tag{22}$$

*where*

$$h_j = (\textit{number of elements of } \{p, b\} \textit{ greater or equal to } j) - 1\,. \tag{23}$$

*Thus the p.d.f. and c.d.f. of W are*

$$f_W(w) = K_j^{p+b-2} \sum_{j=1}^{p+b-2} P_j(w) e^{-\lambda_j w} \qquad (w > 0) \tag{24}$$

*and*

$$F_W(w) = 1 - K_j^{p+b-2} \sum_{j=1}^{p+b-2} P_j^{**}(w) e^{-\lambda_j w} \qquad (w > 0)\,, \tag{25}$$

*with $K_j^{p+b-2}$ given by (5) and $P_j$ and $P_j^{**}$ given by (7) and (18) respectively.*

*The p.d.f. and c.d.f. of $W'$ are thus*

$$f_{W'}(w) = K_j^{p+b-2} \sum_{j=1}^{p+b-2} P_j(-\log w)\, w^{\lambda_j - 1} \qquad (0 < w < 1) \tag{26}$$

*and*

$$F_{W'}(w) = K_j^{p+b-2} \sum_{j=1}^{p+b-2} P_j^{**}(-\log w)\, w^{\lambda_j} \qquad (0 < w < 1)\,. \tag{27}$$

*Here, the coefficients $c_{j,k}$ in the definitions of $P_j$ and $P_j^{**}$ are given by*

$$\begin{aligned} c_{j,r_j}(g,\underline{r}) &= \frac{2^{S-r_j}}{(r_j-1)!\prod_{i=1,i\neq j}^{g}(i-j)^{r_i}} \\ &= \frac{(-1)^{r_j+1}2^{S-r_j}}{(r_j-1)!\prod_{i=1}^{j-2}(i-j)^{r_i}\prod_{i=j+2}^{g}(i-j)^{r_i}} \end{aligned} \tag{28}$$

*and, for $k = 1, \ldots, r_j - 1$,*

$$c_{j,r_j-k}(g,\underline{r}) = \frac{1}{k}\sum_{i=1}^{k}\frac{(r_j-k+i-1)!}{(r_j-k-1)!}R(i-1,j,g,\underline{r})c_{j,r_j-(k-i)}(g,\underline{r})\,, \tag{29}$$

*where*

$$R(n,i,g,\underline{r}) = \sum_{\substack{j=1\\ j\neq i}}^{g} r_j\left(\frac{2}{i-j}\right)^{n+1} \qquad \left(n = 0,\ldots,r_j-1\right), \tag{30}$$

*with $\underline{r}$ as in (9).*

*In (28), $S = r_1 + \cdots + r_g = pb/2$ is the number of exponentials that incorporate the distribution of $W$.*

*Proof* Given the independence of the $p$ Beta random variables, the $h$th moment of $W'$ is

$$E\left(W'^h\right) = \prod_{j=1}^{p}\frac{\Gamma\left(a_j+h\right)\Gamma\left(a_j+(b/2)\right)}{\Gamma\left(a_j+(b/2)+h\right)\Gamma\left(a_j\right)}\,, \tag{31}$$

and thus the characteristic function of $W$ is

$$E\left(e^{itW}\right) = E\left(e^{-it\log W'}\right) = E\left(W'^{-it}\right) = \prod_{j=1}^{p} \frac{\Gamma\left(a_j - it\right)\Gamma\left(a_j + (b/2)\right)}{\Gamma\left(a_j + (b/2) - it\right)\Gamma\left(a_j\right)},$$

where $i = (-1)^{1/2}$ and $t$ is a real constant. But then, given that $p = 2m$ is an even integer, using the duplication formula for the Gamma function

$$\Gamma(2z) = \pi^{-1/2}2^{2z-1}\Gamma(z)\,\Gamma\left(z + \frac{1}{2}\right)$$

we may write,

$$\begin{aligned}\prod_{j=1}^{p}\Gamma\left(a_j - it\right) &= \prod_{j=1}^{m}\Gamma\left(k - \frac{2j-1}{2} - it\right)\Gamma\left(k - \frac{2j}{2} - it\right)\\ &= \prod_{j=1}^{m}\Gamma\left(k - \frac{2j}{2} + \frac{1}{2} - it\right)\Gamma\left(k - \frac{2j}{2} - it\right)\\ &= \pi^{m/2}\prod_{j=1}^{m}\frac{1}{2^{2(k-j)-2it-1}}\Gamma(2k - 2j - 2it)\end{aligned}$$

and similarly

$$\prod_{j=1}^{p}\Gamma\left(a_j + \frac{b}{2}\right) = \pi^{m/2}\prod_{j=1}^{m}\frac{1}{2^{2(k-j)+b-1}}\Gamma(2k - 2j + b)$$

$$\prod_{j=1}^{p}\Gamma\left(a_j + \frac{b}{2} - it\right) = \pi^{m/2}\prod_{j=1}^{m}\frac{1}{2^{2(k-j)+b-2it-1}}\Gamma(2k - 2j + b - 2it)$$

$$\prod_{j=1}^{p}\Gamma\left(a_j\right) = \pi^{m/2}\prod_{j=1}^{m}\frac{1}{2^{2(k-j)-1}}\Gamma(2k - 2j)\,.$$

Then

$$\begin{aligned}E\left(e^{itW}\right) &= \prod_{j=1}^{m}\frac{2^{2(k-j)+b-2it-1}2^{2(k-j)-1}}{2^{2(k-j)-2it-1}2^{2(k-j)+b-1}}\,\frac{\Gamma(2k-2j-2it)\Gamma(2k-2j+b)}{\Gamma(2k-2j+b-2it)\Gamma(2k-2j)}\\ &= \prod_{j=1}^{m}\frac{\Gamma(2k-2j+b)/\Gamma(2k-2j)}{\Gamma(2k-2j+b-2it)/\Gamma(2k-2j-2it)}\end{aligned}$$

which, using for all real or complex $a$ and integer $n$,

$$\frac{\Gamma(a+n)}{\Gamma(a)} = \prod_{i=0}^{n-1}(a+i)\,,$$

gives

$$\begin{aligned}
E\left(e^{itW}\right) &= \prod_{j=1}^{m}\prod_{i=0}^{b-1}(2k-2j+i)(2k-2j+i-2it)^{-1} \\
&= \prod_{j=1}^{m}\prod_{i=0}^{b-1}\left(\frac{2k-2j+i}{2}\right)\left(\frac{2k-2j+i}{2}-it\right)^{-1} \\
&= \prod_{j=1}^{2m+b-2}\left(k+\frac{j-2m-1}{2}\right)^{r_j}\left(k+\frac{j-2m-1}{2}-it\right)^{-r_j}
\end{aligned} \tag{32}$$

where $r_j$ are integers given by (22), with $h_j$ defined as

$$h_j = \begin{cases} 1 & j=1,\ldots,\min(2m,b) \\ 0 & j=1+\min(2m,b),\ldots,\max(2m,b) \\ -1 & j=1+\max(2m,b),\ldots,2m+b-2 \end{cases}$$

or equivalently defined as in (23).

Expression (32) shows that $W$ is the sum of $S = mb = pb/2$ independent exponentials with parameters $k - j + i/2$ $(j = 1,\ldots,m;\, i = 0,\ldots,b)$ or the sum of $2m + b - 2 = p + b - 2$ independent Gamma random variables with parameters $r_j$ and $\lambda_j = k + (j - p - 1)/2$ $(j = 1,\ldots,p + b - 2)$.

This way the p.d.f. of $W$ is a GI Gamma distribution, as defined in (4), with

$$g = p + b - 2$$

$$\lambda_j = k + \frac{j-2m-1}{2}\,.$$

Thus, in order to obtain the p.d.f. and c.d.f. of $W$ in (24) and (25) we just have to replace $g$ by $p+b-2$ and $Z$ by $W$ in (6) and (17). Similarly, the p.d.f. and c.d.f. of $W'$ in (26) and (27) are obtained by replacing $g$ by $p + b - 2$ and $U$ by $W'$ in (19) and (20).

The coefficients $c_{j,k}$, $k = 1,\ldots,r_j$; $j = 1,\ldots,g$, in the polynomials $P_j$ and $P_j^{**}$ are given by (11) through (13), which in this particular case, given that $\lambda_i - \lambda_j = (i - j)/2$, may be written as in (28) through (30). ∎

We may notice that in Theorem 2, $p$ and $b$ are interchangeable as long as they are both even, and that when $b = 1$ the distribution in (26) becomes a Beta distribution.

Theorem 2 may be seen as an extension of the well known result

$$-n\log\{X/(X+Y)\} \sim \chi^2_2$$

(Fujikoshi and Mukaiata, 1993), for two independent random variables $X \sim \chi^2_n$ and $Y \sim \chi^2_2$.

## 4 Application to the Exact Distribution of the (Generalized) Wilks' $\Lambda$

An useful and interesting application of the GI Gamma distribution is in obtaining the distribution of some multivariate statistics. The Wilks' $\Lambda$ statistic is among such statistics.

The Wilks' Lambda (Wilks, 1932, 1935) is a well known statistic used, under a multivariate normal setting, to test in multivariate analysis of variance the existence of overall differences among the level means of a factor, in multivariate regression or canonical analysis to test the equality of the vector of regression parameters or part of it to a given vector or still used to test the independence of $m$ sets of normally distributed variables (Wilks 1932, 1935; Bartlett 1938; Anderson 1958, Chap. 8, 9; Kshirsagar 1972, Chap. 8). Since all the above settings lead to the same test statistic, in order to keep it short we will describe this statistic only under the last setting.

Let $\underline{x}$ be a $p \times 1$ vector of variables with a joint $p$-multivariate normal distribution $N_p(\underline{\mu}, \Sigma)$. Further let $\underline{x}$ be split into $m$ subvectors, the $k$th of which has $p_k$ variables, with $p = p_1 + \cdots + p_m$. Then each subvector $\underline{x}_k$ $(k = 1, \ldots, m)$ has a $p_k$-multivariate normal distribution $N_{p_k}\big(\underline{\mu}_k, \Sigma_{kk}\big)$ and, for a sample of size $n$, the $(2/n)$-th power of the likelihood ratio statistic to test the null hypothesis

$$H_0 : \Sigma = \text{diag}\,(\Sigma_{11}, \ldots, \Sigma_{kk}, \ldots, \Sigma_{mm})\,, \tag{33}$$

i.e., the hypothesis of independence of the $m$ subvectors $\underline{x}_k$, is the Wilks' Lambda,

$$\Lambda = \frac{|A|}{\prod_{k=1}^{m}|A_{kk}|}\,, \tag{34}$$

where $|\cdot|$ stands for the determinant and

$$A = \begin{bmatrix} A_{11} & \cdots & A_{1k} & \cdots & A_{1m} \\ \vdots & & \vdots & & \vdots \\ A_{k1} & \cdots & A_{kk} & \cdots & A_{km} \\ \vdots & & \vdots & & \vdots \\ A_{m1} & \cdots & A_{mk} & \cdots & A_{mm} \end{bmatrix}$$

is either the sample variance-covariance matrix of the $p$ variables in $\underline{x}$ or the Maximum Likelihood Estimator of $\Sigma$.

For $m > 2$ such a test statistic also arises under the setting of the Generalized Canonical Analysis, when the approach proposed by Carroll (1968) is considered (Coelho, 1992, Chap. 4).

A general concise expression for the exact distribution of the (generalized) Wilks' Lambda, without involving any series expansion or unknown coefficients, may then be obtained by direct application of the result in Theorem 2 as long as at most one of the sets of variables has an odd number of variables.

## 4.1 *The General Case of $m \geqslant 2$ Sets of Variables*

For general $m \geqslant 2$ the Wilks' $\Lambda$ statistic in (34) to test the independence of the $m$ sets of variables $\underline{x}_1, \ldots, \underline{x}_m$, may be written as

$$\Lambda = \prod_{k=1}^{m-1} \Lambda_{k(k+1,\ldots m)} \tag{35}$$

where $\Lambda_{k(k+1,\ldots m)}$ stands for the Wilks' $\Lambda$ statistic to test the independence of the set of variables $\underline{x}_k$ and the set formed by the joining of the sets $\left(\underline{x}_{k+1}, \ldots, \underline{x}_m\right)$ (Anderson, 1958, Theorem 9.3.2; Coelho, 1992, Sec. 4.7).

Under the hypothesis of joint multivariate normality of the $m$ sets of variables and the null hypothesis (33), of independence of the $m$ sets of variables, the $m-1$ Wilks' Lambda statistics $\Lambda_{k(k+1,\ldots m)}$ are all independent (Coelho, 1992, Chap. 4). Considering that $\underline{x}_k$ has $p_k$ variables ($k = 1, \ldots, m$), the distribution of $\Lambda_{k(k+1,\ldots,m)}$ is the same as the distribution of $\prod_{j=1}^{p_k} Y_j$ where, for a sample of size $n+1$, with $n \geqslant p_1 + \cdots + p_m$, $Y_j$ are $p_k$ independent Beta random variables with $Y_j \sim B\left((n+1-q_k-j)/2, q_k/2\right)$, using $q_k = p_{k+1} + \cdots + p_m$ (Anderson, 1958, Theorem 9.3.2). Then we have the following Theorem.

**Theorem 3** *For the general case of $m \geqslant 2$ sets of variables, the kth set having $p_k$ variables ($k = 1, \ldots, m$), when at most one of the sets has an odd number of variables, and for a sample of size $n + 1 > p_1 + \cdots + p_m$, the p.d.f. and c.d.f. of*

*$W = -\log \Lambda$, under the null hypothesis (33) are given by*

$$f_W(w) = K_j^g \sum_{j=1}^{g} P_j(w) e^{-\lambda_j w}$$

*and*

$$F_W(w) = 1 - K_j^g \sum_{j=1}^{g} P_j^{**}(w) e^{-\lambda_j w}$$

*respectively, with*

$$g = p - 2 \quad \text{where} \quad p = \sum_{k=1}^{m} p_k \tag{36}$$

*being $K_j^g$ given by (5) and $P_j(w)$ and $P_j^{**}$ defined as in (7) and (18).*
*The p.d.f. and c.d.f. of $\Lambda$ are then given by*

$$f_\Lambda(l) = K_j^g \sum_{j=1}^{g} P_j(-\log l) l^{\lambda_j - 1}$$

*and*

$$F_\Lambda(l) = K_j^g \sum_{j=1}^{g} P_j^{**}(-\log l) l^{\lambda_j} .$$

*The scale parameters are*[1] *$\lambda_j = (n - p + j)/2$ $(j = 1, \ldots, g)$, with $p$ given by (36) above. The shape parameters $r_j$ $(j = 1, \ldots, g)$, are given by*

$$r_j = \begin{cases} h_j & j = 1, 2 \\ r_{j-2} + h_j & j = 3, \ldots, g = p_1 + \cdots + p_m - 2 , \end{cases}$$

*an extension of (22), with the $h_j$'s defined as*

$$h_j = (\textit{number of } p_k \ (k = 1, \ldots, m) \geqslant j) - 1, \quad j = 1, \ldots, g = p - 2 .$$

[1] see author and editors note at the end of the Chapter

*Proof* Given the independence of the $m-1$ Wilks' Lambdas in (35) we have

$$E\left(\Lambda^h\right)=\prod_{k=1}^{m-1} E\left(\Lambda_{k(k+1,\ldots,m)}^h\right).$$

When at most one of the sets has an odd number of variables, without any loss of generality, we may suppose it to be the $m$th set. Then, from the proof of Theorem 2 in Section 3,

$$\begin{aligned}E\left(e^{-it\log\Lambda}\right)=&E\left(\Lambda^{-it}\right)=\prod_{k=1}^{m-1}\prod_{j=1}^{p_k+q_k-2}\left(\frac{n-p_k-q_k+j}{2}\right)^{r_{kj}}\\ &\qquad\qquad\times\left(\frac{n-p_k-q_k+j}{2}-it\right)^{-r_{kj}}\\ =&\prod_{j=1}^{p-2}\left(\frac{n-1-j}{2}\right)^{r_j}\left(\frac{n-1-j}{2}-it\right)^{-r_j}\end{aligned}$$

where $q_k=p_{k+1}+\cdots+p_m$, and $r_{kj}$ $(k=1,\ldots,m-1;\ j=1,\ldots,p_k+q_k-2)$ are defined by

$$r_{kj}=\begin{cases}h_{kj} & j=1,2\\ r_{k,j-2}+h_{kj} & j=3,\ldots,p_k+q_k-2\end{cases} \qquad (37)$$

with

$$h_{kj}=(\text{number of elements in }\{p_k,q_k\}\text{ greater or equal to }j)-1 \qquad (38)$$

and

$$r_j=\sum_{k=1}^{m-1} r_{kj} \qquad (39)$$

with $r_{kj}=0$ if $j>p_k+q_k-2$. Therefore $r_j$ may be defined by (22), where now[2]

$$h_j=\sum_{k=1}^{m-1} h_{kj}=(\text{number of }p_k\ (k=1,\ldots,m)\geqslant j)-1,\quad j=1,\ldots,p-2. \qquad (40)$$

[2] see author and editors note at the end of the Chapter

This shows that the distribution of $-\log \Lambda$ is, for $m \geqslant 2$, the sum of $p-2$ Gamma distributions with scale parameters $\lambda_j = (n-1-j)/2$ $(j=1, \ldots, p-2)$ and shape parameters $r_j$. Thus applying Theorem 2 we obtain the results in this Theorem. ∎

We may note that for $m = 2$ we obtain the distribution of the usual Wilks' Lambda and thus of each of the $m-1$ Wilks' Lambdas $\Lambda_{k(k+1,\ldots,m)}$ $(k = 1, \ldots, m-1)$ on the right hand side of (35). Further, we may also note that the shape parameters $r_j$, in the distribution of the generalized Wilks' Lambda, are the ordered sum of the shape parameters in the distribution of the $m-1$ Wilks' Lambdas $\Lambda_{k(k+1,\ldots,m)}$ $(k = 1, \ldots, m-1)$.

For example, for $m = 3$, with $p_1 = 4$, $p_2 = 4$ and $p_3 = 3$, we have $g = p_1 + p_2 + p_3 - 2 = 9$, and from (35) through (38),

$$\begin{aligned} h &= [h_j] = [2, 2, 2, 1, -1, -1, -1, -1, -1]' \\ \underline{r} &= [r_j] = [2, 2, 4, 3, 3, 2, 2, 1, 1]' \\ r_1 &= [r_{1j}] = [1, 1, 2, 2, 2, 2, 2, 1, 1]' \\ r_2 &= [r_{2j}] = [1, 1, 2, 1, 1]', \end{aligned}$$

where the elements of $\underline{r}_1$ are the shape parameters in the distribution of $\Lambda_{1(2,3)}$, the Wilks' Lambda statistic used to test the independence between the first set and the superset formed by joining the second and third sets of variables, while in $\underline{r}_2$ are the shape parameters in the distribution of $\Lambda_{2(3)}$, the Wilks' Lambda statistic used to test the independence between the second and third sets of variables.

## 5 Discussion

The GI Gamma distribution and the distributions presented in Theorem 2 of Section 3 are of major relevance in obtaining the distribution of several multivariate statistics whose moments are of the form in (31). Based on the above results it is possible to obtain the distributions of these statistics in a general, more concise and manageable form. This will make easier the computation of percentiles and enable us to overcome the problems arising from the use of distributions under a series expansion form.

As an illustration, the null distribution of the Wilks' $\Lambda$ statistic to test the independence of $m$ sets of variables is explicitly obtained in a simple manageable form even for general $m$.

Moreover, the use of the GI Gamma distribution also enables us to get a deeper insight and at the same time have an overall view upon the studies carried out so far on the distributions of a number of multivariate statistics, namely the Wilks' Lambda.

**Acknowledgments** The author thanks and expresses his earnest appreciation for the insightful comments from the editor and a referee.

## References

Anderson, T.W. (1958). *An Introduction to Multivariate Statistical Analysis*, Wiley, New York.

Bartlett, M.S. (1938). Further aspects of the theory of multiple regression. *Proc. Cambridge Philos. Soc.* **34**, 33–40.

Carroll, J.D. (1968). Generalization of canonical correlation analysis to three or more sets of variables. In *Proc., 76 th Annual Convention of the American Psychological Association*, pp. 227–228.

Coelho, C.A. (1992). *Generalized Canonical Analysis*, Ph.D. thesis. The University of Michigan, Ann Arbor, MI.

Fujikoshi, Y., Mukaihata, S. (1993). Approximations for the quantiles of student's $t$ and $F$ distribution and their error bound. *Hiroshima Math. J.* **23**, 557–564.

Gupta, R.D., Richards, D. (1979). Exact distributions of Wilks' $\Lambda$ under the null and non-null (linear) hypotheses. *Statistica* **39**, 333–342.

Kabe, D.G. (1962). On the exact distribution of a class of multivariate test criteria. *Ann. Math. Statist.* **33**, 1197–1200.

Kshirsagar, A.M. (1972). *Multivariate Analysis*. Dekker, New York.

Nandi, S.B. (1977). The exact null distribution of Wilks' criterion. *Sankhyā* **39**, 307–315.

Tretter, M.J., Walster, G.W. (1975). Central and noncentral distributions of Wilks' statistic in Manova as mixtures of incomplete beta functions. *Ann. Statist.* **3**, 467–472.

Wilks, S.S. (1932). Certain generalizations in the analysis of variance. *Biometrika* **24**, 471–494.

Wilks, S.S. (1935). On the independence of $k$ sets of normally distributed statistical variables. *Econometrica* **3**, 309–326.

### *Author and Editors notes:*

[1] the indexation of the scale parameters $\lambda_j$ has actually to be the reverse of the one indicated, that is, it should rather be $\lambda_j = (n-1-j)/2$.

[2] a correction has to be made to the definition of the $h_j$, whose expression is correct, but which are not the sum in $k$ of the $h_{kj}$; the parameters $h_j$ should rather be just defined as

$$h_j = (\text{number of } p_k \ (k=1,\ldots,m) \geqslant j) - 1, \quad j = 1,\ldots,p-2\,.$$

# Addendum to the Paper "The Generalized Integer Gamma Distribution – A Basis for Distributions in Multivariate Statistics"

**Carlos A. Coelho**

**Abstract** A brief remark on the paper "The Generalized Integer Gamma Distribution – A Basis for Distributions in Multivariate Statistics," (1998, *J. Multivariate Anal.* **64**, 86–102) and an additional result concerning the distribution of the product of some particular independent beta random variables, which broadens the scope of the results in that paper, are presented.

**Keywords** Product independent beta variables · Sum independent gamma variables

**AMS 1991 Subject Classifications** 62E15, 62H10

Volume **64**, Number 1 (1998), in Article No. MV971710, in "The Generalized Integer Gamma Distribution—A Basis for Distributions in Multivariate Statistics," by Carlos A. Coelho, pages 86–102: The remark following Theorem 2, although correct, is unnecessarily restrictive when it states that in Theorem 2, $p$ and $b$ are interchangeable as long as they are both even.

---

Reprinted from Journal of Multivariate Analysis, with permission from Elsevier.
First published as Coelho, C. A. Addendum to the paper "The Generalized Integer Gamma Distribution – a Basis for Distributions in Multivariate Analysis". *Journal of Multivariate Analysis* **69**, 281–285 (1999) (https://doi.org/10.1006/jmva.1998.1805)

---

C. A. Coelho (✉)
Departamento de Matemática, Inst. Sup. Agronomia, Universidade Técnica de Lisboa, Lisbon, Portugal

Current Address: Mathematics Department, NOVA School of Science and Technology, NOVA University of Lisbon, Caparica, Portugal
e-mail: cmac@fct.unl.pt



B. C. Arnold et al. (eds.), *Likelihood Ratio Tests and Related Topics in Multivariate Analysis*, Contributions to Statistics,
https://doi.org/10.1007/978-3-031-68008-3_4

As a matter of fact, that paper (Coelho, 1998) should say that $p$ and $b$ are interchangeable as long as at most one of them is odd, that is, if they are both even or one of them is even and the other odd.

This means that the distributions (24) through (27) in Coelho (1998, pp. 94–95) are the distributions (p.d.f. and c.d.f.) of $W$ and $W'$, with

$$W = -\log W' \quad \text{and} \quad W' = \prod_{j=1}^{p} Y_j\,, \tag{1}$$

where

$$Y_j \sim B\left(a_j, \frac{b}{2}\right) \qquad j = 1, \ldots, p \tag{2}$$

are $p$ independent beta random variables with

$$a_j = k - j/2 \;\; (j = 1, \ldots, p) \quad \text{and} \quad k > p/2\,. \tag{3}$$

This is true as long as both $p$ and $b$ are even or if one of them is odd then the other is even.

Although in the proof of Theorem 2 an even $p$ was considered, we note that for even $b$ we have

$$\frac{\Gamma\left(a_j+b/2\right)}{\Gamma\left(a_j\right)} = \prod_{i=0}^{b/2-1} \left(a_j+i\right) \quad \text{and} \quad \frac{\Gamma\left(a_j-it\right)}{\Gamma\left(a_j+b/2-it\right)} = \prod_{i=0}^{b/2-1} \frac{1}{\left(a_j-it+i\right)}$$

so that

$$\begin{aligned} E\left(e^{itW}\right) &= \prod_{j=1}^{p} \prod_{i=0}^{b/2-1} \left(a_j+i\right)\left(a_j-it+i\right)^{-1} \\ &= \prod_{j=1}^{p} \prod_{i=0}^{b/2-1} \left(k-\frac{j}{2}+i\right)\left(k-\frac{j}{2}+i-it\right)^{-1} \\ &= \prod_{j=1}^{p+b-2} \left(k-\frac{p}{2}+\frac{j-1}{2}\right)^{r_j}\left(k-\frac{p}{2}+\frac{j-1}{2}-it\right)^{-r_j}, \end{aligned}$$

which is the same as (32) in the paper, with $r_j$ given by (22) and $h_j$ still defined by

$$h_j = \begin{cases} 1 & j = 1, \ldots, \min(p,b) \\ 0 & j = 1+\min(p,b), \ldots, \max(p,b) \\ -1 & j = 1+\max(p,b), \ldots, p+b-2 \end{cases}$$

so that finally the distributions (24) through (27) are also the distributions of $W = -\log W'$ and $W'$, where $W'$ is the product of any even or odd number $p$ of independent beta random variables with shape parameters $a_j = k - j/2$ $(j = 1, \ldots, p)$ $(k > p/2)$ and integer scale parameter $b/2$, where $b$ is an even integer.

This means that, for example, the p.d.f. in (26) is either the p.d.f. of the product of an even number $p$ of independent beta random variables with parameters $a_j = k - j/2$ $(j = 1, \ldots, p)$ $(k > p/2)$ and $b/2$, where $b$ is any positive integer, or the p.d.f. of the product of any even or odd number of independent beta random variables with parameters $a_j$ and $b/2$, where $b$ is any even positive integer.

The fact that $p$ and $b$ are interchangeable is made more clear if in (3) we make

$$k = l - \frac{b}{2}\left(> \frac{p}{2}\right). \tag{4}$$

In this way, we may write

$$E\left(e^{itW}\right) = \prod_{j=1}^{p+b-2}\left(l - \frac{b}{2} - \frac{p}{2} + \frac{j-1}{2}\right)^{r_j}\left(l - \frac{b}{2} - \frac{p}{2} + \frac{j-1}{2} - it\right)^{-r_j},$$

which shows that the distributions (24) through (27) (Coelho, 1998, pp. 94–95) are also the distributions (p.d.f. and c.d.f.) of $W^*$ and $W^{*\prime}$, with

$$W^* = -\log W^{*\prime} \quad \text{and} \quad W^{*\prime} = \prod_{j=1}^{b} Y_j^*, \tag{5}$$

where

$$Y_j^* \sim B\left(a_j^*, \frac{p}{2}\right) \quad (j = 1, \ldots, b) \tag{6}$$

are $b$ independent beta random variables with

$$a_j^* = k^* - \frac{j}{2} \qquad (j = 1, \ldots, b)$$

where

$$k^* = l - \frac{p}{2}\left(> \frac{b}{2}\right).$$

The condition $k^* > b/2$ comes directly from (4) above, showing that there is no need for any restrictions on the relative magnitude of $p$ and $b$.

In the case where $p$ and $b$ are both odd, the interchangeability of $p$ and $b$ still holds. In this case, following steps similar to those in the Proof of Theorem 2 in Coelho (1998, pp. 95–97), we have for $W$ and $W'$ defined by (1) and (2) above

$$\begin{aligned}
E\left(e^{itW}\right) &= \left[\prod_{j=1}^{p-1} \frac{\Gamma\left(a_j - it\right)\Gamma\left(a_j + (b/2)\right)}{\Gamma\left(a_j + (b/2) - it\right)\Gamma\left(a_j\right)}\right] \frac{\Gamma\left(a_p - it\right)\Gamma\left(a_p + (b/2)\right)}{\Gamma\left(a_p + (b/2) - it\right)\Gamma\left(a_p\right)} \\
&= \left[\prod_{j=1}^{p+b-3} \left(l - \frac{b}{2} - \frac{p}{2} + \frac{j-1}{2}\right)^{r_j} \left(l - \frac{b}{2} - \frac{p}{2} + \frac{j-1}{2} - it\right)^{-r_j}\right] \\
&\quad \times \frac{\Gamma\left(a_p - it\right)\Gamma\left(a_p + (1/2)\right)}{\Gamma\left(a_p\right)\Gamma\left(a_p + (1/2) - it\right)} \prod_{j=0}^{(b-3)/2} \frac{a_p + j}{a_p + j - it} \\
&= \left[\prod_{j=1}^{p+b-3} \left(l - \frac{b}{2} - \frac{p}{2} + \frac{j-1}{2}\right)^{r_j^*} \left(l - \frac{b}{2} - \frac{p}{2} + \frac{j-1}{2} - it\right)^{-r_j^*}\right] \\
&\quad \times \frac{\Gamma\left(l - \frac{b}{2} - \frac{p}{2} - it\right)\Gamma\left(l - \frac{b}{2} - \frac{p}{2} + \frac{1}{2}\right)}{\Gamma\left(l - \frac{b}{2} - \frac{p}{2}\right)\Gamma\left(l - \frac{b}{2} - \frac{p}{2} + \frac{1}{2} - it\right)}
\end{aligned} \tag{7}$$

with

$$r_j^* = \begin{cases} r_j & j \text{ even or } j > \frac{b-1}{2} \\ r_j + 1 & j = 2n-1, n \in \left\{1, \ldots, \frac{b-1}{2}\right\}, \end{cases}$$

where $r_j$ are defined as in Theorem 2. In (7) we considered

$$a_j = k - \frac{j}{2} = l - \frac{b}{2} - \frac{j}{2}$$

and thus

$$a_p = l - \frac{b}{2} - \frac{p}{2}$$

and we used

$$\frac{\Gamma(a+n)}{\Gamma(a)} = \prod_{i=0}^{n-1} (a+i),$$

as well as the fact that since $p$ is odd $p-1$ is even, so that expression (32) in Coelho (1998, p. 96) can be applied.

Expression (7) above shows that the distribution of $W$ is now the same as that of the sum of $p+b-2$ independent random variables, $p+b-3$ of which are gamma random variables with integer shape parameters $r_j^*$ and scale parameters

$$\lambda_l = l - \frac{b}{2} - \frac{p}{2} + \frac{j-1}{2} \qquad (j = 1, \ldots, p+b-3),$$

and the other of which is the logarithm of a beta random variable with parameters $a_p = l - b/2 - p/2$ and $1/2$. This distribution is still also the distribution of $W^*$ defined by (5) and (6) above, but it is no longer a GI gamma distribution and its p.d.f. and c.d.f. are not easy to obtain under a concise manageable form. In order to avoid the use of incomplete gamma or bcta functions one has to use generalized hypergeometric functions. This problem was recently addressed for the particular case of the Wilks $\Lambda$ statistic by Alberto (1998) and is intended to be the subject of a future publication.

When using the Wilks $\Lambda$ statistic to test the independence of two sets of variables, the first set with $p_1$ variables and the second with $p_2$ variables, with a joint $(p_1 + p_2)$-multivariate normal distribution we use a statistic which, according to Anderson (1984) and Kshirsagar (1972), has the same distribution as

$$\prod_{j=1}^{p_1} Y_j \quad \text{or} \quad \prod_{j=1}^{p_2} Y_j^*,$$

where

$$Y_j \sim B\left(\frac{n+1-p_2-j}{2}, \frac{p_2}{2}\right) \quad \text{and} \quad Y_j^* \sim B\left(\frac{n+1-p_1-j}{2}, \frac{p_1}{2}\right).$$

The results presented herein confirm that the distribution of $\prod_{j=1}^{p_1} Y_j$ is the same as the distribution of $\prod_{j=1}^{p_2} Y_j^*$, their p.d.f and c.d.f. being given by (26) and (27) in Coelho (1998, pp. 94–95), when at most one of $p_1$ or $p_2$ is odd. These same results give evidence that no further magnitude relationship needs to hold between $p_1$ and $p_2$.

## References

Alberto, R.P. (1998). *A distribuição exacta da estatística Λ de Wilks*. Master's thesis, Univ. Lusíada, Lisbon.

Anderson, T.W. (1984). *An Introduction to Multivariate Statistical Analysis*, 2nd ed., Wiley, New York.

Coelho, C.A. (1998). The generalized integer gamma distribution – a basis for distributions in multivariate statistics. *J. Multivariate Anal.* **64**, 86–102.

Kshirsagar, A.M. (1972). *Multivariate Analysis*. Dekker, New York.

# The Generalized Near-Integer Gamma Distribution: A Basis for 'Near-Exact' Approximations to the Distribution of Statistics Which Are the Product of an Odd Number of Independent Beta Random Variables

**Carlos A. Coelho**

**Abstract** In this paper the concept of near-exact approximation to a distribution is introduced. Based on this concept it is shown how a random variable whose exponential has a Beta distribution may be closely approximated by a sum of independent Gamma random variables, giving rise to the Generalized Near-Integer (GNI) Gamma distribution. A particular near-exact approximation to the distribution of the logarithm of the product of an odd number of independent Beta random variables is shown to be a GNI Gamma distribution. As an application, a near-exact approximation to the distribution of the generalized Wilks $\Lambda$ statistic is obtained for cases where two or more sets of variables have an odd number of variables. This near-exact approximation gives the exact distribution when there is at most one set with an odd number of variables. In the other cases a near-exact approximation to the distribution of the logarithm of the Wilks Lambda statistic is found to be either a particular Generalized Integer Gamma distribution or a particular GNI Gamma distribution.

---

Reprinted from Journal of Multivariate Analysis, with permission from Elsevier.

First published as Coelho, C. A. The Generalized Near-Integer Gamma distribution – a basis for 'near-exact' approximations to the distributions of statistics which are the product of an odd number of particular independent Beta random variables. *Journal of Multivariate Analysis* **89**, 2, 191–218 (2004). (https://doi.org/10.1016/j.jmva.2003.12.001)

---

C. A. Coelho (✉)
Mathematics Department, Institute of Agronomy, Lisbon University of Technology, Lisboa, Portugal

Current Address: Mathematics Department, NOVA School of Science and Technology, NOVA University of Lisbon, Caparica, Portugal
e-mail: cmac@fct.unl.pt

 
B. C. Arnold et al. (eds.), *Likelihood Ratio Tests and Related Topics in Multivariate Analysis*, Contributions to Statistics,
https://doi.org/10.1007/978-3-031-68008-3_5

**Keywords** Product independent beta variables · Sum independent Gamma variables · Generalized Wilks Lambda · Likelihood ratio statistic

**AMS 2000 Subject Classification** Primary 62E20, Secondary 62H10

## 1 Introduction

It is the aim of this paper to obtain a close approximation to the exact distribution of the product of an odd number of particular independent Beta random variables in a concise manageable form.

Let $Y_j$ $(j = 1, \dots, p)$ represent $p$ independent Beta random variables with parameters $a_j$ and $b/2$, this is, let

$$Y_j \sim B\left(a_j, \frac{b}{2}\right), \qquad j = 1, \dots, p$$

where both $p$ and $b$ are odd integers. We are interested in the distribution of

$$W' = \prod_{j=1}^{p} Y_j \tag{1}$$

or the distribution of

$$W = -\log W' = -\sum_{j=1}^{p} \log Y_j\,, \tag{2}$$

since such distributions are of key importance in Multivariate Statistics, where a number of test statistics used, such as the Wilks $\Lambda$, have a distribution of this type.

However, the exact distributions of $W'$ in (1) and $W$ in (2) have rather complicated forms and expressions, involving either incomplete Beta or Gamma functions or, alternatively, generalized hypergeometric functions, usually rendering the final result not very adequate for further manipulation. Our aim is to obtain a close approximation to this exact distribution by replacing some parts of its characteristic function by adequate approximations with good convergence properties, so chosen that the resulting characteristic function corresponds to a known distribution with manageable c.d.f., as illustrated in the next section. This approach will be based on a distribution that we will call Generalized Near-integer (GNI) Gamma distribution, that is the distribution of the sum of independent Gamma random variables with different rate parameters and shape parameters that are all integer except one. This distribution uses as a basis the Generalized Integer (GI) Gamma distribution (Coelho, 1998), that is the distribution of the sum of independent Gamma random variables with all different rate parameters and integer shape parameters.

Finally we show that a 'near-exact' approximation to the distribution of the logarithm of the generalized Wilks $\Lambda$ statistic (Wilks, 1932, 1935, 1946; Anderson, 1984; Coelho, 1992), when two or more sets of variables have an odd number of variables, is either a particular GI Gamma or a GNI Gamma distribution, according to the cases.

## 2 Some Preliminary Results

Whenever we are able to factor the characteristic function of a random variable into several elementary components and replace one or just a few of those components by an asymptotic approximation, we will call the distribution obtained from that characteristic function, a 'near-exact' approximation. Here 'asymptotic' is used in the sense that the approximation is done in such a way that it will be of a better quality for increasing values of some parameter. This parameter is chosen so that when the distribution under study is used as the distribution of some statistic, increasing values of the parameter are associated with increasing sample sizes.

Using this procedure we obtain a particular type of asymptotic distribution which may be shown to have very good convergence properties (in terms of characteristic function, moments, quantiles and distribution function converging to the exact ones as the 'asymptotic parameter' increases) and which may be seen as lying between the exact and an asymptotic distribution in the usual sense. If we replace all the components of the characteristic function using the same asymptotic result then we will get an asymptotic distribution.

The asymptotic replacement that may be found for one of such components is usually of an even better quality than the one that may be used for the overall characteristic function. Furthermore, if the number of factors of the characteristic function replaced by these asymptotic approximations is kept to a minimum, the resulting distribution will yield a very good approximation to the exact distribution (in terms of moments, quantiles and distribution function itself). One of the aims in obtaining this distribution is that it may be manageable at least in such a way that the computation of quantiles is rendered reasonably easy.

### *2.1 Approximating a LogBeta Distribution with a Sum of Gamma Distributions*

Let us consider the random variable $X$ with a Beta distribution with parameters $a > 0$ and $b > 0$, this is, let

$$X \sim B(a, b)$$

and let $Y = -\log X$. Then $Y$ will have what we call a logBeta distribution with parameters $a$ and $b$, its probability density function (p.d.f.) being

$$f_Y(y) = \frac{1}{B(a,b)}\, e^{-ay}(1-e^{-y})^{b-1}\,.$$

Since we know that the $h$-th moment of $X$ is

$$E\left(X^h\right) = \frac{B(a+h,b)}{B(a,b)} = \frac{\Gamma(a+b)}{\Gamma(a)}\,\frac{\Gamma(a+h)}{\Gamma(a+b+h)} \qquad (a+h>0)$$

the characteristic function of $Y$ is

$$E\left(e^{itY}\right) = E\left(e^{-it\,\log X}\right) = E\left(X^{-it}\right) = \frac{\Gamma(a+b)}{\Gamma(a)}\,\frac{\Gamma(a-it)}{\Gamma(a+b-it)}\,, \tag{3}$$

where $i=\sqrt{-1}$ and $t$ is real. But then, using Result A.2, in Appendix A, we may write

$$E\left(e^{itY}\right) \overset{a}{\simeq} \left(a+\frac{b}{2}-\frac{1}{2}\right)^{b}\left(a+\frac{b}{2}-\frac{1}{2}-it\right)^{-b}\,, \tag{4}$$

which will be a fairly good approximation to the exact characteristic function of $Y$ in (3) for sufficiently large $a$ and $0<b<1$, as stated in Appendix A (and where '$\overset{a}{\simeq}$' has the same meaning as in Appendix A).

It happens that for increasing values of $b>1$, for given $a$ and $t$, the approximation in (4) begins to degrade, but what is interesting is that if $b\geq 1$, we may consider, with $\lfloor\cdot\rfloor$ denoting the floor of the argument,

$$r^* = \lfloor b\rfloor \quad \text{and} \quad r=b-r^*\,, \quad \text{or} \quad b=r^*+r\,, \quad r^*\in\mathbb{N}\,,\ \ 0\leq r<1\,,$$

so that we may always write the characteristic function of $Y$ as

$$\begin{aligned}
E\left(e^{itY}\right) &= \tfrac{\Gamma(a+b)}{\Gamma(a)}\,\tfrac{\Gamma(a-it)}{\Gamma(a+b-it)} = \tfrac{\Gamma(a+r^*+r)}{\Gamma(a)}\,\tfrac{\Gamma(a-it)}{\Gamma(a+r^*+r-it)}\\
&= \tfrac{\Gamma(a+r^*+r)}{\Gamma(a+r^*)}\,\tfrac{\Gamma(a+r^*-it)}{\Gamma(a+r^*+r-it)}\,\tfrac{\Gamma(a+r^*)}{\Gamma(a)}\,\tfrac{\Gamma(a-it)}{\Gamma(a+r^*-it)}\\
&\overset{a+r^*}{\simeq} \left(a+r^*+\tfrac{r}{2}-\tfrac{1}{2}\right)^{r}\left(a+r^*+\tfrac{r}{2}-\tfrac{1}{2}-it\right)^{-r}\\
&\qquad\qquad \times\prod_{j=0}^{r^*-1}(a+j)\,(a+j-it)^{-1}
\end{aligned} \tag{5}$$

where we used Result A.2 in Appendix A, and the fact that being $r^*$ a positive integer, for all real or complex $a$

$$\frac{\Gamma(a+r^*)}{\Gamma(a)} = \prod_{j=0}^{r^*-1}(a+j)\,. \tag{6}$$

Expression (4) shows that if $Y$ has a logBeta distribution with parameters $a$ and $b$, we may obtain as an asymptotic distribution a Gamma distribution with shape parameter $b$ and rate parameter $a + b/2 - 1/2$, by directly applying Result A.2 in Appendix A to its characteristic function. On the other hand (5) shows that, by extracting from the exact distribution of $Y$ the Gamma components with shape parameter equal to one and then applying Result A.2 to the remaining logBeta with parameters $a + r^*$ and $r$, we may obtain what we will call a 'near-exact' approximation to the distribution of $Y$. Such distribution is a sum of $r^*+1$ (with $r^* = \lfloor b \rfloor$) independent Gamma random variables, $r^* = \lfloor b \rfloor$ of which are Exponentials, this is, Gamma random variables with shape parameter 1 and rate parameter $\lambda_j = a - 1 + j$ $(j = 1, \ldots, r^*)$ and a further Gamma random variable with shape parameter $r$ and rate parameter $a + r^* + \frac{r}{2} - \frac{1}{2} = a + b - \frac{r+1}{2}$. This distribution is a particular case of a wider class of distributions, defined in the next section, which will be called Generalized Near-Integer (GNI) Gamma distributions, that are the distributions of the sum of a random variable with a Generalized Integer (GI) Gamma distribution (Coelho, 1998) with an independent Gamma random variable with a non-integer shape parameter.

This latter distribution besides being also asymptotic to the exact distribution, since it yields better approximations to the exact distribution for increasing values of the parameter $a$, which in the distribution of some statistics is an increasing function of the sample size, it also shows high advantage over the asymptotic Gamma distribution as it may be seen through straight computation of either the moments, quantiles or even the distribution function. The approximation it gives also improves for higher values of the parameter $b$, not only because the result is also asymptotic for increasing values of $r^* = \lfloor b \rfloor$ but also because for higher values of this parameter, by following the procedure in (5), the part that is replaced by an asymptotic result will represent a smaller part of the whole.

However, another approach, which will yield yet better results would be to replace the part of the characteristic function we want to approximate asymptotically, and which corresponds to a logBeta distribution, by the characteristic function of a Gamma distribution with the same two first moments, as referred at the end of Appendix A. Although it is too long to show, this approach will also yield an asymptotic result in the sense that the approximation (in terms of the characteristic function, moments, quantiles and distribution function) gets better for increasing values of the parameter $a$, which, as noted above, is an increasing function of the sample size in the distribution of some statistics. Following the approach outlined in the second half of Appendix A, and similar steps to the ones used in the decomposition of the characteristic function of $Y$ in (5) we could write

$$E\left(e^{itY}\right) \overset{a+r^*}{\simeq} \lambda^r(\lambda - it)^{-r} \prod_{j=0}^{r^*-1} (a+j)\,(a+j-it)^{-1}$$

with

$$\lambda = \frac{\Psi(a+b) - \Psi(a+r^*)}{\Psi'(a+r^*) - \Psi'(a+b)} \qquad r = \frac{(\Psi(a+b) - \Psi(a+r^*))^2}{\Psi'(a+r^*) - \Psi'(a+b)}\,,$$

where $\Psi(x) = \frac{d}{dx}\log\Gamma(x)$ is the digamma function and $\Psi'(x) = \frac{d^2}{dx^2}\log\Gamma(x) = \frac{d}{dx}\Psi(x)$ is the trigamma function.

This result which yields an even much better near-exact approximation, with the same two first moments as the exact distribution, was obtained following an insightful and kind lead given by one of the referees, upon a former version of the manuscript.

In the next section the exact p.d.f. and c.d.f. (cumulative distribution function) of a GNI Gamma random variable are obtained in a concise and highly manageable form.

## 3 The Generalized Near-Integer Gamma Distribution

Let $Y$ be a random variable with a GI Gamma distribution (Coelho, 1998) of *depth* $g$, with integer shape parameters $r_1, \ldots, r_g$ and different rate parameters $\lambda_1, .., \lambda_g$, this is, let

$$Y \sim GI\Gamma\left(r_1, \ldots, r_g; \lambda_1, \ldots, \lambda_g\right)$$

and let $X$ be a random variable with a Gamma distribution with a non-integer shape parameter $r > 0$ and rate parameter $\lambda$, this is, let

$$X \sim \Gamma(r, \lambda)\,.$$

Further, let $X$ and $Y$ be independent and let $\lambda \neq \lambda_j$ $(j = 1, \ldots, g)$. We will call the distribution of

$$Z = Y + X$$

a Generalized Near-Integer Gamma distribution of *depth* $g+1$, and we will write

$$Z \sim GNI\Gamma\left(r_1, \ldots, r_g, r; \lambda_1, \ldots, \lambda_g, \lambda\right)\,.$$

The p.d.f. and the c.d.f. of $Z$ are given in the following Theorem.

**Theorem 1** *Let*

$$Z = Y + X$$

*where*

$$Y \sim GI\Gamma\left(r_1, \dots, r_g; \lambda_1, \dots, \lambda_g\right)$$

*and*

$$X \sim \Gamma(r, \lambda) ,$$

*are two independent random variables, with* $\lambda \neq \lambda_j$ $(j = 1, \dots, g)$. *We will say that* $Z$ *has a Generalized Near-Integer (GNI) Gamma distribution. The p.d.f. of* $Z$ *is, in terms of the Kummer confluent hypergeometric function, given by*

$$f_Z(z) = K\,\lambda^r \sum_{j=1}^{g} e^{-\lambda_j z} \sum_{k=1}^{r_j} \left\{ c_{j,k} \frac{\Gamma(k)}{\Gamma(k+r)}\, z^{k+r-1} \; {}_1F_1\big(r, k+r, -(\lambda-\lambda_j)z\big) \right\} \tag{7}$$

*or*

$$f_Z(z) = K\,\lambda^r\, e^{-\lambda z} \sum_{j=1}^{g} \sum_{k=1}^{r_j} \left\{ c_{j,k} \frac{\Gamma(k)}{\Gamma(k+r)}\, z^{k+r-1} \; {}_1F_1\big(k, k+r, (\lambda-\lambda_j)z\big) \right\} \tag{8}$$

*and the c.d.f. by*

$$F_Z(z) = \lambda^r \frac{z^r}{\Gamma(r+1)}\; {}_1F_1(r, r+1, -\lambda z) - K\,\lambda^r \sum_{j=1}^{g} e^{-\lambda_j z} \sum_{k=1}^{r_j} c^*_{j,k} \sum_{i=0}^{k-1} \frac{z^{r+i}\,\lambda_j^i}{\Gamma(r+1+i)}\; {}_1F_1\big(r, r+1+i, -(\lambda-\lambda_j)z\big) \tag{9}$$

*or*

$$F_Z(z) = \lambda^r \frac{z^r}{\Gamma(r+1)}\ {}_1F_1(r, r+1, -\lambda z) - K\,\lambda^r\, e^{-\lambda z}$$
$$\sum_{j=1}^{g}\sum_{k=1}^{r_j} c^*_{j,k} \sum_{i=0}^{k-1} \frac{z^{r+i}\,\lambda_j^i}{\Gamma\,(r+1+i)}\ {}_1F_1\big(i+1, r+1+i, (\lambda-\lambda_j)z\big)\,, \tag{10}$$

*where*

$$K = \prod_{j=1}^{g} \lambda_j^{r_j} \qquad \textit{and} \qquad c^*_{j,k} = \frac{c_{j,k}}{\lambda_j^k}\,\Gamma(k)$$

*with* $c_{j,k}$ *given by* (11)–(13) *in Coelho (1998).*

***Proof*** Given the independence between the two random variables $Y$ and $X$ and the expression for the p.d.f. of the GI Gamma distribution given by (6) and (7) in Coelho (1998) we may write

$$\begin{aligned} f_Z(z) &= \int_0^z K \frac{\lambda^r}{\Gamma(r)} \sum_{j=1}^{g}\sum_{k=1}^{r_j} c_{j,k}\,(z-x)^{k-1}\, e^{-\lambda_j(z-x)} x^{r-1} e^{-\lambda x}\, dx \\ &= K \frac{\lambda^r}{\Gamma(r)} \sum_{j=1}^{g} e^{-\lambda_j z} \sum_{k=1}^{r_j} c_{j,k} \int_0^z (z-x)^{k-1}\, x^{r-1} e^{-(\lambda-\lambda_j)x}\, dx \\ &= K\,\lambda^r \sum_{j=1}^{g} e^{-\lambda_j z} \sum_{k=1}^{r_j} \Bigg\{ c_{j,k}\, \frac{\Gamma(k)}{\Gamma(k+r)}\, z^{k+r-1} \\ &\qquad\qquad {}_1F_1\big(r, k+r, -(\lambda-\lambda_j)z\big)\Bigg\}, \end{aligned} \tag{11}$$

that is expression (7) and where Result B.1 in Appendix B was used in executing the last step. In (11) we may use 13.1.27 in Abramowitz and Stegun (1974) that in this case may be written as

$$e^{-\lambda_j z}\ {}_1F_1\big(a, b, -(\lambda-\lambda_j)z\big) = e^{-\lambda z}\ {}_1F_1\big(b-a, b, (\lambda-\lambda_j)z\big)\,, \tag{12}$$

in order to obtain (8).

Then, to obtain the expression for the c.d.f., one simply has to use Result B.2 in Appendix B, since

$$
\begin{aligned}
F_Z(z) = K\,\lambda^r \sum_{j=1}^{g}\sum_{k=1}^{r_j} \Bigg\{ c_{j,k}\,\frac{\Gamma(k)}{\Gamma(k+r)} \\
\int_0^z e^{-\lambda_j x}\, x^{k+r-1}\;{}_1F_1\big(r, k+r, -(\lambda-\lambda_j)x\big)\, dx \Bigg\} \\
= K\,\lambda^r \sum_{j=1}^{g}\sum_{k=1}^{r_j} c^*_{j,k} \Bigg\{ \frac{z^r}{\Gamma(r+1)}\;{}_1F_1(r, r+1, -\lambda z) \\
-e^{-\lambda_j z} \sum_{i=0}^{k-1} \frac{z^{r+i}\,\lambda_j^i}{\Gamma(r+1+i)}\;{}_1F_1\big(r, r+1+i, -(\lambda-\lambda_j)z\big) \Bigg\},
\end{aligned}
\tag{13}
$$

where

$$
c^*_{j,k} = \frac{c_{j,k}}{\lambda_j^k}\,\Gamma(k)
$$

and where we may use the result right before expression (17) in Coelho (1998) which gives

$$
\sum_{j=1}^{g}\sum_{k=1}^{r_j} \frac{c_{j,k}}{\lambda_j^k}\,\Gamma(k) = K^{-1},
$$

to obtain expression (9). In (13) above we may also use (12) to obtain (10). □

In order to obtain the p.d.f. of $U = e^{-Z}$ one simply has to follow the usual procedure of replacing in (7) $e^{-z}$ by $u$ and multiply by $\left|\frac{dz}{du}(-\log u)\right| = 1/u$, while in order to obtain the c.d.f. of $U = e^{-Z}$ one simply has to use relation (21) in Coelho (1998).

The GNI Gamma distribution may be seen as a further generalization of the GI Gamma distribution since for $r \in \mathbb{N}$ the GNI Gamma distribution reduces to a GI Gamma distribution, as we expected, since in this case we may apply the results in Appendix C. Also, when either $r = 0$ or $\lambda$ tends to positive infinity the GNI Gamma distribution reduces to a GI Gamma distribution since then in either case the random variable with a $\Gamma\,(r, \lambda)$ distribution vanishes. In fact, if $r = 0$, the Kummer functions in (13) evaluate to 1, and if $\lambda$ tends to plus infinity it may be shown that

$$
\lim_{\lambda\to\infty} \lambda^r\;{}_1F_1\big(r, k+r, -(\lambda-\lambda_j)z\big) = \frac{\Gamma(k+r)}{\Gamma(k)}\, z^{-r}\,.
$$

Although the Kummer confluent hypergeometric function is a fairly manageable function which is nowadays handled by a number of software packages, we may note that using (B.3) in Result B.1 in Appendix B both the p.d.f. and the c.d.f. of a GNI Gamma random variable may, for the referred cases, be written in terms of the error function.

# 4 Approximating the Distribution of the Product of an Odd Number of Particular Independent Beta Random Variables

**Theorem 2** *Let $p$ be an odd integer and let*

$$Y_j \sim B\left(a_j, \frac{b}{2}\right) \qquad j = 1, \ldots, p$$

*be independent random variables with Beta distributions, where $b$ is also an odd positive integer and $a_j = k - j/2$ $(j = 1, \ldots, p)$, with $k = a + p/2$, $a \in \mathbb{R}^+$. Further let*

$$W_1' = \prod_{j=1}^{p} Y_j \qquad and \qquad W_1 = -\log W_1' = -\sum_{j=1}^{p} \log Y_j \,.$$

*Let also*

$$Y_j^* \sim B\left(a_j^*, \frac{p}{2}\right) \qquad j = 1, \ldots, b$$

*be independent random variables with Beta distributions, where $a_j^* = k^* - j/2$ $(j = 1, \ldots, b)$, with $k^* = a + b/2$, $a \in \mathbb{R}^+$. Let then*

$$W_2' = \prod_{j=1}^{b} Y_j^* \qquad and \qquad W_2 = -\log W_2' = -\sum_{j=1}^{b} \log Y_j^* \,.$$

*Then a near-exact approximation to the distribution of $W_1$ is a GNI Gamma distribution of* depth *$p + b - 2$ with shape parameters $r_j^*$ given by*

$$r_j^* = \begin{cases} h_j^* & j = 1, 2 \\ r_{j-2}^* + h_j^* & j = 3, \ldots, p + b - 3 \\ 1/2 & j = p + b - 2 \end{cases} \tag{14}$$

*with*

$$h_j^* = \begin{cases} 1 & j = 1, \ldots, \min(p, b) \\ 0 & j = 1 + \min(p, b), \ldots, \max(p, b) \\ -1 & j = 1 + \max(p, b), \ldots, p + b - 3 \end{cases} \tag{15}$$

*or equivalently,*

$$h_j^* = (\textit{number of elements of } \{p, b\} \textit{ greater or equal to } j) - 1 \tag{16}$$

*and rate parameters*

$$\lambda_j = a - \frac{1}{2} + \frac{j}{2} \;\; (j = 1, \ldots, p+b-3), \quad \textit{and} \;\; \lambda_{p+b-2} = a + \frac{p+b}{2} - \frac{5}{4}. \tag{17}$$

*Alternatively, we may define*

$$\lambda_{p+b-2} = \frac{\Psi\left(a_1 + \frac{b}{2}\right) - \Psi\left(a_1 + \frac{b}{2} - \frac{1}{2}\right)}{\Psi'\left(a_1 + \frac{b}{2} - \frac{1}{2}\right) - \Psi'\left(a_1 + \frac{b}{2}\right)} \tag{18}$$

*and*

$$r_{p+b-2} = \frac{\left(\Psi\left(a_1 + \frac{b}{2}\right) - \Psi\left(a_1 + \frac{b}{2} - \frac{1}{2}\right)\right)^2}{\Psi'\left(a_1 + \frac{b}{2} - \frac{1}{2}\right) - \Psi'\left(a_1 + \frac{b}{2}\right)}, \tag{19}$$

*yielding an even better approximation, whose two first moments are the same as the two first exact moments of* $W_1$.

*The corresponding near-exact p.d.f. and c.d.f. of* $W_1$ *are thus given by* (7) *or* (8) *and* (9) *or* (10) *respectively, with* $g = p + b - 3$, $r = r_{p+b-2}$ *and* $\lambda = \lambda_{p+b-2}$. *The corresponding near-exact p.d.f. and c.d.f. of* $W_1'$ *is then obtained simply by making the appropriate change of variable.*

*The distribution of* $W_2$ *is the same as the distribution of* $W_1$ *and thus also the near-exact approximations to their distributions are the same. For this reason also the exact distributions and near-exact approximations to the distributions of* $W_2'$ *and* $W_1'$ *are the same.*

***Proof*** In order to obtain the best possible asymptotic result for the logBeta part of the characteristic function left free, using the result in Appendix A, we have to leave free in the distribution of $W_1$, the logBeta with the largest first parameter and not the one with the smallest, as it was done in Coelho (1999). On this logBeta we have then to implement a procedure similar to the one used in (5), reserving the largest rate parameter for the asymptotic Gamma with shape parameter 1/2. In order to accomplish this objective we may write the characteristic function of $W_1$ as

$$
\begin{aligned}
E\left(e^{itW_1}\right) &= \frac{\Gamma(a_1+b/2)\,\Gamma(a_1-it)}{\Gamma(a_1)\,\Gamma(a_1+b/2-it)}\prod_{j=2}^{p}\frac{\Gamma\left(a_j-it\right)\Gamma\left(a_j+b/2\right)}{\Gamma\left(a_j+b/2-it\right)\Gamma\left(a_j\right)}\\
&= \frac{\Gamma(a_1+b/2)}{\Gamma(a_1+b/2-1/2)}\,\frac{\Gamma(a_1+(b-1)/2)}{\Gamma(a_1)}\\
&\qquad \frac{\Gamma(a_1+b/2-1/2-it)}{\Gamma(a_1+b/2-it)}\,\frac{\Gamma(a_1-it)}{\Gamma(a_1+(b-1)/2-it)}\\
&\qquad \prod_{j=1}^{p-1}\frac{\Gamma\left(a_j-1/2-it\right)\Gamma\left(a_j-1/2+b/2\right)}{\Gamma\left(a_j-1/2+b/2-it\right)\Gamma\left(a_j-1/2\right)}\\
&= \frac{\Gamma\left(a_1+\frac{b}{2}\right)}{\Gamma\left(a_1+\frac{b}{2}-\frac{1}{2}\right)}\,\frac{\Gamma\left(a_1+\frac{b}{2}-\frac{1}{2}-it\right)}{\Gamma\left(a_1+\frac{b}{2}-it\right)}\prod_{j=0}^{\frac{b-1}{2}-1}\frac{a_1+j}{a_1+j-it}\\
&\qquad \prod_{j=1}^{p+b-3}\left(a-\frac{1}{2}+\frac{j}{2}\right)^{r_j}\left(a-\frac{1}{2}+\frac{j}{2}-it\right)^{-r_j}\\
&\overset{a_1+b/2}{\simeq}\left(a_1+\frac{b}{2}-\frac{3}{4}\right)^{1/2}\left(a_1+\frac{b}{2}-\frac{3}{4}-it\right)^{-1/2}\\
&\qquad \prod_{j=0}^{\frac{b-3}{2}}\left(a+\frac{p}{2}-\frac{1}{2}+j\right)\left(a+\frac{p}{2}-\frac{1}{2}+j-it\right)^{-1}\\
&\qquad \prod_{j=1}^{p+b-3}\left(a-\frac{1}{2}+\frac{j}{2}\right)^{r_j}\left(a-\frac{1}{2}+\frac{j}{2}-it\right)^{-r_j}\\
&= \left(a+\frac{p+b}{2}-\frac{5}{4}\right)^{1/2}\left(a+\frac{p+b}{2}-\frac{5}{4}-it\right)^{-1/2}\\
&\qquad \prod_{j=1}^{p+b-3}\left(a-\frac{1}{2}+\frac{j}{2}\right)^{r_j^*}\left(a-\frac{1}{2}+\frac{j}{2}-it\right)^{-r_j^*}
\end{aligned}
\tag{20}
$$

for $r_j^*$ $(j=1,\dots,p+b-3)$ given by (14)–(16) above, and where we used Claim A.1 and also, from the second to the third equality above, (6) and the fact that for $a\in\mathbb{R}^+$ and positive integer $p$ and $b$, where at least one of them is even,

$$
\prod_{j=1}^{p}\frac{\Gamma\left(a+\frac{p}{2}-\frac{j}{2}+\frac{b}{2}\right)}{\Gamma\left(a+\frac{p}{2}-\frac{j}{2}\right)}=\prod_{j=1}^{p+b-2}\left(a+\frac{j}{2}-\frac{1}{2}\right)^{r_j}
\tag{21}
$$

where

$$r_j = \begin{cases} h_j & j = 1, 2 \\ r_{j-2} + h_j & j = 3, \ldots, p + b - 2 \end{cases} \tag{22}$$

with

$$h_j = \begin{cases} 1 & j = 1, \ldots, \min(p, b) \\ 0 & j = 1 + \min(p, b), \ldots, \max(p, b) \\ -1 & j = 1 + \max(p, b), \ldots, p + b - 2 \end{cases} \tag{23}$$

or equivalently,

$$h_j = (\text{number of elements of } \{p, b\} \text{ greater or equal to } j) - 1 \, . \tag{24}$$

Relation (21) was obtained as a by-product in the proof of Theorem 2 in Coelho (1998) and we may notice that for odd $p$ it yields

$$\prod_{j=1}^{p-1} \frac{\Gamma\left(a_j - \frac{1}{2} + \frac{b}{2}\right)}{\Gamma\left(a_j - \frac{1}{2}\right)} = \prod_{j=1}^{p-1} \frac{\Gamma\left(a + \frac{p-1}{2} - \frac{j}{2} + \frac{b}{2}\right)}{\Gamma\left(a + \frac{p-1}{2} - \frac{j}{2}\right)} = \prod_{j=1}^{p+b-3} \left(a + \frac{j}{2} - \frac{1}{2}\right)^{r_j}$$

with $r_j$ given by (22)–(24), with $p$ replaced by $p-1$.

The last expression in (20) shows that a near-exact approximation to the distribution of $W_1$ is a particular GNI Gamma distribution, the sum of $p + b - 2$ independent random variables, $p + b - 3$ of which are Gamma random variables with integer shape parameters $r_j^*$ given by

$$r_j^* = \begin{cases} r_j & j = 1, \ldots, p - 1 \text{ and} \\ & j = p + 2n + 1\, , \quad \left(n = 0, \ldots, \frac{b-5}{2}\right) \\ r_j + 1 & j = p + 2n\, , \quad \left(n = 0, \ldots, \frac{b-3}{2}\right) \\ 1/2 & j = p + b - 2\, , \end{cases}$$

where $r_j$ $(j = 1, \ldots, p + b - 3)$ are given by (22)–(24) above, with $p$ replaced by $p - 1$, or equivalently, with $r_j^*$ given by (14)–(16) and rate parameters $\lambda_j$ $(j = 1, \ldots, p + b - 3)$ given by (17) and a latter random variable with a Gamma distribution with shape parameter $1/2$ and rate parameter $a + (p + b)/2 - 5/4$. Alternatively, instead of using (A.4) in Appendix A to asymptotically approximate the two first factors in the third equality of (20) we could have used the approach in A.2 in Appendix A and (A.15) to obtain an even better approximation, whose two first moments are the same as the exact ones, by taking $\lambda_{p+b-2}$ and $r_{p+b-2}$ given by (18) and (19).

Thus in order to obtain the corresponding near-exact p.d.f. and c.d.f. of $W_1$ one simply has to refer to the results in Theorem 1, being the corresponding near-exact p.d.f. and c.d.f. of $W_1' = e^{-W_1}$ easily obtained by making the appropriate straightforward transformation of variable.

That the characteristic function of $W_2$ is the same as the characteristic function of $W_1$ is most easily shown using, for positive integer $p$ and $b$, the relation

$$\prod_{j=1}^{p} \frac{\Gamma\left(a+\frac{p}{2}-\frac{j}{2}+\frac{b}{2}\right)}{\Gamma\left(a+\frac{p}{2}-\frac{j}{2}\right)} = \prod_{j=1}^{b} \frac{\Gamma\left(a+\frac{b}{2}-\frac{j}{2}+\frac{p}{2}\right)}{\Gamma\left(a+\frac{b}{2}-\frac{j}{2}\right)} \tag{25}$$

that is derived from the results coming out of the proof of Theorem 2 in Coelho (1998) and the results in Coelho (1999).

Then, for example, a direct application of (25) to the characteristic function of $W_2$ leads us immediately to the characteristic function of $W_1$, since

$$\begin{aligned} E\left(e^{itW_2}\right) &= \prod_{j=1}^{b} \frac{\Gamma(a_j^* - it)\,\Gamma(a_j^* + p/2)}{\Gamma(a_j^* + p/2 - it)\,\Gamma(a_j^*)} \\ &= \prod_{j=1}^{b} \frac{\Gamma\left(a+\frac{b}{2}-\frac{j}{2}-it\right)\Gamma\left(a+\frac{b}{2}+\frac{p}{2}-\frac{j}{2}\right)}{\Gamma\left(a+\frac{b}{2}+\frac{p}{2}-\frac{j}{2}-it\right)\Gamma\left(a+\frac{b}{2}-\frac{j}{2}\right)}\,. \end{aligned}$$

Given that the characteristic function of $W_2$ is the same as the characteristic function of $W_1$, $W_2$ also has the same near-exact characteristic functions as $W_1$ and thus the same near-exact approximations to its distribution. Consequently, also $W_2'$ and $W_1'$ have the same exact characteristic function and distribution and thus bear the same near-exact approximations. □

In the cases where either $p$ or $b$ is even or they are both even, the exact distribution of $W_1$ or $W_2$ is known to be a GI Gamma distribution (Coelho, 1998, 1999).

## 5 An Application of the Concept of Near-Exact Approximation and the GNI Gamma Distribution

The generalized Wilks $\Lambda$ statistic is the well-known statistic used to test the fit of the Generalized Canonical Analysis model (Coelho, 1992) or the independence among $m$ sets of jointly normally distributed variables. Let us assume that

$$\underline{x} = [\underline{x}_1', \ldots, \underline{x}_k', \ldots, \underline{x}_m']' \sim N_p\left(\underline{\mu}, \Sigma\right),$$

where

$$\underline{\mu} = \left[\underline{\mu}_1', \ldots, \underline{\mu}_k', \ldots, \underline{\mu}_m'\right]', \quad \Sigma = \begin{bmatrix} \Sigma_{11} & \ldots & \Sigma_{1k} & \ldots & \Sigma_{1m} \\ \Sigma_{21} & \ldots & \Sigma_{2k} & \ldots & \Sigma_{2m} \\ \vdots & & \vdots & & \vdots \\ \Sigma_{k1} & \ldots & \Sigma_{kk} & \ldots & \Sigma_{km} \\ \vdots & & \vdots & & \vdots \\ \Sigma_{m1} & \ldots & \Sigma_{mk} & \ldots & \Sigma_{mm} \end{bmatrix} \tag{26}$$

and

$$p = \sum_{k=1}^{m} p_k$$

is the overall number of variables, $p_k$ being the number of variables in $\underline{x}_k$, the $k$-th set.

The generalized Wilks $\Lambda$ statistic (Wilks, 1932, 1935, 1946; Anderson, 1984; Coelho, 1992)

$$\Lambda = \frac{|V|}{\prod_{k=1}^{m} |V_{kk}|}$$

where $|\cdot|$ stands for the determinant and $V$ is either the MLE of $\Sigma$ or the sample variance-covariance matrix of $\underline{x}$, split in a manner similar to $\Sigma$ in (26) (being thus $V_{kk}$ either the MLE of $\Sigma_{kk}$ or the sample variance-covariance matrix of $\underline{x}_k$) is, for a sample of size $n+1$, the $(2/(n+1))$th power of the likelihood ratio test statistic to test the null hypothesis of independence of the $m$ sets of variables,

$$H_0 : \Sigma = diag\left(\Sigma_{11}, \ldots, \Sigma_{kk}, \ldots, \Sigma_{mm}\right) . \tag{27}$$

It may be shown that the generalized Wilks $\Lambda$ statistic may be written as

$$\Lambda = \prod_{k=1}^{m-1} \Lambda_{k(k+1,\ldots,m)} , \tag{28}$$

where $\Lambda_{k(k+1,\ldots,m)}$ denotes the Wilks $\Lambda$ statistic used to test the independence between the $k$-th set and the set formed by joining sets $k+1$ through $m$. In other words, for $k = 1, \ldots, m-1$, $\Lambda_{k(k+1,\ldots,m)}$ is the Wilks $\Lambda$ statistic used to test the null hypothesis

$$H_0^{(k)} : \left[\Sigma_{k,k+1} \ldots \Sigma_{km}\right] = 0_{p_k \times (p_{k+1}+\ldots+p_m)} \quad (k = 1, \ldots, m-1) .$$

Under the null hypothesis (27) the $m-1$ $\Lambda$ statistics on the right hand side of (28) are independent (Anderson, 1984; Coelho, 1992). Then, from (28) above we may write the $h$-th moment of $\Lambda$ under (27), for a sample of size $n+1$, as

$$E\left(\Lambda^h\right) = \prod_{k=1}^{m-1}\prod_{j=1}^{p_k} \frac{\Gamma\left(\frac{n+1-j}{2}\right)\Gamma\left(\frac{n+1-q_k-j}{2}+h\right)}{\Gamma\left(\frac{n+1-j}{2}+h\right)\Gamma\left(\frac{n+1-q_k-j}{2}\right)}, \tag{29}$$

where $p_k$ is the number of variables in the $k$-th set and $q_k = p_{k+1} + \ldots + p_m$.

In this section we will apply the results obtained so far and the concept of near-exact approximation to the distribution of the generalized Wilks Lambda statistic, when there are two or more sets of variables with an odd number of variables.

Namely when there are more than two sets with an odd number of variables, approximating the exact distribution of the generalized Wilks $\Lambda$ statistic is an interesting problem. However, the concept of near-exact approximation and the form obtained for the near-exact approximation to the distribution of the product of an odd number of particular Beta random variables in Theorem 2 will enable us to obtain a near-exact approximation to the distribution of the generalized Wilks $\Lambda$ statistic under a simple and highly manageable form.

As we will see, when the floor of half the number of sets of variables with an odd number of variables is odd, a near-exact approximation to the distribution of the logarithm of the generalized Wilks $\Lambda$ is a GNI Gamma distribution, and when that number is even it is a GI Gamma distribution.

**Theorem 3** *When among the m sets of variables there are $m-l$ that have an odd number of variables, i.e. there are l sets with an even number of variables, then, for $k^* \in \mathbb{N}_0$, write $m-l=2k^*$ if $m-l$ is even or write $m-l=2k^*+1$ if $m-l$ is odd. This is, let $k^* = \lfloor\frac{m-l}{2}\rfloor$ be the floor of $\frac{m-l}{2}$. Then, under* (27) *and for a sample of size $n+1$, for $k^*$ odd*

$$W \overset{ne}{\sim} GNI\Gamma\left(r_1^*, r_2^*, \ldots, r_{p-2}^*, \tfrac{k^*}{2}; \tfrac{n-p+1}{2}, \tfrac{n-p+2}{2}, \ldots, \tfrac{n-2}{2}, \tfrac{n}{2}-\tfrac{3}{4}\right), \tag{30}$$

*and for $k^*$ even*

$$W \overset{ne}{\sim} GI\Gamma\left(r_1^*, r_2^*, \ldots, r_{p-2}^*, \tfrac{k^*}{2}; \tfrac{n-p+1}{2}, \tfrac{n-p+2}{2}, \ldots, \tfrac{n-2}{2}, \tfrac{n}{2}-\tfrac{3}{4}\right), \tag{31}$$

*where $W = -\log\Lambda$.*

*This means that if $k^*$ is odd, a near-exact approximation to the distribution of $W = -\log\Lambda$ is a GNI Gamma distribution of* depth $p-1$, *where $p = \sum_{k=1}^m p_k$, with rate parameters*

$$\lambda_j = \frac{n-p+j}{2} \quad (j = 1, \ldots, p-2) \qquad \textit{and} \qquad \lambda_{p-1} = \frac{n}{2} - \frac{3}{4} \tag{32}$$

*and shape parameters* $r^*_{p-1} = k^*/2$ *and*

$$r^*_j = \sum_{k=1}^{m-2k^*-1} r_{k,j-p^*_k} + \sum_{\substack{k=m-2k^* \\ \text{step } 2}}^{m-2} r_{k,j-p^*_k} + \sum_{\substack{k=m-2k^*+1 \\ \text{step } 2}}^{m-1} r^*_{k,j-p^*_k} \qquad (j=1,\ldots,p-2) \tag{33}$$

*where* $p^*_k = \sum_{l=1}^{k-1} p_l$,

$$r_{k,j-p^*_k} = 0 \quad if \quad p^*_k \geq j$$
$$r^*_{k,j-p^*_k} = 0 \quad if \quad p^*_k \geq j \;\; or \;\; j = p-2\,,$$

*and where for* $k = 1, \ldots, m-2k^*-1$ *(by steps of 1), and for* $k = m-2k^*, \ldots, m-2$ *(by steps of 2),*

$$r_{kj} = \begin{cases} h_{kj} & j = 1, 2 \\ r_{k,j-2} + h_{kj} & j = 3, \ldots, p_k + q_k - 2 \end{cases} \tag{34}$$

*with*

$$h_{kj} = (\textit{number of elements in } \{p_k, q_k\} \textit{ greater or equal to } j) - 1 \tag{35}$$

*and for* $k = m - 2k^* + 1, \ldots, m-1$ *(by steps of 2)*

$$r^*_{kj} = \begin{cases} h_{kj} & j = 1, 2 \\ r^*_{k,j-2} + h_{kj} & j = 3, \ldots, p_k + q_k - 3 \end{cases} \tag{36}$$

*with* $h_{kj}$ *given by (35) above.*

*The corresponding near-exact p.d.f. and c.d.f. of* $W$ *are thus respectively given by* (7) *or* (8) *and* (9) *or* (10)*, with* $g$ *replaced by* $p-2$*,* $r_j$ *replaced by* $r^*_j$ $(j = 1, \ldots, p-2)$*,* $\lambda_j$ $(j = 1, \ldots, p-2)$ *given by* (32)*,* $r$ *replaced by* $r^*_{p-1}$ *and* $\lambda$ *replaced by* $\lambda_{p-1}$*, being then the corresponding near-exact p.d.f. and c.d.f. of* $\Lambda = e^{-W}$ *easy to obtain from these.*

*Expression* (31) *shows that if* $k^*$ *is even, a near-exact approximation to the distribution of* $W = -\log \Lambda$ *is a GI Gamma distribution of* depth $p-1$ *with similar rate and shape parameters still given by* (32) *and* (33)–(36)*, respectively.*

*Alternatively, and yielding an even better approximation, whose two first moments are the same as the two first exact moments of* $W = -\log \Lambda$, *we may define*

$$\lambda_{p-1} = \frac{\Psi\left(\frac{n}{2}\right) - \Psi\left(\frac{n}{2} - \frac{1}{2}\right)}{\Psi'\left(\frac{n}{2} - \frac{1}{2}\right) - \Psi'\left(\frac{n}{2}\right)} \quad \textit{and} \quad r^*_{p-1} = k^* \frac{\left(\Psi\left(\frac{n}{2}\right) - \Psi\left(\frac{n}{2} - \frac{1}{2}\right)\right)^2}{\Psi'\left(\frac{n}{2} - \frac{1}{2}\right) - \Psi'\left(\frac{n}{2}\right)} . \tag{37}$$

*Expressions* (30) *and* (31) *yield the exact distribution if* $k^* = 0$, *situation in which the Gamma random variable with shape parameter* $k^*/2$ *or* $k^* \frac{\left(\Psi\left(\frac{n}{2}\right) - \Psi\left(\frac{n}{2} - \frac{1}{2}\right)\right)^2}{\Psi'\left(\frac{n}{2} - \frac{1}{2}\right) - \Psi'\left(\frac{n}{2}\right)}$ *vanishes.*

***Proof*** From (29) we know that the characteristic function of $W = -log\ \Lambda$, where $\Lambda$ represents the generalized Wilks $\Lambda$ statistic, may be written, for a sample of size $n+1$, as

$$E\left(e^{itW}\right) = \prod_{k=1}^{m-1} \prod_{j=1}^{p_k} \frac{\Gamma\left(\frac{n+1-j}{2}\right) \Gamma\left(\frac{n+1-q_k-j}{2} - it\right)}{\Gamma\left(\frac{n+1-j}{2} - it\right) \Gamma\left(\frac{n+1-q_k-j}{2}\right)} \tag{38}$$

where $i = (-1)^{1/2}$, $p_k$ is the number of variables in the $k$-th set $(k = 1, \ldots, m)$ and $q_k = p_{k+1} + \ldots + p_m$.

Without any loss of generality, let the sets with an odd number of variables be the last ones among the $m$ sets of variables, say $l+1$ through $m$.

If $m - l$, the number of sets with an odd number of variables, is even, write it as $2k^*$ and if it is odd write it as $1 + 2k^*$, this is, let $k^* = \lfloor \frac{m-l}{2} \rfloor$, with $k^* \in \mathbb{N}_0$.

We will then always have $k^*$ of the products

$$\prod_{j=1}^{p_k} \frac{\Gamma\left(\frac{n+1-j}{2}\right) \Gamma\left(\frac{n+1-q_k-j}{2} - it\right)}{\Gamma\left(\frac{n+1-j}{2} - it\right) \Gamma\left(\frac{n+1-q_k-j}{2}\right)} \tag{39}$$

in (38) where $p_k$ and $q_k$ are both odd. These products have to be handled in the same way the characteristic function of $W_1$ was in the proof of Theorem 2.

It will be for $k = m - 2(k^* - j^*) - 1$, $(j^* = 1, \ldots, k^*)$, that the products in (39) will have both $p_k$ and $q_k$ odd. Then, using (21), for any $k^* \geq 0$, we may write the characteristic function of $W$ as

$$E\left(e^{itW}\right) = \prod_{k=1}^{m-2k^*-1} \underbrace{\prod_{j=1}^{p_k+q_k-2} \left(\frac{n-p_k-q_k+j}{2}\right)^{r_{kj}} \left(\frac{n-p_k-q_k+j}{2}-it\right)^{-r_{kj}}}_{p_k \text{ even}}$$
$$\times \prod_{\substack{k=m-2k^* \\ \text{step } 2}}^{m-2} \underbrace{\prod_{j=1}^{p_k+q_k-2} \left(\frac{n-p_k-q_k+j}{2}\right)^{r_{kj}} \left(\frac{n-p_k-q_k+j}{2}-it\right)^{-r_{kj}}}_{q_k \text{ even}}$$
$$\times \prod_{\substack{k=m-2k^*+1 \\ \text{step } 2}}^{m-1} \underbrace{\prod_{j=1}^{p_k} \frac{\Gamma\left(\frac{n+1-j}{2}\right)\Gamma\left(\frac{n+1-q_k-j}{2}-it\right)}{\Gamma\left(\frac{n+1-j}{2}-it\right)\Gamma\left(\frac{n+1-q_k-j}{2}\right)}}_{\text{both } p_k \text{ and } q_k \text{ odd}}$$

where $r_{kj}$ is given by (34) and (35) and where each product with a lower bound larger than the upper bound evaluates to 1. Then, using the same technique used in the proof of Theorem 2, we may write

$$\begin{aligned}
&\prod_{j=1}^{p_k} \frac{\Gamma\left(\frac{n+1-j}{2}\right)\Gamma\left(\frac{n+1-q_k-j}{2}-it\right)}{\Gamma\left(\frac{n+1-j}{2}-it\right)\Gamma\left(\frac{n+1-q_k-j}{2}\right)} \\
&\quad= \frac{\Gamma\left(\frac{n}{2}\right)\Gamma\left(\frac{n-q_k}{2}-it\right)}{\Gamma\left(\frac{n}{2}-it\right)\Gamma\left(\frac{n-q_k}{2}\right)} \prod_{j=2}^{p_k} \frac{\Gamma\left(\frac{n+1-j}{2}\right)\Gamma\left(\frac{n+1-q_k-j}{2}-it\right)}{\Gamma\left(\frac{n+1-j}{2}-it\right)\Gamma\left(\frac{n+1-q_k-j}{2}\right)} \\
&\quad= \frac{\Gamma\left(\frac{n}{2}\right)\Gamma\left(\frac{n}{2}-\frac{1}{2}-it\right)}{\Gamma\left(\frac{n}{2}-\frac{1}{2}\right)\Gamma\left(\frac{n}{2}-it\right)} \frac{\Gamma\left(\frac{n}{2}-\frac{1}{2}\right)\Gamma\left(\frac{n-q_k}{2}-it\right)}{\Gamma\left(\frac{n}{2}-\frac{1}{2}-it\right)\Gamma\left(\frac{n-q_k}{2}\right)} \\
&\qquad\qquad \times \prod_{j=1}^{p_k-1} \frac{\Gamma\left(\frac{n-j}{2}\right)\Gamma\left(\frac{n-q_k-j}{2}-it\right)}{\Gamma\left(\frac{n-j}{2}-it\right)\Gamma\left(\frac{n-q_k-j}{2}\right)} \\
&\quad\overset{n}{\simeq} \left(\frac{n}{2}-\frac{3}{4}\right)^{1/2}\left(\frac{n}{2}-\frac{3}{4}-it\right)^{-1/2} \prod_{j=0}^{\frac{q_k-1}{2}-1} \left(\frac{n-q_k}{2}+j\right)\left(\frac{n-q_k}{2}+j-it\right)^{-1} \\
&\qquad \times \prod_{j=1}^{p_k+q_k-3} \left(\frac{n-q_k-p_k}{2}+\frac{j}{2}\right)^{r'_{kj}} \left(\frac{n-q_k-p_k}{2}+\frac{j}{2}-it\right)^{-r'_{kj}}
\end{aligned}$$

$$
\begin{aligned}
&= \left(\frac{n}{2}-\frac{3}{4}\right)^{1/2}\left(\frac{n}{2}-\frac{3}{4}-it\right)^{-1/2} \\
&\quad \times \prod_{j=1}^{p_k+q_k-3}\left(\frac{n-q_k-p_k}{2}+\frac{j}{2}\right)^{r^*_{kj}}\left(\frac{n-q_k-p_k}{2}+\frac{j}{2}-it\right)^{-r^*_{kj}}
\end{aligned}
$$

where $r^*_{kj}$ is given by

$$
r^*_{kj} = \begin{cases} r'_{kj} & j=1,\dots,p_k-1 \text{ and} \\ & j=p_k+2n+1\,,\ \left(n=0,\dots,\frac{q_k-5}{2}\right) \\ r'_{kj}+1 & j=p_k+2n\,,\ \left(n=0,\dots,\frac{q_k-3}{2}\right) \end{cases}
$$

with

$$
r'_{kj} = \begin{cases} h'_{kj} & j=1,2 \\ r'_{k,j-2}+h'_{kj} & j=3,\dots,p_k+q_k-3 \end{cases}
$$

where

$$
h'_{kj} = (\text{number of elements in } \{p_k-1,q_k\} \text{ greater or equal to } j)-1\,,
$$

or equivalently, with $r^*_{kj}$ given by (36).

But then we have

$$
\begin{aligned}
E\left(e^{itW}\right) \overset{n}{\simeq} &\prod_{k=1}^{m-2k^*-1}\prod_{j=1}^{p_k+q_k-2}\left(\frac{n-p_k-q_k+j}{2}\right)^{r_{kj}}\left(\frac{n-p_k-q_k+j}{2}-it\right)^{-r_{kj}} \\
&\times \prod_{\substack{k=m-2k^*\\ \text{step } 2}}^{m-2}\prod_{j=1}^{p_k+q_k-2}\left(\frac{n-p_k-q_k+j}{2}\right)^{r_{kj}}\left(\frac{n-p_k-q_k+j}{2}-it\right)^{-r_{kj}} \\
&\times \prod_{\substack{k=m-2k^*+1\\ \text{step } 2}}^{m-1}\prod_{j=1}^{p_k+q_k-3}\left(\frac{n-p_k-q_k+j}{2}\right)^{r^*_{kj}}\left(\frac{n-p_k-q_k+j}{2}-it\right)^{-r^*_{kj}} \\
&\times \left(\frac{n}{2}-\frac{3}{4}\right)^{k^*/2}\left(\frac{n}{2}-\frac{3}{4}-it\right)^{-k^*/2},
\end{aligned}
$$

where $r_{kj}$ is given by (34) and (35), and thus we may finally write

$$E\left(e^{itW}\right) \overset{n}{\simeq} (n/2-3/4)^{k^*/2}\,(n/2-3/4-it)^{-k^*/2} \times \prod_{j=1}^{p-2}\left(\frac{n-p+j}{2}\right)^{r_j^*}\left(\frac{n-p+j}{2}-it\right)^{-r_j^*}$$

with $r_j^*$ given by (33), so that, if $k^*$ is even, a near-exact approximation to the distribution of $W$ is a GI Gamma distribution and if $k^*$ is odd, it is a GNI Gamma distribution, both of *depth* $p-1$, with shape parameters $r_j^*$ $(j=1,\ldots,p-2)$ given by (33) through (36) and $r_{p-1}^* = k^*/2$ and rate parameters $\lambda_j = \frac{n-p+j}{2}$ $(j = 1,\ldots,p-2)$ and $\lambda_{p-1} = \frac{n}{2} - \frac{3}{4}$.

Alternatively, using the approach outlined in A.2 in Appendix A and (A.15), we may define $\lambda_{p-1}$ and $r_{p-1}^*$ through (37), getting an even better approximation, which has the two first moments equal to the exact two first moments of $W = -\log \Lambda$.

It is then easy to see that, if we have at most one of the $m$ sets with an odd number of variables, this is, for $k^* = 0$, we will be in one of the cases studied in Coelho (1998, 1999), so that the resulting GI Gamma distribution of *depth* $p-2$ is the exact distribution of $W = -log\,\Lambda$. □

The corresponding near-exact p.d.f. and c.d.f. for the Wilks $\Lambda$ statistic are then easily derived from the expressions for the near-exact p.d.f. and c.d.f. for $W = -\log \Lambda$, through transformation. For example, the near-exact c.d.f. of $\Lambda$ may be obtained from (9), writing $\lambda$ for $\lambda_{p-1}$ and $r$ for $r_{p-1}^*$, as

$$F_\Lambda(u) = 1 - \lambda^r \frac{(-\log u)^r}{\Gamma(r+1)}\,{}_1F_1\,(r, r+1, \lambda \log u) + K\,\lambda^r \sum_{j=1}^{p-2} u^{\lambda_j} \sum_{k=1}^{r_j^*} c_{j,k}^* \sum_{i=0}^{k-1} \frac{(-\log u)^{r+i}\,\lambda_j^i}{\Gamma\,(r+1+i)}\,{}_1F_1\big(r, r+1+i, (\lambda-\lambda_j)\log u\big)\,.$$

From this c.d.f. near-exact quantiles may be easily computed.

Similar procedures may be applied in order to obtain near-exact approximations to distributions of statistics having moments of the form in (29).

## 6 Conclusions and Final Remarks

The concept of 'near-exact' approximation although being a simple one seems to be quite useful. Indeed, in a number of cases by replacing just a small part of the characteristic function of a random variable by an asymptotic result with very good convergence properties one will be able to get a distribution extremely close to the exact distribution but much easier to handle and yet far better than any of the usual asymptotic distributions.

Further, using such procedure one may usually choose which component parts of the characteristic function one wants to replace. Since a number of times we just have to replace a small number of such components in order to obtain a fairly manageable distribution, we may choose the most adequate ones in terms of the asymptotic result used, giving yet more control over the process and the possibility to obtain very good approximations.

If these replacements are kept to a minimum, just in order to enable the resulting distribution to be easily manageable, and if the asymptotic replacements used are of a high quality, then one may obtain distributions which besides being far more manageable than the exact distribution are very close to this exact distribution, in terms of characteristic function, moments, distribution function and quantiles, and thus highly adequate to be used for example in computing near-exact quantiles.

Using this idea and associated way of handling the characteristic function a near-exact approximation for the distribution of the product of an odd number of particular independent Beta random variables was obtained under a form that although involving the Kummer confluent hypergeometric function is fairly concise and manageable, mainly nowadays that several computer programs may easily handle such functions which have generally good convergence properties. This distribution is a particular case of the GNI Gamma distribution introduced in Section 3. It was also shown that a near-exact approximation to the distribution of the logarithm of the generalized Wilks $\Lambda$ is either a particular GNI or GI Gamma distribution, depending on whether the number of sets of variables with an odd number of variables, expressed as $1+2k^*$ or $2k^*$, yields an odd or even $k^*$.

**Acknowledgments** The author wants to express his earnest appreciation for the insightful comments from the referees which, besides other improvements, allowed to obtain even better near-exact approximations for a logBeta random variable and for the product of an odd number of independent particular Beta random variables.

This research was financially supported by the Portuguese Foundation for Science and Technology (FCT).

## Appendix A. Approximating the Characteristic Function of a LogBeta Random Variable by the Characteristic Function of a Gamma Random Variable

Being $f_a$ and $g_a$ two non-constant functions of $a$, we will use the notation

$$f_a \overset{a}{\simeq} g_a$$

to denote that

i)
$$\lim_{a\to\infty} \frac{f_a}{g_a} = 1\,;$$

and

ii) there is an injective, i.e. one-to-one, function of $a$, $h_a(\cdot)$ such that

$$h_a(g_a) = g \quad \text{and} \quad h_a^{-1}(g) = g_a$$

where $g$ is no longer a function of $a$, with

$$\lim_{a\to\infty} h_a\left(f_a\right) = h_a\left(g_a\right) = g\,.$$

**Claim A.1** Based on the fact that

$$\lim_{a\to\pm\infty} \left(\frac{a}{a+b}\right)^{1/b} (a+b/2+1/2) - a = \frac{b}{2} - \frac{1}{2}\,,$$

we may write,

$$\frac{a}{a+b} \overset{a}{\simeq} \left(\frac{a+b/2-1/2}{a+b/2+1/2}\right)^b . \tag{A.1}$$

The result is indeed exact for $b = 0$, $b = 1$ and $b = -1$, being the approximation, for other fixed values of $b$, good to second order in $a$. For $b > 0$ and a given $a \geq 1/2$, the difference $\frac{a}{a+b} - \left(\frac{a+b/2-1/2}{a+b/2+1/2}\right)^b$ attains its maximum at a value of $b$ which approaches $\gamma$ (the Euler gamma constant) from below as $a$ tends to infinity. This maximum has a value which lowers as $a$ increases and is always much lower than $e^{-2}$. For $b > 1$ the value of that difference decreases towards $-e^{-2}$ as $b$ tends to positive infinity. For $a \leq -1/2$ and $b < 0$, that difference has a symmetrical behavior. If $b$ is a positive half-integer, this is, if $b = l/2$ where $l$ is a positive odd integer, the approximation given by (A.1) is better for $l = 1$ than for any other value of $l$. Thus, although for general $0 < b < 1$ the value $b = 1/2$ is around the point where the approximation given by (A.1) is worse, this approximation is yet fairly good.

For $|a| < 1/2$ the quality of the approximation is not quite good at all, but these cases are not of interest for the contents of the paper. □

*Result A.1* For $z \in \mathbb{C}$,

$$z = \prod_{n=1}^{\infty} \frac{n+z-1}{n+z} \frac{n+1}{n}\,. \tag{A.2}$$

***Proof*** The result may be either straightforwardly worked out from expression (6.1.3) in Abramowitz and Stegun (1974), that is

$$\frac{1}{\Gamma(z)} = z\, e^{\gamma z} \prod_{n=1}^{\infty} \left[ \left(1+\frac{z}{n}\right) e^{-z/n} \right] \tag{A.3}$$

where $\gamma$ is the Euler gamma constant, or observing that being it a telescoping product we have

$$\begin{aligned} \prod_{n=1}^{\infty} \frac{n+z-1}{n+z} \frac{n+1}{n} &= \lim_{N\to\infty} \prod_{n=1}^{N} \frac{n+z-1}{n+z} \frac{n+1}{n} \\ &= \lim_{N\to\infty} \frac{z}{N+z}(N+1) = z\,. \quad \square \end{aligned}$$

Although for $a \in \mathbb{R}^+$ and $b \in \mathbb{N}_0$ the exact value for $\frac{\Gamma(a+b)}{\Gamma(a)}$ is easily expressed as a finite product, for general $b \in \mathbb{R}^+$, the following result is useful for some applications.

*Result A.2* For $a, b \in \mathbb{R}^+$ such that $a + b/2 - 1/2 > 0$,

$$\frac{\Gamma(a+b)}{\Gamma(a)} \overset{a}{\simeq} \left(a+\frac{b}{2}-\frac{1}{2}\right)^{b} \tag{A.4}$$

being the approximation of a better quality for higher values of $a$ and generally smaller values of $b$, in the same terms as the ones expressed in the statement of Claim A.1. Expression (A.4) gives clearly the exact result for $b = 0$, $b = 1$ or $b = -1$.

***Proof*** From (A.3), we may, on one hand, write

$$\frac{\Gamma(a+b)}{\Gamma(a)} = \frac{a}{a+b} e^{-\gamma b} \prod_{n=1}^{\infty} \frac{n+a}{n+a+b} e^{b/n} \tag{A.5}$$

while, on the other hand, making $z = 1$,

$$1 = e^{\gamma} \prod_{n=1}^{\infty} \frac{n+1}{n} e^{-\frac{1}{n}} \quad \text{or} \quad e^{-\gamma} = \prod_{n=1}^{\infty} \frac{n+1}{n} e^{-\frac{1}{n}}\,. \tag{A.6}$$

But then using (A.6) on (A.5) we may write

$$\frac{\Gamma(a+b)}{\Gamma(a)} = \frac{a}{a+b}\prod_{n=1}^{\infty}\frac{n+a}{n+a+b}\left(\frac{n+1}{n}\right)^{b} \tag{A.7}$$

which, either using Result A.1 or simply adjusting the product index may yet be written as

$$\frac{\Gamma(a+b)}{\Gamma(a)} = \prod_{n=1}^{\infty}\frac{n+a-1}{n+a+b-1}\left(\frac{n+1}{n}\right)^{b}.$$

Finally, using first Claim A.1 and then Result A.1, since

$$\frac{n+a-1}{n+a+b-1}\Big/\left(\frac{n+a-1+b/2-1/2}{n+a+b/2-1/2}\right)^{b}$$

converges uniformly to 1 (in $n$), we may write, for $a+b/2>1/2$,

$$\begin{aligned}\frac{\Gamma(a+b)}{\Gamma(a)} &= \prod_{n=1}^{\infty}\left(\frac{n+a-1+b/2-1/2}{n+a+b/2-1/2}\right)^{b}\left(\frac{n+1}{n}\right)^{b}\underbrace{\prod_{n=1}^{\infty}\frac{n+a-1}{n+a+b-1}\Big/\left(\frac{n+a-1+b/2-1/2}{n+a+b/2-1/2}\right)^{b}}_{\xrightarrow[a\to\infty]{}1}\\ &\overset{a}{\simeq}\prod_{n=1}^{\infty}\left(\frac{n+a-1+b/2-1/2}{n+a+b/2-1/2}\right)^{b}\left(\frac{n+1}{n}\right)^{b} = \left(a+\frac{b}{2}-\frac{1}{2}\right)^{b},\end{aligned} \tag{A.8}$$

where the quality of the approximation is ruled by the statement in Claim A.1, being particularly adequate for sufficiently large values of $a$ and $0<b<1$. □

We may notice that from (A.8), by successive derivations, we may obtain another much useful approximation,

$$\Psi^{(k)}(a+b)-\Psi^{(k)}(a)\overset{a}{\simeq}(-1)^{k}\frac{k!\,b}{(a+b/2-1/2)^{k+1}}, \tag{A.9}$$

(where $\Psi^{(k)}(\cdot)$ is the polygamma function of order $k+1$, that is, for $k=1$ it is the digamma function) which, among other interesting applications, may also be used in obtaining (A.12)–(A.14).

We may also notice that instead of using the approximation given by Result A.2 one could think of using a truncation of (A.7). However, although this is possible, the use of Result A.2 is highly advantageous. For example, for $a=5$ and $b=1/2$ in order to achieve an approximation similar to the one given by Result A.2, one would need more than 3400 factors in (A.7) with much consumption of computation time and the build up of a huge algebraic complication in the expressions of the distributions where such result would be used. Also, while for $b = 1/2$ it is common to approximate $\Gamma(a+1/2)/\Gamma(a)$ by $a^{1/2}$, Result A.2 would give $(a-1/4)^{1/2}$. This latter approximation is of a much better quality than $a^{1/2}$, for any value of $a$ such that $a + \frac{b}{2} - \frac{1}{2} = a - \frac{1}{4} > 0$.

On the other hand Result A.2 may be used to obtain, for $i = \sqrt{-1}$ and $t \in \mathbb{R}$,

$$\frac{\Gamma(a+b)}{\Gamma(a)} \frac{\Gamma(a-it)}{\Gamma(a+b-it)} \overset{a}{\simeq} \left(a + \frac{b}{2} - \frac{1}{2}\right)^b \left(a + \frac{b}{2} - \frac{1}{2} - it\right)^{-b},$$

that is used in the approximation of characteristic functions in the paper. We should note that this approximation, owing to the particular structure of the expression to be approximated is yet of a much better quality than the one in (A.4).

The use of this approximation would lead us to approximate a logBeta distribution with parameters $a$ and $b = 1/2$ by a Gamma distribution with shape parameter $1/2$ and rate parameter $a - 1/4$, which indeed in terms of characteristic function, moments, quantiles and distribution function is a much better approximation than a Gamma distribution with shape parameter $1/2$ and rate parameter $a$.

## *A.1 How Does the Proposed Approximation Work, Concerning Moments*

If

$$E\left[\left|X^k\right|\right] < \infty$$

we have, for $i = \sqrt{-1}$,

$$E\left(X^k\right) = \frac{1}{i^k} \left(\frac{d^k}{dt^k} E\left(e^{itX}\right)\right)\Big|_{t=0}.$$

So, from the characteristic function of $Y$ in (3) one may see that the expected value of $Y$, a logBeta random variable with parameters $a$ and $b$, is

$$E(Y) = \frac{1}{i} \left(\frac{d}{dt} E\left(e^{itY}\right)\right)\Big|_{t=0} = \Psi(a+b) - \Psi(a) \tag{A.10}$$

and its second non-centered moment is

$$E(Y^2) = \frac{1}{i^2} \left(\frac{d^2}{dt^2} E\left(e^{itY}\right)\right)\Big|_{t=0} = \Psi'(a) - \Psi'(a+b) + (\Psi(a+b) - \Psi(a))^2$$

so that its variance is

$$Var(Y) = \Psi'(a) - \Psi'(a+b) , \tag{A.11}$$

where $\Psi(x) = \frac{d}{dx}\log\Gamma(x)$ is the digamma function and $\Psi'(x) = \frac{d^2}{dx^2}\log\Gamma(x) = \frac{d}{dx}\Psi(x)$ is the trigamma function.

One may then notice that

$$\lim_{a\to\infty}\frac{[E(Y)]^2}{Var(Y)} = \lim_{a\to\infty}\frac{(\Psi(a+b)-\Psi(a))^2}{\Psi'(a)-\Psi'(a+b)} = b \tag{A.12}$$

and

$$\lim_{a\to\infty}\frac{E(Y)}{Var(Y)} - a = \lim_{a\to\infty}\frac{\Psi(a+b)-\Psi(a)}{\Psi'(a)-\Psi'(a+b)} - a = \frac{b}{2}-\frac{1}{2}, \tag{A.13}$$

so that we may write

$$\frac{E(Y)}{Var(Y)} = \frac{\Psi(a+b)-\Psi(a)}{\Psi'(a)-\Psi'(a+b)} \overset{a}{\simeq} a+\frac{b}{2}-\frac{1}{2}. \tag{A.14}$$

Expressions (A.12) and (A.13) yield the exact result for $b=1$, being the rate of convergence in (A.12) and (A.13) generally better for $0<b<1$ than for $b>1$, and accordingly higher for $b=1/2$ than for any other positive half-integer.

Thus, for large $a$, a Gamma distribution with shape parameter $b$ and rate parameter $\lambda=a+\frac{b}{2}-\frac{1}{2}$ will approximate the logBeta distribution also in the sense that, in agreement with (A.9), it will have its expected value, $b/\left(a+\frac{b}{2}-\frac{1}{2}\right)$, and its variance, $b/\left(a+\frac{b}{2}-\frac{1}{2}\right)^2$, verifying the relations $\Psi(a+b)-\Psi(a)\overset{a}{\simeq}b/\left(a+\frac{b}{2}-\frac{1}{2}\right)$ and $\Psi'(a)-\Psi'(a+b)\overset{a}{\simeq}b/\left(a+\frac{b}{2}-\frac{1}{2}\right)^2$.

## *A.2 A Different Approach in Approximating a LogBeta Distribution by a Gamma Distribution*

The approach proposed so far in this Appendix, of approximating a logBeta distribution with parameters $a$ and $b$ by a Gamma distribution with shape parameter $b$ and rate parameter $a + b/2 - 1/2$, given the results presented, clearly yields an asymptotic approximation for increasing values of $a$, the parameter that, in the distribution of some statistics, will be an increasing function of the sample size, yielding a distribution that gets closer to the logBeta distribution in terms of characteristic function, moments, quantiles and distribution function, for increasing values of the parameter $a$. However, the results about moments just obtained in the second part of this Appendix may suggest (following an insightful lead from one of the referees) that we could think of approximating a logBeta distribution with parameters $a$ and $b$ by a Gamma distribution with rate and shape parameters $\lambda$ and $r$ given by

$$\lambda = \frac{\Psi(a+b) - \Psi(a)}{\Psi'(a) - \Psi'(a+b)}, \qquad r = \frac{(\Psi(a+b) - \Psi(a))^2}{\Psi'(a) - \Psi'(a+b)}. \tag{A.15}$$

As it is easy to verify, this Gamma distribution would have the same two first moments as the original logBeta distribution, this is, the expected value and variance in (A.10) and (A.11).

Although it is not all that easy to prove, since the algebra involving polygamma functions may get a little complicated, the Gamma distribution obtained this way is, for increasing values of the parameter $a$, also an asymptotic distribution for the logBeta distribution with parameters $a$ and $b$, yielding in fact an outstanding approximation since both distributions even have the same two first moments.

## Appendix B. Results Used in Obtaining the Explicit Expressions for the p.d.f. and c.d.f. of the GNI Gamma Distribution

*Result B.1* For $Re(a) > -1$, and $Re(r) > 0$,

$$\int_0^z (z-x)^a\, x^{r-1}\, e^{-kx}\, dx = \frac{\Gamma(a+1)\,\Gamma(r)}{\Gamma(a+r+1)}\, z^{a+r}\; {}_1F_1(r, a+r+1, -kz)\,, \tag{B.1}$$

where

$${}_1F_1(a,b,z) = \sum_{n=0}^{\infty} \frac{(a)_n\, z^n}{(b)_n\, n!}$$

is the Kummer confluent hypergeometric function, with

$$(a)_n = \frac{\Gamma(a+n)}{\Gamma(a)}.$$

For integer $a$ $(\geq 0)$ and $r = b/2$, where $b$ is a positive odd integer we may also write

$$\int_0^z (z-x)^a x^{\frac{b}{2}-1} e^{-kx}\,dx = \frac{1}{k^{a+b/2}} \left[ e^{-kz} \sum_{i=0}^{\max\left(a,\frac{b-1}{2}\right)-1} (-1)^{i+1}\,\chi_{a-1-i}\,k^{i+1/2}\,z^{i+1/2} + (-1)^a\,\phi\left(\sqrt{kz}\right) \sum_{i=0}^{a} (-1)^i\,\delta_i\,k^i\,z^i \right] \tag{B.2}$$

where $\phi(z) = \dfrac{2}{\sqrt{\pi}} \displaystyle\int_0^z e^{-x^2} dx$ is the error function,

$$\delta_i = \binom{a}{i}\,\Gamma\left(a+\frac{b}{2}-i\right), \qquad i = 0, \ldots, a$$

and

$$\chi_i = \sum_{j=i+1}^{a} (-1)^j\,\gamma_{j-i-1,j}\,, \qquad i = \min\left(0, a-\frac{b-1}{2}\right), \ldots, a-1$$

with

$$\gamma_{i,j} = \binom{a}{i} \frac{\Gamma\left(a+\frac{b}{2}-i\right)}{\Gamma\left(a-j+3/2\right)}\,, \qquad i = 0, \ldots, a;\; j = 1, \ldots, a\,.$$

From (B.1) and (B.2) we obtain, for integer $a$ $(\geq 0)$ and odd $b$ $(\geq 1)$ the following relation,

$$ {}_1F_1\left(\tfrac{b}{2}, a+\tfrac{b}{2}+1, -kz\right) = \frac{1}{k^{a+\frac{b}{2}}\,z^{a+\frac{b}{2}}} \frac{\Gamma(a+b/2+1)}{\Gamma(a+1)\,\Gamma(b/2)} \times \left[ e^{-kz} \sum_{i=0}^{\max\left(a,\frac{b-1}{2}\right)-1} (-1)^{i+1}\,\chi_{a-1-i}\,k^{i+1/2}\,z^{i+1/2} + (-1)^a\,\phi\left(\sqrt{kz}\right) \sum_{i=0}^{a} (-1)^i\,\delta_i\,k^i\,z^i \right], \tag{B.3}$$

which, for these cases, expresses the Kummer confluent functions in (B.1) in terms of the error function.

***Proof*** The result in (B.1) may be obtained from expression (13.2.1) in Abramowitz and Stegun (1974), which may be written, for $Re(a) > -1$ and $Re(r) > 0$, as

$$\int_0^1 e^{-zt}\, t^{r-1}(1-t)^a\, dt = \frac{\Gamma(a+1)\,\Gamma(r)}{\Gamma(a+r+1)}\; {}_1F_1(r, a+r+1, -z)\,. \tag{B.4}$$

Making in (B.4) the change of variable $x = \frac{tz}{k}$ we have

$$\begin{aligned}\int_0^{z/k} e^{-kx}\left(\frac{kx}{z}\right)^{r-1}\left(1-\frac{kx}{z}\right)^a \frac{k}{z}\, dx \\ &= \left(\frac{k}{z}\right)^{a+r}\int_0^{z/k} e^{-kx}\, x^{r-1}\left(\frac{z}{k}-x\right)^a dx \\ &= \frac{\Gamma(a+1)\,\Gamma(r)}{\Gamma(a+r+1)}\; {}_1F_1(r, a+r+1, -z)\,,\end{aligned}$$

or (B.1) by just replacing $z$ by $kz$.

The result for $a \in \mathbb{N}_0$ and odd $b$, under the form in (B.2), may be obtained by successive integrations by parts and induction. □

We may note that Result B.1 is still valid for $a = 0$, in which case the Kummer function in (B.1) may be expressed in terms of the incomplete Gamma function, as

$${}_1F_1\,(r, r+1, -kz) = \frac{r}{k^r\, z^r}\,(\Gamma\,(r) - \Gamma\,(r, kz))$$

where

$$\Gamma\,(r, kz) = \int_{kz}^{\infty} t^{r-1}\, e^{-t}\, dt$$

is the incomplete Gamma function. In this case (B.2) and (B.3) still hold.

*Result B.2* For $Re(l-m) < 0$ and integer $k \geq 0$

$$\begin{aligned}\int_0^z e^{-mx}\, x^{k+r-1}\; {}_1F_1\,(r, k+r, lx)\, dx \\ = \frac{\Gamma\,(k+r)}{m^k}\Bigg[&\frac{z^r}{\Gamma\,(r+1)}\; {}_1F_1\,(r, r+1, (l-m)z) \\ &- e^{-mz}\sum_{j=0}^{k-1}\frac{z^{r+j}\, m^j}{\Gamma\,(r+1+j)}\; {}_1F_1\,(r, r+1+j, lz)\Bigg]\,.\end{aligned} \tag{B.5}$$

***Proof*** From 1.14.7 in Prudnikov et al. (1990),

$$\int (lz)^{k+r-2}\,{}_1F_1\,(r, k+r-1, lz)\;dz \;=\; \frac{l^{k+r-2}}{k+r-1}\,z^{k+r-1}\,{}_1F_1\,(r, k+r, lz)\,. \tag{B.6}$$

Then, using integration by parts and (B.6) we have

$$\begin{aligned}\int e^{-mz}\,z^{k+r-2}\,{}_1F_1\,(r, k+r-1, lz)\;dz&\\ = \tfrac{e^{-mz}}{k+r-1}\,z^{k+r-1}\,{}_1F_1\,(r, k+r, lz)&\\ +\frac{m}{k+r-1}\int e^{-mz}\,z^{k+r-1}\,{}_1F_1\,(r, k+r, lz)\;dz&\end{aligned}$$

or

$$\begin{aligned}\int e^{-mz}\,z^{k+r-1}\,{}_1F_1\,(r, k+r, lz)\;dz&\\ = \frac{k+r-1}{m}\int e^{-mz}\,z^{k+r-2}\,{}_1F_1\,(r, k+r-1, lz)\;dz&\\ -\frac{e^{-mz}}{m}\,z^{k+r-1}\,{}_1F_1\,(r, k+r, lz)\;,&\end{aligned} \tag{B.7}$$

which has to be applied recursively $k$ times, so that we get on the right-hand side

$$\int e^{-mz}\,z^{r-1}\,{}_1F_1\,(r, r, lz)\;dz$$

where

$${}_1F_1\,(r, r, lz) \;=\; e^{lz}$$

so that

$$\int e^{-mz}\,z^{r-1}\,{}_1F_1\,(r, r, lz)\;dz \;=\; \int e^{(l-m)z}\,z^{r-1}\,dz$$

which is under the form of Result B.1, for $a = 0$, yielding

$$\int_0^z e^{(l-m)x}\,x^{r-1}\,dx \;=\; \frac{z}{r}\,{}_1F_1\,(r, r+1, (l-m)z)\;.$$

This way, applying (B.7) recursively $k$ times we get

$$\begin{aligned}\int e^{-mz}\, z^{k+r-1}\, {}_1F_1\,(r,k+r,lz)\, dz \\ &= \frac{\prod_{j=1}^{k}(k+r-j)}{m^k}\,\frac{1}{r}\, z^r\, {}_1F_1\,(r,r+1,(l-m)z) \\ &\quad -\frac{e^{-mz}}{m}\sum_{j=1}^{k}\frac{\prod_{i=1}^{j-1}(k+r-i)}{m^{j-1}}\, z^{k+r-j}\, {}_1F_1\,(r,k+r+1-j,lz)\,,\end{aligned} \tag{B.8}$$

where

$$\prod_{j=1}^{k}(k+r-j) = \frac{\Gamma\,(k+r)}{\Gamma\,(r)}\,,$$

yielding after some small rearrangements (B.5), together with the fact that the right-hand side of (B.8) clearly evaluates to zero at $z=0$. □

## Appendix C. Results on the Kummer Hypergeometric Function for Integer Arguments

For any $n, m \in \mathbb{N}$, with $n < m$, we have

$$\begin{aligned}{}_1F_1(n,m,z) = (-1)^n\frac{(m-1)!}{z^{m-1}}\Bigg[&\sum_{k=0}^{m-n-1}\frac{z^k}{\Gamma(n)\,k!}\,\frac{\Gamma(m-1-k)}{\Gamma(m-n-k)} \\ &-e^z\sum_{k=0}^{n-1}\frac{(-z)^k}{\Gamma(n-k)\,k!}\,\frac{\Gamma(m-1-k)}{\Gamma(m-n)}\Bigg],\end{aligned}$$

what yields for example, for $r \in \mathbb{N}$,

$${}_1F_1(r,r+1,-\lambda z) = \frac{r!}{(\lambda z)^r}\left[1-e^{-\lambda z}\sum_{k=0}^{r-1}\frac{(\lambda z)^k}{k!}\right]$$

where

$$1-e^{-\lambda z}\sum_{k=0}^{r-1}\frac{(\lambda z)^k}{k!}$$

is the c.d.f. of a Gamma random variable with integer shape parameter $r \in \mathbb{N}$ and rate parameter $\lambda$.

## References


Abramowitz, M., Stegun, I. A. (1974). *Handbook of Mathematical Functions.* (eds.). 9-th print., Dover, New York.

Anderson, T. W. (1984). *An Introduction to Multivariate Statistical Analysis*, 2nd ed., J. Wiley & Sons, New York.

Coelho, C. A. (1992). *Generalized Canonical Analysis*, Ph.D. thesis. The University of Michigan, Ann Arbor, MI.

Coelho, C. A. (1998). The Generalized Integer Gamma Distribution – A Basis for Distributions in Multivariate Statistics. *J. Multivariate Anal.* **64**, 86–102.

Coelho, C. A. (1999). Addendum to the paper "The Generalized Integer Gamma Distribution – A Basis for Distributions in Multivariate Statistics". *J. Multivariate Anal.* **69**, 281–285.

Prudnikov, A. P., Brychkov, Yu. A. and Marichev, O. I. (1990). *Integrals and Series*, Vol III, Gordon and Breach Sci. Publ., New York (Trans. in English by G. G. Gould).

Wilks, S. S. (1932). Certain generalizations in the analysis of variance. *Biometrika* **24**, 471–494.

Wilks, S. S. (1935). On the independence of *k* sets of normally distributed statistical variables. *Econometrica* **3**, 309–326.

Wilks, S. S. (1946). Sample criteria for testing equality of means, equality of variances, and equality of covariances in a normal multivariate distribution. *Ann. Math. Stat.* **17**, 257–281.

# The Wrapped Gamma Distribution and Wrapped Sums and Linear Combinations of Independent Gamma and Laplace Distributions

**Carlos A. Coelho**

**Abstract** In this paper we first obtain an expression for the probability density function of the wrapped or circular Gamma distribution and then we show how it may be seen, both for integer and non-integer shape parameter, as a mixture of truncated Gamma distributions. Some other properties of the wrapped Gamma distribution are studied and it is shown how this distribution and mixtures of these distributions may be much useful tools in modelling directional data in biology and meteorology. Based on the results obtained, namely the ones concerning mixtures, and on some properties of the distributions of the sum of independent Gamma random variables, the wrapped versions of the distributions of such sums, for both integer and non-integer shape parameters are derived. Also the wrapped sum of independent generalized Laplace distributions is introduced as a particular case of a mixture of wrapped Gamma distributions. Among the particular cases of the distributions introduced there are symmetrical, slightly skewed and highly skewed wrapped distributions as well as the recently introduced wrapped Exponential and Laplace distributions.

**Keywords** Mixtures · Truncated Gamma distributions · Circular data · Circular distributions · Wrapped symmetrical and skew distributions

---

Reprinted from Journal of Statistical Theory and Practice, with permission from Springer Nature. First published as Coelho, C. A. The wrapped Gamma distribution and wrapped sums and linear combinations of independent Gamma and Laplace distributions. Journal of Statistical Theory and Practice, 1, 1–29 (2007). (https://link.springer.com/article/10.1080/15598608.2007.10411821)

---

C. A. Coelho (✉)
Mathematics Department, Faculty of Science and Technology, The New University of Lisbon, Lisbon, Portugal
e-mail: cmac@fct.unl.pt



B. C. Arnold et al. (eds.), *Likelihood Ratio Tests and Related Topics in Multivariate Analysis*, Contributions to Statistics,
https://doi.org/10.1007/978-3-031-68008-3_6

## 1 Introduction

The wrapped Exponential and Laplace distributions have been recently introduced and studied by Jammalamadaka and Kozubowski (2003, 2004) and their adequacy to model directional data in biology was shown. However, the wrapped or circular Gamma distribution and associated distributions have never been adequately studied, although they may be even better candidates to model directional data in several areas of research such as biology and meteorology.

Although wrapping the Exponential distribution is equivalent to only truncate it at $2\pi$, wrapping the Gamma distribution raises other interesting problems and opens interesting views. In this paper, expressions for both the pdf (probability density function) and the cdf (cumulative distribution function) of the wrapped Gamma distribution are obtained under much manageable forms. A couple of alternative representations, based on mixtures, are obtained. These representations show that for both integer and non-integer values of the shape parameter the wrapped Gamma distribution may be seen as a mixture of Gamma distributions truncated at $2\pi$ and they are much useful in obtaining manageable expressions for the cdf of the wrapped Gamma distribution. These mixtures will be finite for integer values of the shape parameter and infinite when the shape parameter assumes a non-integer value.

To establish the notation and nomenclature used, we will say that the r.v. (random variable) $X$ has a Gamma distribution with shape parameter $r$ ( $>$ 0) and rate parameter $\lambda$ ( $>$ 0) if its pdf may be written as

$$f_X(x) = \frac{\lambda^r}{\Gamma(r)} e^{-\lambda x} x^{r-1}, \quad (x > 0).$$

We will represent this fact by

$$X \sim \Gamma(r, \lambda),$$

and we will use $\Gamma(r, \lambda)$ to denote such distribution. If $r$ is an integer we will call the Gamma distribution an integer Gamma distribution.

We will also use

$$\gamma(r; \lambda y) = \int_0^y \lambda^r e^{-\lambda x} x^{r-1} dx \tag{1}$$

to denote the incomplete Gamma function and the notation

$$\gamma^*(r; \lambda y) = \frac{\gamma(r; \lambda y)}{\Gamma(r)}. \tag{2}$$

With this notation, the cdf of the r.v. $X \sim \Gamma(r, \lambda)$ is given by

$$F_X(x) = P(X \leq x) = \gamma^*(r; \lambda x) = \frac{\gamma(r; \lambda x)}{\Gamma(r)}$$

and the pdf of a Gamma distribution with shape parameter $r$ and rate parameter $\lambda$, truncated at $b > 0$, written as

$$f(x) = \frac{\frac{\lambda^r}{\Gamma(r)} e^{-\lambda x} x^{r-1}}{\gamma^*(r; \lambda b)} = \frac{\lambda^r}{\gamma(r; \lambda b)} e^{-\lambda x} x^{r-1} .$$

Hereon, we will denote the fact that the r.v. $X$ has a Gamma distribution with shape parameter $r$ and rate parameter $\lambda$, truncated at $b > 0$, by

$$X \sim \Gamma\Big(r, \lambda; [0, b)\Big) ,$$

and we will use $\Gamma\Big(r, \lambda; [0, b)\Big)$ to denote such a distribution.

Yet, to shorten the notation, we will use: i) $\mathbb{N}$ to denote the set of positive integers and $\mathbb{N}_0$ to denote the set of non-negative integers; ii) $\mathbb{R}$ to denote the set of all reals, with $\mathbb{R}^+$ for the positive, $\mathbb{R}_0^+$ for the non-negative, and $\mathbb{R}_0^-$ for the non-positive reals.

## 2 The Wrapped Gamma Distribution

### 2.1 A First Representation for the Pdf and Some Properties of the Distribution

Consider

$$\Phi(z, s, a) = \sum_{k=0}^{\infty} \frac{z^k}{(a+k)^s} , \tag{3}$$

where any term with $a+k = 0$ is excluded. The function $\Phi(z, s, a)$ is usually called Lerch's transcendental function and it is a generalization of both Riemann's generalized Zeta function and the Polylogarithm function. The generalized Riemann's Zeta function is defined as

$$\xi(s, a) = \sum_{k=0}^{\infty} \frac{1}{(a+k)^s} = \Phi(1, s, a)$$

where any term with $a+k=0$ is also excluded, while the Polylogarithm function is defined as

$$\begin{aligned} PLog(z,s) &= \sum_{k=1}^{\infty} \frac{z^k}{k^s} \\ &= \frac{1}{\Gamma(s)} \int_0^1 \frac{z\,(-\log t)^{s-1}}{1-tz}\, dt \;=\; \Phi(z,s,0)\,. \end{aligned} \tag{4}$$

Let $X \sim \Gamma(r,\lambda)$ and let

$$\theta \equiv \theta(X) = X \ \mathrm{mod}\ 2\pi\,.$$

Then $\theta$ is a wrapped (around the circle) or circular Gamma r.v., that for $t \in [0,2\pi)$ has pdf

$$\begin{aligned} f_\theta(t) &= \sum_{k=0}^{\infty} f_X(t+2k\pi) \\ &= \frac{\lambda^r}{\Gamma(r)} \sum_{k=0}^{\infty} e^{-\lambda(t+2k\pi)}\,(t+2k\pi)^{r-1} \\ &= \frac{\lambda^r}{\Gamma(r)}\, e^{-\lambda t}\,(2\pi)^{r-1} \sum_{k=0}^{\infty} e^{-2\lambda\pi k} \left(\frac{t}{2\pi}+k\right)^{r-1} \\ &= \frac{\lambda^r}{\Gamma(r)}\, e^{-\lambda t}\,(2\pi)^{r-1}\, \Phi\left(e^{-2\lambda\pi}, 1-r, \frac{t}{2\pi}\right). \end{aligned} \tag{5}$$

We will denote the fact that the r.v. $\theta$ has a wrapped Gamma distribution with shape parameter $r$ and rate parameter $\lambda$, with the above pdf, by

$$\theta \sim W\Gamma(r,\lambda)\,.$$

The representation for the pdf of the wrapped Gamma distribution in (5) is rather convenient since Lerch's function is already well implemented in a number of software packages like Mathematica (from Wolfram Research). However, it may present some problems in terms of the determination of a manageable representation for the cdf.

Alternative representations of the pdf of the wrapped Gamma distribution may be obtained directly from the series expression in (5), under the form of mixtures. These representations will shed some more light on the intrinsic structure of the wrapped Gamma distribution and will also enable us to obtain simple representations for the cdf. Details are shown in the next subsection, where it is also shown that if the shape parameter $r$ assumes integer values then negative values of the rate parameter

$\lambda$ are allowed, producing mirrored distributions (around the origin) of the ones with positive rate parameter.

Concerning the arbitrariness of the choice of the origin we should rather write the pdf of the $W\Gamma(r, \lambda)$ distribution as

$$f_\theta(t) = \frac{\lambda^r}{\Gamma(r)} e^{-\lambda(t-a)} (2\pi)^{r-1} \Phi\left(e^{-2\lambda\pi}, 1-r, \frac{t-a}{2\pi}\right)$$

for $t - a \in [0, 2\pi)$ or $t \in [a, a + 2\pi)$, $a$ being the origin. However, for the sake of simplicity, we will use $a = 0$ throughout this manuscript, thus using the notation in (5).

We should note that as either $\lambda \to 0$ or $r \to \infty$, the $W\Gamma(r, \lambda)$ distribution tends to the wrapped uniform distribution, with density

$$f(t) = \frac{1}{2\pi}, \qquad t \in [0, 2\pi).$$

This fact may be analyzed in Figures 2 and 3 and it is derived from the facts that

$$\lim_{r\to+\infty} \frac{\Phi\left(e^{-2\lambda\pi}, 1-r, \frac{t}{2\pi}\right)}{e^{\lambda t}\, \Gamma(r)\, (2\lambda\pi)^{-r}} = 1 \tag{6}$$

and

$$\lim_{\lambda\to0^+} \frac{\Phi\left(e^{-2\lambda\pi}, 1-r, \frac{t}{2\pi}\right)}{\Gamma(r)\, (2\lambda\pi)^{-r}} = 1, \tag{7}$$

so that

$$\lim_{r\to+\infty} f_\theta(t) = \lim_{r\to+\infty} \frac{\lambda^r}{\Gamma(r)} e^{-\lambda t} (2\pi)^{r-1} \Phi\left(e^{-2\lambda\pi}, 1-r, \frac{t}{2\pi}\right) = \frac{1}{2\pi}$$

and also

$$\lim_{\lambda\to0^+} f_\theta(t) = \lim_{\lambda\to0^+} \frac{\lambda^r}{\Gamma(r)} e^{-\lambda t} (2\pi)^{r-1} \Phi\left(e^{-2\lambda\pi}, 1-r, \frac{t}{2\pi}\right) = \frac{1}{2\pi}.$$

The rate of convergence of the limit in (6) is higher for smaller values of $\lambda$, while the rate of convergence of the limit in (7) is somehow erratic for different values of $r$. See Figure 1.

The graphical aspects of the $W\Gamma(r, \lambda)$ distribution are much variable, somehow even more than for its linear counterpart. Contrary to what happens with the linear Gamma distribution, with the $W\Gamma(r, \lambda)$ distribution the lower point or antimode of the pdf is not always placed at the origin (see Figures 2, 3 and 4).

The mode is, for sufficiently large $\lambda$ and $r > 1$ (the largest is $r$, the largest has $\lambda$ to be), placed a little below $t = \frac{r-1}{\lambda}$. For $0 < r \leq 1$ it is placed exactly on the

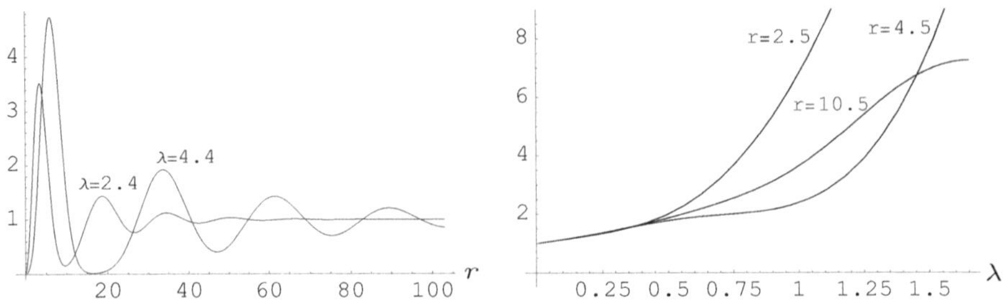


**Fig. 1** Plots of the functions in (6) and (7), to illustrate the rates of convergence.

right of $t = 0$ and the antimode on the left of $t = 2\pi$. For $1 < r \leq 2$ the antimode is placed on the right of $t = 0$. For $r = 2$ the mode is at $t = \frac{1}{\lambda} - \frac{2\pi}{e^{2\lambda\pi}-1}$, while for $r = 3$ the value of $t$ where the mode is placed already has a very complicated expression,

$$t = \frac{1}{\lambda} + \frac{2\lambda\pi\left(e^{2\lambda\pi}-1\right) - \sqrt{\sum_{i=0}^{4}\binom{4}{i}(-1)^i\, e^{2\lambda\pi i} + 4\lambda^2\pi^2\sum_{i=0}^{2}(-1)^{i+1}\binom{2}{i}e^{2\lambda\pi i}}}{\lambda\sum_{i=0}^{2}(-1)^{i+1}\binom{2}{i}\, e^{2\lambda\pi i}}$$

with the antimode at

$$t = \frac{1}{\lambda} + \frac{2\lambda\pi\left(e^{2\lambda\pi}-1\right) + \sqrt{\sum_{i=0}^{4}\binom{4}{i}(-1)^i\, e^{2\lambda\pi i} + 4\lambda^2\pi^2\sum_{i=0}^{2}(-1)^{i+1}\binom{2}{i}e^{2\lambda\pi i}}}{\lambda\sum_{i=0}^{2}(-1)^{i+1}\binom{2}{i}\, e^{2\lambda\pi i}}.$$

The mode and the antimode are anyway, for any value of $r$ and $\lambda$, at the values of $t$ for which $\frac{d}{dt} f_\theta(t)$ does not exist or $\frac{d}{dt} f_\theta(t) = 0$.

Since[1]

$$\frac{d}{dt}\Phi\left(e^{-2\lambda\pi}, 1-r, \frac{t}{2\pi}\right) = \frac{1}{2\pi}\Phi\left(e^{-2\lambda\pi}, 2-r, \frac{t}{2\pi}\right)$$

as it is easy to derive from the definition of Lerch's Phi function, the abscissas for the mode and the antimode are the values of $t \in [0, 2\pi)$ such that $\frac{d}{dt} f_\theta(t)$ does not exist or

$$\frac{d}{dt} f_\theta(t) = 0 \iff \Phi\left(e^{-2\lambda\pi}, 1-r, \frac{t}{2\pi}\right) = \frac{r-1}{2\lambda\pi}\Phi\left(e^{-2\lambda\pi}, 2-r, \frac{t}{2\pi}\right).$$

[1] see author and editors note at the end of the Chapter

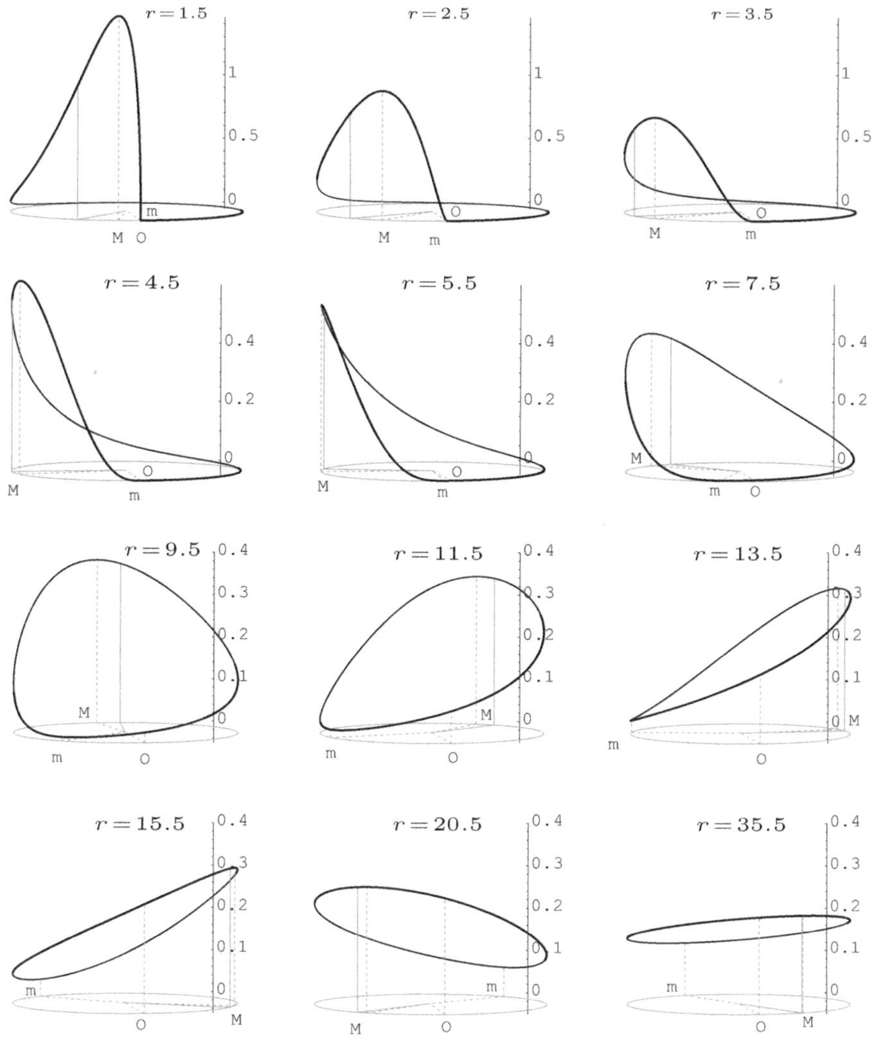


**Fig. 2** Three dimensional views of wrapped Gamma densities for $\lambda = 2.8$ and increasing values of $r$. All plots have the origin placed at the same spot. In each plot are marked the origin (O), the mode (M), the antimode (m) and the mean direction, this one marked with a continuous line (for a definition of the mean direction see section 3). (All plots on the same row have the same vertical scale but plots in different rows have different vertical scales)

For values of $r \leq 1$ the mode is at the point $t = 0$, where $\frac{d}{dt} f(t)$ is not defined.

As it happens with the wrapped Exponential distribution, also the integer wrapped Gamma distribution is defined for $\lambda < 0$, with the pdf for $\lambda < 0$ yielding a symmetrical aspect (relative to the origin) of the one it does have for positive $\lambda$. More precisely, we have the following Theorem.

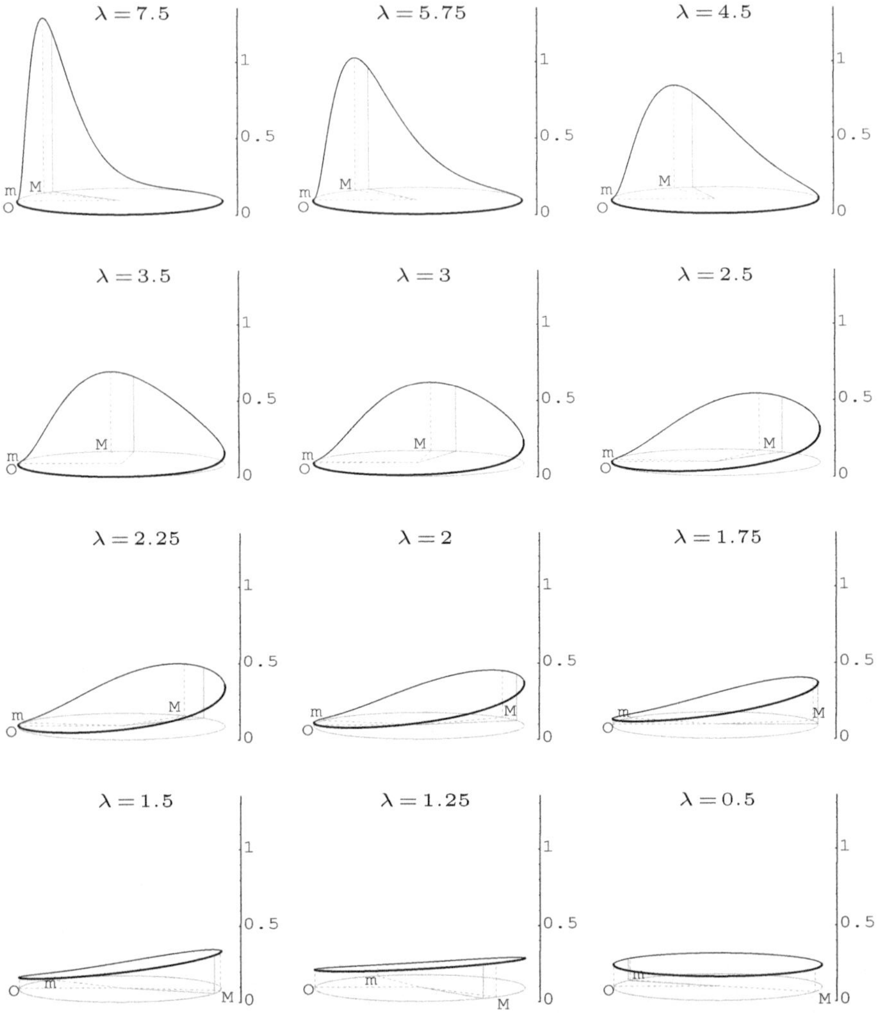


**Fig. 3** Three-dimensional views of wrapped Gamma densities for $r = 6.1$ and increasing values of $\lambda$. All plots have the origin placed at the same spot. In each plot are marked the origin (O), the mode (M), the antimode (m) and the mean direction, this one marked with a continuous line (for a definition of the mean direction see section 3). All plots have the same vertical scale

**Theorem 1** *For positive integer r, if*

$$\theta_1 \sim W\Gamma(r, \lambda) \quad \text{and} \quad \theta_2 \sim W\Gamma(r, -\lambda)$$

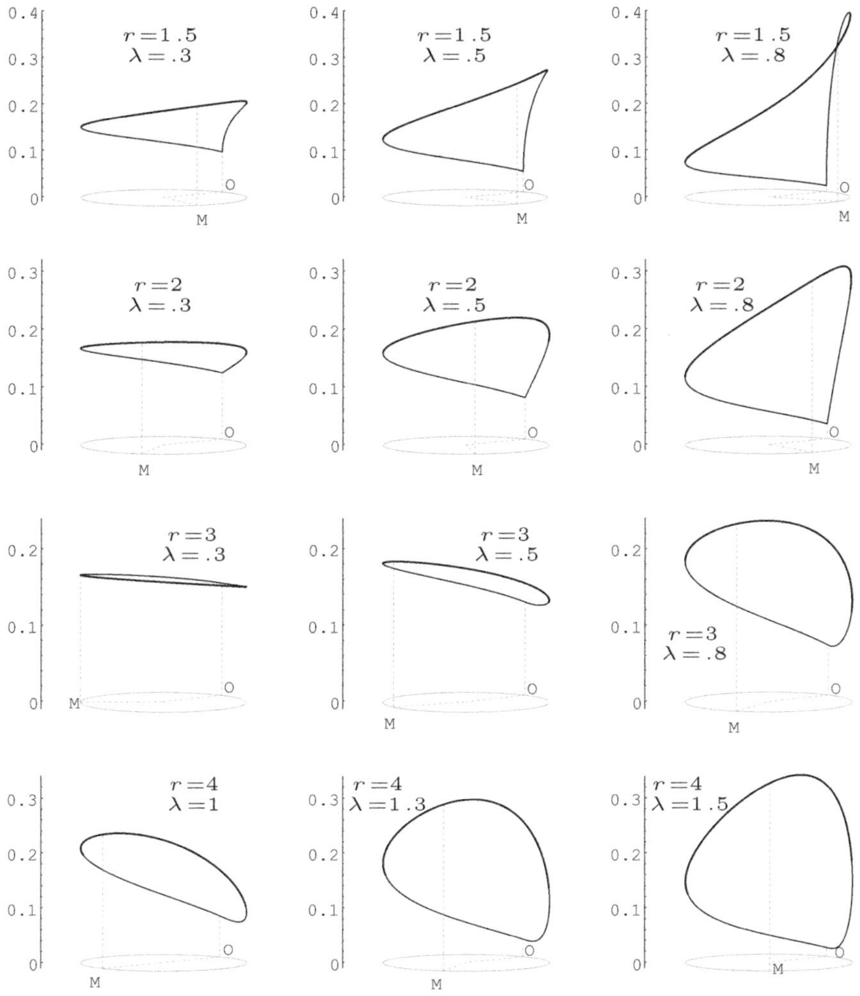


**Fig. 4** Three-dimensional views of wrapped Gamma densities for different values of $\lambda$ and $r$. All plots have the origin placed at the same spot. In each plot are marked the origin (O) and the mean direction (M) (for a definition of the mean direction see section 3). Plots on the same row have the same vertical scale but plots on different rows may have different vertical scales

*then*

$$f_{\theta_1}(t) = f_{\theta_2}(2\pi - t) \qquad \text{or} \qquad f_{\theta_2}(t) = f_{\theta_1}(2\pi - t),$$

*that is, for positive integer r, wrapped Gama distributions with symmetric rate parameters have pdf's that are mirrored about the origin.*

***Proof*** We may note that for positive integer $r$ we may write the pdf of the linear $\Gamma(r, \lambda)$ distribution as

$$
\begin{aligned}
f(x; r, \lambda) &= \frac{\lambda^r}{\Gamma(r)} e^{-\lambda x} x^{r-1} \\
&= -\frac{(-\lambda)^r}{\Gamma(r)} e^{-(-\lambda)(-x)} (-x)^{r-1} \\
&= -f(-x; r, -\lambda)
\end{aligned}
$$

what shows that for positive integer $r$, the symmetric of $f(x; r, \lambda)$ with $\lambda(> 0)$ replaced by $-\lambda$, is a pdf on $\mathbb{R}^-$, mirrored from $f(x; r, \lambda)$.

Using this result we may easily show that for positive integer $r$, we have for the $W\Gamma(r, \lambda)$ distribution (with $\lambda > 0$),

$$
f_\theta(2\pi - t; r, -\lambda) = f_\theta(t; r, \lambda), \quad (t \in [0, 2\pi))
$$

since for $t \in [0, 2\pi)$,

$$
\begin{aligned}
f_\theta(2\pi - t; r, -\lambda) &= \sum_{k=-1}^{-\infty} -f(2\pi - t + 2k\pi; r, -\lambda) = \sum_{k=-1}^{-\infty} -f(-t + 2\pi(k+1); r, -\lambda) \\
&= \sum_{k=0}^{-\infty} -f(-t + 2\pi k; r, -\lambda) = \sum_{k=0}^{\infty} -f(-t - 2\pi k; r, -\lambda) \\
&= \sum_{k=0}^{\infty} f(t + 2\pi k; r, \lambda) = f_\theta(t; r, \lambda),
\end{aligned}
$$

where the equality from the row preceeding the last to the last row is only valid for $r \in \mathbb{N}$. ■

The definition of Lerch's Phi function in most packages, as it happens with the software Mathematica, usually has this duality built in, so that if we implement the pdf of the $W\Gamma(r, \lambda)$ distribution through the use of Lerch's Phi function, and if then we take, for $r \in \mathbb{N}$ and $\lambda > 0$, $f_\theta(2\pi - t; r, -\lambda)$, we obtain the same values as for $f_\theta(t; r, \lambda)$.

Also for wrapped or circular distributions the usual properties of mixtures hold. Namely:

i) if $f_i(t)$ $(i = 0, 1, \dots)$ are pdf's of wrapped distributions and if $\sum_{i=0}^{n} p_i = 1$, with $0 < p_i < 1$ and where $n$ may be either finite or infinite, then

$$
f(t) = \sum_{i=0}^{n} p_i\, f_i(t)
$$

is the pdf of a wrapped distribution;

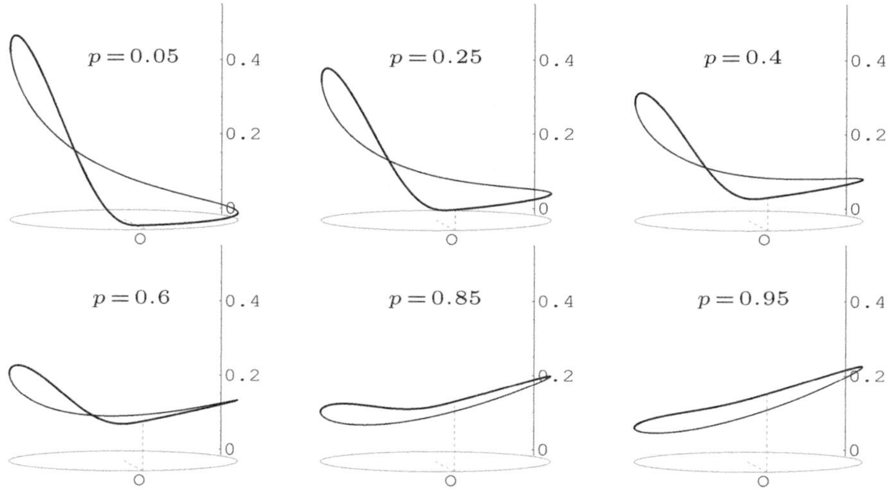


**Fig. 5** Three-dimensional views of the densities of the mixture of two wrapped Gamma distributions, $p * W\Gamma(2.5, 14.1) + (1 - p) * W\Gamma(2.5, 4.5)$, for different values of $p$. All plots have the origin placed at the same spot and all plots have the same vertical scale

ii) if $\Phi_i(p)$ $(i = 0, 1, \ldots)$ are the c.f.'s (characteristic functions) of the wrapped distributions with pdf's $f_i(t)$ in i), then

$$\Phi(p) = \sum_{i=0}^{n} p_i \, \Phi_i(p)$$

is the c.f. of the wrapped mixture distribution with pdf $f(t)$ in i) and thus the cdf's and moments also mix.

Mixtures of wrapped Gamma distributions are extremely flexible and are a powerful modeling tool for circular data, adequate to model both symmetric and asymmetric circular data with more than one mode, since mixtures of $k$ wrapped Gamma distributions may have from 1 to $k$ modes. Some plots of densities of mixtures of wrapped Gamma distributions are in Figures 5, 6 and 7.

Linearized plots of the pdf's may here be useful to better analyze the location and number of modes, although the circular plots may be also quite elucidative.

In Figures 5 and 6 we may see how simple changes in the weight may dramatically affect the aspect of the density of the mixture of two wrapped Gamma distributions, while in Figures 7 and 8 we may see how mixtures of wrapped Gamma distributions may exhibit multimodal densities.

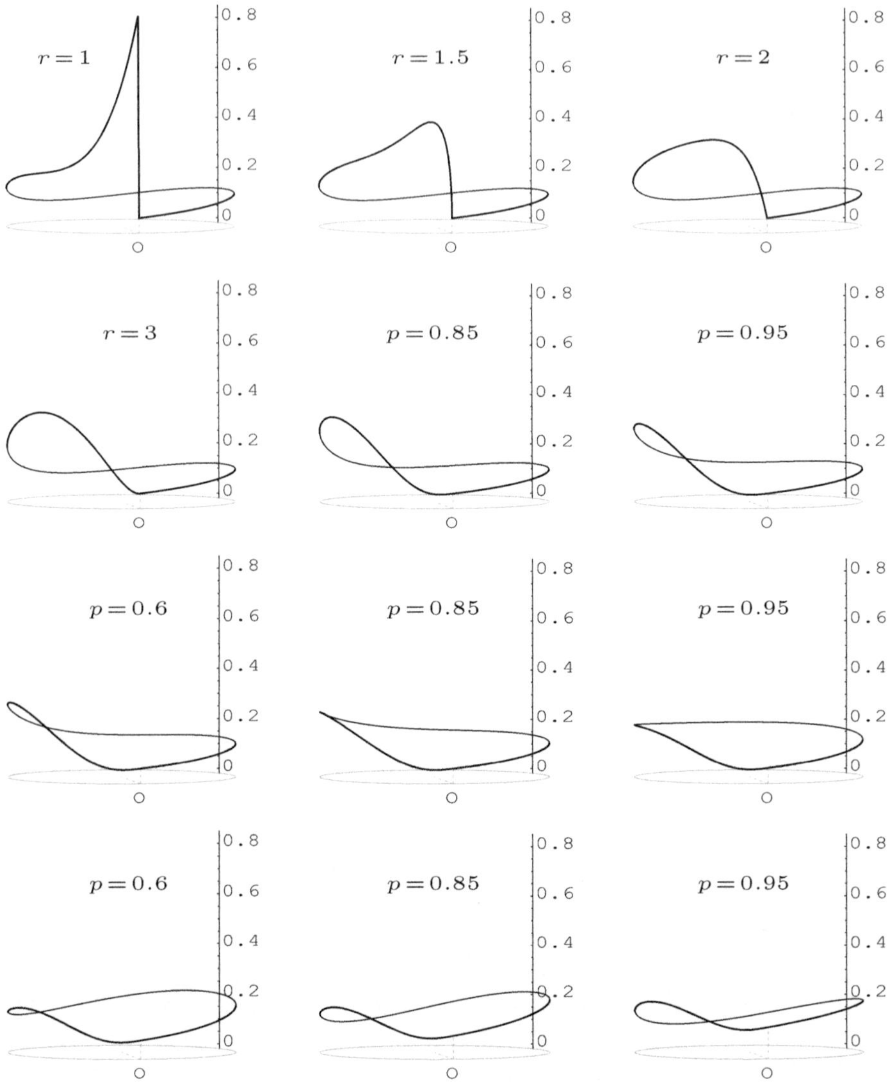


**Fig. 6** Three-dimensional views of the densities of the mixture of three wrapped Gamma distributions, $3/7 * W\Gamma(2.8, 13.5) + 2/7 * W\Gamma(2.8, 4.5) + 2/7 * W\Gamma(2.8, r)$, for different values of $r$. All plots have the origin placed at the same spot and all plots have the same vertical scale

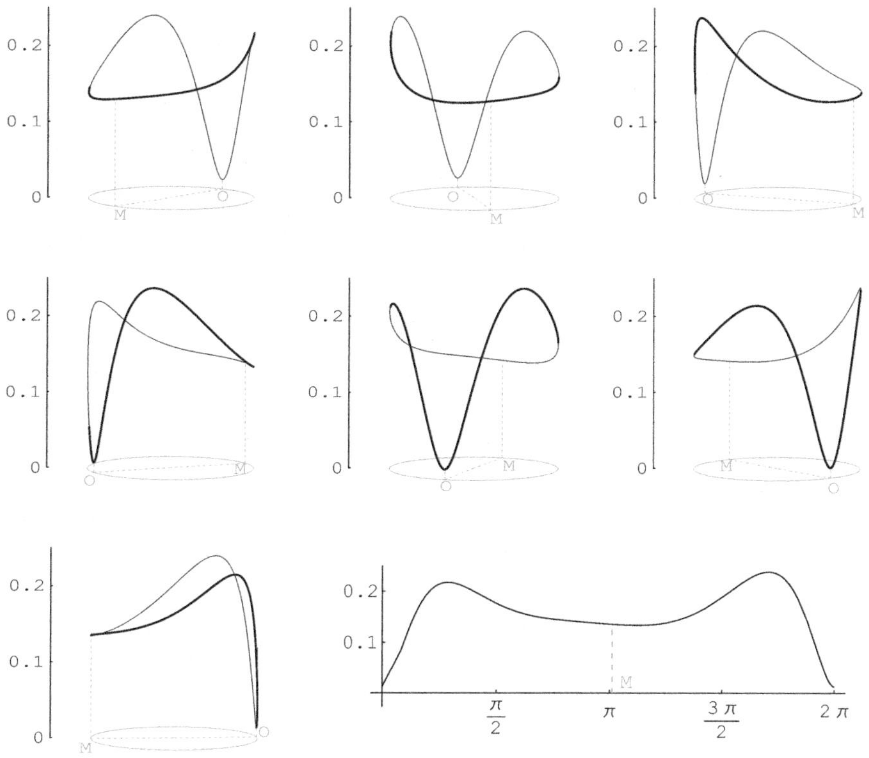


**Fig. 7** Three dimensional views of the density of a mixture of three wrapped Gamma distributions: $1/3 * W\Gamma(-2.3, 3) + 1/3 * W\Gamma(2, 7) + 1/3 * W\Gamma(2.3, 3)$ and linear plot of the same density. In each plot are marked the origin (O) and the mean direction (M)

## 2.2 *Alternative Representations for the pdf of the Wrapped Gamma Distribution*

The following two theorems establish two results which show how the wrapped Gamma distribution may be seen as a mixture of truncated linear Gamma distributions both for integer and non-integer $r$.

**Theorem 2** *For non-integer $r$ the $W\Gamma(r, \lambda)$ distribution is a mixture with weights*

$$\begin{aligned} p_i &= \gamma^*(i+1; 2\lambda\pi)\,(-1)^i\,\frac{\lambda^{r-i-1}}{\Gamma(r)}\frac{\Gamma(1-r+i)}{\Gamma(1-r)} \\ &\quad \times PLog\left(e^{-2\lambda\pi}, i+1-r\right),\; i=0, 1, \ldots \end{aligned} \tag{8}$$

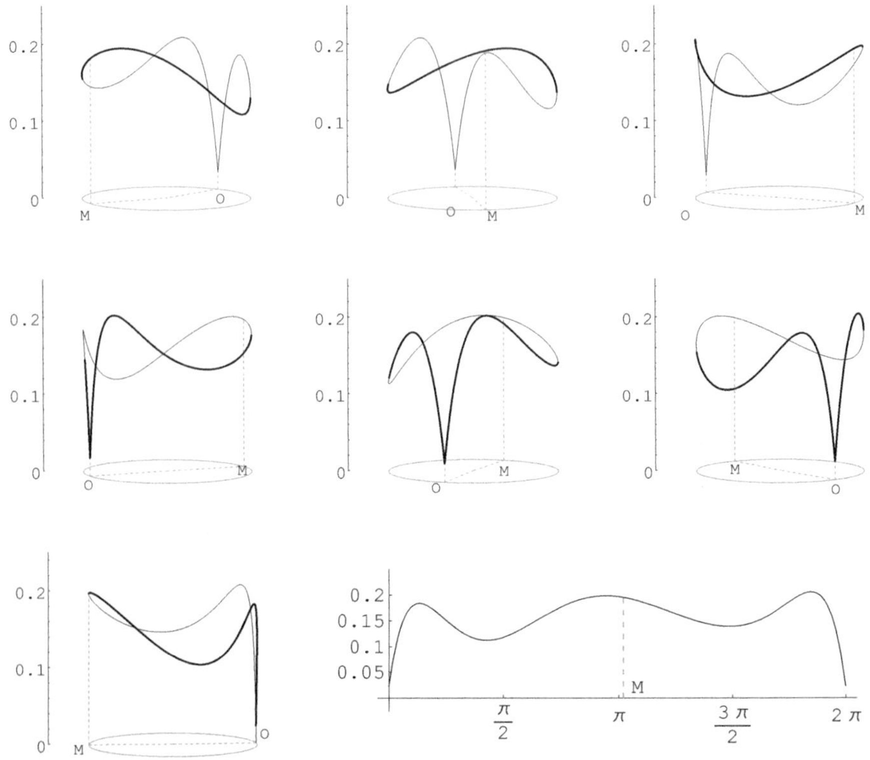


**Fig. 8** Three dimensional views of the density of a mixture of three wrapped Gamma distributions: $.2 * W\Gamma(-2.3, 2) + .6 * W\Gamma(2, 7) + .2 * W\Gamma(2.3, 2)$ and linear plot of the same density. In each plot are marked the origin (O) and the mean direction (M)

*of infinitely many* $\Gamma(i+1, \lambda; [0, 2\pi))$ *distributions* $(i = 0, 1, \ldots)$ *with an additional* $\Gamma(r, \lambda; [0, 2\pi))$ *distribution with weight*

$$p^* = \gamma^*(r, 2\lambda\pi)\,. \tag{9}$$

***Proof*** Since for $t \in [0, 2\pi)$ and $k \geq 1$ we have $\left|\frac{t}{2k\pi}\right| < 1$, we may write, for $c \notin \mathbb{N}_0$,

$$\left(1 + \frac{t}{2k\pi}\right)^c = {}_1F_0\left(-c; -\frac{t}{2k\pi}\right) = \sum_{i=0}^{\infty} \frac{\Gamma(-c+i)}{\Gamma(-c)} \frac{(-1)^i}{i!} \left(\frac{t}{2k\pi}\right)^i .$$

Then, given the uniform convergence of the series involved, we may write, directly from the definition of $f_\theta(t)$ and from (5),[2]

$$\begin{aligned} f_\theta(t) &= \frac{\lambda^r}{\Gamma(r)} e^{-\lambda t} \sum_{k=0}^{\infty} e^{-2\pi\lambda k} (t + 2k\pi)^{r-1} \\ &= \frac{\lambda^r}{\Gamma(r)} e^{-\lambda t} t^{r-1} + \frac{\lambda^r}{\Gamma(r)} e^{-\lambda t} \sum_{k=1}^{\infty} e^{-2\pi\lambda k} \left(\frac{t}{2k\pi} + 1\right)^{r-1} (2k\pi)^{r-1} \\ &= \frac{\lambda^r}{\Gamma(r)} e^{-\lambda t} t^{r-1} \frac{\Gamma(r)}{\gamma(r; 2\lambda\pi)} \frac{\gamma(r; 2\lambda\pi)}{\Gamma(r)} \end{aligned}$$

$$\begin{aligned} &+ \frac{\lambda^r}{\Gamma(r)} e^{-\lambda t} \sum_{k=1}^{\infty} e^{-2\pi\lambda k} (2k\pi)^{r-1} \sum_{i=0}^{\infty} (-1)^i \left(\frac{t}{2k\pi}\right)^i \frac{\Gamma(1-r+i)}{\Gamma(1-r)} \frac{1}{i!} \\ &= \frac{\lambda^r}{\gamma(r; 2\lambda\pi)} e^{-\lambda t} t^{r-1} \frac{\gamma(r; 2\lambda\pi)}{\Gamma(r)} + \sum_{i=0}^{\infty} \frac{\lambda^{i+1}}{\gamma(i+1; 2\lambda\pi)} e^{-\lambda t} t^i \\ &\quad \frac{\gamma(i+1; 2\lambda\pi)}{\Gamma(i+1)} (-1)^i \frac{\lambda^{r-i-1}}{\Gamma(r)} \frac{\Gamma(1-r+i)}{\Gamma(1-r)} \sum_{k=1}^{\infty} e^{-2\pi\lambda k} (2k\pi)^{r-1-i} \\ &= \underbrace{\frac{\lambda^r}{\gamma(r; 2\lambda\pi)} e^{-\lambda t} t^{r-1}}_{\text{pdf of } \Gamma\left(r,\lambda;[0,2\pi)\right)} \underbrace{\gamma^*(r; 2\lambda\pi)}_{p^*} + \sum_{i=0}^{\infty} \underbrace{\frac{\lambda^{i+1}}{\gamma(i+1; 2\lambda\pi)} e^{-\lambda t} t^i}_{\text{pdf of } \Gamma\left(i+1,\lambda;[0,2\pi)\right)} \\ &\quad \underbrace{\gamma^*(i+1; 2\lambda\pi) (-1)^i \frac{\lambda^{r-i-1}}{\Gamma(r)} \frac{\Gamma(1-r+i)}{\Gamma(1-r)} PLog\left(e^{-2\lambda\pi}, i+1-r\right)}_{p_i} . \ \blacksquare \end{aligned} \tag{10}$$

We should note that indeed for the weights $p_i$ in (8), and $p^*$ in (9),

$$\sum_{i=0}^{\infty} p_i = 1 - p^* , \qquad \text{or} \qquad p^* + \sum_{i=0}^{\infty} p_i = 1 .$$

For negative integer $s$ the Lerch and Polylogarithm functions in (3) and (4) have simpler finite representations, which are useful to develop. Such representations will be used to obtain a finite mixture representation for the $W\Gamma(r, \lambda)$ distribution for $r \in \mathbb{N}$.

We will define, for positive integers $n$ and $i$, with $1 < i < n$,

$$t_{n,i} = (n - i + 1)\, t_{n-1,i-1} + i\, t_{n-1,i}$$

with

$$t_{n,n} = t_{n,1} = 1$$

[2] see author and editors note at the end of the Chapter

and

$$t_{n,i} = 0 \quad \text{for} \quad i > n \ \text{ or } \ i < 0\,.$$

We may note that as a consequence we have, as it is really desirable,

$$t_{n,0} = \begin{cases} 1\,, & n = 0 \\ 0\,, & n > 0 \end{cases} \qquad \text{and} \quad t_{n,i} = 0 \ \text{ for } \ n < 0\,.$$

Values of $t_{n,i}$ for $n = 1, \ldots, 10$ (and $1 \le i \le n$) are tabulated in Appendix A.

Then we may write, for positive integer $n$, and $z \in \mathbb{R}$

$$\begin{aligned} \Phi^*(z, n) &= \sum_{k=0}^{\infty} z^k\, k^n = \sum_{k=1}^{\infty} z^k\, k^n = \Phi(z, -n, 0) = PLog(z, -n) \\ &= \frac{\sum_{i=1}^{n} t_{n,i}\, z^{-i}}{\left(z^{-1} - 1\right)^{n+1}} = \frac{\sum_{i=1}^{n} t_{n,i}\, z^{n-i+1}}{(1 - z)^{n+1}} \end{aligned} \tag{11}$$

and for $n = 0$

$$\begin{aligned} \Phi^*(z, 0) &= \sum_{k=0}^{\infty} z^k\, k^0 = 1 + \sum_{k=1}^{\infty} z^k\, k^0 = 1 + \Phi(z, 0, 0) = 1 + PLog(z, 0) \\ &= \frac{z^{-1}}{z^{-1} - 1} = \frac{1}{1 - z}\,, \end{aligned}$$

so that we may write, for $n \in \mathbb{N}_0$,

$$\Phi^*(z, n) = \sum_{k=0}^{\infty} z^k\, k^n = \frac{t_{n,0} + \sum_{i=1}^{n} t_{n,i}\, z^{n+1-i}}{(1 - z)^{n+1}}\,. \tag{12}$$

**Theorem 3** *For $r \in \mathbb{N}$ the $W\Gamma(r, \lambda)$ distribution is a mixture with weights*

$$p_h = \gamma^*(h + 1; 2\lambda\pi)\, \frac{(2\pi\lambda)^{r-h-1}}{(r - h - 1)!}\, \Phi^*\left(e^{-2\lambda\pi}, r - h - 1\right), \qquad h = 0, \ldots, r - 1 \tag{13}$$

*of $r$ distributions $\Gamma\Big(h + 1, \lambda, [0, 2\pi)\Big)$, for $h = 0, \ldots, r - 1$.*

***Proof*** For $r \in \mathbb{N}$ we may write

$$\begin{aligned}
f_\theta(t) &= \frac{\lambda^r}{\Gamma(r)}\, e^{-\lambda t} \sum_{k=0}^{\infty} e^{-2\lambda\pi k}\, (t+2\pi k)^{r-1} \\
&= \frac{\lambda^r}{\Gamma(r)}\, e^{-\lambda t} \sum_{k=0}^{\infty} e^{-2\lambda\pi k} \sum_{h=0}^{r-1} \binom{r-1}{h}\, t^h\, (2\pi k)^{r-1-h} \\
&= \sum_{h=0}^{r-1} \frac{\lambda^{h+1}}{\Gamma(h+1)}\, e^{-\lambda t}\, t^h\, \frac{\lambda^{r-h-1}}{(r-h-1)!}\, (2\pi)^{r-1-h} \sum_{k=0}^{\infty} e^{-2\lambda\pi k}\, k^{r-1-h} \\
&= \sum_{h=0}^{r-1} \underbrace{\frac{\lambda^{h+1}}{\gamma(h+1;\, 2\lambda\pi)}\, e^{-\lambda t}\, t^h}_{\text{pdf of } \Gamma\left(h+1,\lambda;[0,2\pi)\right)} \\
&\qquad\qquad \underbrace{\gamma^*(h+1;\, 2\lambda\pi)\, \frac{(2\pi\lambda)^{r-h-1}}{(r-h-1)!}\, \Phi^*\left(e^{-2\lambda\pi}, r-h-1\right)}_{p_h}
\end{aligned} \tag{14}$$

where $\sum_{h=0}^{r-1} p_h = 1$, $\Phi^*(\cdot,\cdot)$ is given by (12) above, and where, for non-negative integer $h$,

$$\gamma^*(h+1;\, 2\lambda\pi) = \frac{\gamma(h+1;\, 2\lambda\pi)}{\Gamma(h+1)} = 1 - e^{-2\lambda\pi} \sum_{j=0}^{h} \frac{(2\lambda\pi)^j}{j!}. \quad \blacksquare$$

For the particular case $r = 1$ we have the wrapped Exponential distribution of Jammalamadaka and Kozubowski (2003, 2004).

We should note that the mixture expression obtained for non-integer $r$ still holds for integer $r$.

The mixture representations of the $W\Gamma(r, \lambda)$ distribution are useful for a number of reasons. Besides showing us an interesting result, they allow us to easily derive expressions for the cdf, both for integer and non-integer $r$. Using the incomplete Gamma function defined in (1) and (2) and the mixture representation in (10) we may write the cdf of the $W\Gamma(r, \lambda)$ distribution for non-integer $r$ as

$$F_\theta(t) = p^*\, \frac{\gamma(r;\, \lambda t)}{\gamma(r;\, 2\pi\lambda)} + \sum_{i=0}^{\infty} p_i\, \frac{\gamma(i+1;\, \lambda t)}{\gamma(i+1;\, 2\pi\lambda)} \tag{15}$$

for $p_i$ $(i = 0, 1, \ldots)$ given by (8) and $p^*$ given by (9), while for $r \in \mathbb{N}$ we may write, from (14), the cdf of the $W\Gamma(r, \lambda)$ distribution as

$$F_\theta(t) = \sum_{h=0}^{r-1} p_h \frac{\gamma(h+1; \lambda t)}{\gamma(h+1; 2\pi\lambda)} \tag{16}$$

for $p_h$ $(h = 0, \ldots, r-1)$ given in (13).

## 3 Trigonometric Moments and Related Parameters

The c.f. (characteristic function) of the $W\Gamma(r, \lambda)$ distribution is

$$\Phi(p) = \lambda^r(\lambda - ip)^{-r}, \qquad p = \pm 1, \pm 2, \ldots,$$

where $i = (-1)^{1/2}$.

Using the fact that for $a > 0$ and $b, r \in \mathbb{R}$, we may write

$$(a - ib)^{-r} = \left(a^2 + b^2\right)^{-r/2} e^{ir \,\arctan(b/a)}$$

we have for $\lambda > 0$,

$$\Phi(p) = \lambda^r \left(\lambda^2 + p^2\right)^{-r/2} e^{ir \,\arctan(p/\lambda)} = \rho_p \, e^{i\mu_p}$$

with

$$\rho_p = \lambda^r \left(\lambda^2 + p^2\right)^{-r/2}$$

and

$$\mu_p = r \,\arctan(p/\lambda),$$

where $\arctan(\cdot)$ is the usual inverse function of the tangent function, with counter-domain $]-\pi/2, \pi/2[$. This way, the non-central trigonometric moments are

$$\begin{aligned}\alpha_p &= \rho_p \cos(\mu_p) = \lambda^r \left(\lambda^2 + p^2\right)^{-r/2} \cos\left(r \,\arctan(p/\lambda)\right)\\ \beta_p &= \rho_p \sin(\mu_p) = \lambda^r \left(\lambda^2 + p^2\right)^{-r/2} \sin\left(r \,\arctan(p/\lambda)\right)\end{aligned}$$

with

$$\Phi(p) = \alpha_p + i\,\beta_p, \quad p = \pm 1, \pm 2, \ldots,$$

and the central trigonometric moments are given by

$$\overline{\alpha}_p = \rho_p \cos(\mu_p - p\mu_1) = \lambda^r \left(\lambda^2 + p^2\right)^{-r/2} \cos\left(r \arctan(p/\lambda) - pr \arctan(1/\lambda)\right)$$
$$\overline{\beta}_p = \rho_p \sin(\mu_p - p\mu_1) = \lambda^r \left(\lambda^2 + p^2\right)^{-r/2} \sin\left(r \arctan(p/\lambda - pr \arctan(1/\lambda)\right) .$$

The resultant length and mean direction are thus respectively

$$\rho = \rho_1 = \lambda^r \left(\lambda^2 + 1\right)^{-r/2}$$

and

$$\mu_1 = r \arctan(1/\lambda) .$$

The circular variance is

$$V_0 = 1 - \rho = 1 - \lambda^r \left(\lambda^2 + 1\right)^{-r/2}$$

and the circular standard deviation is

$$\sigma_0 = \sqrt{-2\log \rho} = \sqrt{-r \log\left(1 + \frac{1}{\lambda^2}\right)} ,$$

with the skewness and kurtosis not having any interesting expressions, being just defined through the usual expressions, respectively

$$\gamma_1^0 = \overline{\beta}_2 / V_0^{3/2}$$

and

$$\gamma_2^0 = \frac{\overline{\alpha}_2 - \rho^4}{(1-\rho)^2} .$$

For $r \in \mathbb{N}$ and $\lambda < 0$ we have to use the fact that for $a < 0$, $b > 0$ and $r \in \mathbb{R}$, we may write

$$(a - ib)^{-r} = \left(a^2 + b^2\right)^{-r/2} e^{ir(\pi + \arctan(b/a))}$$

so that in this case

$$\Phi(p) = \lambda^r \left(\lambda^2 + p^2\right)^{-r/2} e^{ir(\pi + \arctan(p/\lambda))}$$

with $\rho_p$ being the same as for $\lambda > 0$ and

$$\mu_p = r\,(\pi + \arctan(p/\lambda))$$

with the concomitant changes in the mean direction and all other moments and parameters depending on $\mu_p$.

All these results confirm, for $r = 1$, the results from Jammalamadaka and Kozubowski (2003, 2004), except for the mean direction with $\lambda < 0$, where there seems to be a small mistake in Jammalamadaka and Kozubowski (2003, 2004).

## 4 The Wrapped Sum and Linear Combination of Gamma Distributions and the Wrapped Generalized Integer Gamma Distribution

It may be shown that wrapped sums of independent Gamma distributions are particular types of mixtures of wrapped Gamma distributions.

In fact both for integer and non-integer shape parameters, the sum of independent Gamma distributions is a mixture of Gamma distributions. Thus its wrapped version is a mixture of wrapped Gamma distributions. The same applies to the wrapped version of linear combinations of independent Gamma distributions with all positive coefficients.

Let

$$X_i \sim \Gamma(r_i, \lambda_i)\,, \qquad i = 1, \ldots, n$$

be $n$ independent Gamma r.v.'s, and consider the r.v.

$$Z = \sum_{i=1}^{n} a_i\, X_i\,, \quad a_i \in \mathbb{R}^+,\ \forall i\,.$$

Then we may see the r.v. $Z$ as the sum of $n$ independent Gamma r.v.'s

$$Y_i = a_i\, X_i \sim \Gamma(r_i, \mu_i)$$

where $\mu_i = \lambda_i / a_i$ ($> 0$).

In the general case, Moschopoulos (1985) has shown that the distribution of $Z$ is an infinite mixture of $\Gamma(r + k, \mu)$ ($k = 0, 1, \ldots$) distributions, where

$$r = \sum_{i=1}^{n} r_i \quad \text{and} \quad \mu = \max_{1 \le i \le n} \mu_i \tag{17}$$

with weights

$$w_k = \delta_k \prod_{i=1}^{n} \left(\frac{\mu_i}{\mu}\right)^{r_i} \quad (k = 0, 1, \ldots) \tag{18}$$

where $\delta_0 = 1$ and for $k = 1, 2, \ldots$

$$\delta_k = \frac{1}{k} \sum_{j=1}^{k} \gamma_j \, \delta_{k-j} \tag{19}$$

with

$$\gamma_j = \sum_{i=1}^{n} r_i \left(1 - \frac{\mu_i}{\mu}\right)^j, \quad j = 1, 2, \ldots . \tag{20}$$

Thus the wrapped distribution of $Z$, that is, the distribution of the r.v.

$$\theta(Z) = Z \mod 2\pi$$

is an infinite mixture, with weights $w_k$ $(k = 0, 1, \ldots)$ of $W\Gamma(r+k, \mu)$ distributions, each one of which is itself an infinite mixture of truncated Gamma distributions at $2\pi$, unless $r \in \mathbb{N}$, in which case each of the $W\Gamma(r + k, \mu)$ $(k = 0, 1, \ldots)$ distributions is a finite mixture of $r + k$ truncated Gamma distributions.

In the general case, that is, for $r_i \in \mathbb{R}^+ (i = 1, \ldots, n)$, we may write the pdf of $\theta(Z)$ as

$$f_{\theta(Z)}(t) = \sum_{k=0}^{\infty} w_k \frac{\mu^{r+k}}{\Gamma(r+k)} e^{-\mu t} (2\pi)^{r+k-1} \Phi\left(e^{-2\mu\pi}, 1 - r - k, \frac{t}{2\pi}\right) \tag{21}$$

with $w_k$ given by (18) through (20) above, and $\mu$ and $r$ given by (17).

The corresponding cdf is then easy to derive from (21) above, using the incomplete Gamma function and the mixture structure of the distribution. For non-integer $r$, using (15) it may be written as

$$F_{\theta(Z)}(t) = \sum_{k=0}^{\infty} w_k \left( p_k^* \frac{\gamma(r+k; \mu t)}{\gamma(r+k; 2\pi\mu)} + \sum_{i=0}^{\infty} p_{ik} \frac{\gamma(i+1; \mu t)}{\gamma(i+1; 2\pi\mu)} \right)$$

with $w_k$, $r$ and $\mu$ defined as in (21) above and, from (9), with

$$p_k^* = \gamma^*(r + k; 2\mu\pi)$$

and, from (8), for $i, k = 0, 1, \ldots,$

$$p_{ik} = \gamma^*(i+1; 2\mu\pi)\,(-1)^i\,\frac{\mu^{r+k-i-1}}{\Gamma(r+k)}\,\frac{\Gamma(1-r-k+i)}{\Gamma(1-r-k)}\,(2\pi)^{r+k-1-i}\; PLog\left(e^{-2\mu\pi}, i+1-r-k\right),$$

while for integer $r$, taking (16) as a reference, we may write

$$F_{\theta(Z)}(t) = \sum_{k=0}^{\infty} w_k \sum_{h=0}^{r+k-1} p_{hk}\,\frac{\gamma(h+1;\mu t)}{\gamma(h+1; 2\pi\mu)}$$

with, from (13), for $h = 0, \ldots, r+k-1$ and $k = 0, 1, \ldots,$ and taking $\Phi^*(\cdot,\cdot)$ defined as in (11),

$$p_{hk} = \gamma^*(h+1; 2\mu\pi)\,\frac{(2\pi\mu)^{r+k-h-1}}{(r+k-h-1)!}\,\Phi^*\left(e^{-2\mu\pi}, r+k-h-1\right).$$

However, if all the $r_i$'s are integer, the distribution of $\theta(Z)$ may be shown to have a finite representation, since in this case the distribution of $Z$ is what Coelho (1998) called a GIG (Generalized Integer Gamma) distribution of depth $n$, with pdf

$$f_Y(y) = K \sum_{j=1}^{n} P_j(y)\, e^{-\mu_j y}$$

and cdf

$$F_Y(y) = 1 - K \sum_{j=1}^{n} P_j^*(y)\, e^{-\mu_j y}$$

where

$$K = \prod_{j=1}^{n} \mu_j^{r_j}, \qquad P_j(y) = \sum_{k=1}^{r_j} c_{jk}\, y^{k-1}$$

and

$$P_j^*(y) = \sum_{k=1}^{r_j} c_{jk}\,(k-1)! \sum_{i=0}^{k-1} \frac{y^i}{i!\,\mu_i^{k-i}}$$

with

$$c_{j,r_j} = \frac{1}{(r_j-1)!} \prod_{\substack{i=1\\ i\neq j}}^{n} (\mu_i - \mu_j)^{-r_i}, \qquad j = 1, \ldots, n,$$

and

$$c_{j,r_j-k} = \frac{1}{k} \sum_{i=1}^{k} \frac{(r_j - k + i - 1)!}{(r_j - k - 1)!} R(i, j, n)\, c_{j,r_j-(k-i)}\,, \quad \begin{array}{l}(k = 1, \ldots, r_j - 1)\\ (j = 1, \ldots, n)\end{array}$$

where

$$R(i, j, n) = \sum_{\substack{k=1\\k\neq j}}^{n} r_k \left(\mu_j - \mu_k\right)^{-i} \quad (i = 1, \ldots, r_j - 1)\,,$$

and where we assumed all $\mu_i$ $(i = 1, \ldots, n)$ to be different. In case $m$ of them $(2 \leq m \leq n)$ are equal, we have to add the corresponding shape parameters, and then consider the $n - m + 1$ different rate parameters and corresponding shape parameters to obtain a GIG distribution of depth $n - m + 1$.

The important point is that we may indeed look at the GIG distribution of $Z$ as a mixture of integer Gamma distributions, since we may write the pdf of $Z$ as

$$\begin{aligned} f_Z(z) &= K \sum_{j=1}^{n} \sum_{k=1}^{r_j} c_{jk}\, z^{k-1}\, e^{-\mu_j z} \\ &= \sum_{j=1}^{n} \sum_{k=1}^{r_j} \underbrace{\frac{K}{\mu_j^k}\, \Gamma(k)\, c_{jk}}_{p^*_{kj}}\, \frac{\mu_j^k}{\Gamma(k)}\, e^{-\mu_j z}\, z^{k-1} \end{aligned}$$

what shows that it is a mixture of $\sum_{j=1}^{n} r_j$ integer $\Gamma(k, \mu_j)$ distributions with weights

$$p^*_{kj} = \frac{K}{\mu_j^k}\, \Gamma(k)\, c_{jk}\,, \qquad k = 1, \ldots, r_j;\ j = 1, \ldots, n$$

with $\sum_{j=1}^{n} \sum_{k=1}^{r_j} p^*_{kj} = 1$.

As such, the wrapped GIG distribution of depth $n$ is a mixture with weights $p^*_{kj}$ of $\sum_{j=1}^{n} r_j$ integer $W\Gamma(k, \mu_j)$ distributions $(k = 1, \ldots, r_j;\ j = 1, \ldots, n)$, its pdf being given by

$$\begin{aligned} f_{\theta(Z)}(t) &= \sum_{j=1}^{n} \sum_{k=1}^{r_j} p^*_{kj}\, \frac{\mu_j^k}{\Gamma(k)}\, e^{-\mu_j t}\, (2\pi)^{k-1}\, \Phi\left(e^{-2\mu_j \pi}, 1 - k, \frac{t}{2\pi}\right) \\ &= K \sum_{j=1}^{n} \sum_{k=1}^{r_j} c_{jk}\, e^{-\mu_j t}\, (2\pi)^{k-1}\, \Phi\left(e^{-2\mu_j \pi}, 1 - k, \frac{t}{2\pi}\right) \end{aligned}$$

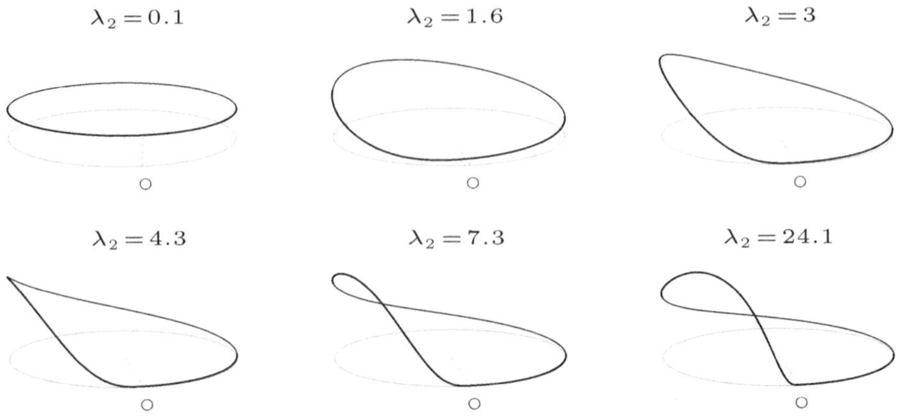


**Fig. 9** Three-dimensional views of densities of wrapped GIG distributions of depth 2 with shape parameters $r_1 = 2$, $r_2 = 3$ and rate parameters $\lambda_1 = 1.5$ and different values of $\lambda_2$. All plots have the origin placed at the same spot and all plots have the same vertical scale

or

$$f_{\theta(Z)}(t) = K \sum_{j=1}^{n} \sum_{k=1}^{r_j} \frac{\Gamma(k)}{\mu_j^k} c_{jk} \sum_{h=0}^{k-1} p_h \frac{\mu_j^{h+1}}{\Gamma(h+1)} \frac{e^{-\mu_j t}\, t^h}{\gamma^*(h+1; 2\mu_j \pi)}$$

and the cdf by

$$F_{\theta(Z)}(t) = K \sum_{j=1}^{n} \sum_{k=1}^{r_j} \frac{\Gamma(k)}{\mu_j^k} c_{jk} \sum_{h=0}^{k-1} p_h \frac{\gamma(h+1; \mu_j t)}{\gamma(h+1; 2\mu_j \pi)},$$

where $p_h$ $(h = 0, \ldots, k-1)$ is given by (13), with $\lambda$ replaced by $\mu_j$ and $r$ replaced by $k$.

We may then also easily consider mixtures of wrapped GIG distributions and of wrapped sums of Gammas.

On Figure 9 we may see the effect of the change in one of the rate parameters on the shape of the density of wrapped GIG distributions of depth 2, while on Figure 10 we may see the effect of the change in one of the shape parameters.

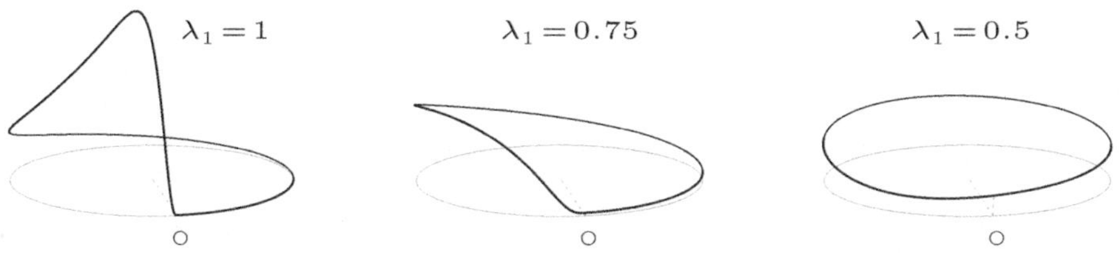


**Fig. 10** Three-dimensional views of densities of wrapped GIG distributions of depth 2 with shape parameters $r_1 = 2$, $r_2 = 3$ and rate parameters $\lambda_2 = 24.1$ and different values of $\lambda_1$. All plots have the origin placed at the same spot and all plots have the same vertical scale

# 5 The Wrapped Sum of Independent Generalized Laplace Distributions or the Wrapped Difference of Two Sums of Independent Exponentials

One important particular type of mixture of wrapped Gamma distributions is the mixture of two wrapped Exponential distributions, that is, two wrapped integer Gamma distributions with shape parameter equal to 1, weights equal to 1/2 and symmetric rate parameters. Such a distribution is a wrapped Laplace distribution (see Jammalamadaka and Kozubowski (2003, 2004)). Indeed one way of looking at the linear Laplace distribution is to see it as the difference of two independent Exponential distributions with the same rate parameter, what also turns out to be a mixture with weights 1/2 of two Exponential distributions with symmetrical rate parameters, one with support $\mathbb{R}_0^+$ and the other with support $\mathbb{R}_0^-$.

Indeed we may even define the generalized linear Laplace distribution with rate parameters $\lambda_1$ and $\lambda_2$, as the distribution with pdf

$$f(x) = \begin{cases} \frac{\lambda_1}{\lambda_1+\lambda_2}\,\lambda_2\, e^{\lambda_2 x}\,, & x \leq 0 \\ \frac{\lambda_2}{\lambda_1+\lambda_2}\,\lambda_1\, e^{-\lambda_1 x}\,, & x \geq 0 \end{cases}$$

$$= \frac{\lambda_1}{\lambda_1+\lambda_2}\,\lambda_2\, e^{\lambda_2 x}\, I_{]-\infty,0]}(x) + \frac{\lambda_2}{\lambda_1+\lambda_2}\,\lambda_1\, e^{-\lambda_1 x}\, I_{[0,+\infty[}(x)$$

where

$$I_{[a,b[}(x) = \begin{cases} 1 \text{ if } x \in [a,b[ \\ 0 \text{ if } x \notin [a,b[ \end{cases}$$

is the indicator function for the set $[a, b[$. This shows that the generalized Laplace distribution may be seen as a mixture with weights $\frac{\lambda_2}{\lambda_1+\lambda_2}$ and $\frac{\lambda_1}{\lambda_1+\lambda_2}$ respectively of an Exponential distribution with rate parameter $\lambda_1$ with support $\mathbb{R}_0^+$ and a 'negative' Exponential distribution with rate parameter $\lambda_2$ with support $\mathbb{R}_0^-$. For $\lambda_1 = \lambda_2$ this distribution yields the usual Laplace distribution.

When wrapping the generalized Laplace distribution the mixture structure and the weights are kept, with the wrapped version of the generalized Laplace distribution with pdf [3]

$$
\begin{aligned}
f(t) &= \frac{\lambda_2}{\lambda_1+\lambda_2}\,\frac{\lambda_1\,e^{-\lambda_1 t}}{1-e^{-2\lambda_1\pi}} - \frac{\lambda_1}{\lambda_1+\lambda_2}\,\frac{\lambda_2\,e^{-\lambda_2 t}}{1-e^{2\lambda_2\pi}}\\
&= \frac{\lambda_2}{\lambda_1+\lambda_2}\,\frac{\lambda_1\,e^{-\lambda_1 t}}{1-e^{-2\lambda_1\pi}} + \frac{\lambda_1}{\lambda_1+\lambda_2}\,\frac{(-\lambda_2)\,e^{\lambda_2 t}}{1-e^{2\lambda_2\pi}}\,,
\end{aligned}
$$

a mixture with weights $\frac{\lambda_2}{\lambda_1+\lambda_2}$ and $\frac{\lambda_1}{\lambda_1+\lambda_2}$ respectively of a wrapped Exponential with rate $\lambda_1$ and another wrapped Exponential distribution with rate $-\lambda_2$.

The linear generalized Laplace distribution is quite simple and flexible since it may exhibit both symmetrical and asymmetrical forms and its wrapped version was studied by Jammalamadaka and Kozubowski (2003, 2004). However, these distributions have the limitation of having a sharp top always placed at the origin. It would be interesting to be able to develop somehow similar distributions, able to exhibit both symmetrical and asymmetrical forms but which could have either sharp or smooth tops, not necessarily placed always at the origin, namely when the distribution would assume an asymmetric form.

The answer to such a goal is the development of the wrapped sum of independent generalized Laplace distributions. However, since another way to look at the linear Laplace distributions is also as the difference of two independent Exponential distributions, exactly the Exponential distributions that appear in the mixture structure, we may alternatively think of the wrapped difference of two independent sums of independent Exponential distributions, which if each of these two sums is the sum of an equal number of Exponential distributions, may then be seen as the sum of independent generalized Laplace distributions. Also in the case where the two sums of independent Exponential distributions are sums of an equal number of distributions all the formulation becomes a bit simpler. We have therefore chosen this approach.

Let $Z$ be a linear combination of $n$ independent Exponential distributions, with all positive coefficients, that is, let

$$
Z = \sum_{i=1}^{n} a_i\, X_i
$$

[3] see author and editors note at the end of the Chapter

where $a_i > 0\ (i = 1, \ldots, n)$ and $X_i \sim Exp(\lambda_i)$, so that

$$Z = \sum_{i=1}^{n} Y_i$$

where $Y_i \sim Exp(\mu_i)$ with $\mu_i = \lambda_i/a_i(> 0)$. Then the distribution of $Z$ is just a GIG distribution of depth $n$ (see the previous section), with all shape parameters equal to 1 and rate parameters $\mu_i\ (i = 1, \ldots, n)$, which has pdf

$$f_Z(z) = K \sum_{j=1}^{n} c_j\, e^{-\mu_j y}$$

where

$$K = \prod_{j=1}^{n} \mu_j \qquad \text{and} \qquad c_j = \prod_{\substack{k=1\\k\neq j}}^{n} \frac{1}{\mu_j - \mu_k} \quad (j = 1, \ldots, n)\,.$$

We will denote the fact that $Z$ has this distribution by

$$Z \sim SE(\mu_j,\ j \in \{1, \ldots, n\})\,.$$

Let then

$$Z_1 \sim SE(\mu_j,\ j \in \{1, \ldots, n\}) \quad \text{and} \quad Z_2 \sim SE(\nu_j,\ j \in \{1, \ldots, n\})$$

be two independent r.v.'s and let

$$W = Z_1 - Z_2\,.$$

The pdf's of $Z_1$ and $Z_2$ may then be respectively written as

$$f_{Z_1}(z_1) = K_1 \sum_{j=1}^{n} c_j\, e^{-\mu_j z_1} \quad \text{and} \quad f_{Z_2}(z_2) = K_2 \sum_{j=1}^{n} d_j\, e^{-\nu_j z_2}$$

where

$$K_1 = \prod_{j=1}^{n} \mu_j\,, \quad K_2 = \prod_{j=1}^{n} \nu_j$$

and, for $j = 1, \ldots, p$,

$$c_j = \prod_{\substack{k=1\\k\neq j}}^{n} \frac{1}{\mu_j - \mu_k}\,, \quad d_j = \prod_{\substack{k=1\\k\neq j}}^{n} \frac{1}{\nu_j - \nu_k}$$

so that the p.d.f. of $W$ will be given by

$$\begin{aligned} f_W(w) &= \int_{\max(w,0)}^{+\infty} K_1 K_2 \left(\sum_{j=1}^{n} c_j\, e^{-\mu_j z_1}\right)\left(\sum_{j=1}^{n} d_j\, e^{-\nu_j (z_1 - w)}\right) dz_1 \\ &= K_1 K_2 \sum_{j=1}^{n}\sum_{k=1}^{n} e^{\nu_k w}\, c_j\, d_k \int_{\max(z_1,0)}^{+\infty} e^{-(\mu_j+\nu_k) z_1}\, dz_1 \end{aligned}$$

or,

$$\begin{aligned} f_W(w) &= \begin{cases} K_1 K_2 \sum_{j=1}^{n} H_{1j}\, c_j\, e^{-\mu_j w} & w \geq 0 \\ K_1 K_2 \sum_{j=1}^{n} H_{2j}\, d_j\, e^{\nu_j w} & w \leq 0 \end{cases} \\ &= K_1 K_2 \sum_{j=1}^{n} H_{2j}\, d_j\, e^{\nu_j w}\, I_{]-\infty,0]}(w) \\ &\qquad\qquad + K_1 K_2 \sum_{j=1}^{n} H_{1j}\, c_j\, e^{-\mu_j w}\, I_{[0,+\infty[}(w) \end{aligned}$$

where

$$H_{1j} = \sum_{h=1}^{n} \frac{d_h}{\mu_j + \nu_h} \quad \text{and} \quad H_{2j} = \sum_{h=1}^{n} \frac{c_h}{\mu_h + \nu_j}\,.$$

Therefore, the wrapped version of this sum of independent generalized Laplace distributions or difference of two sums of independent Exponentials, that is, the distribution of

$$\theta(W) = W \mod 2\pi$$

is a mixture of $2n$ wrapped Exponentials, $n$ of which are $W\Gamma(\mu_j, 1)$ distributions, with weights $K_1 K_2\, H_{1j}\, c_j$ $(j = 1, \ldots, n)$, and the other $n$ are $W\Gamma(-\nu_j, 1)$

distributions, with weights $K_1 K_2 H_{2j} d_j$ $(j = 1, \ldots, n)$, with pdf

$$f_{\theta(W)}(t) = K_1 K_2 \sum_{j=1}^{n} \left( H_{1j} c_j \frac{\mu_j e^{-\mu_j t}}{1 - e^{-2\mu_j \pi}} + H_{2j} d_j \frac{-\nu_j e^{\nu_j t}}{1 - e^{2\nu_j \pi}} \right)$$

and cdf

$$F_{\theta(W)}(t) = K_1 K_2 \sum_{j=1}^{n} \left( H_{1j} c_j \frac{1 - e^{-2\mu_j t}}{1 - e^{-2\mu_j \pi}} + H_{2j} d_j \frac{1 - e^{2\nu_j t}}{1 - e^{2\nu_j \pi}} \right).$$

We will denote this distribution by $WSL(\mu_1, \ldots, \mu_n; \nu_1, \ldots, \nu_n)$. For $n = 1$ this distribution reduces to the wrapped generalized Laplace distribution.

This distribution has a symmetric shape if the set of rate parameters $\mu_j$ and the set of rate parameters $\nu_j$ $(j = 1, \ldots, n)$ are the same, the order of the rate parameters $\mu_j$ and $\nu_j$ $(j = 1, \ldots, n)$ in each set being irrelevant. We should also note that for the distribution above all the $\mu_j$ are supposed to be different as well as all the $\nu_j$, since this is what is required in the GIG distribution, from which this distribution was derived. Two wrapped sums of independent generalized Laplace distributions are symmetrical, relative to the origin, if the two sets of parameters $\mu_j$ and $\nu_J$ are interchanged. Also, this distribution tends to the uniform distribution if at least one of the $\mu_j$'s and one of the $\nu_j$'s tend to zero. Both the sum of generalized Laplace distributions and the GIG or the sum of independent Gammas are unimodal, both in their linear and wrapped forms. However, mixtures of these distributions are a powerful tool in modeling circular data, since a criterious choice of parameters may easily yield multimodal distributions. The trigonometric moments for these distributions, and related parameters, may be easily obtained from the mixture structure and the results for the single wrapped Gamma distribution.

In Figures 11 and 12 we may see some plots of densities of wrapped sums of generalized Laplace distributions.

## 6 The Wrapped Non-central Gamma Distribution

We will say that the r.v. $X$ has a non-central Gamma distribution with shape parameter $r$, rate parameter $\lambda$ and non-centrality parameter $\delta$ if the r.v. $X$ has pdf

$$\begin{aligned} f_X(x) &= \frac{\lambda^r}{\Gamma(r)} e^{-\lambda x} x^{r-1} e^{-\delta/2} {}_0F_1(r; \lambda x \delta/2) \\ &= \sum_{i=0}^{\infty} \frac{e^{-\delta/2} (\delta/2)^i}{i!} \frac{\lambda^{r+i}}{\Gamma(r+i)} e^{-\lambda x} x^{r+i-1}, \end{aligned}$$

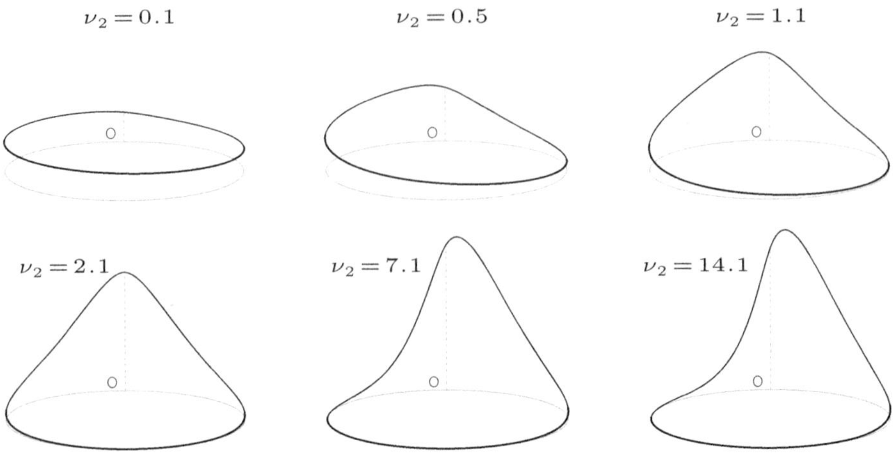


**Fig. 11** Three-dimensional views of densities of $WSL(2.9, 3.3; 4.5, \nu_2)$ for different values of $\nu_2$. All plots have the origin placed at the same spot and all plots have the same vertical scale

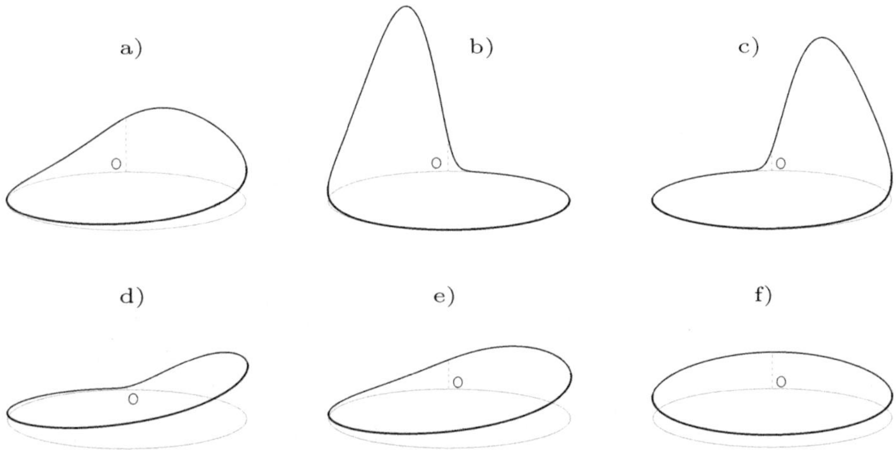


**Fig. 12** Three-dimensional views of densities of: a) $WSL(2.9, 3.3, 1.1; 4.5, 3.5, 5.6)$, b) $WSL(32.9, 33.3, 35.1; 4.5, 3.5, 5.6)$, c) $WSL(2.9, 3.3, 5.1; 34.5, 33.5, 35.6)$, d) $WSL(1.9, 1.3, 1.1; 34.5, 33.5, 35.6)$, e) $WSL(1.9, 1.3, 1.1; 4.5, 3.5, 5.6)$, f) $WSL(1.9, 1.3, 1.1; 1.25, 1.5, 1.6)$. All plots have the origin placed at the same spot and all plots have the same vertical scale

that is, if the distribution of $X$ is a mixture, with Poisson weights with parameter $\delta/2$, of infinitely many $\Gamma(r + i, \lambda)$ distributions ($i = 0, 1, \ldots$).

This way, the wrapped form of the non-central Gamma distribution will be a mixture with the same Poisson weights of infinitely many $W\Gamma(r+i, \lambda)$ distributions

$(i = 0, 1, \ldots)$, with pdf

$$f_\theta(t) = e^{-(\lambda t+\delta/2)} \sum_{i=0}^{\infty} \frac{(\delta/2)^i}{i!} \frac{\lambda^{r+i}}{\Gamma(r+i)} (2\pi)^{r+i-1} \Phi\left(e^{-2\lambda\pi}, 1-r-i, \frac{t}{2\pi}\right)$$

where

$$\theta \equiv \theta(X) = X \bmod 2\pi$$

and where, for non-integer $r$ we may write

$$\begin{aligned} f_\theta(t) = & \sum_{j=0}^{\infty} \frac{e^{-\delta/2}(\delta/2)^j}{j!} \left( p_j^* \frac{\lambda^{r+j}}{\gamma(r+j; 2\lambda\pi)} e^{-\lambda t} t^{r+j-1} \right. \\ & \left. + \sum_{i=0}^{\infty} p_{i,j} \frac{\lambda^{i+1}}{\gamma(i+1; 2\lambda\pi)} e^{-\lambda t} t^i \right) \\ = & e^{-\delta/2+\lambda t} \sum_{j=0}^{\infty} \frac{(\delta/2)^j}{j!} \left( \frac{t^{r+j-1}}{\Gamma(r+j)} + \sum_{i=0}^{\infty} \frac{t^i(-1)^i}{\Gamma(i+1)} \frac{(2\pi\lambda)^{r+j-h-1}}{(r+j-h-1)!} \right. \\ & \left. \Phi^*\left(e^{-2\lambda\pi}, r+j-h-1\right) \right) \end{aligned}$$

where

$$p_j^* = \gamma^*(r+j, 2\lambda\pi) \tag{22}$$

and

$$p_{i,j} = \gamma^*(i+1, 2\lambda\pi)(-1)^i \frac{\lambda^{r+j-i-1}}{\Gamma(r)} \frac{\Gamma(1-r-j+i)}{\Gamma(1-r-j)} PLog\left(e^{-2\lambda\pi}, i+1-r-j\right) \tag{23}$$

while for $r \in \mathbb{N}$,

$$\begin{aligned} f_\theta(t) = & \sum_{j=0}^{\infty} \frac{e^{-\delta/2}(\delta/2)^j}{j!} \sum_{h=0}^{r-1} p_{h,j}^* \frac{\lambda^{h+1}}{\gamma(h+1, 2\lambda\pi)} e^{-\lambda t} t^h \\ = & e^{-(\delta/2+\lambda t)} \sum_{j=0}^{\infty} \frac{(\delta/2)^j}{j!} \sum_{h=0}^{r-1} \frac{\lambda^{h+1}}{\Gamma(h+1)} \frac{(2\lambda\pi)^{r+j-h-1}}{(r+j-h-1)!} \\ & \Phi^*\left(e^{-2\lambda\pi}, r+j-h-1\right) \end{aligned}$$

where

$$p^*_{h,j} = \gamma^*(h+1, 2\lambda\pi)\,\frac{(2\lambda\pi)^{r+j-h-1}}{(r+j-h-1)!}\,\Phi^*\!\left(e^{-2\lambda\pi}, r+j-h-1\right). \tag{24}$$

The corresponding cdf's are given by

$$F_\theta(t) = \sum_{j=0}^{\infty} \frac{e^{-\delta/2}(\delta/2)^j}{j!}\left(p^*_j\,\frac{\gamma(r+j,\lambda t)}{\gamma(r+j,2\lambda\pi)} + \sum_{i=0}^{\infty} p_{i,j}\,\frac{\gamma(i+1,\lambda t)}{\gamma(i+1,2\lambda\pi)}\right)$$

for non-integer $r$, with $p_{i,j}$ $(i = 0, 1, \ldots)$ given by (23) and $p^*_j$ $(j = 0, 1, \ldots)$ given by (22), and

$$F_\theta(t) = \sum_{j=0}^{\infty} \frac{e^{-\delta/2}(\delta/2)^j}{j!} \sum_{h=0}^{r-1} p^*_{h,j}\,\frac{\gamma(h+1,\lambda t)}{\gamma(h+1,2\lambda\pi)}$$

for $r \in \mathbb{N}$, where $p^*_{h,j}$ $(h = 0, \ldots, r-1;\ j = 0, 1, \ldots)$ are given by (24).

We may note that, interestingly enough, this latter non-central distribution depends on $r$ only through the weights $p^*_{h,j}$, while for non-integer $r$ the non-central distribution has all of its components but the first depending on $r$ only through the weights $p_{i,j}$, with the first component being a function of $r$ both through the weight $p^*_j$ and also through the distribution part itself.

We should also note that when $\delta \to \infty$ the wrapped non-central Gamma distribution tends to the wrapped Uniform distribution, since then the components in the mixture with higher values for the shape parameter will have higher weights in the mixture, and as we saw in section 2, when the shape parameter tends to infinity the wrapped Gamma distribution tends to the Uniform distribution. In Figure 13 we may see a few three-dimensional plots of densities of wrapped non-central Gamma distributions, where this fact may be analyzed.

## 7 Final Remarks

Once the mixture structure of the wrapped Gamma distribution was unveiled, it became easy to obtain expressions for its cdf and also once considered the mixture structure of the other linear distributions considered, as the distribution of linear combinations of independent Gamma distributions, the distribution of the sum of independent generalized Laplace distributions and the non-central Gamma distribution, it was easy to obtain expressions for both the pdf and cdf of their wrapped versions.

The great advantage of the wrapped sum of generalized Laplace distributions is that these distributions may exhibit both symmetrical and fairly asymmetrical

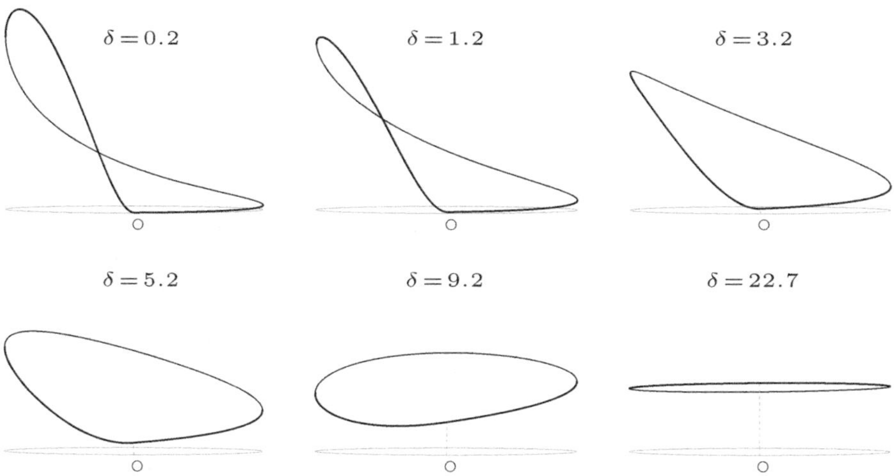


**Fig. 13** Three-dimensional views of densities of wrapped non-central Gamma densities with shape parameter $r = 3$, rate parameter $\lambda = 1.9$ and different values of the non-centrality parameter $\delta$. All plots have the origin placed at the same spot and all plots have the same vertical scale

shapes with either a sharp or smooth top. The sharp top only occurs for $n = 1$, that is for the single generalized Laplace distribution. Although these distributions are always unimodal, we may use mixtures of them to obtain very flexible multimodal distributions.

The wrapped sum or linear combination of independent Gamma distributions, the wrapped GIG distribution, the wrapped distribution of the sum of independent generalized Laplace distributions and the wrapped non-central Gamma distribution are all very flexible distributions. However, they are all unimodal. Of course when one needs to handle multimodal circular data we may think about using mixtures of these distributions. However, then they may become a bit hard to handle from an estimation point of view.

Mere mixtures of wrapped Gamma distributions seem to be quite a powerful tool, since they may have from one to as many modes as the number of Gamma distributions in the mixture. However, in cases where we need some degree of symmetry, the wrapped sum of generalized Laplace distributions will be a distribution to consider. Of course one may use mixtures of different distributions, as for example a mixture of a wrapped sum of generalized Laplace distributions with a wrapped Gamma distribution.

Modules programmed in Mathematica© (from Wolfram Research, Inc.) to compute and plot the pdf's and cdf's in this paper are available from the author.

# Appendix A

| $n$ | | | | | | | | | | |
|---|---|---|---|---|---|---|---|---|---|---|
| $n=1$ | 1 | | | | | | | | | |
| 2 | 1 | 1 | | | | | | | | |
| 3 | 1 | 4 | 1 | | | | | | | |
| 4 | 1 | 11 | 11 | 1 | | | | | | |
| 5 | 1 | 26 | 66 | 26 | 1 | | | | | |
| 6 | 1 | 57 | 302 | 302 | 57 | 1 | | | | |
| 7 | 1 | 120 | 1191 | 2416 | 1191 | 120 | 1 | | | |
| 8 | 1 | 247 | 4293 | 15619 | 15619 | 4293 | 247 | 1 | | |
| 9 | 1 | 502 | 14608 | 88234 | 156190 | 88234 | 14608 | 502 | 1 | |
| 10 | 1 | 1013 | 47840 | 455192 | 1310354 | 1310354 | 455192 | 47840 | 1013 | 1 |

# References


Coelho, C. A. (1998). The generalized integer Gamma distribution - a basis for distributions in multivariate analysis. *J. Multivariate Anal.* **64**, 86–102.

Jammalamadaka, S. R. and Kozubowski, T. J. (2003). A new family of circular models: the wrapped Laplace distributions. *Adv. Appl. Stat.* **3**, 77–103.

Jammalamadaka, S. R. and Kozubowski, T. J. (2004). New families of wrapped distributions for modeling skew circular data. *Comm. Statist. Theory Methods* **33**, 2059–2074.

Moschopoulos, P. G. (1985). The distribution of the sum of independent Gamma random variables. *Ann. Inst. Statist. Math.* **37**, 541–544.


### *Author and Editors notes:*

[1] on the right hand side of the equal sign the fraction should rather read $\dfrac{r-1}{2\pi}$

[2] on the last row of expression (10) the expression for the weight $p_i$ should rather be

$$\gamma^*(i+1;\,2\lambda\pi)(2\pi)^{r-1-i}(-1)^i\,\frac{\lambda^{r-i-1}}{\Gamma(r)}\,\frac{\Gamma(1-r+i)}{\Gamma(1-r)}\,PLog\left(e^{-2\lambda\pi},i+1-r\right)$$

[3] in the first epression of $f(t)$ there is a small typo in the exponent of the second exponential that appears in a numerator, where the minus sign before $\lambda_2$ should not be there, that is, this first expression for $f(t)$ should rather be

$$f(t)=\frac{\lambda_2}{\lambda_1+\lambda_2}\,\frac{\lambda_1\,e^{-\lambda_1 t}}{1-e^{-2\lambda_1\pi}}-\frac{\lambda_1}{\lambda_1+\lambda_2}\,\frac{\lambda_2\,e^{\lambda_2 t}}{1-e^{2\lambda_2\pi}}$$

# On the Distribution of the Product and Ratio of Independent Generalized Gamma-Ratio Random Variables

**Carlos A. Coelho and João T. Mexia**

**Abstract** Using a decomposition of the characteristic function of the logarithm of the product of independent generalized gamma-ratio random variables (r.v.'s), we obtain explicit expressions for both the probability density and cumulative distribution functions of the product of independent r.v.'s with generalized $F$ or generalized gamma-ratio distributions in the form of particular mixtures of generalized Pareto and inverted Pareto distributions. The expressions obtained do not involve any unsolved integrals and are convenient for computer implementation. By considering power parameters which are not required to be positive, we were able to obtain, as particular cases, not only the distributions for the product of folded T and folded Cauchy r.v.'s but also for the ratio of two independent products of generalized gamma-ratio r.v.'s. Theoretical applications of the results as well as simulations are presented.

**Keywords** Particular mixtures · Pareto and inverted Pareto distributions · GIG distribution · sum of exponentials · Difference of exponentials · Folded T · Folded Cauchy · Beta prime · Beta second kind

**AMS (2000) Subject Classification** Primary 62E15; Secondary 62E10

---

Republished from Sankhya, with authorization from the Indian Statistical Institute, Calcutta.
First published as Coelho, C. A., Mexia, J. T.: On the distribution of the product and ratio of independent generalized Gamma-ratio random variables. Sankhya, Ser. A, 69, 2, 221–255 (2007). https://www.jstor.org/stable/25664553

---

C. A. Coelho (✉) · J. T. Mexia
The New University of Lisbon, Lisbon, Portugal

Current Address: Department of Mathematics, NOVA School of Science and Technology, NOVA University of Lisbon, Caparica, Portugal
e-mail: cmac@fct.unl.pt; jtm@fct.unl.pt

 
B. C. Arnold et al. (eds.), *Likelihood Ratio Tests and Related Topics in Multivariate Analysis*, Contributions to Statistics,
https://doi.org/10.1007/978-3-031-68008-3_7

# 1 Introduction

The problem of obtaining an explicit expression, without involving any unsolved integrals, for both the probability density function (p.d.f.) and cumulative distribution function (c.d.f.) of the product of independent Generalized Gamma-Ratio random variables (r.v.'s) or Generalized F r.v.'s, as they are also called, is a challenging one, especially since the characteristic function is not readily available for such r.v.'s. In this paper we determine the distribution of the product of independent Generalized Gamma-Ratio (GGR) random variables under the form of particular mixtures of Pareto and inverted Pareto distributions. Expressions for the p.d.f. of such a product were obtained by Shah and Rathie (1974) in terms of Fox's $H$ function. Also Pham-Gia and Turkkan (2002) obtained expressions for the p.d.f. of the product of only two independent generalized $F$ r.v.'s in terms of the Lauricella hypergeometric $D$-function of two variables. However, even nowadays when good softwares for symbolic and numeric computation are available Fox's $H$ function and the Lauricella function are not readily computable, being usually computed in terms of the integrals that define them. Although Pham-Gia and Turkkan (2002) developed an efficient computer code to compute the Lauricella hypergeometric function, their approach is not extensible to the product of more than two r.v.'s and these authors point out that when we consider the product of more than two GGR r.v.'s, it seems that "frequently, however, no closed form solution for these operations can be obtained and one has to resort to approximate approaches, including simulation". The same authors, when referring to the results in Shah and Rathie (1974) say that "these results, although very convenient notationwise, are difficult to be programmed on a computer and hence are difficult to be used in applications". Yet the same authors say that "they are, however, essential when the number of variables in the product, or quotient, is larger than 2". In this paper, our aim is to obtain explicit simpler expressions for the p.d.f. and c.d.f. of the product of independent Generalized Gamma-Ratio r.v.'s which may be readily implemented computationally and that, given its structure, may also give us ready access to asymptotic and near-exact distributions. Given the approach followed, not only the distributions for the non-central case are readily at hand but in addition the distribution for the product of any non-null power of GGR r.v.'s is also available.

As particular immediate cases, we have the product of independent generalized second kind Beta or Beta prime, folded $T$ and folded Cauchy r.v.'s and yet, of course, $F$ r.v.'s.

Given the fact that the characteristic functions for the GGR random variables are not readily available and given the fact that we are dealing with a product of random variables, it is more convenient to carry out our work through the decomposition of the characteristic function of the logarithm of the product of the GGR random variables.

Another novelty is that, although usually only positive power parameters are considered for the GGR distributions, actually nothing forces those parameters to be positive. In fact, to consider a negative power parameter in the GGR distribution is clearly equivalent to considering the reciprocal of that given random variable with the symmetrical positive power parameter. Given the way the problem is approached, even negative power parameters may be easily considered in the GGR distributions and also the distribution of the ratio of two GGR random variables or of the ratio of two products of GGR random variables are particular cases of the results obtained in the paper.

Products of several independent GGR random variables arise in several areas of application in statistics as it is shown in Section 4 and are also related to a test statistic used in the multivariate linear functional model (Provost, 1986).

## 2 Some Preliminary Results

### *2.1 The Generalized Gamma-Ratio (GGR) Distribution*

In order to establish some of the notation, nomenclature and a result used, we will start by defining what we mean by a Generalized Gamma-Ratio (GGR) distribution. Let

$$X_1 \sim \Gamma(r_1, \lambda_1) \quad \text{and} \quad X_2 \sim \Gamma(r_2, \lambda_2)$$

be two independent r.v.'s with Gamma distributions with shape parameters $r_1$ and $r_2$ and rate parameters $\lambda_1$ and $\lambda_2$, that is, for example, $X_1$ has p.d.f.

$$f_{X_1}(x) = \frac{\lambda_1^{r_1}}{\Gamma(r_1)} \exp(-\lambda_1 x)\, x^{r_1-1}, \quad r_1, \lambda_1 > 0;\ x > 0\,.$$

Let then

$$Y_1 = X_1^{1/\beta}, \quad Y_2 = X_2^{1/\beta}, \quad \beta \in \mathbb{R}\backslash\{0\}$$

and

$$Z = Y_1/Y_2\,.$$

We will say that $Y_1$ and $Y_2$ have Generalized Gamma distributions and that $Z$ has a GGR distribution. Using standard methods, we have the p.d.f.'s of $Y_i$ $(i = 1, 2)$ and $Z$ given by

$$f_{Y_i}(y_i) = \frac{|\beta|\, \lambda_i^{r_i}}{\Gamma(r_i)} \exp\left(-\lambda_i y_i^{\beta}\right) y_i^{\beta r_i - 1}, \quad y_i > 0\,, \quad (i = 1, 2)$$

and

$$f_Z(z) = \frac{|\beta|\, k^{r_1}}{B(r_1, r_2)} \left(1 + k z^{\beta}\right)^{-r_1-r_2} z^{\beta r_1 - 1}, \quad z > 0,$$

where $k = \lambda_1/\lambda_2$ and $B(\cdot, \cdot)$ is the Beta function.

We will denote the fact that $Z$ has the GGR distribution with parameters $k, r_1, r_2$ and $\beta$ by

$$Z \sim GGR(k, r_1, r_2, \beta).$$

The $h$-th moment of $Z$ is easily derived as

$$E\left(Z^h\right) = k^{-h/\beta} \frac{\Gamma(r_1 + h/\beta)}{\Gamma(r_1)} \frac{\Gamma(r_2 - h/\beta)}{\Gamma(r_2)}, \quad (-r_1 < h/\beta < r_2). \tag{1}$$

If $\beta = 1$, $k = m/n$, $r_1 = m/2$ and $r_2 = n/2$, with $m, n \in \mathbb{N}$, then $Z$ has an $F$ distribution with $m$ and $n$ degrees of freedom. This is the reason why the distribution of $Z$ is also called a Generalized $F$ distribution (Shah and Rathie, 1974).

## 2.2 The Generalized Integer Gamma (GIG) Distribution

In this subsection and in the two following ones, we will establish some distributions that will be used in the next section. Let

$$X_j \sim \Gamma(r_j, \lambda_j), \quad j = 1, \ldots, p$$

be $p$ independent r.v.'s with Gamma distributions with shape parameters $r_j \in \mathbb{N}$ and rate parameters $\lambda_j > 0$ $(j = 1, \ldots, p)$. We will say that then the r.v.

$$Y = \sum_{j=1}^{p} X_j$$

has a GIG distribution of depth $p$, with shape parameters $r_j$ and rate parameters $\lambda_j$, $(j = 1, \ldots, p)$, and we will denote this fact by

$$Y \sim GIG(r_j, \lambda_j;\; p).$$

The p.d.f. and the c.d.f. of $Y$ are given by (Coelho, 1998)

$$f_Y(y) = K \sum_{j=1}^{p} P_j(y) \exp(-\lambda_j y) \tag{2}$$

and

$$F_Y(y) = 1 - K \sum_{j=1}^{p} P_j^*(y) \exp(-\lambda_j y), \tag{3}$$

where

$$K = \prod_{j=1}^{p} \lambda_j^{r_j}, \qquad P_j(y) = \sum_{k=1}^{r_j} c_{j,k}\, y^{k-1} \tag{4}$$

and

$$P_j^*(y) = \sum_{k=1}^{r_j} c_{j,k}\, (k-1)! \sum_{i=0}^{k-1} \frac{y^i}{i!\, \lambda_i^{k-i}},$$

with

$$c_{j,r_j} = \frac{1}{(r_j-1)!} \prod_{\substack{i=1 \\ i \neq j}}^{p} (\lambda_i - \lambda_j)^{-r_i}, \qquad j = 1, \ldots, p, \tag{5}$$

and

$$c_{j,r_j-k} = \frac{1}{k} \sum_{i=1}^{k} \frac{(r_j-k+i-1)!}{(r_j-k-1)!} R(i,j,p)\, c_{j,r_j-(k-i)}, \quad \begin{array}{l}(k = 1, \ldots, r_j-1) \\ (j = 1, \ldots, p),\end{array} \tag{6}$$

where

$$R(i,j,p) = \sum_{\substack{k=1 \\ k \neq j}}^{p} r_k \left(\lambda_j - \lambda_k\right)^{-i} \quad (i = 1, \ldots, r_j - 1). \tag{7}$$

### 2.3 The Distribution of the Sum of Random Variables with Exponential Distribution and the Distribution of the Difference of Two of These Random Variables

Let

$$X_i \sim Exp(\lambda_i) \sim \Gamma(1, \lambda_i), \qquad i = 1, \ldots, p,$$

be $p$ independent Exponential r.v.'s with rate parameters $\lambda_i$ $(i = 1, \ldots, p)$, and let

$$Y = \sum_{i=1}^{p} X_i .$$

The distribution of the r.v. $Y$ is a particular case of the GIG distribution (Coelho, 1998) of depth $p$, with all the shape parameters equal to 1, whose p.d.f. may be written as

$$f_Y(y) = K \sum_{j=1}^{p} c_{j,1} \exp(-\lambda_j y), \tag{8}$$

where

$$K = \prod_{j=1}^{p} \lambda_j \qquad \text{and} \qquad c_{j,1} = \prod_{\substack{k=1\\k\neq j}}^{p} \frac{1}{\lambda_k - \lambda_j} \quad (j = 1, \ldots, p). \tag{9}$$

We will denote the fact that $Y$ has this distribution by

$$Y \sim SE(\lambda_j;\, p) \sim GIG(1, \lambda_j;\, p). \tag{10}$$

**Theorem 1** *Let*

$$Y_1 \sim SE(\lambda_j;\, p) \quad \text{and} \quad Y_2 \sim SE(\nu_j;\, p) \tag{11}$$

*be two independent r.v.'s and let*

$$Z = Y_1 - Y_2 .$$

*The p.d.f. and c.d.f. of Z are respectively given by*

$$f_Z(z) = \begin{cases} K_1 K_2 \sum_{j=1}^{p} H_{2j}\, d_{j,1} \exp(\nu_j z)\,, & z \le 0 \\ K_1 K_2 \sum_{j=1}^{p} H_{1j}\, c_{j,1} \exp(-\lambda_j z)\,, & z \ge 0 \end{cases} \tag{12}$$

*and* [1]

$$F_Z(z) = \begin{cases} K_1 K_2 \sum_{j=1}^{p} H_{2j} \frac{d_{j,1}}{\nu_j} \exp(\nu_j z)\,, & z \le 0 \\ K_1 K_2 \sum_{j=1}^{p} \left( H_{2j} \frac{d_{j,1}}{\nu_j} + H_{1j} \frac{c_{j,1}}{\lambda_j} \left(1 - \exp(-\lambda_j z)\right)\right), & z \ge 0 \end{cases} \tag{13}$$

*respectively, where*

$$K_1 = \prod_{j=1}^{p} \lambda_j\,, \quad K_2 = \prod_{j=1}^{p} \nu_j \tag{14}$$

*and, for* $j = 1, \ldots, p$,

$$c_{j,1} = \prod_{\substack{k=1 \\ k \neq j}}^{p} \frac{1}{\lambda_k - \lambda_j}\,, \quad d_{j,1} = \prod_{\substack{k=1 \\ k \neq j}}^{p} \frac{1}{\nu_k - \nu_j}\,, \tag{15}$$

$$H_{1j} = \sum_{h=1}^{p} \frac{d_{h,1}}{\lambda_j + \nu_h} \quad \textit{and} \quad H_{2j} = \sum_{h=1}^{p} \frac{c_{h,1}}{\lambda_h + \nu_j}\,. \tag{16}$$

*Proof* The p.d.f.'s of $Y_1$ and $Y_2$ may then be respectively written as

$$f_{Y_1}(y_1) = K_1 \sum_{j=1}^{p} c_{j,1} \exp(-\lambda_j y_1) \quad \text{and} \quad f_{Y_2}(y_2) = K_2 \sum_{j=1}^{p} d_{j,1} \exp(-\nu_j y_2) \tag{17}$$

[1] see the author and editors note at the end of the chapter

respectively, with $K_1$ and $K_2$ given by (14), $c_{j,1}$ and $d_{j,1}$ given by (15) and $H_{1j}$ and $H_{2j}$ given by (16), so that the p.d.f. of $Z$ will be given by

$$
\begin{aligned}
f_Z(z) &= \int_{\max(z,0)}^{+\infty} K_1 K_2 \left(\sum_{j=1}^{p} c_{j,1} \exp(-\lambda_j\, y_1)\right)\left(\sum_{j=1}^{p} d_{j,1} \exp\{-\nu_j(y_1-z)\}\right) dy_1 \\
&= K_1 K_2 \sum_{j=1}^{p}\sum_{k=1}^{p} \exp(\nu_k z)\, c_{j,1}\, d_{k,1} \int_{\max(z,0)}^{+\infty} \exp\{-(\lambda_j+\nu_k)y_1\}\, dy_1\,,
\end{aligned}
$$

where

$$
\int_{\max(z,0)}^{\infty} \exp\{-(\lambda_j+\nu_k)y_1\}\, dy_1 = \begin{cases} \frac{1}{\lambda_j+\nu_k}\,, & z \le 0 \\ \frac{\exp\{-(\lambda_j+\nu_k)z\}}{\lambda_j+\nu_k}\,, & z \ge 0 \end{cases}
$$

so that, after some small rearrangements, we have $f_Z(z)$ given by (12). The c.d.f. of $Z$ is then easily derived from (12) as given by (13). □

**Corollary 1** *If we take, for some $k > 0$,*

$$
W = k^{-1}\exp(Z)
$$

*then, the p.d.f. and the c.d.f. of $W$ will be given by*

$$
f_W(w) = \begin{cases} K_1 K_2 \displaystyle\sum_{j=1}^{p} H_{1j}\, c_{j,1}\, (kw)^{-\lambda_j} \frac{1}{w}\,, & w \ge k^{-1} \\ K_1 K_2 \displaystyle\sum_{j=1}^{p} H_{2j}\, d_{j,1}\, (kw)^{\nu_j} \frac{1}{w}\,, & 0 < w \le k^{-1}\,. \end{cases} \tag{18}
$$

*and* [2]

$$
F_W(w) = \begin{cases} K_1 K_2 \displaystyle\sum_{j=1}^{p} \left(H_{2j}\frac{d_{j,1}}{\nu_j} + H_{1j}\frac{c_{j,1}}{\lambda_j}\left(1-(kw)^{-\lambda_j}\right)\right), & w \ge k^{-1} \\ K_1 K_2 \displaystyle\sum_{j=1}^{p} H_{2j}\frac{d_{j,1}}{\nu_j}(kw)^{\nu_j}\,, & 0 < w \le k^{-1}\,, \end{cases} \tag{19}
$$

*respectively.* □

[2] see the author and editors note at the end of the chapter

The distribution of $Z$ is also the distribution of the sum of $p$ independent r.v.'s with the distribution of the difference of two independent r.v.'s with exponential distribution, either with similar or different parameters, i.e., the distribution of the sum of $p$ independent r.v.'s with Laplace or generalized Laplace distributions.

We should note that although, namely in (14), it may seem that it would not be reasonable to take $p \to \infty$, as a matter of fact in both (12) and (18) taking $p \to \infty$ will yield genuine p.d.f.'s (see Appendix).

Taking into account that if the r.v. $X$ has an Exponential distribution with rate parameter $\lambda$, with p.d.f.

$$f_X(x) = \lambda \exp(-\lambda x), \qquad x > 0;\ \lambda > 0,$$

for $k > 0$, the r.v. $Y = k * \exp(X)$ has a Pareto distribution with rate parameter $\lambda$ and lower bound parameter $k$, with p.d.f.

$$f_Y(y) = \lambda \left(\frac{y}{k}\right)^{-\lambda} \frac{1}{y}, \qquad y \geq k;\ \lambda > 0\,.$$

We will say that the r.v. $X_1 = -X$ with p.d.f.

$$f_{X_1}(x) = \lambda \exp(\lambda x), \qquad x < 0;\ \lambda > 0$$

has a symmetrical Exponential distribution with rate parameter $\lambda$ and that the r.v. $Y_1 = 1/Y = (1/k)\exp(X_1) = (1/k)\exp(-X)$, with p.d.f.

$$f_{Y_1}(y) = \lambda \left(\frac{y}{k^{-1}}\right)^{\lambda} \frac{1}{y}, \qquad 0 < y \leq \frac{1}{k};\ \lambda > 0\,,$$

has an inverted Pareto distribution with rate parameter $\lambda$ and lower bound parameter $k^{-1}$.

We may then note that while the distribution of $Z$, for $z \geq 0$, may be seen as a particular mixture of Exponential distributions with rate parameters $\lambda_j$ $(j = 1, \ldots, p)$, with weights

$$p_j = K_1 K_2 H_{1j} \frac{c_j}{\lambda_j}, \qquad j = 1, \ldots, p\,,$$

such that

$$\sum_{j=1}^{p} p_j = P[Z \geq 0]\,;$$

and for $z \leq 0$ as a particular mixture of symmetrical Exponential distributions with rate parameters $\nu_j$, with weights

$$q_j = K_1 K_2 H_{2j} \frac{d_j}{\nu_j}, \qquad j = 1, \ldots, p,$$

such that

$$\sum_{j=1}^{p} q_j = P[Z \leq 0],$$

the distribution of $W$ may, for $w \geq k^{-1}$, be seen as a particular mixture of Pareto distributions with rate parameters $\lambda_j$ $(j = 1, \ldots, p)$ and lower bound parameters $k^{-1}$, with weights $p_j$ $(j = 1, \ldots, p)$, such that

$$\sum_{j=1}^{p} p_j = P[W \geq k^{-1}],$$

and for $w \leq k^{-1}$, it may be seen as a mixture of inverted Pareto distributions with rate parameters $\nu_j$ $(j = 1, \ldots, p)$ and lower bound parameters $k$, with weights $q_j$ $(j = 1, \ldots, p)$, such that

$$\sum_{j=1}^{p} q_j = P[W \leq k^{-1}].$$

## 2.4 *The Distribution of the Difference of Two GIG Distributions*

**Theorem 2** *Let, for $j = 1, \ldots, p_1$ and $l = 1, \ldots, p_2$,*

$$Y_1 \sim GIG(r_{1j}, \lambda_j; p_1), \quad \text{and} \quad Y_2 \sim GIG(r_{2l}, \nu_l; p_2),$$

*be two independent r.v.'s, and let*

$$Z = Y_1 - Y_2.$$

*Then, taking $K_1$ and $c_{jk}$ defined in a similar manner to $K$ and $c_{jk}$ in (4) and (5)–(7) respectively, and $K_2$ and $d_{lh}$ defined in a corresponding manner, replacing $p_1$ by*

*$p_2$, $r_{1j}$ by $r_{2l}$ and $\lambda_j$ by $\nu_l$, the p.d.f. of Z is given by* [3]

$$f_Z(z) = \begin{cases} K_1 K_2 \sum_{j=1}^{p_1} P_{1j}^{**}(z) \exp(-\lambda_j z), & z \geq 0, \\ K_1 K_2 \sum_{j=1}^{p_1} P_{2j}^{**}(z) \exp(\nu_j z), & z \leq 0, \end{cases} \tag{20}$$

*where*

$$P_{1j}^{**}(z) = \sum_{k=1}^{r_{1j}} c_{jk} \sum_{l=1}^{p_2} \sum_{h=1}^{r_{2l}} d_{lh} \sum_{i=0}^{k-1} \binom{k-1}{i} z^{k-1-i} \frac{(h+i-1)!}{(\lambda_j+\nu_l)^{h+i}} \tag{21}$$

*and*

$$P_{2l}^{**}(z) = \sum_{h=1}^{r_{2l}} d_{lh} \sum_{j=1}^{p_1} \sum_{k=1}^{r_{1j}} c_{jk} \sum_{i=0}^{h-1} \binom{h-1}{i} (-z)^{h-1-i} \frac{(k+i-1)!}{(\lambda_j+\nu_l)^{k+i}}\,. \tag{22}$$

*and the c.d.f. by*

$$F_Z(z) = \begin{cases} 1 - K_1 K_2 \sum_{j=1}^{p_1} P_{1j}^{***}(z) \exp(-\lambda_j\, z)\,, & z \geq 0 \\ K_1 K_2 \sum_{l=1}^{p_2} P_{2l}^{***}(z) \exp(\nu_l\, z)\,, & z \leq 0 \end{cases} \tag{23}$$

*with*

$$P_{1j}^{***}(z) = \sum_{k=1}^{r_{1j}} c_{jk} \sum_{l=1}^{p_2} \sum_{h=1}^{r_{2l}} d_{lh} \sum_{i=0}^{k-1} \binom{k-1}{i} \frac{(h+i-1)!}{(\lambda_j+\nu_l)^{h+i}} \sum_{t=0}^{k-1-i} \frac{(k-1-i)!}{t!} \frac{z^t}{\lambda_j^{k-i-t}} \tag{24}$$

*and*

$$P_{2l}^{***}(z) = \sum_{h=1}^{r_{2l}} d_{lh} \sum_{j=1}^{p_1} \sum_{k=1}^{r_{1j}} c_{jk} \sum_{i=0}^{h-1} \binom{h-1}{i} \frac{(k+i-1)!}{(\lambda_j+\nu_l)^{k+i}} \sum_{t=0}^{h-1-i} \frac{(h-1-i)!}{t!} \frac{(-z)^t}{\nu_l^{h-i-t}}\,. \tag{25}$$

[3] see the author and editors note at the end of the Chapter, concerning expressions (20), (22), (23) and (25)

*Proof* Considering (2) and taking $K_1$ and $c_{jk}$ defined in a similar manner to $K$ and $c_{jk}$ in (4) and (5)–(7) respectively, and $K_2$ and $d_{lh}$ defined in a corresponding manner, the p.d.f. of $Z$ is given by

$$
\begin{aligned}
f_Z(z) &= K_1 K_2 \int_{\max(z,0)}^{+\infty} \left\{ \sum_{j=1}^{p_1} \left( \sum_{k=1}^{r_{ij}} c_{jk}\, y_1^{k-1} \right) \exp(-\lambda_j y_1) \right\} \\
&\quad \times \left\{ \sum_{l=1}^{p_2} \left( \sum_{h=1}^{r_{2l}} d_{lh} (y_1 - z)^{h-1} \right) \exp\{-\nu_l (y_1 - z)\} \right\} dy_1 \\
&= K_1 K_2 \int_{\max(z,0)}^{+\infty} \sum_{j=1}^{p_1} \sum_{l=1}^{p_2} \left( \sum_{k=1}^{r_{1j}} c_{jk}\, y_1^{k-1} \right) \left( \sum_{h=1}^{r_{2l}} d_{lh} (y_1 - z)^{h-1} \right) \\
&\quad \times \exp\{-(\lambda_j + \nu_l) y_1\} \exp(\nu_l z)\, dy_1 \\
&= K_1 K_2 \sum_{j=1}^{p_1} \sum_{l=1}^{p_2} \int_{\max(z,0)}^{+\infty} \sum_{k=1}^{r_{1j}} \sum_{h=1}^{r_{2l}} c_{jk}\, d_{lh}\, y_1^{k-1} (y_1 - z)^{h-1} \\
&\quad \times \exp\{-(\lambda_j + \nu_l) y_1\} \exp(\nu_l z)\, dy_1 \\
&= K_1 K_2 \sum_{j=1}^{p_1} \sum_{l=1}^{p_2} \exp(\nu_l z) \sum_{k=1}^{r_{1j}} \sum_{h=1}^{r_{2l}} c_{jk}\, d_{lh} \sum_{i=0}^{h-1} \binom{h-1}{i} (-z)^{h-1-i} \\
&\quad \int_{\max(z,0)}^{+\infty} \exp\{-(\lambda_j + \nu_l) y_1\}\, y_1^{k+i-1}\, dy_1,
\end{aligned}
\tag{26}
$$

where, for $m > 0$ and $k \in \mathbb{N}_0$,

$$
\int_{\max(z,0)}^{+\infty} \exp(-my)\, y^k\, dy = \begin{cases} \exp(-mz) \displaystyle\sum_{i=0}^{k} \frac{k!}{i!} \frac{z^i}{m^{k-i+1}}, & z \geq 0 \\ \dfrac{k!}{m^{k+1}}, & z \leq 0 \end{cases}
\tag{27}
$$

(which is, for $z > 0$, a particular version of an incomplete Gamma function) yielding (20) with $P_{1j}^{**}(z)$ given by

$$
\begin{aligned}
P_{1j}^{**}(z) = \sum_{k=1}^{r_{1j}} c_{jk} \sum_{\ell=1}^{p_2} \sum_{h=1}^{r_{2\ell}} d_{\ell h} \sum_{i=0}^{h-1} \binom{h-1}{i} (-z)^{h-1-i} \\
\sum_{t=0}^{k+i-1} \frac{(k+i-1)!}{t!} \frac{z^t}{(\lambda_j + \nu_l)^{k+i-t}}.
\end{aligned}
\tag{28}
$$

It is however interesting and useful to observe that, given that the distribution of $Y_1 - Y_2$ and $Y_2 - Y_1$ are symmetrical, and that we may in (26) integrate in order to $y_2$ instead of $y_1$, we may obtain the p.d.f. of $Z$ given by a similar expression to the one in (20), with $P_{1j}^{**}(z)$ given by (21) and [4]

$$P_{2j}^{**}(z) = \sum_{k=1}^{r_{1j}} c_{jk} \sum_{\ell=1}^{p_2} \sum_{h=1}^{r_{2\ell}} d_{\ell h} \sum_{i=0}^{k-1} \binom{k-1}{i} z^{k-1-i} \sum_{t=0}^{h+i-1} \frac{(h+i-1)!}{t!} \frac{z^t}{(\lambda_j + \nu_\ell)^{h+i-t}} . \tag{29}$$

We may further note that indeed (for $z \in \mathbb{R}$ and $h, k \in \mathbb{N}$),

$$\sum_{i=0}^{h-1} \binom{h-1}{i} (-z)^{h-1-i} \sum_{t=0}^{k+i-1} \frac{(k+i-1)!}{t!} \frac{z^t}{(\lambda_j+\nu_l)^{k+i-t}} = \sum_{i=0}^{k-1} \binom{k-1}{i} z^{k-1-i} \frac{(h+i-1)!}{(\lambda_j+\nu_l)^{h+i}} .$$

This way, in order to obtain a simpler expression for the p.d.f. of $Z$, we may consider the p.d.f. in (20) with $P_{1j}^{**}(z)$ given by (21) and $P_{2j}^{**}(z)$ given by (22).

Then, from (20) through (22) it is easy to derive the c.d.f. of $Z$ as given by (23) through (25). $\square$

**Corollary 2** *If we consider, for some $k > 0$, the r.v.*

$$W = k^{-1} \exp(Z)$$

*we have* [5]

$$f_W(w) = \begin{cases} K_1 K_2 \displaystyle\sum_{j=1}^{p_1} P_{1j}^{**}(\log(kw))(kw)^{-\lambda_j} \frac{1}{w}, & w \ge k^{-1}, \\ K_1 K_2 \displaystyle\sum_{j=1}^{p_1} P_{2j}^{**}(\log(kw))(kw)^{\nu_j} \frac{1}{w}, & 0 < w \le k^{-1}, \end{cases} \tag{30}$$

[4] see the author and editors note at the end of the Chapter concerning expression (29)

[5] see the author and editors note at the end of the Chapter concerning expressions (30) and (31)

*and*

$$F_W(w) = \begin{cases} K_1 K_2 \sum_{j=1}^{p_1} P_{1j}^{***}(\log(kw))(kw)^{-\lambda_j}, \; w \geq k^{-1}, \\ K_1 K_2 \sum_{j=1}^{p_1} P_{2j}^{***}(\log(kw))(kw)^{\nu_j}, \quad 0 < w \leq k^{-1}. \; \square \end{cases} \tag{31}$$

As we did with (12) and (18), also in (20) through (31) we may take both $p_1 \to \infty$ and $p_2 \to \infty$, which still yields genuine distributions (see Appendix).

# 3 The Distribution of the Product of *m* Independent Random Variables with GGR Distributions

## *3.1 The Case with All Distinct Shape Parameters for the Numerator and Denominator*

In this section we will obtain explicit and concise expressions for the p.d.f. and c.d.f. of the r.v.

$$W = \prod_{j=1}^{m} X_j \,,$$

where the $X_j$'s are independent r.v.'s with all distinct shape parameters. The p.d.f. and c.d.f. of $W$ are stated in the following Theorem.

**Theorem 3** *Let*

$$X_j \sim GGR(k_j, r_{1j}, r_{2j}, \beta_j)\,, \quad j = 1, \ldots, m$$

*be m independent r.v.'s where the power parameters* $\beta_j$ $(j = 1, \ldots, m)$ *are not necessarily all positive. Let*

$$\beta_j^* = |\beta_j|\,, \quad s_{1j} = \begin{cases} r_{1j} \text{ if } \beta_j > 0 \\ r_{2j} \text{ if } \beta_j < 0 \end{cases} \quad \text{and} \quad s_{2j} = \begin{cases} r_{2j} \text{ if } \beta_j > 0 \\ r_{1j} \text{ if } \beta_j < 0 \end{cases}. \tag{32}$$

*Then, if $\beta_j^* s_{1j} \neq \beta_k^* s_{1k}$ and $\beta_j^* s_{2j} \neq \beta_k^* s_{2k}$, $\forall j \neq k,\ j,k \in \{1,\dots,m\}$*[6]*, the p.d.f. and c.d.f. of the r.v.*

$$W = \prod_{j=1}^{m} X_j\,, \tag{33}$$

*are respectively given by*[7]

$$f_W(w) = \begin{cases} \lim\limits_{n\to\infty} K_1 K_2 \sum\limits_{j=1}^{m}\sum\limits_{h=0}^{n} H_{2jh}\, d_{hj}\, (w/K^*)^{s_{1jh}^*}\, \dfrac{1}{w}, & 0 < w \leq K^*, \\ \lim\limits_{n\to\infty} K_1 K_2 \sum\limits_{j=1}^{m}\sum\limits_{h=0}^{n} H_{1jh}\, c_{hj}\, (w/K^*)^{-s_{2jh}^*}\, \dfrac{1}{w}, & w \geq K^*, \end{cases} \tag{34}$$

*and*[8]

$$F_W(w) = \begin{cases} \lim\limits_{n\to\infty} K_1 K_2 \sum\limits_{j=1}^{m}\sum\limits_{h=0}^{n} H_{2jh} \dfrac{d_{hj}}{s_{1jh}^*} \left(\dfrac{w}{K^*}\right)^{s_{1jh}^*}, & 0 < w \leq K^* \\ \lim\limits_{n\to\infty} K_1 K_2 \sum\limits_{j=1}^{m}\sum\limits_{h=0}^{n} \left\{ H_{2jh} \dfrac{d_{hj}}{s_{1jh}^*} + H_{1jh} \dfrac{c_{hj}}{s_{2jh}^*} \left(1 - \left(\dfrac{w}{K^*}\right)^{-s_{2jh}^*}\right)\right\}, & w \geq K^* \end{cases} \tag{35}$$

*where*

$$K^* = \prod_{j=1}^{m} k_j^{-1/\beta_j}\,, \tag{36}$$

$$K_1 = \prod_{j=1}^{m}\prod_{h=0}^{n} s_{1jh}^*\,, \quad K_2 = \prod_{j=1}^{m}\prod_{h=0}^{n} s_{2jh}^*\,, \tag{37}$$

---

[6] see the author and editors note at the end of the Chapter

[7] see the author and editors note at the end of the Chapter concerning expressions (34), (35), (38) and (39)

[8] see the author and editors note at the end of the Chapter

*and, for* $j = 1, \dots, m$ *and* $h = 0, \dots, n$,

$$c_{hj} = \prod_{\substack{\eta=1 \\ (\{\eta \neq j \wedge \nu \neq h\})}}^{m} \prod_{\nu=0}^{n} \frac{1}{s^*_{1\eta\nu} - s^*_{1jh}}, \qquad d_{hj} = \prod_{\substack{\eta=1 \\ (\{\eta \neq j \wedge \nu \neq h\})}}^{m} \prod_{\nu=0}^{n} \frac{1}{s^*_{2\eta\nu} - s^*_{2jh}}, \tag{38}$$

*and*

$$H_{1jh} = \sum_{k=1}^{m} \sum_{l=0}^{n} \frac{d_{kl}}{s^*_{1jh} + s^*_{2kl}}, \qquad H_{2jh} = \sum_{k=1}^{m} \sum_{l=0}^{n} \frac{c_{kl}}{s^*_{2jh} + s^*_{1kl}}, \tag{39}$$

*with*

$$s^*_{1jh} = \beta^*_j (s_{1j} + h) \quad and \quad s^*_{2jh} = \beta^*_j (s_{2j} + h) \quad (j = 1, \dots, m; h = 0, \dots, n). \tag{40}$$

*Proof* From (1) we have

$$E\left(X_j^h\right) = k_j^{-h/\beta_j} \frac{\Gamma(r_{1j} + h/\beta_j)\,\Gamma(r_{2j} - h/\beta_j)}{\Gamma(r_{1j})\,\Gamma(r_{2j})}$$

so that if we take

$$Y_j = k_j^{1/\beta_j} X_j,$$

then we have

$$E\left(Y_j^h\right) = \frac{\Gamma(r_{1j} + h/\beta_j)\,\Gamma(r_{2j} - h/\beta_j)}{\Gamma(r_{1j})\,\Gamma(r_{2j})}.$$

Thus, if we define

$$W' = \prod_{j=1}^{m} Y_j,$$

then we have

$$W = \prod_{j=1}^{m} X_j = \underbrace{\left( \prod_{j=1}^{m} k_j^{-1/\beta_j} \right)}_{=K^*} \prod_{j=1}^{m} Y_j = K^* W',$$

where we may take

$$Z = \log W' = \sum_{j=1}^{m} \log Y_j ,$$

so that the characteristic function $\Phi_Z(t)$ is given, under the form of a Mellin transform, for $i = \sqrt{-1}$, by

$$\begin{aligned}\Phi_Z(t) &= \prod_{j=1}^{m} \Phi_{\log Y_j}(t) = \prod_{j=1}^{m} E\left(Y_j^{it}\right) \\ &= \prod_{j=1}^{m} \frac{\Gamma(r_{1j} + it/\beta_j)\,\Gamma(r_{2j} - it/\beta_j)}{\Gamma(r_{1j})\,\Gamma(r_{2j})} .\end{aligned} \tag{41}$$

Using in (41) the relation

$$\Gamma(z) = \frac{\exp(-\gamma z)}{z} \prod_{i=1}^{\infty} \left( \frac{i}{i+z} \exp(z/i) \right) ,$$

which is valid for any $z \in \mathbb{C}\backslash\{0, -1, -2, \ldots\}$ (Ahlfors, 1979, § 2.4, pp. 198; Lang, 1999, Chap. XV, § 2, pp. 413), where $\gamma$ is the Euler gamma constant, we may, after some simplifications, write

$$\Phi_Z(t)=\prod_{j=1}^{m}\prod_{k=0}^{\infty} \beta_j(r_{1j}+k)\big(\beta_j(r_{1j}+k) + it\big)^{-1} \beta_j(r_{2j}+k)\big(\beta_j(r_{2j}+k)-it\big)^{-1}. \tag{42}$$

Since the $\beta_j$ $(j = 1, \ldots, m)$ are not necessarily all positive, we take $\beta_j^*$, $s_{1j}$ and $s_{2j}$ as given by (32), so that we may write

$$\Phi_Z(t)=\prod_{j=1}^{m}\prod_{k=0}^{\infty} \beta_j^*(s_{1j}+k)\Big(\beta_j^*(s_{1j}+k) + it\Big)^{-1} \beta_j^*(s_{2j}+k)\Big(\beta_j^*(s_{2j}+k)-it\Big)^{-1}, \tag{43}$$

where now $\beta_j^*(s_{1j} + k) > 0$ and $\beta_j^*(s_{2j} + k) > 0$ for all $j = 1, \ldots, m$ and $k = 0, 1, \ldots$. Expression (43) shows that, if $\beta_j^* s_{1j} \neq \beta_k^* s_{1k}$ and $\beta_j^* s_{2j} \neq \beta_k^* s_{2k}$, $\forall j \neq k,\ j, k \in \{1, \ldots, m\}$, the distribution of $Z$ is the same as the distribution of a sum of infinitely many independent r.v.'s distributed as the difference of two independent Exponential distributions, with parameters $\beta_j^*(s_{1j} + k)$ and $\beta_j^*(s_{2j} + k)$ $(j = 1, \ldots, m; k = 0, 1, \ldots)$. Alternatively, $Z$ is distributed as the difference of two independent r.v.'s, each one having the distribution of the sum of infinitely many independent Exponential r.v.'s.

Thus, since

$$W = K^* \exp(Z)\,, \tag{44}$$

with $K^*$ given by (36), taking $s^*_{1jh} = \beta^*_j(s_{1j}+h)$ and $s^*_{2jh} = \beta^*_j(s_{2j}+h)$ and using (18) and (19) in Corollary 1 as a basis, we may write the p.d.f. and c.d.f. of $W$ as in (34) through (40). □

To simplify the notation we may replace the pair of indices $(k, j)$ by $h = km + j$, setting

$$s^*_{ih} = \beta^*_j(s_{ij}+k)\,, \quad \text{for } i = 1, 2; \;\; j = 1, \dots, m; \;\; k = 0, \dots, n\,. \tag{45}$$

We may now define $K_1$, $K_2$, $c_j$, $d_j$, $H_{1j}$ and $H_{2j}$ $(j = 1, \dots, m(n+1))$, (with $n \to \infty$) in a way similar to the one used in Subsection 1.2, that is,

$$K_1 = \prod_{j=1}^{m(n+1)} s^*_{1j}\,, \qquad K_2 = \prod_{j=1}^{m(n+1)} s^*_{2j}\,,$$

and, for $j = 1, \dots,$

$$c_j = \prod_{\substack{k=1\\k\neq j}}^{m(n+1)} \frac{1}{s^*_{1j} - s^*_{1k}}\,, \qquad d_j = \prod_{\substack{k=1\\k\neq j}}^{m(n+1)} \frac{1}{s^*_{2j} - s^*_{2k}}$$

and

$$H_{1j} = \sum_{h=1}^{m(n+1)} \frac{d_h}{s^*_{1j} + s^*_{2h}}\,, \qquad H_{2j} = \sum_{h=1}^{m(n+1)} \frac{c_h}{s^*_{1h} + s^*_{2j}}\,,$$

so that we may, for example, write the c.d.f. of the r.v. $W$ as [9]

$$F_W(w) = \begin{cases} \lim\limits_{n\to\infty} K_1K_2 \sum\limits_{j=1}^{m(n+1)} H_{2j}\dfrac{d_j}{s^*_{1j}}\left(\dfrac{w}{K^*}\right)^{s^*_{1j}}, & 0 < w \le K^*,\\[2ex] \lim\limits_{n\to\infty} K_1K_2 \sum\limits_{j=1}^{m(n+1)} \left[H_{2j}\dfrac{d_j}{s^*_{1j}} + H_{1j}\dfrac{c_j}{s^*_{2j}}\left\{1-\left(\dfrac{w}{K^*}\right)^{-s^*_{2j}}\right\}\right], & w \ge K^*. \end{cases} \tag{46}$$

[9] see the author and editors note at the end of the Chapter concerning expression (46)

## 3.2 The General Case

In the general case, we will allow for the possibility that some of the parameters $\tilde{s}_{1j} = \beta_j^* s_{1j}$, $j \in \{i, \ldots, m\}$, will be equal and that also some of the parameters $\tilde{s}_{2j} = \beta_j^* s_{2j}$, $j \in \{i, \ldots, m\}$, will be equal. More precisely, and without any loss of generality, let us suppose that $\tau_1$ of the $m$ parameters $\tilde{s}_{1j}$ are equal to $\tilde{s}_{11}$, $\tau_2$ are equal to $\tilde{s}_{12}$, and so on, and that $\tau_{p_1}$ are equal to $\tilde{s}_{1p_1}$, with $p_1 < m$ and

$$\sum_{j=1}^{p_1} \tau_j = m\,,$$

and that, similarly, $\eta_1$ of the $m$ parameters $\tilde{s}_{2j}$ are equal to $\tilde{s}_{21}$, $\eta_2$ are equal to $\tilde{s}_{22}$, and so on, and that $\eta_{p_2}$ are equal to $\tilde{s}_{2p_2}$, with $p_2 < m$ and

$$\sum_{j=1}^{p_2} \eta_j = m\,.$$

Then the characteristic function in (41) may be written as

$$\begin{aligned}
\Phi_Z(t) &= \prod_{j=1}^{p_1}\prod_{k=0}^{\infty}\left(\tilde{s}_{1j}+\beta_j^*k\right)^{\tau_j}\left(\tilde{s}_{1j}+\beta_j^*k-it\right)^{-\tau_j} \\
&\qquad\qquad \prod_{j=1}^{p_2}\prod_{k=0}^{\infty}\left(\tilde{s}_{2j}+\beta_j^*k\right)^{\eta_j}\left(\tilde{s}_{2j}+\beta_j^*k-it\right)^{-\eta_j} \\
&= \lim_{n\to\infty}\prod_{j=1}^{p_1}\prod_{k=0}^{n}\left(\tilde{s}_{1j}+\beta_j^*k\right)^{\tau_j}\left(\tilde{s}_{1j}+\beta_j^*k-it\right)^{-\tau_j} \\
&\qquad\qquad \prod_{j=1}^{p_2}\prod_{k=0}^{n}\left(\tilde{s}_{2j}+\beta_j^*k\right)^{\eta_j}\left(\tilde{s}_{2j}+\beta_j^*k-it\right)^{-\eta_j}\,,
\end{aligned}$$

that is the characteristic function of the difference of two independent r.v.'s with GIG distributions, the first one, that is, the one with positive sign, with depth $p_1\times(n+1)$ (with $n \to \infty$), with rate parameters and associated shape parameters

$$\underbrace{\tilde{s}_{11}+\beta_1^*k}_{k=0,\ldots,n}\,,\;\ldots\;,\;\underbrace{\tilde{s}_{1j}+\beta_j^*k}_{k=0,\ldots,n}\,,\;\ldots\;,\;\underbrace{\tilde{s}_{1p_1}+\beta_{p_1}^*k}_{k=0,\ldots,n}$$

$$\underbrace{\tau_1,\ldots,\tau_1}_{n+1}\,,\;\ldots\;,\;\underbrace{\tau_j,\ldots,\tau_j}_{n+1}\,,\;\ldots\;,\;\underbrace{\tau_{p_1},\ldots,\tau_{p_1}}_{n+1}$$

and the second one, that is, the one with negative sign, with depth $p_2 \times (n+1)$ (with $n \to \infty$), with shape parameters and associated rate parameters

$$\underbrace{\tilde{s}_{21} + \beta_1^* k}_{k=0,\ldots,n}, \ \ldots\ , \underbrace{\tilde{s}_{2j} + \beta_j^* k}_{k=0,\ldots,n}, \ \ldots\ , \underbrace{\tilde{s}_{2p_1} + \beta_{p_2}^* k}_{k=0,\ldots,n}$$

$$\underbrace{\eta_1, \ldots, \eta_1}_{n+1}, \ \ldots\ , \underbrace{\eta_j, \ldots, \eta_j}_{n+1}, \ \ldots\ , \underbrace{\eta_{p_2}, \ldots, \eta_{p_2}}_{n+1}.$$

Let us consider the vectors

$$\underline{\tau}^* = \Big[\underbrace{\tau_1, \ldots, \tau_{p_1}, \tau_1, \ldots, \tau_{p_1}, \ \ldots\ , \tau_1, \ldots, \tau_{p_1}}_{n+1 \text{ times}}\Big]',$$

and

$$\underline{\eta}^* = \Big[\underbrace{\eta_1, \ldots, \eta_{p_2}, \eta_1, \ldots, \eta_{p_2}, \ \ldots\ , \eta_1, \ldots, \eta_{p_2}}_{n+1 \text{ times}}\Big]',$$

where, for $j = 1, \ldots, p_1(n+1)$ and $l = 1, \ldots, p_2(n+1)$, with $j = kp_1 + h$ and $l = kp_2 + i$, for $k = 0, \ldots, n$, $h = 1, \ldots, p_1$ and $i = 1, \ldots, p_2$,

$$\tau_j^* = \tau_h, \qquad \text{and} \qquad \eta_l^* = \eta_i,$$

and, similarly to the vectors $\underline{s}_1^*$ and $\underline{s}_2^*$ considered in the previous subsection, the vectors

$$\underline{\tilde{s}}_1^* = \left[\tilde{s}_{11}, \ldots, \tilde{s}_{1p_1}, \tilde{s}_{11} + \beta_1^*, \ldots, \tilde{s}_{1p_1} + \beta_{p_1}^*, \ldots, \tilde{s}_{11} + \beta_1^* n, \ldots, \tilde{s}_{1p_1} + \beta_{p_1}^* n\right]'$$

and

$$\underline{\tilde{s}}_2^* = \left[\tilde{s}_{21}, \ldots, \tilde{s}_{2p_1}, \tilde{s}_{21} + \beta_1^*, \ldots, \tilde{s}_{2p_1} + \beta_{p_1}^*, \ldots, \tilde{s}_{21} + \beta_1^* n, \ldots, \tilde{s}_{2p_1} + \beta_{p_1}^* n\right]'$$

where, once again, for $j = 1, \ldots, p_1(n+1)$ and $l = 1, \ldots, p_2(n+1)$, with $j, l, k$, $h$ and $i$ defined as above,

$$\tilde{s}_{1j}^* = \tilde{s}_{1h} + \beta_h^* k \qquad \text{and} \qquad \tilde{s}_{2l}^* = \tilde{s}_{2i} + \beta_i^* k\,.$$

The p.d.f. and c.d.f. of $W = K^* \exp(Z)$, for $K^*$ defined as in (44), may then be derived respectively from (30) and (31) in Corollary 2, taking into account the shape and rate parameters mentioned above, as

$$f_W(w) = \begin{cases} \lim\limits_{n\to\infty} K_1 K_2 \sum\limits_{j=1}^{p_2(n+1)} P_{2j}^{**}\left(\log\left(\frac{w}{K^*}\right)\right)\left(\frac{w}{K^*}\right)^{\tilde{s}_{2j}^*}\frac{1}{w}, & 0<w\leq K^* \\ \lim\limits_{n\to\infty} K_1 K_2 \sum\limits_{j=1}^{p_1(n+1)} P_{1j}^{**}\left(\log\left(\frac{w}{K^*}\right)\right)\left(\frac{w}{K^*}\right)^{-\tilde{s}_{1j}^*}\frac{1}{w}, & w\geq K^* \end{cases} \tag{47}$$

and [10]

$$F_W(w) = \begin{cases} \lim\limits_{n\to\infty} K_1 K_2 \sum\limits_{j=1}^{p_2(n+1)} P_{2j}^{***}\left(\log\left(\frac{w}{K^*}\right)\right)\left(\frac{w}{K^*}\right)^{\tilde{s}_{2j}^*}, & 0<w\leq K^*, \\ \lim\limits_{n\to\infty} K_1 K_2 \sum\limits_{j=1}^{p_1(n+1)} P_{1j}^{***}\left(\log\left(\frac{w}{K^*}\right)\right)\left(\frac{w}{K^*}\right)^{-\tilde{s}_{1j}^*}, & w\geq K^*, \end{cases} \tag{48}$$

where now $K_1$, $K_2$, $P_{1j}^{***}(\cdot)$ and $P_{2j}^{***}(\cdot)$ are defined as in Theorem 2 and Corollary 2, with $p_1$ replaced by $p_1(n+1)$, $p_2$ replaced by $p_2(n+1)$, $\lambda_j$ replaced by $\tilde{s}_{1j}^*$, $\nu_l$ replaced by $\tilde{s}_{2j}^*$, $r_{1j}$ replaced by $\tau_j^*$ and $r_{2l}$ replaced by $\eta_l^*$.

# 4 Applications

## *4.1 Testing the (Mutual) Independence of a Set of Variables*

Let us suppose that the random vector $\underline{X}$ $(p\times 1)$ has a $p$-multivariate Normal distribution,

$$\underline{X} \sim N_p(\underline{\mu}, \Sigma)\,,$$

and we want to test the hypothesis that the $p$ random variables in $\underline{X}$ are mutually independent. This hypothesis may be written as

$$\Sigma = diag(\sigma_1^2, \ldots, \sigma_p^2)\,. \tag{49}$$

[10] see the author and editors note at the end of the Chapter concerning expression (48)

If we have a sample of size $N$ ($> p$), the power $2/N$ of the likelihood ratio test statistic is (Anderson, 1984, Chap. 9, sec. 9.2; Coelho, 1992, 1998, 2004)

$$\Lambda = \frac{|V|}{\prod_{j=1}^{p} V_{jj}}, \tag{50}$$

where $V$ is either the sample variance-covariance matrix or the maximum likelihood estimator of $\Sigma$, and $V_{jj}$ is the $j$-th diagonal element of $V$. With $V_j^*$ the corresponding matrix for the last $p - j$ variables (a lower right submatrix of $V$), we have (see Anderson (1984, Chap. 9, Thm. 9.3.2), Coelho (1992, Chap. 4))

$$\Lambda = \prod_{j=1}^{p-1} \Lambda_{j(j+1,\ldots,p)} \tag{51}$$

with

$$\Lambda_{j(j+1,\ldots,p)} = \frac{|V_j^*|}{V_{jj}\,|V_{j+1}^*|}$$

as the statistic for testing the independence between the $j$-th variable and those with indexes $j + 1, \ldots, p$. Moreover, when $H_0$ holds, $\Lambda_{j(j+1,\ldots,p)}$ will have a Beta distribution with parameters $(n^* - p + j)/2$ and $(p - j)/2$, where $n^* = N - 1$ (Anderson, 1984, Chap. 9, sec. 9.3).

When $H_0$ holds,

$$\frac{\Lambda_{j(j+1,\ldots,p)}}{1 - \Lambda_{j(j+1,\ldots,p)}} \sim GGR\left(1, \frac{n^* - p + j}{2}, \frac{p - j}{2}, 1\right)$$

and

$$\frac{1 - \Lambda_{j(j+1,\ldots,p)}}{\Lambda_{j(j+1,\ldots,p)}} \sim GGR\left(1, \frac{p - j}{2}, \frac{n^* - p + j}{2}, 1\right).$$

We may thus replace $\Lambda$ as test statistic by

$$G^* = \prod_{j=1}^{p-1} \frac{\Lambda_{j(j+1,\ldots,p)}}{1 - \Lambda_{j(j+1,\ldots,p)}} \quad \text{or} \quad G^{**} = \prod_{j=1}^{p-1} \frac{1 - \Lambda_{j(j+1,\ldots,p)}}{\Lambda_{j(j+1,\ldots,p)}},$$

both of which, under $H_0$ in (49), have the distribution of the product of independent Gamma ratio random variables with parameters $k_j = \beta_j = 1$ and $r_{1j} = (n^* - p + j)/2$, $r_{2j} = (p - j)/2$, and $r_{1j} = (p - j)/2$, $r_{2j} = (n^* - p + j)/2$, $(j = 1, \ldots, p - 1)$, respectively.

## 4.2 *Random Effects Models with Balanced Cross-nesting*

Let us assume that there are $L$ groups with $u_1, \dots, u_L$ factors. The first factors in the groups will have $a_\ell(1)$ levels and, if $u_\ell > 1$, the $h$-th factor in the $\ell$-th group will have $a_\ell(h)$ levels nested inside each level of the preceding factor ($h = 2, \dots, u_\ell$). There will be $c_\ell(h) = \prod_{k=1}^{h} a_\ell(h)$ level combinations of the first $h$ factors in the $\ell$-th group, each nesting $b_\ell(h) = c_\ell(u_\ell)/c_\ell(h)$ level combinations of the remaining factors ($h = 1, \dots, u_\ell$; $\ell = 1, \dots, L$). We also take $c_\ell(-1) = c_\ell(0) = 1$, $\ell = 1, \dots, L$.

Since when a factor nests another they do not interact, the variance components will correspond to sets of factors belonging to the different groups. These sets of factors are indicated by the vectors $\underline{h}$ with components $h_\ell = 0, \dots, u_\ell$, $\ell = 1, \dots, L$. If $h_\ell = 0$, no factor is taken from the $\ell$-th group, otherwise $h_\ell$ will be the index of the factor taken in that group. Let $\Gamma$ represent the set of vectors $\underline{h}$ and $\sigma^2(\underline{h})$ the variance component for the set of factors corresponding to $\underline{h}$.

If $r$ observations are taken for each treatment the sum $S(\underline{h})$ of squares for the effects or interactions indicated by $\underline{h}$ will be (see Fonseca et al. (2003)) the product of

$$\gamma(\underline{h}) = \sigma^2 + \sum_{\underline{k}:\underline{h}\leq\underline{k}} b(\underline{k})\,\sigma^2(\underline{k})\,, \quad \underline{h} \in \Gamma$$

where $b(\underline{k}) = r\prod_{\ell=1}^{k} b_\ell(k_\ell)$, and a central chi-square with $g(\underline{k}) = \prod_{\ell=1}^{L}\{c_\ell(k_\ell) - c_\ell(k_\ell - 1)\}$ degrees of freedom.

When only one component, say $h_\ell$, of $\underline{h}$, is less than the corresponding bound $u_\ell$, let $\underline{h}^+$ be the vector obtained increasing that component by 1. Then

$$\gamma(\underline{h}) = \gamma(\underline{h}^+) + b(\underline{h})\,\sigma^2(\underline{h})$$

and we may use

$$\mathcal{F}(\underline{h}) = \frac{g(\underline{h}^+)}{g(\underline{h})}\,\frac{S(\underline{h})}{S(\underline{h}^+)}$$

to test

$$H_0(\underline{h}) : \sigma^2(\underline{h}) = 0\,.$$

We point out that $\mathcal{F}(\underline{h})$ will be the product of $1 + \sigma^2(\underline{h})/\gamma(\underline{h}^+)$ and a variable having a central $F$ distribution with $g(\underline{h})$ and $g(\underline{h}^+)$ degrees of freedom.

The $S(\underline{h})$, $h \in \Gamma$, are independent (Fonseca et al., 2003). Thus if the pairs $(\underline{h}_j, \underline{h}_j^+)$, $j = 1, \ldots, d$, are formed by distinct vectors we may use

$$\overline{\mathcal{F}} = \prod_{j=1}^{d} \mathcal{F}(\underline{h}_j)$$

to test

$$\overline{H}_0 : \sigma^2(\underline{h}_1) = \cdots = \sigma^2(\underline{h}_d) = 0 .$$

This test statistic, when $\overline{H}_0$ holds, will be the product of $d$ independent variables with central $F$ distributions.

## 4.3 Combining Independent F Tests in Meta-Analysis

An emergent problem in Meta-Analysis is how to combine independent $F$ tests (Khuri et al., 1998). One possible way to do so is by considering an overall test statistic, for the whole meta-analysis being considered, which will be the product of the elementary $F$ statistics for each single analysis. By using this procedure, we may also obtain the distribution of the overall statistic under the alternative hypothesis, which will be the distribution of the product of non-central $F$ distributions. This distribution may be derived from the distributions obtained in this manuscript and will be the object of a future work.

## 4.4 A Test for Equality of Two Generalized Variances

Let us suppose that

$$\underline{X}_1 \sim N_p(\underline{\mu}_1, \Sigma_1) \qquad \text{and} \qquad \underline{X}_2 \sim N_p(\underline{\mu}_2, \Sigma_2) ,$$

and that we want to test the hypothesis

$$H_0 : |\Sigma_1| = |\Sigma_2| , \tag{52}$$

based on two independent samples of sizes $n_1$ and $n_2$ of $\underline{X}_1$ and $\underline{X}_2$ respectively.

Taking the work of Cao (2006, Chap. 4, sec. 4.1, 4.2) as a basis, we may use the statistic

$$T = \frac{|S_1|}{|S_2|} , \tag{53}$$

where $S_1$ and $S_2$ are either the sample variance-covariance matrices or the maximum likelihood estimators of $\Sigma_1$ and $\Sigma_2$, based on the two independent random samples of sizes $n_1$ and $n_2$ respectively of $\underline{X}_1$ and $\underline{X}_2$. Alternatively, $S_1$ and $S_2$ may yet be taken as the matrices of squares and cross-products of the deviations from the sample means. A test of size $\alpha$ for $H_0$ in (52) may then be constructed by rejecting $H_0$ in (52) if, for the statistic $T$ in (53),

$$T_{calc} > T_{1-\alpha/2} \quad \text{or} \quad T_{calc} < T_{\alpha/2}\,,$$

where $T_{1-\alpha/2}$ and $T_{\alpha/2}$ represent respectively the $1-\alpha/2$ and the $\alpha/2$ quantiles of the statistic $T$ in (53).

Taking $S_1$ and $S_2$ as the maximum likelihood estimators of $\Sigma_1$ and $\Sigma_2$, we will have

$$S_1 \sim W_p\left(n_1-1, \frac{1}{n_1}\Sigma_1\right) \quad \text{and} \quad S_2 \sim W_p\left(n_2-1, \frac{1}{n_2}\Sigma_2\right),$$

with

$$|S_1| \sim \left|\frac{1}{n_1}\Sigma_1\right| \prod_{j=1}^{p} W_j \quad \text{and} \quad |S_2| \sim \left|\frac{1}{n_2}\Sigma_2\right| \prod_{j=1}^{p} Y_j\,,$$

where

$$W_j \sim \chi^2_{n_1-j} \quad \text{and} \quad Y_j \sim \chi^2_{n_2-j} \quad (j=1,\ldots,p)$$

are all independent. Thus, under $H_0$ in (52), the distribution of the statistic $T$ in (53) is the same as the distribution of

$$\frac{\left(\frac{1}{n_1}\right)^p \prod_{j=1}^{p} W_j}{\left(\frac{1}{n_2}\right)^p \prod_{j=1}^{p} Y_j} = \frac{\prod_{j=1}^{p} W_j^*}{\prod_{j=1}^{p} Y_j^*} = \prod_{j=1}^{p} Z_j$$

where

$$W_j^* \sim \Gamma\left(\frac{n_1-j}{2}, \frac{n_1}{2}\right) \quad \text{and} \quad Y_j^* \sim \Gamma\left(\frac{n_2-j}{2}, \frac{n_2}{2}\right) \quad (j=1,\ldots,p),$$

i.e., under $H_0$ in (52), the exact distribution of the statistic $T$ in (53) is the product of $p$ r.v.'s

$$Z_j \sim GGR\left(\frac{n_1}{n_2}, \frac{n_1-j}{2}, \frac{n_2-j}{2}, 1\right) \quad (j=1,\ldots,p)\,.$$

# 5 A Few Simulations and Numerical Studies for the Distributions Obtained

We will consider in this section a few numerical studies, together with the results from simulations in order to show the correct behaviour of the distributions obtained and the correctness of the corresponding expressions [11]. To show the agreement between the simulated data and the distributions obtained, simultaneous plots of the histograms of relative frequencies (simple and cumulative) and the exact p.d.f.'s and c.d.f.'s are shown.

The simulations were carried out by generating five vectors, each with $10^6$ randomly generated values of the distribution under study. These vectors were then ordered and their average used to compute the simulated quantiles and plot the relative frequency histograms.

We should note that although for any value of $n \in \mathbb{N}$, the expressions for the p.d.f.'s and c.d.f.'s in (34), (35), (47) and (48) yield proper p.d.f.'s and c.d.f.'s, we need to consider sufficiently high values of $n \in \mathbb{N}$ in order that the p.d.f.'s and c.d.f.'s in (34) and (35) or (47) and (48) converge to the exact distribution of the desired product of GGR r.v.'s.

All simulations and computations were done with the software Mathematica, from Wolfram Research, version 5.2.

## 5.1 The Test of Independence in a Set of p Variables

The case presented in this subsection corresponds to the example of application discussed in Subsection 4.1.

For $p = 5$ and $N = 10$, according to what was exposed in Subsection 4.1, if we want to test the independence of the $p = 5$ variables, we may, in this case, use the test statistic

$$G^* = G_1 G_2 G_3 G_4 , \tag{54}$$

where

$$\begin{array}{ll} G_1 \sim GGR(1, 5/2, 2, 1) , & G_2 \sim GGR(1, 6/2, 3/2, 1) , \\ G_3 \sim GGR(1, 7/2, 1, 1) , & G_4 \sim GGR(1, 8/2, 1/2, 1) , \end{array} \tag{55}$$

[11] see the author and editors note at the end of the Chapter

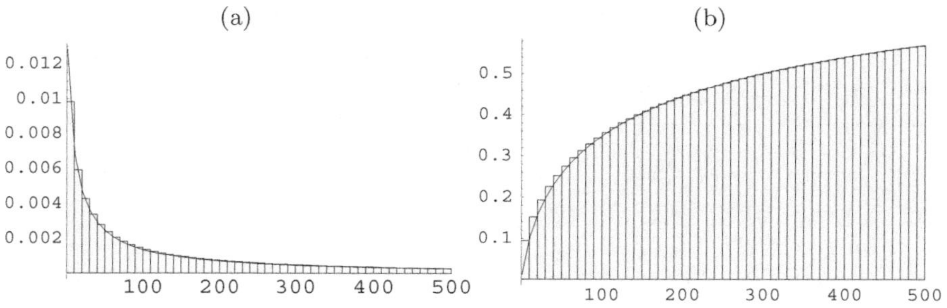


**Fig. 1** Overlaid plots of the histograms of relative frequencies generated from simulated data and the exact p.d.f. and c.d.f. for $n = 100$ for the statistic in (54): (a) histogram of relative frequencies and exact p.d.f.; (b) histogram of cumulative relative frequencies and exact c.d.f.

so that, using the result that states that if

$$X_1 \sim \chi^2_m \qquad \text{and} \qquad X_2 \sim \chi^2_{m-1}$$

are two independent r.v.'s, then

$$X_1 X_2 \sim \frac{\left(\chi^2_{2m-2}\right)^2}{4}$$

we may also write

$$G^* = F_1^* F_2^* , \tag{56}$$

where

$$F_1^* \sim GGR(1, 10/2, 2/2, 1/2) \quad \text{and} \quad F_2^* \sim GGR(1, 14/2, 6/2, 1/2) . \tag{57}$$

In order to obtain the exact distribution of the statistic $G^*$ we may then either use (54) and (55) or (56) and (57).

In Figure 1 we may observe the overlaid plots of the histograms of relative frequencies generated from the simulated data, obtained from (55), and the exact p.d.f. and c.d.f. for the statistic $G^*$ in (54), for $n = 100$, where we may observe the very long tail that the distribution has in this case.

In Table 1 we have the simulated 0.90, 0.95 and 0.99 quantiles for the statistic $G^*$ in (54), together with the quantiles computed from different truncations of the exact distributions corresponding to (54)–(55) and (56)–(57).

We may note that although, as expected, both exact distributions have quantiles that converge to the simulated ones as $n \to \infty$, the performance of the truncations with $n = 50$ and $n = 100$ is rather poor, with the truncations of the exact distribution based on (56) and (57) exhibiting a much worse behaviour. The exact distribution

**Table 1** Quantiles $q$ (simulated and from different truncations of the exact distribution) for the statistic $G^*$ in (54)

| | | From the exact distribution in (54), (55) with $n =$ | | | From the exact distribution in (56), (57) with $n =$ | | |
|---|---|---|---|---|---|---|---|
| $q$ | simulat. | 50 | 100 | 200 | 50 | 100 | 200 |
| 0.90 | 23748 | 19887 | 21729 | 22754 | 16767 | 19853 | 21720 |
| 0.95 | 107386 | 89567 | 98030 | 102742 | 75270 | 89409 | 97985 |
| 0.99 | 2935411 | 2469174 | 2706682 | 2839113 | 2068931 | 2464760 | 2705447 |

**Table 2** Relative error for quantiles obtained from different truncations of the exact distributions of the statistic $G^*$ in (54) and (56)

| | value of $n$ in the exact dist. in (54) and (55) | | | value of $n$ in the exact dist. in (54) and (55) | | |
|---|---|---|---|---|---|---|
| Quant. | 50 | 100 | 200 | 50 | 100 | 200 |
| 0.90 | −0.16257 | −0.08500 | −0.04187 | −0.29396 | −0.16402 | −0.08541 |
| 0.95 | −0.16594 | −0.08713 | −0.04325 | −0.29908 | −0.16741 | −0.08755 |
| 0.99 | −0.15883 | −0.07792 | −0.03281 | −0.29518 | −0.16034 | −0.07834 |

of $G^*$ based on (56) and (57) is indeed simpler than the one based on (54) and (55) since it only involves two r.v.'s instead of four, and as such it is also much easier and quicker to compute. But, on the other hand, it seems to need twice the number of terms to attain a similar accuracy as the one exhibited by the distribution based on (54) and (55). Although this worse performance of the exact distribution based on the product of two GGR r.v.'s may come out as a surprise, in the end it seems that the result points towards somewhat similar number of terms being necessary while one uses either one of the two exact distributions.

Similar conclusions may be drawn from Table 2, where we may analyse the values of the relative errors for the quantiles obtained from the truncations of the exact distributions used in Table 1. Now that we do not have the exact quantiles, the relative errors are computed relatively to the simulated quantiles as $(approximate - simulated)/simulated$. Once again it is interesting to note the almost linear relation between the values of the relative error and the value of $n$ used for both exact distributions.

The alternative approach consists in using the statistic

$$G^{**} = G_1^* G_2^* G_3^* G_4^* \tag{58}$$

where

$$\begin{array}{ll} G_1^* \sim GGR(1, 2, 5/2, 1)\,, & G_2^* \sim GGR(1, 3/2, 6/2, 1)\,, \\ G_3^* \sim GGR(1, 1, 7/2, 1)\,, & G_4^* \sim GGR(1, 1/2, 8/2, 1) \end{array} \tag{59}$$

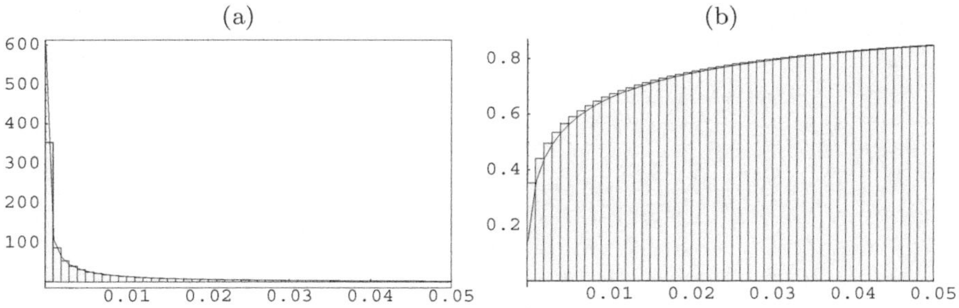


**Fig. 2** Overlaid plots of the histograms of relative frequencies generated from simulated data and the exact p.d.f. and c.d.f. for $n = 100$ for the statistic in (58): (a) histogram of relative frequencies and exact p.d.f.; (b) histogram of cumulative relative frequencies and exact c.d.f.

**Table 3** Quantiles $q$ (simulated and from different truncations of the exact distribution) for the statistic $G^*$ in (58)

| | | From the exact distribution in (58), (59) with $n =$ | | | From the exact distribution in (60), (61) with $n =$ | | |
|---|---|---|---|---|---|---|---|
| $q$ | simulat. | 50 | 100 | 200 | 50 | 100 | 200 |
| 0.90 | 0.09090 | 0.10169 | 0.09619 | 0.09345 | 0.11315 | 0.10180 | 0.09622 |
| 0.95 | 0.21654 | 0.23908 | 0.22307 | 0.21714 | 0.26257 | 0.23925 | 0.21853 |
| 0.99 | 1.03876 | 1.08365 | 1.08246 | 1.06167 | 1.20176 | 1.12303 | 1.08266 |

or,

$$G^{**} = F_1^{**} F_2^{**} \tag{60}$$

where

$$F_1^{**} \sim GGR(1, 2/2, 10/2, 1/2) \quad \text{and} \quad F_2^{**} \sim GGR(1, 6/2, 14/2, 1/2)\,. \tag{61}$$

In Figure 2 we have the overlaid plots of the histograms of relative frequencies generated from the simulated data, obtained from (59), and the exact p.d.f. and c.d.f. for the statistic $G^*$ in (58), for $n = 100$.

In Table 3 we have the simulated 0.90, 0.95 and 0.99 quantiles and corresponding quantiles from different truncations of the exact distributions derived from (59) and (61), and in Table 4 the relative errors of these quantiles, computed taking as reference the values of the simulated quantiles.

We may observe in Table 4 that the relative errors are, for a given value of $n$, generally smaller when using the statistic $G^{**}$, with a noticeable small value of the relative error for the 0.95 quantiles for $n = 200$.

**Table 4** Relative error for quantiles obtained from different truncations of the exact distributions of the statistic $G^{**}$ in (58) and (60)

| | value of $n$ in the exact dist. in (58) and (59) | | | value of $n$ in the exact dist. in (60) and (61) | | |
|---|---|---|---|---|---|---|
| Quant. | 50 | 100 | 200 | 50 | 100 | 200 |
| 0.90 | 0.11876 | 0.05827 | 0.02804 | 0.24483 | 0.11998 | 0.05857 |
| 0.95 | 0.10409 | 0.03015 | 0.00278 | 0.21255 | 0.10487 | 0.00916 |
| 0.99 | 0.04322 | 0.04207 | 0.02206 | 0.15691 | 0.08113 | 0.04226 |

## 5.2 *Simultaneous Tests for Variance Components in Random Effects Models with Balanced Cross-Nesting*

Let us assume we have three groups of factors, the first and third of which have a single factor and the second with five factors sequentially nested. Let us further suppose that all factors have two levels and random effects. When we want to test the simultaneous nullity of the variance components associated with the factor in the first group and with its interaction with the second and fourth factors in the second group, we have to use the statistic

$$F^* = F_1 F_2 F_3 , \tag{62}$$

where

$$F_1 = \frac{1}{1}\,\frac{S\left(\begin{bmatrix}1\\0\end{bmatrix}\right)}{S\left(\begin{bmatrix}1\\1\end{bmatrix}\right)} \sim F_{1,1} \sim GGR\left(1, \tfrac{1}{2}, \tfrac{1}{2}, 1\right)$$

$$F_2 = \frac{4}{2}\,\frac{S\left(\begin{bmatrix}1\\2\end{bmatrix}\right)}{S\left(\begin{bmatrix}1\\3\end{bmatrix}\right)} \sim F_{2,4} \sim GGR\left(\tfrac{2}{4}, \tfrac{2}{2}, \tfrac{4}{2}, 1\right)$$

$$F_3 = \frac{16}{8}\,\frac{S\left(\begin{bmatrix}1\\4\end{bmatrix}\right)}{S\left(\begin{bmatrix}1\\5\end{bmatrix}\right)} \sim F_{8,16} \sim GGR\left(\tfrac{8}{16}, \tfrac{8}{2}, \tfrac{16}{2}, 1\right) .$$

Figure 3 shows the good graphical agreement between the relative frequency histograms generated from the simulated data and the exact p.d.f. and c.d.f., while in Tables 5 and 6 we may see once again that in order to be able to achieve a reasonable good accuracy for the computation of quantiles we may need quite a large number of terms in the exact distribution.

In Table 6 we report the values of the relative errors for the quantiles obtained from the truncations of the exact distribution used in Table 5. Once again we may note the almost linear decay of the values of the relative errors for increasing values of $n$. Only for the quantile 0.95 this rate seems to be slightly higher, what may

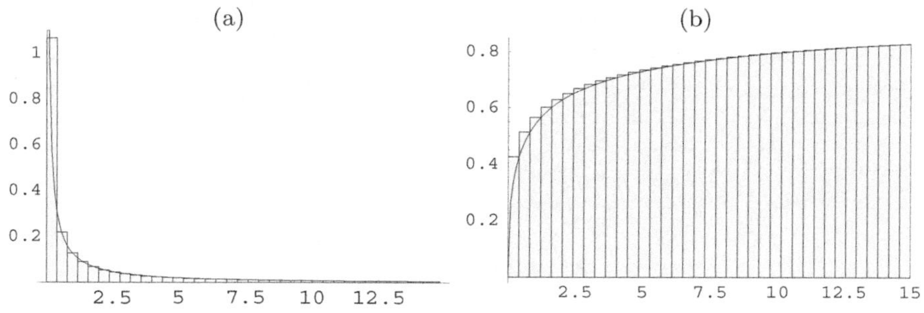


**Fig. 3** Overlaid plots of the histograms of relative frequencies generated from simulated data and the exact p.d.f. and c.d.f. for $n = 300$ for the statistic in (62): (a) histogram of relative frequencies and exact p.d.f.; (b) histogram of cumulative relative frequencies and exact c.d.f.

**Table 5** Quantiles (simulated and from different truncations of the exact distribution) for the statistic $F^*$ in (62)

| | | from the exact distribution with $n =$ | | | |
|---|---|---|---|---|---|
| Quant. | simulat. | 50 | 100 | 200 | 300 |
| 0.90 | 48.112 | 51.448 | 49.880 | 49.042 | 48.754 |
| 0.95 | 203.590 | 216.070 | 209.760 | 206.380 | 205.218 |
| 0.99 | 5169.479 | 5505.706 | 5348.353 | 5263.981 | 5234.950 |

**Table 6** Relative error for quantiles obtained from different truncations of the exact distribution of the statistic $F^*$ in (62)

| | value of $n$ in the exact distribution | | | |
|---|---|---|---|---|
| Quant. | 50 | 100 | 200 | 300 |
| 0.90 | 0.06934 | 0.03674 | 0.01933 | 0.01335 |
| 0.95 | 0.06130 | 0.03031 | 0.01371 | 0.00800 |
| 0.99 | 0.06504 | 0.03460 | 0.01828 | 0.01267 |

indeed be due to the fact that the value of the simulated quantile, relative to which the relative errors are computed, may be a bit higher than what it should be.

## 6 Relations with Other Distributions

We should emphasize that the results obtained may be easily and directly generalized to the case where we are interested in the distribution of the r.v.

$$Z = \prod_{j=1}^{n} \gamma_j \, Y_j^{\alpha_j},$$

where $\gamma_j \in \mathbb{R}^+$, $\alpha_j \in \mathbb{R}\backslash\{0\}$ and $Y_j$ are independent r.v.'s with GGR distributions, since it is straightforward to show that if

$$Y_j \sim GGR(k_j, r_{1j}, r_{2j}, \beta_j), \tag{63}$$

then

$$\gamma_j Y_j^{\alpha_j} \sim GGR\left(\frac{k_j}{\gamma_j^{\beta_j/\alpha_j}}, r_{1j}, r_{2j}, \frac{\beta_j}{\alpha_j}\right).$$

The distributions obtained in Section 3 may also be seen as the distributions of the product of independent Beta prime r.v.'s. Recall that if the r.v. $X$ has a standard Beta distribution, the distribution of either $(1-X)/X$ or $X/(1-X)$ is then usually called a standard Beta prime or Beta second kind distribution. However, we should note that the distributions of either $(1-X)/X$ or $X/(1-X)$ are then only particular GGR distributions. Actually, it is easy to show that if

$$Y_j \sim GGR(k_j, r_{1j}, r_{2j}, \beta_j)$$

with $k_j = 1$ and $\beta_j = 1$, then the r.v.'s $1/(1+Y_j)$ and $Y_j/(1+Y_j)$ have standard Beta distributions with parameters $r_{2j}$ and $r_{1j}$ or $r_{1j}$ and $r_{2j}$, respectively, while for general $k_j > 0$ and general $\beta_j \in \mathbb{R}\backslash\{0\}$ the r.v. $X_j = Y_j/(1+Y_j)$ has what we call a generalized Beta distribution with p.d.f.[12]

$$f_{X_j}(x) = \frac{|\beta_j|\, k_j^{r_{1j}}}{B\left(r_{1j}, r_{2j}\right)} \left(1 + k_j \frac{x}{1-x}\right)^{-r_{1j}-r_{2j}} (1-x)^{-\beta_j r_{1j}-1} x^{\beta_j r_{1j}-1},$$

which clearly reduces to the standard Beta p.d.f. for $k_j = 1$ and $\beta_j = 1$, while the r.v. $1 - X_j = 1/(1+Y_j)$ has of course a similar p.d.f. with $r_{1j}$ and $r_{2j}$ swapped. For this reason, the distribution of $Y_j$ is also called, for $k_j = 1$ and $\beta_j = 1$, a Beta prime distribution and for general $k_j$ and $\beta_j$ a generalized Beta prime distribution. Thus, for the distribution in (63), we may also say that $Y_j$ has a generalized Beta prime distribution and thus, the distributions obtained in Section 3 are also the distributions of the product of independent generalized Beta prime r.v.'s.

Clearly, if in (63) we have $r_{1j} = m_j/2$, $r_{2j} = n_j/2$, with $m_j, n_j \in \mathbb{N}$, $k = m_j/n_j$ and $\beta_j = 1$, the r.v.'s $Y_j$ will have $F$ distributions with $m_j$ and $n_j$ degrees of freedom, and thus the results in Section 3 will then give the distribution of the product of independent $F$ distributed r.v.'s.

Also, if in (63) we have $r_{1j} = 1/2$, $r_{2j} = n_j/2$, $k_j = 1/n_j$ and $\beta_j = 2$, with $n_j \in \mathbb{N}$, we have $Y_j$ with the so-called folded $T$ distribution, which is the distribution of a r.v. $Y_j = |T_j|$, where $T_j$ has a Student $T$ distribution with $n_j$

[12] see the author and editors note at the end of the Chapter

degrees of freedom. If instead we take $r_{2j} = 1/2$ and $k_j = 1$, we will have $Y_j$ with the so-called folded Cauchy distribution, that is the distribution of the absolute value of a r.v. with a standard Cauchy distribution, or a Student $T$ distribution with only 1 degree of freedom. In both cases, once again, the results in Section 3 may be readily applied.

Since, contrary to what is commonly done, we considered the power parameters as real and not necessarily positive (only non-null), by allowing them to be negative, the results obtained may be directly extended to the distribution of the ratio of two independent GGR random variables or the distribution of the ratio of two independent products of independent GGR random variables. In order to obtain the distribution of the ratio of two independent GGR random variables one simply has to consider $m = 2$ in (33), taking then for the random variable in the denominator the symmetric of its power parameter. The distribution for the ratio of two independent products of independent GGR random variables may then be obtained by considering the distribution of the product of the whole set of random variables, taking the power parameters for the random variables in the denominator with a minus sign.

As a by-product, in Subsections 2.3 and 2.4, we also obtain closed form representations, not involving any infinite series or unsolved integrals, for the distribution of the difference of two independent sums of a finite number of Exponential random variables with all different rate parameters or for the distribution of the difference of two independent sums of Gamma random variables with all different rate parameters and integer shape parameters, under the form of particular mixtures of either Exponential or Gamma distributions, according to the case. Also, if we consider the exponential of a Gamma random variable with integer shape parameter as a generalized Pareto distribution, then the distribution of the random variable $W$ in Subsection 2.3 is the distribution of the ratio of two independent products of Pareto distributions, while the distribution of the same random variable in Subsection 2.4 is the distribution of the ratio of two independent products of generalized Pareto distributions, expressed as a particular mixture of Pareto and inverted Pareto distributions.

## 7 Conclusions and Final Remarks

Given the form of the exact distributions obtained for the product of GGR r.v.'s and the problems related to the need for a very large number of terms in the series in order to obtain the desired precision, the development of near-exact distributions seems to be a much desirable goal. Such near-exact distributions are indeed not too hard to obtain, being not treated in this paper due to length limitations. While the simple truncation of the series obtained gives indeed mainly unsatisfactory results, namely in terms of c.d.f. and quantiles, owing to the fact that in this case the weights do not add up to 1, preventing this way the c.d.f. from reaching this value of 1, much better results may be obtained if we consider near-exact distributions, based on the concept of keeping the major part of the exact characteristic function unchanged and approximating the remaining part by an asymptotic result (Coelho, 2003, 2004). Given the form obtained for the exact distributions, such an approach would lead us to consider, for example, the truncation of the infinite series obtained, coupled with one or two more terms which would at the same time enable the weights to add up to 1 and the first few moments to match the exact ones.

Yet, the extension of the results presented in this paper to non-central distributions is almost straightforward, as it only requires consideration of the mixtures with Poisson weights that would account for the non-centrality. Once again, this topic is not addressed in this paper as the paper is already quite long.

## Appendix

The facts that in (2)–(3), and thus also in (4)–(7) and (8)–(19), we may take $p \to \infty$, and also in (26)–(31) we may take $p_1 \to \infty$ and $p_2 \to \infty$, with no need to worry about the convergence of the series involved, as well as about the convergence of the infinite products in (42) and (43), are justified as follows. If

$$X_i \sim \Gamma(r_j, \lambda_j), \qquad i = 1, 2, \ldots,$$

then the characteristic function of

$$Z = \sum_{j=1}^{\infty} X_j \tag{64}$$

is

$$\prod_{j=1}^{\infty} \lambda_j^{r_j} (\lambda_j - it)^{-r_j} = \lim_{p\to\infty} \prod_{j=1}^{p} \lambda_j^{r_j} (\lambda_j - it)^{-r_j},$$

with

$$\left| \prod_{j=1}^{\infty} \lambda_j^{r_j} (\lambda_j - it)^{-r_j} \right| = \left| \lim_{p\to\infty} \prod_{j=1}^{p} \lambda_j^{r_j} (\lambda_j - it)^{-r_j} \right| \leq 1,$$

not only because $\prod_{j=1}^{\infty} \lambda_j^{r_j} (\lambda_j - it)^{-r_j}$ is a characteristic function but also because

$$\left| \prod_{j=1}^{\infty} \lambda_j^{r_j} (\lambda_j - it)^{-r_j} \right| \leq \left| \prod_{j=1}^{p_1} \lambda_j^{r_j} (\lambda_j - it)^{-r_j} \right| \leq 1 \tag{65}$$

for any $p_1 < \infty$, also because for any $j$, we clearly have

$$\left|\lambda_j^{r_j}(\lambda_j - it)^{-r_j}\right| \leq 1 .$$

But then, let

$$f_{Z^*}(z; \underline{r}, \underline{\lambda}, p) = \frac{1}{2\pi}\int_{-\infty}^{\infty} e^{-itz}\prod_{j=1}^{p}\lambda_j^{r_j}(\lambda_j - it)^{-r_j}\,dt$$

represent the p.d.f. of the random variable

$$Z^* = \sum_{j=1}^{p} X_j ,$$

and let

$$f_Z(z; \infty) = \lim_{p\to\infty} f_{Z^*}(z; \underline{r}, \underline{\lambda}; p)$$

represent the p.d.f. of the random variable $Z$ in (64). The fact that $f_Z(z; \infty)$ is a legitimate p.d.f. follows directly from (65) and the dominated convergence theorem, since thus

$$\begin{aligned} f_Z(z; \infty) &= \lim_{p\to\infty}\frac{1}{2\pi}\int_{-\infty}^{\infty} e^{-itz}\prod_{j=1}^{p}\lambda_j^{r_j}(\lambda_j - it)^{-r_j}\,dt \\ &= \frac{1}{2\pi}\int_{-\infty}^{\infty} e^{-itz}\prod_{j=1}^{\infty}\lambda_j^{r_j}(\lambda_j - it)^{-r_j}\,dt \end{aligned}$$

(where the integral is uniformly convergent given the result in (65)).

If all the shape parameters $r_j$ $(j = 1, 2, \ldots)$ are integer, the p.d.f $f_Z(z; \infty)$ is the p.d.f. of a GIG distribution with infinite depth, what shows that in expressions (14) through (20) we may take $p \to \infty$.

Similar arguments could then be used for the distribution of the difference of two sums of infinitely many Exponential random variables considered after expression (23), or the distribution of the difference of two sums of infinitely many Gamma random variables all with integer shape parameters, considered at the end of Section 2, both of these distributions being the distribution of the difference of two GIG distributions with infinite depth.

**Acknowledgments** The authors want to thank the remarks and suggestions made by the referees and editors which improved the presentation of the paper. The reviewers suggested the need for Sections 4 and 5, which help in understanding the scope, usefulness and functioning of the distributions presented in the paper. The authors also want to thank Professor Thomas Mathew of the University of Maryland, who, as the adviser, brought to our attention and made available to us the material in Cao (2006).

## References

Ahlfors, L.V. (1979). *Complex Analysis – An Introduction to the Theory of Analytic functions of One Complex Variable*, 3rd edition, McGraw-Hill, London.

Anderson, T.W. (1984). *An Introduction to Multivariate Statistical Analysis*, 2nd edition, Wiley, New York.

Cao, L. (2006). *The Assessment of Multivariate Bioequivalence*. Ph.D. Thesis, Department of Mathematics and Statistics, The University of Maryland, Baltimore County, MD.

Coelho, C.A. (1992). *The Generalized Canonical Analysis*. Ph.D. Thesis, The University of Michigan, Ann Arbor, MI.

Coelho, C.A. (1998). The Generalized Integer Gamma distribution – a basis for distributions in Multivariate Statistics. *J. Multivariate Anal.*, **64**, 86–102.

Coelho, C.A. (2003). A Generalized Integer Gamma distribution as an asymptotic replacement for a Logbeta distribution & applications. *Amer. J. Math. Management Sci.*, **23**, 383–399.

Coelho, C.A. (2004). The Generalized Near-Integer Gamma distribution: a basis for 'near-exact' approximations to the distribution of statistics which are the product of an odd number of independent Beta random variables. *J. Multivariate Anal.*, **89**, 191–218.

Coelho, C.A., Mexia J.T. (2006). On the exact and near-exact distributions of statistics used in generalized $F$ tests. Technical Report DM/FCT http://www.dm.fct.unl.pt/sites/www.dm.fct.unl.pt/files/preprints/2006/23_06.pdf (2006)

Fonseca, M., Mexia, J.T., Zmyslony, R. (2003). Estimators and tests for variance components in cross nested orthogonal designs. *Discuss. Math., Probab. Stat.*, **23**, 175–201.

Khuri, A.I., Mathew, T., Sinha, B.K. (1998). *Statistical Tests for Mixed Linear Models*, Wiley, New York.

Lang, S. (199). *Complex Analysis*, 4th edition, Springer-Verlag, New York.

Pham-Gia, T., Turkkan, N. (2002). Operations on the generalized-$F$ variables and applications. *Statistics*, **36**, 195–209.

Provost, S.B. (1986). On the distribution of some test statistics connected with the multivariate linear functional relationship model. *Commun. Statist. Theory Methods*, **15**, 1285–1298.

Shah, M.C., Rathie, P.N. (1974). Exact distribution of product of generalized $F$-variates. *Canad. J. Statist.*, **2**, 13–24.

### *Author and Editors notes:*

[1] It seemed useful to bring to the attention of the reader that, for example for the c.d.f. in (13), we have

$$K_1 K_2 \sum_{j=1}^{p} \left( H_{2j} \frac{d_{j,1}}{\nu_j} + H_{1j} \frac{c_{j,1}}{\lambda_j} \right) = 1 ,$$

so that the c.d.f. in (13) may be written as

$$F_Z(z) = \begin{cases} K_1 K_2 \sum_{j=1}^{p} H_{2j} \frac{d_{j,1}}{\nu_j} \exp(\nu_j z), & z \le 0 \\ 1 - K_1 K_2 \sum_{j=1}^{p} H_{1j} \frac{c_{j,1}}{\lambda_j} \exp(-\lambda_j z), & z \ge 0 \end{cases}$$

and the c.d.f. in (19) as

$$F_W(w) = \begin{cases} 1 - K_1 K_2 \sum_{j=1}^{p} H_{1j} \frac{c_{j,1}}{\lambda_j} (kw)^{-\lambda_j}, & w \ge k^{-1} \\ K_1 K_2 \sum_{j=1}^{p} H_{2j} \frac{d_{j,1}}{\nu_j} (kw)^{\nu_j}, & 0 < w \le k^{-1}. \end{cases}$$

Also in (35), the c.d.f. may be written as

$$F_W(w) = \begin{cases} \lim_{n\to\infty} K_1 K_2 \sum_{j=1}^{m} \sum_{h=0}^{n} H_{2jh} \frac{d_{hj}}{s^*_{2jh}} \left(\frac{w}{K^*}\right)^{s^*_{2jh}} 0, < w \le K^* \\ 1 - \lim_{n\to\infty} K_1 K_2 \sum_{j=1}^{m} \sum_{h=0}^{n} H_{1jh} \frac{c_{hj}}{s^*_{1jh}} \left(\frac{w}{K^*}\right)^{-s^*_{1jh}}, \ w \ge K^* \end{cases}$$

and the c.d.f. in (46) as

$$F_W(w) = \begin{cases} \lim_{n\to\infty} K_1 K_2 \sum_{j=1}^{m(n+1)} H_{2j} \frac{d_j}{s^*_{2j}} \left(\frac{w}{K^*}\right)^{s^*_{2j}}, 0 < w \le K^* \\ 1 - \lim_{n\to\infty} K_1 K_2 \sum_{j=1}^{m(n+1)} H_{1j} \frac{c_j}{s^*_{1j}} \left(\frac{w}{K^*}\right)^{-s^*_{1j}}, \ w \ge K^*. \end{cases}$$

In fact, the distribution in Theorem 1 is a particular case of the distribution in Theorem 2, since it is indeed the distribution of the difference of two independent GIG r.v.'s, with $p_1 = p_2 = p$ and $r_{1j} = r_{2j} = 1$ $(j = 1, \ldots, p)$, and as such the expressions for the p.d.f. and c.d.f. in Theorem 1 may be obtained from the expressions in Theorem 2, by taking in this Theorem, $p_1 = p_2 = p$ and $r_{1j} = r_{2j} = 1$ $(j = 1, \ldots, p)$. By doing this, the expression for the c.d.f. in (13) will immediately come out in the form stated at the beginning of this note.

[2] see [1] above

[3] Expressions (20), (22), (23) and (25) have some typos; these should rather read

$$f_Z(z) = \begin{cases} K_1 K_2 \displaystyle\sum_{j=1}^{p_1} P^{**}_{1j}(z)\,\exp(-\lambda_j\, z)\,, & z \geq 0 \\ K_1 K_2 \displaystyle\sum_{l=1}^{p_2} P^{**}_{2l}(z)\,\exp(\nu_l\, z)\,, & z \leq 0 \end{cases} \tag{20}$$

$$P^{**}_{2l}(z) = \sum_{h=1}^{r_{2l}} d_{lh} \sum_{j=1}^{p_1}\sum_{k=1}^{r_{1j}} c_{jk} \sum_{i=0}^{h-1} \binom{h-1}{i} (-z)^{h-1-i} \frac{(k+i-1)!}{(\lambda_j+\nu_l)^{k+i}}\,. \tag{22}$$

$$F_Z(z) = \begin{cases} 1 - K_1 K_2 \displaystyle\sum_{j=1}^{p_1} P^{***}_{1j}(z)\,\exp(-\lambda_j\, z)\,, & z \geq 0 \\ K_1 K_2 \displaystyle\sum_{l=1}^{p_2} P^{***}_{2l}(z)\,\exp(\nu_l\, z)\,, & z \leq 0 \end{cases} \tag{23}$$

$$P^{***}_{2l}(z) = \sum_{h=1}^{r_{2l}} d_{lh} \sum_{j=1}^{p_1}\sum_{k=1}^{r_{1j}} c_{jk} \sum_{i=0}^{h-1} \binom{h-1}{i} \frac{(k+i-1)!}{(\lambda_j+\nu_l)^{k+i}} \sum_{t=0}^{h-1-i} \frac{(h-1-i)!}{t!} \frac{(-z)^t}{\nu_l^{h-i-t}}\,. \tag{25}$$

[4] a number of typos have to be corrected in expression (29); this expression should rather read

$$P^{**}_{2l}(z) = \sum_{h=1}^{r_{2l}} d_{lh} \sum_{j=1}^{p_1}\sum_{k=1}^{r_{1j}} c_{jk} \sum_{i=0}^{k-1} \binom{k-1}{i} z^{k-1-i} \sum_{t=0}^{h+i-1} \frac{(h+i-1)!}{t!} \frac{(-z)^t}{(\lambda_j+\nu_l)^{h+i-t}}\,. \tag{29}$$

[5] a number of typos also have to be corrected in expressions (30) and (31); these should rather read

$$f_W(w) = \begin{cases} K_1 K_2 \sum\limits_{j=1}^{p_1} P_{1j}^{**}(\log(kw))\,(kw)^{-\lambda_j}\,\frac{1}{w}, & w \ge k^{-1} \\ K_1 K_2 \sum\limits_{l=1}^{p_2} P_{2l}^{**}(\log(kw))\,(kw)^{\nu_l}\,\frac{1}{w}, & 0 < w \le k^{-1} \end{cases} \tag{30}$$

and

$$F_W(w) = \begin{cases} 1 - K_1 K_2 \sum\limits_{j=1}^{p_1} P_{1j}^{***}(\log(kw))\,(kw)^{-\lambda_j}, & w \ge k^{-1} \\ K_1 K_2 \sum\limits_{l=1}^{p_2} P_{2l}^{***}(\log(kw))\,(kw)^{\nu_l}, & 0 < w \le k^{-1}. \end{cases} \tag{31}$$

[6] indeed we need all $s_{1j}$ $(j = 1, \dots, m)$, on one hand, and all $s_{2j}$ $(j = 1, \dots, m)$, on the other hand, to not differ by integers, since what we really need is that all $s_{1jh}^*$ in (40), on one hand, and all $s_{2jh}^*$ in (40), on the other hand, to be different.

[7] a number of typos have to be corrected in expressions (34), (35), (38) and (39); these should rather read

$$f_W(w) = \begin{cases} \lim\limits_{n\to\infty} K_1 K_2 \sum\limits_{j=1}^{m}\sum\limits_{h=0}^{n} H_{2jh}\, d_{hj}\, \left(w/K^*\right)^{s_{2jh}^*}\,\frac{1}{w}, & 0 < w \le K^* \\ \lim\limits_{n\to\infty} K_1 K_2 \sum\limits_{j=1}^{m}\sum\limits_{h=0}^{n} H_{1jh}\, c_{hj}\, \left(w/K^*\right)^{-s_{1jh}^*}\,\frac{1}{w}, & w \ge K^* \end{cases} \tag{34}$$

$$F_W(w) = \begin{cases} \lim\limits_{n\to\infty} K_1 K_2 \sum\limits_{j=1}^{m}\sum\limits_{h=0}^{n} H_{2jh}\, \frac{d_{hj}}{s_{2jh}^*} \left(\frac{w}{K^*}\right)^{s_{2jh}^*}, & 0 < w \le K^* \\ \lim\limits_{n\to\infty} K_1 K_2 \sum\limits_{j=1}^{m}\sum\limits_{h=0}^{n} \left(H_{2jh}\, \frac{d_{hj}}{s_{2jh}^*} + H_{1jh}\, \frac{c_{hj}}{s_{1jh}^*} \left(1 - \left(\frac{w}{K^*}\right)^{-s_{1jh}^*}\right)\right), & w \ge K^* \end{cases} \tag{35}$$

[8] see [1]

[9] a number of typos have to be corrected in expression (46), which should rather read

$$F_W(w) = \begin{cases} \lim_{n\to\infty} K_1 K_2 \sum_{j=1}^{m(n+1)} H_{2j} \frac{d_j}{s_{2j}^*} \left(\frac{w}{K^*}\right)^{s_{2j}^*}, & 0 < w \le K^* \\ \lim_{n\to\infty} K_1 K_2 \sum_{j=1}^{m(n+1)} \left(H_{2j} \frac{d_j}{s_{2j}^*} + H_{1j} \frac{c_j}{s_{1j}^*} \left(1 - \left(\frac{w}{K^*}\right)^{-s_{1j}^*}\right)\right), & w \ge K^*. \end{cases} \tag{46}$$

[10] a typo has to be corrected in expression (48); this expression should rather look like

$$F_W(w) = \begin{cases} \lim_{n\to\infty} K_1 K_2 \sum_{j=1}^{p_2(n+1)} P_{2j}^{***} \left(\log\left(\frac{w}{K^*}\right)\right) \left(\frac{w}{K^*}\right)^{\tilde{s}_{2j}^*}, & 0 < w \le K^* \\ 1 - \lim_{n\to\infty} K_1 K_2 \sum_{j=1}^{p_1(n+1)} P_{1j}^{***} \left(\log\left(\frac{w}{K^*}\right)\right) \left(\frac{w}{K^*}\right)^{-\tilde{s}_{1j}^*}, & w \ge K^* \end{cases} \tag{48}$$

[11] We should note that despite the fact that there were a number of typos in some of the expressions in Sections 2 and 3, in the original paper, the computational modules used in this section were using the correct formulations.

[12] a typo has to be corrected in this expression; it should rather be

$$f_{X_j}(x) = \frac{|\beta_j|\, k_j^{r_{1j}}}{B(r_{1j}, r_{2j})} \left(1 + k_j \left(\frac{x}{1-x}\right)^{\beta_j}\right)^{-r_{1j}-r_{2j}} (1-x)^{-\beta_j r_{1j}-1}\, x^{\beta_j r_{1j}-1}$$

# The Advantage of Decomposing Elaborate Hypotheses on Covariance Matrices into Conditionally Independent Hypotheses in Building Near-Exact Distributions for the Test Statistics

**Carlos A. Coelho and Filipe J. Marques**

**Abstract** The aim of this paper is to show how the decomposition of elaborate hypotheses on the structure of covariance matrices into conditionally independent simpler hypotheses, by inducing the factorization of the overall test statistic into a product of several independent simpler test statistics, may be used to obtain near-exact distributions for the overall test statistics, even in situations where asymptotic distributions are not available in the literature and adequately fit ones are not easy to obtain.

**Keywords** Near-exact distributions · Conditionally independent hypotheses · Likelihood ratio test statistics · Generalized sphericity tests · Generalized Near-Integer Gamma distribution · Mixtures

## 1 Introduction

The concept of a near-exact distribution has already been introduced in a number of papers (Coelho, 2003, 2004; Coelho and Mexia, 2006; Alberto and Coelho, 2007; Grilo and Coelho, 2007a,b; Coelho and Marques, 2007a; Marques and Coelho,

---

Reprinted from Linear Algebra and Its Applications, with permission from Elsevier.
First published as Coelho, C. A., Marques, F. J. (2009). The advantage of decomposing elaborate hypotheses on covariance matrices into conditionally independent hypotheses in building near-exact distributions for the test statistics. *Linear Algebra and its Applications*, **430**, 2592–2606. (https://www.sciencedirect.com/science/article/pii/S0024379508003480)

---

C. A. Coelho (✉) · F. J. Marques
Mathematics Department, Faculty of Sciences and Technology, The New University of Lisbon, Quinta da Torre, Caparica, Portugal
e-mail: cmac@fct.unl.pt; fjm@fct.unl.pt

 
B. C. Arnold et al. (eds.), *Likelihood Ratio Tests and Related Topics in Multivariate Analysis*, Contributions to Statistics,
https://doi.org/10.1007/978-3-031-68008-3_8

2008a). In a nutshell, near-exact distributions are new asymptotic distributions that lay closer to the exact distribution than the usual asymptotic distributions. They correspond to what we call near-exact c.f.s (characteristic functions) which are c.f.s obtained from the exact c.f. by leaving most of it unchanged and replacing the remaining smaller part by an asymptotic result (which is intended to be asymptotic both in terms of sample size and overall number of variables, in a manner that will be more precisely stated in this section). This replacement is done in such a way that the resulting c.f. corresponds to a distribution with a manageable c.d.f. (cumulative distribution function) so that the computation of near-exact quantiles is rendered feasible, easy and precise.

There are mainly two ways in which near-exact distributions, or rather, near-exact c.f.s may be obtained. One is when the exact distribution is an infinite mixture, in which case we may cut short the series that corresponds to the exact c.f. at a given number of terms and replace the remainder by the c.f. of just one, or a mixture of two or three distributions of the same kind of the ones in the infinite mixture. The parameters in these replacing parts of the c.f. are computed in such a way that the first few derivatives at the origin match the derivatives of the part being replaced. More precisely, if the exact c.f. of the statistic $W$ may be written as

$$\Phi_W(t) = \sum_{i=0}^{\infty} p_i \, \Phi_i(t) \,,$$

where $p_i > 0$ ($i = 0, 1, \dots$) and $\sum_{i=0}^{\infty} p_i = 1$, and where $\Phi_i(t)$ are c.f.s, then we will use instead of $\Phi_W(t)$,

$$\Phi_W^*(t) = \sum_{i=0}^{n^*} p_i \, \Phi_i(t) + \theta \, \Phi_2(t) \,,$$

where $\theta = 1 - \sum_{i=0}^{n^*} p_i$ and $\Phi_2(t)$ is usually either a c.f. of the type of $\Phi_i(t)$ or the c.f. of the mixture of two or three such c.f.s, defined in such a way that

$$\left. \frac{d}{dt^h} \theta \, \Phi_2(t) \right|_{t=0} = \left. \frac{d}{dt^h} \sum_{i=n^*+1}^{\infty} p_i \, \Phi_i(t) \right|_{t=0} ,$$

for $h \in H$ (where usually we will have $H = \{1, 2\}$, $H = \{1, \dots, 4\}$ or $H = \{1, \dots, 6\}$).

Another way to obtain a near-exact c.f. is through the factorization of the exact c.f. If we may write

$$\Phi_W(t) = \Phi_1(t) \, \Phi_2(t) \,,$$

where we recognize $\Phi_1(t)$ as corresponding to a known and manageable distribution and $\Phi_2(t)$ as corresponding to a non-manageable distribution, then we will use as a near-exact c.f. for $W$ the c.f.

$$\Phi_W^*(t) = \Phi_1(t)\,\Phi_2^*(t)\,, \tag{1}$$

where, for $h \in H$ (usually with $H=\{1,2\}$, $H=\{1,\ldots,4\}$ or $H=\{1,\ldots,6\}$),

$$\left.\frac{d}{dt^h}\,\Phi_2^*(t)\right|_{t=0} = \left.\frac{d}{dt^h}\,\Phi_2(t)\right|_{t=0} \tag{2}$$

in such a way that, if we write $\Phi_2^*(t)$ as a function of the sample size (say $n$) and the overall number of variables involved (say $p$), that is, if we write $\Phi_2^*(t) \equiv \Phi_2^*(t;n,p)$, we want

$$\lim_{n\to\infty}\int_{-\infty}^{+\infty}\left|\frac{\Phi_2(t)-\Phi_2^*(t;n,p)}{t}\right| dt = 0$$

and

$$\lim_{p\to\infty}\int_{-\infty}^{+\infty}\left|\frac{\Phi_2(t)-\Phi_2^*(t;n,p)}{t}\right| dt = 0$$

with

$$\begin{aligned}\int_{-\infty}^{+\infty}\left|\frac{\Phi_2(t)-\Phi_2^*(t;n,p)}{t}\right| dt &\geq \int_{-\infty}^{+\infty}|\Phi_1(t)|\left|\frac{\Phi_2(t)-\Phi_2^*(t;n,p)}{t}\right| dt\\ &\geq \max_w \left|F_W(w)-F_W^*(w;n,p)\right|\,,\end{aligned} \tag{3}$$

where $F_W(\cdot)$ is the exact c.d.f. of $W$ and $F_W^*(\cdot\,;n,p)$ is the near-exact c.d.f. of $W$, corresponding to the c.f. $\Phi_W^*(t)$ in (1). The relation in (3) will be used at the end of section 3 in the definition of a measure of proximity between distributions (see expression (6)).

When dealing with l.r.t. (likelihood ratio test) statistics, mainly those more commonly used in Multivariate Statistics, $\Phi_1(t)$ is the c.f. of a sum of independent Logbeta r.v.s (random variables), that are r.v.s whose exponential has a Beta distribution. This c.f. may be alternatively written as the c.f. of a GIG (Generalized Integer Gamma) distribution, that is, the distribution of the sum of a given number of independent Gamma distributions, all with different rate parameters and integer shape parameters (see Coelho (1998)), while $\Phi_2(t)$ is the c.f. of a sum of other independent Logbeta r.v.s, which is not possible to be written under the form of the c.f. of a GIG distribution and which will be asymptotically replaced by the c.f. of a single Gamma distribution or the c.f. of a mixture of two or three Gamma distributions with the same rate parameters. This replacement was already used in Marques and Coelho (2008a) and is indeed well justified since a single Logbeta distribution may be represented under the form of an infinite mixture of Exponential distributions (see Coelho et al. (2006)). A sum of independent Logbeta random variables may thus be represented under the form of an infinite mixture of sums of Exponential distributions, which are themselves mixtures of Exponential or Gamma distributions.

The near-exact distributions we are interested in this paper are exactly the ones of this second kind.

## 2 How May the Decomposition of a Complex Hypothesis into More Elementary Hypotheses Help in Building Near-Exact Distributions for the Overall Test Statistic?

Let us suppose we have $\Lambda$, the l.r.t. statistic to test a given null hypothesis $H_0$, and that we want to write, for $W = -\log \Lambda$,

$$\Phi_W(t) = \Phi_1(t)\, \Phi_2(t)$$

(the reason why we usually want to handle the c.f. of $W = -\log \Lambda$ instead of the c.f. of $\Lambda$ is that while the moments of $\Lambda$ may be relatively easy to obtain and they commonly exist for any positive integer order, and even for any order not necessarily integer just above a given negative value, the expression for the c.f. of $\Lambda$ may be too hard to obtain and to handle; while on the other hand once obtained a near-exact distribution for $W$ it will then be easy to obtain the corresponding near-exact distribution for $\Lambda = e^{-W}$).

It may happen that this factorization may be too hard to obtain from scratch, given the complexity of $\Lambda$ itself.

But, let us suppose we may write

$$\Lambda = \prod_{j=1}^{m} \Lambda_j\,, \tag{4}$$

where $\Lambda_j$ is the l.r.t. statistic to test $H_{0j|1,\dots,j-1}$, the $j$-th nested conditionally independent null hypothesis we may split $H_0$ into. That is, we are assuming we may write

$$H_0 \equiv H_{0m|1,\dots,m-1} \circ \dots \circ H_{03|1,2} \circ H_{02|1} \circ H_{01}\,,$$

which means that to test $H_0$ is equivalent to test sequentially the $m$ independent hypotheses $H_{0j|1,\dots,j-1}$ $(j = 1, \dots, m)$, starting with $H_{01}$ and testing then $H_{02|1}$ and then $H_{03|1,2}$, and so on, and where to test $H_{0j|1,\dots,j-1}$ is to test $H_{0j}$ assuming that $H_{01}$ through $H_{0,j-1}$ are true. The hypotheses $H_{0j|1,\dots,j-1}$ $(j = 1, \dots, m)$ are all independent in the sense that under the null hypothesis $H_0$ it is possible to prove that the $\Lambda_j$ in (4) are independent. This is the case with all the likelihood ratio statistics used and mentioned in this paper (see Basu (1955); Hogg and Craig (1956)), although in some non-regular cases independence may not hold as Hogg (1961) shows. Let us further suppose that, for

$$W_j = -\log \Lambda_j \qquad (j = 1, \dots, m),$$

we have available the factorizations

$$\Phi_{W_j}(t) = \Phi_{j1}(t)\,\Phi_{j2}(t)\,. \tag{5}$$

Then, under $H_0$, given the independence of the $\Lambda_j$ and thus also of the $W_j$, we may easily write

$$\begin{aligned}\Phi_W(t) &= \prod_{j=1}^{m} \Phi_{W_j}(t) = \underbrace{\left\{\prod_{j_1=1}^{m} \Phi_{j_1}(t)\right\}}_{\Phi_1(t)} \underbrace{\left\{\prod_{j_2=1}^{m} \Phi_{j_2}(t)\right\}}_{\Phi_2(t)} \\ &= \Phi_1(t)\,\Phi_2(t)\,,\end{aligned}$$

where for the most common l.r.t. statistics used in Multivariate Statistics the c.f.s $\Phi_{j_1}(t)$ may be obtained under the form of c.f.s of GIG distributions (Coelho, 2003, 2004; Marques and Coelho, 2008a; Coelho and Marques, 2007a) and all the $\Phi_{j_2}(t)$ are c.f.s of the sum of independent Logbeta r.v.s. Then, $\Phi_1(t)$ itself will be the c.f. of a GIG distribution, and $\Phi_2(t)$ will be itself the c.f. of the sum of independent Logbeta r.v.s, being thus adequately asymptotically replaced by the c.f. of a single Gamma distribution or the c.f. of the mixture of two or three Gamma distributions, verifying (2) (see Marques and Coelho (2008a)) and yielding this way either a GNIG (Generalized Near-Integer Gamma) distribution or a mixture of two or three GNIG distributions (the GNIG distribution is the distribution of the sum of a r.v. with a GIG distribution with an independent r.v. with a Gamma distribution with a non-integer shape parameter – for details on this distribution see Coelho (2004)).

## 3 Examples of Application

Since decompositions of the c.f.s of the type in (5) are already available for the Wilks $\Lambda$ statistic, or the l.r.t. statistic to test the independence of several sets of variables (see Coelho (2003, 2004)), and also for the l.r.t. statistic to test sphericity (see Marques and Coelho (2008a)) and for the l.r.t. statistic to test the equality of several variance-covariance matrices (see Coelho and Marques (2007a)), we may think of l.r.t.s whose statistic may be factorized as in (4) and where the $\Lambda_j$ are the above mentioned statistics.

This was indeed what was somehow done when obtaining either the exact distribution for the generalized Wilks $\Lambda$ statistic, under the form of a GIG distribution, when at most one of the sets of variables has an odd number of variables (see Coelho (1998)) or when obtaining a near-exact distribution for the same statistic, under the form of a GNIG distribution, for the general case when two or more sets have an odd number of variables (see Coelho (2003, 2004)).

In the subsections ahead we will use the following notation:

- $\Lambda_1(q, p; N_j)$, with $\Lambda_1(q, p; N_j) = \dfrac{\prod_{j=1}^{q} |A_j|^{N_j/2}}{|A|^{N/2}} \dfrac{N^{Np/2}}{\prod_{j=1}^{q} N_j^{N_j p/2}}$, to denote the l.r.t. statistic used to test $H_{01} : \Sigma_1 = \cdots = \Sigma_q$, based on samples of size $N_j$ $(j = 1, \ldots, q)$ from $N_p(\underline{\mu}_j, \Sigma_j)$, with $A_j = \hat{\Sigma}_j$, $A = A_1 + \cdots + A_q$ and $N = N_1 + \cdots + N_q$;
- $\Lambda_2(N; p_1, \ldots, p_k)$, with $\Lambda_2(N; p_1, \ldots, p_k) = \left(\dfrac{|A|}{\prod_{i=1}^{k} |A_{ii}|}\right)^{N/2}$, to denote the l.r.t. statistic used to test $H_{02} : \Sigma = diag(\Sigma_{11}, \ldots, \Sigma_{ii}, \ldots, \Sigma_{kk})$, based on a sample of size $N$ from $N_p(\underline{\mu}, \Sigma)$, with $A = \hat{\Sigma}$, $A_{ii} = \hat{\Sigma}_{ii}$ and

$$\underline{\mu} = \left[\underline{\mu}_1, \ldots, \underline{\mu}_i, \ldots, \underline{\mu}_k\right]', \quad \Sigma = \begin{bmatrix} \Sigma_{11} & \ldots & \Sigma_{1i} & \ldots & \Sigma_{1k} \\ \vdots & \ddots & \vdots & & \vdots \\ \Sigma_{i1} & \ldots & \Sigma_{ii} & \ldots & \Sigma_{ik} \\ \vdots & & \vdots & \ddots & \vdots \\ \Sigma_{k1} & \ldots & \Sigma_{ki} & \ldots & \Sigma_{kk} \end{bmatrix};$$

- $\Lambda_3(p; N)$, with $\Lambda_3(p; N) = \left(\dfrac{|A|}{\left(tr\,\frac{1}{p} A\right)^p}\right)^{N/2}$, to denote the l.r.t. statistic used to test $H_{03} : \Sigma = \sigma^2 I_p$, based on a sample of size $N$ from $N_p(\underline{\mu}, \Sigma)$, with $A = \hat{\Sigma}$.

To evaluate the proximity of the approximations obtained, to the exact distribution, we will use the measure

$$\Delta = \frac{1}{2\pi} \int_{-\infty}^{+\infty} \left| \frac{\Phi_W(t) - \Phi_W^*(t)}{t} \right| dt\,, \tag{6}$$

where $\Phi_W(t)$ and $\Phi_W^*(t)$ represent respectively the exact and the approximate (near-exact) c.f. of $W$, the logarithm of the l.r.t. statistic under study.

This measure of proximity has already been used in Grilo and Coelho (2007a) and Marques and Coelho (2008a) and its values are an upper bound on the absolute value of the difference between the corresponding c.d.f.s, with

$$\Delta \geq \max_{x \in S_W} \left| F_W(x) - F_W^*(x) \right| , \tag{7}$$

where $S_W$ is the support of the r.v. $W$ and $F_W(x)$ and $F_W^*(x)$ represent respectively the exact c.d.f. of $W$ and the approximate c.d.f. corresponding to $\Phi_W^*(t)$. This measure is related with the Berry-Esseen bound (see Berry (1941); Esseen (1945); Hwang (1998) and Loève (1977, Chap. VI, Sec. 21)).

### *3.1 The Test of Equality of Several Multivariate Normal Distributions*

It may be the best well known example of the situation we are trying to illustrate, since we may write in this case (Anderson, 2003; Kshirsagar, 1972; Muirhead, 1986)

$$H_0 : \underline{\mu}_1 = \cdots = \underline{\mu}_q \, , \quad \Sigma_1 = \cdots = \Sigma_q \tag{8}$$

with

$$H_0 \equiv H_{02|1} \circ H_{01} ,$$

where

$$H_{01} : \Sigma_1 = \cdots = \Sigma_q$$

and

$$H_{02|1} : \underline{\mu}_1 = \cdots = \underline{\mu}_q \quad \text{given that } \Sigma_1 = \cdots = \Sigma_q \, ,$$

so that, using the notation defined at the beginning of this section, we may write the l.r.t. statistic to test $H_0$ in (8) as

$$\Lambda = \Lambda_2(N; p, q-1) \, \Lambda_1(q, p; N_j) \qquad (\text{with } N = N_1 + \cdots + N_q) \, , \tag{9}$$

where $p$ is the order of the matrices $\Sigma_1, \ldots, \Sigma_q$ and where, under $H_0$ in (8), $\Lambda_2(N; p, q-1)$ and $\Lambda_1(q, p; N_j)$ are independent.

Since factorizations of the type in (5) for the c.f. of $-\log \Lambda_2(N; p, q-1)$ and $-\log \Lambda_1(q, p; N_j)$ are available (Coelho, 2003, 2004; Coelho and Marques, 2007a), the process of obtaining near-exact distributions for the l.r.t. statistic for this test may be quite easily implemented by using the approach proposed in this paper.

For $N_1 = \cdots = N_q = n$, and $\alpha = 1, \ldots, \lfloor p/q \rfloor$, the exact c.f. of $W = -\log \Lambda$, where $\Lambda$ is the l.r.t. statistic in (9), may be written either as (see Anderson (2003); Kshirsagar (1972); Muirhead (1986))

$$\begin{aligned} \Phi_W(t) = {} & \prod_{j=1}^{p} \frac{\Gamma\left(\frac{nq-j}{2}\right)\Gamma\left(\frac{(n-1)q+1-j}{2} - \frac{nq}{2}\mathrm{i}t\right)}{\Gamma\left(\frac{nq-j}{2} - \frac{nq}{2}\mathrm{i}t\right)\Gamma\left(\frac{(n-1)q+1-j}{2}\right)} \\ & \times \prod_{j=1}^{p}\prod_{k=1}^{q} \frac{\Gamma\left(\frac{n}{2} + \frac{1-j}{2q} + \frac{k-1}{q}\right)\Gamma\left(\frac{n+1-j}{2} - \frac{n}{2}\mathrm{i}t\right)}{\Gamma\left(\frac{n}{2} + \frac{1-j}{2q} + \frac{k-1}{q} - \frac{n}{2}\mathrm{i}t\right)\Gamma\left(\frac{n+1-j}{2}\right)} \end{aligned} \tag{10}$$

or as

$$\begin{aligned} \Phi_W(t) = {} & \left.\prod_{\substack{j=2\\ j\neq\alpha q}}^{p+q-1} \left(\frac{n-j/q}{n}\right)^{r_j}\left(\frac{n-j/q}{n} - \mathrm{i}t\right)^{-r_j} \times \prod_{j=2}^{p} \left(\frac{n-j}{n}\right)^{u_j}\left(\frac{n-j}{n} - \mathrm{i}t\right)^{-u_j} \right\} \Phi_1(t) \\ & \left.\begin{aligned} & \times \left(\frac{\Gamma\left(\frac{nq-1}{2}\right)\Gamma\left(\frac{nq-2}{2} - \frac{nq}{2}\mathrm{i}t\right)}{\Gamma\left(\frac{nq-1}{2} - \frac{nq}{2}\mathrm{i}t\right)\Gamma\left(\frac{nq-2}{2}\right)}\right)^{\mathrm{mod}(p,2)} \\ & \times \prod_{j=1}^{\lfloor p/2 \rfloor}\prod_{k=1}^{q} \frac{\Gamma\left(a_j + b_{jk}\right)\Gamma\left(a_j + b^*_{jk} - n\mathrm{i}t\right)}{\Gamma\left(a_j + b^*_{jk}\right)\Gamma\left(a_j + b_{jk} - n\mathrm{i}t\right)} \\ & \times \left(\prod_{k=1}^{q} \frac{\Gamma(a_p + b_{pk})}{\Gamma(a_p + b^*_{pk})} \frac{\Gamma\left(a_p + b^*_{pk} - \frac{n}{2}\mathrm{i}t\right)}{\Gamma\left(a_p + b_{pk} - \frac{n}{2}\mathrm{i}t\right)}\right)^{\mathrm{mod}(p,2)} \end{aligned}\right\} \Phi_2(t) \end{aligned} \tag{11}$$

for suitably defined shape parameters $r_j$ ($j = 2, \ldots, p+q-1$; $j \neq \alpha q$) and $u_j$ ($j = 2, \ldots, p$) (see Marques and Coelho (2008b), also for parameters $a_j$, $b_{jk}$, $a_p$, $b_{pk}$, $b^*_{jk}$ and $b^*_{pk}$).

But then, according to the explanation in the previous Sections 1 and 2, we may replace $\Phi_2(t)$ in (11) by

$$\Phi_2^*(t) = \sum_{k=1}^{h/2} p_k\, \lambda^{s_k}(\lambda - \mathrm{i}t)^{-s_k}\,, \tag{12}$$

with

$$\sum_{k=1}^{h/2} p_k = 1\,, \text{ and } 0 < p_k \leq 1;\; k = 1, \ldots, h/2\,, \tag{13}$$

where $h = 2$, $h = 4$ or $h = 6$, is the number of derivatives of $\Phi_2(t)$ at $t = 0$ that will be equated by the corresponding derivatives of $\Phi_2^*(t)$ at $t = 0$. This way we obtain

$$\Phi_W(t) = \left.\begin{array}{l} \displaystyle\prod_{\substack{j=2\\ j\neq\alpha q}}^{p+q-1} \left(\frac{n-j/q}{n}\right)^{r_j} \left(\frac{n-j/q}{n} - \mathrm{i}t\right)^{-r_j} \\ \displaystyle\times \prod_{j=2}^{p} \left(\frac{n-j}{n}\right)^{u_j} \left(\frac{n-j}{n} - \mathrm{i}t\right)^{-u_j} \end{array}\right\} \Phi_1(t) \tag{14}$$
$$\times\; \Phi_2^*(t)$$

as a near-exact c.f. for $W$, yielding thus near-exact distributions for $W$ which are, according to the number of moments of $\Phi_2(t)$ being equated and the concomitant definition of $\Phi_2^*(t)$, either a GNIG distribution or a mixture of two or three GNIG distributions of depth $2p + q - 2 - \lfloor p/q \rfloor$, with integer shape parameters $r_j$ ($j = 2, \ldots, p + q - 1;\; j \neq \alpha q$) and $u_j$ ($j = 2, \ldots, p$) and non-integer shape parameter $s_k$, for the $k$-th mixture term ($k = 1, \ldots, h/2$). These near-exact distributions will be respectively designated in the Tables by the acronyms GNIG, M2GNIG and M3GNIG.

Once obtained a well-fit near-exact distribution for $W$ under the form of a GNIG distribution or a mixture of GNIG distributions, the corresponding c.d.f.s are easily available (see Coelho (2004) and also Marques and Coelho (2008a,b)), being the corresponding near-exact c.d.f.s for $\Lambda$ very easily derived by simple exponentiation of the r.v. $W$. From these c.d.f.s the computation of near-exact quantiles is quick and easy, when using one of the several available softwares able to handle extended precision computations.

The computation of near-exact quantiles is not carried out in this subsection since this would only be a matter of interest in order to compare them with the exact ones, but the computation of these quantiles with enough precision (at least twelve to sixteen decimal places) would be a difficult and time consuming task, almost rendered useless, given the very good approximations that the near-exact distributions developed above exhibit, which is possible to be assessed through the computation of the values of the measure $\Delta$ using the c.f.s in (10) and (14) for different values of $n$, $p$ and $q$. To establish a comparison, we have also used the asymptotic Box-style distribution for $W = -\log \Lambda$ in Anderson (2003, sec. 10.5). Results are shown in Tables 1–6.

**Table 1** Values of $\Delta$ for the asymptotic and near-exact distributions for the l.r.t. statistic of equality of $q$ multivariate Normal distributions, for $n=20$, $p=10$

| | $q=4$ | $q=5$ | $q=6$ | $q=7$ |
|---|---|---|---|---|
| asymp. | $3.61\times10^{-2}$ | $3.49\times10^{-2}$ | $3.64\times10^{-2}$ | $3.92\times10^{-2}$ |
| GNIG | $1.19\times10^{-6}$ | $6.13\times10^{-7}$ | $4.61\times10^{-7}$ | $2.89\times10^{-7}$ |
| M2GNIG | $3.39\times10^{-10}$ | $8.16\times10^{-11}$ | $5.72\times10^{-11}$ | $2.69\times10^{-11}$ |
| M3GNIG | $1.08\times10^{-13}$ | $2.56\times10^{-15}$ | $6.82\times10^{-15}$ | $2.86\times10^{-15}$ |

**Table 2** Values of $\Delta$ for the asymptotic and near-exact distributions for the l.r.t. statistic of equality of $q$ multivariate Normal distributions, for $n=30$, $p=10$

| | $q=4$ | $q=5$ | $q=6$ | $q=7$ |
|---|---|---|---|---|
| asymp. | $1.06\times10^{-2}$ | $8.98\times10^{-3}$ | $8.30\times10^{-3}$ | $8.10\times10^{-3}$ |
| GNIG | $5.95\times10^{-7}$ | $3.16\times10^{-7}$ | $2.37\times10^{-7}$ | $1.50\times10^{-7}$ |
| M2GNIG | $8.08\times10^{-11}$ | $1.94\times10^{-11}$ | $1.46\times10^{-11}$ | $6.94\times10^{-12}$ |
| M3GNIG | $1.26\times10^{-14}$ | $1.88\times10^{-15}$ | $9.62\times10^{-16}$ | $4.60\times10^{-16}$ |

**Table 3** Values of $\Delta$ for the asymptotic and near-exact distributions for the l.r.t. statistic of equality of $q$ multivariate Normal distributions, for $n=50$, $p=10$

| | $q=4$ | $q=5$ | $q=6$ | $q=7$ |
|---|---|---|---|---|
| asymp. | $2.94\times10^{-3}$ | $2.34\times10^{-3}$ | $1.98\times10^{-3}$ | $1.77\times10^{-3}$ |
| GNIG | $2.25\times10^{-7}$ | $1.22\times10^{-7}$ | $9.17\times10^{-8}$ | $5.86\times10^{-8}$ |
| M2GNIG | $1.13\times10^{-11}$ | $2.50\times10^{-12}$ | $2.15\times10^{-12}$ | $1.04\times10^{-12}$ |
| M3GNIG | $9.04\times10^{-16}$ | $7.62\times10^{-16}$ | $9.53\times10^{-17}$ | $5.42\times10^{-17}$ |

**Table 4** Values of $\Delta$ for the asymptotic and near-exact distributions for the l.r.t. statistic of equality of $q$ multivariate Normal distributions, for $n=12$, $p=10$

| | $q=4$ | $q=5$ | $q=6$ | $q=7$ |
|---|---|---|---|---|
| asymp. | $4.03\times10^{-1}$ | $4.73\times10^{-1}$ | $5.44\times10^{-1}$ | $6.12\times10^{-1}$ |
| GNIG | $1.47\times10^{-6}$ | $6.83\times10^{-7}$ | $5.11\times10^{-7}$ | $3.13\times10^{-7}$ |
| M2GNIG | $6.85\times10^{-10}$ | $1.44\times10^{-10}$ | $9.47\times10^{-11}$ | $4.33\times10^{-11}$ |
| M3GNIG | $4.27\times10^{-13}$ | $2.70\times10^{-14}$ | $1.90\times10^{-14}$ | $6.92\times10^{-15}$ |

The results in Tables 1–6 show not only the extreme closeness to the exact distribution that all the near-exact distributions exhibit, with, as expected, the M3GNIG distribution displaying always the best fit and the single GNIG distribution a less good fit, although always much better than the asymptotic distribution. Moreover, the near-exact distributions display not only an asymptotic behaviour for increasing sample sizes but also an asymptotic behaviour for increasing values of $p$ and $q$, which is a much desirable feature. Apart from a couple of values which came out a little too good for the M3GNIG distribution, as for example for $n=20$, $p=10$ and $q=5$ and for $n=13$, $p=11$ and $q=5$, the evolution of the values of the measure

**Table 5** Values of $\Delta$ for the asymptotic and near-exact distributions for the l.r.t. statistic of equality of $q$ multivariate Normal distributions, for $n=14$, $p=12$

| | $q=4$ | $q=5$ | $q=6$ | $q=7$ |
|---|---|---|---|---|
| asymp. | $5.33\times10^{-1}$ | $6.27\times10^{-1}$ | $7.17\times10^{-1}$ | $7.99\times10^{-1}$ |
| GNIG | $8.58\times10^{-7}$ | $4.64\times10^{-7}$ | $2.63\times10^{-7}$ | $2.03\times10^{-7}$ |
| M2GNIG | $2.45\times10^{-10}$ | $8.71\times10^{-11}$ | $2.71\times10^{-11}$ | $1.87\times10^{-11}$ |
| M3GNIG | $7.85\times10^{-14}$ | $2.02\times10^{-14}$ | $2.21\times10^{-15}$ | $1.79\times10^{-15}$ |

**Table 6** Values of $\Delta$ for the asymptotic and near-exact distributions for the l.r.t. statistic of equality of $q$ multivariate Normal distributions, for $n=19$, $p=17$

| | $q=4$ | $q=5$ | $q=6$ | $q=7$ |
|---|---|---|---|---|
| asymp. | $8.45\times10^{-1}$ | $9.84\times10^{-1}$ | $1.11\times10^{0}$ | $1.22\times10^{0}$ |
| GNIG | $4.44\times10^{-7}$ | $2.64\times10^{-7}$ | $2.03\times10^{-7}$ | $1.09\times10^{-7}$ |
| M2GNIG | $8.23\times10^{-11}$ | $3.38\times10^{-11}$ | $2.23\times10^{-11}$ | $7.87\times10^{-12}$ |
| M3GNIG | $1.68\times10^{-14}$ | $5.26\times10^{-15}$ | $3.08\times10^{-15}$ | $7.55\times10^{-16}$ |

$\Delta$ is pretty steady. Indeed we should bear in mind that for either even $p$ or odd $q$ we have the exact distribution for the part of the l.r.t. statistic which is a Wilks Lambda statistic, given under the form of a GIG distribution.

Also, even for very small sample sizes, that is, for sample sizes which are very close to the number of variables being used, all the near-exact distributions show a very good performance, still with a clear asymptotic behaviour for increasing values of $p$ and $q$, even in situations where the asymptotic distribution breaks down (indeed not being a distribution any more).

## 3.2 *The Sphericity Test*

The sphericity test itself, whose null hypothesis may be written as

$$H_0 : \Sigma = \sigma^2 I_p \tag{15}$$

may be seen as another example of a l.r.t. whose null hypothesis may be written as the composition of two conditionally independent null hypotheses (see Ch. 10, subsec. 10.7.3 in Anderson (2003)), since we may indeed write $H_0$ in (15) once again as

$$H_0 \equiv H_{02|1} \circ H_{01}$$

where

$$H_{01} : \Sigma = diag(\sigma_1^2, \ldots, \sigma_p^2)$$

and

$$H_{02|1} : \sigma_1^2 = \cdots = \sigma_p^2 \\ \text{given that } \Sigma = diag(\sigma_1^2, \ldots, \sigma_p^2) ,$$

so that, using the notation defined at the beginning of this section, we may write the l.r.t. statistic to test $H_0$ in (15) as

$$\Lambda = \Lambda_2(n; \underbrace{1, \ldots, 1}_{p}) \, \Lambda_1(p, 1; n) ,$$

where, under $H_0$ in (15), $\Lambda_2(n; 1, \ldots, 1)$ and $\Lambda_1(p, 1; n)$ are independent (see for example Anderson (2003, Subsection 10.7.3)).

Although we may question the usefulness of this approach, moreover given that very accurate near-exact distributions have already been obtained for the sphericity l.r.t. statistic in Marques and Coelho (2008a), the same authors were able to obtain even more accurate near-exact distributions for moderately large values of $p$ for the same statistic through the use of this approach (Coelho and Marques, 2007b). These near-exact distributions, besides having a simpler formulation, are even more accurate than the previously developed ones namely for larger number of variables ($p \geq 10$).

The l.r.t. statistic to test $H_0$ in (15) is, for a sample of size $n$ (Mauchly, 1940; Anderson, 2003),

$$\Lambda = \left( \frac{|A|}{\left( tr \, \frac{1}{p} \, A \right)^p} \right)^{n/2} , \tag{16}$$

where $A$ is the maximum likelihood estimator of $\Sigma$.

The exact c.f. of $W = -\log \Lambda$ may be written as (see Anderson (2003); Kshirsagar (1972); Muirhead (1986))

$$\Phi_W(t) = \prod_{j=2}^{p} \frac{\Gamma\left(\frac{n-1}{2} + \frac{j-1}{p}\right) \Gamma\left(\frac{n-j}{2} - \frac{n}{2}\,\mathrm{i}t\right)}{\Gamma\left(\frac{n-1}{2} + \frac{j-1}{p} - \frac{n}{2}\,\mathrm{i}t\right) \Gamma\left(\frac{n-j}{2}\right)} \tag{17}$$

or, alternatively, using the approach outlined in Section 2, as (see Coelho and Marques (2007b))

$$\Phi_W(t) = \underbrace{\prod_{j=2}^{p} \left(\frac{n-j}{n}\right)^{r_j} \left(\frac{n-j}{n} - \mathrm{i}t\right)^{-r_j}}_{\Phi_1(t)} \\ \underbrace{\times \left\{ \prod_{j=1}^{p-k^*} \frac{\Gamma\left(\frac{n-1}{2}+\frac{j-1}{p}\right)\Gamma\left(\frac{n-1}{2}-\frac{n}{2}\mathrm{i}t\right)}{\Gamma\left(\frac{n-1}{2}\right)\Gamma\left(\frac{n-1}{2}+\frac{j-1}{p}-\frac{n}{2}\mathrm{i}t\right)} \right\} \times \left\{ \prod_{j=p-k^*+1}^{p} \frac{\Gamma\left(\frac{n-1}{2}+\frac{j-1}{p}\right)\Gamma\left(\frac{n}{2}-\frac{n}{2}\mathrm{i}t\right)}{\Gamma\left(\frac{n}{2}\right)\Gamma\left(\frac{n-1}{2}+\frac{j-1}{p}-\frac{n}{2}\mathrm{i}t\right)} \right\}}_{\Phi_2(t)}, \tag{18}$$

where $k^* = \lfloor p/2 \rfloor$ and $r_j = \lfloor \frac{p-j+2}{2} \rfloor$ $(j = 2, \ldots, p)$.

Then, if in (18) we replace $\Phi_2(t)$ by a c.f. $\Phi_2^*(t)$ of the type of the one in (12), using the specification in (13), we will obtain as near-exact distributions for $W$ a GNIG distribution or a mixture of two or three GNIG distributions of depth $p$ with integer shape parameters $r_j$ and non-integer shape parameter $s_k$, for the $k$-th mixture term $(k = 1, \ldots, h/2)$. As in the previous subsection, these near-exact distributions will be respectively designated, namely in the Tables, by the acronyms GNIGnew, M2GNIGnew and M3GNIGnew (being the acronyms GNIG, M2GNIG and M3GNIG used in Tables 8 and 9 for the corresponding near-exact distributions obtained for the same statistic in Marques and Coelho (2008a), without using the decomposition approach proposed in this paper). We will compare these near-exact distributions with the exact distribution as well as in terms of quantiles as through the use of the measure $\Delta$ in (6).

Consul (1967) obtained representations for the exact p.d.f. of the $(2/n)$-th power of the l.r.t. statistic in (16) for $p = 3, 4$ and 6 through the use of the Gauss hypergeometric function (as a side note, there is only a minor typo in expression (3.14), where the constant term should be $K(n)/9!$ instead of $2\,K(n)/9!$), leading to expressions that yield easy computations when using one of the common symbolic softwares with extended precision. However, a common expression for general $p$, was only obtained by the same author, in Consul (1969), where he makes use of the Meijer $G$-function. But, since the Meijer $G$-function is indeed only an alternative representation for a Mellin-Barnes type of integral, its computation is still quite heavy, even when using an extended precision software. Such representations for the p.d.f. only allow for the computation of quantiles for the $(2/n)$-th power of the l.r.t. statistic with the required precision (that is, up to sixteen decimal places of precision, in order to be able to compare them with the near-exact ones) for moderate values of $p$ $(p \leq 10)$.

**Table 7** Quantiles 0.05 and 0.01 for the $(2/n)$-th power of the l.r.t. statistic in (16)

| | quantile | | | |
|---|---|---|---|---|
| | 0.05 | $\delta$ | 0.01 | $\delta$ |
| $n = 7,\ p = 5$ | | | | |
| exact | 0.0012622014344466 | | 0.0002183943236075 | |
| Box | 0.0018598053836777 | 3.2 | 0.0004370905088500 | 3.7 |
| GNIGnew | 0.0012621657432522 | 7.4 | 0.0002183808232977 | 7.9 |
| M2GNIGnew | 0.0012622015298382 | 10.0 | 0.0002183943197572 | 11.4 |
| M3GNIGnew | 0.0012622014349577 | 12.3 | 0.0002183943234605 | 12.8 |
| $n = 13,\ p = 6$ | | | | |
| exact | 0.0365289320027240 | | 0.0187996936542321 | |
| Box | 0.0371562902726480 | 3.2 | 0.0193651668489626 | 3.2 |
| GNIGnew | 0.0365288722277811 | 7.2 | 0.0187995858186483 | 7.0 |
| M2GNIGnew | 0.0365289321616112 | 9.8 | 0.0187996937407390 | 10.1 |
| M3GNIGnew | 0.0365289320026219 | 13.0 | 0.0187996936542503 | 13.7 |
| $n = 9,\ p = 7$ | | | | |
| exact | 0.0001472413497881 | | 0.0000242384697502 | |
| Box | 0.0003324977520427 | 3.7 | 0.0000879672231885 | 4.2 |
| GNIGnew | 0.0001472404530786 | 7.2 | 0.0000242381703782 | 7.0 |
| M2GNIGnew | 0.0001472413507666 | 9.8 | 0.0000242384696727 | 10.1 |
| M3GNIGnew | 0.0001472413497887 | 13.0 | 0.0000242384697498 | 13.7 |
| $n = 14,\ p = 8$ | | | | |
| exact | 0.0056126167409998 | | 0.0024755558596933 | |
| Box | 0.0060583304947613 | 3.4 | 0.0027924174017203 | 3.5 |
| GNIGnew | 0.0056126113751765 | 8.3 | 0.0024755489107065 | 8.2 |
| M2GNIGnew | 0.0056126167475432 | 11.2 | 0.0024755558612765 | 11.8 |
| M3GNIGnew | 0.0056126167409703 | 13.5 | 0.0024755558596937 | 15.4 |
| $n = 16,\ p = 10$ | | | | |
| exact | 0.0012140286105857 | | 0.0005064717865027 | |
| Box | 0.0014135955350563 | 3.7 | 0.0006334878610630 | 3.9 |
| GNIGnew | 0.0012140280320840 | 9.2 | 0.0005064711210440 | 9.2 |
| M2GNIGnew | 0.0012140286109532 | 12.4 | 0.0005064717870345 | 12.3 |
| M3GNIGnew | 0.0012140286105719 | 13.9 | 0.0005064717869780 | 12.3 |

We have computed exact quantiles for the $(2/n)$-th power of the l.r.t. statistic in (16) for $p = 5, 6, 7, 8, 10$ and given values of $n$ and we compare them with the near-exact quantiles. The results are presented in Table 7, where a measure

$$\delta = -\log_{10} |exact\ quantile - approx.\ quantile|$$

**Table 8** Values of $\Delta$ for the near-exact distributions for the l.r.t. test statistic of sphericity for several values of $n$ and $p$

| | $p = 10, n = 12$ | $p = 12, n = 14$ | $p = 14, n = 16$ |
|---|---|---|---|
| GNIG | $1.03\times10^{-6}$ | $5.02\times10^{-7}$ | $2.30\times10^{-7}$ |
| M2GNIG | $2.79\times10^{-10}$ | $6.18\times10^{-11}$ | $1.86\times10^{-11}$ |
| M3GNIG | $2.20\times10^{-13}$ | $1.74\times10^{-14}$ | $4.82\times10^{-15}$ |
| GNIGnew | $1.20\times10^{-7}$ | $5.09\times10^{-8}$ | $2.48\times10^{-8}$ |
| M2GNIGnew | $6.59\times10^{-11}$ | $1.60\times10^{-11}$ | $4.86\times10^{-12}$ |
| M3GNIGnew | $1.23\times10^{-14}$ | $2.60\times10^{-15}$ | $6.11\times10^{-16}$ |

**Table 9** Values of $\Delta$ for the near-exact distributions for the l.r.t. test statistic of sphericity for several values of $n$ and $p$

| | $p = 10, n = 50$ | $p = 12, n = 52$ | $p = 14, n = 54$ |
|---|---|---|---|
| GNIG | $1.94\times10^{-7}$ | $1.41\times10^{-7}$ | $8.81\times10^{-8}$ |
| M2GNIG | $1.25\times10^{-11}$ | $3.84\times10^{-12}$ | $2.61\times10^{-12}$ |
| M3GNIG | $1.06\times10^{-14}$ | $2.54\times10^{-15}$ | $1.26\times10^{-15}$ |
| GNIGnew | $2.66\times10^{-8}$ | $1.62\times10^{-8}$ | $1.06\times10^{-8}$ |
| M2GNIGnew | $8.90\times10^{-12}$ | $3.73\times10^{-12}$ | $1.76\times10^{-12}$ |
| M3GNIGnew | $2.69\times10^{-15}$ | $7.82\times10^{-16}$ | $2.71\times10^{-16}$ |

is used to have a more clear and extended assessment of the number of decimal places matched by the approximate quantiles. In order to establish a comparison, we use Box's asymptotic approximation (see Anderson (2003), subsec. 10.7.4).

We may see how all the near-exact distributions show a very good behaviour for both quantiles, when compared with the Box asymptotic approximation, mainly for the smaller sample sizes (that is, for the smaller differences between $n$ and $p$). Only for $p = 10$ there seems to be a decline in the performance of the M3GNIG near-exact distribution, but the smaller than expected value for $\delta$ may be indeed due to a lack of precision in the computation of the exact quantile.

In Tables 8 and 9 we may examine the values of the measure $\Delta$ in (6), computed using the exact c.f. in (17) and the near-exact distributions GNIG, M2GNIG and M3GNIG in Marques and Coelho (2008a), and the near-exact distributions GNIGnew, M2GNIGnew and M3GNIGnew, whose c.f.s are derived from (18) by replacing $\Phi_2(t)$ by c.f.s of the type of $\Phi_2^*(t)$ in (12).

We may see how the near-exact distributions developed using the decomposition technique described in this paper have an even better behaviour than the previously developed and already well-fit near-exact distributions, both for small and large sample sizes, displaying a good asymptotic behaviour both for increasing sample size and increasing number of variables.

### 3.3 Generalized Sphericity Tests

Other tests to which the authors intend to apply the approach proposed in this paper are extended versions of the sphericity test, which we may call the 'multi-sample scalar-block sphericity test' and the 'multi-sample matrix-block sphericity test'.

For the multi-sample scalar-block sphericity test we have

$$H_0 : \Sigma_1 = \cdots = \Sigma_q = \begin{bmatrix} \sigma_1^2 I_{p_1} & & 0 \\ & \ddots & \\ 0 & & \sigma_k^2 I_{p_k} \end{bmatrix} \tag{19}$$

with

$$H_0 \equiv H_{03|1,2} \circ H_{02|1} \circ H_{01}$$

where

$$H_{01} : \Sigma_1 = \cdots = \Sigma_q ,$$

$$\begin{aligned} H_{02|1} : \Sigma &= diag(\Sigma_{11}, \ldots, \Sigma_{kk}) \\ &\text{given that } \Sigma_1 = \cdots = \Sigma_q = \Sigma , \end{aligned}$$

where $\Sigma$ is of dimensions $p \times p$, while $\Sigma_{ii}$ $(i = 1, \ldots, k)$ of dimensions $p_i \times p_i$ is the $i$-th diagonal block of $\Sigma$, with $p = \sum_{i=1}^{k} p_i$, and

$$\begin{aligned} H_{03|1,2} : \Sigma_{ii} &= \sigma_i^2 I_{p_i} , \quad \text{for } i = 1, \ldots, k \\ &\text{given that } \Sigma = diag(\Sigma_{11}, \ldots, \Sigma_{kk}) \text{ and} \\ &\text{given that } \Sigma_1 = \cdots = \Sigma_q = \Sigma , \end{aligned}$$

so that we may write the l.r.t. statistic to test $H_0$ in (19), for $N = N_1 + \cdots + N_q$, as

$$\Lambda = \Lambda_1(q, p; N_j)\, \Lambda_2(N; p_1, \ldots, p_k) \prod_{i=1}^{k} \Lambda_3(p_i; N) . \tag{20}$$

Since, under $H_0$ in (19), it is possible to prove the independence of all the statistics in (20) and since factorizations of the type in (5) are available for the c.f.s of the logarithms of all the statistics in (20) (Coelho, 2003, 2004; Coelho and Marques, 2007a; Marques and Coelho, 2008a), the process of obtaining near-exact distributions for the l.r.t. statistic for this test may be implemented by using the approach proposed in this paper.

Concerning the multi-sample matrix-block sphericity test, its null hypothesis may be written as

$$H_0 : \Sigma_1 = \cdots = \Sigma_q = \Delta \otimes I_k \left( = \begin{bmatrix} \Delta & & 0 \\ & \ddots & \\ 0 & & \Delta \end{bmatrix} \right) \tag{21}$$

with

$$H_0 \equiv H_{03|1,2} \circ H_{02|1} \circ H_{01} ,$$

where

$$H_{01} : \Sigma_1 = \cdots = \Sigma_q \, (= \Sigma) ,$$

$$\begin{aligned} H_{02|1} : \Sigma &= diag(\Sigma_{11}, \ldots , \Sigma_{kk}) \\ &\text{given that } \Sigma_1 = \cdots = \Sigma_q \, (= \Sigma) , \end{aligned}$$

where $\Sigma$ is of dimensions $p \times p$, while $\Sigma_{ii}$ $(i = 1, \ldots , k)$ of dimensions $p^* \times p^*$ is the $i$-th diagonal block of $\Sigma$, with $p = kp^*$, and

$$\begin{aligned} H_{03|1,2} : \Sigma_{11} &= \cdots = \Sigma_{kk} (= \Delta) \\ &\text{given that } \Sigma = diag(\Sigma_{11}, \ldots , \Sigma_{kk}) \text{ and} \\ &\text{given that } \Sigma_1 = \cdots = \Sigma_q = \Sigma , \end{aligned}$$

so that we may write the l.r.t. statistic to test $H_0$ in (21), for $N = N_1 + \cdots + N_q$, as

$$\Lambda = \Lambda_1(q, p; N_j) \, \Lambda_2(N; \underbrace{p^*, \ldots , p^*}_{k}) \, \Lambda_1(k, p^*; N) . \tag{22}$$

Since, under $H_0$ in (21), it is possible to prove the independence of all the statistics in (22) and since factorizations of the type in (5) are available for the c.f.s of the logarithms of all the statistics in (22) (Coelho, 2003, 2004; Marques and Coelho, 2008a), the process of obtaining near-exact distributions for the l.r.t. statistic for this test may once again be implemented by using the approach proposed in this paper.

## 4 Final Remarks

The use of the decomposition approach developed in this paper not only enables us to build very accurate and manageable near-exact approximations to the exact distribution of the overall test statistics but also concomitantly enables us to easily overcome the problems of controlling statistical errors (in particular the error of the first kind), which arise when we have to test sequentially the partial hypotheses. Now we may easily compute near-exact quantiles which enable us to carry the overall test in just one step, avoiding this way the problems brought to our attention by Hogg (1961) and avoiding also the need for any corrections of the first kind error level, like Sidak's correction (Sidak, 1967, 1968).

We may even plan to go beyond the tests presented in the previous section, namely those in Section 3, with virtually no limit, by adequately nesting the three elementary multivariate likelihood ratio tests (referred at the beginning of the previous section). In every case we will still be able to obtain near-exact distributions for the overall test statistics, in situations where the construction of well-fit asymptotic distributions, using the usual techniques, is too complicated or even virtually impossible.

**Acknowledgments** This research was financially supported by the Portuguese Foundation for Science and Technology, through the Center for Mathematics and its Applications (CMA) from the New University of Lisbon.

The authors would like also to express their appreciation for the insightful comments and suggestions made by the referees.

## References

Alberto, R.P., Coelho, C. A. (2007). Study of the quality of several asymptotic and near-exact approximations based on moments for the distribution of the Wilks Lambda statistic, *J. Statist. Plann. Inference* **137**, 1612–1626.

Anderson, T.W. (2003). *An Introduction to Multivariate Statistical Analysis*, 3rd ed.. J. Wiley & Sons, New York.

Berry, A. (1941). The accuracy of the Gaussian approximation to the sum of independent variates, *Trans. Amer. Math. Soc.* **49**, 122–136.

Basu, D. (1955). On statistics independent of a complete sufficient statistics, *Sankhyā* **15**, 377–380.

Coelho, C.A. (1998). The Generalized Integer Gamma distribution – a basis for distributions in Multivariate Statistics, *J. Multivariate Anal.* **64**, 86–102.

Coelho, C.A. (2003). The Generalized Integer Gamma distribution as an asymptotic replacement for the Logbeta random variable – Applications, *Amer. J. Math. Management Sci.* **23**, 383–399.

Coelho, C.A. (2004). The Generalized Near-Integer Gamma distribution: a basis for 'near-exact' approximations to the distributions of statistics which are the product of an odd number of independent Beta random variables, *J. Multivariate Anal.* **89**, 191–218.

Coelho, C.A., Alberto, R.P., Grilo, L.M. (2006). A mixture of Generalized Integer Gamma distributions as the exact distribution of the product of an odd number of independent Beta random variables, *J. Interdiscip. Math.* **9**, 229–248.

Coelho, C.A.:, Marques, F.J. (2007a). Near-exact distributions for the likelihood ratio test statistic for testing equality of several variance-covariance matrices. The New University of Lisbon, Mathematics Department, Technical Report #12/2007 (submitted for publication). (http://www.dm.fct.unl.pt/sites/www.dm.fct.unl.pt/files/preprints/2009/9_09.pdf)

Coelho, C.A., Marques, F.J. (2007b). Near-exact distributions for the independence and sphericity likelihood ratio test statistics (submitted for publication). The New University of Lisbon, Mathematics Department, Technical Report #20/2007. (https://www.dm.fct.unl.pt/sites/www.dm.fct.unl.pt/files/preprints/2007/20_07.pdf)

Coelho, C.A., Mexia, J.T. (2006). On the exact and near-exact distributions of statistics used in generalized $F$ tests. The New University of Lisbon, Mathematics Department, Technical Report #23/2006. (http://www.dm.fct.unl.pt/sites/www.dm.fct.unl.pt/files/preprints/2006/23_06.pdf)

Consul, P.C. (1967). On the exact distributions of the criterion W for testing sphericity in a p-variate Normal distribution, *Ann. Math. Statist.* **38**, 1170–1174.

Consul, P.C. (1969). The exact distributions of likelihood criteria for different hypotheses. In: *Multivariate Analysis II*, Proc. 2nd Intl. Symp., Dayton, Ohio, pp. 171–181, Academic Press, New York.

Esseen, C.-G. (1945). Fourier analysis of distribution functions. A Mathematical Study of the Laplace-Gaussian Law, *Acta Math.* **77**, 1–125.

Grilo, L.M., Coelho, C.A. (2007a). Development and study of two near-exact approximations to the distribution of the product of an odd number of independent Beta random variables, *J. Statist. Plann. Inference* **137**, 1560–1575.

Grilo, L.M., Coelho, C.A. (2007b). The exact and near-exact distributions for the Wilks Lambda statistic used in the test of independence of two sets of variables. The New University of Lisbon, Mathematics Department, Technical Report #15/2007 (submitted for publication). (http://www.dm.fct.unl.pt/sites/www.dm.fct.unl.pt/files/preprints/2007/15_07.pdf)

Hogg, R.V. (1961). On the resolution of statistical hypotheses. *J. Amer. Statist. Assoc.* **56**, 978–989.

Hogg, R.V., Craig, A.T. (1956). Sufficient statistics in elementary distribution theory, *Sankhyā* **17**, 209–216.

Hwang, H.-K. (1998). On convergence rates in the central limit theorems for combinatorial structures. *European J. Combin.* **19**, 329–343.

Kshirsagar, A.M. (1972). *Multivariate Analysis*. Marcel Dekker, New York.

Loève, M. (1977). *Probability Theory*, Vol. I, 4th ed.. Springer-Verlag, New York.

Marques, F.J., Coelho, C.A. (2008a). Near-exact distributions for the sphericity likelihood ratio test statistic. *J. Statist. Plann. Inference* **138**, 726–741.

Marques, F.J., Coelho, C.A. (2008b). A general near-exact distribution theory for the most common likelihood ratio test statistics used in Multivariate Statistics. The New University of Lisbon, Mathematics Department, Technical Report #5/2008 (submitted for publication). (https://www.dm.fct.unl.pt/sites/www.dm.fct.unl.pt/files/preprints/2008/5_08.pdf)

Mauchly, J.W. (1940). Significance test for sphericity of a normal $n$-variate distribution. *Ann. Math. Statist.* **11**, 204–209.

Muirhead, R.J. (1986). *Aspects of Multivariate Statistical Theory*. J. Wiley & Sons, New York.

Sidak, Z. (1967). Rectangular confidence regions for the means of multivariate normal distributions. *J. Amer. Statist. Assoc.* **62**, 626–633.

Sidak, Z. (1968). On multivariate normal probabilities of rectangles: Their dependence on correlations. *Ann. Math. Statist.* **39**, 1425–1434.

# Near-Exact Distributions for Certain Likelihood Ratio Test Statistics

**Carlos A. Coelho, Barry C. Arnold, and Filipe J. Marques**

**Abstract** In this paper we will show how, using an expansion of a Logbeta distribution as an infinite mixture of Gamma distributions we are able to obtain near-exact distributions for the negative logarithm of the l.r.t. (likelihood ratio test) statistics used in Multivariate Analysis to test the independence of several sets of variables, the equality of several mean vectors, sphericity and the equality of several variance-covariance matrices which will match as many of the exact moments as we wish and for which we will be able to have an *a priori* upper-bound for the difference between their c.d.f. and the exact c.d.f.. These near-exact distributions also display very good performance, with an agreement with the exact distribution which may virtually be taken as far as we wish and which it is not possible to obtain with the usual asymptotic distributions. Furthermore, based on the results presented it will be easy to build near-exact distributions for any l.r.t. statistics which may be built as the product of the above l.r.t. statistics.

**Keywords** Mixtures of Gamma distributions · Independence test · Equality of mean vectors test · Sphericity test · Test of equality of covariance matrices · Generalized Near-Integer Gamma distribution · Mixtures

---

Reprinted from Journal of Statistical Theory and Practice, with permission from Springer Nature. First published as Coelho, C.A., Arnold, B.C., Marques, F.J. (2010). Near-exact distributions for certain likelihood ratio test statistics. *Journal of Statistical Theory and Practice* **4**, 711–725. https://link.springer.com/article/10.1080/15598608.2010.10412014

---

C. A. Coelho (✉) · F. J. Marques
Departamento de Matemática, Faculdade de Ciências e Tecnologia, Universidade Nova de Lisboa, Lisbon, Portugal
e-mail: cmac@fct.unl.pt; fjm@fct.unl.pt

B. C. Arnold
Department of Statistics, University of California, Riverside, Riverside, CA, USA
e-mail: barry.arnold@ucr.edu



B. C. Arnold et al. (eds.), *Likelihood Ratio Tests and Related Topics in Multivariate Analysis*, Contributions to Statistics,
https://doi.org/10.1007/978-3-031-68008-3_9

**AMS Subject Classification** 62H10; 62E20; 62H05; 62H15

# 1 Introduction

Marques, Coelho and Arnold (2010) have shown that it is possible to obtain a common structure for both the exact and the near-exact distributions of the l.r.t. (likelihood ratio test) statistics used in Multivariate analysis under multivariate elliptically contoured or left orthogonal-invariant distributions to test

(i) the independence of several sets of variables
(ii) the equality of several mean vectors
(iii) sphericity
(iv) the equality of several variance-covariance matrices,

as well as for any l.r.t. statistic which may be obtained as the product of any number of independent such l.r.t. statistics.

In this paper we will show how, based on a new approach to the common structure for the exact distribution of the above l.r.t. statistics we may develop even better performing near-exact distributions which also exhibit a common structure for all the l.r.t. addressed. We will first show how, based on the results in Tricomi and Erdélyi (1951), we may write any Logbeta distribution, that is the distribution of the negative logarithm of a Beta distributed r.v. (random variable) as an infinite mixture of Gamma distributions. Then we will show how, based on this result, we may obtain near-exact distributions for any of the l.r.t. statistics above, which may, by construction, match as many of the exact moments as we wish and for which we will be able to have *a priori* upper-bounds on the difference between their c.d.f.'s and the exact c.d.f.. Indeed, this approach may be used for any l.r.t. statistic which may be built as the product of any number of independent such l.r.t. statistics (Coelho and Marques, 2009).

# 2 The Logbeta Distribution as an Infinite Mixture of Gammas

From the two first expressions in section 5 of Tricomi and Erdélyi (1951) and also expressions (11) and (14) in the same paper, we may write

$$\frac{\Gamma(a-it)}{\Gamma(a+b-it)} = \sum_{j=0}^{\infty} p_j(b)(a-it)^{-b-j} \tag{1}$$

where

$$p_j(b) = \frac{1}{j}\sum_{m=0}^{j-1}\left(\frac{\Gamma(1-b-m)}{\Gamma(-b-j)(j-m+1)!} + (-1)^{j+m}b^{j-m+1}\right)p_m(b),$$
$$j = 1, 2, \ldots, \tag{2}$$

with $p_0(b) = 1$.

Then, since the c.f. (characteristic function) of $Y = -\log X$, where $X \sim \text{Beta}(a, b)$, is given by

$$\Phi_Y(t) = \frac{\Gamma(a+b)}{\Gamma(a)}\frac{\Gamma(a-it)}{\Gamma(a+b-it)}, \tag{3}$$

using (1), we may write

$$\Phi_Y(t) = \sum_{j=0}^{\infty} \underbrace{\frac{\Gamma(a+b)}{\Gamma(a)}\frac{p_j(b)}{a^{b+j}}}_{p_j^*(a,b)} a^{b+j}(a-it)^{-(b+j)}, \tag{4}$$

which is the c.f. of an infinite mixture of $\Gamma(b+j, a)$ distributions, with weights $p_j^*(a, b)$, with $p_j(b)$ given by (2).

## 3 The Exact Distribution of the l.r.t. Statistics to Test Independence and Equality of Mean Vectors as Infinite Mixtures of GNIG Distributions

From (3) and (4) above we may write,

$$\frac{\Gamma\left(\frac{n-1}{2}\right)\Gamma\left(\frac{n-2}{2}-\frac{n}{2}\text{i}t\right)}{\Gamma\left(\frac{n-2}{2}\right)\Gamma\left(\frac{n-1}{2}-\frac{n}{2}\text{i}t\right)}$$
$$= \sum_{\ell=0}^{\infty} p_\ell^*\left(\tfrac{n-2}{2}, \tfrac{1}{2}\right)\left(\frac{n-2}{n}\right)^{\frac{1}{2}+\ell}\left(\frac{n-2}{n}-\text{i}t\right)^{-\left(\frac{1}{2}+\ell\right)}$$

with

$$p_\ell^*\left(\tfrac{n-2}{2}, \tfrac{1}{2}\right) = \frac{\Gamma\left(\frac{n-1}{2}\right)p_\ell\left(\frac{1}{2}\right)}{\Gamma\left(\frac{n-2}{2}\right)\left(\frac{n-2}{2}\right)^{\frac{1}{2}+\ell}}, \quad \ell = 0, 1, \ldots \tag{5}$$

where $p_\ell\left(\frac{1}{2}\right)$ $(\ell = 1, 2, \ldots)$ are given by (2) above, with $p_0\left(\frac{1}{2}\right) = 1$, so that, for any $k^* \in \mathbb{N}$, we may write

$$\begin{aligned}
&\left(\frac{\Gamma\left(\frac{n-1}{2}\right)\Gamma\left(\frac{n-2}{2}-\frac{n}{2}\text{i}t\right)}{\Gamma\left(\frac{n-2}{2}\right)\Gamma\left(\frac{n-1}{2}-\frac{n}{2}\text{i}t\right)}\right)^{k^*}\\
&\qquad=\left(\sum_{\ell=0}^{\infty} p_\ell^*\left(\tfrac{n-2}{2},\tfrac{1}{2}\right)\left(\frac{n-2}{n}\right)^{\ell+\frac{1}{2}}\left(\frac{n-2}{n}-\text{i}t\right)^{-\left(\ell+\frac{1}{2}\right)}\right)^{k^*}\\
&\qquad=\sum_{\ell=0}^{\infty}\pi_\ell\left(\frac{n-2}{n}\right)^{\ell+\frac{k^*}{2}}\left(\frac{n-2}{n}-\text{i}t\right)^{-\left(\ell+\frac{k^*}{2}\right)}
\end{aligned}$$

with

$$\pi_\ell = \sum_{\text{all } L_\ell}\prod_{\ell^*\in L_\ell} p_{\ell^*}^*\left(\tfrac{n-2}{2},\tfrac{1}{2}\right), \tag{6}$$

where $L_\ell$ is a set of $k^*$ values $\ell^* \in \mathbb{N}_0$ such that their sum is equal to $\ell$.

But then, from the results in section 5 of Coelho (2004), considering now a sample of size $n$, we may write the c.f. of $W_1$, the negative logarithm of the l.r.t. statistic to test the independence of $m$ sets of variables with an elliptically contoured or left orthogonal-invariant distribution, under the null hypothesis of independence of the $m$ sets of variables, for a sample of size $n$, and where $\ell^*(\leq m)$ of the sets have an odd number of variables, as

$$\Phi_{W_1}(t) = \left\{\prod_{j=3}^{p}\left(\frac{n-j}{n}\right)^{r_j}\left(\frac{n-j}{n}-\text{i}t\right)^{-r_j}\right\}\left(\frac{\Gamma\left(\frac{n-1}{2}\right)\Gamma\left(\frac{n-2}{2}-\frac{n}{2}\text{i}t\right)}{\Gamma\left(\frac{n-2}{2}\right)\Gamma\left(\frac{n-1}{2}-\frac{n}{2}\text{i}t\right)}\right)^{k^*} \tag{7}$$

$$\begin{aligned}
&=\left\{\prod_{j=3}^{p}\left(\frac{n-j}{n}\right)^{r_j}\left(\frac{n-j}{n}-\text{i}t\right)^{-r_j}\right\}\\
&\qquad\times\left\{\sum_{\ell=0}^{\infty}\pi_\ell\left(\frac{n-2}{n}\right)^{\frac{k^*}{2}+\ell}\left(\frac{n-2}{n}-\text{i}t\right)^{-\left(\frac{k^*}{2}+\ell\right)}\right\}
\end{aligned} \tag{8}$$

$$=\sum_{\ell=0}^{\infty}\pi_\ell\left\{\prod_{j=2}^{p}\left(\frac{n-j}{n}\right)^{r_j^*}\left(\frac{n-j}{n}-\text{i}t\right)^{-r_j^*}\right\} \tag{9}$$

where

$$k^* = \left\lfloor \frac{\ell^*}{2} \right\rfloor, \tag{10}$$

$p = \sum_{k=1}^{m} p_k$, where $p_k$ is the number of variables in the $k$-th set ($k = 1, \ldots, m$), and

$$r_j^* = \begin{cases} \frac{k^*}{2} + \ell, & j = 2 \\ r_j, & j = 3, \ldots, p \end{cases} \tag{11}$$

where

$$r_j = \begin{cases} h_{j-2} + (-1)^j k^*, & j = 3, 4 \\ r_{j-2} + h_{j-2}, & j = 5, \ldots, p \end{cases} \tag{12}$$

with

$$h_j = (\#\text{ of } p_k (k = 1, \ldots, m) \geq j) - 1, \quad j = 1, \ldots, p-2. \tag{13}$$

This expression for the c.f. of $W_1$ shows that the exact distribution of $W_1$ is an infinite mixture of GNIG (Generalized Near-Integer Gamma) distributions of depth $p-1$, with weights $\pi_\ell$ ($\ell = 0, 1, \ldots$), each GNIG distribution having shape parameters $r_j^*$ and rate parameters $\frac{n-j}{n}$ ($j = 2, \ldots, p$) (see Coelho (2004) for further details on the GNIG distribution, its p.d.f. and c.d.f.).

We should note that in (8) and (9), for $k^* = 0$ we have $\pi_0 = 1$ and all other $\pi_\ell = 0$ ($\ell = 1, 2, \ldots$) and thus both (8) and (9) in fact reduce to the product which is the first term in (7), since in this case we will also have $r_2^* = 0$.

We will denote by

$$f^{GNIG}\left(\cdot \mid s_1, \ldots, s_{d-1}; s_d; \lambda_1, \ldots, \lambda_{d-1}; \lambda_d; d\right)$$

and

$$F^{GNIG}\left(\cdot \mid s_1, \ldots, s_{d-1}; s_d; \lambda_1, \ldots, \lambda_{d-1}; \lambda_d; d\right)$$

the p.d.f. and the c.d.f. of a GNIG distribution of depth $d$, with integer shape parameters $s_1, \ldots, s_{d-1}$, non-integer shape parameter $s_d$ and corresponding rate parameters $\lambda_1, \ldots, \lambda_{d-1}, \lambda_d$.

The exact p.d.f. and c.d.f. of $W_1$ and of $\Lambda_1 = e^{-W_1}$, when there are two or more sets with an odd number of variables, that is, when $k^* \geq 1$, are thus given by the following Theorem.

**Theorem 1** *The exact p.d.f. and c.d.f. of* $W_1 = -\log \Lambda_1$, *where* $\Lambda_1$ *is the l.r.t. statistic to test the independence of* $m$ *sets of variables with a joint elliptically* *contoured or left orthogonal-invariant distribution, based on a sample of size n, under the null hypothesis of independence, are respectively*

$$f_{W_1}(w) = \sum_{\ell=0}^{\infty} \pi_\ell f^{GNIG}\left(w \mid r_3^*, \ldots, r_p^*; r_2^*; \frac{n-3}{n}, \ldots, \frac{n-p}{n}; \frac{n-2}{n}; p-1\right), \quad w > 0$$

*and*

$$F_{W_1}(w) = \sum_{\ell=0}^{\infty} \pi_\ell F^{GNIG}\left(w \mid r_3^*, \ldots, r_p^*; r_2^*; \frac{n-3}{n}, \ldots, \frac{n-p}{n}; \frac{n-2}{n}; p-1\right), \quad w > 0$$

*while the exact p.d.f. and c.d.f. of $\Lambda_1$ are given by*

$$f_{\Lambda_1}(z) = \sum_{\ell=0}^{\infty} \pi_\ell f^{GNIG}\Bigg(-\log z \mid r_3^*, \ldots, r_p^*; r_2^*; \frac{n-3}{n}, \ldots, \frac{n-p}{n}; \frac{n-2}{n}; p-1\Bigg)\frac{1}{z}, \quad 0 < z < 1$$

*and*

$$F_{\Lambda_1}(w) = \sum_{\ell=0}^{\infty} \pi_\ell \Bigg(1 - F^{GNIG}\Bigg(-\log z \mid r_3^*, \ldots, r_p^*; r_2^*; \frac{n-3}{n}, \ldots, \frac{n-p}{n}; \frac{n-2}{n}; p-1\Bigg)\Bigg), \quad 0 < z < 1$$

*where the $\pi_\ell$ are given by (6) and the $r_j^*$ by (11).*

We should note that when $k^*$ is even, $r_2^*$ is an integer, and thus, in this case, the GNIG distributions in the previous theorem reduce to GIG (Generalized Integer Gamma) distributions (Coelho, 1998, 2004).

When at most one of the sets of variables has an odd number of variables, $k^* = 0$ and then the exact distribution of $W_1$ is just a GIG distribution (Coelho, 1998).

The distribution of the l.r.t. statistic to test the equality of $q$ mean vectors from $q$ $p$-multivariate elliptically contoured or left orthogonal-invariant distributions, based on $q$ independent random samples, the $k$-th of which has size $n_k$ ($k = 1, \ldots, q$), is only a particular case of the distribution of $\Lambda_1$ above, for $m = 2$, $p_1 = p$, $p_2 = q-1$ and $n = \sum_{k=1}^{q} n_k$. If we call this statistic $\Lambda_2$ and take $W_2 = -\log \Lambda_2$, the exact c.f. of $W_2$ may be written as

$$\Phi_{W_2}(t) = \left\{ \prod_{j=3}^{p+q-1} \left(\frac{n-j}{n}\right)^{r_j} \left(\frac{n-j}{n} - \mathrm{i}t\right)^{-r_j} \right\} \times \left( \frac{\Gamma\left(\frac{n-1}{2}\right)\Gamma\left(\frac{n-2}{2} - \frac{n}{2}\mathrm{i}t\right)}{\Gamma\left(\frac{n-2}{2}\right)\Gamma\left(\frac{n-1}{2} - \frac{n}{2}\mathrm{i}t\right)} \right)^{k^*} \tag{14}$$

$$= \left\{ \prod_{j=3}^{p+q-1} \left(\frac{n-j}{n}\right)^{r_j} \left(\frac{n-j}{n} - \mathrm{i}t\right)^{-r_j} \right\} \times \left( \sum_{\ell=0}^{\infty} p_\ell^*\left(\tfrac{n-2}{2}, \tfrac{1}{2}\right) \left(\frac{n-2}{n}\right)^{\ell+\frac{1}{2}} \left(\frac{n-2}{n} - \mathrm{i}t\right)^{-\left(\ell+\frac{1}{2}\right)} \right)^{k^*}$$

where the $p_\ell^*\left(\frac{n-2}{2}, \frac{1}{2}\right)$ are given by (5),

$$r_j = \begin{cases} h_1 - k^* & j = 3 \\ h_2 + k^* & j = 4 \\ r_{j-2} + h_{j-2} & j = 5, \ldots, p+q-1 \end{cases} \tag{15}$$

with

$$h_j = (\#\text{ of elements of } \{p, q-1\} \geq j) - 1, \quad j = 1, \ldots, p+q-3, \tag{16}$$

and

$$k^* = \begin{cases} 1 & \text{if } p \text{ is odd and } q \text{ is even} \\ 0 & \text{all other cases} \end{cases} \tag{17}$$

so that if $p$ is even or $q$ is odd, the exact distribution of $W_2$ is a GIG distribution, while if $p$ is odd and $q$ is even, the exact c.f. of $W_2$ may be written as

$$\Phi_{W_2}(t) = \sum_{\ell=0}^{\infty} p_\ell^*\left(\tfrac{n-2}{2}, \tfrac{1}{2}\right) \prod_{j=2}^{p+q-1} \left(\frac{n-j}{j}\right)^{r_j^*} \left(\frac{n-j}{n} - \mathrm{i}t\right)^{-r_j^*}, \tag{18}$$

where

$$r_j^* = \begin{cases} k^*\left(\frac{1}{2}+\ell\right) & j=2 \\ r_j & j=3,\ldots,p+q-1 \end{cases} \tag{19}$$

for $r_j$ given by (15)–(16) above.

We should note that the c.f. in (18) reduces, for $k^* = 0$, to the product which is the first term in (14), since in this case we will have $r_2^* = 0$ and thus the product in (18) reduces to the product in (14), which not being any more a function of $\ell$, may move out of the summation which then clearly adds up to 1.

We have thus the following Theorem.

**Theorem 2** *The exact p.d.f. and c.d.f. of* $W_2 = -\log \Lambda_2$ *and* $\Lambda_2$*, where* $\Lambda_2$ *is the l.r.t. statistic to test the equality of q p-multivariate mean vectors from q p-multivariate elliptically contoured or left-orthogonal invariant distributions, based on q independent samples, the k-th of which with size* $n_k (k = 1, \ldots, q)$*, when p is odd and q is even, under the null hypothesis of equality of the mean vectors, are, for* $n = \sum_{k=1}^{q} n_k$*, respectively given by*

$$f_{W_2}(w) = \sum_{\ell=0}^{\infty} p_\ell^*\left(\tfrac{n-2}{2}, \tfrac{1}{2}\right) f^{GNIG}\left(w \mid r_3^*, \ldots, r_{p+q-1}^*; r_2^*; \frac{n-3}{n}, \ldots, \frac{n-(p+q-1)}{n}; \frac{n-2}{n}; p+q-2\right), \quad w > 0,$$

$$F_{W_2}(w) = \sum_{\ell=0}^{\infty} p_\ell^*\left(\tfrac{n-2}{2}, \tfrac{1}{2}\right) F^{GNIG}\left(w \mid r_3^*, \ldots, r_{p+q-1}^*; r_2^*; \frac{n-3}{n}, \ldots, \frac{n-(p+q-1)}{n}; \frac{n-2}{n}; p+q-2\right), \quad w > 0,$$

$$f_{\Lambda_2}(z) = \sum_{\ell=0}^{\infty} p_\ell^*\left(\tfrac{n-2}{2}, \tfrac{1}{2}\right) f^{GNIG}\left(-\log z \mid r_3^*, \ldots, r_{p+q-1}^*; r_2^*; \frac{n-3}{n}, \ldots, \frac{n-(p+q-1)}{n}; \frac{n-2}{n}; p+q-2\right) \frac{1}{z}, \quad 0 < z < 1$$

*and*

$$F_{\Lambda_2}(w) = \sum_{\ell=0}^{\infty} p_\ell^*\left(\tfrac{n-2}{2}, \tfrac{1}{2}\right) \left(1 - F^{GNIG}\left(-\log z \mid r_3^*, \ldots, r_{p+q-1}^*; r_2^*; \frac{n-3}{n}, \ldots, \frac{n-(p+q-1)}{n}; \frac{n-2}{n}; p+q-2\right)\right), \quad 0 < z < 1,$$

*where the* $p_\ell^*\left(\frac{n-2}{2}, \frac{1}{2}\right)$ *are given by (5) and the* $r_j^*$ *by (19).*

# 4 On the Exact Distributions of the l.r.t. Statistics to Test Sphericity and Equality of Covariance Matrices

We may try to deal with the exact distributions of the l.r.t. statistics to test sphericity and equality of covariance matrices using a similar approach to the one used in the previous section.

## 4.1 On the Exact Distribution of the l.r.t. Statistic to Test Sphericity

From Lemma 4 in section 3 of Coelho and Marques (2010) we see that the part of the c.f. of $W_3 = -\log \Lambda_3$, where $\Lambda_3$ is the l.r.t. statistic to test sphericity in a set of $p$ variates with a joint elliptically contoured or left-orthogonal invariant distribution, denoted by $\Phi_{2,W_3}(t)$ is, for $k^* = \lfloor p/2 \rfloor$, the c.f. of the sum of $p - k^* - 1$ independent Logbeta$\left(\frac{n-1}{2}, \frac{j-1}{p}\right)$ $(j = 2, \ldots, p - k^*)$ r.v.'s, multiplied by $n/2$, with other $k^*$ independent Logbeta $\left(\frac{n}{2}, \frac{j-1}{p} - \frac{1}{2}\right)$ $(j = p - k^* + 1, \ldots, p)$ r.v.'s, multiplied by $n/2$, which, from (3) and (4) is the sum of $p - k^* - 1$ independent infinite mixtures of $\Gamma\left(\frac{j-1}{p} + \ell, \frac{n-1}{n}\right)$ $(j = 2, \ldots, p - k^*; \ell = 0, 1, \ldots)$ distributions, with other $k^*$ independent infinite mixtures of $\Gamma\left(\frac{j-1}{p} - \frac{1}{2} + \ell, \frac{n}{n}\right)$ $(j = p - k^* + 1, \ldots, p;\ \ell = 0, 1, \ldots)$ distributions. In the case in which all the rate parameters are the same, say equal to $\lambda$, this sum of mixtures would simplify to only one infinite mixture of $\Gamma\left(\frac{p-2}{4} + \ell, \lambda\right)$ $(\ell = 0, 1, \ldots)$ distributions for even $p$ or one infinite mixture of $\Gamma\left(\frac{p-1}{4} + \ell, \lambda\right)$ $(\ell = 0, 1, \ldots)$ distributions for odd $p$, since for even $p$ we have

$$\sum_{j=1}^{p-k^*} \frac{j-1}{p} + \sum_{j=p-k^*+1}^{p} \left(\frac{j-1}{p} - \frac{1}{2}\right) = \left\{\sum_{j=1}^{p} \frac{j-1}{p}\right\} - \frac{1}{2}\frac{p}{2} = \frac{p-1}{2} - \frac{p}{4}$$
$$= \frac{p-2}{4}$$

while for odd $p$ we have

$$\sum_{j=1}^{p-k^*} \frac{j-1}{p} + \sum_{j=p-k^*+1}^{p} \left(\frac{j-1}{p} - \frac{1}{2}\right) = \left\{\sum_{j=1}^{p} \frac{j-1}{p}\right\} - \frac{1}{2}\frac{p-1}{2}$$
$$= \frac{p-1}{2} - \frac{1}{2}\frac{p-1}{2} = \frac{p-1}{4}.$$

But then in order to approximate this sum of independent infinite mixtures of Gamma distributions by a single infinite mixture of Gamma distributions, we propose an infinite mixture of Gamma distributions with shape parameters $r+\ell$ ($\ell = 0, 1, \ldots$), with

$$r = \begin{cases} \frac{p-2}{4} & \text{for even } p \\ \frac{p-1}{4} & \text{for odd } p \end{cases}$$

and all with rate parameter $\lambda$ with the same value obtained for $\lambda$ in (3.3) in Marques, Coelho and Arnold (2010), where near-exact distributions are obtained in the form of finite mixtures of Gamma distributions with the same rate parameter.

## 4.2 *On the Exact Distribution of the l.r.t. Statistic to Test Equality of Covariance Matrices*

By using a similar procedure and similar arguments to the ones used in the previous subsection, we may approximate the exact distribution of $W_4 = -\log \Lambda_4$, where $\Lambda_4$ is the l.r.t. statistic to test the equality of $q$ covariance matrices from $p$-variate elliptically contoured or left orthogonal-invariant distributions, based on $q$ independent samples of size $n$, by replacing the part of the c.f. of $W_4$ denoted in expression (A.9) in Marques, Coelho and Arnold (2010) by $\Phi_{2,W_4}(t)$, which, using (3) and (4), may be seen as the c.f. of the sum of $q \times \lfloor p/2 \rfloor$ independent infinite mixtures of $\Gamma\left(\frac{k-2j}{q} - \left\lfloor \frac{k-2j}{q} \right\rfloor + \ell, \frac{n-1}{n} - \frac{1}{n}\left\lfloor \frac{k-2j}{q} \right\rfloor\right)$ ($k = 1, \ldots, q$; $j = 1, \ldots, \lfloor p/2 \rfloor$; $\ell = 0, 1, \ldots$) distributions, plus, for odd $p$, the sum of $q$ independent infinite mixtures of $\Gamma\left(\frac{2k-p-1}{2q} - \left\lfloor \frac{2k-p-1}{2q} \right\rfloor + \ell, \frac{n-1}{n} - \frac{2}{n}\left\lfloor \frac{2k-p-1}{2q} \right\rfloor\right)$ ($k = 1, \ldots, q$; $\ell = 0, 1, \ldots$) distributions, by the c.f. of an infinite mixture of $\Gamma(r+\ell, \lambda)$ ($\ell = 0, 1, \ldots$) distributions, where

$$\begin{aligned} r &= \left\{ \sum_{j=1}^{\lfloor p/2 \rfloor} \sum_{k=1}^{q} \frac{k-2j}{q} - \left\lfloor \frac{k-2j}{q} \right\rfloor \right\} + \left( \frac{2k-p-1}{2q} - \left\lfloor \frac{2k-p-1}{2q} \right\rfloor \right)^{\mathrm{mod}(p,2)} \\ &= \left\lfloor \frac{p+1}{2} \right\rfloor \frac{q-1}{2}, \end{aligned}$$

and $\lambda$ is once again defined as in the previous subsection.

## 5 Near-Exact Distributions

Although the approximations proposed in the previous section for the l.r.t. statistics to test sphericity and equality of covariance matrices provide crude approximations to the exact distributions of those statistics and their logarithms, such approximations are in fact not as accurate as one would wish. Anyway, even if they were judged to be sufficiently accurate, for practical applications we would still have to face the problem of the truncation of the series corresponding to their p.d.f.'s and c.d.f.'s as also happens with the exact distributions obtained for the l.r.t. statistics to test the independence of several sets of variables or the equality of several mean vectors in section 3.

The development of near-exact distributions, based on such exact distributions or approximations, which will take the form of finite mixtures of GNIG distributions for the negative logarithm of such l.r.t. statistics, will solve such a problem and provide us with very accurate approximations to the exact distributions.

The near-exact distributions proposed for the four basic l.r.t. statistics addressed in this paper, which are: (i) the l.r.t. statistic to test the independence of several sets of variables; (ii) the l.r.t. statistic to test the equality of several mean vectors; (iii) the l.r.t. statistic to test for sphericity; and (iv) the l.r.t. statistic to test the equality of several covariance matrices, have c.f.'s which may be written, for the negative logarithm of those statistics (generally denoted here by $W$), as[1]

$$\Phi^*_W(t) = \sum_{k=0}^{m^*} p_k \left\{ \lambda^{b+j}(\lambda - \mathrm{i}t)^{-(b+j)} \prod_{j=2}^{p} \left(\frac{n-j}{n}\right)^{r_j} \left(\frac{n-1}{n} - \mathrm{i}t\right)^{-r_j} \right\} \tag{20}$$

which yields for $W$, and for $\Lambda$, the l.r.t. statistic itself, near-exact distributions with p.d.f.'s and c.d.f.'s respectively given by

$$f_W(w) = \sum_{k=0}^{m^*} p_k f^{GNIG}\left(w \mid r_2, \ldots, r_p; r_k^+; \frac{n-2}{n}, \ldots, \frac{n-p}{n}; \lambda; p\right), \quad w > 0$$

$$F_W(w) = \sum_{k=0}^{m^*} p_k F^{GNIG}\left(w \mid r_2, \ldots, r_p, r_k^+; \frac{n-2}{n}, \ldots, \frac{n-p}{n}, \lambda; p\right), \quad w > 0$$

$$f_\Lambda(z) = \sum_{k=0}^{m^*} p_k f^{GNIG}\left(-\log z \mid r_2, \ldots, r_p; r_k^+; \frac{n-2}{n}, \ldots, \frac{n-p}{n}; \lambda; p\right)\frac{1}{z}, \quad 0 < z < 1$$

$$F_\Lambda(z) = \sum_{k=0}^{m^*} p_k \left(1 - F^{GNIG}\left(-\log z \mid r_2, \ldots, r_p; r_k^+; \frac{n-2}{n}, \ldots, \frac{n-p}{n}; \lambda; p\right)\right), \quad 0 < z < 1$$

[1] see author and editors note at the end of the Chapter

where $p$ is the overall number of variables involved, except in the case of the test of equality of $q$ $p$-variate mean vectors, where it is to be taken as $p+q-1$, the $r_j$ are given by:

(i) expressions (12) and (13), with $r_2 = 0$, for the l.r.t. statistic to test the independence of $m$ sets of elliptically contoured or left-orthogonal invariant distributed r.v.'s,
(ii) expressions (15) and (16), with $r_2 = 0$, for the l.r.t. statistic to test the equality of mean vectors of $q$ populations with elliptically or left-orthogonal invariant distributions,
(iii) $\left[\frac{p-j+2}{2}\right]$, for the l.r.t. statistic to test sphericity in a set of elliptically contoured or left-orthogonal invariant r.v.'s (Coelho and Marques, 2010),
(iv) expressions (A.12) – (A.16) in Marques, Coelho and Arnold (2010) for the l.r.t. statistic to test the equality of covariance matrices of $q$ elliptically contoured or left-orthogonal invariant sets of r.v.'s,

$r_k^+$ is given by

(i) $\frac{k^*}{2}+k$, with $k^*$ in (10), for the l.r.t. statistic to test independence of $m$ sets of variables,
(ii) $k^*\left(\frac{1}{2}+k\right)$, with $k^*$ in (17), for the l.r.t. statistic to test the equality of $q$ $p$-variate mean vectors,
(iii) $\frac{p-2}{4}+k$ for even $p$ or $\frac{p-1}{4}+k$ for odd $p$, for the l.r.t. statistic to test sphericity,
(iv) $\left\lfloor\frac{p+1}{2}\right\rfloor\frac{q-1}{2}+k$, for the l.r.t. statistic to test the equality of $q$ covariance matrices,

and $\lambda$ assumes the value

(i),(ii) $\frac{n-2}{n}$ for the l.r.t. statistics to test independence of several sets of variables and the equality of several mean vectors,
(iii),(iv) of $\lambda$ in (3.3) of Marques, Coelho and Arnold (2010) when we match 4 or 6 exact moments, for the l.r.t. statistics to test sphericity and to test the equality of several covariance matrices,

and where the depth of the GNIG distributions is indeed $p-1$ for the l.r.t. of independence, where $r_2 = 0$ and $p+q-2$ for the l.r.t. of equality of mean vectors, where $r_2 = 0$.

We should note here that when $r_k^+$ are integer, the above near-exact distributions are indeed mixtures of GIG distributions.

In the above expressions, $m^*$ represents the number of exact moments equated to obtain the near-exact distribution; the weights $p_k$ $(k=0,\dots,m^*-1)$ are determined by equating the first $m^*$ moments from (20) and the first $m^*$ exact moments of $W$, that is, in such a way that

$$\left.\frac{\partial^h}{\partial t^h}\Phi_W(t)\right|_{t=0} = \left.\frac{\partial^h}{\partial t^h}\Phi_W^*(t)\right|_{t=0}, \qquad h=1,\dots,m^*,$$

where $\Phi_W(t)$ represents the exact c.f. of $W$ and $\Phi^*_W(t)$ the near-exact c.f. of $W$ in (20). The weight $p_{m^*}$ is obtained through the relation

$$p_{m^*} = 1 - \sum_{k=0}^{m^*-1} p_k \,.$$

These near-exact distributions are thus, if $m^*$ exact moments are matched, mixtures of $m^* + 1$ GNIG distributions with shape parameters $r_2, \ldots, r_p, r_k^+$ and rate parameters $\frac{n-2}{n}, \ldots, \frac{n-p}{n}, \lambda$.

The number of exact moments matched by these distributions may be virtually as many as one wishes.

If the number of exact moments matched is equal to two, four or six, the near-exact distributions developed in this paper will indeed have a slightly more complicated structure than the ones in Marques, Coelho and Arnold (2010) that match the same number of exact moments, since while these distributions would correspond respectively to a single GNIG distribution or to a mixture of two or three GNIG distributions, we now have a mixture of respectively three, five or seven GNIG distributions. However, the near-exact distributions proposed in the present paper enjoy several advantages, among which are the following:

- they may equate virtually as many of the exact moments as one wishes,
- for a given number of exact moments matched, they always show a better performance and even better asymptotic characteristics than the ones previously proposed, with even better performances for very small sample sizes,
- the system of equations to be solved in order to determine the weights $p_k$ ($k = 0, \ldots, m^* - 1$) is linear and as such very simple to be solved,
- it is possible to obtain an *a priori* upper-bound on the probability error, that is, an upper-bound on the difference between the near-exact and the exact c.d.f..

In Tables 1–4 we may analyze the values for the measure

$$\Delta = \frac{1}{2\pi} \int_{-\infty}^{+\infty} \left| \frac{\Phi_W(t) - \Phi^*_W(t)}{t} \right| dt$$

where $\Phi_W(t)$ represents the exact c.f. of $W$ and $\Phi^*_W(t)$ the near-exact c.f. of $W$ in (20), with

$$\max_{w \in \mathbb{R}^+} \left| F_W(w) - F^*_W(w) \right| \leq \Delta$$

and

$$\max_{z \in ]0,1[} \left| F_\Lambda(z) - F^*_\Lambda(z) \right| \leq \Delta \,,$$

where $F_W(\cdot)$ and $F^*_W(\cdot)$ represent respectively the exact c.d.f. of $W$ and the near-exact c.d.f. corresponding to $\Phi^*_W(t)$ in (20) and $F_\Lambda(\cdot)$ and $F^*_\Lambda(\cdot)$ represent respectively the exact c.d.f. of $\Lambda$ and the near-exact c.d.f. corresponding to $\Phi^*_W(t)$ in (20). See Coelho and Marques (2009) and Coelho and Mexia (2010) for details on the use of the measure $\Delta$ as a measure of proximity between distributions.

Tables 1–4 display values of the measure $\Delta$ for the near-exact distributions proposed in this paper for the cases in which 4, 6, 10 and 15 exact moments are matched, as well as for those proposed in Marques, Coelho and Arnold (2010) in which 4 and 6 exact moments were matched. These are designated by M2GNIG and M3GNIG respectively (since they are mixtures of respectively 2 and 3 GNIG distributions).

All the computations relative to the l.r.t. statistics to test sphericity and the equality of several covariance matrices, for the newly developed near-exact distributions use a value of $\lambda$ which is equal to the value obtained for $\lambda$ in (3.3) of Marques, Coelho and Arnold (2010) when equating 4 exact moments.

From Table 1 we may see that the new near-exact distributions besides showing even better performance for very small sample sizes than the already very good performance of the formerly proposed ones, also will show an improved asymptotic behavior, in terms of sample size and number of variables involved, since they display a more accentuated decrease in the values of the measure $\Delta$ even for small increases in the sample size or in the number of variables involved.

By analyzing Table 2 we may see how also for the l.r.t. statistic used to test the equality of $q$ $p$-multivariate mean vectors, assuming equality of the covariance matrices, the near-exact distributions proposed in this paper exhibit improved behavior for increasing sample size, increasing dimension ($p$) and also for increasing values of $q$, the number of mean vectors being tested. Although this also happens for the former near-exact distributions, namely for the M2GNIG and M3GNIG distributions, this behavior is now more accentuated for the newly proposed ones. We may also notice that, as expected, the larger the number of exact moments matched, the more accentuated is the asymptotic behavior of the new near-exact distributions, since the decrease in the values of the measure $\Delta$ is even more substantial for increasing values of $n$, $p$ and $q$ when we match a larger number of exact moments.

From Table 3 we may see that, as with the other l.r.t. statistics, for the l.r.t. statistic used to test sphericity, the near-exact distributions proposed in this paper provide improved approximations both for increasing sample sizes as well as for increasing dimension. Once again, use of the distributions that match more exact moments results in even better approximations and more marked improvement as sample size and/or dimension increases.

By analyzing the values of the measure $\Delta$ in Table 4 we see that, for the l.r.t. statistic to test the equality of $q$ $p$-variate covariance matrices, the newly proposed near-exact distributions once again display improved behavior and provide excellent approximations to the exact distributions, as also happened for all the other l.r.t. statistics addressed in this paper. Once again the approximations proposed in this paper show better behavior than the previously proposed approximations, with

**Table 1** Values of the measure $\Delta$ for several near-exact distributions for the l.r.t. statistic to test the independence of $m$ groups of variables, the $k$-th of which with $p_k$ variables ($k = 1, \ldots, m$)

| | | | former near-exact dists. | | new near-exact dists. (# of exact moments matched) | | | |
|---|---|---|---|---|---|---|---|---|
| $m$ | $p_k$ | $n$ | M2GNIG | M3GNIG | 4 | 6 | 10 | 15 |
| 3 | {3,5,6} | 16 | $2.16\times10^{-11}$ | $3.81\times10^{-15}$ | $3.84\times10^{-13}$ | $1.24\times10^{-16}$ | $1.09\times10^{-21}$ | $5.88\times10^{-27}$ |
| | | 50 | $1.33\times10^{-11}$ | $5.53\times10^{-15}$ | $2.04\times10^{-14}$ | $2.93\times10^{-18}$ | $5.81\times10^{-25}$ | $9.96\times10^{-32}$ |
| | | 100 | $2.21\times10^{-12}$ | $1.21\times10^{-15}$ | $7.95\times10^{-16}$ | $3.17\times10^{-20}$ | $5.76\times10^{-28}$ | $2.04\times10^{-36}$ |
| 3 | {5,7,10} | 24 | $3.09\times10^{-13}$ | $9.10\times10^{-18}$ | $2.41\times10^{-15}$ | $1.48\times10^{-19}$ | $2.74\times10^{-27}$ | $2.65\times10^{-34}$ |
| | | 50 | $8.16\times10^{-13}$ | $1.11\times10^{-16}$ | $1.33\times10^{-15}$ | $6.73\times10^{-20}$ | $1.80\times10^{-27}$ | $2.97\times10^{-35}$ |
| | | 100 | $1.86\times10^{-13}$ | $2.04\times10^{-17}$ | $7.16\times10^{-17}$ | $1.15\times10^{-21}$ | $3.69\times10^{-30}$ | $1.75\times10^{-39}$ |
| 3 | {7,11,15} | 35 | $7.50\times10^{-15}$ | $1.18\times10^{-19}$ | $2.65\times10^{-17}$ | $2.98\times10^{-22}$ | $2.92\times10^{-31}$ | $2.08\times10^{-40}$ |
| | | 50 | $4.14\times10^{-14}$ | $1.71\times10^{-18}$ | $6.90\times10^{-17}$ | $1.09\times10^{-21}$ | $2.90\times10^{-30}$ | $2.87\times10^{-39}$ |
| | | 100 | $1.78\times10^{-14}$ | $7.68\times10^{-19}$ | $7.04\times10^{-18}$ | $4.56\times10^{-23}$ | $2.46\times10^{-32}$ | $1.34\times10^{-42}$ |
| 4 | {3,5,7,7} | 24 | $5.83\times10^{-13}$ | $1.60\times10^{-17}$ | $4.25\times10^{-15}$ | $2.68\times10^{-19}$ | $2.13\times10^{-27}$ | $2.90\times10^{-34}$ |
| | | 50 | $1.32\times10^{-12}$ | $1.56\times10^{-16}$ | $1.97\times10^{-15}$ | $9.54\times10^{-20}$ | $2.21\times10^{-27}$ | $2.67\times10^{-35}$ |
| | | 100 | $2.89\times10^{-13}$ | $5.37\times10^{-17}$ | $1.01\times10^{-16}$ | $1.52\times10^{-21}$ | $4.07\times10^{-30}$ | $1.37\times10^{-39}$ |
| 4 | {7,7,9,10} | 35 | $6.74\times10^{-15}$ | $1.01\times10^{-19}$ | $2.39\times10^{-17}$ | $2.55\times10^{-22}$ | $2.24\times10^{-31}$ | $1.36\times10^{-40}$ |
| | | 50 | $3.25\times10^{-14}$ | $1.22\times10^{-18}$ | $5.42\times10^{-17}$ | $7.76\times10^{-22}$ | $1.70\times10^{-30}$ | $1.30\times10^{-39}$ |
| | | 100 | $1.28\times10^{-14}$ | $4.85\times10^{-19}$ | $5.08\times10^{-18}$ | $2.89\times10^{-23}$ | $1.21\times10^{-32}$ | $4.76\times10^{-43}$ |
| 5 | {3,5,5,5,4} | 24 | $5.45\times10^{-13}$ | $1.45\times10^{-17}$ | $3.98\times10^{-15}$ | $2.43\times10^{-19}$ | $1.79\times10^{-27}$ | $2.19\times10^{-34}$ |
| | | 50 | $1.12\times10^{-12}$ | $1.23\times10^{-16}$ | $1.67\times10^{-15}$ | $7.57\times10^{-20}$ | $1.54\times10^{-27}$ | $1.57\times10^{-35}$ |
| | | 100 | $2.38\times10^{-13}$ | $4.09\times10^{-17}$ | $8.32\times10^{-17}$ | $1.16\times10^{-21}$ | $2.69\times10^{-30}$ | $7.56\times10^{-40}$ |
| 5 | {5,5,7,7,9} | 35 | $1.36\times10^{-14}$ | $1.92\times10^{-19}$ | $4.43\times10^{-17}$ | $4.95\times10^{-22}$ | $4.76\times10^{-31}$ | $2.49\times10^{-40}$ |
| | | 50 | $6.23\times10^{-14}$ | $2.16\times10^{-18}$ | $9.47\times10^{-17}$ | $1.38\times10^{-21}$ | $2.98\times10^{-30}$ | $1.96\times10^{-39}$ |
| | | 100 | $2.36\times10^{-14}$ | $1.62\times10^{-18}$ | $8.43\times10^{-18}$ | $4.80\times10^{-23}$ | $1.88\times10^{-32}$ | $6.11\times10^{-43}$ |
| 5 | {9,9,11,11,8} | 50 | $4.70\times10^{-16}$ | $2.46\times10^{-21}$ | $7.19\times10^{-19}$ | $1.60\times10^{-24}$ | $8.83\times10^{-35}$ | $6.89\times10^{-46}$ |
| | | 100 | $2.25\times10^{-15}$ | $6.03\times10^{-20}$ | $8.10\times10^{-19}$ | $1.82\times10^{-24}$ | $1.12\times10^{-34}$ | $3.68\times10^{-46}$ |
| | | 200 | $5.39\times10^{-16}$ | $5.20\times10^{-21}$ | $4.69\times10^{-20}$ | $3.39\times10^{-26}$ | $2.25\times10^{-37}$ | $1.93\times10^{-50}$ |
| 6 | {11,11,15, | 100 | $9.06\times10^{-18}$ | $2.25\times10^{-23}$ | $2.77\times10^{-21}$ | $6.28\times10^{-28}$ | $3.57\times10^{-40}$ | $3.01\times10^{-54}$ |
| | 15,19,21} | 200 | $1.66\times10^{-17}$ | $5.30\times10^{-24}$ | $1.21\times10^{-21}$ | $1.99\times10^{-28}$ | $6.07\times10^{-41}$ | $9.08\times10^{-56}$ |

**Table 2** Values of the measure $\Delta$ for several near-exact distributions for the l.r.t. statistic to test the equality of $q$ $p$-variate mean vectors, for odd $p$ and even $q$

| | | | former near-exact dists. | | new near-exact dists. (# of exact moments matched) | | | |
|---|---|---|---|---|---|---|---|---|
| $p$ | $q$ | $n$ | M2GNIG | M3GNIG | 4 | 6 | 10 | 15 |
| 5 | 4 | 30 | $2.82\times10^{-9}$ | $5.61\times10^{-12}$ | $9.16\times10^{-12}$ | $1.15\times10^{-14}$ | $5.88\times10^{-20}$ | $1.16\times10^{-24}$ |
| | | 50 | $7.06\times10^{-10}$ | $1.36\times10^{-12}$ | $7.69\times10^{-13}$ | $3.75\times10^{-16}$ | $6.51\times10^{-22}$ | $1.06\times10^{-27}$ |
| | | 100 | $9.62\times10^{-11}$ | $1.17\times10^{-13}$ | $2.46\times10^{-14}$ | $3.06\times10^{-18}$ | $4.12\times10^{-25}$ | $1.17\times10^{-32}$ |
| 5 | 10 | 70 | $1.29\times10^{-11}$ | $6.51\times10^{-15}$ | $9.36\times10^{-15}$ | $9.72\times10^{-19}$ | $1.10\times10^{-25}$ | $5.46\times10^{-33}$ |
| | | 100 | $5.02\times10^{-12}$ | $2.02\times10^{-15}$ | $1.73\times10^{-15}$ | $9.16\times10^{-20}$ | $2.82\times10^{-27}$ | $1.79\times10^{-35}$ |
| 7 | 4 | 35 | $6.64\times10^{-10}$ | $9.36\times10^{-13}$ | $1.76\times10^{-12}$ | $1.28\times10^{-15}$ | $3.71\times10^{-21}$ | $2.56\times10^{-26}$ |
| | | 50 | $2.67\times10^{-10}$ | $3.59\times10^{-13}$ | $3.30\times10^{-13}$ | $1.25\times10^{-16}$ | $1.43\times10^{-22}$ | $1.57\times10^{-28}$ |
| | | 100 | $3.88\times10^{-11}$ | $3.41\times10^{-14}$ | $1.12\times10^{-14}$ | $1.11\times10^{-18}$ | $1.02\times10^{-25}$ | $2.01\times10^{-33}$ |
| 7 | 10 | 100 | $2.02\times10^{-12}$ | $5.72\times10^{-16}$ | $7.31\times10^{-16}$ | $2.83\times10^{-20}$ | $4.91\times10^{-28}$ | $1.65\times10^{-36}$ |
| | | 200 | $2.98\times10^{-13}$ | $4.91\times10^{-17}$ | $2.61\times10^{-17}$ | $2.65\times10^{-22}$ | $3.25\times10^{-31}$ | $1.78\times10^{-41}$ |
| 15 | 4 | 65 | $1.30\times10^{-11}$ | $6.40\times10^{-15}$ | $1.11\times10^{-14}$ | $1.25\times10^{-18}$ | $1.68\times10^{-25}$ | $1.17\times10^{-32}$ |
| | | 100 | $4.52\times10^{-12}$ | $1.76\times10^{-15}$ | $1.57\times10^{-15}$ | $8.04\times10^{-20}$ | $2.35\times10^{-27}$ | $1.42\times10^{-35}$ |
| | | 200 | $6.81\times10^{-13}$ | $1.55\times10^{-16}$ | $5.70\times10^{-17}$ | $7.70\times10^{-22}$ | $1.59\times10^{-30}$ | $1.55\times10^{-40}$ |
| 15 | 10 | 200 | $3.94\times10^{-14}$ | $2.93\times10^{-18}$ | $3.67\times10^{-18}$ | $1.79\times10^{-23}$ | $5.45\times10^{-33}$ | $6.01\times10^{-44}$ |
| | | 300 | $1.28\times10^{-14}$ | $6.76\times10^{-19}$ | $5.23\times10^{-19}$ | $1.17\times10^{-24}$ | $7.50\times10^{-35}$ | $7.23\times10^{-47}$ |
| 25 | 4 | 120 | $6.00\times10^{-13}$ | $1.13\times10^{-16}$ | $1.53\times10^{-16}$ | $3.26\times10^{-21}$ | $1.80\times10^{-29}$ | $1.26\times10^{-38}$ |
| | | 200 | $1.64\times10^{-13}$ | $3.60\times10^{-17}$ | $1.47\times10^{-17}$ | $1.22\times10^{-22}$ | $1.02\times10^{-31}$ | $3.68\times10^{-42}$ |
| 25 | 10 | 260 | $4.78\times10^{-15}$ | $1.65\times10^{-19}$ | $2.67\times10^{-19}$ | $4.65\times10^{-25}$ | $1.86\times10^{-35}$ | $1.26\times10^{-47}$ |
| | | 300 | $3.25\times10^{-15}$ | $9.98\times10^{-20}$ | $1.36\times10^{-19}$ | $1.81\times10^{-25}$ | $2.21\times10^{-36}$ | $1.22\times10^{-48}$ |
| | | 500 | $7.82\times10^{-16}$ | $1.54\times10^{-20}$ | $1.17\times10^{-20}$ | $5.81\times10^{-27}$ | $1.90\times10^{-38}$ | $2.38\times10^{-52}$ |

**Table 3** Values of the measure $\Delta$ for several near-exact distributions for the l.r.t. statistic to test sphericity

| | | former near-exact dists. | | new near-exact dists. (# of exact moments matched) | | | |
|---|---|---|---|---|---|---|---|
| $p$ | $n$ | M2GNIG | M3GNIG | 4 | 6 | 10 | 15 |
| 3 | 5 | $2.27\times10^{-6}$ | $5.78\times10^{-7}$ | $6.47\times10^{-8}$ | $1.20\times10^{-8}$ | $1.54\times10^{-10}$ | $2.06\times10^{-12}$ |
| | 15 | $2.76\times10^{-7}$ | $1.24\times10^{-9}$ | $1.46\times10^{-9}$ | $7.40\times10^{-12}$ | $5.96\times10^{-15}$ | $3.59\times10^{-18}$ |
| 6 | 8 | $3.57\times10^{-9}$ | $3.74\times10^{-12}$ | $4.79\times10^{-11}$ | $6.76\times10^{-14}$ | $1.64\times10^{-17}$ | $2.01\times10^{-21}$ |
| | 18 | $1.14\times10^{-9}$ | $5.05\times10^{-12}$ | $2.62\times10^{-12}$ | $1.67\times10^{-15}$ | $2.42\times10^{-21}$ | $2.53\times10^{-26}$ |
| | 30 | $3.03\times10^{-10}$ | $8.19\times10^{-13}$ | $2.37\times10^{-13}$ | $5.99\times10^{-17}$ | $3.12\times10^{-23}$ | $1.84\times10^{-29}$ |
| 10 | 12 | $6.59\times10^{-11}$ | $1.23\times10^{-14}$ | $4.26\times10^{-13}$ | $1.15\times10^{-16}$ | $2.24\times10^{-22}$ | $7.89\times10^{-28}$ |
| | 22 | $6.13\times10^{-11}$ | $2.56\times10^{-14}$ | $1.09\times10^{-13}$ | $1.80\times10^{-17}$ | $3.13\times10^{-24}$ | $2.12\times10^{-30}$ |
| | 35 | $2.22\times10^{-11}$ | $8.30\times10^{-15}$ | $1.51\times10^{-14}$ | $1.16\times10^{-18}$ | $7.09\times10^{-26}$ | $3.93\times10^{-33}$ |
| | 50 | $8.90\times10^{-12}$ | $2.69\times10^{-15}$ | $2.91\times10^{-15}$ | $1.16\times10^{-19}$ | $2.15\times10^{-27}$ | $1.69\times10^{-35}$ |
| 15 | 17 | $5.34\times10^{-12}$ | $9.62\times10^{-17}$ | $3.97\times10^{-14}$ | $4.52\times10^{-18}$ | $8.84\times10^{-24}$ | $7.78\times10^{-30}$ |
| | 27 | $9.72\times10^{-12}$ | $1.17\times10^{-15}$ | $2.92\times10^{-14}$ | $3.95\times10^{-18}$ | $5.44\times10^{-25}$ | $2.26\times10^{-31}$ |
| | 50 | $3.03\times10^{-12}$ | $4.34\times10^{-16}$ | $2.64\times10^{-15}$ | $1.57\times10^{-19}$ | $6.79\times10^{-27}$ | $3.90\times10^{-34}$ |
| | 100 | $5.04\times10^{-13}$ | $8.02\times10^{-18}$ | $1.09\times10^{-16}$ | $1.88\times10^{-21}$ | $9.26\times10^{-30}$ | $1.30\times10^{-38}$ |
| 20 | 22 | $3.14\times10^{-13}$ | $1.80\times10^{-17}$ | $5.96\times10^{-16}$ | $9.86\times10^{-21}$ | $1.83\times10^{-29}$ | $1.01\times10^{-37}$ |
| | 35 | $7.39\times10^{-13}$ | $1.30\times10^{-17}$ | $5.38\times10^{-16}$ | $8.09\times10^{-21}$ | $2.14\times10^{-29}$ | $3.62\times10^{-38}$ |
| | 50 | $4.19\times10^{-13}$ | $4.12\times10^{-18}$ | $1.48\times10^{-16}$ | $1.34\times10^{-21}$ | $1.46\times10^{-30}$ | $4.95\times10^{-40}$ |
| | 100 | $8.07\times10^{-14}$ | $1.87\times10^{-19}$ | $6.98\times10^{-18}$ | $1.88\times10^{-23}$ | $1.99\times10^{-33}$ | $2.05\times10^{-44}$ |
| 25 | 27 | $8.79\times10^{-14}$ | $2.48\times10^{-18}$ | $2.23\times10^{-16}$ | $3.92\times10^{-21}$ | $1.30\times10^{-29}$ | $7.19\times10^{-38}$ |
| | 50 | $2.12\times10^{-13}$ | $1.16\times10^{-17}$ | $1.56\times10^{-16}$ | $2.54\times10^{-21}$ | $1.03\times10^{-29}$ | $3.64\times10^{-38}$ |
| | 100 | $4.89\times10^{-14}$ | $1.95\times10^{-18}$ | $8.97\times10^{-18}$ | $4.85\times10^{-23}$ | $2.99\times10^{-32}$ | $3.66\times10^{-42}$ |

**Table 4** Values of the measure $\Delta$ for several near-exact distributions for the l.r.t. statistic to test the equality of $q$ covariance matrices of dimension $p$

| | | | former near-exact dists. | | new near-exact dists. (# of exact moments matched) | | | |
|---|---|---|---|---|---|---|---|---|
| $p$ | $q$ | $n$ | M2GNIG | M3GNIG | 4 | 6 | 10 | 15 |
| 5 | 3 | 7 | $2.24\times10^{-7}$ | $7.42\times10^{-10}$ | $5.23\times10^{-9}$ | $1.98\times10^{-11}$ | $2.39\times10^{-14}$ | $1.15\times10^{-16}$ |
| | | 50 | $3.85\times10^{-11}$ | $2.11\times10^{-12}$ | $3.64\times10^{-13}$ | $7.62\times10^{-17}$ | $8.60\times10^{-23}$ | $1.50\times10^{-29}$ |
| | | 100 | $1.03\times10^{-11}$ | $1.29\times10^{-13}$ | $1.23\times10^{-14}$ | $4.41\times10^{-19}$ | $2.05\times10^{-26}$ | $4.93\times10^{-34}$ |
| 5 | 7 | 7 | $9.86\times10^{-8}$ | $5.42\times10^{-10}$ | $2.13\times10^{-8}$ | $9.79\times10^{-11}$ | $4.71\times10^{-15}$ | $6.07\times10^{-20}$ |
| | | 50 | $1.25\times10^{-10}$ | $1.04\times10^{-13}$ | $5.51\times10^{-13}$ | $4.43\times10^{-17}$ | $1.11\times10^{-24}$ | $9.17\times10^{-32}$ |
| | | 100 | $1.25\times10^{-11}$ | $2.32\times10^{-14}$ | $1.49\times10^{-14}$ | $2.97\times10^{-19}$ | $5.79\times10^{-28}$ | $8.55\times10^{-37}$ |
| 7 | 5 | 9 | $5.17\times10^{-9}$ | $6.54\times10^{-12}$ | $3.33\times10^{-10}$ | $2.51\times10^{-13}$ | $1.61\times10^{-18}$ | $5.80\times10^{-24}$ |
| | | 50 | $2.39\times10^{-11}$ | $8.90\times10^{-15}$ | $3.22\times10^{-14}$ | $1.55\times10^{-18}$ | $4.00\times10^{-26}$ | $6.34\times10^{-33}$ |
| | | 100 | $1.83\times10^{-12}$ | $7.32\times10^{-15}$ | $5.43\times10^{-16}$ | $1.00\times10^{-20}$ | $2.94\times10^{-29}$ | $1.99\times10^{-37}$ |
| 7 | 10 | 9 | $2.35\times10^{-9}$ | $2.53\times10^{-12}$ | $3.82\times10^{-10}$ | $3.18\times10^{-13}$ | $4.23\times10^{-19}$ | $2.84\times10^{-26}$ |
| | | 50 | $1.88\times10^{-11}$ | $2.72\times10^{-14}$ | $1.24\times10^{-13}$ | $4.70\times10^{-18}$ | $1.39\times10^{-26}$ | $4.68\times10^{-35}$ |
| | | 100 | $1.83\times10^{-12}$ | $1.63\times10^{-14}$ | $3.44\times10^{-15}$ | $3.28\times10^{-20}$ | $6.58\times10^{-30}$ | $4.00\times10^{-40}$ |
| 10 | 5 | 12 | $1.64\times10^{-10}$ | $3.25\times10^{-14}$ | $3.88\times10^{-12}$ | $5.14\times10^{-16}$ | $8.45\times10^{-24}$ | $2.41\times10^{-31}$ |
| | | 50 | $3.59\times10^{-12}$ | $1.34\times10^{-14}$ | $3.68\times10^{-15}$ | $1.96\times10^{-19}$ | $3.01\times10^{-28}$ | $3.68\times10^{-39}$ |
| | | 100 | $1.67\times10^{-13}$ | $9.61\times10^{-16}$ | $2.46\times10^{-16}$ | $2.31\times10^{-21}$ | $1.97\times10^{-31}$ | $1.08\times10^{-43}$ |
| 10 | 15 | 12 | $7.13\times10^{-12}$ | — | $1.84\times10^{-13}$ | $5.58\times10^{-18}$ | $5.73\times10^{-27}$ | $1.56\times10^{-36}$ |
| | | 50 | $2.81\times10^{-13}$ | — | $3.51\times10^{-16}$ | $1.24\times10^{-21}$ | $3.48\times10^{-33}$ | $5.51\times10^{-43}$ |
| | | 100 | $1.97\times10^{-14}$ | $1.03\times10^{-20}$ | $5.23\times10^{-18}$ | $3.29\times10^{-24}$ | $8.09\times10^{-36}$ | $5.82\times10^{-46}$ |

improved performance both for increasing sample sizes as well as for increasing dimension ($p$) and also for increasing number of matrices involved ($q$).

Also, given the oscillating sign nature of the weights $p_k$ in (20), for all the new near-exact distributions, we may take the absolute value of the last of these weights, that is $p_{m^*-1}$, as a rough upper-bound on the difference between the near-exact and exact c.d.f.'s. This means that before we determine the value for $\Delta$ in each single case, which may indeed take some time, we may obtain from the absolute value of $p_{m^*-1}$ a rough upper-bound on this value, which may come in quite handy.

## 6 Conclusions

Although, for a given number of exact moments matched, the near-exact distributions proposed in this paper have a somewhat more complicated structure than the ones proposed in Marques, Coelho and Arnold (2010), since they will correspond to a mixture of a larger number of GNIG distributions, they are still quite manageable when using an adequate software and they show smaller values of the discrepancy measure $\Delta$, indicating in this way that an excellent approximation to the exact distribution has been obtained.

Moreover, the computation of the parameters involved in these near-exact distributions, which have to be computed by matching some of the exact moments, shows absolutely no problems since the systems of equations to be solved are always linear and as such very easy to solve. This would not be the case with the formerly proposed near-exact distributions when the number of exact moments matched is larger than 4.

With the proposed technique we may essentially get near-exact distributions that match as many of the exact moments as we wish and which will be as close to the exact distribution as we wish. As we may see from the tables in the previous section, in many situations, for such near-exact distributions, even when a quite small number of exact moments is matched, the discrepancy measure $\Delta$ is remarkably small. In practice, even for applications requiring very good approximations, near-exact p-values and quantiles may be used instead of the ones obtained from the exact distributions, which are considerably less manageable.

Moreover, the technique used to build the near-exact distributions proposed in this paper may be readily extended to any l.r.t. statistic which may be built as the product of any number of independent l.r.t. statistics of the type studied in this paper, just by applying the methods and ideas in Coelho and Marques (2009).

For practical purposes, the newly proposed near-exact distributions are considerably easier to implement than the former ones. Of course one will equate a larger number of exact moments when more precision is needed and a smaller number, obtaining this way an even more manageable distribution, when less precision is needed. MATHEMATICA modules, together with user instructions, for the implementation and computation of the near-exact distributions in this paper are available from the first author.

**Acknowledgments** This research was financially supported by the Portuguese Foundation for Science and Technology, through CMA – 'Centro de Matemática e Aplicações' of Universidade Nova de Lisboa (Financiamento Base 2010 ISFL-1-297 from FCT/MCTES/PT).

## References

Coelho, C.A. (1998). The Generalized Integer Gamma distribution – a basis for distribution in Multivariate Statistics. *J. Multivariate Anal.* **64**, 86–102.

Coelho, C.A. (2004). The generalized near-integer Gamma distribution, a basis for "near-exact" approximations to the distributions of statistics which are the product of an odd number of independent Beta random variables. *J. Multivariate Anal.* **89**, 191–218.

Coelho, C.A., Marques, F.J. (2009). The advantage of decomposing elaborate hypotheses on covariance matrices into conditionally independent hypotheses in building near-exact distributions for the test statistics. *Linear Algebra Appl.* **430**, 2592–2606.

Coelho, C.A., Marques, F.J. (2010). Near-exact distributions for the independence and sphericity likelihood ratio test statistics. *J. Multivariate Anal.* **101**, 583–593.

Coelho, C.A., Mexia, J.T. (2010). *Product and Ratio of Generalized Gamma-Ratio Random Variables: Exact and Near-exact Distributions - Applications.* LAP – Lambert Academic Publishing AG & Co. KG, Saarbrücken, Germany (ISBN: 978-3-8383-5846-8)

Marques, F.J., Coelho, C.A., Arnold, B.C. (2010). A general near-exact distribution theory for the most common likelihood ratio test statistics used in Multivariate Analysis. *Test* (in print, https://doi.org/10.1007/s11749-010-0193-3)

Tricomi, F.G., Erdélyi, A. (1051). The asymptotic expansion of a ratio of Gamma functions. *Pacific J. Math.* **1**, 133–142.

### *Author and Editors note:*

[1] there are 3 small typos in (20), two of them in the two first exponents, where $b+j$ should rather be $b+k$ and in the second ratio where $n-1$ should be $n-j$, that is, this expression should rather read

$$\Phi^*_W(t) = \sum_{k=0}^{m^*} p_k \left\{ \lambda^{b+k}(\lambda - \mathrm{i}t)^{-(b+k)} \prod_{j=2}^{p} \left(\frac{n-j}{n}\right)^{r_j} \left(\frac{n-j}{n} - \mathrm{i}t\right)^{-r_j} \right\} \tag{20}$$

# A General Near-Exact Distribution Theory for the Most Common Likelihood Ratio Test Statistics Used in Multivariate Analysis

**Filipe J. Marques, Carlos A. Coelho, and Barry C. Arnold**

**Abstract** In this paper we first show how the exact distributions of the most common likelihood ratio test (l.r.t.) statistics, that is, the ones used to test the independence of several sets of variables, the equality of several variance-covariance matrices, sphericity and the equality of several mean vectors, may be expressed as the distribution of the product of independent Beta random variables or the product of a given number of independent random variables whose logarithm has a Gamma distribution times a given number of independent Beta random variables. Then, we show how near-exact distributions for the logarithms of these statistics may be expressed as Generalized Near-Integer Gamma distributions or mixtures of these distributions, whose rate parameters associated with the integer shape parameters, for samples of size $n$, all have the form $(n - j)/n$ for $j = 2, \ldots, p$, where for three of the statistics, $p$ is the number of variables involved, while for the other one, it is the sum of the number of variables involved and the number of mean vectors being tested. What is interesting is that the similarities exhibited by these statistics are even more striking in terms of near-exact distributions than in terms of exact distributions. Moreover all the l.r.t. statistics that may be built as products of these basic statistics also inherit a similar structure for their near-exact distributions. To

Reprinted from TEST, with permission from Springer Nature.
First published as Marques, F.J., Coelho, C.A., Arnold, B.C.: A general near-exact distribution theory for the most common likelihood ratio test statistics used in Multivariate Analysis. TEST, 20, 1, 180–203 (2011). https://link.springer.com/article/10.1007/s11749-010-0193-3

F. J. Marques · C. A. Coelho (✉)
Departamento de Matemática, Faculdade de Ciências e Tecnologia, Universidade Nova de Lisboa, Lisbon, Portugal
e-mail: fjm@fct.unl.pt; cmac@fct.unl.pt

B. C. Arnold
Statistics Department, University of California, Riverside, Riverside, CA, USA
e-mail: barry.arnold@ucr.edu

 
B. C. Arnold et al. (eds.), *Likelihood Ratio Tests and Related Topics in Multivariate Analysis*, Contributions to Statistics,
https://doi.org/10.1007/978-3-031-68008-3_10

illustrate this fact, an application is made to the l.r.t. statistic to test the equality of several multivariate Normal distributions.

**Keywords** Wilks Lambda statistic · Independence test · Sphericity test · Generalized Integer Gamma distribution · Generalized Near-Integer Gamma distribution · Mixtures

**Mathematics Subject Classification (2000)** 62H10 · 62E20 · 62H05 · 62H15

## 1 Introduction

When using likelihood ratio tests in Multivariate Analysis, and in particular when using the tests addressed in this paper, the practitioner will face many situations where the exact distributions are extremely difficult if not impossible to handle, while at the same time the available asymptotic distributions may not yield the desired accuracy. We should note that even the representations of the exact distributions in forms that use generalized functions like the Meijer or the $H$ functions will not be adequate since such functions are not computable with the necessary accuracy even with current computers and the available extended precision software. Thus, in cases like these, the development of near-exact distributions is a welcome addition to the researcher's toolbox.

Near-exact distributions are asymptotic distributions which besides being much closer to the exact distribution than common asymptotic distributions are, and besides displaying an asymptotic behavior in the usual sense, that is, for increasing sample sizes, also display an asymptotic behavior for increasing number of variables and the number of matrices or distributions involved (when more than one matrix or distribution are involved), which is a very desirable property. Yet, these distributions also have a very good behavior for very small sample sizes, that is, sample sizes which scarcely exceed the number of variables involved. Moreover, the near-exact distributions also display an improved asymptotic behavior, that is, a faster rate of convergence towards the exact distribution, than common asymptotic distributions do.

This is achieved through the use of a new concept of approximating the c.f. of a distribution, when developing the near-exact distributions. By first identifying the component parts of the c.f. which may be considered together to yield a manageable distribution and then identifying a good asymptotic replacement for the remaining smaller part of that c.f. which, after that replacement, yields a c.f. corresponding to a known manageable distribution, from which quantiles and $p$-values may be easily computed, we are able to obtain what we call a near-exact c.f.

It is a well-known fact that the distributions of the l.r.t. (likelihood ratio test) statistics used in Multivariate Statistics (under normality assumptions) to test the independence of several sets of variables and to test the equality of several mean vectors may be expressed as distributions of products of independent Beta r.v.'s (random variables) raised to the power equal to one-half the sample-size (see, for example, Anderson (2003, Theorems 9.3.1 and 8.4.4 and Lemma 8.4.2)). However, this is not the case for the l.r.t. statistics to test sphericity and the equality of several covariance matrices, whose distributions are usually represented as the distribution of a product of independent Beta r.v.'s, raised to the power equal to one-half the sample-size, multiplied by the complement of some of them, also raised to some power (see, for example, Theorem 10.4.2 and subsection 10.7.3 in Anderson (2003)).

We will show, in Sect. 2.1 of this paper, that also for these two statistics, it is possible to represent their exact distribution as the distribution of the product of independent Beta r.v.'s, raised to the power equal to one-half the sample size.

Based on these results, we will then present for each of the four l.r.t. statistics (that is, the l.r.t. statistics to test independence of several groups of variables, the equality of several mean vectors, the equality of several covariance matrices, and sphericity) expressions for their exact distribution represented as the distribution of the product of a given number (closely related with the number of variables being used) of independent r.v.'s whose logarithms have Gamma distributions with integer shape parameters and a given number (a function of the parity of the number of variables involved) of independent Beta r.v.'s raised to the power equal to one-half the sample size.

This way of looking at the exact distribution of these statistics will enable us to establish a common strategy concerning the development of near-exact distributions for these statistics, as well as a common formulation for such near-exact distributions. Based on previously published results (as well as others submitted for publication) (Coelho, 2004; Marques and Coelho, 2008; Coelho and Marques, 2009b, 2010), we will show how, in terms of near-exact distributions, the similarities among the distributions of these statistics are even more striking.

If we think that the usual normality assumptions under which the l.r.t. statistics and their distributions are derived are too restrictive, we should bear in mind the results in Anderson et al. (1986), Anderson and Fang (1990), Anderson (2003), Jensen and Good (1981) and Kariya (1981) which show that such statistics are still the l.r.t. statistics for the same hypotheses when the underlying distributions for the samples are multivariate elliptically contoured or left orthogonal-invariant and that under the null hypotheses such statistics have the same distributions as in the normal case, in this way making the results presented somewhat more general.

We will also show how the results derived may be applied to more complex tests, where the null hypothesis may be decomposed into a number of conditionally independent hypotheses, by illustrating with the test for equality of several multivariate Normal distributions. More precisely, we illustrate with the test statistic for the equality of several multivariate Normal distributions how the decompositions of the c.f. of the negative logarithm of the l.r.t. statistics in this paper may be used to extend similar decompositions to the c.f. of the negative logarithm of l.r.t. statistics which may themselves be factored into the product of two or more independent l.r.t. statistics, corresponding to the l.r.t. statistics for the conditionally independent null hypotheses into which the original null hypothesis may be decomposed (Coelho and

Marques, 2009a, 2010). This is a useful approach which may be used to build near-exact distributions for more elaborate statistics which cannot be handled by other methods of building approximate distributions.

In this manner, in Sect. 2 we will describe the exact distribution of the l.r.t. statistics addressed in this paper both as the distribution of the product of powers of independent Beta distributed r.v.'s and as the distribution of products of a given number of independent r.v.'s whose logarithms have Gamma distributions with integer shape parameters and a given number of powers of independent Beta r.v.'s. Then in Sect. 3 we will use these latter representations to develop near-exact distributions for those statistics, and we will use these representations and the method used to develop a general near-exact distribution theory for the l.r.t. statistics. Finally in Sect. 4 we illustrate with the l.r.t. statistic to test the equality of several multivariate Normal distributions how we may further use the results obtained to develop near-exact distributions for other l.r.t. statistics that may be expressed as the product of independent l.r.t. statistics. Finally in Sect. 5 we present some numerical studies.

In order to make the near-exact distributions based on the unified approach developed in this paper widely available and easy to use, the authors provide modules for the computation of the near-exact c.f.'s, p.d.f's, c.d.f.'s, quantiles and $p$-values for the negative logarithm of the l.r.t. statistics addressed in this paper (web-page: https://sites.google.com/site/nearexactdistributions/).

## 2 Exact Distributions of the l.r.t. Statistics

### *2.1 As the Distribution of the Product of Independent Beta r.v.'s*

As Hsieh (1979) remarks, "In multivariate analysis under the normality assumptions, many likelihood ratio criteria are distributed as

$$K \prod_{i=1}^{m} Z_{1i}^{a_i} (1 - Z_{1i})^{b_i} \prod_{j=1}^{m'} Z_{2j}^{c_j} \tag{1}$$

for some constants $K$, $m$, $m'$, $a_i$, $b_i$ and $c_j$ when the associated null hypotheses are true, where $Z_{ij}$ are independently distributed beta variates", and where we may have $m = 0$ or $m' = 0$ for some l.r.t. statistics, in which case the corresponding product is taken as equal to 1. This is exactly the case for all the l.r.t. statistics we address in this paper.

However, since it is more useful for our ultimate purpose of developing near-exact distributions, we will show that for all four l.r.t. statistics addressed in this paper, as well as for all other l.r.t. statistics built as products of these statistics, their exact distribution, when the associated null hypothesis is true, may be written as the distribution of

$$\prod_{j=1}^{m}\prod_{k=1}^{m'}\left(Y_{jk}\right)^{c} \tag{2}$$

for some $m, m' \in \mathbb{N}$, and $c > 0$, in general a simple function of the sample size, where the $Y_{jk}$'s are independently distributed Beta r.v.'s. In (2) we may have either $m = 1$ or $m' = 1$. Although the change from (1) above may seem slight, it is an important one given that the existence of the terms $(1 - Z_{1i})$ in (1) cause some trouble when trying to obtain alternative representations for the exact distribution of the corresponding l.r.t. statistic, as we will need to do. Representations in the form of (2) will be much convenient for that goal, which ultimately will lead us to be able to develop a uniform strategy leading to the development of near-exact distributions with a common structure for all l.r.t. statistics addressed in this paper.

By working directly with the c.f. of the negative logarithm of the l.r.t. statistic it is possible to establish the following results.

(i) The l.r.t. statistic to test, under the normality assumption, the null hypothesis

$$\begin{aligned} H_{01} : \Sigma &= diag(\Sigma_{11}, \dots, \Sigma_{kk}, \dots, \Sigma_{mm}) \\ &\Longleftrightarrow \Sigma_{ij} = 0_{p_i \times p_j}\,, \quad i \neq j;\ i, j \in \{1, \dots, m\}\,, \end{aligned} \tag{3}$$

that is, the null hypothesis of independence of $m$ sets of variables, the $k$-th of which has $p_k$ variables ($k = 1, \dots, m$), based on a sample of size $n$, is

$$\Lambda_1 = \left(\frac{|A|}{\prod_{k=1}^{m}|A_{kk}|}\right)^{n/2}, \tag{4}$$

where $A$ is the m.l.e. (maximum likelihood estimator) of $\Sigma$ and $A_{kk}$ is the m.l.e. of $\Sigma_{kk}$ ($k = 1, \dots, m$) (for details see Appendix A.1). For $\Lambda_1$ we may write (see Anderson (2003, Theorems 9.3.2 and 9.3.3) and Coelho (1992, 1998))

$$\Lambda_1 \stackrel{d}{=} \prod_{k=1}^{m-1}\prod_{j=1}^{p_k}\left(Y_{jk}\right)^{n/2}, \qquad \text{with} \qquad Y_{jk} \sim Beta\left(\frac{n - q_k - j}{2}, \frac{q_k}{2}\right) \tag{5}$$

where the $Y_{jk}$ are independent, $q_k = p_{k+1} + \cdots + p_m$, and "$\stackrel{d}{=}$" means "equal in distribution".

(ii) The l.r.t. statistic to test the null hypothesis

$$H_{02} : \underline{\mu}_1 = \cdots = \underline{\mu}_q\,, \tag{6}$$

based on $q$ independent samples from $q$ multivariate Normal distributions $N_p(\underline{\mu}_j, \Sigma)$ $(j = 1, \ldots, q)$, the sample from the $j$-th population being of size $n_j$, may be written as

$$\Lambda_2 = \left(\frac{|A|}{|A+B|}\right)^{n/2}, \tag{7}$$

where $n = n_1 + \cdots + n_q$, and $A$ and $B$ are two independent $p \times p$ Wishart matrices, both with parameter matrix $\Sigma$ and respectively $n-q$ and $q-1$ degrees of freedom, and both central, under $H_{02}$ in (6) (see Kshirsagar (1972, Chap. 10, Sect. 10.1)).

The distribution of $\Lambda_2$ under $H_{02}$ in (6) is thus the same as the distribution of $\Lambda_1$ in (4) for $m = 2$, $p_1 = p$ and $p_2 = q-1$. Therefore we may write

$$\Lambda_2 \stackrel{d}{=} \prod_{j=1}^{p} (Y_j)^{n/2} \quad \text{with} \quad Y_j \sim Beta\left(\frac{n-q+1-j}{2}, \frac{q-1}{2}\right), \tag{8}$$

where the $Y_j$ are independent.

(iii) If we assume that

$$\underline{X} \sim N_p(\underline{\mu}, \Sigma)$$

and we want to test the null hypothesis of sphericity,

$$H_{03} : \Sigma = \sigma^2 I_p \quad (\sigma^2 \text{ unspecified}), \tag{9}$$

based on a sample of size $n$ from $\underline{X}$, the l.r.t. statistic is (Mauchly, 1940; Anderson, 2003)

$$\Lambda_3 = \left(\frac{|A|}{\left(tr\,\frac{1}{p}\,A\right)^p}\right)^{n/2}, \tag{10}$$

where $A$ is the m.l.e. of $\Sigma$.

It is then interesting and useful to see that it is possible to establish a similar result to the one obtained for the statistics in (i) and (ii) for the exact distribution of $\Lambda_3$. Indeed, from (A.6) in Appendix A we may write

$$\Lambda_3 \stackrel{d}{=} \prod_{j=2}^{p} (Y_j)^{n/2} \quad \text{with} \quad Y_j \sim Beta\left(\frac{n-j}{2}, \frac{j-1}{p} + \frac{j-1}{2}\right), \tag{11}$$

where the $Y_j$ are independent.

(iv) If we assume that

$$\underline{X}_k \sim N_p(\underline{\mu}_k, \Sigma_k)\,, \qquad k = 1, \ldots, q\,,$$

and we want to test the null hypothesis

$$H_{04} : \Sigma_1 = \cdots = \Sigma_q\,, \tag{12}$$

that is, the null hypothesis of equality of the $q$ variance-covariance matrices, we may write the l.r.t. statistic to test $H_{04}$ in (12), for independent samples of size $n$ from each $\underline{X}_k$ $(k = 1, \ldots, q)$, as

$$\Lambda_4 = \left( q^{pq} \frac{\prod_{k=1}^{q} |A_k|}{|A|^q} \right)^{n/2}, \tag{13}$$

where $A_k$ is the m.l.e. of $\Sigma_k$ $(k = 1, \ldots, q)$, and $A = A_1 + \cdots + A_q$.

It is then interesting and helpful to see that, similarly to the results in (i), (ii) and (iii), we may write, from (A.8) in Appendix A,

$$\Lambda_4 \stackrel{d}{=} \prod_{j=1}^{p} \prod_{k=1}^{q} \left(Y_{jk}\right)^{n/2} \;\text{ with }\; Y_{jk} \sim Beta\left( \frac{n-j}{2}, \frac{j(q-1)+2k-1-q}{2q} \right), \tag{14}$$

where the $Y_{jk}$ are independent.

## 2.2 Bridging to the Near-Exact Distributions

From the alternative expressions for the c.f.'s of the negative logarithms of the l.r.t. statistics in Appendix A we may obtain alternative representations for the exact distribution of these statistics, which will be the basis for the development of a uniform strategy and analogous near-exact distributions for these statistics.

(i) From the second expression in (A.1) in Appendix A we see that we may write for $\Lambda_1$ in (4), the l.r.t statistic to test the independence of $m$ sets of variables, with an overall number of variables equal to $p$,

$$\Lambda_1 \stackrel{d}{=} \left\{ \prod_{j=3}^{p} e^{Z_j} \right\} \left\{ \prod_{j=1}^{k^*} (Y_j)^{n/2} \right\}, \tag{15}$$

with

$$Z_j \sim \Gamma\left(r_j, \frac{n-j}{n}\right) \quad \text{and} \quad Y_j \sim Beta\left(\frac{n-2}{2}, \frac{1}{2}\right),$$

where all $Z_j$ and $Y_j$ are independent, with $r_j$ given by (A.2)–(A.3) and $k^* = \lfloor \ell/2 \rfloor$, where $\ell$ is the number of sets of variables with an odd number of variables. See Appendix A.1 for details and Appendix B for the notation on the Gamma distribution.

This result also shows that if $\ell$ is at most 1, the exact distribution of $W_1 = -\log \Lambda_1$ is a GIG (Generalized Integer Gamma) distribution of depth $p-2$ with shape parameters $r_j$ and rate parameters $\frac{n-j}{n}$ $(j = 3, \ldots, p)$, since in this case the second product in (15) vanishes.

(ii) Concerning the l.r.t. statistic used to test the equality of $q$ $p$-variate mean vectors, $\Lambda_2$ in (7), we may write, from Appendix A.2,

$$\Lambda_2 \stackrel{d}{=} \left\{ \prod_{j=3}^{p+q-1} e^{Z_j} \right\} \times \left(Y^{n/2}\right)^{k^*} \tag{16}$$

with

$$Z_j \sim \Gamma\left(r_j, \frac{n-j}{n}\right) \quad \text{and} \quad Y \sim Beta\left(\frac{n-2}{2}, \frac{1}{2}\right),$$

where $k^*$ is given by (A.4), $Z_j$ $(j = 3, \ldots, p+q-1)$ and $Y$ are a set of $p+q-2$ independent r.v.'s, with $r_j$ given by (A.2)–(A.3) for $m = 2$, $p_1 = p$ and $p_2 = q-1$. See Appendix A.2 for details.

This result also shows that in case $p$ is even or $q$ is odd, the exact distribution of $W_2 = -\log \Lambda_2$ is a GIG distribution of depth $p+q-3$ with shape parameters $r_j$ and rate parameters $\frac{n-j}{n}$ $(j = 3, \ldots, p+q-1)$, since in this case the r.v. $Y$ in (16) vanishes.

(iii) If we consider the l.r.t. statistic $\Lambda_3$ in (10) to test sphericity, from expression (A.6) in Appendix A.3, we may write

$$\Lambda_3 \stackrel{d}{=} \left\{ \prod_{j=2}^{p} e^{Z_j} \right\} \left\{ \prod_{j=1}^{p-k^*} (Y_j)^{n/2} \right\} \left\{ \prod_{j=p-k^*+1}^{p} \left(Y_j^*\right)^{n/2} \right\}, \tag{17}$$

where

$$Z_j \sim \Gamma\left(r_j, \frac{n-j}{n}\right),$$
$$Y_j \sim Beta\left(\frac{n-1}{2}, \frac{j-1}{p}\right) \quad \text{and}$$

for $r_j$ given by (A.7), $k^* = \lfloor p/2 \rfloor$ and where all the $Z_j$, $Y_j$ and $Y_j^*$ r.v.'s are independent.

(iv) Finally, for the l.r.t. statistic $\Lambda_4$ in (13), used to test the equality of $q$ variance-covariance matrices, from expression (A.9) in Appendix A.4, we may write

$$\Lambda_4 \stackrel{d}{=} \left\{ \prod_{j=2}^{p} e^{Z_j} \right\} \left\{ \prod_{j=1}^{\lfloor p/2 \rfloor} \prod_{k=1}^{q} (Y_{jk})^n \right\} \left\{ \prod_{k=1}^{q} (Y_k^*)^{n/2} \right\}^{p \perp\!\!\!\perp 2}, \tag{18}$$

where $\perp\!\!\!\perp$ is defined in (A.5),

$$Z_j \sim \Gamma\left(r_j, \frac{n-j}{n}\right), \quad Y_{jk} \sim Beta\left(a_j + b_{jk}^*, b_{jk} - b_{jk}^*\right)$$

and

$$Y_k^* \sim Beta(a_p + b_{pk}^*, b_{pk} - b_{pk}^*)$$

for $r_j$ given by (A.12) through (A.16), $a_j$, $b_{jk}$, $a_p$, $a_{pk}$, $b_{jk}^*$ and $b_{pk}^*$ are given by (A.10) and (A.11) and where the r.v.'s $Z_j$, $Y_{jk}$ and $Y_k^*$ are all independent, and where the r.v.'s $Y_k^*$ are only involved if $p$ is odd.

## 3 Near-Exact Distributions for the l.r.t. Statistics

### *3.1 Preamble*

One question that may be asked is: what do near-exact approximations bring new to the distribution approximation techniques? We may systemize the answer in the following points:

- they are built using a completely new technique that can be used for all test statistics whose c.f., or the one of its logarithm, may be factorized;
- they are much more accurate, that is, much closer to the exact distribution, than the usual asymptotic approximations;
- they have good asymptotic properties not only in terms of the sample size but also when other parameters are considered, such as the number of variables or matrices involved;
- they may be implemented using the computational resources available nowadays, and therefore they may be used for a large number of practical problems;
- when the null hypothesis of the test can be decomposed into a sequence of two or more null hypotheses, near-exact approximations for the l.r.t. statistic may be developed from the factorization induced on the characteristic function of the logarithm of the l.r.t. statistic — an example of this advantage will be given in

Sect. 4 for the l.r.t statistic to test the equality of several $p$-multivariate Normal populations.

The near-exact distributions are developed using a technique that combines the factorization of the c.f. of the test statistic or of its logarithm with a matching moments technique used to build asymptotic approximations for a convenient smaller factor of that c.f. It is true that these near-exact distributions are a little more elaborate than most asymptotic approximations available and that its use usually requires the help of a computer. However, they are far more manageable than the exact distributions, which, except in extremely simple cases, are not computable not even using modern symbolic softwares and powerful computers. Moreover, we would like to raise one question: is it possible to use most of the already known asymptotic and saddle-point approximations without the help of a computer?

Based on the results provided in Sect. 2.2, we will now develop a general near-exact approximation technique for the most common l.r.t statistics used in multivariate statistics.

Like the asymptotic distributions, near-exact distributions for a given statistic may usually be obtained in several different forms. The near-exact distributions we will focus our attention on in this paper are near-exact distributions that are derived from a factorization of the exact c.f. of the logarithm of the statistic under study, $W_k = -\log \Lambda_k$ (where $k \in \{1, \ldots, 4\}$ denotes the l.r.t. statistic in subsection $k$ of Sect. 2) of the form

$$\Phi_{W_k}(t) = \Phi_{1,W_k}(t)\, \Phi_{2,W_k}(t)\,, \tag{19}$$

where $\Phi_{1,W_k}(t)$ is the part of $\Phi_{W_k}(t)$ that will be left unchanged, and $\Phi_{2,W_k}(t)$ is the part of $\Phi_{W_k}(t)$ that will be replaced by $\Phi^*_{2,W_k}(t)$, giving rise to

$$\Phi^*_{W_k}(t) = \Phi_{1,W_k}(t)\, \Phi^*_{2,W_k}(t)\,, \tag{20}$$

which is intended to be asymptotic in terms of all parameters in the distribution of $W_k$, in the sense that

$$\lim_{\gamma \to \infty} \int_{-\infty}^{\infty} \left| \frac{\Phi_{W_k}(t) - \Phi^*_{W_k}(t)}{t} \right| dt = 0\,,$$

where $\gamma$ represents any parameter in the distribution of $W_k$. In (19), $\Phi_{1,W_k}(t)$ is the c.f. of a sum of independent Logbeta random variables (multiplied by $n/2$, for samples of size $n$) which may also be expressed as a c.f. of the sum of independent Gamma random variables, all with integer shape parameters, while $\Phi_{2,W_k}(t)$ is the c.f. of another sum of independent Logbeta random variables (also multiplied by $n/2$) which cannot be expressed as a sum of independent Gamma random variables with integer shape parameters.

In (20), $\Phi_{1,W_k}(t)$ will be written in the form of the c.f. of the sum of a given number of independent Gamma r.v.'s, all with integer shape parameters, while

$\Phi^*_{2,W_k}(t)$ will be either the c.f. of a single Gamma distribution or of a mixture of two or three Gamma distributions with the same rate parameter, depending on the number of exact moments we want to match. That is,

$$\Phi^*_{2,W_k}(t) = \sum_{\nu=1}^{h/2} \theta_\nu \, \lambda^{s_\nu} \, (\lambda - \mathrm{i}t)^{-s_\nu} \tag{21}$$

with

$$\theta_{h/2} = 1 - \sum_{\nu=1}^{h/2-1} \theta_\nu \,, \tag{22}$$

and

$$\left. \frac{d^j}{dt^j} \, \Phi^*_{2,W_k}(t) \right|_{t=0} = \left. \frac{d^j}{dt^j} \, \Phi_{2,W_k}(t) \right|_{t=0} , \quad j = 1, \ldots, h \,, \tag{23}$$

for $h = 2, 4$ or $6$, according to the case of $\Phi^*_{2,W_k}(t)$ being the c.f. of a single Gamma distribution, or the c.f. of a mixture of 2 or 3 Gamma distributions with the same rate parameter.

If $\Phi^*_{2,W_k}(t)$ is the c.f. of a single Gamma distribution that equates the two first derivatives of $\Phi_{2,W_k}(t)$ at $t = 0$, there is a simple analytical solution for the problem of equating moments, with the rate and shape parameters of $\Phi^*_{2,W_k}(t)$ being given by

$$\lambda = \frac{m_1}{m_2 - m_1^2} \quad \text{and} \quad s_1 = \frac{m_1}{m_2 - m_1^2} \,,$$

where

$$m_1 = \frac{1}{i} \left. \frac{d}{dt} \, \Phi_{2,W_k}(t) \right|_{t=0} \quad \text{and} \quad m_2 = - \left. \frac{d^2}{dt^2} \, \Phi_{2,W_k}(t) \right|_{t=0} .$$

But, if $\Phi^*_{2,W_k}(t)$ is the c.f. of a mixture of two Gamma distributions it is possible to prove (through rather long and tedious calculations) that there is always one unique analytic real solution, or rather, a pair of conjugate real solutions, with the values for the two shape parameters and corresponding weights interchanged, and if $\Phi^*_{2,W_k}(t)$ is the c.f. of a mixture of three Gamma distributions, it is believed that there is also always one only real solution with all positive parameters, or rather, a six-tuple of conjugate solutions, although this is not easy to prove analytically. In any case, for situations in which $\Phi^*_{2,W_k}(t)$ is the c.f. of a mixture of 2 or 3 Gamma distributions (respectively for $h = 4$ or $h = 6$) we advocate the numerical solution of the system of equations (23).

As already remarked in Marques and Coelho (2008), the replacement of $\Phi_{2,W_k}(t)$ by $\Phi^*_{2,W_k}(t)$, that is, the replacement of a sum of independent Logbeta random variables (multiplied by a constant) by a single Gamma distribution or a mixture of 2 or 3 Gamma distributions, matching the first 2, 4 or 6 exact moments can be expected to yield a good approximation, since, as it is shown in Coelho et al. (2006), a single Logbeta distribution may be represented either in the form of an infinite mixture of Exponential or GIG distributions, and thus a sum of independent Logbeta random variables may thus be represented in the form of an infinite mixture of sums of Exponential or GIG distributions, which are themselves GIG distributions, while the GIG distribution may itself be viewed as a mixture of Gamma distributions (Coelho, 2007).

$\Phi_{1,W_k}(t)$ in (20) will have, for samples of size $n$, the form

$$\Phi_{1,W_k}(t) = \prod_{j=2}^{p} \left(\frac{n-j}{n}\right)^{r_{j,k}} \left(\frac{n-j}{n} - \mathrm{i}t\right)^{-r_{j,k}}, \tag{24}$$

where $p$ is the overall number of variables involved, and the $r_{j,k}$ are the integer shape parameters of the Gamma r.v.'s involved and will be obtained under specific forms, according to the l.r.t. statistic under study. This permits writing the near-exact c.f. of the logarithm of the l.r.t. statistics under consideration in the form

$$\Phi^*_{2,W_k}(t) \times \prod_{j=2}^{p} \left(\frac{n-j}{n}\right)^{r_{j,k}} \left(\frac{n-j}{n} - \mathrm{i}t\right)^{-r_{j,k}}, \tag{25}$$

where $\Phi^*_{2,W_k}(t)$ is either the c.f. of a Gamma distribution or the c.f. of a mixture of 2 or 3 Gamma distributions. The near-exact distributions obtained in this way thus correspond to a GNIG (Generalized Near-Integer Gamma) distribution of depth at most $p+1$ (see Appendix B) or a mixture of two or three GNIG distributions of depth at most $p+1$, which have very manageable expressions, allowing for an easy computation of very accurate near-exact quantiles.

## 3.2 *Expressions for the Near-Exact p.d.f.'s and c.d.f.'s of the l.r.t. Statistics*

Expressions for the near-exact p.d.f.'s and c.d.f.'s of the l.r.t. statistics $\Lambda_k$ in Sect. 2 (where $k$ denotes the Subsect. $k$ of Sect. 2 where $\Lambda_k$ is studied) may thus be obtained, using the notation in Appendix B, under the form

$$f_{\Lambda_k}(\ell) = \sum_{\nu=1}^{h/2} \theta_\nu \, f^{GNIG}\left(-\log \ell | r_2, \dots, r_{f_k}, s_\nu; \frac{n-2}{n}, \dots, \frac{n-f_k}{n}, \lambda; f_k\right) \frac{1}{\ell}$$

and

$$F_{\Lambda_k}(\ell)=1-\sum_{\nu=1}^{h/2}\theta_\nu\, F^{GNIG}\left(-\log\ell|r_2,\ldots,r_{f_k},s_\nu;\frac{n-2}{n},\ldots,\frac{n-f_k}{n},\lambda;f_k\right),$$

where the weights $\theta_\nu$ are subject to the relation (22), and where

$$f_k=\begin{cases} p\,, & k=1,3,4\\ p+q-1\,, & k=2\,,\end{cases} \tag{26}$$

and

$$\begin{array}{ll} r_2,\ldots,r_p & \text{given by (A.2)–(A.3) for } \Lambda_1,\\ r_2,\ldots,r_{p+q-1} & \text{given by (A.2)–(A.3) for } \Lambda_2,\\ r_2,\ldots,r_p & \text{given by (A.7) for } \Lambda_3,\\ r_2,\ldots,r_p & \text{given by (A.12)–(A.16) for } \Lambda_4. \end{array} \tag{27}$$

We should briefly note here that, as referred to in Sect. 2.2, for $k=1$ when there is at most one set of variables with an odd number of variables and for $k=2$ when $p$ is even or $q$ is odd, we have the exact distribution of $\Lambda_k$ available in the form of a GIG distribution (Coelho, 1998), with p.d.f. and c.d.f. respectively given by

$$f_{\Lambda_k}(\ell)=f^{GIG}\left(-\log\,\ell|r_2,\ldots,r_{f_k};\frac{n-2}{n},\ldots,\frac{n-f_k}{n};f_k-1\right)\frac{1}{\ell}$$

and

$$F_{\Lambda_k}(\ell)=1-F^{GIG}\left(-\log\,\ell|r_2,\ldots,r_{f_k};\frac{n-2}{n},\ldots,\frac{n-f_k}{n};f_k-1\right)$$

with $f_k$ given by (26) above and $r_2,\ldots,r_{f_k}$ given by (27).

In order to make these near-exact distributions widely available and easy to use, modules for the computation of the near-exact c.f., p.d.f., c.d.f., quantiles and $p$-values based on this unified approach are provided by the authors on the web-page https://sites.google.com/site/nearexactdistributions/.

## 4 The l.r.t. Statistic to Test the Equality of Several $p$-Multivariate Normal Distributions

Let us suppose we wish to test whether $q$ multivariate Normal distributions $N_p(\underline{\mu}_j, \Sigma_j)$ $(j = 1, \ldots, q)$ are equal, that is, we want to test the null hypothesis

$$H_{05} : \underline{\mu}_1 = \cdots = \underline{\mu}_q,\ \Sigma_1 = \cdots = \Sigma_q\,, \tag{28}$$

based on $q$ independent samples, each of size $n_j$ $(j = 1, \ldots, q)$.

We will decompose, as it is indeed usual, the above null hypothesis as

$$H_{05} :\ H_{02|04} \,^{\mathrm{o}}\, H_{04}$$

where "o" is to be read as "after" or "composed with", and where

$$\begin{aligned} &H_{04} : \Sigma_1 = \cdots = \Sigma_q \\ &H_{02|04} : \underline{\mu}_1 = \cdots = \underline{\mu}_q \ \text{ given that } \ \Sigma_1 = \cdots = \Sigma_q (= \Sigma)\,, \end{aligned} \tag{29}$$

and use the induced factorization on the test statistic to obtain near-exact distributions for the overall test statistic. Now the overall l.r.t. statistic to test $H_{05}$ in (28) may be written as

$$\Lambda_5 = \Lambda_2\,\Lambda_4,$$

where $\Lambda_4$ is the l.r.t. statistic treated in Appendix A.4, used here to test $H_{04}$ in (29) and $\Lambda_2$ is the l.r.t. statistic in Appendix A.2, used to test $H_{02|04}$ in (29), and where $\Lambda_2$ and $\Lambda_4$ are independent under $H_{05}$ in (28).

If we take $n_j = n$ for all $j = 1, \ldots, q$, we may then write, for $\alpha \in \mathbb{N}$, the exact c.f. of $W_5 = -\log \Lambda_5$ as

$$\Phi_{W_5}(t) = \underbrace{\left\{ \prod_{\substack{j=2 \\ j\neq \alpha q}}^{p+q-1} \left(\frac{n-j/q}{n}\right)^{r_j} \left(\frac{n-j/q}{n} - \mathrm{i}t\right)^{-r_j} \right\} \times \left\{ \prod_{j=2}^{p} \left(\frac{n-j}{n}\right)^{u_j} \left(\frac{n-j}{n} - \mathrm{i}t\right)^{-u_j} \right\}}_{\Phi_{1,W_5}(t)}$$

$$\times \left\{ \frac{\Gamma\left(\frac{nq-1}{2}\right) \Gamma\left(\frac{nq-2}{2} - \frac{nq}{2}\mathrm{i}t\right)}{\Gamma\left(\frac{nq-1}{2} - \frac{nq}{2}\mathrm{i}t\right) \Gamma\left(\frac{nq-2}{2}\right)} \right\}^{k^*}$$

$$\underbrace{\times \left\{ \prod_{k=1}^{q} \frac{\Gamma(a_p + b_{pk})}{\Gamma(a_p + b^*_{pk})} \frac{\Gamma\left(a_p + b^*_{pk} - \frac{n}{2}\mathrm{i}t\right)}{\Gamma\left(a_p + b_{pk} - \frac{n}{2}\mathrm{i}t\right)} \right\}^{p \perp\!\!\!\perp 2} \times \left\{ \prod_{j=1}^{\lfloor p/2 \rfloor} \prod_{k=1}^{q} \frac{\Gamma\left(a_j + b_{jk}\right) \Gamma\left(a_j + b^*_{jk} - n\mathrm{i}t\right)}{\Gamma\left(a_j + b^*_{jk}\right) \Gamma\left(a_j + b_{jk} - n\mathrm{i}t\right)} \right\}}_{\Phi_{2,W_5}(t)}$$

where $k^*$ is given by (A.4), $a_j$, $a_p$, $b_{jk}$, $b^*_{jk}$, $b_{pk}$ and $b^*_{pk}$ are given by (A.10) and (A.11), the $r_j$ are given by (A.2) and (A.3), for $m = 2$, with $p_1 = p$ and $p_2 = q - 1$, and

$$u_j = \begin{cases} r^*_j\,, & j \neq \alpha q \\ r^*_j + r_{\alpha q}\,, & j = \alpha q\,,\ \alpha = 1, \ldots, \left\lfloor \frac{p}{q} \right\rfloor, \end{cases}$$

where $r^*_j$ are given by (A.13) and the $r_{\alpha q}$ are given by (A.2) and (A.3), for $m = 2$, with $p_1 = p$ and $p_2 = q - 1$.

Thus, following the exposition in the previous section, near-exact distributions for $\Lambda_5$ may be obtained with p.d.f.'s and c.d.f.'s corresponding to mixtures of GNIG distributions of depth $2p + q - 3 - \left\lfloor \frac{p}{q} \right\rfloor$, respectively given by

$$f_{\Lambda_5}(\ell) = \sum_{\nu=1}^{h/2} \theta_\nu\, f^{GNIG}\Bigg( -\log \ell \,|\, \underbrace{r_3, \ldots, r_j, \ldots, r_{p+q-1}}_{j \neq \alpha q,\, \alpha \in \mathbb{N}}, u_2, \ldots, u_p, s_\nu;$$
$$\underbrace{\frac{n-3/q}{n}, \ldots, \frac{n-j/q}{n}, \ldots, \frac{n-(p+q-1)/q}{n}}_{j \neq \alpha q,\, \alpha \in \mathbb{N}}, \frac{n-2}{n}, \ldots, \frac{n-p}{n}, \lambda;$$
$$2p + q - 3 - \left\lfloor \frac{p}{q} \right\rfloor \Bigg) \frac{1}{\ell}$$

and

$$F_{\Lambda_5}(\ell) = 1 - \sum_{\nu=1}^{h/2} \theta_\nu \, F^{GNIG}\Big( -\log \ell \,|\, \underbrace{r_3, \ldots, r_j, \ldots, r_{p+q-1}}_{j\neq \alpha q,\, \alpha\in\mathbb{N}}, u_2, \ldots, u_p, s_\nu; \underbrace{\frac{n-3/q}{n}, \ldots, \frac{n-j/q}{n}, \ldots, \frac{n-(p+q-1)/q}{n}}_{j\neq \alpha q,\, \alpha\in\mathbb{N}}, \frac{n-2}{n}, \ldots, \frac{n-p}{n}, \lambda; 2p+q-3-\left\lfloor \frac{p}{q} \right\rfloor \Big).$$

## 5 Numerical Studies

According to the discussion in Sect. 3.1, we will use the measure

$$\Delta = \frac{1}{2\pi} \int_{-\infty}^{+\infty} \left| \frac{\Phi_{W_k}(t) - \Phi^*_{W_k}(t)}{t} \right| dt\,,$$

where $\Phi_{W_k}(t)$ and $\Phi^*_{W_k}(t)$ represent respectively the exact and the near-exact c.f. of the r.v.'s $W_k = -\log \Lambda_k$, to assess the closeness of the near-exact distributions to the exact distribution. The measure $\Delta$ is related to the Berry-Esseen upper bound (Berry, 1941; Essen, 1945; Loève, 1977, Chap. VI, Sect. 21; Hwang, 1998), with

$$\Delta \geq \max_{w\in S} \left| F_{W_k}(w) - F^*_{W_k}(w) \right|\,, \quad \text{and} \quad \Delta \underset{\gamma\to\infty}{\longrightarrow} 0 \Longleftrightarrow \lim_{\gamma\to\infty} \frac{\Phi^*_{W_k}(t)}{\Phi_{W_k}(t)} = 1\,,$$

where $F_{W_k}(w)$ and $F^*_{W_k}(w)$ are the exact and near-exact c.d.f. of $W_k$, and $S$ represents the support of $W_k$.

Indeed, given the way near-exact distributions are built, when we combine two or more tests to obtain a new test, we will even gain in precision, whenever we have the exact distribution available for one of the elementary test statistics the overall test statistic is factorized into. For example, if we have an l.r.t. statistic for the overall test whose logarithm may be written as $W = W_k + W_{k'}$, where $W_k$ and $W_{k'}$ are independent, we will only approximate the c.f. of $W_{k'}$ by its near-exact c.f., $\Phi^*_{W_{k'}}(t)$, obtaining

$$\Phi^*_W(t) = \Phi_{W_k}(t)\, \Phi^*_{W_{k'}}(t) \Phi_{W_k}(t)\, \Phi_{W_{k'}}(t) = \Phi_W(t)\,,$$

as a near-exact approximation of $\Phi_W(t)$, we have

$$\begin{aligned}\Delta_W &= \frac{1}{2\pi}\int_{-\infty}^{+\infty}\left|\frac{\Phi_W(t)-\Phi_W^*(t)}{t}\right| dt\\ &= \frac{1}{2\pi}\int_{-\infty}^{+\infty}\left|\frac{\Phi_{W_k}(t)\Phi_{W_{k'}}(t)-\Phi_{W_k}(t)\Phi_{W_{k'}}^*(t)}{t}\right| dt\\ &= \frac{1}{2\pi}\int_{-\infty}^{+\infty}\underbrace{\left|\Phi_{W_k}(t)\right|}_{\leq 1}\left|\frac{\Phi_{W_{k'}}(t)-\Phi_{W_{k'}}^*(t)}{t}\right| dt\\ &\leq \frac{1}{2\pi}\int_{-\infty}^{+\infty}\left|\frac{\Phi_{W_{k'}}(t)-\Phi_{W_{k'}}^*(t)}{t}\right| dt \;=\; \Delta_{W_{k'}}\,.\end{aligned} \tag{30}$$

In the following tables we will use $W_5$ to denote the logarithm of the l.r.t. statistic to test the equality of several (more precisely $q$) $p$-multivariate Normal distributions, that is, the null hypothesis $H_{05}$ in (28), and $W_2$ and $W_4$ to denote the logarithm of the l.r.t. statistics to test the null hypotheses $H_{02|04}$ and $H_{04}$ in (29), respectively.

We will denote by GNIG, M2GNIG and M3GNIG the near-exact distributions corresponding to a GNIG or a mixture of two or three GNIG distributions, respectively.

Tables 1 and 2 refer to cases where we do not have the exact distribution available in a manageable form for the statistic $W_2$. From the results in these tables we may see how the values of $\Delta$ for $W_5$ are in most cases only slightly higher than the values of $\Delta$ for $W_4$, while in the other cases, that is, for $p=5$ and $q=12$ the values of $\Delta$ are actually slightly smaller for $W_5$ than for $W_4$. This also occurs when $p=11$ and $q=12$ for the larger sample size, for all near-exact distributions, and for $p=11$ and $q=4$, only for the M3GNIG distribution.

Tables 3 and 4 refer to cases where we have the exact distribution of $W_2$ available under the form of a GIG distribution. In these cases, the way the near-

**Table 1** Values of $\Delta$ for $p=5$ and $q=4, 12$

| | $p=5$ | | | | | |
|---|---|---|---|---|---|---|
| | $q=4$ | | | $q=12$ | | |
| | $W_2$ | $W_4$ | $W_5$ | $W_2$ | $W_4$ | $W_5$ |
| $n=p+2$ | | | | | | |
| GNIG | $9.8\times10^{-7}$ | $4.2\times10^{-5}$ | $4.7\times10^{-5}$ | $1.4\times10^{-8}$ | $1.6\times10^{-5}$ | $1.5\times10^{-5}$ |
| M2GNIG | $3.4\times10^{-9}$ | $1.8\times10^{-7}$ | $2.5\times10^{-7}$ | $4.5\times10^{-12}$ | $3.7\times10^{-8}$ | $3.2\times10^{-8}$ |
| M3GNIG | $6.4\times10^{-12}$ | $1.1\times10^{-9}$ | $1.8\times10^{-9}$ | $1.6\times10^{-15}$ | $1.4\times10^{-10}$ | $1.1\times10^{-10}$ |
| $n=p+20$ | | | | | | |
| GNIG | $8.5\times10^{-8}$ | $4.0\times10^{-6}$ | $4.4\times10^{-6}$ | $1.2\times10^{-9}$ | $1.5\times10^{-6}$ | $1.2\times10^{-6}$ |
| M2GNIG | $9.6\times10^{-11}$ | $1.5\times10^{-9}$ | $1.7\times10^{-9}$ | $1.3\times10^{-13}$ | $6.2\times10^{-10}$ | $3.9\times10^{-10}$ |
| M3GNIG | $1.2\times10^{-13}$ | $8.3\times10^{-13}$ | $2.8\times10^{-12}$ | $1.7\times10^{-17}$ | $5.6\times10^{-13}$ | $2.5\times10^{-13}$ |

**Table 2** Values of $\Delta$ for $p = 11$ and $q = 4, 12$

| | $p = 11$ | | | | | |
|---|---|---|---|---|---|---|
| | $q = 4$ | | | $q = 12$ | | |
| | $W_2$ | $W_4$ | $W_5$ | $W_2$ | $W_4$ | $W_5$ |
| $n = p + 2$ | | | | | | |
| GNIG | $7.3\times10^{-8}$ | $2.5\times10^{-6}$ | $2.7\times10^{-6}$ | $1.2\times10^{-9}$ | $2.6\times10^{-7}$ | $2.7\times10^{-7}$ |
| M2GNIG | $6.0\times10^{-11}$ | $1.9\times10^{-9}$ | $2.2\times10^{-9}$ | $1.1\times10^{-13}$ | $3.0\times10^{-11}$ | $3.2\times10^{-11}$ |
| M3GNIG | $4.7\times10^{-14}$ | $2.3\times10^{-12}$ | $2.7\times10^{-12}$ | $1.8\times10^{-18}$ | $4.5\times10^{-15}$ | $4.7\times10^{-15}$ |
| $n = p + 20$ | | | | | | |
| GNIG | $1.5\times10^{-8}$ | $1.1\times10^{-6}$ | $1.2\times10^{-6}$ | $2.4\times10^{-10}$ | $1.6\times10^{-7}$ | $1.5\times10^{-7}$ |
| M2GNIG | $6.1\times10^{-12}$ | $3.4\times10^{-10}$ | $3.4\times10^{-10}$ | $9.3\times10^{-15}$ | $9.6\times10^{-12}$ | $8.4\times10^{-12}$ |
| M3GNIG | $2.9\times10^{-15}$ | $1.8\times10^{-13}$ | $1.2\times10^{-13}$ | $4.5\times10^{-19}$ | $9.8\times10^{-16}$ | $5.8\times10^{-16}$ |

exact distributions for $W_5$ were built enables these distributions to exhibit smaller values of $\Delta$ than the corresponding near-exact distributions for $W_4$, as expected, from the result in (30).

Analyzing the overall set of tables we may see that the proposed near-exact distributions exhibit desirable asymptotic behavior for all of the statistics involved, in the sense that they converge to the exact distribution not only for increasing sample sizes but also for increasing values of $p$ (the number of variables involved) and $q$ (the number of Normal distributions involved for $W_5$ or covariance matrices for $W_4$ or mean vectors for $W_2$). Only for the cases of $p = 10$ and $q = 12$ or $p = 11$ and $q = 5$, it appears that the distribution M3GNIG for some reason exhibits unusually small values of $\Delta$ for the smaller sample sizes.

## 6 Conclusions

Although other authors have already shown or observed that all the l.r.t. statistics considered in this paper have the same distribution as the product of a given number of independent r.v.'s, which are either Beta r.v.'s or functions of Beta r.v.'s (see for example, Anderson (2003), Hsieh (1979)), we have shown that indeed all those statistics have the same distribution as the product of independent Beta r.v.'s with different parameters, according to the statistic considered.

On the other hand, we have also shown that not only can one obtain near-exact distributions for all the "basic" l.r.t. statistics used in Multivariate Analysis in a uniform manner, but also the same general strategy permits the development of near-exact distributions for more elaborate l.r.t. statistics, retaining a similar structure and formulation.

The development of near-exact distributions as manageable approximations that are very close to the exact distribution of these l.r.t. statistics seems most useful since

**Table 3** Values of $\Delta$ for $p = 4, 10$ and $q = 4, 12$

| | $p=4$ | | | | $p=10$ | | | |
|---|---|---|---|---|---|---|---|---|
| | $q=4$ | | $q=12$ | | $q=4$ | | $q=12$ | |
| | $W_4$ | $W_5$ | $W_4$ | $W_5$ | $W_4$ | $W_5$ | $W_4$ | $W_5$ |
| $n=p+2$ | | | | | | | | |
| GNIG | $1.2\times10^{-5}$ | $8.9\times10^{-6}$ | $8.7\times10^{-6}$ | $7.1\times10^{-6}$ | $1.6\times10^{-6}$ | $1.5\times10^{-6}$ | $1.3\times10^{-7}$ | $1.3\times10^{-7}$ |
| M2GNIG | $1.7\times10^{-8}$ | $1.1\times10^{-8}$ | $1.1\times10^{-8}$ | $8.0\times10^{-9}$ | $7.8\times10^{-10}$ | $6.9\times10^{-10}$ | $6.6\times10^{-12}$ | $5.9\times10^{-12}$ |
| M3GNIG | $3.8\times10^{-10}$ | $1.9\times10^{-10}$ | $2.5\times10^{-11}$ | $1.5\times10^{-11}$ | $5.2\times10^{-13}$ | $4.3\times10^{-13}$ | $6.9\times10^{-17}$ | $5.8\times10^{-17}$ |
| $n=p+20$ | | | | | | | | |
| GNIG | $1.3\times10^{-6}$ | $7.9\times10^{-7}$ | $7.5\times10^{-7}$ | $4.8\times10^{-7}$ | $7.2\times10^{-7}$ | $6.0\times10^{-7}$ | $8.3\times10^{-8}$ | $6.9\times10^{-8}$ |
| M2GNIG | $8.2\times10^{-10}$ | $3.8\times10^{-10}$ | $1.8\times10^{-10}$ | $8.6\times10^{-11}$ | $1.1\times10^{-10}$ | $8.1\times10^{-11}$ | $1.2\times10^{-12}$ | $8.6\times10^{-13}$ |
| M3GNIG | $1.1\times10^{-12}$ | $3.7\times10^{-13}$ | $1.3\times10^{-13}$ | $4.8\times10^{-14}$ | $2.0\times10^{-14}$ | $1.3\times10^{-14}$ | $3.4\times10^{-16}$ | $2.2\times10^{-16}$ |

**Table 4** Values of $\Delta$ for $p = 4, 10$ and $q = 4, 12$

| | $p=4$ | | | | $p=11$ | | | |
|---|---|---|---|---|---|---|---|---|
| | $q=4$ | | $q=12$ | | $q=5$ | | $q=11$ | |
| | $W_4$ | $W_5$ | $W_4$ | $W_5$ | $W_4$ | $W_5$ | $W_4$ | $W_5$ |
| $n=p+2$ | | | | | | | | |
| GNIG | $3.7\times10^{-5}$ | $3.1\times10^{-5}$ | $1.8\times10^{-5}$ | $1.5\times10^{-5}$ | $5.6\times10^{-7}$ | $5.3\times10^{-7}$ | $2.7\times10^{-7}$ | $2.6\times10^{-7}$ |
| M2GNIG | $1.5\times10^{-7}$ | $1.1\times10^{-7}$ | $4.4\times10^{-8}$ | $3.3\times10^{-8}$ | $8.7\times10^{-11}$ | $7.8\times10^{-11}$ | $3.3\times10^{-11}$ | $3.0\times10^{-11}$ |
| M3GNIG | $9.4\times10^{-10}$ | $6.0\times10^{-10}$ | $1.7\times10^{-10}$ | $1.2\times10^{-10}$ | $3.6\times10^{-15}$ | $3.1\times10^{-15}$ | $5.0\times10^{-15}$ | $4.3\times10^{-15}$ |
| $n=p+20$ | | | | | | | | |
| GNIG | $3.5\times10^{-6}$ | $2.4\times10^{-6}$ | $1.6\times10^{-6}$ | $1.1\times10^{-6}$ | $3.6\times10^{-7}$ | $3.0\times10^{-7}$ | $1.7\times10^{-7}$ | $1.4\times10^{-7}$ |
| M2GNIG | $1.7\times10^{-9}$ | $9.2\times10^{-10}$ | $7.3\times10^{-10}$ | $4.0\times10^{-10}$ | $3.3\times10^{-11}$ | $3.0\times10^{-11}$ | $1.0\times10^{-11}$ | $7.7\times10^{-12}$ |
| M3GNIG | $2.3\times10^{-12}$ | $9.5\times10^{-13}$ | $7.0\times10^{-13}$ | $3.0\times10^{-13}$ | $1.6\times10^{-14}$ | $1.1\times10^{-14}$ | $1.1\times10^{-15}$ | $7.1\times10^{-16}$ |

their exact distributions do not have manageable closed forms (see for example, Aslam and Rocke (2005)).

Two good features of the near-exact distributions are the facts that, as opposed to the usual asymptotic distributions, they show a very good fit even for small samples and their closeness to the exact distribution even improves when the dimension increases.

**Acknowledgments** This research was financially supported by the Portuguese Foundation for Science and Technology, through the Center for Mathematics and its Applications (CMA) from the New University of Lisbon (Financiamento Base 2009 ISFL-1-297 from FCT/MCTES/PT).

The authors thank the Associate Editor and the referees for many helpful comments and suggestions which markedly improved the readability of the paper.

# Appendix A: On the Characteristic Functions of the l.r.t. Statistics

## A.1 The l.r.t. Statistic to Test Independence Among m Sets of Variates

Assume that

$$\underline{X} = \left[\underline{X}_1', \ldots, \underline{X}_k', \ldots, \underline{X}_m'\right]' \sim N_p(\underline{\mu}, \Sigma),$$

where

$$\underline{\mu} = \left[\underline{\mu}_1', \ldots, \underline{\mu}_k', \ldots, \underline{\mu}_m'\right]' \quad \text{and} \quad \Sigma = \begin{bmatrix} \Sigma_{11} & \cdots & \Sigma_{1k} & \cdots & \Sigma_{1m} \\ \vdots & \ddots & \vdots & & \vdots \\ \Sigma_{k1} & \cdots & \Sigma_{kk} & \cdots & \Sigma_{km} \\ \vdots & & \vdots & \ddots & \vdots \\ \Sigma_{m1} & \cdots & \Sigma_{mk} & \cdots & \Sigma_{mm} \end{bmatrix},$$

with

$$\underline{X}_k \sim N_{p_k}(\underline{\mu}_k, \Sigma_{kk}) \quad (k = 1, \ldots, m), \qquad \text{and} \qquad p = \sum_{k=1}^{m} p_k .$$

Then, the l.r.t. statistic to test (3), based on a sample of size $n$ from $\underline{X}$, is given by (4).

According to Coelho (1992, 2004), we may write the exact c.f. (characteristic function) of $W_1 = -\log \Lambda_1$, under $H_{01}$ in (3), for $a_{jk} = \frac{n-q_k-j}{2}$, $b_k = \frac{q_k}{2}$ and $q_k = p_{k+1} + \cdots + p_m$, as

$$\begin{aligned}\Phi_{W_1}(t) &= \prod_{k=1}^{m-1}\prod_{j=1}^{p_k} \frac{\Gamma(a_{jk}+b_k)\,\Gamma\left(a_{jk}-\frac{n}{2}\mathrm{i}t\right)}{\Gamma(a_{jk})\,\Gamma\left(a_{jk}+b_k-\frac{n}{2}\mathrm{i}t\right)} \\ &= \underbrace{\left\{\prod_{j=2}^{p}\left(\frac{n-j}{n}\right)^{r_j}\left(\frac{n-j}{n}-\mathrm{i}t\right)^{-r_j}\right\}}_{\Phi_{1,W_1}(t)} \\ &\qquad \times \underbrace{\left(\frac{\Gamma\left(\frac{n-1}{2}\right)\Gamma\left(\frac{n-1}{2}-\frac{1}{2}-\frac{n}{2}\mathrm{i}t\right)}{\Gamma\left(\frac{n-1}{2}-\frac{1}{2}\right)\Gamma\left(\frac{n-1}{2}-\frac{n}{2}\mathrm{i}t\right)}\right)^{k^*}}_{\Phi_{2,W_1}(t)}\end{aligned} \tag{A.1}$$

where $k^* = \left\lfloor \frac{\ell}{2} \right\rfloor$, with $\ell$ denoting the number of $\underline{X}_k$'s with an odd number of variates, and

$$r_j = \begin{cases} 0\,, & j = 2 \\ h_j + (-1)^j k^*\,, & j = 3, 4 \\ r_{j-2} + h_{j-2}\,, & j = 5, \ldots, p\,, \end{cases} \tag{A.2}$$

where

$$h_j = (\#\text{ of } p_k\ (k = 1, \ldots, m) \geq j) - 1\,, \quad j = 1, \ldots, p-2\,. \tag{A.3}$$

The first expression in (A.1) is the c.f. of the sum of $\sum_{k=1}^{m-1} p_k$ independent Logbeta$(a_{jk}, b_k)$ r.v.'s multiplied by $n/2$, thus giving rise to (5), while the second expression in (A.1) is the c.f. of the sum of $p-2$ independent Gamma r.v.'s with rate parameters $\frac{n-j}{n}$ $(j = 3, \ldots, p)$ and shape parameters $r_j$ given by (A.2)–(A.3), plus $k^*$ i.i.d. Logbeta$\left(\frac{n-2}{2}, \frac{1}{2}\right)$ r.v.'s, multiplied by $n/2$, this way giving rise to (15).

## *A.2 The l.r.t. Statistic to Test the Equality of Several Mean Vectors*

Let us suppose we have $q$ independent samples from the $q$ multivariate Normal distributions $N_p(\underline{\mu}_j, \Sigma)$ $(j = 1, \ldots, q)$, the sample from the $j$-th population being of size $n_j$. Then the l.r.t. to test the null hypothesis in (6) may be written as in (7) (Kshirsagar, 1972, Ch. 10, sec. 10.1).

The distribution of $\Lambda_2$ under $H_{02}$ in (6) is thus the same as the distribution of $\Lambda_1$ in the previous section for $m = 2$, taking $p_1 = p$ and $p_2 = q - 1$. Thus, we have the c.f. of $W_2 = -\log \Lambda_2$ given by

$$\begin{aligned}\Phi_{W_2}(t) &= \prod_{j=1}^{p} \frac{\Gamma\left(\frac{n-j}{2}\right)\Gamma\left(\frac{n-q+1-j}{2} - \frac{n}{2}\mathrm{i}t\right)}{\Gamma\left(\frac{n-j}{2} - \frac{n}{2}\mathrm{i}t\right)\Gamma\left(\frac{n-q+1-j}{2}\right)} \\ &= \underbrace{\left\{\prod_{j=2}^{p+q-1}\left(\frac{n-j}{n}\right)^{r_j}\left(\frac{n-j}{n} - \mathrm{i}t\right)^{-r_j}\right\}}_{\Phi_{1,W_2}(t)} \\ &\qquad\times \underbrace{\left(\frac{\Gamma\left(\frac{n-1}{2}\right)\Gamma\left(\frac{n-2}{2} - \frac{n}{2}\mathrm{i}t\right)}{\Gamma\left(\frac{n-1}{2} - \frac{n}{2}\mathrm{i}t\right)\Gamma\left(\frac{n-2}{2}\right)}\right)^{k^*}}_{\Phi_{2,W_2}(t)}\end{aligned}$$

where

$$k^* = (p \perp\!\!\!\perp 2)((q-1) \perp\!\!\!\perp 2)\,, \tag{A.4}$$

with

$$a \perp\!\!\!\perp 2 = \left\lfloor \frac{a+1}{2} \right\rfloor - \left\lfloor \frac{a}{2} \right\rfloor = \begin{cases} 0 \text{ for } a \text{ even} \\ 1 \text{ for } a \text{ odd}\,, \end{cases} \tag{A.5}$$

the indicator function of odd $a$, and $r_j$ given by (A.2)–(A.3), for $m = 2$, with $p_1 = p$ and $p_2 = q - 1$, which shows that the exact distribution of $\Lambda_2$ may be given by (8) or (16).

### *A.3 The l.r.t. Statistic for Sphericity*

If we assume that

$$\underline{X} \sim N_p(\underline{\mu}, \Sigma)$$

and we want to test the null hypothesis in (9), based on a sample of size $n$ from $\underline{X}$, the l.r.t. statistic is (Mauchly, 1940; Anderson, 2003) given by (10).

From Mauchly (1940), Anderson (2003) and Coelho and Marques (2010), the c.f. of $W_3 = -\log\, \Lambda_3$ may be written as

$$\Phi_{W_3}(t) = \prod_{j=2}^{p} \frac{\Gamma\left(\frac{n-1}{2}+\frac{j-1}{p}\right)\,\Gamma\left(\frac{n-j}{2}-\frac{n}{2}\mathrm{it}\right)}{\Gamma\left(\frac{n-1}{2}+\frac{j-1}{p}-\frac{n}{2}\mathrm{it}\right)\,\Gamma\left(\frac{n-j}{2}\right)}$$

$$= \underbrace{\left\{\prod_{j=2}^{p}\left(\frac{n-j}{n}\right)^{r_j}\left(\frac{n-j}{n}-\mathrm{it}\right)^{-r_j}\right\}}_{\Phi_{1,W_3}(t)}$$

$$\underbrace{\times \left\{\prod_{j=1}^{p-k^*} \frac{\Gamma\left(\frac{n-1}{2}+\frac{j-1}{p}\right)\,\Gamma\left(\frac{n-1}{2}-\frac{n}{2}\mathrm{it}\right)}{\Gamma\left(\frac{n-1}{2}\right)\,\Gamma\left(\frac{n-1}{2}+\frac{j-1}{p}-\frac{n}{2}\mathrm{it}\right)}\right\} \times \left\{\prod_{j=p-k^*+1}^{p} \frac{\Gamma\left(\frac{n-1}{2}+\frac{j-1}{p}\right)\,\Gamma\left(\frac{n}{2}-\frac{n}{2}\mathrm{it}\right)}{\Gamma\left(\frac{n}{2}\right)\,\Gamma\left(\frac{n-1}{2}+\frac{j-1}{p}-\frac{n}{2}\mathrm{it}\right)}\right\}}_{\Phi_{2,W_3}(t)}, \qquad \text{(A.6)}$$

where $k^* = \lfloor p/2 \rfloor$ and

$$r_j = \left\lfloor \frac{p-j+2}{2} \right\rfloor \qquad j = 2, \ldots, p, \qquad \text{(A.7)}$$

which shows that the distribution of $\Lambda_3$ may be given by (11) and (17).

### *A.4 The l.r.t. Statistic to Test the Equality of Several Variance-Covariance Matrices*

If we assume that

$$\underline{X}_k \sim N_p(\underline{\mu}_k, \Sigma_k), \qquad k = 1, \ldots, q,$$

and we want to test the null hypothesis in (12), that is, the null hypothesis of equality of the $q$ variance-covariance matrices, we may write the l.r.t. statistic to test $H_{04}$ in (12), for $q$ independent samples of size $n$ from each $\underline{X}_k$ ($k = 1, \ldots, q$), as in (13).

From expression (1.7) in Coelho and Marques (2009b) and using the multiplication formula for the Gamma function, we may write the c.f. of $W_4 = -\log \Lambda_4$, under $H_{04}$ in (12), either as

$$\Phi_{W_4}(t) = \prod_{j=1}^{p}\prod_{k=1}^{q} \frac{\Gamma\left(\frac{n-1}{2} - \frac{j-1}{2q} + \frac{k-1}{q}\right)\,\Gamma\left(\frac{n-j}{2} - \frac{n}{2}\mathrm{i}t\right)}{\Gamma\left(\frac{n-1}{2} - \frac{j-1}{2q} + \frac{k-1}{q} - \frac{n}{2}\mathrm{i}t\right)\,\Gamma\left(\frac{n-j}{2}\right)}, \tag{A.8}$$

where we should note that for $j = k = 1$, the term in the product yields the value 1, or alternatively as

$$\Phi_{W_4}(t) = \underbrace{\left\{\prod_{j=2}^{p}\left(\frac{n-j}{n}\right)^{r_j}\left(\frac{n-j}{n} - \mathrm{i}t\right)^{-r_j}\right\}}_{\Phi_{1,W_4}(t)}$$
$$\underbrace{\times\left\{\prod_{j=1}^{\lfloor p/2\rfloor}\prod_{k=1}^{q}\frac{\Gamma(a_j+b_{jk})}{\Gamma(a_j+b^*_{jk})}\,\frac{\Gamma(a_j+b^*_{jk}-n\mathrm{i}t)}{\Gamma(a_j+b_{jk}-n\mathrm{i}t)}\right\} \times\left\{\prod_{k=1}^{q}\frac{\Gamma(a_p+b_{pk})}{\Gamma(a_p+b^*_{pk})}\,\frac{\Gamma\left(a_p+b^*_{pk}-\frac{n}{2}\mathrm{i}t\right)}{\Gamma\left(a_p+b_{pk}-\frac{n}{2}\mathrm{i}t\right)}\right\}^{p \perp\!\!\!\perp 2}}_{\Phi_{2,W_4}(t)} \tag{A.9}$$

where

$$a_j = n - 2j\,, \qquad b_{jk} = 2j - 1 + \frac{k-2j}{q}\,, \qquad b^*_{jk} = \lfloor b_{jk}\rfloor\,, \tag{A.10}$$

$$a_p = \frac{n-p}{2}\,, \qquad b_{pk} = \frac{pq-q-p+2k-1}{2q}\,, \qquad b^*_{pk} = \lfloor b_{pk}\rfloor\,, \tag{A.11}$$

and

$$r_j = \begin{cases} r^*_{j-1} & \text{for } j = 2, \ldots, p, \\ & \text{except for } j = p - 2\alpha_1 \\ r^*_{j-1} + (p \perp\!\!\!\perp 2)(\alpha_2 - \alpha_1)\left(q - \frac{p-1}{2} + q\left\lfloor \frac{p}{2q}\right\rfloor\right) & \text{for } j = p - 2\alpha_1 \end{cases} \tag{A.12}$$

with

$$r_j^* = \begin{cases} \gamma_j & \text{for } j = 1, \dots, \alpha+1 \\ q\left(\lfloor\frac{p}{2}\rfloor - \lfloor\frac{j}{2}\rfloor\right) & \text{for } j = \alpha+2, \dots, \min(p-2\alpha_1, p-1) \\ & \text{and } j = 2+p-2\alpha_1, \dots, 2\lfloor\frac{p}{2}\rfloor-1, \text{ by steps of 2} \\ q\left(\lfloor\frac{p+1}{2}\rfloor - \lfloor\frac{j}{2}\rfloor\right) & \text{for } j = 1+p-2\alpha_1, \dots, p-1, \text{ by steps of 2}, \end{cases} \tag{A.13}$$

and

$$\alpha = \left\lfloor \frac{p-1}{q} \right\rfloor, \quad \alpha_1 = \left\lfloor \frac{q-1}{q}\frac{p-1}{2} \right\rfloor, \quad \alpha_2 = \left\lfloor \frac{q-1}{q}\frac{p+1}{2} \right\rfloor, \tag{A.14}$$

where, for $j = 1, \dots, \alpha$,

$$\gamma_j = \left\lfloor\frac{q}{2}\right\rfloor \left((j-1)q - 2\left((q+1) \perp\!\!\!\perp 2\right)\left\lfloor\frac{j}{2}\right\rfloor\right) + \left\lfloor\frac{q}{2}\right\rfloor \left\lfloor\frac{q+j \perp\!\!\!\perp 2}{2}\right\rfloor \tag{A.15}$$

and

$$\begin{aligned}\gamma_{\alpha+1} = -\left(\left\lfloor\frac{p}{2}\right\rfloor - \alpha\left\lfloor\frac{q}{2}\right\rfloor\right)^2 + q\left(\left\lfloor\frac{p}{2}\right\rfloor - \left\lfloor\frac{\alpha+1}{2}\right\rfloor\right) \\ +(q \perp\!\!\!\perp 2)\left(\alpha\left\lfloor\frac{p}{2}\right\rfloor + \frac{\alpha \perp\!\!\!\perp 2}{4} - \frac{\alpha^2}{4} - \alpha^2\left\lfloor\frac{q}{2}\right\rfloor\right).\end{aligned} \tag{A.16}$$

Expression (A.8) shows that we may write (14) while (A.9) shows that we may write (18).

## Appendix B: The Gamma, GIG (Generalized Integer Gamma) and GNIG (Generalized Near-Integer Gamma) Distributions

We will use this Appendix to establish some notation concerning distributions used in the paper, as well as to give the expressions for the p.d.f.'s (probability density functions) and c.d.f.'s (cumulative distribution functions) of the GIG (Generalized Integer Gamma) and GNIG (Generalized Near-Integer Gamma) distributions.

Let

$$X_j \sim \Gamma(r_j, \lambda_j), \quad j = 1, \dots, p,$$

be $p$ independent r.v.'s with Gamma distributions with shape parameters $r_j \in \mathbb{N}$ and rate parameters $\lambda_j > 0$, with $\lambda_j \neq \lambda_{j'}$, for all $j, j' \in \{1, \dots, p\}$. We will say that then the r.v. $Y = \sum_{j=1}^{p} X_j$ has a GIG (Generalized Integer Gamma) distribution of depth $p$, with shape parameters $r_j$ and rate parameters $\lambda_j$ $(j = 1, \dots, p)$, and we will denote this fact by

$$Y \sim GIG(r_j, \lambda_j;\ p)\,.$$

The p.d.f. and c.d.f. (cumulative distribution function) of $Y$ are respectively given by (Coelho, 1998)

$$f^{GIG}(y|r_1, \dots, r_p; \lambda_1, \dots, \lambda_p; p) = K \sum_{j=1}^{p} e^{-\lambda_j y} \sum_{k=1}^{r_j} c_{j,k}\, y^{k-1}\,, \qquad (y > 0)$$

and

$$F^{GIG}(y|r_1, \dots, r_j; \lambda_1, \dots, \lambda_p; p) = 1 - K \sum_{j=1}^{p} e^{-\lambda_j y} \sum_{k=1}^{r_j} c_{j,k} (k-1)! \sum_{i=0}^{k-1} \frac{1}{i!} \frac{y^i}{\lambda_j^{k-i}}\,, \qquad (y > 0)$$

where $K$ is given by (5) and $c_{j,k}$ by (11)-(13) in Coelho (1998).

The GNIG (Generalized Near-Integer Gamma) distribution of depth $p + 1$ (Coelho, 2004) is the distribution of the r.v.

$$Z = Y_1 + Y_2$$

where $Y_1$ and $Y_2$ are independent, $Y_1$ having a GIG distribution of depth $p$ and $Y_2$ with a Gamma distribution with a non-integer shape parameter $r$ and a rate parameter $\lambda \neq \lambda_j$ $(j = 1, \dots, p)$. The p.d.f. (probability density function) of $Z$ is given by

$$f^{GNIG}(z|r_1, \dots, r_p, r;\ \lambda_1, \dots, \lambda_p, \lambda;\ p+1) = K\lambda^r \sum_{j=1}^{p} e^{-\lambda_j z} \sum_{k=1}^{r_j} \left\{ c_{j,k} \frac{\Gamma(k)}{\Gamma(k+r)} z^{k+r-1}\, {}_1F_1(r, k+r, -(\lambda-\lambda_j)z) \right\}, \quad (z>0)$$

and the c.d.f. (cumulative distribution function) is given by

$$F^{GNIG}(z|r_1, \dots, r_p, r;\ \lambda_1, \dots, \lambda_p, \lambda;\ p+1) = \frac{\lambda^r z^r}{\Gamma(r+1)}\, {}_1F_1(r, r+1, -\lambda z) - K\lambda^r \sum_{j=1}^{p} e^{-\lambda_j z} \sum_{k=1}^{r_j} c^*_{j,k} \sum_{i=0}^{k-1} \frac{z^{r+i} \lambda_j^i}{\Gamma(r+1+i)}\, {}_1F_1(r, r+1+i, -(\lambda-\lambda_j)z), \quad (z>0)$$

where

$$c^*_{j,k} = \frac{c_{j,k}}{\lambda_j^k}\Gamma(k)$$

with $c_{j,k}$ given by (11) through (13) in Coelho (1998). In the above expressions ${}_1F_1(a, b; z)$ is the Kummer confluent hypergeometric function. This function typically has very good convergence properties and is nowadays easily handled by a number of software packages.

## References


Anderson, T.W. (2003). *An Introduction to Multivariate Statistical Analysis*, 3rd edn. Wiley, New York.

Anderson, T.W., Fang, K.-T. (1990). Inference in multivariate elliptically contoured distributions based on maximum likelihood. In: Fang K.-T., Anderson T.W. (eds.) *Statistical Inference in Elliptically Contoured and Related Distributions*, pp. 201–216.

Anderson, T.W., Fang, K.-T., Hsu, H. (1986). Maximum-likelihood estimates and likelihood-ratio criteria for multivariate elliptically contoured distributions. *Canad. J. Statist.* **14**, 55–59.

Aslam, S., Rocke, D.M. (2005). A robust testing procedure for the equality of covariance matrices. *Comput. Statist. Data Anal.* **49**, 863–874.

Berry, A. (1941). The accuracy of the Gaussian approximation to the sum of independent variates. *Trans. Amer. Math. Soc.* **49**, 122–136.

Coelho, C.A. (1992). *Generalized Canonical Analysis*. Ph.D. Thesis, The University of Michigan, Ann Arbor, MI.

Coelho, C.A. (1998). The Generalized Integer Gamma distribution – a basis for distributions in Multivariate Statistics. *J. Multivariate Anal.* **64**, 86–102.

Coelho, C.A. (2004). The Generalized Near-Integer Gamma distribution: a basis for 'near-exact' approximations to the distributions of statistics which are the product of an odd number of independent Beta random variables. *J. Multivariate Anal.* **89**, 191–218.

Coelho, C.A. (2007). The wrapped Gamma distribution and wrapped sums and linear combinations of independent Gamma and Laplace distributions. *J. Stat. Theory Pract.* **1**, 1–29.

Coelho, C.A., Marques, F.J. (2009a). The advantage of decomposing elaborate hypotheses on covariance matrices into conditionally independent hypotheses in building near-exact distributions for the test statistics. *Linear Algebra Appl.* **430**, 2592–2606.

Coelho, C.A., Marques, F.J. (2009b). Near-exact distributions for the likelihood ratio test statistic for testing equality of several variance-covariance matrices (revisited). The New University of Lisbon, Mathematics Department, Technical Report #9/2009 (submitted for publication) (http://www.dm.fct.unl.pt/sites/www.dm.fct.unl.pt/files/preprints/2009/9_09.pdf)

Coelho, C.A., Marques, F.J. (2010). Near-exact distributions for the independence and sphericity likelihood ratio test statistics. *J. Multivariate Anal.* **101**, 583–593.

Coelho, C.A., Alberto, R.P., Grilo, L.M. (2006). A mixture of Generalized Integer Gamma distributions as the exact distribution of the product of an odd number of independent Beta random variables: applications. *J. Interdiscip. Math.* **9**, 229–248.

Esseen, C.-G. (1945). Fourier analysis of distribution functions. A mathematical study of the Laplace-Gaussian Law. *Acta Math.* **77**, 1–125.

Hsieh, H.K. (1979). On asymptotic optimality of likelihood ratio tests for multivariate normal distributions. *Ann. Statist.* **7**, 592–598.

Hwang, H.-K. (1998). On convergence rates in the central limit theorems for combinatorial structures. *European J. Combin.* **19**, 329–343.

Jensen, D.R., Good, I.J. (1981). Invariant distributions associated with matrix laws under structural symmetry. *J. R. Statist. Soc.* Ser. B **43**, 327–332.

Kariya, T. (1981). Robustness of multivariate tests. *Ann. Statist.* **9**, 1267–1275.

Kshirsagar, A.M. (1972). *Multivariate Analysis*. Marcel Dekker, New York.

Loève, M. (1977). *Probability Theory*, 4th edn, Vol. I. Springer, New York.

Marques, F.J., Coelho, C.A. (2008). Near-exact distributions for the sphericity likelihood ratio test statistic. *J. Statist. Plann. Inference* **138**, 726–741.

Mauchly, J.W. (1940). Significance test for sphericity of a normal $n$-variate distribution. *Ann. Math. Stat.* **11**, 204–209.

# The Distribution of the Product of Powers of Independent Uniform Random Variables – A Simple but Useful Tool to Address and Better Understand the Structure of Some Distributions

**Barry C. Arnold, Carlos A. Coelho, and Filipe J. Marques**

**Abstract** What is the distribution of the product of given powers of independent uniform (0,1) random variables? Is this distribution useful? Is this distribution commonly used in some contexts? Is this distribution somehow related to the distribution of the product of other random variables? Are there some test statistics with this distribution? This paper will give the answers to the above questions. It will be seen that the answer to the last four questions above is: yes! We will show how particular choices of the numbers of variables involved and their powers will result in interesting and useful distributions and how these distributions may help us to shed some new light on some well-known distributions and also how it may help us to address, in a much simpler way, some distributions usually considered to be rather complicated as is the case with the exact distribution of a number of statistics used in Multivariate Analysis, including some whose exact distribution up until now is not available in a concise and manageable form.

**Keywords** Product of independent Beta random variables · Generalized Integer Gamma distribution · Exponentiated Generalized Integer Gamma distribution ·

---

Reprinted from Journal of Multivariate Analysis, with permission from Elsevier.
First published as Arnold, B.C., Coelho, C.A., Marques, F.J. (2013). The distribution of the product of powers of independent Uniform random variables – a simple but useful tool to address and better understand the structure of some distributions, *Journal of Multivariate Analysis* **113**, 19–36. (https://doi.org/10.1016/j.jmva.2011.04.006)

---

B. C. Arnold
Statistics Department, University of California, Riverside, Riverside, CA, USA
e-mail: bcarnold@ucr.edu

C. A. Coelho (✉) · F. J. Marques
Departamento de Matemática, Universidade Nova de Lisboa, Faculdade de Ciências e Tecnologia, Quinta da Torre, Caparica, Portugal
e-mail: cmac@fct.unl.pt; fjm@fct.unl.pt



B. C. Arnold et al. (eds.), *Likelihood Ratio Tests and Related Topics in Multivariate Analysis*, Contributions to Statistics,
https://doi.org/10.1007/978-3-031-68008-3_11

Wilks Lambda statistic · Equality of mean vectors · Testing circularity · Testing equality of mean vectors and circularity

**AMS Subject Classifications** 62E15, 62E10, 62H05, 62H10, 62H15

# 1 Introduction

This paper intends to provide simpler expressions for the exact distributions of several likelihood ratio test (l.r.t.) statistics used in Multivariate Analysis. Simpler derivations of these expressions are obtained through the use of only one base distribution and a useful unified procedure.

The distributions we are interested in studying are the distributions of r.v.'s (random variables) of the form

$$V = \prod_{j=1}^{p} \prod_{i=1}^{r_j} U_{ij}^{\frac{1}{\delta_{ij}}} , \quad \text{for } \delta_{ij} > 0 , \tag{1}$$

where the $U_{ij}$'s are i.i.d. Unif(0, 1) r.v.'s, for $j = 1, \ldots, p$; $i = 1, \ldots, r_j$.

In this paper we will show how particular choices of $\delta_{ij}$, and also sometimes of the $r_j$, in (1), will result in $V$ having interesting and useful distributions, whose p.d.f.'s (probability density functions) and c.d.f.'s (cumulative distribution functions) have concise and manageable representations.

More precisely, we may show that $V$ in (1) can be identified with the distribution of (i) the exponential of the sum of independent Exponential r.v.'s or the exponential of the sum of independent integer Gamma r.v.'s (Section 2), (ii) a product of independent Pareto or generalized Pareto r.v.'s (Section 3), and (iii) a product of independent Beta r.v.'s (Section 4). As a consequence of this last result, we are able to obtain the exact distribution of several l.r.t. statistics used in Multivariate Analysis whose exact distributions up until now are only available in less convenient forms, such as those used to test

- independence of groups of variables, both in the real and complex cases,
- equality of several mean vectors both in the real and complex cases,
- circularity, in the case when the dimension is odd,
- circularity and equality of means, for any dimension.

What the choices for $\delta_{ij}$ and $r_j$ in (1) have to be in order to obtain these distributions will be explained in Sections 2 – 5 ahead.

In terms of notation, we will use $\mathbb{N}$ to denote the set of positive integers and $\mathbb{R}$ to denote the set of reals. We will also use " $\sim$" to denote "is distributed as" and " $\stackrel{st}{\sim}$ " to denote "has the same distribution as" or "is stochastically equivalent to".

## 2 The General Case – The Distribution of $V$ as an EGIG Distribution

Since for $U_{ij} \sim \text{Unif}(0, 1)$ we have

$$E\left(U_{ij}^h\right) = \frac{1}{1+h},$$

and given the independence of the r.v.'s $U_{ij}$ in (1), we may write

$$E\left(V^h\right) = \prod_{j=1}^{p}\prod_{i=1}^{r_j} E\left(U_{ij}^{\frac{h}{\delta_{ij}}}\right) = \prod_{j=1}^{p}\prod_{i=1}^{r_j} \delta_{ij}(\delta_{ij}+h)^{-1} \quad (h > -\min_{i,j}\delta_{ij}). \qquad (2)$$

This result will enable us to easily obtain the expressions for the $h$-th moment of $V$ in any of the cases either in this section as well as in Sections 3–5 ahead and in this way to obtain in Section 5 alternative expressions to those more commonly used for the moments of the l.r.t. statistics studied, namely the Wilks $\Lambda$ statistic.

From (2) above it is easy to derive the c.f. of $W = -\log V$, for $\mathrm{i} = \sqrt{-1}$, as

$$\Phi_W(t) = E\left(e^{\mathrm{i}tW}\right) = E\left(V^{-\mathrm{i}t}\right) = \prod_{j=1}^{p}\prod_{i=1}^{r_j} \delta_{ij}(\delta_{ij}-\mathrm{i}t)^{-1} \quad (t \in \mathbb{R}).$$

If in (1) we take $\delta_{ij} > 0$ to be all different for $i = 1, \ldots, r_j$ and $j = 1, \ldots, p$, then (see Appendices A.2 and B) $V$ has an EGIG (Exponential Generalized Integer Gamma) distribution of depth $\sum_{j=1}^{p} r_j$ with all shape parameters equal to 1 and rate parameters $\delta_{ij}$.

On the other hand, if $\delta_{ij} = \delta > 0$ for all $i = 1, \ldots, r_j$ and all $j = 1, \ldots, p$, $V$ has the distribution of the exponential of the negative of an integer Gamma r.v. with shape parameter $\sum_{j=1}^{p} r_j$ and rate parameter $\delta$.

In the more general case where we just consider $\delta_{ij} > 0$, allowing some of them to be equal, we will have to call say $\delta_1^*, \ldots, \delta_h^*$ the set of different values that all the $\delta_{ij}$ take, and call $r_\ell^*$ the number of times the value $\delta_\ell^*$ $(\ell = 1, \ldots, h)$ occurs. Then $V$ has an EGIG distribution of depth $h$, with shape parameters $r_\ell^*$ and rate parameters $\delta_\ell^*$ $(\ell = 1, \ldots, h)$.

Some other cases that may occur are:

- if we take $\delta_{ij} = \delta_j > 0$ for $i = 1, \ldots, r_j$; $j = 1, \ldots, p$ and all the $\delta_j$ to be different for $j = 1, \ldots, p$; in this case $V$ will have an EGIG distribution of depth $p$ with shape parameters $r_j$ and rate parameters $\delta_j$;
- if we take $p = 1$, $r_1 = 1$ and $\delta_{11} = \delta$, then $V$ has a Beta$(\delta, 1)$ distribution which is the distribution of the exponential of the negative of an Exponential$(\delta)$ r.v..

# 3 The Distribution of $V$ in (1) as the Distribution of the Product of Pareto r.v.'s

Essentially we will have to consider in this subsection exactly the same cases that were considered in the previous section, with the only difference that now we have all $\delta_{ij} < 0$, so that

$$U_{ij}^{\frac{1}{\delta_{ij}}} \sim \text{Pareto}(-\delta_{ij})\,.$$

This way, if we take in (1) $r_j = 1$ for all $j = 1, \ldots, p$ and $\delta_{ij} = \delta_j < 0$, all different for $j = 1, \ldots, p$, then $V$ in (1) will have the distribution of the product of $p$ independent standard Pareto r.v.'s. This distribution is [1], from the previous section, an EGIG distribution of depth $p$ with all shape parameters equal to 1 and rate parameters $-\delta_j < 0$.[2]

On the other hand, if all $\delta_j$ are equal, with $\delta_j = \delta$ $(j = 1, \ldots, p)$, then $V$ has a generalized Pareto distribution (see Appendix A.2) with shape parameter $p$ and rate parameter $-\delta$.

If in the general case, with $r_j \in \mathbb{N}$, we take $\delta_{ij} = \delta_j < 0$ for $j = 1, \ldots, p$, then $V$ has the distribution of the product of $p$ independent generalized Pareto distributions with shape parameters $r_j$ and rate parameters $-\delta_j$, which is thus [3] an EGIG distribution of depth $p$ with shape parameters $r_j$ and rate parameters $-\delta_j$ $(j = 1, \ldots, p)$.

If on the other hand we take in (1) all $\delta_{ij} < 0$ to be different for all $i = 1, \ldots, r_j$ and $j = 1, \ldots, p$, in this case $V$ has the distribution of the product of $\sum_{j=1}^{p} r_j$ independent standard Pareto distributions with rate parameters $-\delta_{ij}$, which is [4] an EGIG distribution of depth $\sum_{j=1}^{p} r_j$ with all shape parameters equal to 1 and rate parameters $-\delta_{ij}$, and if only $h$ of the $\delta_{ij}$ are different, say with values $\delta_1^*, \ldots, \delta_h^*$, where $r_\ell^*$ is the number of times $\delta_\ell^*$ $(\ell = 1, \ldots, h)$ occurs, the distribution of $V$ will be the distribution of the product of $h$ independent generalized Pareto r.v.'s with shape parameters $r_\ell^*$ and rate parameters $-\delta_\ell^*$, which is thus [5] an EGIG distribution of depth $h$ with shape parameters $r_\ell^*$ and rate parameters $-\delta_\ell^*$ $(\ell = 1, \ldots, h)$.

In case we want the distribution of the product of non-standard Pareto r.v.'s all we have to do is to collect say in the constant $c$ the product of all the product-shift parameters and then consider the distribution of the r.v. $V^* = cV$, which is easily obtained by transformation from the distribution of $V$.

---

[1] see author and editors notes at the end of the Chapter

[2] see author and editors notes at the end of the Chapter

[3] see author and editors notes at the end of the Chapter

[4] see author and editors notes at the end of the Chapter

[5] see author and editors notes at the end of the Chapter

For the cases in which the $\delta_j$'s are all equal or all different, these results agree with those in Arnold (1983).

# 4 When Does $V$ Have the Distribution of a Beta r.v. or the Distribution of the Product of Beta r.v.'s

## 4.1 $V$ in (1) as a Beta or a Power of a Beta r.v.

It is easy to see that for any choice of $\delta_{ij} > 0$ in (1) the r.v.'s $U_{ij}^{\frac{1}{\delta_{ij}}}$ will have a $Beta(\delta_{ij}, 1)$ distribution (see Appendix A.2), so that if in (1) we take $p = 1$ and $r_1 = 1$, $V$ will have a $Beta(\delta_{11}, 1)$ distribution.

But our question is: apart from this trivial case, are there other cases for which $V$ in (1) has a Beta distribution? The answer is: Yes! And some of them may lead us to quite interesting results, as will be seen below.

So, our question now is: for what choices of $\delta_{ij}$ and $r_j$ in (1) will the r.v. $V$ have a Beta distribution?

It is not hard to see that if in (1) we take $\delta_{ij} = \delta_j = a + j - 1$ for some $a > 0$ and $r_j = 1$ for all $j = 1, \ldots, p$, we will have

$$V_j = U_{1j}^{\frac{1}{\delta_j}} \sim Beta(a + j - 1, 1)$$

so that

$$V = \prod_{j=1}^{p} V_j \sim Beta(a, p), \tag{3}$$

since, using the relation

$$\frac{\Gamma(a+n)}{\Gamma(a)} = \prod_{i=1}^{n}(a+i-1), \qquad \text{for any } a > 0,\ n \in \mathbb{N}, \tag{4}$$

we may write the c.f. of $W = -\log V$ as

$$\begin{aligned}\Phi_W(t) &= \prod_{j=1}^{p}(a+j-1)\,(a+j-1-\mathrm{i}t)^{-1}\\ &= \frac{\Gamma(a+p)}{\Gamma(a)}\,\frac{\Gamma(a-\mathrm{i}t)}{\Gamma(a+p-\mathrm{i}t)}\end{aligned}$$

which is the c.f. of a $Logbeta(a, p)$ r.v. and as such shows that $V$ has the distribution in (3) above. We may note that in this case the distribution of $V$ still is an EGIG distribution of depth $p$ with all shape parameters equal to 1 and rate parameters $a + j - 1$ ($j = 1, \ldots, p$), which is also a $Beta(a, p)$ distribution.

Moreover, if in (1) we take $\delta_{ij} = \frac{a+j-1}{p} + i - 1$, for some $a > 0$, and $r_j = r$ for all $j = 1, \ldots, p$, we will have

$$V_j = \prod_{i=1}^{r} U_{ij}^{\frac{1}{\delta_{ij}}} \sim Beta\left(\frac{a+j-1}{p}, r\right)$$

and

$$V = \prod_{j=1}^{p} V_j$$

will have the distribution of the $p$-th power of a $Beta(a, pr)$ r.v., with p.d.f.

$$f_V(v) = \frac{1}{p\, B(a, pr)}\, v^{\frac{a}{p}-1} \left(1 - v^{1/p}\right)^{pr-1} \qquad (0 < v < 1)\,,$$

or,

$$V^{1/p} \sim Beta(a, pr) \tag{5}$$

since in this case, using (4) above and the Gamma multiplication formula,

$$\Gamma(nz) = (2\pi)^{\frac{1}{2}(1-n)}\, n^{nz-\frac{1}{2}} \prod_{k=0}^{n-1} \Gamma\left(z + \frac{k}{n}\right), \tag{6}$$

we may write the c.f. of $W = -\log V$ as

$$\begin{aligned}
\Phi_W(t) &= \prod_{j=1}^{p}\prod_{i=1}^{r} \left(\frac{a+j-1}{p} + i - 1\right)\left(\frac{a+j-1}{p} + i - 1 - \mathrm{i}t\right)^{-1} \\
&= \prod_{j=1}^{p} \frac{\Gamma\left(\frac{a+j-1}{p} + r\right)}{\Gamma\left(\frac{a+j-1}{p}\right)}\, \frac{\Gamma\left(\frac{a+j-1}{p} - \mathrm{i}t\right)}{\Gamma\left(\frac{a+j-1}{p} + r - \mathrm{i}t\right)} \\
&= \frac{\Gamma\,(a + pr)}{\Gamma\,(a)}\, \frac{\Gamma\,(a - \mathrm{i}pt)}{\Gamma\,(a + pr - \mathrm{i}pt)}
\end{aligned}$$

which is the c.f. of a $Logbeta(a, pr)$ r.v., multiplied by $p$, what implies that $V$ has the distribution in (5) (see Appendix A.1).

We should notice that in this case the distribution of $V$ still is an EGIG distribution of depth $pr$ with all shape parameters equal to 1 and rate parameters $\frac{a+j-1}{p}+i-1$, $(i=1,\ldots,r;\ j=1,\ldots,p)$.

Then it is easy to see that if we take $\delta_{ij}=a+j-1+p(i-1)$ $(i=1,\ldots,r;\ j=1,\ldots,p)$, we have

$$V \sim Beta(a, pr)\,,$$

with

$$V=\prod_{j=1}^{p} V_j \quad \text{where} \quad V_j^{1/p} \sim Beta(a+j-1, r) \quad \text{are } p \text{ independent r.v.'s.}$$

In these two last distributions all the Beta r.v.'s in the product, that is, all the $V_j$, had a second parameter that was not only an integer but also it was always the same, for all $j=1,\ldots,p$. The question now is: is it possible to make $V$ in (1) to have the distribution of the product of independent Beta r.v.'s whose second parameter is not the same for all $j=1,\ldots,p$ or not even an integer? The answer is indeed 'Yes!'. But if so, how? The answer to this question will be given in the next section.

## 4.2 *V in (1) as the Product of Independent Beta r.v.'s*

### 4.2.1 A First Very Simple Case

Indeed if in (1) we simply take $\delta_{ij}=a_j+i-1$, for some $a_j>0$, then

$$V_j=\prod_{i=1}^{r_j} U_{ij}^{\frac{1}{\delta_{ij}}} \sim Beta(a_j, r_j)$$

since in this case the c.f. of $W_j=-\log V_j$, using (4), may be, for $\mathrm{i}=\sqrt{-1}$, written as

$$\Phi_{W_j}(t)=\prod_{i=1}^{r_j}(a_j+i-1)\,(a_j+i-1-\mathrm{i}t)^{-1}=\frac{\Gamma(a_j+r_j)}{\Gamma(a_j)}\,\frac{\Gamma(a_j-\mathrm{i}t)}{\Gamma(a_j+r_j-\mathrm{i}t)}$$

which is the c.f. of the negative logarithm of a $Beta(a_j, r_j)$ r.v., and thus in this case $V$ will have the same distribution as

$$\prod_{j=1}^{p} V_j \quad \text{where} \quad V_j \sim Beta(a_j, r_j) \quad (a_j>0;\ r_j\in\mathbb{N}) \quad \text{are } p \text{ independent r.v.'s.}$$

This distribution is, from Section 2, an EGIG distribution, with depth $\sum_{j=1}^{p} r_j$, with all shape parameters equal to 1 and rate parameters $a_j+i-1$, if all the $a_j+i-1$ are different for $i = 1, \ldots, r_j$; $j = 1, \ldots, p$. One may contemplate many other different situations according to the relations among the $a_j$, but the distribution of $V$ in (1) will always be an EGIG distribution with a given depth, as explained in Section 2.

### 4.2.2 A First More Elaborate Case

However, there remains the question: is it possible for $V$ in (1) to have the same distribution as a product of independent Beta r.v.'s whose parameters are not integers?

The answer, once again is 'Yes!', but only if some restrictions are placed on the parameters of the Beta r.v.'s.

More precisely, in order to be able to argue that $V$ in (1) has the distribution of a product of independent Beta r.v.'s whose second parameter is not an integer, those Beta r.v.'s in the product need to have a first parameter that increases or decreases by steps of $\frac{1}{k}$ where $k \in \mathbb{N}$, and a second parameter that is then a multiple of $\frac{1}{k}$. In addition, the number of these Beta r.v.'s has to be a multiple of $k$.

Let us take in (1), for some $k \in \mathbb{N}$, and either some $u$, $vk \in \mathbb{N}$ or some $v$, $uk \in \mathbb{N}$,

$$p = (u + v - 1)k\,. \tag{7}$$

Then let

$$r_j = \begin{cases} h_j & j = 1, \ldots, k \\ h_j + r_{j-k} & j = k+1, \ldots, p \end{cases} \tag{8}$$

with

$$h_j = (\#\text{ of elements in } \{uk, vk\} \leq j) - 1 \tag{9}$$

and

$$\delta_{ij} = a - 1 - \frac{j}{k}\,, \qquad (i = 1, \ldots, r_j;\ j = 1, \ldots, p) \tag{10}$$

for some $a > u + v$.

Then we will be able to show that the distribution of $V$ in (1) is the same as the distribution of

(i) (a) $V_{1a}=\prod_{j=1}^{vk} Y_j$ where $Y_j \sim Beta\left(a-u-\frac{j}{k},\, u\right)$ or
$Y_j \sim Beta\left(a-u-v+\frac{j-1}{k},\, u\right)$
are $vk$ independent r.v.'s;

(b) $V_{1b}=\prod_{j=1}^{uk} Y_j^{*}$ where $Y_j^{*} \sim Beta\left(a-v-\frac{j}{k},\, v\right)$ or
$Y_j^{*} \sim Beta\left(a-u-v+\frac{j-1}{k},\, v\right)$
are $uk$ independent r.v.'s;

(ii) (a) $V_{2a}=\prod_{j=1}^{u}\prod_{i=1}^{k} Y_{ij}$ where $Y_{ij} \sim Beta\left(a-v-j+1-\frac{i}{k},\, v\right)$ or
$Y_{ij} \sim Beta\left(a-u-v+j-1+\frac{i-1}{k},\, v\right)$
are $uk$ independent r.v.'s,
(if $u,\, vk \in \mathbb{N}$, even if $v \notin \mathbb{N}$);

(b) $V_{2b}=\prod_{j=1}^{v}\prod_{i=1}^{k} Y_{ij}^{*}$ where $Y_{ij}^{*} \sim Beta\left(a-u-j+1-\frac{i}{k},\, u\right)$ or
$Y_{ij}^{*} \sim Beta\left(a-u-v+j-1+\frac{i-1}{k},\, v\right)$ [6]
are $vk$ independent r.v.'s,
(if $v,\, uk \in \mathbb{N}$, even if $u \notin \mathbb{N}$);

(iii) (a) $V_{3a}=\prod_{j=1}^{u} Y_j^{**}$ where $\left(Y_j^{**}\right)^{1/k} \sim Beta\big(k(a-v-j),\, vk\big)$ or
$\left(Y_j^{**}\right)^{1/k} \sim Beta\big(k(a-u-v+j-1),\, vk\big)$
are $u$ independent r.v.'s,
(if $u,\, vk \in \mathbb{N}$, even if $v \notin \mathbb{N}$);

(b) $V_{3b}=\prod_{j=1}^{v} Y_j^{***}$ where $\left(Y_j^{***}\right)^{1/k} \sim Beta\big(k(a-u-j),\, uk\big)$ or
$\left(Y_j^{***}\right)^{1/k} \sim Beta\big(k(a-u-v+j-1),\, uk\big)$
are $v$ independent r.v.'s,
(if $v,\, uk \in \mathbb{N}$, even if $u \notin \mathbb{N}$);

(iv) (a) $V_{4a}=\prod_{j=1}^{\lfloor vk \rfloor}\prod_{i=1}^{u} Y_{ij}^{**}$ where $Y_{ij}^{**} \sim Beta\left(a-\frac{j}{k}-i,\, 1\right)$ or
$Y_{ij}^{**} \sim Beta\left(a-u-v-1+\frac{j-1}{k}+i,\, 1\right)$
are $u\lfloor vk \rfloor$ independent r.v.'s
(if $u \in \mathbb{N}$, even if neither $vk \notin \mathbb{N}$ nor $v \notin \mathbb{N}$);[7]

---

[6] see author and editors notes at the end of the Chapter

[7] see author and editors notes at the end of the Chapter

(b) $V_{4b} = \prod_{j=1}^{\lfloor uk \rfloor} \prod_{i=1}^{v} Y_{ij}^{***}$ where $Y_{ij}^{***} \sim Beta\left(a - \frac{j}{k} - i, 1\right)$ or
$Y_{ij}^{***} \sim Beta\left(a - u - v - 1 + \frac{j-1}{k} + i, 1\right)$
are $v\lfloor uk \rfloor$ independent r.v.'s
(if $v \in \mathbb{N}$, even if neither $uk \notin \mathbb{N}$ nor $u \notin \mathbb{N}$).[8]

Parts (a) and (b) of (i) – (iv) above show the interchangeability of $u$ and $v$, in case they are both integers, in which case all the above eight distributions are the same as the distribution of $V$ in (1), for $r_j$ and $\delta_{ij}$ given by (8)–(10).

A complete proof of the equivalence of the distribution of $V$ in (1) with the distributions in (i) – (iv) above, may be easily achieved through the use of c.f.'s, and we will do it only for the first distribution on each of the items (i) – (iv). The second distributions in (i) – (iv) are obtained only by reversing the indexes.

We will take $W = -\log V$ and $W_{ij} = -\log V_{ij}$ $(i = 1, \ldots, 4;\ j = a, b)$, for $V_{ij}$ in (i) – (iv) above. The equivalence of the distributions in (i)(a), (iv)(a) and (1) may then be shown by writing

$$\Phi_{W_{1a}}(t) = \prod_{j=1}^{vk} \frac{\Gamma\left(a - \frac{j}{k}\right)}{\Gamma\left(a - u - \frac{j}{k}\right)} \frac{\Gamma\left(a - u - \frac{j}{k} - \mathrm{i}t\right)}{\Gamma\left(a - \frac{j}{k} - \mathrm{i}t\right)} \tag{11}$$

$$= \prod_{j=1}^{vk} \prod_{i=1}^{u} \left(a - u - \frac{j}{k} + i - 1\right) \left(a - u - \frac{j}{k} + i - 1 - \mathrm{i}t\right)^{-1} \tag{12}$$

$$= \prod_{j=1}^{vk} \prod_{i=1}^{u} \left(a - \frac{j}{k} - i\right) \left(a - \frac{j}{k} - i - \mathrm{i}t\right)^{-1} = \Phi_{W_{4a}}(t) \tag{13}$$

$$= \prod_{j=1}^{vk+uk-k} \prod_{i=1}^{r_j} \left(a - 1 - \frac{j}{k}\right) \left(a - 1 - \frac{j}{k} - \mathrm{i}t\right)^{-1}$$

$$= \prod_{j=1}^{p} \left(a - 1 - \frac{j}{k}\right)^{r_j} \left(a - 1 - \frac{j}{k} - \mathrm{i}t\right)^{-r_j} = \Phi_W(t) \tag{14}$$

where we go from (11) to (12) by using (4), from (12) to (13) just by reversing the index $i$, and from (13) to (14) by just identifying and counting the different values for $\left(a - \frac{j}{k} - i\right)$, using a counting technique similar to the one used in Wald and Brookner (1949) and Coelho (2006) and then by using the definition of $p$ in (7) and $r_j$ in (8)–(9) and the fact that $\left(a - 1 - \frac{j}{k}\right)$ is not a function of $i$.

[8] see author and editors notes at the end of the Chapter

The second set of equivalences, that is, of the distributions in (iv)(a), (iii)(a) and (ii)(a), may be shown by writing

$$\Phi_{W_{4a}}(t) = \prod_{j=1}^{vk}\prod_{i=1}^{u}\left(a-\frac{j}{k}-i\right)\left(a-\frac{j}{k}-i-\mathrm{i}t\right)^{-1}$$

$$=\prod_{j=1}^{vk}\prod_{i=1}^{u}\left(a-v+\frac{j-1}{k}-i\right)\left(a-v+\frac{j-1}{k}-i-\mathrm{i}t\right)^{-1} \tag{15}$$

$$=\prod_{j=1}^{vk}\prod_{i=1}^{u}(k(a-v-i)+j-1)\,(k(a-v-i)+j-1-\mathrm{i}tk)^{-1} \tag{16}$$

$$=\prod_{i=1}^{u}\frac{\Gamma\,(k(a-v-i)+vk)}{\Gamma\,(k(a-v-i))}\,\frac{\Gamma\,(k(a-v-i)-\mathrm{i}tk)}{\Gamma\,(k(a-n-i)+vk-\mathrm{i}tk)} = \Phi_{W_{3a}}(t) \tag{17}$$

$$=\prod_{i=1}^{u}\prod_{j=1}^{k}\frac{\Gamma\left(a-i+\frac{j-1}{k}\right)}{\Gamma\left(a-v-i+\frac{j-1}{k}\right)}\,\frac{\Gamma\left(a-v-i+\frac{j-1}{k}-\mathrm{i}t\right)}{\Gamma\left(a-i+\frac{j-1}{k}-\mathrm{i}t\right)} \tag{18}$$

$$=\prod_{j=1}^{u}\prod_{i=1}^{k}\frac{\Gamma\,(a-j+1-\frac{i}{k})}{\Gamma\,(a-v-j+1-\frac{i}{k})}\,\frac{\Gamma\,(a-v-j+1-\frac{i}{k}-\mathrm{i}t)}{\Gamma\,(a-j+1-\frac{i}{k}-\mathrm{i}t)} = \Phi_{W_{2a}}(t), \tag{19}$$

where from (15) to (16) we have multiplied and divided by $k$ and from (16) to (17) we have used (4). Then, from (17) to (18) we have used the reverse version of (6) and from (18) to (19) we have just switched $i$ and $j$ and then reversed the index $i$.

The distributions in (iv) are particularly useful not only in bridging between the distribution of $V$ in (1) and the distributions in (i) but also in order to help establishing the equivalence between versions (a) and (b) of the distributions in (i)–(iv).

Indeed directly from (13) we may write

$$\Phi_{W_{4a}}(t) = \prod_{j=1}^{vk}\prod_{i=1}^{u}\left(a-\frac{j}{k}-i\right)\left(a-\frac{j}{k}-i-\mathrm{i}t\right)^{-1}$$
$$=\prod_{j=1}^{v}\prod_{i=1}^{uk}\left(a-j-\frac{i}{k}\right)\left(a-j-\frac{i}{k}-\mathrm{i}t\right)^{-1} = \Phi_{W_{4b}}(t),$$

and then, by using similar techniques to the ones used above we will be able to establish the equivalence, on one hand with (i)(b) and on the other, with (iii)(b) and (ii)(b).

We should note that in order to have the equivalence between the distributions in (i)(a) and (i)(b) we only need to have $uk, vk \in \mathbb{N}$, although to have these distributions to be equivalent to the distribution of $V$ in (1) we indeed need to have either $vk, u \in \mathbb{N}$ or $uk, v \in \mathbb{N}$. Of course if we have $u, v \in \mathbb{N}$, always considering $k \in \mathbb{N}$, we will then also have the equivalence between the distribution of $V$ in (1) and the distributions in (i)–(iv). However, we should note that if we just have $uk, v \in \mathbb{N}$, with $u \notin \mathbb{N}$, then we will only have the equivalence between the distribution of $V$ in (1) and the distributions in (i)(a), (i)(b), (ii)(b), (iii)(b) and (iv)(b), while if we have $vk, u \in \mathbb{N}$, with $v \notin \mathbb{N}$, we will have the equivalence of the distribution of $V$ in (1) and the distributions in (i)(a), (i)(b), (ii)(a), (iii)(a) and (iv)(a).

The distribution of $V$ in (1) is in this case an EGIG distribution of depth $p$, with shape parameters $r_j$ and rate parameters $\delta_{ij} = \delta_j = a - 1 - \frac{j}{k}$ $(j = 1, \ldots, p)$.

### 4.2.3 A Second More Elaborate Case

$V$ in (1) will also have the distribution of a compound product of independent Beta r.v.'s, most of them having non-integer parameters, which now are function of the indices of the products, if we take

$$\delta_{ij} = a + \frac{s-j}{k}, \qquad (i = 1, \ldots, r_j;\ j = 1, \ldots, p+s)$$

with $p = mk$ for some $m, k \in \mathbb{N}$, and if we define, for some $s \in \mathbb{N}_0$,

$$r_j = \begin{cases} m & j = 1, \ldots, s \\ m + 1 + \left\lfloor \frac{s-j}{k} \right\rfloor & j = s+1, \ldots, p+s\,(= mk+s)\,. \end{cases} \tag{20}$$

In this case the distribution of $V$ in (1), which, from section 2, will be an EGIG distribution of depth $p+s\,(= mk+s)$, with shape parameters $r_j$ and rate parameters $\delta_{ij} = a + \frac{s-j}{k}$ $(j = 1, \ldots, p+s\,(= mk+s))$, is the same as the distribution of

$$V^* = \prod_{\ell=1}^{k} \prod_{j=1+(\ell-1)m}^{\ell m} Y_{j\ell}\,, \quad \text{where} \quad Y_{j\ell} \sim Beta\left(a - \tfrac{j}{k}, \tfrac{j+\ell+s-1}{k}\right) \tag{21}$$

are $p\,(= mk)$ independent r.v.'s.

The proof of this result may be easily obtained, once again, through the use of c.f.'s. If we take $W^* = -\log V^*$ and $W = -\log V$, for $V$ in (1) and $V^*$ in (21), and use the relations in (6) and (4), and yet the relation

$$\prod_{\ell=1}^{k} \Gamma\left(a+\frac{\ell+s-1}{k}\right) = \left\{\prod_{h=0}^{s-1}\left(a+\frac{h}{k}\right)\right\}\left\{\prod_{\ell=1}^{k}\Gamma\left(a+\frac{\ell-1}{k}\right)\right\}$$
$$= \left\{\prod_{h=0}^{s-1}\left(a+\frac{h}{k}\right)\right\} \Gamma(ak)\,(2\pi)^{(k-1)/2}\,k^{\frac{1}{2}-ak},$$

where any product with an upper limit smaller than the lower limit is equal to 1, we have

$$\Phi_{W^*}(t) = \prod_{\ell=1}^{k}\prod_{j=1+(\ell-1)m}^{\ell m} \frac{\Gamma\left(a+\frac{\ell+s-1}{k}\right)\Gamma\left(a-\frac{j}{k}-\mathrm{i}t\right)}{\Gamma\left(a-\frac{j}{k}\right)\Gamma\left(a+\frac{\ell+s-1}{k}-\mathrm{i}t\right)}$$

$$= \prod_{j=1}^{m}\left\{\prod_{h=0}^{s-1}\left(a+\frac{h}{k}\right)\left(a+\frac{h}{k}-\mathrm{i}t\right)^{-1}\right\}\frac{\Gamma(ak)\,\Gamma(ak-kj-ki t)}{\Gamma(ak-kj)\,\Gamma(ak-ki t)}$$

$$= \left\{\prod_{h=1}^{s}\left(a+\frac{s-h}{k}\right)^{m}\left(a+\frac{s-h}{k}-\mathrm{i}t\right)^{-m}\right\}$$
$$\times\left\{\prod_{j=1}^{m}\frac{\Gamma(ak)\,\Gamma(ak-kj-ki t)}{\Gamma(ak-kj)\,\Gamma(ak-ki t)}\right\}$$

$$= \left\{\prod_{h=1}^{s}\left(a+\frac{s-h}{k}\right)^{m}\left(a+\frac{s-h}{k}-\mathrm{i}t\right)^{-m}\right\}$$
$$\times\left\{\prod_{j=1}^{m}\prod_{\ell=0}^{kj-1}(ak-kj+\ell)\,(ak-kj+\ell-ki t)^{-1}\right\}$$

$$= \left\{\prod_{h=1}^{s}\left(a+\frac{s-h}{k}\right)^{m}\left(a+\frac{s-h}{k}-\mathrm{i}t\right)^{-m}\right\}$$
$$\times\left\{\prod_{j=1}^{m}\prod_{\ell=0}^{kj-1}\left(a-j+\frac{\ell}{k}\right)\left(a-j+\frac{\ell}{k}-\mathrm{i}t\right)^{-1}\right\}$$

$$= \left\{\prod_{h=1}^{s}\left(a+\frac{s-h}{k}\right)^{m}\left(a+\frac{s-h}{k}-\mathrm{i}t\right)^{-m}\right\}$$

$$\times \left\{ \prod_{j=1}^{mk} \left(a - \frac{j}{k}\right)^{r_j^*} \left(a - \frac{j}{k} - \mathrm{i}t\right)^{-r_j^*} \right\}$$

$$= \left\{ \prod_{h=1}^{s} \left(a + \frac{s-h}{k}\right)^{m} \left(a + \frac{s-h}{k} - \mathrm{i}t\right)^{-m} \right\}$$

$$\times \left\{ \prod_{j=s+1}^{mk+s} \left(a + \frac{s-j}{k}\right)^{r_j} \left(a + \frac{s-j}{k} - \mathrm{i}t\right)^{-r_j} \right\}$$

$$= \prod_{j=1}^{mk+s} \left(a + \frac{s-j}{k}\right)^{r_j} \left(a + \frac{s-j}{k} - \mathrm{i}t\right)^{-r_j} = \Phi_W(t),$$

for

$$r_j^* = m + 1 + \left\lfloor \frac{-j}{k} \right\rfloor, \quad j = 1, \ldots, mk,$$

and $r_j$ given by (20).

## 5 The Distribution of $V$ in (1) as the Exact Distribution of Several l.r.t. Statistics Used in Multivariate Analysis

The question now is: are there any statistics which have a distribution which fits into any of the cases studied in the previous section? The answer, once again, is: Yes!

Indeed, as we will show in this section, there are a number of statistics used in Multivariate Analysis whose distributions have the form of the distributions studied in the previous section. Among these, we have:

i) the Wilks Lambda statistic to test independence among several sets of variables

   a) when a real multivariate elliptical contoured distribution is assumed as the underlying model and at most one of the sets has an odd number of variables,
   b) when a complex multivariate Normal distribution is assumed as the underlying model, in this case for any number of variables in the groups and any number of groups of variables,

ii) the l.r.t. statistic to test the equality of several mean vectors

   a) when a real multivariate elliptical contoured distribution is assumed as the underlying model and either the number of variables involved is even or the number of means vectors involved is odd,

b) when a complex multivariate Normal distribution is assumed as the underlying model, in this case for any number of variables and any number of mean vectors involved,

iii) the l.r.t. statistic to test circularity of the covariance matrix under multivariate normality, for an odd number of variables involved,
iv) the l.r.t. statistic to test simultaneously that means are equal and the covariance matrix is circular, for any number of variables involved.

We will address in detail the exact distribution of these statistics, namely devising for which values of $r_j$ and $\delta_{ij}$ they may be written in the form in (1).

## 5.1 *The Distribution of V in (1) as the Exact Distribution of the Wilks Λ Statistic*

Let us assume

$$\underline{X} = \left[\underline{X}'_1, \underline{X}'_2, \dots, \underline{X}'_m\right]' \sim N_p(\underline{\mu}, \Sigma) \tag{22}$$

where

$$\underline{\mu} = \left[\underline{\mu}'_1, \underline{\mu}'_2, \dots, \underline{\mu}'_m\right]' \quad \text{and} \quad \Sigma = \begin{bmatrix} \Sigma_{11} & \Sigma_{12} & \dots & \Sigma_{1m} \\ \Sigma_{21} & \Sigma_{22} & \dots & \Sigma_{2m} \\ \vdots & \vdots & \ddots & \vdots \\ \Sigma_{m1} & \Sigma_{m2} & \dots & \Sigma_{mm} \end{bmatrix}$$

with

$$\underline{\mu}_h = E(\underline{X}_h) \text{ and } \Sigma_{hh} = Var(\underline{X}_h), \ \Sigma_{hh'} = Cov(\underline{X}_h, \underline{X}_{h'}), \ h,h' \in \{1, \dots, m\},$$

and that we wish to test the null hypothesis

$$H_0 : \ \Sigma_{hh'} = 0 \ \text{ for all } h \neq h' \in \{1, \dots, m\}, \tag{23}$$

that is the null hypothesis of independence of the $m$ sets of variables $\underline{X}_h$ $(h = 1, \dots, m)$.

We will suppose that each $\underline{X}_h$ has $p_h$ variables, and we will take

$$p = \sum_{h=1}^{m} p_h \, . \tag{24}$$

We consider both the real and complex cases.

### 5.1.1 The Real Case

Let us suppose that the distribution in (22) is a real $p$-multivariate Normal distribution and that we have a sample of size $n$ from the distribution of $\underline{X}$. Then, the $2/n$-th power of the l.r.t. statistic to test $H_0$ in (23) is the Wilks $\Lambda$ statistic,

$$\Lambda = \frac{|S|}{\prod_{h=1}^{m} |S_{hh}|} \tag{25}$$

where $S$ is the MLE of $\Sigma$ and $S_{hh}$ the MLE of $\Sigma_{hh}$ $(h = 1, \ldots, m)$.

In this subsection we will show how when we assume for $\underline{X}$ a real $p$-multivariate Normal distribution and at most one of the $p_h$'s is odd, the exact distribution of the statistic $\Lambda$ in (25) may be written in the form in (1), yielding an EGIG distribution as the exact distribution of $\Lambda$, in these cases.

Note that this distribution, under $H_0$ in (23) is still valid if we assume for $\underline{X}$ an underlying elliptically contoured or left orthogonal-invariant distribution (Anderson et al., 1986; Anderson and Fang, 1990; Jensen and Good, 1981; Kariya, 1981).

For $m \geq 2$, it may be shown (Anderson, 2003, Chap. 9) that we may write

$$\Lambda = \prod_{h=1}^{m-1} \Lambda_{h,(h+1,\ldots,m)}\,, \tag{26}$$

where $\Lambda_{h,(h+1,\ldots,m)}$ $(h = 1, \ldots, m-1)$ is the l.r.t. statistic to test

$$H_{0h}: \; \Sigma_{h\ell} = 0, \; \ell = h+1, \ldots, m \qquad (h = 1, \ldots, m-1).$$

Under $H_0$ in (23), the $m-1$ statistics $\Lambda_{h,(h+1,\ldots,m)}$ $(h = 1, \ldots, m)$ are independent, with

$$\Lambda_{h,(h+1,\ldots,m)} \overset{\text{st}}{\sim} \prod_{j=1}^{p_h} Y_j \overset{\text{st}}{\sim} \prod_{j=1}^{q_h} Y_j^*\,, \quad \text{where} \quad \begin{array}{l} Y_j \sim Beta\left(\frac{n-q_h-j}{c}, \frac{q_h}{c}\right) \text{ or} \\ Y_j \sim Beta\left(\frac{n-1-q_{h-1}+j}{c}, \frac{q_h}{c}\right) \\ \text{and} \\ Y_j^* \sim Beta\left(\frac{n-p_h-j}{c}, \frac{p_h}{c}\right) \text{ or} \\ Y_j^* \sim Beta\left(\frac{n-1-p_{h-1}+j}{c}, \frac{p_h}{c}\right) \\ \text{are } p_h \text{ or } q_h \text{ independent r.v.'s,} \end{array} \tag{27}$$

where $c = 2$ and $q_h = p_{h+1} + \cdots + p_m$.

Therefore, if at most one of the $p_h$'s is odd, without any loss of generality, let it be $p_m$, so that from (26) and (27) and the equivalence between the distributions in (i)(a) and (b) and (1) in Section 4.2.2, the distributions in (27) above, are, for $k = 2$,

$a = n/2$, $u = q_h/2$ and $v = p_h/2$, or $uk = q_h$ and $vk = p_h$, so that if either $p_h$ or $q_h$ is even, that is, if either $u$ or $v \in \mathbb{N}$, equivalent to the distribution of

$$\prod_{j=1}^{p_h+q_h-c} \prod_{i=1}^{r_{hj}} U_{ij}^{\frac{1}{(n-j)/c-1}}, \tag{28}$$

where $c = 2$, the $U_{ij}$ are i.i.d. Unif(0, 1) and the $r_{hj}$, for $h = 1, \ldots, m-1$, are defined in the same manner as the $r_j$'s in (8) and (9) with $k = 2$ and $vk$ and $uk$ replaced respectively by $p_h$ and $q_h$.

But then, from (26) we may write

$$\Lambda \overset{\text{st}}{\sim} \prod_{k=1}^{m-1} \prod_{j=1}^{p_k+q_k-c} \prod_{i=1}^{r_{kj}} U_{ij}^{\frac{1}{(n-j)/c-1}}, \tag{29}$$

which, by taking $p$ as defined in (24), actually also with $p = p_1 + q_1$, allows us to write, for $c = 2$, the c.f. of $W = -\log \Lambda$ as

$$\begin{aligned}\Phi_W(t) &= \prod_{h=1}^{m-1} \prod_{j=1}^{p_h+q_h-c} \left(\frac{n-j}{c} - 1\right)^{r_{hj}} \left(\frac{n-j}{c} - 1 - \mathrm{i}t\right)^{-r_{hj}} \\ &= \prod_{j=1}^{p-c} \left(\frac{n-j}{c} - 1\right)^{r_j} \left(\frac{n-j}{c} - 1 - \mathrm{i}t\right)^{-r_j},\end{aligned} \tag{30}$$

with

$$r_j = \sum_{h=1}^{m-1} r_{hj}^* \tag{31}$$

in which, for $c = 2$ and $h = 1, \ldots, m-1$,

$$r_{hj}^* = \begin{cases} r_{hj} & j = 1, \ldots, p_h + q_h - c \\ 0 & j = p_h + q_h - (c-1), \ldots, p - c\,. \end{cases} \tag{32}$$

Consequently we may write, once again for $c = 2$,

$$\Lambda \overset{\text{st}}{\sim} \prod_{j=1}^{p-c} \prod_{i=1}^{r_j} U_{ij}^{\frac{1}{(n-j)/c-1}} \tag{33}$$

which shows that the exact distribution of $\Lambda$ is in this case an EGIG distribution of depth $p-2$ with shape parameters $r_j$ and rate parameters $\delta_j = \frac{n-j}{2} - 1$ ($j = 1, \ldots, p-2$), and thus with p.d.f. and c.d.f. given by (see (B.1) and (B.2) in Appendix B)

$$f_\Lambda(\ell) = f^{EGIG}\left(\ell \,\middle|\, r_j; \frac{n-j}{c} - 1; j = 1, \ldots, p-2\right) \tag{34}$$

and

$$F_\Lambda(\ell) = F^{EGIG}\left(\ell \,\middle|\, r_j; \frac{n-j}{c} - 1; j = 1, \ldots, p-2\right), \tag{35}$$

for $c = 2$, $0 < \ell < 1$, $p = \sum_{h=1}^{m} p_h$, and $r_j$ given by (31) and (32) above. We should remark that these results exactly agree with those in Coelho (1998).

While from (26) and (27) and the expression for the $s$-th moment of a Beta r.v., we have the usual expression for the $s$-th moment of the Wilks $\Lambda$ statistic given by

$$E\left(\Lambda^s\right) = \prod_{h=1}^{m-1} \prod_{j=1}^{p_h} \frac{\Gamma\left(\frac{n-j}{c}\right) \Gamma\left(\frac{n-q_h-j}{c} + s\right)}{\Gamma\left(\frac{n-q_h-j}{c}\right) \Gamma\left(\frac{n-j}{c} + s\right)}, \quad \left(s > -\tfrac{n-p}{c}\right), \tag{36}$$

for $c = 2$, from (33) and (2), we see that for the case in which at most one of the $m$ sets of variables has an odd number of variables, the $s$-th moment of $\Lambda$ can be also expressed explicitly as (see also Appendix B for the moments of the EGIG distribution)

$$E\left(\Lambda^s\right) = \prod_{j=1}^{p-c} \prod_{\ell=1}^{r_j} \frac{1}{\frac{c}{n-j-c}s + 1} = \prod_{j=1}^{p-c} \left(\frac{n-c-j}{n-c-j+cs}\right)^{r_j}, \quad \left(s > -\tfrac{n-p}{c}\right) \tag{37}$$

for $c = 2$ and $r_j$ given by (31) and (32).

We should note that if at most one of the $p_h$'s is odd, then we will always have either $p_h$ or $q_h$ even for $h = 1, \ldots, m-1$ in (27) so that, according to the observations in Section 4.2.2, the representations in (28) and (29) hold. If two or more of the $p_h$'s are odd, we will have at least one of the $\Lambda_{h,(h+1,\ldots,m)}$ with both $p_h$ and $q_h$ odd, more precisely, if there are $\ell$ sets with an odd number of variables, there will be $\lfloor \ell/2 \rfloor$ of the $\Lambda_{h,(h+1,\ldots,m)}$ with both $p_h$ and $q_h$ odd, so that in this case neither (28) nor (29) are valid.

### 5.1.2 The Complex Case

Let us suppose that the distribution in (22) is a complex $p$-multivariate Normal distribution with expected value $\underline{\mu}$ and variance-covariance matrix $\Sigma$ (Wooding

(1956); Goodman (1963a), see also (Anderson, 2003, problem 2.64)). Then, for a sample of size $n$, the $1/n$-th power of the l.r.t. statistic to test the independence of the $m$ sets of variables $\underline{X}_h$ $(h = 1, \ldots, m)$ will be the statistic in (25) (see Wooding (1956); Goodman (1963a); see also (Anderson, 2003, problem 3.11) for references concerning the maximum likelihood estimators in the complex case). Now, under $H_0$ in (23), using a similar procedure to the one used for the real case and the results in Goodman (1963b) on the distribution of the determinant of a complex Wishart matrix, it is possible to show that, for the statistics $\Lambda_{h(h+1,\ldots,m)}$ $(h = 1, \ldots, m-1)$ in (26), a similar relation to the one in (27) still holds, now with $c = 1$.

Therefore, for any values of $p_h$ $(h = 1, \ldots, m)$, from (26) and (27) and the equivalence between the distributions in (i)(a) and (b) and (1) in Section 4.2.2, the distributions in (27) are, for $k = 1$, $a = n$, $u = q_h$ and $v = p_h$, so that $u \in \mathbb{N}$ and $v \in \mathbb{N}$, equivalent to the distribution in (28), where the $r_{hj}$, for $h = 1, \ldots, m-1$, are now defined in the same manner as the $r_j$'s in (8) and (9) with $k = 1$ and $vk = v$ and $uk = u$ replaced respectively by $p_h$ and $q_h$.

But then we have (29), now for $c = 1$, which allows us to write the c.f. of $W = -\log \Lambda$ as in (30), with $c = 1$ and $r_j$ given by (31), in which the $r^*_{hj}$ are defined in a similar manner as in (32), now with $c = 1$.

Consequently, for the complex case we have (33) for $c = 1$, which shows that the exact distribution of $\Lambda$ is in the complex case, for any $p_h$ $(h=1, \ldots, m)$ an EGIG distribution of depth $p-1$ with shape parameters $r_j$ given by (31) and (32), for $c = 1$, and rate parameters $\delta_j = n - 1 - j$ $(j = 1, \ldots, p-1)$, and thus with p.d.f. and c.d.f. respectively given by (34) and (35), for $c = 1$.

While from (26) and (27) and the expression for the $s$-th moment of a Beta r.v., we have the expression for the $s$-th moment of the Wilks $\Lambda$ statistic for the complex case, given by (36) for $c = 1$, from (33) and (2), we see that for the complex case the $s$-th moment of $\Lambda$ can also be expressed explicitly as in (37) for $c = 1$ and $r_j$ given by (31) and (32), also for $c = 1$.

We remark that while the expression for the $s$-th moment of $\Lambda$ as given by (36) for $c = 1$, agrees with the expressions used by Krishnaiah et al. (1976) and Fang et al. (1982), the result obtained, concerning the expression for the exact distribution of $\Lambda$ in the complex case, with p.d.f. and c.d.f. given by (34) and (35) for $c = 1$, for any $p_h$ $(h = 1, \ldots, m)$, is considerably simpler than any former existing results (see for example Krishnaiah et al. (1976); Fang et al. (1982)).

## 5.2 *The Distribution of V in (1) as the Exact Distribution of the l.r.t. Statistic to Test Equality of Several Mean Vectors*

Let us suppose that

$$\underline{X}_h \sim N_p\left(\underline{\mu}_h, \Sigma\right), \quad h = 1, \ldots, q \tag{38}$$

and let us suppose that we have a sample of size $n_h$ from the distribution of $\underline{X}_h$ $(h = 1, \ldots, q)$. Let us further suppose that the $q$ samples are independent, the sample from $\underline{X}_h$ being arranged in the matrix $X_h$ of dimensions $n_h \times p$, and that we want to test the null hypothesis

$$H_0 : \underline{\mu}_1 = \cdots = \underline{\mu}_q \,. \tag{39}$$

Consider the matrices

$$A = \sum_{h=1}^{q} \left( X_h - E_{n_h 1} \overline{\underline{X}}_h' \right)' \left( X_h - E_{n_h 1} \overline{\underline{X}}_h' \right)$$

and

$$B = \sum_{h=1}^{q} n_h \left( \overline{\underline{X}}_h - \overline{\underline{X}} \right) \left( \overline{\underline{X}}_h - \overline{\underline{X}} \right)'$$

where, for $n = \sum_{h=1}^{q} n_h$,

$$\overline{\underline{X}}_h = \frac{1}{n_h} X_h' E_{n_h 1} \qquad \text{and} \qquad \overline{\underline{X}} = \frac{1}{n} n_h \overline{\underline{X}}_h \,,$$

with $E_{nm}$ denoting a matrix of 1's with dimensions $n \times m$. Consider then the statistic

$$\Lambda_2 = \frac{|A|}{|A + B|} \,. \tag{40}$$

Here also, we consider both the real and the complex cases.

### 5.2.1 The Real Case

Let us suppose that the $p$-multivariate Normal distributions in (38) are real valued. Then, for $n = \sum_{h=1}^{q} n_h$, the $2/n$-th power of the l.r.t. statistic to test $H_0$ in (39) is the statistic $\Lambda_2$ in (40), where under $H_0$ in (39), $A$ and $B$ are two independent Wishart matrices, with

$$A \sim W_p(n - q, \Sigma)$$

and

$$B \sim W_p(q - 1, \Sigma) \,,$$

so that, (see Anderson (2003, sec. 8.8, 8.9)), under $H_0$ in (39),

$$\Lambda_2 \overset{\text{st}}{\sim} \prod_{j=1}^{p} Y_j \overset{\text{st}}{\sim} \prod_{j=1}^{q-1} Y_j^*, \quad \text{where} \quad \begin{array}{l} Y_j \sim Beta\left(\frac{n-(q-1)-j}{c}, \frac{q-1}{c}\right) \text{ or} \\ Y_j \sim Beta\left(\frac{n-p-q+j}{c}, \frac{q-1}{c}\right) \\ \text{and } Y_j^* \sim Beta\left(\frac{n-p-j}{c}, \frac{p}{c}\right) \text{ or} \\ Y_j^* \sim Beta\left(\frac{n-q-p+j}{c}, \frac{p}{c}\right) \\ \text{are } p \text{ and } q-1 \text{ independent r.v.'s,} \end{array} \tag{41}$$

for $c = 2$. Therefore, if $p$ is even, or $q$ is odd, from the equivalence between the distributions in (i)(a) and (b) and (1) in Section 4.2.2, the distributions in (41) above, are, for $k = 2$, $a = n/2$, $u = (q-1)/2$ and $v = p/2$, or $uk = q-1$ and $vk = p$, equivalent to the distribution of

$$\prod_{j=1}^{p+q-c-1} \prod_{i=1}^{r_j} U_{ij}^{\frac{1}{(n-j)/c-1}}, \tag{42}$$

for $c = 2$, where the $U_{ij}$ are i.i.d. Unif(0, 1) and the $r_j$ are defined in the same manner as the $r_j$'s in (8) and (9) with $k = 2$ and $vk$ and $uk$ replaced respectively by $p$ and $q-1$, so that if either $p$ or $q-1$ are even, then either $u \in \mathbb{N}$ or $v \in \mathbb{N}$.

This shows that the exact distribution of $\Lambda_2$ is in this case an EGIG distribution of depth $p+q-3$ with shape parameters $r_j$ and rate parameters $\delta_j = \frac{n-j}{2} - 1$ ($j = 1, \ldots, p+q-3$), and thus with p.d.f. and c.d.f. given, for $c = 2$, by (see (B.1) and (B.2) in Appendix B)

$$f_{\Lambda_2}(\ell) = f^{EGIG}\left(\ell \,\middle|\, r_j; \frac{n-j}{c} - 1; j = 1, \ldots, p+q-c-1\right) \tag{43}$$

and

$$F_{\Lambda_2}(\ell) = F^{EGIG}\left(\ell \,\middle|\, r_j; \frac{n-j}{c} - 1; j = 1, \ldots, p+q-c-1\right), \tag{44}$$

for $0 < \ell < 1$ and $r_j$ given by (8) and (9) with $k = 2$ and $vk$ and $uk$ replaced respectively by $p$ and $q-1$.

We remark that these results exactly agree with those in Coelho (1998) and Marques et al. (2010).

### 5.2.2 The Complex Case

Let us suppose that the distribution in (38) is a complex $p$-multivariate Normal distribution with expected value $\underline{\mu}_h$ and variance-covariance matrix $\Sigma$ (Wooding (1956); Goodman (1963a); see also (Anderson, 2003, problem 2.64)). Then, for $n = \sum_{h=1}^{q} n_h$, the $1/n$-th power of the l.r.t. statistic to test $H_0$ in (39) will be the statistic in (40) (see Wooding (1956); Goodman (1963b); Khatri (1965) and also (Anderson, 2003, problem 3.11) for references concerning the maximum likelihood estimators in the complex case). Now, using a similar procedure to the one used for the real case, it is possible to show that, under $H_0$ in (39), $A$ and $B$ are two independent complex Wishart matrices, with

$$A \sim CW_p(n-q, \Sigma) \qquad \text{and} \qquad B \sim CW_p(q-1, \Sigma)\,,$$

so that, (see Goodman (1963b)), under $H_0$ in (39),

$$2^p\,|A| \overset{\text{st}}{\sim} |\Sigma| \prod_{j=1}^{p} W_j \qquad \text{and} \qquad 2^p\,|A+B| \overset{\text{st}}{\sim} |\Sigma| \prod_{j=1}^{p} Z_j\,,$$

where,

$$W_j \sim \chi^2_{2(n+1-q-j)} \qquad \text{and} \qquad Z_j \sim \chi^2_{2(n-j)}\,, \quad j = 1, \ldots, p\,,$$

actually with

$$Z_j = W_j + W_j^*$$

where $W_j$ and $W_j^*$ are independent, with

$$W_j^* \sim \chi^2_{2(q-1)}\,, \quad j = 1, \ldots, p\,,$$

so that in this case we have (41), now for $c = 1$. Therefore, for any $p$ and any $q$, from (26) and (41), for $c = 1$, and the equivalence between the distributions in (i)(a) and (b) and (1) in Section 4.2.2, the distributions in (41) are now, for $k = 1$, $a = n$, $u = q - 1$ and $v = p$, or $uk = q - 1$ and $vk = p$, equivalent to the distribution in (42) for $c = 1$, where the $r_j$ are defined in the same manner as the $r_j$'s in (8) and (9) with $k = 1$ and $vk$ and $uk$ replaced respectively by $p$ and $q - 1$.

This shows that the exact distribution of $\Lambda_2$ is in this case an EGIG distribution of depth $p + q - 2$ with shape parameters $r_j$ and rate parameters $\delta_j = n - j - 1$ $(j = 1, \ldots, p + q - 2)$, and thus with p.d.f. and c.d.f. given respectively by (43) and (44) for $c = 1$, with $r_j$ given by (8) and (9) with $k = 1$ and $vk$ and $uk$ replaced respectively by $p$ and $q - 1$.

We should remark that the results in this subsection give decisively much simpler expressions for the exact p.d.f. and c.d.f. of the l.r.t. statistic to test the equality of several mean vectors than former existing results as it is the case with the ones in Khatri (1965).

## 5.3 *The Distribution of V in (1) as the Exact Distribution of the l.r.t. Statistic Involved in Tests of Circularity*

### 5.3.1 The Distribution of $V$ in (1) as the Exact Distribution of the l.r.t. Statistic to Test Circularity of the Covariance Matrix

A symmetric covariance matrix $\Sigma = \left[\sigma_{ij}\right]$ $(i, j = 1, \ldots, p)$, is said to be circular if

$$\sigma_{ij} = \sigma^2 \rho_{j-i}\,, \quad \text{with} \quad \rho_0 = 1;\ \rho_k = \rho_{-k} = \rho_{p-k} \ \ (k = 1, \ldots, p-1)\,.$$

Let us assume that

$$\underline{X} \sim N_p(\underline{\mu}, \Sigma)\,,$$

and that, based on a sample of size $n$, we want to test the hypothesis of circularity of the covariance matrix.

Then, for $m = \lfloor p/2 \rfloor$, according to Olkin and Press (1969), the $2/n$-th power of the l.r.t. statistic, call it $\Lambda_3$, will have the same distribution as

$$\left\{\prod_{j=1}^{m} Y_j\right\}\left\{\prod_{j=m+1}^{p-1} Y_j^*\right\}, \quad \text{where} \quad \begin{array}{l} Y_j \sim Beta\left(\frac{n-1-j}{2}, \frac{j}{2}\right) \text{ and} \\ Y_j^* \sim Beta\left(\frac{n-1-j}{2}, \frac{j+1}{2}\right) \\ \text{are } p-1 \text{ independent r.v.'s.} \end{array} \tag{45}$$

But then, according to the results in Section 4.2.3, for odd $p$ we only have to take $k = 2$, $s = 0$, $m = \frac{p-1}{2}$ and $a = \frac{n-1}{2}$ in (21) to see that the exact distribution of $\Lambda_3$ is in this case an EGIG distribution of depth $p-1$, with rate parameters $\frac{n-1-j}{2}$ and shape parameters

$$r_j = \frac{p+1}{2} + \left\lfloor -\frac{j}{2} \right\rfloor, \quad j = 1, \ldots, p-1\,,$$

with p.d.f. and c.d.f. respectively given by (see (B.1) and (B.2) in Appendix B)

$$f_{\Lambda_3}(\ell) = f^{EGIG}\left(\ell \,\middle|\, \frac{p+1}{2} + \left\lfloor -\frac{j}{2} \right\rfloor; \frac{n-1-j}{2}; j = 1, \ldots, p-1\right)$$

and

$$F_{\Lambda_3}(\ell) = F^{EGIG}\left(\ell \,\Big|\, \frac{p+1}{2} + \left\lfloor -\frac{j}{2} \right\rfloor ; \frac{n-1-j}{2} ; j = 1, \ldots, p-1\right),$$

for $0 < \ell < 1$.

We should remark that this result represents an improvement over the results in Olkin and Press (1969), where only an asymptotic distribution is given for $\Lambda_3$, and is far simpler and more manageable than the results in Nagar et al. (2004), involving no infinite sums.

However, for even $p$ we cannot apply the results in Section 4.2.3.

### 5.3.2 The Distribution of $V$ in (1) as the Exact Distribution of the l.r.t. Statistic for the Simultaneous Test of Circularity of the Covariance Matrix and Equality of the Means

Let, once again,

$$\underline{X} \sim N_p(\underline{\mu}, \Sigma),$$

and let us assume that, based on a sample of size $n$, we want to test simultaneously the hypothesis of circularity of the covariance matrix and the hypothesis of equality of all the means in $\underline{\mu}$.

Then, for $m = \lfloor p/2 \rfloor$, according to Olkin and Press (1969), the $2/n$-th power of the l.r.t. statistic, call it $\Lambda_4$, will have the same distribution as

$$\left\{\prod_{j=1}^{m} Y_j\right\}\left\{\prod_{j=m+1}^{p-1} Y_j^*\right\}, \quad \text{where} \quad \begin{array}{l} Y_j \sim Beta\left(\frac{n-1-j}{2}, \frac{j+1}{2}\right) \text{ and} \\ Y_j^* \sim Beta\left(\frac{n-1-j}{2}, \frac{j+2}{2}\right) \\ \text{are } p-1 \text{ independent r.v.'s.} \end{array} \tag{46}$$

Then, for odd $p$, since we have $\frac{p-1}{2}$ variables in each of the products in (46), we just have to take $k = 2$, $s = 1$, $a = \frac{n-1}{2}$ and $m = \frac{p-1}{2}$ in (21) in order to be able to see that the exact distribution of $\Lambda_4$ in this case is an EGIG distribution of depth $p\ (= p-1+s)$ with rate parameters $\frac{n-j}{2}$ $(j = 1, \ldots, p)$ and shape parameters

$$r_j = \begin{cases} \frac{p-1}{2} & j = 1 \\ \frac{p+1}{2} + \left\lfloor \frac{1-j}{2} \right\rfloor & j = 2, \ldots, p\,. \end{cases} \tag{47}$$

Thus, for odd $p$, the exact p.d.f. and c.d.f. of $\Lambda_4$ are thus given by

$$f_{\Lambda_4}(\ell) = f^{EGIG}\left(\ell \,\middle|\, r_j; \frac{n-j}{2}; j=1,\ldots,p\right) \tag{48}$$

and

$$F_{\Lambda_4}(\ell) = F^{EGIG}\left(\ell \,\middle|\, r_j; \frac{n-j}{2}; j=1,\ldots,p\right), \tag{49}$$

for $0 < \ell < 1$ and $r_j$ given by (47).

For even $p$, in order to have an equal number of terms in each product, more precisely $\frac{p}{2}-1$, we may see that the distribution of $\Lambda_4$ is the same as the distribution of

$$Y_0^{**}\left\{\prod_{j=1}^{m-1} Y_j^{**}\right\}\left\{\prod_{j=m}^{p-2} Y_j^{***}\right\}, \quad \text{where} \quad \begin{array}{l} Y_j^{**} \sim Beta\left(\frac{n-2-j}{2}, \frac{j+2}{2}\right) \text{ and} \\ Y_j^{***} \sim Beta\left(\frac{n-2-j}{2}, \frac{j+3}{2}\right), \\ \text{are } p-1 \text{ independent r.v.'s,} \end{array} \tag{50}$$

so that taking $k=2$, $s=2$, $a=\frac{n-2}{2}$ and $m=\frac{p}{2}$ in (21) we see that the distribution of

$$\left\{\prod_{j=1}^{m-1} Y_j^{**}\right\}\left\{\prod_{j=m}^{p-2} Y_j^{***}\right\}$$

in (50), is an EGIG distribution of depth $p\,(=p-2+s)$ with rate parameters $\frac{n-2-j}{2}$ $(j=-1,\ldots,p-2)$ and shape parameters

$$r_j = \begin{cases} \frac{p}{2}-1 & j=-1,0 \\ \frac{p}{2}+\left\lfloor -\frac{j}{2}\right\rfloor & j=1,\ldots,p-2\,. \end{cases}$$

But then, accounting for $Y_0^{**}$, we have that the exact distribution of $\Lambda_4$ is, for even $p$, an EGIG distribution of depth $p$ with shape parameters $\frac{n-j}{2}$ $(j=1,\ldots,p)$ and rate parameters

$$r_j = \begin{cases} \frac{p}{2}-1 & j=1 \\ \frac{p}{2} & j=2 \\ \frac{p}{2}+\left\lfloor \frac{2-j}{2}\right\rfloor & j=3,\ldots,p\,. \end{cases} \tag{51}$$

Thus, for even $p$, the exact p.d.f. and c.d.f. of $\Lambda_4$ are given by (48) and (49), with $r_j$ given by (51) above.

These results on the exact distribution of the l.r.t. statistic to test simultaneously the circularity of the covariance matrix and the equality of means show a significant improvement over the results presented in Olkin and Press (1969), where only asymptotic distributions are presented and discussed.

## 6 Finite Representations for the Meijer G Function in Some Particular Cases

From the definition of the Meijer G function (see for example expression (9) in Springer and Thompson (1970) for a sharp compact definition) and from the results in Section 4.2, it is easy to see that, besides the rather simple cases in Theorem 8 and the succeeding Corollary, at the end of Section 5 of Springer and Thompson (1970) and Theorems 4.4.1 and 4.4.2 in Springer (1979), there are other more elaborate and encompassing cases, in which the arguments of the Meijer G function do not even need to be integers, for which there are finite simple representations of this function. We summarize in the following Theorem the results obtained in Section 4.2, concerning finite representations under the form of EGIG p.d.f.'s for some particular cases of the Meijer G function.

**Theorem 1** *For*

(i) $a_j > 0$ *and* $b_j \in \mathbb{N}$, $j = 1, \dots, m$,
(ii) $a_j = a - v - \frac{j}{k}$ *and* $b_j = v$, $j = 1, \dots, m (= uk)$,
*for some* $a > u + v$, *with* $u, k \in \mathbb{N}$ *and* $vk \in \mathbb{N}$ *(not necessarily with* $v \in \mathbb{N}$*),*
(iii) $a_j = a - \frac{j}{k}$ *and* $b_j = \frac{j+s+\left\lfloor \frac{j-1}{m} \right\rfloor}{k}$, $j = 1, \dots, m (= uk)$,
*for some* $a > u$ *and* $s \in \mathbb{N}_0$, *with* $u, k \in \mathbb{N}$,

$$G^{m\,0}_{m\,0}\left(z \,\middle|\, \begin{matrix} a_1+b_1-1, \dots, a_m+b_m-1 \\ b_1-1, \dots, b_m-1 \end{matrix}\right) = \frac{1}{K}\, f^{\text{EGIG}}(z|r_j; \lambda_j; j = 1, \dots, p) \tag{52}$$

*where* $f^{\text{EGIG}}(\cdot\,|\,\cdot;\ \cdot;\ \cdot)$ *is defined in (B.1) in Appendix B,*

$$K = \prod_{j=1}^{m} \frac{\Gamma(a_j + b_j)}{\Gamma(a_j)}$$

*and* $r_j$ *and* $\lambda_j$, *for* (i)–(iii) *above are respectively given by*

(i) $\lambda_1, \dots, \lambda_p$ *are the* $p$ *different values of*

$$a_j + i - 1, \quad i = 1, \dots, b_j;\ j = 1, \dots, m$$

*and* $r_j$ $(j = 1, \dots, p)$ *is the number of times* $\lambda_j$ *occurs in the* $a_j$+$i$−1 *sequence,*

(ii) $\lambda_j = a - 1 - \frac{j}{k}$, $j = 1, \ldots, p(= (u + v - 1)k)$, *and, for* $h_j$ *given by (9),*

$$r_j = \begin{cases} h_j & j = 1, \ldots, k \\ h_j + r_{j-k} & j = k + 1, \ldots, p \end{cases}$$

(iii) $\lambda_j = a + \frac{s-j}{k}$, $j = 1, \ldots, p\,(= uk + s)$ *and*

$$r_j = \begin{cases} m & j = 1, \ldots, s \\ m + 1 + \left\lfloor \frac{s-j}{k} \right\rfloor & j = s + 1, \ldots, p\,(= uk + s)\,. \end{cases}$$

***Proof*** Omitted, since the result is directly derived from the definition of the Meijer G function and the results in Section "*V* in (1) as the Product of Independent Beta r.v.'s". □

From the computing times in Tables 1–3 we may see how much faster to compute are the EGIG finite sum representations of the Meijer G function on the right hand side of (52) when compared with the usual Mathematica Meijer G function implementation.

In Tables 1–3 we have the computing times (in seconds) obtained with an Intel(R) Core(TM)2 Duo CPU P8700 at 2.53 GHz machine with 4 GB of RAM, running on Windows 7 (64-bit), at full speed and with Mathematica as the only running application. All computing times reported are the average of 3 runs. We

**Table 1** Comparison of computing times (in seconds) for the implementation of the Meijer G function in Mathematica and the EGIG implementation in (i) of Theorem 1 for $a_j = \left\{\frac{23}{17}, \frac{18}{17}, \frac{4}{3}, \frac{17}{15}, \frac{13}{10}, \frac{7}{6}, \frac{19}{17}, \frac{12}{13}, \frac{5}{4}, \frac{7}{6}, \frac{13}{12}, \frac{9}{5}, \frac{5}{3}, \frac{6}{5}, \frac{25}{11}\right\}$ $(j = 1, \ldots, p)$ and $b_j = \{5, 3, 7, 5, 3, 4, 8, 6, 7, 5, 7, 6, 4, 3, 9\}$ $(j = 1, \ldots, p)$ for several values of $p$ (all values were computed for $z = 2/1000$ and $a = 55.3$, with $d$ precision digits, in order to assure 16 decimal places in the result).

| | | computing times | | | | | |
|---|---|---|---|---|---|---|---|
| | | Mathematica ver. 7 | | Mathematica ver. 5.2 | | ratios | |
| $p$ | $d$ | EGIG (1) | Meijer-G (2) | EGIG (3) | Meijer-G (4) | (2)/(1) | (4)/(3) |
| 5 | 23 | < 0.01 | 4.84 | 0.02 | 3.84 | >484 | 192 |
| 6 | 27 | 0.02 | 12.93 | 0.02 | 10.43 | 647 | 522 |
| 7 | 30 | 0.02 | 59.08 | 0.02 | 47.97 | 2954 | 2399 |
| 8 | 35 | 0.03 | 149.50 | 0.02 | 128.11 | 4983 | 6406 |
| 9 | 41 | 0.03 | –/– | 0.05 | 5381.21 | — | 107624 |
| 10 | 47 | 0.05 | –/– | 0.05 | –/– | — | — |
| 12 | 58 | 0.06 | –/– | 0.08 | –/– | — | — |
| 15 | 73 | 0.09 | –/– | 0.09 | –/– | — | — |

–/– indicates a computing time which exceeded 2 hours

**Table 2** Comparison of computing times (in seconds) for the implementation of the Meijer G function in Mathematica and the EGIG implementation in (ii) of Theorem 1 for different values of $u$, $k$ and $v$, and for $z = 0.5$ (all values were computed for $a = 15.3$, with $d$ precision digits, in order to assure 16 decimal places in the result).

| | | | | computing times | | | | | |
|---|---|---|---|---|---|---|---|---|---|
| | | | | Mathematica ver. 7 | | Mathematica ver. 5.2 | | ratios | |
| $u$ | $k$ | $v$ | $d$ | EGIG (1) | Meijer-G (2) | EGIG (3) | Meijer-G (4) | (2)/(1) | (4)/(3) |
| 6 | 3 | $7/k$ | 70 | 0.02 | 2.62 | 0.02 | 2.42 | 131 | 121 |
| 6 | 4 | $7/k$ | 72 | 0.02 | 1.51 | 0.02 | 1.45 | 76 | 73 |
| 6 | 5 | $7/k$ | 72 | 0.03 | 184.85 | 0.02 | 13.46 | 6162 | 673 |
| 6 | 6 | $7/k$ | 72 | 0.02 | 3.88 | 0.02 | 3.82 | 194 | 191 |
| 6 | 7 | $7/k$ | 72 | 0.03 | 0.03 | 0.02 | 2.81 | 1 | 141 |
| 6 | 8 | $7/k$ | 72 | 0.02 | 1.95 | 0.03 | 1.98 | 98 | 66 |
| 6 | 9 | $7/k$ | 72 | 0.03 | 4.52 | 0.03 | 4.34 | 151 | 145 |
| 6 | 10 | $7/k$ | 72 | 0.03 | 56.77 | 0.02 | 14.74 | 1892 | 737 |
| 6 | 11 | $7/k$ | 72 | 0.05 | 4.82 | 0.05 | 4.60 | 96 | 92 |
| 6 | 12 | $7/k$ | 72 | 0.05 | 41.82 | 0.05 | 18.02 | 836 | 360 |
| 6 | 13 | $7/k$ | 72 | 0.06 | 6.44 | 0.05 | 5.60 | 107 | 112 |
| 6 | 14 | $7/k$ | 72 | 0.06 | 6.52 | 0.06 | 5.57 | 109 | 93 |
| 7 | 7 | $5/k$ | 57 | 0.02 | 1.61 | 0.02 | 1.56 | 81 | 78 |
| 7 | 7 | $6/k$ | 69 | 0.03 | 2.70 | 0.03 | 2.68 | 90 | 89 |
| 7 | 7 | $7/k$ | 79 | 0.03 | 0.05 | 0.03 | 4.48 | 2 | 149 |
| 7 | 7 | $8/k$ | 92 | 0.05 | 27.88 | 0.05 | 21.75 | 558 | 435 |
| 7 | 7 | $9/k$ | 106 | 0.05 | 34.10 | 0.03 | 27.28 | 682 | 909 |
| 7 | 7 | $10/k$ | 119 | 0.05 | 47.35 | 0.06 | 34.12 | 947 | 569 |
| 7 | 7 | $11/k$ | 132 | 0.06 | 75.57 | 0.06 | 46.40 | 1260 | 773 |
| 7 | 7 | $12/k$ | 145 | 0.06 | 114.08 | 0.06 | 61.54 | 1901 | 1026 |
| 7 | 7 | $13/k$ | 159 | 0.08 | 164.94 | 0.08 | 81.96 | 2062 | 1025 |
| 7 | 7 | $14/k$ | 173 | 0.08 | 0.42 | 0.08 | 187.59 | 5 | 2345 |
| 7 | 7 | $15/k$ | 187 | 0.08 | 343.09 | 0.11 | 316.15 | 4289 | 2874 |
| 7 | 7 | $16/k$ | 202 | 0.11 | 477.05 | 0.11 | 445.60 | 4337 | 4051 |
| 10 | 10 | $5/k$ | 74 | 0.09 | 1097.61 | 0.08 | 130.67 | 12196 | 1633 |
| 10 | 10 | $6/k$ | 92 | 0.09 | 323.30 | 0.08 | 100.70 | 3592 | 1259 |
| 10 | 10 | $7/k$ | 108 | 0.09 | 683.72 | 0.09 | 174.28 | 7597 | 1936 |
| 10 | 10 | $8/k$ | 127 | 0.09 | 1946.64 | 0.11 | 240.62 | 21629 | 2187 |
| 10 | 10 | $9/k$ | 143 | 0.12 | 47.92 | 0.09 | 44.48 | 399 | 494 |
| 10 | 10 | $10/k$ | 162 | 0.11 | 0.12 | 0.12 | 0.33 | 1 | 3 |
| 10 | 10 | $11/k$ | 181 | 0.14 | –/– | 0.14 | 2539.56 | — | 18140 |
| 7 | 14 | $9/k$ | 107 | 0.09 | 26.82 | 0.09 | 17.24 | 298 | 192 |
| 7 | 17 | $9/k$ | 107 | 0.13 | 25.48 | 0.11 | 23.01 | 196 | 209 |
| 7 | 22 | $9/k$ | 108 | 0.17 | 29.09 | 0.16 | 25.35 | 171 | 158 |
| 7 | 27 | $9/k$ | 109 | 0.23 | 30.08 | 0.20 | 25.94 | 131 | 130 |
| 7 | 27 | $19/k$ | 250 | 0.34 | 723.97 | 0.41 | 665.75 | 2129 | 1624 |

–/– indicates a computing time which exceeded 2 hours

**Table 3** Comparison of computing times (in seconds) for the implementation of the Meijer G function in Mathematica and the EGIG implementation in (iii) of Theorem 1 for different values of $m$, $k$ and $s$, and for the given value of $z$ (all values were computed for $a = 95.3$, with $d$ precision digits, in order to assure 16 decimal places in the result).

| | | | | | computing times | | | | | |
|---|---|---|---|---|---|---|---|---|---|---|
| | | | | | Mathematica ver. 7 | | Mathematica ver. 5.2 | | ratios | |
| $m$ | $k$ | $s$ | $d$ | $z$ | EGIG(1) | Meijer-G(2) | EGIG(3) | Meijer-G(4) | (2)/(1) | (4)/(3) |
| 3 | 3 | 3 | 76 | 0.9 | 0.02 | 0.22 | 0.02 | 0.89 | 11 | 45 |
| 4 | 4 | 4 | 152 | 0.9 | 0.03 | 0.52 | 0.05 | 25.93 | 17 | 519 |
| 5 | 5 | 5 | 276 | 0.9 | 0.08 | 1.48 | 0.13 | –//– | 19 | — |
| 6 | 6 | 6 | 462 | 0.9 | 0.27 | 6.10 | 0.53 | –//– | 23 | — |
| 7 | 7 | 7 | 721 | 0.9 | 0.72 | 22.32 | 3.28 | –//– | 31 | — |
| 8 | 8 | 8 | 653 | 0.2 | 0.98 | 85.86 | 1.70 | –//– | 88 | — |
| 9 | 9 | 9 | 867 | 0.1 | 2.09 | 395.18 | 6.29 | –//– | 189 | — |
| 10 | 10 | 10 | 897 | 0.001 | 2.96 | 1900.65 | 13.01 | –//– | 642 | — |
| 12 | 12 | 12 | 1503 | 0.0001 | 45.74 | –/– | 61.01 | –//– | — | — |

–/– indicates a computing time which exceeded 2 hours

–//– indicates that the available memory was exhausted before the computation was done

should remark here that while the computing times for the EGIG implementation were highly stable, most of times coinciding to the hundredth of the second, the computing times for the Mathematica implementation of the Meijer G function were highly variable.

In each table $d$ indicates the number of precision digits that was necessary to use in order to obtain a result with 16 decimal places. In all cases the numerical results from the EGIG implementation and the results obtained from the Mathematica implementation of the Meijer G function were exactly the same.

All reported computing times are for a first run. For subsequent runs of exactly the same command, that is, a second run with exactly all the same parameters and variable values, one may indeed obtain much lower computing times but this is of no meaning when one tries to compute or plot a function.

We may see that for all the three cases considered in Theorem 1 the computing times of the Mathematica implementation of the Meijer G function are usually from some tens or a few hundreds to a few thousand or even some hundreds of thousand times higher than the computing times for the EGIG implementation, which commonly takes less than a tenth of a second to compute.

Only for case (ii) in Theorem 1, when $v/k$ is an integer, is the Mathematica version 7 implementation of the Meijer function able to almost match the computing times of the EGIG implementation. Something that version 5.2 of the same software is not able to do.

It is interesting to note how version 5.2 of Mathematica seems to have somehow lower computing times for case (ii) in Theorem 1 than version 7, with a switch for the other two cases considered in the same Theorem.

In the overall, for the cases considered in this paper and summarized in Theorem 1, there is an enormous advantage in using the EGIG implementation, even from a strictly computation point of view.

## 7 Conclusions

Interestingly, the distribution of $V$ in (1) yields, for different values of the $\delta_{ij}$'s, several other distributions, including the exact distribution of the l.r.t. statistics used to test the independence of several sets of variables, the equality of several mean vectors, circularity of the variance-covariance matrix and equality of means in the mean vector and circularity of the variance-covariance matrix, simultaneously. The difficulties encountered in writing down concise and manageable expressions for the exact distribution of the l.r.t. statistic to test independence of several sets of variables when more than one set of variables has an odd number of variables, or for the l.r.t. statistic to test the equality of several mean vectors when the number of variables is odd and at the same time the number of mean vectors being tested is even, or yet for the l.r.t. statistic to test circularity of the covariance matrix when the dimension is even, are related with the fact that in these cases we are not able to represent such distributions as a product of powers of Unif(0, 1) r.v.'s.

The distribution of the product of powers of independent Unif(0, 1) r.v.'s arises thus as a useful tool to approach in an unified way the distributions of several random variables, which would be more challenging to deal with if they were not recognized as being amenable to such a representation. The distribution of the product of powers of Unif(0, 1) r.v.'s appears as the common link and common elementary structure among these distributions.

**Acknowledgments** This research was financially supported in part by the Portuguese Foundation for Science and Technology, through CMA (Centro de Matemática e Aplicações) of Universidade Nova de Lisboa (Financiamento Base 2010 ISFL-1-297 from FCT/MCTES/PT).

# Appendix A. Some Simple Relations Among Distributions

## A.1. The Logbeta Distribution and Its Characteristic Function

In a number of situations throughout the paper we will be dealing with Beta r.v.'s and their logarithms, or rather, their negative logarithms. We know that

$$X \sim \text{Beta}(a,b) \quad \Longrightarrow \quad E\left(X^h\right) = \frac{\Gamma(a+b)}{\Gamma(a)} \frac{\Gamma(a+h)}{\Gamma(a+b+h)}, \quad (h > -a),$$

so that if we take $Y = -\log X$, we have, for $\mathrm{i} = \sqrt{-1}$,

$$\Phi_Y(t) = E\left(e^{\mathrm{i}tY}\right) = E\left(X^{-\mathrm{i}t}\right) = \frac{\Gamma(a+b)}{\Gamma(a)} \frac{\Gamma(a-\mathrm{i}t)}{\Gamma(a+b-\mathrm{i}t)}, \quad (t \in \mathbb{R}) \qquad \text{(A.1)}$$

and we will say that the r.v. $Y$ has a Logbeta distribution with parameters $a$ and $b$. The identification of $Y$ as a Logbeta r.v., or, equivalently, of $X = e^{-Y}$ as a Beta r.v. from the observation of the fact that the c.f. of $Y$ is of the type in (A.1) is widely used throughout the paper.

It is then clear that a sum of independent Logbeta r.v.'s is thus the negative logarithm of a product of independent Beta r.v.'s, with the same parameters.

## A.2. Establishing Some Simple Concepts and Notation

We know that

$$Y_i \sim Exp(\delta), \quad i = 1, \dots, r, \quad \text{i.i.d.} \Longrightarrow \quad T = \sum_{i=1}^{r} Y_i \sim \Gamma(r, \delta).$$

The r.v. $T$ has what we call an integer Gamma distribution, with integer shape parameter $r$ and rate parameter $\delta$, with p.d.f. and c.f. respectively given by

$$f_T(t) = \frac{\delta^r}{\Gamma(r)} e^{-\delta t}\, t^{r-1} \quad (t > 0; r, \delta > 0)$$

and

$$\Phi_T(t) = E\left(e^{\mathrm{i}tT}\right) = \delta^r (\delta - \mathrm{i}t)^{-r} \quad (\text{for } \mathrm{i} = \sqrt{-1}).$$

Then

$$W^* = e^T = e^{(\sum_{i=1}^r Y_i)} = \prod_{i=1}^r W_i \,,$$

where $W_i = e^{Y_i} \sim Pareto(1, \delta)$ are i.i.d. standard Pareto r.v.'s, has what we call a generalized Pareto distribution with shape parameter $r$ and rate parameter $\delta$. The r.v. $W^{*-1}$ will have what we call an inverted generalized Pareto distribution. See Appendix B on the GIG distribution, that is the distribution of the sum of independent integer Gamma r.v.'s, and on the EGIG distribution which is the distribution of the product of independent inverted generalized Pareto r.v.'s.

The $h$-th moment of $W^*$ may be obtained either from the moments of $W_i$ or from the c.f. of $T$,

$$\begin{aligned} E\left((W^*)^h\right) &= \prod_{i=1}^r E\left(W_i^h\right) = \prod_{i=1}^r \frac{\delta}{\delta - h} = \delta^r (\delta - h)^{-r} \\ &= E\left(\left(e^T\right)^h\right) = E\left(e^{hT}\right) = \delta^r (\delta - h)^{-r} \quad (h < \delta)\,. \end{aligned}$$

## Appendix B. The GIG and EGIG Distributions

We will say that a r.v. $Y$ has a GIG (Generalized Integer Gamma) distribution of depth $p$, with shape parameters $r_j \in \mathbb{N}$ and rate parameters $\delta_j$ $(j = 1, \ldots, p)$, if

$$Y = \sum_{j=1}^p Y_j$$

where

$$Y_j \sim \Gamma(r_j, \delta_j)\,, \quad r_j \in \mathbb{N},\ \delta_j > 0,\ j = 1, \ldots, p$$

are $p$ independent integer Gamma or Erlang r.v.'s, with $\delta_j \neq \delta_{j'}$ for all $j \neq j'$, with $j, j' \in \{1, \ldots, p\}$.

The r.v. $Y$ has c.f.

$$\Phi_Y(t) = \prod_{j=1}^p \delta_j^{r_j} (\delta_j - \mathrm{i}t)^{-r_j}$$

and p.d.f. and c.d.f. given by (see Coelho (1998)),

$$f^{GIG}(y\,|\,r_j,\delta_j;\,p) \;=\; K\sum_{j=1}^{p} P_j(y)\,e^{-\delta_j y}\,, \qquad (y>0)$$

and

$$F^{GIG}(y\,|\,r_j,\delta_j;\,p) \;=\; 1-K\sum_{j=1}^{p} P_j^*(y)\,e^{-\delta_j y}\,, \qquad (y>0)$$

where $K=\prod_{j=1}^{p}\delta_j^{r_j}$,

$$P_j(y)=\sum_{k=1}^{r_j} c_{j,k}\,y^{k-1} \quad \text{and} \quad P_j^*(y)=\sum_{k=1}^{r_j} c_{j,k}\,(k-1)!\sum_{i=0}^{k-1}\frac{y^i}{i!\,\delta_j^{k-i}}$$

with

$$c_{j,r_j}=\frac{1}{(r_j-1)!}\prod_{\substack{i=1\\ i\neq j}}^{p}(\delta_i-\delta_j)^{-r_i}\,, \qquad j=1,\ldots,p\,,$$

and, for $k=1,\ldots,r_j-1;\ j=1,\ldots,p$,

$$c_{j,r_j-k}=\frac{1}{k}\sum_{i=1}^{k}\frac{(r_j-k+i-1)!}{(r_j-k-1)!}\,R(i,j,p)\,c_{j,r_j-(k-i)}\,,$$

where

$$R(i,j,p)=\sum_{\substack{k=1\\ k\neq j}}^{p} r_k\left(\delta_j-\delta_k\right)^{-i} \qquad (i=1,\ldots,r_j-1)\,.$$

We will then say that the r.v. $X=e^{-Y}$ has an EGIG (Exponentiated Generalized Integer Gamma) distribution of depth $p$, with shape parameters $r_j$ and rate parameters $\delta_j$, with p.d.f. and c.d.f. respectively given by

$$f^{EGIG}(x\,|\,r_j,\delta_j;\,p)=f^{GIG}(-\log x\,|\,r_j,\delta_j;\,p)\,\frac{1}{x} \qquad (0<x<1) \tag{B.1}$$

and

$$F^{EGIG}(x\,|\,r_j,\delta_j;\,p)=1-F^{GIG}(-\log x\,|\,r_j,\delta_j;\,p) \qquad (0<x<1)\,. \tag{B.2}$$

Note that the $h$-th moment of $X$ is, for $h > -\min_{1\le j\le p} \delta_j$, given by

$$E(X^h) = E\left(e^{-hY}\right) = \prod_{j=1}^{p} E\left(e^{-hY_j}\right) = \prod_{j=1}^{p} \delta_j^{r_j} \left(\delta_j + h\right)^{-r_j} .$$

We may note that one of the advantages of using the GIG distribution in the form above is that even in the case of depth 1 the expression for the c.d.f. yields an expression for the c.d.f. of an integer Gamma or Erlang r.v. without the use of the incomplete Gamma function, since in this case the c.d.f. may be expressed as a finite summation.

## References

Anderson, T. W. (2003). *An Introduction to Multivariate Statistical Analysis*, 3rd ed., J. Wiley & Sons, New York.

Anderson, T. W., Fang, K.-T. and Hsu, H. (1986). Maximum-likelihood estimates and likelihood-ratio criteria for multivariate elliptically contoured distributions. *Canad. J. Statist.* **14**, 55–59.

Anderson, T. W. and Fang, K.-T. (1990). Inference in multivariate elliptically contoured distributions based on maximum likelihood. *Statistical Inference in Elliptically Contoured and Related Distributions*, K.-T. Fang, T. W. Anderson, eds., 201–216.

Arnold, B. C. (1983). *Pareto Distributions*, International Cooperative Publ. House, Fairland, MD, 1983.

Coelho, C. A. (1998). The Generalized Integer Gamma distribution – a basis for distributions in Multivariate Statistics. *J. Multivariate Anal.* **64**, 86–102.

Coelho, C. A. (2006). The exact and near-exact distributions of the product of independent Beta random variables whose second parameter is rational. *J. Combin. Inform. System Sci.* **31**, 21–44.

Fang, C., Krishnaiah, P. R., Nagarsenker, B. N. (1982). Asymptotic distributions of the likelihood ratio test statistics for covariance structures of the complex multivariate normal distributions. *J. Multivariate Anal.* **12**, 597–611.

Goodman, N. R. (1963a). Statistical analysis based on a certain multivariate complex Gaussian distribution (An Introduction). *Ann. Math. Statist.* **34**, 152–177.

Goodman, N. R. (1963b). The distribution of the determinant of a complex Wishart distributed matrix. *Ann. Math. Statist.* **34**, 178–180.

Jensen, D. R. and Good, I. J. (1981). Invariant distributions associated with matrix laws under structural symmetry. *J. R. Stat. Soc.*, Ser. B **43**, 327–332.

Kariya, T. (1981). Robustness of multivariate tests. *Ann. Statist.* **9**, 1267–1275.

Khatri, C. G. (1965). Classical statistical analysis based on a certain multivariate complex Gaussian distribution. *Ann. Math. Statist.* **36**, 98–114.

Krishnaiah, P. R., Lee, J. C., Chang, T. C. (1976). The distributions of the likelihood ratio statistics for tests of certain covariance structures of complex multivariate Normal populations. *Biometrika* **63**, 543–549.

Marques, F. J., Coelho, C. A., Arnold, B. C. (2010). A general near-exact distribution theory for the most common likelihood ratio test statistics used in Multivariate Analysis. *TEST* (in print, https://doi.org/10.1007/s11749-010-0193-3).

Nagar, D. K., Chen, J., Gupta, A. K. (2004). Distribution and percentage points of the likelihood ratio statistic for testing circular symmetry. *Comput. Statist. Data Anal.* **47**, 79–89.

Olkin, I., Press, S. J. (1969). Testing and estimation for a circular stationary model. *Ann. Math. Statist.* **40**, 1358–1373.

Springer, M. D. and Thompson, W. E. (1970). The Distribution of products of Beta, Gamma and Gaussian random variables. *SIAM Journal on Applied Mathematics* **18**, 721-737.

Springer, M. D. (1979). *The Algebra of Random Variables*, J. Wiley & Sons, New York.

Wald, A. and Brookner, R. J. (1941). On the distribution of Wilks' statistic for testing the independence of several groups of variates. *Ann. Math. Statist.* **12**, 137–152.

Wooding, R. A. (1956). The multivariate distribution of complex Normal variables. *Biometrika* **43**, 212–215.

### *Author and Editors notes:*

[1] there is a typo in the original paper where "This distribution" should be replaced by "The distribution of $V^{-1}$"

[2] there is a small typo in the original paper: where is written $-\delta_j < 0$, should clearly be $-\delta_j > 0$, since, as clearly stated, in this subsection we are considering $\delta_j < 0$

[3] there is a typo in the original paper where "which is thus" should be replaced by "whose reciprocal has"

[4] there is a typo in the original paper where "which is" should be replaced by "and $V^{-1}$ has"

[5] there is a typo in the original paper where "which is thus" should be replaced by "whose reciprocal has"

[6] the second parameter in the Beta distribution of $Y_{ij}^{**}$ should be $v$ instead of $u$

[7] this sentence should rather read "if $u, vk \in \mathbb{N}$, even if $v \notin \mathbb{N}$, or $u \in \mathbb{N}$, even if $vk \notin \mathbb{N}$ and $v \notin \mathbb{N}$ for the first distribution of the $Y_{ij}^{**}$

[8] this sentence should rather read "if $v, uk \in \mathbb{N}$, even if $u \notin \mathbb{N}$, or $v \in \mathbb{N}$, even if $uk \notin \mathbb{N}$ and $u \notin \mathbb{N}$ for the first distribution of the $Y_{ij}^{***}$

# The Exact and Near-Exact Distributions of the Main Likelihood Ratio Test Statistics Used in the Complex Multivariate Normal Setting

Carlos A. Coelho, Barry C. Arnold, and Filipe J. Marques

**Abstract** In this paper the authors show how it is possible to establish a common structure for the exact distribution of the main likelihood ratio test (LRT) statistics used in the complex multivariate normal setting. In contrast to what happens when dealing with real random variables, for complex random variables it is shown that it is possible to obtain closed-form expressions for the exact distributions of the LRT statistics to test independence, equality of mean vectors and the equality of an expected value matrix to a given matrix. For the LRT statistics to test sphericity and the equality of covariance matrices, cases where the exact distribution has a non-manageable expression, easy to implement and very accurate near-exact distributions are developed. Numerical studies show how these near-exact distributions outperform by far any other available approximations. As an example of application of the results obtained, the authors develop a near-exact approximation for the distribution of the LRT statistic to test the equality of several complex normal distributions.

**Mathematics Subject Classification** 62H05, 62H10, 62E15, 62H15, 62E20

Reprinted from TEST, with permission from Springer Nature.

First published as Coelho, C.A., Arnold, B.C., Marques, F.J. (2015). The exact and near-exact distributions of the main likelihood ratio test statistics used in the complex multivariate normal setting, Test **24**, 386–416. (https://doi.org/10.1007/s11749-014-0418-y)

C. A. Coelho · F. J. Marques
Departamento de Matemática, Faculdade de Ciências e Tecnologia, Centro de Matemática e Aplicações (CMA-FCT/UNL), Universidade Nova de Lisboa, Caparica, Portugal
e-mail: cmac@fct.unl.pt; fjm@fct.unl.pt

B. C. Arnold (✉)
Statistics Department, University of California, Riverside, CA, USA
e-mail: barry.arnold@ucr.edu

 
B. C. Arnold et al. (eds.), *Likelihood Ratio Tests and Related Topics in Multivariate Analysis*, Contributions to Statistics,
https://doi.org/10.1007/978-3-031-68008-3_12

**Keywords** Covariance matrix · Equality of covariance matrices · Equality of mean vectors · Fourier transforms · Generalized integer gamma (GIG) distribution · Generalized near-integer gamma (GNIG) distribution · Independence · Mixtures · Expected value matrix · Sphericity · Statistical distributions (distribution functions)

## 1 Introduction

The implementation of many important likelihood ratio tests (LRTs) used in multivariate analysis is often hindered because of difficulties in handling the exact distribution, which is most of the time too complicated to be used in practice. On the other hand, the commonly available asymptotic distributions do not deliver the necessary accuracy, particularly for small samples and/or large numbers of variables involved.

In this paper the authors show how, in the complex multivariate normal setting, it is possible to obtain simple closed form expressions for the exact distribution of the LRT statistics to test (i) independence of sets of variables, (ii) equality of mean vectors and (iii) the equality of an expected value matrix to a given matrix, and very sharp near-exact distributions for the LRT statistics to test (iv) sphericity and (v) equality of covariance matrices.

These results are obtained, based on the fact that the LRT statistics in (i)–(v) have, under the null hypothesis, a common structure for their exact distributions, which for some $p \in \mathbb{N}$ and $u \in \mathbb{N}_0$, may be written as

$$\Lambda \overset{\text{st}}{\sim} \left(\prod_{j=1}^{p-1} e^{-Z_j}\right) \times \left(\prod_{k=1}^{u} Y_k\right) . \tag{1}$$

In (1), $\Lambda$ represents the LRT statistic, '$\overset{\text{st}}{\sim}$' is to be read as 'is stochastically equivalent to', and, for $r_j \in \mathbb{N}$ $(j = 1, \dots, p-1)$ and $a_k, b_k > 0$ $(k = 1, \dots, u)$,

$$Z_j \sim \Gamma\left(r_j, \frac{n-1-j}{n}\right) \quad \text{and} \quad Y_k \sim Beta(a_k, b_k) , \tag{2}$$

are independent r.v.'s (random variables). In (2), $\Gamma(r_j, \lambda_j)$ denotes a gamma distribution with shape parameter $r_j$ and rate parameter $\lambda_j$ and $n$ is the sample size. In (1), $p$ represents the number of variables involved, for the tests in (i), (iv) and (v), or the sum of the number of variables plus the number of vectors minus 1, for the test in (ii), or the sum of the number of variables involved plus the number of columns in the matrix $\mu$ in (22), for the test in (iii). Also, in (1), we have $u = 0$ for the LRT statistics in (i), (ii) and (iii).

Krishnaiah et al. (1976) used the first four moments of the LRT statistics to test independence, equality of covariance matrices and sphericity to approximate the distributions of certain powers of these LRT statistics by Pearson type I distributions, while Fang et al. (1982) use infinite mixtures of beta distributions to asymptotically approximate the distributions of the first two of those LRT statistics. Khatri (1965) obtained the exact distribution of the LRT statistic to test if an expected value matrix is null also in terms of an infinite mixture of beta distributions, while Gupta (1971) obtained closed form expressions for the distribution of the same test statistic for the cases of two and three variables, stating the possibility of extending such representations to a general number of variables, although without obtaining explicit expressions for the parameters in the distribution. Pillai and Nagarsenker (1971) and Nagarsenker and Das (1975) address the exact distribution of the LRT statistic for sphericity through contour integration, obtaining representations in terms of Meijer G functions and infinite series. Authors such as Nagarsenker and Nagarsenker (1981) and Nagar et al. (1985) studied a block sphericity test, which has as a particular case the common sphericity test, obtaining the exact distribution for the corresponding LRT statistic in the form of series representation, which in the case of the first authors has the form of an infinite mixture of beta distributions for the LRT statistic itself, or the form of an infinite mixture of chi-square distributions for its logarithm, while the second authors start with representations based on Meijer G and hypergeometric functions, to end-up with highly complicated series representations. Gupta (1976) and Gupta and Rathie (1983) address the non-null distribution of Wilks $\Lambda$ statistic for MANOVA, obtaining in the first reference explicit expressions for the probability density function (pdf) of $\Lambda$ only for $p = 2$ and $p = 3$ and only the general form of the expression for $p > 3$ and in the second reference the distribution for general $p$, in the form of Meijer G functions and what the authors claim to be computable series forms, which anyway in the null case remain too intricate. Somehow similar representations were obtained by Pillai and Jouris (1971) also for the LRT statistic for MANOVA and the LRT statistics to test equality of two covariance matrices and independence of two groups of variables. Mathai (1973) refers the non-computability of the representations involving Meijer G functions, which is still a reality, and obtains series representations for the distribution of the LRT for the equality of two covariance matrices and for sphericity, which anyway have very elaborate expressions.

In this paper the authors obtain:

- very simple expressions for the exact pdf and cumulative distribution function (cdf) of the statistics in (i), (ii) and (iii), avoiding the need for the approximations in Krishnaiah et al. (1976); Fang et al. (1982); Khatri (1965) and the complex expressions in Khatri (1965); Gupta (1971); Tang and Gupta (1986); Gupta (1976); Gupta and Rathie (1983); Pillai and Jouris (1971); Mathai (1973) even for the null case,
- very accurate manageable near-exact distributions for the statistics in (iv) and (v), which are far more precise and manageable than the approximations and expressions in Krishnaiah et al. (1976); Fang et al. (1982); Tang and Gupta (1986); Pillai and Jouris (1971); Mathai (1973); Pillai and Nagarsenker (1971);

Nagar et al. (1985); Nagarsenker and Das (1975); Nagarsenker and Nagarsenker (1981).

In Section 2 the exact distributions for the LRT statistics in (i), (ii) and (iii) are derived and it is shown that they have simple closed form expressions, which do not involve any infinite sums. In Sections 3 and 4 the exact distribution of the LRT statistics in (iv) and (v) is obtained in the form of (1). Then, in Section 5, near-exact distributions for these statistics are developed, and in Section 6 some numerical studies are carried out, confirming the very good performance of these near-exact distributions. Based on the results obtained, near-exact distributions for the LRT statistic to test equality of several complex normal distributions are developed in Section 7, and an example of application of these near-exact distributions is provided in the Online Resource.

Near-exact distributions are asymptotic distributions developed under a new concept. Usually based on factorizations of the characteristic function (c.f.) of the test statistic under study, or of its logarithm, they are built by leaving unchanged one part of this c.f. and replacing asymptotically the remaining part, in such a way that the resulting c.f. corresponds to a manageable distribution. These distributions, when correctly built for statistics used in Multivariate Analysis, are not only asymptotic for increasing sample sizes but also have a marked asymptotic behavior for increasing number of variables and number of populations involved. All this amounts to obtaining asymptotic distributions that lie much closer to the exact distribution than common asymptotic distributions (Marques et al., 2011).

In this paper the definition of the complex multivariate normal distribution that will be used is that found in Wooding (1956), Goodman (1963a), Brillinger (2001, sec. 4.2) and Anderson (2003, prob. 2.64): the random vector $\underline{X}$ $(p\times 1)$ has a complex multivariate normal distribution, with expected value $\underline{\mu}$ and Hermitian variance-covariance matrix $\Sigma$ if the pdf of $\underline{X}$ is

$$f_{\underline{X}}(\underline{x}) = \pi^{-p}|\Sigma|^{-1}\, e^{-\overline{(\underline{x}-\underline{\mu})}'\Sigma^{-1}(\underline{x}-\underline{\mu})}\,,$$

where $\overline{(\underline{x}-\underline{\mu})}$ denotes the complex conjugate of $(\underline{x}-\underline{\mu})$ and the prime denotes the transpose, with

$$\Sigma = 2\Theta + 2\mathrm{i}\Phi,$$

where $\Theta = Cov(Re(\underline{X})) = Cov(Im(\underline{X}))$ and $\Phi = Cov(Re(\underline{X}), Im(\underline{X}))$ are, respectively, a positive-definite and a skew-symmetric matrix. In this case we will write

$$\underline{X} \sim CN_p(\underline{\mu}, \Sigma)\,. \tag{3}$$

The complex multivariate normal distribution has applications in a wide range of areas, from crystallography (Pannu et al., 2003) to spectral analysis in time series (Brillinger and Krishnaiah, 1982; Brillinger, 2001; Shumway and Stoffer, 2006;

Goodman, 1963a; Turin, 1960; Krishnaiah et al., 1976) and to studies on satellite images, signal processing, communications and information theory (Conradsen, 2012; Conradsen et al., 2003; Lehmann et al., 2007).

# 2 The Exact Distribution of the LRT Statistics to Test Independence, the Equality of Mean Vectors and the Equality of an Expected Value Matrix to a Given Matrix

In this section it is shown how, in contrast to what happens when dealing with real r.v.'s (Marques et al., 2011), in the complex case it is possible to obtain the exact distribution for the negative logarithm of the LRT statistics to test independence, the equality of mean vectors or the equality of an expected value matrix to a given matrix as generalized integer gamma (GIG) distributions (Coelho, 1998), for any number of variables or mean vectors involved, this way obtaining the exact distribution of the LRT test statistics in the form in (1), with $u = 0$. The simplicity of the expressions obtained for the exact pdf's and cdf's is quite striking when compared with the expressions obtained by other authors (Fang et al., 1982; Gupta and Rathie, 1983; Khatri, 1965; Krishnaiah et al., 1976; Pillai and Jouris, 1971), and although bearing some resemblance with the expressions in Tang and Gupta (1986), they have, in contrast to these, the advantage of involving only finite sums. Even when compared with the quite close representations in Gupta (1971); Mathai (1973), Gupta (1976), they have the advantage of being more general and with coefficients which have much simpler forms. This is especially the case with respect to the representation in Gupta (1976), while in Gupta (1971) the author did not obtain explicit expressions for such coefficients.

## *2.1 The LRT Statistic to Test Independence Among Sets of Variables*

Suppose that the random vector $\underline{X}$ in (3) is split into $m$ subvectors $\underline{X}_k$ $(k = 1, \dots, m)$ and that we intend to test the independence of these $m$ subvectors. The partition of the vector $\underline{X}$ will induce the following structure for $\Sigma$

$$\Sigma = \begin{bmatrix} \Sigma_{11} & \cdots & \Sigma_{1k} & \cdots & \Sigma_{1m} \\ \vdots & \ddots & \vdots & & \vdots \\ \Sigma_{k1} & \cdots & \Sigma_{kk} & \cdots & \Sigma_{km} \\ \vdots & & \vdots & \ddots & \vdots \\ \Sigma_{m1} & \cdots & \Sigma_{mk} & \cdots & \Sigma_{mm} \end{bmatrix},$$

where $\Sigma_{kk} = Var(\underline{X}_k)$ $(k=1, ..., m)$ and $\Sigma_{ij} = Cov(\underline{X}_i, \underline{X}_j)$ $(i, j \in \{1, ..., m\})$. The hypothesis of mutual independence of the random subvectors $\underline{X}_k$, may thus be written as

$$H_0 : \Sigma = diag(\Sigma_{11}, \dots, \Sigma_{kk}, \dots, \Sigma_{mm}) \equiv \Sigma_{ij} = 0, \text{ for all } i \neq j \,. \tag{4}$$

Let us further suppose that, for each $k$, $\underline{X}_k$ has $p_k$ variables, with $p = \sum_{k=1}^{m} p_k$. Then, for a sample of size $n$, the LRT statistic to test $H_0$ in (4) is

$$\Lambda_1 = \left( \frac{|A|}{\prod_{k=1}^{m} |A_{kk}|} \right)^n , \tag{5}$$

where $A$ is the maximum likelihood estimator (MLE) of $\Sigma$ and $A_{kk}$ its $k$-th diagonal block $(k = 1, \dots, m)$, with

$$A = \frac{1}{n} \left( \overline{X - \frac{1}{n} E_{nn} X} \right)' \left( X - \frac{1}{n} E_{nn} X \right), \tag{6}$$

where once again the bar denotes the complex conjugate, $X$ is the $n \times p$ sample matrix and $E_{np}$ is an $n \times p$ matrix of ones (see Goodman (1963a) and (Anderson, 2003, problem 3.11) for references concerning the MLE of $\Sigma$ for the multivariate complex normal distribution).

Following a similar procedure to that used in the real case it is possible to show that, under $H_0$ in (4), for $h > p/n - 1$,

$$E\left(\Lambda_1^h\right) = \prod_{k=1}^{m-1} \prod_{j=1}^{p_k} \frac{\Gamma(n-j)\, \Gamma(n-q_k-j+nh)}{\Gamma(n-q_k-j)\, \Gamma(n-j+nh)} = \prod_{k=1}^{m-1} \prod_{j=1}^{p_k} E\left(Y_{jk}^{nh}\right), \tag{7}$$

where $Y_{jk} \sim Beta\,(n - q_k - j, q_k)$ are a set of $p = \sum_{k=1}^{m} p_k$ independent r.v.'s (for a proof, see Appendix B, section B.1 in the Online Resource, or section B.5.1 for an alternative proof based on the results in Jensen (1988, Thm. 5)). Note that (7) matches the expressions for the moments of $\Lambda_1$ in Krishnaiah et al. (1976) and Fang et al. (1982).

But then, since $0 < \Lambda_1 < 1$, its distribution is well-defined by the set of its moments and consequently

$$\Lambda_1 \overset{\text{st}}{\sim} \prod_{k=1}^{m-1} \prod_{j=1}^{p_k} \left(Y_{jk}\right)^n .$$

Therefore, the c.f. of

$$W_1 = -\log \Lambda_1 , \tag{8}$$

may be written as

$$\Phi_{W_1}(t) = E\left(e^{\mathrm{i}tW_1}\right) = E\left(\Lambda_1^{-\mathrm{i}t}\right) = \prod_{k=1}^{m-1}\prod_{j=1}^{p_k} \frac{\Gamma(n-j)}{\Gamma(n-q_k-j)} \frac{\Gamma(n-q_k-j-n\mathrm{i}t)}{\Gamma(n-j-n\mathrm{i}t)}, \tag{9}$$

giving rise to Theorem 1.

**Theorem 1** *The exact distribution of $W_1$ in (8) is a GIG distribution of depth $p-1$ with pdf (see Marques et al. (2011, App. B) for the notation on the GIG pdf and cdf)*

$$f_{W_1}(w) = f^{GIG}\left(w \,\middle|\, r_1,\ldots,r_{p-1}; \frac{n-2}{n},\ldots,\frac{n-p-2}{n}; p-1\right)$$

*and cdf*

$$F_{W_1}(w) = F^{GIG}\left(w \,\middle|\, r_1,\ldots,r_{p-1}; \frac{n-2}{n},\ldots,\frac{n-p-2}{n}; p-1\right),$$

*where $w > 0$ and*

$$r_j = \begin{cases} h_j & j = 1 \\ h_j + r_{j-1} & j = 2,\ldots,p-1 \end{cases} \tag{10}$$

*with, $h_j =$ (number of $p_k$ greater or equal to $j$) $-1$, for $j = 1,\ldots,p-1$.*

***Proof*** From (9), using, for any complex number $z$ and positive integer $n$, the relation

$$\frac{\Gamma(z+n)}{\Gamma(z)} = \prod_{\nu=0}^{n-1}(z+\nu), \tag{11}$$

it follows that

$$\begin{aligned}\Phi_{W_1}(t) &= \prod_{k=1}^{m-1}\prod_{j=1}^{p_k}\prod_{\nu=0}^{q_k-1} \frac{n-q_k-j+\nu}{n-q_k-j+\nu-n\mathrm{i}t} \\ &= \prod_{j=1}^{p-1}\left(\frac{n-1-j}{n}\right)^{r_j}\left(\frac{n-1-j}{n}-\mathrm{i}t\right)^{-r_j},\end{aligned}$$

which is the c.f. of the sum of $p-1$ independent gamma r.v.'s with shape parameters $r_j$, given by (10) in the body of the Theorem, and rate parameters $\frac{n-1-j}{n}$ ($j = 1,\ldots,p-1$), which is a GIG distribution of depth $p-1$, with shape parameters $r_j$ and rate parameters $\frac{n-1-j}{n}$. □

Then Corollary 1 gives the exact pdf and cdf of $\Lambda_1 = e^{-W_1}$.

**Corollary 1** *The exact pdf and cdf of the statistic* $\Lambda_1 = e^{-W_1}$ *in (5) are, for* $0 < \ell \leq 1$,

$$f_{\Lambda_1}(\ell) = f^{GIG}\left(-\log \ell \,\middle|\, r_1,\ldots,r_{p-1};\, \frac{n-2}{n},\ldots,\frac{n-p-2}{n};\, p-1\right)\frac{1}{\ell}$$

*and*

$$F_{\Lambda_1}(\ell) = 1 - F^{GIG}\left(-\log \ell \,\middle|\, r_1,\ldots,r_{p-1};\, \frac{n-2}{n},\ldots,\frac{n-p-2}{n};\, p-1\right),$$

*for* $r_j$ *given by (10).*

The exact distribution of $\Lambda_1$ is thus of the form (1), with $u = 0$.

## 2.2 The LRT Statistic to Test the Equality of Several Mean Vectors

Let us suppose that

$$\underline{X}_j \sim CN_p(\underline{\mu}_j, \Sigma_j), \quad j = 1,\ldots,q$$

and that, assuming $\Sigma_1 = \cdots = \Sigma_q (= \Sigma)$, the null hypothesis to be tested is

$$H_0 : \underline{\mu}_1 = \cdots = \underline{\mu}_q\,, \tag{12}$$

based on $q$ independent samples, the $j$-th of which is from $\underline{X}_j$, with size $n_j$. Also, suppose that the $j$-th sample is stored in the $n_j \times p$ matrix $X_j$.

Then, the LRT statistic used to test $H_0$ in (12) is given by

$$\Lambda_2 = \left(\frac{|A|}{|A+B|}\right)^n \tag{13}$$

where $n = \sum_{j=1}^{q} n_j$,

$$A = \sum_{j=1}^{q} \left(\overline{X_j - E_{n_j 1}\underline{\widetilde{X}}_j'}\right)' \left(X_j - E_{n_j 1}\underline{\widetilde{X}}_j'\right) \quad \text{and}$$

$$B = \sum_{j=1}^{q} n_j \left(\underline{\widetilde{X}}_j - \underline{\widetilde{X}}\right)\left(\overline{\underline{\widetilde{X}}_j - \underline{\widetilde{X}}}\right)' \tag{14}$$

where once again the bar denotes the complex conjugate and where

$$\underset{\sim}{\underline{X}}_j = \frac{1}{n_j} X'_j E_{n_j 1}\,,$$

is the vector of sample means from the $j$-th sample, and

$$\underset{\sim}{\underline{X}} = \frac{1}{n}\sum_{j=1}^{q} n_j \underset{\sim}{\underline{X}}_j\,.$$

The $p\times p$ matrix $A$ has what is called a complex Wishart distribution (Goodman, 1963a,b) with $n-q$ degrees of freedom and parameter matrix $\Sigma$. This fact is denoted by $A \sim CW_p(n-q, \Sigma)$. Under $H_0$ in (12), $B \sim CW_p(q-1, \Sigma)$, and then, given the independence, for normal r.v.'s, of the MLE's of the mean and variance, the matrices $A$ and $B$ are independent and thus, under $H_0$ in (12),

$$A + B \sim CW_p(n-1, \Sigma)\,.$$

It then follows that (see Goodman (1963b)),

$$2^p\,|A| \overset{\text{st}}{\sim} |\Sigma| \prod_{j=1}^{p} W_j \qquad \text{and} \qquad 2^p\,|A+B| \overset{\text{st}}{\sim} |\Sigma| \prod_{j=1}^{p} Z_j\,,$$

where $W_j$ $(j = 1, \ldots, p)$ and $Z_j$ $(j = 1, \ldots, p)$ are two independent sets of $p$ independent r.v.'s, with

$$W_j \sim \chi^2_{2(n-q-j+1)} \quad \text{and} \quad Z_j = W_j + W_j^* \sim \chi^2_{2(n-1-j+1)}\,,$$

where each

$$W_j^* \sim \chi^2_{2(q-1)}\,, \qquad j = 1, \ldots, p\,,$$

is independent of $W_j$ $(j = 1, \ldots, p)$.

Thus, under $H_0$ in (12),

$$\Lambda_2 \overset{\text{st}}{\sim} \prod_{j=1}^{p} (Y_j)^n \quad \text{where} \quad Y_j \sim Beta\,(n-q-j+1, q-1)\,, \text{ are } p \text{ independent r.v.'s.} \tag{15}$$

But then, for $h > (p+q)/n - 1$,

$$E\left(\Lambda_2^h\right) = \prod_{j=1}^{p} \frac{\Gamma(n-j)}{\Gamma(n-q-j+1)} \, \frac{\Gamma(n-q-j+1+nh)}{\Gamma(n-j+nh)}\,. \tag{16}$$

This result may also be obtained in an alternative way as in Appendix B.2 in the Online Resource.

From (16), the c.f. of

$$W_2 = -\log \Lambda_2 \tag{17}$$

may be written as

$$\Phi_{W_2}(t) = E\left(e^{\mathrm{i}t W_2}\right) = E\left(\Lambda_2^{-\mathrm{i}t}\right) = \prod_{j=1}^{p} \frac{\Gamma(n-j)}{\Gamma(n-q-j+1)} \frac{\Gamma(n-q-j+1-n\mathrm{i}t)}{\Gamma(n-j-n\mathrm{i}t)}, \tag{18}$$

from which Theorem 2 may be established.

**Theorem 2** *The exact distribution of $W_2$ in (17) is a GIG distribution of depth $p+q-2$ with pdf*

$$f_{W_2}(w) = f^{GIG}\left(w \,\middle|\, r_1, \ldots, r_{p+q-2}; \frac{n-2}{n}, \ldots, \frac{n-p-q+1}{n}; p+q-2\right)$$

*and cdf*

$$F_{W_2}(w) = F^{GIG}\left(w \,\middle|\, r_1, \ldots, r_{p+q-2}; \frac{n-2}{n}, \ldots, \frac{n-p-q+1}{n}; p+q-2\right),$$

*where $w > 0$ and*

$$r_j = \begin{cases} h_j & j = 1 \\ h_j + r_{j-1} & j = 2, \ldots, p+q-2 \end{cases} \tag{19}$$

*with,*

$$h_j = (\#\textit{ of elements in } \{p, q-1\} \textit{ greater or equal to } j) - 1,$$

*for $j = 1, \ldots, p+q-2$.*

***Proof*** From (18), using the relation in (11), the c.f. of $W_2$ may be written as

$$\begin{aligned} \Phi_{W_2}(t) &= \prod_{j=1}^{p} \prod_{\ell=0}^{q-2} \frac{n-q-j+1+\ell}{n-q-j+1+\ell-n\mathrm{i}t} \\ &= \prod_{j=1}^{p+q-2} \left(\frac{n-1-j}{n}\right)^{r_j} \left(\frac{n-1-j}{n} - \mathrm{i}t\right)^{-r_j} \end{aligned} \tag{20}$$

with $r_j$ given by (19) in the body of the Theorem. From (20) it is possible to conclude that the distribution of $W_2$ is indeed a GIG distribution of depth $p+q-2$, with shape parameters $r_j$ and rate parameters $\frac{n-1-j}{n}$ $(j=1,\ldots,p+q-2)$. □

Then Corollary 2 gives the exact pdf and cdf of $\Lambda_2 = e^{-W_2}$.

**Corollary 2** *The exact pdf and cdf of the statistic* $\Lambda_2 = e^{-W_2}$ *in (13) are, for* $0 < \ell \leq 1$,

$$f_{\Lambda_2}(\ell) = f^{GIG}\left(-\log \ell \,\middle|\, r_1,\ldots,r_{p+q-2};\, \frac{n-2}{n},\ldots,\frac{n-p-q+1}{n};\, p+q-2\right)\frac{1}{\ell}$$

*and*

$$F_{\Lambda_2}(\ell) = 1 - F^{GIG}\left(-\log \ell \,\middle|\, r_1,\ldots,r_{p+q-2};\, \frac{n-2}{n},\ldots,\frac{n-p-q+1}{n};\, p+q-2\right),$$

*for* $r_j$ *given by (19).*

Once again, the distribution of $\Lambda_2$ is of the form (1) with $u=0$.

## 2.3 The LRT Statistic to Test if an Expected Value Matrix Is Equal to a Given Matrix

A generalization of the test for an expected value matrix developed in Khatri (1965, Subsection 3.2) will be addressed in this subsection. Let $Z$ $(p{\times}n)$ be a matrix with a complex multivariate normal distribution with expected value $\boldsymbol{\mu} M$, where $\boldsymbol{\mu}$ is a $p{\times}q$ complex matrix and $M$ is $q{\times}n$ of rank $q$ $(\leq n)$, and variance $I_n \otimes \Sigma$, that is, with $var(vec(Z)) = I_n \otimes \Sigma$. This fact is denoted by

$$Z_{p\times n} \sim CN_{p\times n}(\boldsymbol{\mu} M, I_n \otimes \Sigma)\,. \tag{21}$$

Let us then suppose that the hypothesis to be tested, for a given matrix $\Xi$, is

$$H_0 : \boldsymbol{\mu}_{(p\times q)} = \Xi_{(p\times q)}\,. \tag{22}$$

Then, according to Khatri (1965), the LRT statistic to test $H_0$ is

$$\Lambda_3 = \left(\frac{|\Psi|}{\left|\Psi + \frac{1}{n}(\beta-\Xi)(M\overline{M}^{'})(\overline{\beta-\Xi})'\right|}\right)^n, \tag{23}$$

where

$$\Psi = \frac{1}{n} Z\left( I_n - \overline{M}'\left(M\overline{M}'\right)^{-1} M\right)\overline{Z}' \quad \text{and} \quad \beta = Z\overline{M}'\left(M\overline{M}'\right)^{-1} \tag{24}$$

are, respectively, the MLE's of $\Sigma$ and $\mu$, and as such independent.

But then, since $\left(I_n - \overline{M}'\left(M\overline{M}'\right)^{-1} M\right)$ is the projector on the null space of the columns of $M$, it is idempotent with

$$rank\left(I_n - \overline{M}'\left(M\overline{M}'\right)^{-1} M\right) = tr\left(I_n - \overline{M}'\left(M\overline{M}'\right)^{-1} M\right) = n - q\,.$$

Then, given the distribution of $Z$ in (21),

$$\Psi \sim CW_p\left(n-q, \frac{1}{n}\Sigma\right). \tag{25}$$

From (21),

$$\beta = Z\overline{M}'\left(M\overline{M}'\right)^{-1} \sim CN_{p\times q}\left(\boldsymbol{\mu}, \left(M\overline{M}'\right)^{-1}\otimes\Sigma\right),$$

so that

$$(\beta - \Xi)\left(M\overline{M}'\right)^{1/2} \sim CN_{p\times q}\left((\boldsymbol{\mu} - \Xi)\left(M\overline{M}'\right)^{1/2}, I_q\otimes\Sigma\right),$$

where, under $H_0$ in (22), $(\boldsymbol{\mu} - \Xi)\left(M\overline{M}'\right)^{1/2} = 0$. Consequently under $H_0$ in (22),

$$(\beta - \Xi)\left(M\overline{M}'\right)\left(\overline{\beta - \Xi}\right)' \sim CW_p(q, \Sigma),$$

independent of $\Psi$, so that, under $H_0$ in (22),

$$\Psi + \frac{1}{n}(\beta - \Xi)(M\overline{M}')(\overline{\beta - \Xi})' \sim CW_p\left(n, \frac{1}{n}\Sigma\right).$$

Thus, following similar steps to those in Sect. 2.2 (see also Appendix B.2 in the Online Resource), under $H_0$ in (22) for $n > q + p - 1$,

$$\Lambda_3 \overset{st}{\sim} \prod_{j=1}^{p} (Y_j)^n \quad \text{where} \quad Y_j \sim Beta\,(n - q - j + 1, q) \tag{26}$$
$$\text{are } p \text{ independent r.v.'s}$$

so that, for $h > (q + p - 1)/n - 1$,

$$E\left(\Lambda_3^h\right) = \prod_{j=1}^{p} \frac{\Gamma(n+1-j)}{\Gamma(n+1-q-j)} \frac{\Gamma(n+1-q-j+nh)}{\Gamma(n+1-j+nh)}.$$

Thus, the c.f. of $W_3 = -\log \Lambda_3$ is

$$\begin{aligned}
\Phi_{W_3}(t) &= E\left(e^{itW_3}\right) = E\left(\Lambda_3^{-it}\right) \\
&= \prod_{j=1}^{p} \frac{\Gamma(n+1-j)\,\Gamma(n+1-q-j-nit)}{\Gamma(n+1-q-j)\,\Gamma(n+1-j-nit)} \qquad (27)\\
&= \prod_{j=1}^{p}\prod_{\ell=0}^{q-1} \frac{n+1-q-j+\ell}{n+1-q-j+\ell-nit} \\
&= \prod_{j=1}^{p+q-1} (n-j)^{r_j}\,(n-j-nit)^{-r_j} = \prod_{j=1}^{p+q-1} \left(\frac{n-j}{n}\right)^{r_j} \left(\frac{n-j}{n}-it\right)^{-r_j} \qquad (28)
\end{aligned}$$

for $r_j$ given by (19) in Theorem 2, with $q$ replaced by $q + 1$.

Comparing (26) with (5.2.2) and (5.2.3) in Khatri (1965) and the expression in (27) with (5.3.1) and (5.3.3) in the same reference, it can be seen that there is a small mistake in Khatri's paper, where $q$ has to be subtracted from the first argument of the beta r.v.'s in (5.2.2) and (5.2.3) and also from the arguments of all gamma functions in (5.3.1) of Khatri (1965).

From (28) Theorem 3 and Corollary 3 may be established.

**Theorem 3** *The exact distribution of $W_3 = -\log \Lambda_3$ is a GIG distribution of depth $p + q - 1$ with pdf*

$$f_{W_3}(w) = f^{GIG}\left(w \,\middle|\, r_1, \ldots, r_{p+q-1}; \frac{n-1}{n}, \ldots, \frac{n-p-q+1}{n}; p+q-1\right)$$

*and cdf*

$$F_{W_3}(w) = F^{GIG}\left(w \,\middle|\, r_1, \ldots, r_{p+q-1}; \frac{n-1}{n}, \ldots, \frac{n-p-q+1}{n}; p+q-1\right),$$

*where $w > 0$ and $r_j$ $(j = 1, \ldots, p)$ are given by (19) in Theorem 2, with $q$ replaced by $q + 1$.*

**Corollary 3** *The exact pdf and cdf of the statistic $\Lambda_3 = e^{-W_3}$ in (23) are, for $0 < \ell \le 1$,*

$$f_{\Lambda_3}(\ell) = f^{GIG}\left(-\log \ell \,\middle|\, r_1, \ldots, r_{p+q-1}; \frac{n-1}{n}, \ldots, \frac{n-p-q+1}{n}; p+q-1\right) \frac{1}{\ell}$$

*and*

$$F_{\Lambda_3}(\ell) = 1 - F^{GIG}\left(-\log \ell \,\middle|\, r_1, \ldots, r_{p+q-1}; \frac{n-1}{n}, \ldots, \frac{n-p-q+1}{n}; p+q-1\right),$$

*for $r_j$ given by (19), with $q$ replaced by $q+1$.*

This test may be easily extended to test hypothesis of the type

$$H_0 : \boldsymbol{\mu} D = 0\,, \tag{29}$$

where $D$ is a $q \times q$ non-random matrix.

By the invariance property of the MLEs, which entails the invariance of the LRT statistics, the LRT statistic to test $H_0$ in (29) may be obtained just by replacing $\beta = \hat{\boldsymbol{\mu}}$ in the expression of $\Lambda_3$ by $\beta^* = \widehat{\boldsymbol{\mu} D} = \hat{\boldsymbol{\mu}} D$. Since in many cases $D$ will not be full-rank, the distribution of the matrix $B^*$, where

$$B^* = \frac{1}{n}\beta^*(M\overline{M}')\overline{\beta}^{*\prime} \qquad \text{with} \qquad \beta^* = Z\overline{M}'(M\overline{M}')^{-1}D \tag{30}$$

is now a complex Wishart distribution with a number of degrees of freedom equal to

$$rank\left(\overline{M}'\overline{D}'(M\overline{M}')^{-1}DM\right)$$

which, since $M\overline{M}'$ is full-rank, will be equal to $rank(D)$.

The LRT statistic to test $H_0$ in (29) is thus

$$\Lambda_3^* = \left(\frac{|\Psi|}{|\Psi + B^*|}\right)^n, \tag{31}$$

with $\Psi \sim CW_p\left(n-q, \frac{1}{n}\Sigma\right)$, as in (25), and $B^*$ as in (30). If $rank(D) = q^*(\leq q)$, then

$$B^* \sim CW_p\left(q^*, \frac{1}{n}\Sigma\right),$$

independent of $\Psi$, so that $\Psi + B^* \sim CW_p\left(n-q+q^*, \frac{1}{n}\Sigma\right)$ and as such (see Appendix B.2 in the Online Resource)

$$\Lambda_3^* \overset{\text{st}}{\sim} \prod_{j=1}^{p}\left(Y_j^*\right)^n \quad \text{where} \quad Y_j^* \sim Beta\,(n-q-j+1, q^*)$$
$$\text{are } p \text{ independent r.v.'s}\,.$$

Then, with an appropriate choice of the matrices $Z$, $M$ and $D$, this test may be used to implement the test of equality of mean vectors in Sect. 2.2. Indeed, if $Z$ is the sample matrix, with

$$\underset{(p\times n)}{Z} = \begin{bmatrix} Z_{111}\dots Z_{11n_1} & Z_{211}\dots Z_{21n_2} & \dots & Z_{k11}\dots Z_{k1n_k} & \dots & Z_{q11}\dots Z_{q1n_q} \\ \vdots\quad\vdots & \vdots\quad\vdots & & \vdots\quad\vdots & & \vdots\quad\vdots \\ Z_{1j1}\dots Z_{1jn_1} & Z_{2j1}\dots Z_{2jn_2} & \dots & Z_{kj1}\dots Z_{kjn_k} & \dots & Z_{qj1}\dots Z_{qjn_q} \\ \vdots\quad\vdots & \vdots\quad\vdots & & \vdots\quad\vdots & & \vdots\quad\vdots \\ Z_{1p1}\dots Z_{1pn_1} & Z_{2p1}\dots Z_{2pn_2} & \dots & Z_{kp1}\dots Z_{kpn_k} & \dots & Z_{qp1}\dots Z_{qpn_q} \end{bmatrix},$$

where $Z_{kj\ell}$ represents the $\ell$-th observation for the $j$-th variable in the $k$-th population, for $k = 1, \dots, q$, $j = 1, \dots, p$ and $\ell = 1, \dots, n_k$, the matrix $M$ the design matrix, with

$$\underset{(q\times n)}{M} = \begin{bmatrix} \overbrace{1\;\;1\;\dots\;1}^{n_1} & \overbrace{0\;\;0\;\dots\;0}^{n_2} & \dots & \overbrace{0\;\;0\;\dots\;0}^{n_k} & \dots & \overbrace{0\;\;0\;\dots\;0}^{n_q} \\ 0\;\;0\;\dots\;0 & 1\;\;1\;\dots\;1 & \dots & 0\;\;0\;\dots\;0 & \dots & 0\;\;0\;\dots\;0 \\ \vdots\;\;\vdots\quad\;\;\vdots & \vdots\;\;\vdots\quad\;\;\vdots & & \vdots\;\;\vdots\quad\;\;\vdots & & \vdots\;\;\vdots\quad\;\;\vdots \\ 0\;\;0\;\dots\;0 & 0\;\;0\;\dots\;0 & \dots & 1\;\;1\;\dots\;1 & \dots & 0\;\;0\;\dots\;0 \\ \vdots\;\;\vdots\quad\;\;\vdots & \vdots\;\;\vdots\quad\;\;\vdots & & \vdots\;\;\vdots\quad\;\;\vdots & & \vdots\;\;\vdots\quad\;\;\vdots \\ 0\;\;0\;\dots\;0 & 0\;\;0\;\dots\;0 & \dots & 0\;\;0\;\dots\;0 & \dots & 1\;\;1\;\dots\;1 \end{bmatrix}$$

and

$$\underset{(q\times q)}{D} = I_q - \frac{1}{n}MM'E_{qq} = \begin{bmatrix} 1-\frac{n_1}{n} & -\frac{n_1}{n} & \dots & -\frac{n_1}{n} \\ -\frac{n_2}{n} & 1-\frac{n_2}{n} & \dots & -\frac{n_2}{n} \\ \vdots & \vdots & \ddots & \vdots \\ -\frac{n_q}{n} & -\frac{n_q}{n} & \dots & 1-\frac{n_q}{n} \end{bmatrix},$$

clearly with $rank(D) = q - 1$, then the matrix $B^*$ in (30) is exactly the same as matrix $B$ in (14), while the matrix $\Psi$ in (24) is the same as matrix $A$ in (14) and the distribution of the LRT statistic $\Lambda_3^*$ in (31) the same as that of $\Lambda_2$ in Sect. 2.2.

## 3 The Exact Distribution of the LRT Statistic to Test Sphericity of the Covariance Matrix

Suppose that $\underline{X} \sim CN_p(\underline{\mu}, \Sigma)$, and that the hypothesis to be tested is

$$H_0 : \Sigma = \sigma^2 I_p \text{ (for some unspecified } \sigma^2 > 0). \tag{32}$$

Then, for a sample of size $n$, the LRT statistic to test $H_0$ in (32), is,

$$\Lambda_4 = \left( \frac{|A|}{\left(tr\, \frac{1}{p} A\right)^p} \right)^n , \tag{33}$$

where $A$ is the MLE of $\Sigma$ (see (6) and the note after this expression for references on the MLE of $\Sigma$). Under $H_0$ in (32), the $h$-th moment of $\Lambda_4$ is

$$E\left(\Lambda_4^h\right) = \prod_{j=1}^{p-1} \frac{\Gamma\left(n-1+\frac{j}{p}\right)}{\Gamma(n-j-1)} \frac{\Gamma(n-j-1+nh)}{\Gamma\left(n-1+\frac{j}{p}+nh\right)} = \prod_{j=1}^{p-1} E\left(Y_j\right)^{nh} , \tag{34}$$

for $h > \frac{p}{n} - 1$, and where $Y_j \sim Beta\left(n-j-1, \frac{j}{p}+j\right)$ are $p-1$ independent r.v.'s. See sections B.3 and B.5.2 of Appendix B in the Online Resource for details on the derivation of the expression for $E\left(\Lambda_4^h\right)$.

Since the whole set of moments of $\Lambda_4$ defines its distribution,

$$\Lambda_4 \overset{\text{st}}{\sim} \prod_{j=1}^{p-1} \left(Y_j\right)^n .$$

Then, for $W_4 = -\log \Lambda_4$, since the expression in (34) is well defined for $h$ in a neighborhood of zero, it follows that (see Appendix C in the Online Resource for details),

$$\begin{aligned}\Phi_{W_4}(t) = E\left(e^{\text{it}\, W_4}\right) = E\left(\Lambda_4^{-\text{it}}\right) &= \prod_{j=1}^{p-1} \frac{\Gamma\left(n-1+\frac{j}{p}\right)}{\Gamma(n-j-1)} \frac{\Gamma(n-j-1-nit)}{\Gamma\left(n-1+\frac{j}{p}-nit\right)} \\ &= \underbrace{\left\{ \prod_{j=1}^{p-1} \left(\frac{n-j-1}{n}\right)^{p-j} \left(\frac{n-j-1}{n} - \text{it}\right)^{-(p-j)} \right\}}_{\Phi_{1,W_4}(t)} \\ &\qquad \times \underbrace{\left\{ \prod_{j=1}^{p-1} \frac{\Gamma\left(n-1+\frac{j}{p}\right) \Gamma(n-1-nit)}{\Gamma(n-1)\, \Gamma\left(n-1+\frac{j}{p}-nit\right)} \right\}}_{\Phi_{2,W_4}(t)}\end{aligned} \tag{35}$$

which shows that the exact distribution of $W_4$ is the same as the distribution of the sum of a GIG distributed r.v. with depth $p-1$, with shape parameters $r_j = p-j$ and rate parameters $\frac{n-j-1}{n}$ $(j = 1, \ldots, p-1)$, with $p-1$ independent

*Logbeta* $\left(n-1, \frac{j}{p}\right)$ r.v.'s ($j = 1, \ldots, p-1$). This shows that the distribution of $\Lambda_4$ is thus of the form (1), with $u = p-1$. The possibility of expressing the exact distribution of $\Lambda_4$ in this form will enable us to develop a very accurate near-exact distribution for $\Lambda_4$ in Sect. 5.1.

## 4 The Exact Distribution of the LRT Statistic to Test Equality of Several Covariance Matrices

In this section the authors will only address the case of equal sample sizes. The case of unequal sample sizes, given its complexity, will only be addressed in the next section.

Hence, suppose that $\underline{X}_k \sim CN_p(\underline{\mu}_k, \Sigma_k)$, $(k = 1, \ldots, q)$, and that, based on $q$ independent samples, each of size $n$, the hypothesis to be tested is

$$H_0 : \Sigma_1 = \cdots = \Sigma_q \,. \tag{36}$$

Then, the LRT statistic is

$$\Lambda_5 = \left(q^{pq}\,\frac{\prod_{k=1}^{q}|A_k|}{|A|^q}\right)^n , \tag{37}$$

where $A_k$ is $n$ times the MLE of $\Sigma_k$ $(k = 1, \ldots, q)$, and $A = A_1 + \cdots + A_q$ (see (6) and the note after for references on the MLE's of $\Sigma_k$). Under $H_0$ in (36), the $h$-th moment of $\Lambda_5$ is

$$E\left(\Lambda_5^h\right) = \prod_{j=1}^{p}\prod_{k=1}^{q}\frac{\Gamma(n+nh-j)}{\Gamma(n-j)}\,\frac{\Gamma\left(n-1+\frac{k-j}{q}\right)}{\Gamma\left(n-1+\frac{k-j}{q}+nh\right)} = \prod_{j=1}^{p}\prod_{k=1}^{q}E\left(Y_{jk}^{nh}\right) \tag{38}$$

which may be obtained from expression (70) in Appendix B.4 in the Online Resource (see also subsection B.5.3 in Appendix B in the Online Resource for an alternative proof based on the results in Jensen (1988, Thm. 5)), and where

$$Y_{jk} \sim Beta\left(n-j, j-1+\frac{k-j}{q}\right)$$

are $pq - 1$ independent r.v.'s. Since the distribution of $\Lambda_5$ is determined by the set of its moments,

$$\Lambda_5 \overset{\text{st}}{\sim} \prod_{\substack{j=1 \\ \text{(except for} \\ j=k=1)}}^{p} \prod_{k=1}^{q} \left(Y_{jk}\right)^n .$$

Then, taking $W_5 = -\log \Lambda_5$, we have (see Appendix D in the Online Resource for details)

$$\begin{aligned}
\Phi_{W_5}(t) &= E\left(e^{-\mathrm{i}tW_5}\right) = E\left(\Lambda_5^{-\mathrm{i}t}\right) = \prod_{j=1}^{p}\prod_{k=1}^{q} \frac{\Gamma\left(n-1+\frac{k-j}{q}\right)\Gamma\left(n-j-nit\right)}{\Gamma\left(n-1+\frac{k-j}{q}-nit\right)\Gamma\left(n-j\right)} \\
&= \underbrace{\left\{\prod_{j=1}^{p-1}\left(\frac{n-1-j}{n}\right)^{r_j}\left(\frac{n-1-j}{n}-\mathrm{i}t\right)^{-r_j}\right\}}_{\Phi_{1,W_5}(t)} \\
&\quad \times \underbrace{\left\{\prod_{j=1}^{p}\prod_{k=1}^{q}\frac{\Gamma\left(n-1+\frac{k-j}{q}\right)\Gamma\left(n-1+\left\lfloor\frac{k-j}{q}\right\rfloor-nit\right)}{\Gamma\left(n-1+\left\lfloor\frac{k-j}{q}\right\rfloor\right)\Gamma\left(n-1+\frac{k-j}{q}-nit\right)}\right\}}_{\Phi_{2,W_5}(t)}
\end{aligned} \tag{39}$$

for

$$r_j = \begin{cases} q(q-1)\left(j-\frac{1}{2}\right) & j=1,\ldots,\left\lceil\frac{p-1}{q}\right\rceil-1 \\ \frac{1}{2}\left(p-p^2+2jpq+q-3jq-q^2(j-1)^2\right) & j=\left\lceil\frac{p-1}{q}\right\rceil \\ q(p-j) & j=\left\lceil\frac{p-1}{q}\right\rceil+1,\ldots,p-1\,, \end{cases} \tag{40}$$

which shows that the exact distribution of $\Lambda_5$ is of the form (1), with $u = pq$. As a consequence, as was the case for the statistic in the previous section, very accurate near-exact distributions can be developed for $\Lambda_5$. See Sect. 5.2.

## 5 Near-Exact Distributions

Given the complexity of the exact distributions of the statistics in Sects. 3 and 4, or rather, of the expressions that might be obtained for their exact pdf's and cdf's and the concomitant issues related with their manageability, the development of near-exact distributions for such statistics arises as a sensible goal.

Let then, as a general notation, $W$ stand for the negative logarithm of the LRT statistics in Sects. 3 and 4. The near-exact distributions developed in this section, for $W$, assume the form of mixtures of generalized near-integer gamma (GNIG) distributions (see Marques et al. (2011, App. B) for the complete notation of the pdf and cdf of theis distribution), which, for some $m^* \in \mathbb{N}$, will have pdf's and cdf's respectively of the form

$$f(w) = \sum_{\nu=0}^{m^*} \pi_\nu \, f^{GNIG}\left(w \,\middle|\, r_1, \ldots, r_{p-1}, r+\nu; \frac{n-2}{n}, \ldots, \frac{n-p}{n}, \lambda; p\right) \tag{41}$$

and

$$F(w) = \sum_{\nu=0}^{m^*} \pi_\nu \, F^{GNIG}\left(w \,\middle|\, r_1, \ldots, r_{p-1}, r+\nu; \frac{n-2}{n}, \ldots, \frac{n-p}{n}, \lambda; p\right), \tag{42}$$

where $w > 0$ represents a possible value of $W$ and where, for the statistic in Sect. 3,

$$r = \frac{p-1}{2}, \quad \lambda = \frac{n-1}{n} \quad \text{and} \quad r_j = p - j\,, \tag{43}$$

for $j = 1, \ldots, p-1$, while for the statistic in Sect. 4, $r_j$ are given by (40),

$$r = p\,\frac{q-1}{2} \quad \text{and} \quad \lambda = \lambda^*\,,$$

where $\lambda^*$ is the rate parameter in

$$\Phi^{**}_{2,W_5}(t) = p_1(\lambda^*)^{s_1}(\lambda^* - \mathrm{i}t)^{-s_1} + (1-p_1)\left(\lambda^*\right)^{s_2}\left(\lambda^* - \mathrm{i}t\right)^{-s_2}, \tag{44}$$

which is determined together with $s_1$, $s_2$ and $p_1$ in such a way that the first 4 derivatives of $\Phi^{**}_{2,W_5}(t)$ and $\Phi_{2,W_5}(t)$ in (39) at $t = 0$ are the same.

These near-exact distributions are built by leaving $\Phi_{1,W_4}(t)$ and $\Phi_{1,W_5}(t)$, respectively in (35) and (39), unchanged and then asymptotically approximating the c.f.'s $\Phi_{2,W_4}(t)$ and $\Phi_{2,W_5}(t)$ in the same expressions by c.f.'s of finite mixtures of gamma distributions. This replacement is done based on the results in Sects. 5 and 6 of Tricomi and Erdélyi (1951), which show that any $Logbeta(a, b)$ distribution may, for non-integer $b$, be asymptotically approximated by an infinite mixture of $\Gamma(b+\nu, a)$ $(\nu = 0, 1, \ldots)$ distributions.

This yields as near-exact distributions for $W$, finite mixtures of sums of $p$ independent gamma r.v.'s, $p-1$ of which have integer shape parameters $r_j$ while the $p$th one has shape parameter $r$. For non-integer $r$, these are finite mixtures of GNIG distributions of depth $p$, but since for odd $p$, $r = \frac{p-1}{2}$ and for odd $q$, $r = p\frac{q-1}{2}$ are integers, for $\Lambda_4$ with odd $p$ and for $\Lambda_5$ with odd $q$, the components of the mixtures are indeed GIG distributions (see Marques et al. (2011, App. B) for a complete notation of the pdf and cdf of these distributions), although for generality of notation the components of the mixtures in (41) and (42) are denoted as GNIG distributions.

The near-exact distributions built in this way are asymptotic not only for increasing sample sizes but also for increasing number of variables and populations involved, as is shown by the numerical studies carried out in the next section. These studies also show the extreme closeness of these near-exact distributions to the corresponding exact distributions even for very small sample sizes. Their parameters being very simple to determine, their implementation is very easy with the help of adequate software.

Further details on the construction of these near-exact distributions are given in the subsections ahead.

## 5.1 Near-Exact Distribution for the LRT Statistic in Sect. 3

Using the results in Sects. 5 and 6 of Tricomi and Erdélyi (1951), as a first step, $\Phi_{2,W_4}(t)$ in (35), which is the c.f. of the sum of $p-1$ independent $Logbeta\left(n-1,\frac{j}{p}\right)$ r.v.'s ($j=1,\dots,p-1$), multiplied by $n$, is replaced by the c.f. of the sum of $p-1$ independent infinite mixtures of $\Gamma\left(\frac{j}{p}+\nu, n-1\right)$ distributions ($\nu=0,1,\dots$), multiplied by $n$, which is the c.f. of the sum of $p-1$ independent infinite mixtures of $\Gamma\left(\frac{j}{p}+\nu,\frac{n-1}{n}\right)$ distributions ($\nu=0,1,\dots$), and which in turn, and given that the rate parameters of the gamma distributions are not functions of either $j$ or $\nu$, is an infinite mixture of $\Gamma\left(\left(\sum_{j=1}^{p-1}\frac{j}{p}\right)+\nu,\frac{n-1}{n}\right)$ distributions, where $\sum_{j=1}^{p-1}\frac{j}{p}=\frac{p-1}{2}$. This way we obtain a representation of the distribution which bears some resemblance with the representations in Nagar et al. (1985); Nagarsenker and Das (1975); Tang and Gupta (1986), but with coefficients which are much simpler to compute. This representation has the advantage of allowing for the easy development of very accurate near-exact approximations. Moreover, our representation does not have any parameters that are not well-defined, in contrast to the representation in Tang and Gupta (1986).

Therefore, in order to obtain a near-exact distribution for $W_4$, $\Phi_{2,W_4}(t)$ in (35) will be replaced by the c.f. of a finite mixture of $\Gamma\left(\frac{p-1}{2}+\nu,\frac{n-1}{n}\right)$ distributions, for $\nu=0,\dots,m^*$,

$$\Phi^*_{2,W_4}(t)=\sum_{\nu=0}^{m^*}\pi_\nu\left(\frac{n-1}{n}\right)^{\frac{p-1}{2}+\nu}\left(\frac{n-1}{n}-\mathrm{i}t\right)^{-\left(\frac{p-1}{2}+\nu\right)}, \tag{45}$$

where the weights $\pi_0, \ldots, \pi_{m^*-1}$ are determined as the numerical solution of the system of $m^*$ equations

$$\frac{\partial^h}{\partial t^h} \Phi^*_{2,W_4}(t)\Bigg|_{t=0} = \frac{\partial^h}{\partial t^h} \Phi_{2,W_4}(t)\Bigg|_{t=0}, \quad h = 1, \ldots, m^* \tag{46}$$

with $\pi_{m^*} = 1 - \sum_{\nu=0}^{m^*-1} \pi_\nu$. In this way these near-exact distributions have, by construction, the first $m^*$ moments equal to the first $m^*$ exact moments of $W_4$. Hence,

$$\Phi^*_{W_4}(t) = \Phi_{1,W_4}(t)\, \Phi^*_{2,W_4}(t),$$

with $\Phi_{1,W_4}(t)$ given by (35) and $\Phi^*_{2,W_4}(t)$ given by (45), will be used as a near-exact c.f. for $W_4$, to which corresponds the pdf and the cdf in (41) and (42), with $r$ and $\lambda$ given by (43).

Then, Theorem 4 and Corollary 4 may be stated.

**Theorem 4** *Let $W_4 = -\log \Lambda_4$, where $\Lambda_4$ is the LRT statistic in (33). Then, for some $m^* \in \mathbb{N}$, distributions with pdf and cdf given by (41) and (42), with $r$, $\lambda$ and $r_j$ given by (43) and $\pi_0, \ldots, \pi_{m^*-1}$ determined from (46), with $\Phi^*_{2,W_4}(t)$ given by (45) and $\pi_{m^*} = 1 - \sum_{\nu=0}^{m^*-1} \pi_\nu$, are near-exact distributions for $W_4$.*

From Theorem 4, Corollary 4 is then readily obtained.

**Corollary 4** *Let $\Lambda_4$ be the LRT statistic in (33). Then distributions with pdf and cdf, respectively given by $f(-\log \ell)\frac{1}{\ell}$ and $1 - F(-\log \ell)$ where $f(\cdot)$ and $F(\cdot)$ are given by (41) and (42), with $0 < \ell \leq 1$ representing a possible value for $\Lambda_4$ and $r_j$ $(j = 1, \ldots, p-1)$, $r$, $\lambda$ and $\pi_\nu$ $(\nu = 0, \ldots, m^*)$ defined as in the previous Theorem, are near-exact distributions for $\Lambda_4$.*

## 5.2 *Near-Exact Distribution for the LRT Statistic in Sect. 4*

### The Case of Equal Sample Sizes

For the statistic $\Lambda_5$ in Sect. 4, in the case where all $q$ samples have size $n$, we may asymptotically replace $\Phi_{2,W_5}(t)$ in (39), which is the c.f. of the sum of $pq - \min(p, q)$ independent $Logbeta\left(n - 1 + \left\lfloor \frac{k-j}{q} \right\rfloor, \frac{k-j}{q} - \left\lfloor \frac{k-j}{q} \right\rfloor\right)$ r.v.'s $(j = 1, \ldots, p;\, k = 1, \ldots, q;\, j \neq k)$, multiplied by $n$, by the c.f. of the sum of $pq - \min(p, q)$ independent infinite mixtures of $\Gamma\left(\frac{k-j}{q} - \left\lfloor \frac{k-j}{q} \right\rfloor + \nu, \frac{1}{n}\left(n - 1 + \left\lfloor \frac{k-j}{q} \right\rfloor\right)\right)$ distributions $(\nu = 0, 1, \ldots)$. If it were the case that the rate parameters in these gamma distributions were functions of neither $j$ nor $k$, as happened in the previous subsection, then this sum of mixtures would yield a simple mixture of gamma distributions. However, since now the rate parameters of the gamma distributions in the sum of mixtures are functions of both $j$ and $k$, it renders it difficult to add the different mixtures of gamma distributions. For this reason it is used as an asymptotic replacement for $\Phi_{2,W_5}(t)$ in (39) the c.f. of a finite mixture of $m^*+1$ $\Gamma\left(\sum_{j=1}^{p}\sum_{k=1}^{q}\frac{k-j}{q}-\left\lfloor\frac{k-j}{q}\right\rfloor+\nu,\lambda^*\right)$ distributions ($\nu=0,\ldots,m^*$), which is

$$\Phi^*_{2,W_5}(t)=\sum_{\nu=0}^{m^*}\pi_\nu(\lambda^*)^{r+\nu}(\lambda^*-\mathrm{i}t)^{-(r+\nu)}\,, \tag{47}$$

with

$$r=\sum_{j=1}^{p}\sum_{k=1}^{q}\frac{k-j}{q}-\left\lfloor\frac{k-j}{q}\right\rfloor=p\,\frac{q-1}{2}\,, \tag{48}$$

and where $\lambda^*$ is the common rate parameter of a mixture of two gamma distributions whose first four moments match the first four moments of the sum of independent logbeta r.v.'s whose c.f. is $\Phi_{2,W_5}(t)$ in (39), that is, where $\lambda^*$ is the rate parameter in $\Phi^{**}_{2,W_5}(t)$ in (44), which is determined together with $s_1$, $s_2$ and $p_1$ in such a way that,

$$\left.\frac{\partial^h}{\partial t^h}\,\Phi^{**}_{2,W_5}(t)\right|_{t=0}=\left.\frac{\partial^h}{\partial t^h}\,\Phi_{2,W_5}(t)\right|_{t=0}\,,\quad h=1,\ldots,4, \tag{49}$$

and where the weights $\pi_\nu$ ($\nu=0,\ldots,m^*-1$), are then determined in such a way that

$$\left.\frac{\partial^h}{\partial t^h}\,\Phi^{*}_{2,W_5}(t)\right|_{t=0}=\left.\frac{\partial^h}{\partial t^h}\,\Phi_{2,W_5}(t)\right|_{t=0}\,,\quad h=1,\ldots,m^*, \tag{50}$$

with $\pi_{m^*}=1-\sum_{\nu=0}^{m^*-1}\pi_\nu$.

Therefore,

$$\Phi^*_{W_5}(t)=\Phi_{1,W_5}(t)\,\Phi^*_{2,W_5}(t)\,,$$

with $\Phi_{1,W_5}(t)$ given by (39) and $\Phi^*_{2,W_5}(t)$ given by (47), will be used as a near-exact c.f. for $W_5$, and as such one may thus enunciate the results summarized in Theorem 5 and Corollary 5.

**Theorem 5** *Let $W_5=-\log\Lambda_5$, where $\Lambda_5$ is the LRT statistic in (37). Then, for some $m^*\in\mathbb{N}$, distributions with pdf and cdf given by (41) and (42), with $r_j$ given by (40), $r$ given by (48) and $\lambda=\lambda^*$ obtained by solving (49) in order to $\lambda^*$, $s_1$, $s_2$*

*and* $p_1$, *and* $\pi_0, \ldots, \pi_{m^*-1}$ *determined from (50), with* $\Phi^*_{2,W_5}(t)$ *given by (47) and* $\pi_{m^*} = 1 - \sum_{\nu=0}^{m^*-1} \pi_\nu$, *are near-exact distributions for* $W_5$.

**Corollary 5** *Let* $\Lambda_5$ *be the LRT statistic in (37). Then distributions with pdf and cdf, respectively given by* $f(-\log \ell)\frac{1}{\ell}$ *and* $1 - F(-\log \ell)$ *where* $f(\cdot)$ *and* $F(\cdot)$ *are given by (41) and (42), with* $0 < \ell \leq 1$ *representing the running value for* $\Lambda_5$ *and* $r_j$ $(j = 1, \ldots, p-1)$, $r$, $\lambda$ *and* $\pi_\nu$ $(\nu = 0, \ldots, m^*)$ *defined as in the previous Theorem, are near-exact distributions for* $\Lambda_5$.

## The Case of Unequal Sample Sizes

When the samples have different sizes, with the $k$-th sample having size $n_k$, the LRT statistic $\Lambda_5$ is now

$$\Lambda_5 = \frac{N^{Np}}{\prod_{k=1}^{q} n_k^{n_k p}} \frac{\prod_{k=1}^{q} |A_k|^{n_k}}{|A|^N}, \tag{51}$$

where $N = \sum_{k=1}^{q} n_k$, and $A_k$ $(k = 1, \ldots, q)$ is equal to $n_k$ times the MLE of $\Sigma_k$ and $A = A_1 + \cdots + A_q$.

The c.f. of $W_5 = -\log \Lambda_5$ may be obtained from (69) in the Online Resource section, replacing $h$ by $-\mathrm{i}t$, as

$$\begin{aligned}\Phi_{W_5}(t) &= E\left(\Lambda_5^{-\mathrm{i}t}\right)\\ &= \frac{N^{-Np\mathrm{i}t}}{\prod_{k=1}^{q} n_k^{-n_k p\mathrm{i}t}} \prod_{j=1}^{p} \left\{ \frac{\Gamma(N-q+1-j)}{\Gamma(N-q+1-j-N\mathrm{i}t)} \prod_{k=1}^{q} \frac{\Gamma(n_k - j - n_k \mathrm{i}t)}{\Gamma(n_k - j)} \right\}.\end{aligned}$$

It happens that the exact distribution of $\Lambda_5$ in (51) is quite complicated and it is not even possible to give it a structure of the form (1).

However, using a procedure similar to the one used in Coelho and Marques (2012), the c.f. of $W_5$ may be written as

$$\Phi_{W_5}(t) = \Phi_{1,W_5}(t) \frac{\Phi_{W_5}(t)}{\Phi_{1,W_5}(t)},$$

where $\Phi_{1,W_5}(t)$ is given by (39), now with $n = N/q$. Then, in order to build a near-exact approximation for $W_5$, $\Phi_{1,W_5}(t)$ will be left unchanged and $\frac{\Phi_{W_5}(t)}{\Phi_{1,W_5}(t)}$ will be replaced by $\Phi^*_{2,W_5}(t)$, given by (47), once again with $\lambda^*$ defined in a similar manner to the one used in the previous subsection, with $r$ either equal to $s_1$ in (44) or given by (48). As it will be shown in the next section, while the first choice for

$r$ will yield near-exact distributions that are asymptotic for increasing sample sizes as well as for increasing values of $p$ and $q$, the second choice may give slightly better approximations for small values of $p$ and even better approximations for large sample sizes. However, these latter near-exact distributions will no longer be asymptotic for increasing values of $p$ or $q$, but only for increasing sample sizes.

As such, in terms of near-exact distributions, for the case of unequal sample sizes, similar results to the ones in Theorem 5 and Corollary 5 still hold, with the due changes in the parameters, namely with all the occurrences of $n$ changed to $N/q$.

## 6 Numerical Studies

In order to assess the proximity of the near-exact distributions developed in Sect. 5 to the exact distribution and their performance in different situations, we will use the measure

$$\Delta = \frac{1}{2\pi} \int_{-\infty}^{+\infty} \left| \frac{\Phi_W(t) - \Phi_W^*(t)}{t} \right| dt\,,$$

which yields

$$\max_{0<\ell\leq 1} \left| F_\Lambda(\ell) - F_\Lambda^*(\ell) \right| = \max_{w>0} \left| F_W(w) - F_W^*(w) \right| \leq \Delta\,,$$

where $W$ represents generally either $W_4$ or $W_5$, $\Lambda = e^{-W}$, $\Phi_W(t)$ represents the exact c.f. of $W$ and $\Phi_W^*(t)$ its approximate c.f., usually a near-exact c.f., but occasionally an asymptotic or other c.f., $F_W(\cdot)$ and $F_\Lambda(\cdot)$ the exact cdf's of $W$ and $\Lambda$, corresponding to $\Phi_W(t)$, and $F_W^*(\cdot)$ and $F_\Lambda^*(\cdot)$ the cdf's corresponding to $\Phi_W^*(t)$. Some further details on $\Delta$ and its relation with the Berry-Esseen bound are discussed in Coelho and Marques (2012). It should be noted that, quite clearly, a given value of $\Delta$ for any approximation to the exact distribution of either $W_4$ or $W_5$ will be the same as the corresponding value of the measure for the corresponding approximation to the distribution of respectively either $\Lambda_4 = e^{-W_4}$ or $\Lambda_5 = e^{-W_5}$.

All computations are done with Mathematica®, version 7.0.0.

### 6.1 Numerical Studies on the Approximations for $\Lambda_4$

In Table 1 may be analyzed the values of the measure $\Delta$ for the statistic $\Lambda_4$, for values of $p$, the number of variables involved, ranging from 3 to 50 and sample sizes $n$ which exceed $p$ by 2, 12, 50 and 100. In order to compare the performance of the near-exact distributions with other available approximations in the literature, the

Box type asymptotic distribution (Box, 1949), used in Nagarsenker and Nagarsenker (1981) and the Pearson type I distribution used in Krishnaiah et al. (1976) have also been considered. The Pearson type I distribution was fitted by matching the first four exact moments of $\Lambda_4^{1/nb}$. In each case it is used for $b$ the positive integer value which would give a better fit, this way obtaining indeed much better approximations than the ones obtained in Krishnaiah et al. (1976). These values of $b$ are specified in Table 1, inside square brackets, right after the value of $\Delta$ for the Pearson type I approximation. It may be noted that this Pearson type I approximation, with the slight change introduced, has a much better performance than truncations, even with a rather large number of terms, of any of the expansions in Nagarsenker and Das (1975).

The results in Table 1 show that although the Pearson type I distribution, with the improvement introduced of finding the integer value of $b$ which gives the best fit to the exact distribution, has a good performance, with an asymptotic behavior not only for increasing sample sizes but also for increasing values of $p$, it still is no match even for the near-exact distribution which matches only 4 moments. Besides, finding the integer value of $b$ which gives the best fit for the Pearson type I distribution is not an easy task, and it was not possible to fit any such distribution for $p = 50$ and $n = 52$. The Box type asymptotic distribution has the poorest behavior of all the approximations. Its performance being worse for larger values of $p$, with values of $\Delta$ greater than 1, which shows that in these cases the asymptotic distribution yielded by the approximation is not a true distribution. Moreover, it produces values of $\Delta$ quite close to 1 for a number of other cases.

The near-exact distributions always show a very good performance, with very low values of $\Delta$. They exhibit a very good performance even for the smaller sample sizes, always showing an asymptotic behavior both for increasing sample sizes as well as for increasing values of $p$, and, of course, with the near-exact distributions which match a larger number of exact moments showing an increasingly better performance.

## 6.2 Numerical Studies on the Approximations for $\Lambda_5$

For the case of equal sample sizes, are shown in Table 2 the values of the measure $\Delta$ for the statistic $\Lambda_5$, for different values of $p$, $q$ and $n$, respectively the number of variables involved, the number of covariance matrices involved and the common sample size of the $q$ independent samples. In order to compare the performance of the near-exact distributions with other available approximations in the literature, the mixture of two beta distributions in Fang et al. (1982) and the Pearson type I distribution used by Krishnaiah et al. (1976) are used. These two distributions were fitted by matching respectively the first two and four exact moments of $\Lambda_5^{1/nb}$, for the choice of $b \in \mathbb{N}$ which gives the lowest value of $\Delta$. These values of $b$ are specified in Table 2, inside square brackets, right after the value of $\Delta$ for each of

**Table 1** Values of the measure $\Delta$ for the near-exact and asymptotic distributions for the LRT statistic to test sphericity

| | | Near-exact distribution<br># of exact moments matched | | | | | |
|---|---|---|---|---|---|---|---|
| $p$ | $n$ | 4 | 6 | 10 | 15 | Box | Pearson type I |
| 3 | 5 | $1.09\times10^{-8}$ | $2.74\times10^{-11}$ | $3.57\times10^{-13}$ | $9.28\times10^{-16}$ | $5.75\times10^{-2}$ | $2.17\times10^{-3}$ [1] |
| | 15 | $6.27\times10^{-11}$ | $1.21\times10^{-13}$ | $1.39\times10^{-18}$ | $4.27\times10^{-23}$ | $9.12\times10^{-3}$ | $2.78\times10^{-5}$ [1] |
| | 53 | $2.21\times10^{-11}$ | $3.42\times10^{-14}$ | $1.19\times10^{-18}$ | $7.73\times10^{-23}$ | $2.88\times10^{-3}$ | $6.85\times10^{-7}$ [1] |
| | 103 | $1.08\times10^{-12}$ | $6.43\times10^{-16}$ | $1.16\times10^{-21}$ | $2.71\times10^{-27}$ | $1.52\times10^{-3}$ | $9.58\times10^{-8}$ [1] |
| 10 | 12 | $3.54\times10^{-16}$ | $2.18\times10^{-20}$ | $2.61\times10^{-28}$ | $2.39\times10^{-37}$ | $8.07\times10^{-1}$ | $3.56\times10^{-4}$ [3] |
| | 22 | $2.19\times10^{-15}$ | $7.50\times10^{-20}$ | $2.26\times10^{-27}$ | $1.64\times10^{-35}$ | $3.52\times10^{-1}$ | $2.17\times10^{-6}$ [5] |
| | 60 | $1.21\times10^{-15}$ | $3.02\times10^{-20}$ | $7.29\times10^{-28}$ | $2.04\times10^{-36}$ | $1.16\times10^{-1}$ | $7.88\times10^{-8}$ [5] |
| | 110 | $9.24\times10^{-17}$ | $7.55\times10^{-22}$ | $3.06\times10^{-30}$ | $2.87\times10^{-40}$ | $6,17\times10^{-2}$ | $1.19\times10^{-8}$ [5] |
| 50 | 52 | $6.91\times10^{-18}$ | $3.35\times10^{-25}$ | $8.87\times10^{-35}$ | $8.46\times10^{-47}$ | $6.48\times10^{0}$ | —— |
| | 62 | $2.57\times10^{-16}$ | $6.21\times10^{-23}$ | $1.88\times10^{-32}$ | $4.66\times10^{-43}$ | $4.85\times10^{0}$ | $1.27\times10^{-6}$ [16] |
| | 100 | $1.56\times10^{-16}$ | $3.48\times10^{-23}$ | $3.92\times10^{-33}$ | $8.04\times10^{-46}$ | $2.48\times10^{0}$ | $2.91\times10^{-8}$ [24] |
| | 150 | $8.77\times10^{-18}$ | $6.80\times10^{-25}$ | $5.13\times10^{-36}$ | $3.71\times10^{-48}$ | $1.57\times10^{0}$ | $3.97\times10^{-9}$ [27] |

**Table 2** Values of the measure $\Delta$ for the near-exact and asymptotic distributions for the LRT statistic to test the equality of $q$ covariance matrices for equal sample sizes

| | | | Near-exact distribution | | | | | |
|---|---|---|---|---|---|---|---|---|
| | | | # of exact moments matched | | | | Mixture of beta dists. | Pearson type I |
| $p$ | $q$ | $n$ | 4 | 6 | 10 | 15 | | |
| 5 | 3 | 7 | $1.42\times10^{-8}$ | $7.99\times10^{-11}$ | $5.30\times10^{-15}$ | $8.89\times10^{-20}$ | $9.63\times10^{-5}$ [6] | $1.36\times10^{-5}$ [6] |
| | | 50 | $1.75\times10^{-13}$ | $1.43\times10^{-17}$ | $1.41\times10^{-25}$ | $2.15\times10^{-34}$ | $7.67\times10^{-6}$ [4] | $4.78\times10^{-9}$ [7] |
| | | 100 | $4.25\times10^{-15}$ | $8.67\times10^{-20}$ | $4.90\times10^{-29}$ | $2.29\times10^{-39}$ | $2.68\times10^{-7}$ [8] | $5.54\times10^{-10}$ [7] |
| | 10 | 7 | $3.19\times10^{-10}$ | $1.97\times10^{-13}$ | $1.08\times10^{-19}$ | $1.91\times10^{-27}$ | $1.29\times10^{-4}$ [26] | $3.24\times10^{-6}$ [22] |
| | | 50 | $1.85\times10^{-14}$ | $2.77\times10^{-19}$ | $8.62\times10^{-29}$ | $8.70\times10^{-39}$ | $6.12\times10^{-8}$ [30] | $1.56\times10^{-9}$ [28] |
| | | 100 | $5.28\times10^{-16}$ | $1.98\times10^{-21}$ | $3.93\times10^{-32}$ | $9.28\times10^{-44}$ | $2.18\times10^{-8}$ [30] | $1.50\times10^{-10}$ [29] |
| | 15 | 7 | $2.25\times10^{-10}$ | $1.22\times10^{-13}$ | $5.54\times10^{-20}$ | $1.28\times10^{-27}$ | $6.80\times10^{-6}$ [37] | $1.60\times10^{-6}$ [34] |
| | | 50 | $1.47\times10^{-14}$ | $1.88\times10^{-19}$ | $4.49\times10^{-29}$ | $1.48\times10^{-40}$ | $5.16\times10^{-8}$ [46] | $7.25\times10^{-10}$ [44] |
| | | 100 | $4.26\times10^{-16}$ | $1.35\times10^{-21}$ | $1.97\times10^{-32}$ | $5.84\times10^{-46}$ | $6.09\times10^{-9}$ [45] | $3.85\times10^{-11}$ [44] |
| 15 | 3 | 17 | $5.70\times10^{-11}$ | $2.47\times10^{-14}$ | $8,54\times10^{-21}$ | $1.84\times10^{-28}$ | $4.02\times10^{-4}$ [16] | $1.67\times10^{-5}$ [12] |
| | | 50 | $7.30\times10^{-13}$ | $6.13\times10^{-17}$ | $7.45\times10^{-25}$ | $3.16\times10^{-34}$ | $3.07\times10^{-6}$ [23] | $1.90\times10^{-8}$ [19] |
| | | 100 | $1.22\times10^{-14}$ | $2.54\times10^{-19}$ | $1.88\times10^{-28}$ | $4.57\times10^{-39}$ | $4.69\times10^{-7}$ [23] | $1.26\times10^{-9}$ [20] |
| | 10 | 17 | $7.78\times10^{-14}$ | $1.89\times10^{-18}$ | $1.58\times10^{-27}$ | $1.30\times10^{-38}$ | $3.99\times10^{-4}$ [65] | $2.88\times10^{-6}$ [43] |
| | | 50 | $1.68\times10^{-15}$ | $1.18\times10^{-20}$ | $8.00\times10^{-31}$ | $3.46\times10^{-43}$ | $1.29\times10^{-6}$ [83] | $2.78\times10^{-9}$ [79] |
| | | 100 | $3.64\times10^{-17}$ | $7.31\times10^{-23}$ | $3.86\times10^{-34}$ | $8.19\times10^{-48}$ | $8.30\times10^{-8}$ [86] | $2.27\times10^{-10}$ [83] |
| | 15 | 17 | $5.05\times10^{-15}$ | $1.99\times10^{-20}$ | $3.35\times10^{-30}$ | $6.42\times10^{-42}$ | $6.35\times10^{-5}$ [95] | $1.76\times10^{-6}$ [63] |
| | | 50 | $5.11\times10^{-17}$ | $5.68\times10^{-22}$ | $1.36\times10^{-32}$ | $6.31\times10^{-46}$ | $1.27\times10^{-7}$ [117] | $1.64\times10^{-9}$ [119] |
| | | 100 | $6.63\times10^{-18}$ | $9.61\times10^{-24}$ | $1.34\times10^{-35}$ | $2.13\times10^{-50}$ | $6.96\times10^{-7}$ [118] | $1.25\times10^{-10}$ [127] |

these approximations. The mixture of two beta distributions used is a mixture of two beta distributions with the same first parameter and a second parameter given by expressions (4.7) and (4.9) in Fang et al. (1982). Then the first weight in the mixture and the first parameter in the beta distributions were determined by equating the two first exact moments of $\Lambda_5^{1/nb}$ and the two first moments of the mixture. The mixture of two beta distributions in Fang et al. (1982) is indeed not implemented exactly in this way. The decision to follow this alternative implementation was due to the facts that not only the definition of the parameter $s$ in (4.9) and (4.10) in Fang et al. (1982) yields a conflicting definition for $A_2$ in the same reference, but also to the fact that by defining the parameters as suggested above, the approximation has indeed a much better performance.

From the results in Table 2 is possible to see how the Pearson type I approximation always outperforms the mixture of two beta distributions, mainly for the larger sample sizes. However, the Pearson type I distribution is always itself largely outperformed by the near-exact distribution which matches only 4 moments. The near-exact distributions once again show not only a marked asymptotic behavior for increasing sample sizes but also for increasing values of both $p$ and $q$, as well as a consistent extremely good performance for very small sample sizes, avoiding the cumbersomeness of the determination of the best value for the parameter $b$ in the Pearson type I and in the mixture of betas approximations.

In the case of different sample sizes, it was not feasible to fit the Pearson type I approximation to any case. Actually the authors in Krishnaiah et al. (1976) have only used this approximation for the case of equal sample sizes. However, it was possible to fit the mixture of two beta distributions to powers of $\Lambda_5^{1/b}$. But, when trying to find the best integer value of $b$ many local minima were found, rendering the process of determining the best value of $b$ a very frustrating and time consuming task. This, together with the fact that the near-exact distributions have a very good performance also for this different sample sizes case, for all sample sizes and all values of $p$ and $q$, leads to the conclusion that it is indeed much preferable to use the near-exact approximations, which seem to be the only approximations with a very good and consistent performance.

The near-exact distributions which use for $\lambda^*$ and $r$ respectively the values of $\lambda^*$ and $s_1$ in (44), obtained by solving the system of equations in (49) exhibit a very good performance even for the smaller sample sizes and a clear asymptotic behavior not only for increasing sample sizes but also for increasing values of $p$ and $q$. The near-exact distributions which match only four exact moments always outperform the mixture of two beta distributions except for the smaller sample sizes for $p = 5$, with the measures for this mixture exhibiting a somewhat erratic behavior. Values of $\Delta$ for these near-exact distributions as well as for the mixture of two beta distributions, with the indication, inside square brackets, of the value of $b$ used, may be analyzed in Table 3 in the supplementary material in the Online Resource.

The near-exact distributions which use the same value of $\lambda^*$ as the ones above, but use for $r$ the value given by (48) give better values of $\Delta$ for larger sample sizes and also for smaller values of $p$, but they loose their asymptotic character for increasing values of $p$, the number of variables involved. See Table 4 in the supplementary material in the Online Resource for the values of $\Delta$ for these near-exact distributions.

## 7 An Application of the Results Obtained — The LRT Statistic to Test the Equality of Several Multivariate Complex Normal Distributions

In this section the authors show how the results obtained may be combined, as proposed in Coelho and Marques (2009), allowing a smooth path towards the development of near-exact distributions for LRT statistics for more elaborate hypotheses. As an application, the authors will show how very sharp near-exact distributions may be obtained for the LRT statistic to test the equality of several multivariate complex Normal distributions, based on the results in the previous sections. Numerical results show the extreme closeness of the near-exact distribution obtained to the exact distribution of the LRT statistic.

Let us consider $q$ independent samples of size $n$, the $k$-th of which obtained from $CN_p(\underline{\mu}_k, \Sigma_k)$ $(k = 1, \ldots, q)$, and that we want to test the hypothesis of equality of the $q$ $CN_p(\underline{\mu}_k, \Sigma_k)$ distributions,

$$H_0 : \underline{\mu}_1 = \cdots = \underline{\mu}_k = \cdots = \underline{\mu}_q,\ \Sigma_1 = \cdots = \Sigma_k = \cdots = \Sigma_q\,. \tag{52}$$

The null hypothesis $H_0$ in (52) may be written as

$$H_0 \equiv H_{02|05} \circ H_{05}$$

where "$\circ$" is to be read as "after" or "composed with", and where

$$H_{05} : \Sigma_1 = \cdots = \Sigma_k = \cdots = \Sigma_q \tag{53}$$

and

$$H_{02|05} : \underline{\mu}_1 = \cdots = \underline{\mu}_k = \cdots = \underline{\mu}_q, \quad \text{assuming } H_{05} \tag{54}$$

which are the null hypotheses in Sects. 4 and 2.2.

According to Lemma 10.3.1 in Anderson (2003), the LRT statistic to test $H_0$ in (52) will be the product of the LRT statistics to test $H_{05}$ in (53) and $H_{02|05}$ in (54), so that this statistic will be

$$\Lambda = \underbrace{\left(q^{pq}\,\frac{\prod_{k=1}^{q}|A_k|}{|A|^q}\right)^{n}}_{\Lambda_5}\;\underbrace{\frac{|A|^{nq}}{|A+B|^{nq}}}_{\Lambda_{2|5}} = \left(q^{pq}\,\frac{\prod_{k=1}^{q}|A_k|}{|A+B|^q}\right)^{n}, \tag{55}$$

where $\Lambda_5$ is the LRT statistic to test $H_{05}$ in (53) (see Sect. 4), $\Lambda_{2|5}$ is the LRT statistic to test $H_{02|05}$ (see Sect. 2.2), $A = A_1 + \cdots + A_k + \cdots + A_q$ and $B$ is given by (14), with $n_j = n$, with $A_k$ being equal to $n$ times the MLE of $\Sigma_k$.

Given the independence of $\Lambda_5$ and $\Lambda_{2|5}$ (see Anderson (2003, Lemma 10.4.1)), the $h$-th moment of $\Lambda$ may be obtained as

$$E\left(\Lambda^h\right) = E\left(\Lambda_5^h\right)E\left(\Lambda_{2|5}^h\right),$$

where $E(\Lambda_5^h)$ is the $h$-th moment of $\Lambda_5$ and as such, given by (38), and $E(\Lambda_{2|5}^h)$ is the $h$-th moment of $\Lambda_{2|5}$, and as such given by (16), using $nq$ in place of $n$. In this context and assuming $H_0$ in (52), $\Lambda_5$ and $\Lambda_{2|5}$ are independent, given that $\Lambda_5$ is independent of $A = A_1 + \cdots + A_k + \cdots + A_q$, which may be shown using a similar procedure to the one in Anderson (2003, Sec. 10.4).

The c.f. of $W = -\log\Lambda$ may be written, from (39) and (20), where in this last expression $n$ has to be replaced by $nq$, as

$$\begin{aligned}\Phi_W(t) = \underbrace{\left\{\prod_{j=1}^{p-1}\left(\frac{n-1-j}{n}\right)^{s_j}\left(\frac{n-1-j}{n}-\mathrm{i}t\right)^{-s_j}\right\} \times\left\{\prod_{\substack{j=1\\ j\neq\alpha q-1}}^{p+q-2}\left(\frac{n-(j+1)/q}{n}\right)^{r_j}\left(\frac{n-(j+1)/q}{n}-\mathrm{i}t\right)^{-r_j}\right\}}_{\Phi_{1,W}(t)}\\ \times\underbrace{\left\{\prod_{j=1}^{p}\prod_{k=1}^{q}\frac{\Gamma\left(n-1+\frac{k-j}{q}\right)\Gamma\left(n-1+\left\lfloor\frac{k-j}{q}\right\rfloor-n\mathrm{i}t\right)}{\Gamma\left(n-1+\left\lfloor\frac{k-j}{q}\right\rfloor\right)\Gamma\left(n-1+\frac{k-j}{q}-n\mathrm{i}t\right)}\right\}}_{\Phi_{2,W}(t)},\end{aligned}$$

where

$$s_j = \begin{cases} r_j^* & j = 1,\ldots,p-1,\ j\neq\alpha q-1\\ r_j^* + r_{\alpha q-1} & j = \alpha q-1\,,\end{cases}$$

for $\alpha=2,\ldots,\left\lfloor\frac{p-1}{q}\right\rfloor$, with $r_j$ $(j=1,\ldots,p+q-2)$ given by (19) and $r_j^*$ $(j=1,\ldots,p-1)$ equal to $r_j$ in (40).

Near-exact distributions for $W$ and $\Lambda$ are thus obtained by leaving $\Phi_{1,W}(t)$ unchanged and replacing $\Phi_{2,W}(t)$ by $\Phi^*_{2,W}(t)$ in (47), with $r$ given by (48) and $\lambda^*$ determined as described in Sect. 5.2. These near-exact distributions will have pdf's and cdf's which for $\Lambda$ are given by

$$f_\Lambda(\ell)=\sum_{\nu=0}^{m^*}\pi_\nu\, f^{GNIG}\Bigg(-\log\ell\;\Bigg|\;\underbrace{r_1,\ldots,r_j,\ldots,r_{p+q-2}}_{j\neq\alpha q-1,\,\alpha=2,\ldots,\lfloor(p-1)/q\rfloor},s_1,\ldots,s_{p-1},r+\nu;\;\underbrace{\frac{n-2/q}{n},\ldots,\frac{n-(j+1)/q}{n},\ldots,\frac{n-(p+q-2)/q}{n}}_{j\neq\alpha q-1,\;\alpha=2,\ldots,\lfloor\frac{p-1}{q}\rfloor},\frac{n-2}{n},\ldots,\frac{n-p}{n},\lambda^*;\;2p+q-2-\left\lfloor\frac{p-1}{q}\right\rfloor\Bigg)\frac{1}{\ell}$$

and

$$F_\Lambda(\ell)=1-\sum_{\nu=0}^{m^*}\pi_\nu\, F^{GNIG}\Bigg(-\log\ell\;\Bigg|\;\underbrace{r_1,\ldots,r_j,\ldots,r_{p+q-2}}_{j\neq\alpha q-1,\,\alpha=2,\ldots,\lfloor(p-1)/q\rfloor},s_1,\ldots,s_{p-1},r+\nu;\;\underbrace{\frac{n-2/q}{n},\ldots,\frac{n-(j+1)/q}{n},\ldots,\frac{n-(p+q-2)/q}{n}}_{j\neq\alpha q-1,\;\alpha=2,\ldots,\lfloor\frac{p-1}{q}\rfloor},\frac{n-2}{n},\ldots,\frac{n-p}{n},\lambda^*;\;2p+q-2-\left\lfloor\frac{p-1}{q}\right\rfloor\Bigg),\tag{56}$$

where the GNIG components of the mixture may indeed be replaced by GIG components in case the shape parameter $r$, given by (48), is an integer.

These near-exact distributions yield very sharp approximations to the exact distribution, which are asymptotic not only for increasing sample sizes but also for increasing values of the number of variables in the distributions ($p$) and the number of distributions involved ($q$), as it may be seen from the values of the measure $\Delta$ in Table 5 in the Online Resource.

An example of application of this test is provided in the supplementary material in the Online Resource.

## 8 Conclusions

The authors were able to show that the main LRT statistics used in multivariate analysis in the complex multivariate normal setting all have a distribution which can be written in the form in (1). This enabled a much deeper insight into the true structure of such distributions and made possible to obtain very simple expressions for the exact distributions of the LRT statistics to test the independence of several groups of variables, the equality of several mean vectors and the equality of an expected value matrix to a given matrix and to develop very well-fitting near-exact approximations for the LRT statistics to test sphericity and the equality of covariance matrices.

Although (1) may seem to bear some resemblance to the form obtained for the real multivariate normal distribution in Marques et al. (2011), there are major differences between the complex and real cases: in the complex case (i) it is possible to obtain the exact distribution for the statistics discussed in Sect. 2 in a closed and very manageable form, thus avoiding the necessity for approximations as in Krishnaiah et al. (1976); Fang et al. (1982); Khatri (1965), or the complex expressions in Khatri (1965); Gupta (1971); Tang and Gupta (1986); Gupta (1976); Gupta and Rathie (1983); Pillai and Jouris (1971); Mathai (1973), even for the null case, and (ii) for the statistics in Sects. 3 and 4, the shape parameters of the gamma r.v.'s in the exact and near-exact distributions have much simpler expressions.

The near-exact approximations show very good performances even for very small sample sizes, displaying an asymptotic behavior not only for increasing sample sizes but also for increasing number of variables and matrices involved, outperforming by far any other available approximations, such as those in Krishnaiah et al. (1976) and Fang et al. (1982). Using one of the symbolic softwares available their implementation is very simple and their use enables us to avoid the cumbersome problems associated with the determination of the best value for the parameter $b$ in the Pearson approximation. Moreover, the numerical determination of their parameters is well defined and poses no numerical problems. Benefiting from all these features, the near-exact approximations are, beyond any doubt, the recommended approximation for the statistics $\Lambda_4$ and $\Lambda_5$.

The common structure of the distributions of the LRT statistics addressed, established in (1), also enables the smooth development of near-exact distributions for LRT statistics for more elaborate hypotheses. In Sect. 7, this feature is used to address the test of equality of multivariate complex Normal distributions and to develop near-exact distributions for its LRT statistic. Numerical studies confirm the extreme closeness between the near-exact distributions obtained and the exact distribution.

Modules for the near-exact distributions developed in this paper are available in the web-site: https://sites.google.com/site/nearexactdistributions/complex-normal, and a short users-guide for these modules is made available in Appendix A.

**Acknowledgments** Research partially supported by National Funds through FCT – Fundação para a Ciência e a Tecnologia, project PEst-OE/MAT/UI0297/2014 (CMA/ UNL). The authors would like to express their gratitude to two anonymous Referees, the Associate Editor and also the Editor-in-Chief, whose suggestions contributed to a more solid and self-contained paper.

# Appendix A: Short User's Guide for the Computational Modules

## A.1 Modules for the Exact Distributions of Statistics in Sect. 2

Concerning the modules for the exact distributions of the three LRT statistics addressed in Sect. 2, there are five modules made available for each statistic $\Lambda_i$. These modules are: CFWi, PDFWi, CDFWi, PDFLi and CDFLi, where $i$ is to be replaced by either 1, 2 or 3. These modules give as output, respectively, the value for the c.f. of $W_i = -\log\, \Lambda_i$ $(i = 1, 2, 3)$, the value for its pdf, the value for its cdf, the value for the pdf of $\Lambda_i$ and the value for the cdf of $\Lambda_i$, computed at the value for the running variable.

In the case of the LRT statistic $\Lambda_1$, the arguments of these five modules are, in this order: (i) a list whose components are the number of variables of each group, (ii) the sample size, (iii) the running variable for the function. For $\Lambda_2$, the arguments are, in this order: (i) the number of variables (equal for each population), (ii) the number of mean vectors being tested, (iii) the overall sample size, (iv) the running variable for the function. In the case of $\Lambda_3$, the arguments are: (i) the number of variables, (ii) is either the rank of the matrix $D$, in case this matrix is used, or the number of columns of the matrix $\mu$.

The modules used to compute the pdf and cdf of $W_i = -\log\, \Lambda_i$ make use of the modules for the pdf and cdf of the GIG distribution, while the modules for the pdf and cdf of $\Lambda_i$ make use of the corresponding modules for $W_i$, by using the usual technique of r.v. transformation.

To obtain the computational modules and for a few examples of their use download the Mathematica® file `Mathematica_Modules_for_Exact_distr_stats_Sec_2.nb`, readily available from the web-page https://sites.google.com/site/nearexactdistributions/complex-normal.

## A.2 Modules for the Near-Exact Distributions of Statistics in Sects. 3 and 4

For the statistics $\Lambda_4$ and $\Lambda_5$, in Sects. 3 and 4 and whose near-exact distributions are developed in Sect. 5, there are also five modules made available for each statistic. These modules are: NECFWi, NEPDFWi, NECDFWi, NEPDFLi and NECDFLi, where $i$ is to be replaced by either 4 or 5. These modules give as output, respectively, the value for the near-exact c.f. of $W_i = -\log\, \Lambda_i$ $(i = 4, 5)$, the value for its near-exact pdf and cdf and the value for the near-exact pdf and cdf of $\Lambda_i$, computed at the value for the running variable.

In the case of the LRT statistic $\Lambda_4$, the arguments of the five modules are, in this order: (i) the number of variables, (ii) the sample size, (iii) the number of exact moments to be matched, (iv) the running variable for the function and (v) an optional argument which is the number of precision digits used to compute the exact moments to be matched by the near-exact distribution and which has a default value of 100. For $\Lambda_5$, the arguments are, in this order: (i) the number of variables, (ii) the number of covariance matrices being tested, (iii) the sample size, (iv) the number of exact moments to be matched, (v) the running variable for the function and (vi) the same optional argument as in (v) for $\Lambda_4$.

The modules used to compute the near-exact pdf and cdf of $W_i = -\log \Lambda_i$ ($i = 4, 5$), make use of the modules for the pdf and cdf of the GIG and GNIG distributions, while the modules for the near-exact pdf and cdf of $\Lambda_i$ make use of the corresponding modules for $W_i$, by using the usual technique of r.v. transformation.

To obtain the computational modules and for a few examples of their use download the Mathematica® file `Mathematica_Modules_for_Near-Exact_distr_stats_Sec_3_4.nb`, readily available from the web-page https://sites.google.com/site/nearexactdistributions/complex-normal.

## References

Anderson, T.W. (2003). *An Introduction to Multivariate Statistical Analysis*, 3rd edn. Wiley, New York.

Box, G.E.P. (1949). A general distribution theory for a class of likelihood criteria. *Biometrika* **36**, 317–346

Brillinger, D.R. (2001). *Time Series: Data Analysis and Theory*. SIAM, Philadelphia.

Brillinger D.R., Krishnaiah P.R. (1982) (Eds.). *Handbook of Statistics*, Vol. 2: *Time Series in Frequency Domain*. North Holland, Amsterdam.

Coelho, C.A. (1998). The generalized integer Gamma distribution – A basis for distributions in multivariate statistics. *J. Multivariate Anal.* **64**, 86–102.

Coelho, C.A., Marques, F.J. (2009). The advantage of decomposing elaborate hypotheses on covariance matrices into conditionally independent hypotheses in building near-exact distributions for the test statistics. *Linear Algebra Appl.* **430**, 2592–2606.

Coelho, C.A., Marques, F.J. (2012). Near-exact distributions for the likelihood ratio test statistic to test equality of several variance-covariance matrices in elliptically contoured distributions. *Comput. Statist.* **27**, 627–659.

Conradsen, K. (2012). Multivariate analysis of polarimetric SAR images", *LINSTAT'12 - Intern. Conf. Trends and Perspectives in Lin. Statist. Infer. and IWMS'12 21st Intern. Workshop on Matrices and Statistics - Abstract Book*, 87–88.

Conradsen, K., Nielsen, A.A., Schou, J., Skiver, H. (2003). A test statistic in the complex Wishart distribution and its application to change detection in polarimetric SAR data. *IEEE Trans. Geosci. Remote Sens.* **41**, 4–19.

Fang, C., Krishnaiah, P.R., Nagarsenker, B.N. (1982) Asymptotic distributions of the likelihood ratio test statistics for covariance structures of the complex multivariate normal distributions. *J. Multivariate Anal.* **12**, 597–611.

Goodman N.R. (1963a). Statistical analysis based on a certain multivariate complex Gaussian distribution (An introduction). *Ann. Math. Stat.* **34**, 152–177.

Goodman N.R. (1963b). The Distribution of the Determinant of a Complex Wishart Distributed Matrix. *Ann. Math. Stat.* **34**, 178–180.

Gupta A.K. (1971). Distribution of Wilks' likelihood-ratio criterion in the complex case. *Ann. Inst. Statist. Math.* **23**, 77–87.

Gupta A.K. (1976). Nonnull distribution of Wilks' statistic for MANOVA in the complex case. *Comm. Statist. Simulation Comput.* **5**, 177–188.

Gupta A.K., Rathie P.N. (1983). Nonnull distributions of Wilks' Lambda in the complex case. *Statistica* **43**, 445–450.

James AT (1964). Distributions of matrix variates and latent roots derived from normal samples. Ann Math Stat 35:475–501

Jensen S.T. (1988). Covariance hypotheses which are linear in both the covariance and the inverse covariance. *Ann. Statist.* **16**, 302–322.

Khatri C.G. (1965). Classical Statistical Analysis Based on a Certain Multivariate Complex Gaussian Distribution. *Ann. Math. Stat.* **36**, 98–114.

Krishnaiah P.R., Lee J.C., Chang T.C. (1976). The Distributions of the Likelihood Ratio Statistics for Tests of Certain Covariance Structures of Complex Multivariate Normal Populations. *Biometrika* **63**, 543–549.

Lehmann N.H., Fishler E., Haimovich A.M., Blum R.S., Chizhik D., Cimini L.J., Valenzuela R.A. (2007). Evaluation of Transmit Diversity in MIMO-Radar Direction Finding. *IEEE Trans. Signal Process* **55**, 2215–2225.

Marques F.J., Coelho C.A., Arnold B.C. (2011). A general near-exact distribution theory for the most common likelihood ratio test statistics used in Multivariate Analysis. *Test* **20**, 180–203.

Mathai A.M. (1973). A few remarks about some recent articles on the exact distributions of multivariate test criteria: I. *Ann. Inst. Statist. Math.* **25**, 557–566.

Muirhead RJ (2005). Aspects of Multivariate Statistical Theory. Wiley, New York

Nagar D.K., Jain S.K., Gupta A.K. (1985). Distribution of LRC for testing sphericity of a complex multivariate Gaussian model. *Int. J. Math. Math. Sci.* **8**, 555–562.

Nagarsenker B.N., Das M.M. (1975). Exact distribution of sphericity criterion in the complex case and its percentage points. *Comm. Statist.* **4**, 363–374.

Nagarsenker B.N., Nagarsenker P.B. (1981). Distribution of the likelihood ratio statistic for testing sphericity structure for a complex normal covariance matrix. *Sankhyā*, Ser. B **43**, 352–359.

Pannu N.S., Coy A.J., Read J. (2003). Application of the complex multivariate normal distribution to crystallographic methods with insights into multiple isomorphous replacement phasing. *Acta Crystallogr.*, Sec. D **59**, 1801–1808.

Pillai K.C.S., Jouris G.M. (1971). Some distribution problems in the multivariate complex Gaussian case. *Ann. Math. Stat.* **42**, 517–525.

Pillai K.C.S., Nagarsenker B.N. (1971). On the distribution of the sphericity test criterion in classical and complex normal populations having unknown covariance matrices. *Ann. Math. Stat.* **42**, 764–767.

Shumway R.H., Stoffer D.S. (2006). *Time Series Analysis and Its Applications: with R examples*, 2nd ed., New York: Springer Verlag

Tang J., Gupta A.K. (1986). Exact distribution of certain general test statistics in Multivariate Analysis. *Aust. J. Statist.* **28**, 107–114.

Tricomi F.G., Erdélyi A. (1951). The asymptotic expansion of a ratio of Gamma functions. *Pacific J. Math.* **1**, 133–142.

Turin G.L. (1960). The characteristic function of Hermitian quadratic forms in complex Normal variables. *Biometrika* **47**, 199–201.

Wooding R.A. (1956). The multivariate distribution of complex normal variables. *Biometrika* **43**, 212–215.

## Supplementary Material

The material here presented makes an integral part of the original paper but was taken out from the main paper because of length restrictions.

## Tables Referenced in Sects. 6 and 7 of the Main Paper

**Table 3** Values of the measure $\Delta$ for the near-exact and asymptotic distributions for the LRT statistic to test the equality of $q$ covariance matrices, for different sample sizes $n_k$

| | | | Near-exact distribution | | | | Mixture of | |
|---|---|---|---|---|---|---|---|---|
| | | | # of exact moments matched | | | | | |
| $p$ | $q$ | $n_k$(*) | 4 | 6 | 10 | 15 | beta dists. | |
| 5 | 5 | 7(3)19 | $9.21\times10^{-5}$ | $9.27\times10^{-6}$ | $1.86\times10^{-7}$ | $3.28\times10^{-9}$ | $2.83\times10^{-5}$ | [68] |
| | 10 | 7(3)34 | $4.10\times10^{-5}$ | $3.09\times10^{-6}$ | $4.59\times10^{-8}$ | $7.10\times10^{-10}$ | $1.66\times10^{-4}$ | [212] |
| | 15 | 7(3)49 | $2.45\times10^{-5}$ | $1.56\times10^{-6}$ | $1.77\times10^{-8}$ | $1.94\times10^{-10}$ | $9.93\times10^{-5}$ | [400] |
| 15 | 5 | 17(3)29 | $6.60\times10^{-5}$ | $5.50\times10^{-6}$ | $1.06\times10^{-7}$ | $2.32\times10^{-9}$ | $1.25\times10^{-4}$ | [527] |
| | 10 | 17(3)44 | $2.81\times10^{-5}$ | $1.84\times10^{-6}$ | $2.23\times10^{-8}$ | $2.60\times10^{-10}$ | $6.78\times10^{-5}$ | [1233] |
| | 15 | 17(3)59 | $1.69\times10^{-5}$ | $9.50\times10^{-7}$ | $8.24\times10^{-9}$ | $6.32\times10^{-11}$ | $2.22\times10^{-4}$ | [2245] |
| 5 | 5 | 37(3)49 | $4.85\times10^{-9}$ | $2.97\times10^{-10}$ | $3.40\times10^{-12}$ | $3.96\times10^{-14}$ | $8.68\times10^{-9}$ | [573] |
| | 10 | 37(3)64 | $3.74\times10^{-9}$ | $1.04\times10^{-10}$ | $2.74\times10^{-13}$ | $5.56\times10^{-16}$ | $7.23\times10^{-5}$ | [352] |
| | 15 | 37(3)79 | $3.21\times10^{-9}$ | $5.78\times10^{-11}$ | $6.66\times10^{-14}$ | $4.98\times10^{-17}$ | $8.38\times10^{-6}$ | [550] |
| 15 | 5 | 47(3)59 | $4.22\times10^{-9}$ | $1.80\times10^{-11}$ | $2.05\times10^{-15}$ | $1.13\times10^{-19}$ | $1.55\times10^{-6}$ | [2023] |
| | 10 | 47(3)74 | $6.79\times10^{-9}$ | $1.38\times10^{-11}$ | $1.96\times10^{-16}$ | $1.29\times10^{-21}$ | $2.09\times10^{-7}$ | [4050] |
| | 15 | 47(3)89 | $7.41\times10^{-9}$ | $1.16\times10^{-11}$ | $3.51\times10^{-17}$ | $1.09\times10^{-23}$ | $8.91\times10^{-7}$ | [7764] |

(*) the notation 'i(s)f' for sample sizes, indicates the initial value, the step and the 'final' value, so that for example, 7(3)19 stands for values of $n_1, \ldots, n_5$ equal to 7,10,13,16,19.

**Table 4** Values of the measure $\Delta$ for a second version of near-exact distributions for the LRT statistic to test the equality of $q$ covariance matrices, for different sample sizes $n_k$

| | | | # of exact moments matched by the near-exact dist. | | | |
|---|---|---|---|---|---|---|
| $p$ | $q$ | $n_k$(*) | 4 | 6 | 10 | 15 |
| 5 | 5 | 7(3)19 | $2.94\times10^{-5}$ | $1.75\times10^{-6}$ | $7.20\times10^{-8}$ | $1.67\times10^{-9}$ |
| | 10 | 7(3)34 | $2.23\times10^{-5}$ | $5.41\times10^{-7}$ | $1.63\times10^{-8}$ | $2.66\times10^{-10}$ |
| | 15 | 7(3)49 | $1.59\times10^{-5}$ | $2.51\times10^{-7}$ | $5.65\times10^{-9}$ | $6.41\times10^{-11}$ |
| 5 | 5 | 37(3)49 | $1.76\times10^{-10}$ | $1.65\times10^{-13}$ | $3.27\times10^{-19}$ | $6.24\times10^{-26}$ |
| | 10 | 37(3)64 | $2.16\times10^{-10}$ | $1.40\times10^{-13}$ | $1.12\times10^{-19}$ | $6.24\times10^{-27}$ |
| | 15 | 37(3)79 | $2.34\times10^{-10}$ | $1.36\times10^{-13}$ | $7.19\times10^{-20}$ | $2.04\times10^{-27}$ |
| 15 | 5 | 47(3)59 | $1.80\times10^{-9}$ | $2.96\times10^{-12}$ | $1.65\times10^{-17}$ | $1.13\times10^{-23}$ |
| | 10 | 47(3)74 | $1.67\times10^{-9}$ | $3.35\times10^{-12}$ | $1.33\times10^{-17}$ | $3.58\times10^{-24}$ |
| | 15 | 47(3)89 | $2.92\times10^{-9}$ | $2.68\times10^{-12}$ | $5.50\times10^{-19}$ | $1.42\times10^{-24}$ |

(*) please see the note after Table 3.

**Table 5** Values of the measure $\Delta$ for the near-exact distributions for the LRT statistic to test the equality of $q$ multivariate complex Normal distributions

| | | | Near-exact distribution — # of exact moments matched | | | |
|---|---|---|---|---|---|---|
| $p$ | $q$ | $n$ | 4 | 6 | 10 | 15 |
| 5 | 3 | 7 | $9.36\times10^{-9}$ | $4.41\times10^{-11}$ | $2.02\times10^{-15}$ | $2.15\times10^{-20}$ |
| | | 50 | $8.17\times10^{-14}$ | $5.07\times10^{-18}$ | $3.00\times10^{-26}$ | $2.56\times10^{-35}$ |
| | | 100 | $1.95\times10^{-15}$ | $2.99\times10^{-20}$ | $1.00\times10^{-29}$ | $2.62\times10^{-40}$ |
| | 10 | 7 | $2.27\times10^{-10}$ | $1.22\times10^{-13}$ | $4.96\times10^{-20}$ | $5.93\times10^{-28}$ |
| | | 50 | $8.53\times10^{-15}$ | $9.42\times10^{-20}$ | $1.61\times10^{-29}$ | $7.88\times10^{-40}$ |
| | | 100 | $2.36\times10^{-16}$ | $6.46\times10^{-22}$ | $6.93\times10^{-33}$ | $7.76\times10^{-45}$ |
| | 15 | 7 | $1.61\times10^{-10}$ | $7.62\times10^{-14}$ | $2.62\times10^{-20}$ | $4.21\times10^{-28}$ |
| | | 50 | $6.75\times10^{-15}$ | $6.35\times10^{-20}$ | $8.29\times10^{-30}$ | $1.30\times10^{-41}$ |
| | | 100 | $1.90\times10^{-16}$ | $4.39\times10^{-22}$ | $3.43\times10^{-33}$ | $4.66\times10^{-47}$ |
| 15 | 3 | 17 | $5.20\times10^{-11}$ | $2.17\times10^{-14}$ | $6.90\times10^{-21}$ | $1.33\times10^{-28}$ |
| | | 50 | $5.66\times10^{-13}$ | $4.29\times10^{-17}$ | $4.26\times10^{-25}$ | $1.41\times10^{-34}$ |
| | | 100 | $9.20\times10^{-15}$ | $1.71\times10^{-19}$ | $1.01\times10^{-28}$ | $1.87\times10^{-39}$ |
| | 10 | 17 | $7.18\times10^{-14}$ | $1.69\times10^{-18}$ | $1.32\times10^{-27}$ | $9.97\times10^{-39}$ |
| | | 50 | $1.31\times10^{-15}$ | $8.37\times10^{-21}$ | $4.64\times10^{-31}$ | $1.57\times10^{-43}$ |
| | | 100 | $2.75\times10^{-17}$ | $4.93\times10^{-23}$ | $2.08\times10^{-34}$ | $3.34\times10^{-48}$ |
| | 15 | 17 | $4.67\times10^{-15}$ | $1.78\times10^{-20}$ | $2.81\times10^{-30}$ | $4.95\times10^{-42}$ |
| | | 50 | $3.99\times10^{-17}$ | $4.02\times10^{-22}$ | $7.92\times10^{-33}$ | $2.86\times10^{-46}$ |
| | | 100 | $5.00\times10^{-18}$ | $6.48\times10^{-24}$ | $7.24\times10^{-36}$ | $8.70\times10^{-51}$ |

## Example of Application of the Test of Equality of Multivariate Complex Normal Distributions

Suppose that there are two independent samples, both with size $n = 6$ from 4-variate complex Normal distributions and that we want to test whether or not we may consider that they come from the same multivariate complex Normal distribution. More precisely, suppose we have one sample of size 6 from a $CN_4(\underline{\mu}_1, \Sigma_1)$ distribution and another independent sample, also of size 6, from a $CN_4(\underline{\mu}_2, \Sigma_2)$ distribution, and we want to test the null hypothesis

$$H_0 : \underline{\mu}_1 = \underline{\mu}_2,\ \Sigma_1 = \Sigma_2 . \tag{57}$$

The sample mean vectors and covariance matrices, or rather, the sum of squares and sum of products matrices, obtained from those two samples, rounded to three decimal places, were the ones in Appendix E, listed there respectively as $\underline{\widetilde{\mathcal{X}}}_1$, $\underline{\widetilde{\mathcal{X}}}_2$, $\mathcal{A}_1$ and $\mathcal{A}_2$. The LRT statistic to test our null hypothesis in (57), will be, from (55),

$$\Lambda = \left(2^{4\times 2}\,\frac{|A_1|\,|A_2|}{|A+B|^2}\right)^6 , \tag{58}$$

where $A = A_1 + A_2$ and $B$ is given by (14), with $q = 2$, $n_1 = n_2 = 6$ and $\underline{\widetilde{X}} = (\underline{\widetilde{X}}_1 + \underline{\widetilde{X}}_2)/2$. Using in (58) the estimates $\underline{\widetilde{\mathcal{X}}}_1$, $\underline{\widetilde{\mathcal{X}}}_2$, $\mathcal{A}_1$, $\mathcal{A}_2$ and $\mathcal{B}$ in Appendix E, one obtains for $\Lambda$ in (58) a computed value of $1.7036793418 \times 10^{-10}$. This computed value is larger than the 0.05 quantile of $7.4047854401 \times 10^{-16}$ for the near-exact distribution for $\Lambda$ that matches 4 exact moments, and as such we should not reject the null hypothesis in (57), for a test of level $\alpha = 0.05$. Otherwise, we may compute the near-exact p-value that corresponds to this computed value. This p-value is 0.46023.

The cdf of this near-exact distribution that matches the first four exact moments is given by (56) for $m^* = 4$, $p = 4$, $q = 2$ and $n = 6$. This amounts to a set of shape parameters

$$\underbrace{\{r_1, \ldots, r_j, \ldots, r_{p+q-2}\}}_{j\neq \alpha q-1, \alpha=2,\ldots,\lfloor (p-1)/q\rfloor} = \underbrace{\{r_1, \ldots, r_4\}}_{j\neq 3} = \{r_1, r_2, r_4\} = \{1, 1, 1\}$$

together with

$$\{s_1, \ldots, s_{p-1}\} = \{s_1, \ldots, s_3\} = \{2, 3, 2\} ,$$

and $r = 2$, as given by (48). Note that from (19)

$$h_j = \{1, 0, 0, 0\} , \qquad \text{and} \qquad r_j = \{1, 1, 1, 1\} , \qquad j = 1, \ldots, p+q-2 ,$$

and from (40),

$$r_j^* = \{1, 3, 2\} .$$

The corresponding rate parameters are

$$\left\{\frac{n-j/q}{n}\right\}_{\substack{j=2,\ldots,p+q-2\\ j\neq \alpha q-1,\alpha=2,\ldots,\left\lfloor\frac{p-1}{q}\right\rfloor}} = \left\{\frac{5}{6}, \frac{3}{4}, \frac{7}{12}\right\}, \quad \left\{\frac{n-j}{n}\right\}_{j=2,\ldots,p} = \left\{\frac{2}{3}, \frac{1}{2}, \frac{1}{3}\right\}$$

and

$$\begin{aligned}\lambda^* = 0.&613816493458495161624292651922869958750441634431890923277611221882\\&657141748414538325819685026992069150216299282055641652747056331026\\&40196895312277918233345329089096913090468302893850344808 97\,,\end{aligned}$$

this last one obtained by solving the system of equations (49), as described in subsection 5.2.1. This way, for example, the cdf for this near-exact distribution of $\Lambda$, would be written, using the notation in (56), as

$$F_\Lambda(\ell) = 1 - \sum_{\nu=0}^{4} \pi_\nu \, F^{GNIG}\left(-\log \ell \,\middle|\, 1, 1, 1, 2, 3, 2, r+\nu; \frac{5}{6}, \frac{3}{4}, \frac{7}{12}, \frac{2}{3}, \frac{1}{2}, \frac{1}{3}, \lambda^*; 7\right),$$

where the weights $\pi_\nu$, obtained by solving the system of equations (50), are listed in Appendix E, with $\pi_4 = 1 - \sum_{\nu=0}^{3} \pi_\nu$. Note that, given the fact that the shape parameter $r$, given by (48) is an integer, the near-exact distribution obtained is indeed a mixture of GIG distributions.

As may be seen one needs to keep a good number of decimal places for $\lambda^*$ and also the weights $\pi_\nu$ in order to obtain the desired precision in the computation of quantiles.

Also, this near-exact distribution has a value of $1.94 \times 10^{-7}$, for the measure $\Delta$, which means that its cdf never gives values which are away from the exact one by more than that amount.

# Appendixes

## B On the Distributions of the LRT Statistics

### *B.1 On the Distribution of $\Lambda_1$*

If $A^*$ is a $p \times p$ matrix with a complex Wishart distribution with $n$ degrees of freedom and parameter matrix $\Sigma$, that is, if $A^* \sim CW_p(n, \Sigma)$, then its pdf is given by (see Goodman (1963a); James (1964, Sec. 8))

$$f_{A^*}(\mathcal{A}) = \frac{e^{-tr(\Sigma^{-1}\mathcal{A})}|\mathcal{A}|^{n-p}}{\Gamma_p(n)\,|\Sigma|^n}\,, \tag{59}$$

where

$$\Gamma_p(a) = \pi^{p(p-1)/2}\prod_{j=1}^{p}\Gamma(a-j+1) \tag{60}$$

is the complex multivariate gamma function of order $p$.

The $h$-th moment of $|A^*|$ may be easily obtained from (59), for $h > p+1-n$, as

$$\begin{aligned} E\left(|A^*|^h\right) &= \int_{\mathcal{A}>0}|\mathcal{A}|^h\, f_{A^*}(\mathcal{A})\,d\mathcal{A} = \int_{\mathcal{A}>0}|\mathcal{A}|^h\,\frac{e^{-tr(\Sigma^{-1}\mathcal{A})}|\mathcal{A}|^{n-p}}{\Gamma_p(n)\,|\Sigma|^n}\,d\mathcal{A}\\ &= \frac{\Gamma_p(n+h)\,|\Sigma|^{n+h}}{\Gamma_p(n)\,|\Sigma|^n}\underbrace{\int_{\mathcal{A}>0}\underbrace{\frac{e^{-tr(\Sigma^{-1}\mathcal{A})}|\mathcal{A}|^{n+h-p-1}}{\Gamma_p(n+h)\,|\Sigma|^{n+h}}}_{\text{pdf of } CW_p(n+h,\,\Sigma)}d\mathcal{A}}_{=1}\\ &= |\Sigma|^h\,\frac{\Gamma_p(n+h)}{\Gamma_p(n)} \end{aligned} \tag{61}$$

where $\int_{\mathcal{A}>0}$ indicates that the integral is taken over the space of all positive-definite hermitian matrices.

Let then $\Lambda_1$ be given by (5), where, for a sample of size $n$, $A \sim CW_p(n-1, \frac{1}{n}\Sigma)$. Then, using a procedure in all similar to the one used in the real case (see Muirhead (2005, subsect. 11.2.2)), under $H_0$ in (4), given the independence of the matrices $A_{kk}$, the $h$-th moment of $\Lambda_1$ will be

$$\begin{aligned} E\left(\Lambda_1^h\right) &= \int_{\mathcal{A}>0}\frac{|\mathcal{A}|^{nh}}{\prod_{k=1}^{m}|\mathcal{A}_{kk}|^{nh}}\,\frac{e^{-tr(n\Sigma^{-1}\mathcal{A})}|\mathcal{A}|^{n-1-p}}{\Gamma_p(n-1)\,|\frac{1}{n}\Sigma|^{n-1}}\,d\mathcal{A}\\ &= \frac{\Gamma_p(n-1+nh)}{\Gamma_p(n-1)}\,|\tfrac{1}{n}\Sigma|^{nh}\\ &\qquad \int_{\mathcal{A}>0}\frac{1}{\prod_{k=1}^{m}|\mathcal{A}_{kk}|^{nh}}\underbrace{\frac{e^{-tr(n\Sigma^{-1}\mathcal{A})}|\mathcal{A}|^{n-1+nh-p}}{\Gamma_p(n-1+nh)\,|\frac{1}{n}\Sigma|^{n-1+nh}}}_{\text{pdf of } CW_p(n-1+nh,\,\frac{1}{n}\Sigma)}d\mathcal{A}\\ &= \frac{\Gamma_p(n-1+nh)}{\Gamma_p(n-1)}\,|\tfrac{1}{n}\Sigma|^{nh}\prod_{k=1}^{m}E\left(|A_{kk}|^{-nh}\right) \end{aligned} \tag{62}$$

for $A_{kk} \sim CW(n-1+nh, \frac{1}{n}\Sigma_{kk})$. But then, from (61) and from the fact that under $H_0$ in (4), $|\Sigma| = \prod_{k=1}^{m} |\Sigma_{kk}|$, using (60), we may write

$$E\left(\Lambda_1^h\right) = \frac{\Gamma_p(n-1+nh)}{\Gamma_p(n-1)} |\tfrac{1}{n}\Sigma|^{nh} \times \prod_{k=1}^{m} \frac{\Gamma_{p_k}(n-1+nh-nh)}{\Gamma_{p_k}(n-1+nh)} |\tfrac{1}{n}\Sigma_{kk}|^{-nh} \tag{63}$$

$$= \frac{\Gamma_p(n-1+nh)}{\Gamma_p(n-1)} \prod_{k=1}^{m} \frac{\Gamma_{p_k}(n-1)}{\Gamma_{p_k}(n-1+nh)} \tag{64}$$

$$= \left\{\prod_{j=1}^{p} \frac{\Gamma(n+nh-j)}{\Gamma(n-j)}\right\} \left\{\prod_{k=1}^{m}\prod_{j=1}^{p_k} \frac{\Gamma(n-j)}{\Gamma(n+nh-j)}\right\}$$

$$= \left\{\prod_{j=p_1+1}^{p} \frac{\Gamma(n+nh-j)}{\Gamma(n-j)}\right\} \left\{\prod_{k=2}^{m}\prod_{j=1}^{p_k} \frac{\Gamma(n-j)}{\Gamma(n+nh-j)}\right\}$$

$$= \left\{\prod_{k=2}^{m}\prod_{j=1}^{p_k} \frac{\Gamma(n+nhp_k^*-j)}{\Gamma(n-p_k^*-j)}\right\} \left\{\prod_{k=2}^{m}\prod_{j=1}^{p_k} \frac{\Gamma(n-j)}{\Gamma(n+nh-j)}\right\}$$

for $p_k^* = p_1 + \ldots + p_{k-1}$. Then, by reversing the indexing of the sets, $p_1 \to p_m$, $p_2 \to p_{m-1}, \ldots, p_{m-1} \to p_2$, $p_m \to p_1$, it is possible to write, for $q_k = p_{k+1} + \cdots + p_m$, $E\left(\Lambda_1^h\right)$ as in (7).

See subsection B.5 of this Appendix for an alternative proof.

## B.2 On the Distribution of $\frac{|A|}{|A+B|}$ when A and B Are Independent Complex Wishart Matrices

In this section it is shown an alternative way to obtain the distribution of

$$\Lambda_2 = \left(\frac{|A|}{|A+B|}\right)^n$$

when $A$ and $B$ have independent complex Wishart distributions.

Let the matrix $A$ be a $p\times p$ matrix with a complex Wishart distribution with $n-q$ degrees of freedom and parameter matrix $\Sigma$ and $B$ an independent $p\times p$ complex Wishart matrix with $q$ degrees of freedom and parameter matrix $\Sigma$, that is, let

$$A \sim CW_p(n-q, \Sigma) \quad \text{and} \quad B \sim CW_p(q-1, \Sigma) \tag{65}$$

be two independent matrices.

It is not hard to obtain the $h$-th moment of $\Lambda_2$ directly from the joint distribution of $A$ and $B$. It is known that if $A^* \sim CW_p(n, \Sigma)$, then its pdf is given by (59).

As such, assuming that $A$ and $B$ are independent, with the distributions in (65), and using (61) and (60), it follows that, for $h > (p+q)/n - 1$,

$$
\begin{aligned}
E\left(\Lambda_2^h\right) &= E\left(\frac{|A|^{nh}}{|A+B|^{nh}}\right)\\
&= \int_{\mathcal{A}>0}\int_{\mathcal{B}>0} \frac{|\mathcal{A}|^{nh}}{|\mathcal{A}+\mathcal{B}|^{nh}} \frac{e^{-tr(\Sigma^{-1}\mathcal{A})}|\mathcal{A}|^{n-q-p}}{\Gamma_p(n-q)\,|\Sigma|^{n-q}} \frac{e^{-tr(\Sigma^{-1}\mathcal{B})}|\mathcal{B}|^{q-p-1}}{\Gamma_p(q-1)\,|\Sigma|^{q-1}} \, d\mathcal{B}\, d\mathcal{A}\\
&= |\Sigma|^{nh}\frac{\Gamma_p(n-q+nh)}{\Gamma_p(n-q)} \int_{\mathcal{A}>0}\int_{\mathcal{B}>0} \frac{1}{|\mathcal{A}+\mathcal{B}|^{nh}}\\
&\qquad \underbrace{\frac{e^{-tr(\Sigma^{-1}\mathcal{A})}|\mathcal{A}|^{n-q+nh-p}}{\Gamma_p(n-q+nh)\,|\Sigma|^{n-q+nh}}}_{\text{pdf of } CW_p(n-q+nh,\,\Sigma)} \underbrace{\frac{e^{-tr(\Sigma^{-1}\mathcal{B})}|\mathcal{B}|^{q-p-1}}{\Gamma_p(q-1)\,|\Sigma|^{q-1}}}_{\text{pdf of } CW_p(q-1,\,\Sigma)} \, d\mathcal{B}\, d\mathcal{A}\\
&= |\Sigma|^{nh}\frac{\Gamma_p(n-q+nh)}{\Gamma_p(n-q)}\, E\left(|A+B|^{-nh}\right)
\end{aligned}
$$

where $A+B \sim CW_p(n+nh-1, \Sigma)$, so that, from (61) we may write

$$
\begin{aligned}
E\left(\Lambda_2^h\right) &= |\Sigma|^{nh}\frac{\Gamma_p(n-q+nh)}{\Gamma_p(n-q)}\,|\Sigma|^{-nh}\frac{\Gamma_p(n-1+nh-nh)}{\Gamma_p(n-1+nh)}\\
&= \prod_{j=1}^{p}\frac{\Gamma(n-q+nh-j+1)}{\Gamma(n-q-j+1)}\frac{\Gamma(n-j)}{\Gamma(n+nh-j)} = \prod_{j=1}^{p} E\left(Y_j^h\right) \quad (h>(p+q)/n-1)
\end{aligned}
\tag{66}
$$

where the r.v.'s $Y_j$ are the r.v.'s in (15). Since $0 < \Lambda_2 < 1$, the set of moments in (66) completely identifies the distribution of $\Lambda_2$, and as such it is possible to infer the result in (15).

### *B.3 On the Distribution of $\Lambda_4$*

Let $\Lambda_4$ be the statistic in (33). Then,

$$
\Lambda_4 = \left(\frac{|A|}{\left(tr\frac{1}{p}A\right)^p}\right)^n = p^{np}\frac{|A|^n}{(tr\,A)^{np}},
$$

and, from the fact that $A \sim CW_p(n-1, \Sigma)$, we have, for $h > \frac{p}{n} - 1$,

$$
\begin{aligned}
E\left(\Lambda_4^h\right) &= p^{nph} \int_{\mathcal{A}>0} \frac{|A|^{nh}}{(tr\,\mathcal{A})^{nph}} \frac{e^{-tr(\Sigma^{-1}\mathcal{A})}\,|\mathcal{A}|^{n-1-p}}{\Gamma_p(n-1)\,|\Sigma|^{n-1}}\, d\mathcal{A} \\
&= p^{nph} \frac{\Gamma_p(n-1+nh)}{\Gamma_p(n-1)} |\Sigma|^{nh} \int_{\mathcal{A}>0} \frac{1}{(tr\,\mathcal{A})^{nph}} \\
&\qquad \underbrace{\frac{e^{-tr(\Sigma^{-1}\mathcal{A})}\,|\mathcal{A}|^{n-1-p+nh}}{\Gamma_p(n-1+nh)\,|\Sigma|^{n-1+nh}}}_{\text{pdf of } CW_p(n-1+nh,\,\Sigma)} d\mathcal{A} \\
&= p^{nph} \frac{\Gamma_p(n-1+nh)}{\Gamma_p(n-1)} |\Sigma|^{nh}\, E\Big((tr\,A)^{-nph}\Big)\,,
\end{aligned}
$$

for $A \sim CW_p(n-1+nh, \Sigma)$.

But then, since under $H_0$ in (32) the diagonal elements of $A$, $a_{jj}$ $(j=1,\ldots,p)$, are independent, one has $tr\,A = \sum_{j=1}^{p} a_{jj} \sim CW_1((n-1)p+nph, \sigma^2)$, which upon applying (61), since $tr\,A = |tr\,A|$, yields

$$
E\left(\Lambda_4^h\right) = p^{nph} \frac{\Gamma_p(n-1+nh)}{\Gamma_p(n-1)} |\Sigma|^{nh} (\sigma^2)^{-nph} \frac{\Gamma((n-1)p+nph-nph)}{\Gamma((n-1)p+nph)}\,,
$$

which, on noticing that under $H_0$ in (32) we have $|\Sigma| = (\sigma^2)^p$, agrees with (4.2) in Pillai and Nagarsenker (1971), taking into account that this expression was designed to yield the non-null moments and that the statistic used by these authors is the power $(1/n)$ of $\Lambda_4$. Then, applying (60), yields in turn

$$
E\left(\Lambda_4^h\right) = p^{nph} \prod_{j=1}^{p} \frac{\Gamma(n-j+nh)}{\Gamma(n-j)} \frac{\Gamma((n-1)p+nph-nph)}{\Gamma((n-1)p+nph)}\,, \tag{67}
$$

which matches expression (2.10) in Krishnaiah et al. (1976), taking into account that also in this paper the statistic used is the power $(1/n)$ of $\Lambda_4$.

Then, by using on (67) the gamma multiplication formula

$$
\Gamma(nz) = (2\pi)^{\frac{1}{2}(1-n)}\, n^{nz-\frac{1}{2}} \prod_{k=0}^{n-1} \Gamma\left(z+\frac{k}{n}\right)\,, \tag{68}
$$

one obtains

$$E\left(\Lambda_4^h\right) = p^{nph}\prod_{j=1}^{p}\frac{\Gamma(n-j+nh)}{\Gamma(n-j)}\, p^{-nph}\prod_{j=0}^{p-1}\frac{\Gamma\left(n-1+\frac{j}{p}\right)}{\Gamma\left(n-1+\frac{j}{p}+nh\right)}$$
$$= \prod_{j=2}^{p}\frac{\Gamma(n-j+nh)}{\Gamma(n-j)}\prod_{j=1}^{p-1}\frac{\Gamma\left(n-1+\frac{j}{p}\right)}{\Gamma\left(n-1+\frac{j}{p}+nh\right)},$$

which yields (34).

See subsection B.5 of this Appendix for an alternative proof.

## B.4 On the Distribution of $\Lambda_5$

The $h$-th moment of $\Lambda_5$ in (51), for different $n_k$, may be obtained from the expression of the pdf of the complex Wishart distribution in (59) and the fact that the $q$ matrices $A_k$ have independent $CW_p(n_k-1, \Sigma_k)$ distributions, where, under $H_0$ in (36), $\Sigma_k = \Sigma$ $(k=1,\dots,q)$. From this expression it is then possible to derive the expression for the moments of $\Lambda_5$ in (37), for $n_k = n$ $(k=1,\dots,q)$. For $N = n_1+\cdots+n_q$,

$$E\left(\Lambda_5^h\right) = \frac{N^{Nph}}{\prod_{k=1}^{q} n_k^{n_k ph}}\int_{\mathcal{A}_1>0}\cdots\int_{\mathcal{A}_q>0}\frac{\prod_{k=1}^{q}|\mathcal{A}_k|^{n_k h}}{|\mathcal{A}|^{Nh}} \times\prod_{k=1}^{q}\frac{e^{-tr(\Sigma^{-1}\mathcal{A}_k)}\,|\mathcal{A}_k|^{n_k-1-p}}{\Gamma_p(n_k-1)\,|\Sigma|^{n_k-1}}\prod_{k=1}^{q}d\mathcal{A}_k$$
$$= \frac{N^{Nph}}{\prod_{k=1}^{q} n_k^{n_k ph}}\prod_{k=1}^{q}\frac{\Gamma_p(n_k+n_k h-1)}{\Gamma_p(n_k-1)}|\Sigma|^{n_k h}$$
$$\int_{\mathcal{A}_1>0}\cdots\int_{\mathcal{A}_q>0}\frac{1}{|\mathcal{A}|^{Nh}}\prod_{k=1}^{q}\underbrace{\frac{e^{-tr(\Sigma^{-1}\mathcal{A}_k)}\,|\mathcal{A}_k|^{n_k+n_k h-1-p}}{\Gamma_p(n_k+n_k h-1)\,|\Sigma|^{n_k+n_k h-1}}}_{\text{pdf of } CW_p(n_k+n_k h-1,\,\Sigma)}\prod_{k=1}^{q}d\mathcal{A}_k$$
$$= \frac{N^{Nph}}{\prod_{k=1}^{q} n_k^{n_k ph}}\left\{\prod_{k=1}^{q}\frac{\Gamma_p(n_k+n_k h-1)}{\Gamma_p(n_k-1)}|\Sigma|^{n_k h}\right\} E\left(|A|^{-Nh}\right)$$

for $A=\sum_{k=1}^{q}A_k \sim CW_p(N-q+Nh, \Sigma)$, so that, from (61), we may write

$$E\left(\Lambda_5^h\right) = \frac{N^{Nph}}{\prod_{k=1}^{q} n_k^{n_k ph}}\left\{\prod_{k=1}^{q}\frac{\Gamma_p(n_k+n_k h-1)}{\Gamma_p(n_k-1)}|\Sigma|^{n_k h}\right\}|\Sigma|^{-Nh} \times\frac{\Gamma_p(N-q+Nh-Nh)}{\Gamma_p(N-q+Nh)},$$

which on using (60), gives

$$E\left(\Lambda_5^h\right) = \frac{N^{Nph}}{\prod_{k=1}^{q} n_k^{n_k ph}} \prod_{j=1}^{p} \left\{ \frac{\Gamma(N-q+1-j)}{\Gamma(N-q+1-j+Nh)} \prod_{k=1}^{q} \frac{\Gamma(n_k - j + n_k h)}{\Gamma(n_k - j)} \right\}, \tag{69}$$

which matches (2.14) in Krishnaiah et al. (1976), taking $d = 1$, $N_i = n_k$ and taking into account that these authors use the 'modified' statistic, so that one has to replace in (2.13) in that reference all the degrees of freedom of the Wishart distributions by the corresponding sample sizes.

Taking $n_k = n$ $(k = 1, \ldots, q)$ in (69), and applying the gamma multiplication formula in (68), we obtain

$$\begin{aligned} E\left(\Lambda_5^h\right) = q^{pqnh} &\left\{ \prod_{k=1}^{q} \prod_{j=1}^{p} \frac{\Gamma(n+nh-j)}{\Gamma(n-j)} \right\} \\ &\times \left\{ \prod_{j=1}^{p} q^{-nqh} \prod_{k=1}^{q} \frac{\Gamma\left(n-1-\frac{j-1}{q}+\frac{k-1}{q}\right)}{\Gamma\left(n-1-\frac{j-1}{q}+\frac{k-1}{q}+nh\right)} \right\} \end{aligned} \tag{70}$$

which, in turn, may assume the form in (38).

For an alternative proof of this result see subsection B.5 of this Appendix.

## B.5 *Alternative Proofs for the Expressions of the Moments of the LRT Statistics in Sections 2.1, 3 and 4*

Let $\text{MLE}_0$ denote the MLE under the null hypothesis of the parameter being tested. Theorem 5 in Jensen (1988) is an interesting result which allows us to state that any of the LRT statistics studied in this paper are, under the corresponding null hypothesis, independent of the $\text{MLE}_0$ of the parameter or parameters being tested. This is a very useful result in obtaining the distributions of these LRT statistics in alternative ways to those used in the previous sections of this Appendix. These are very useful in allowing us to view such statistics from different perspectives.

## On the Moments of $\Lambda_1$

Concerning the LRT statistic $\Lambda_1$ in (5), to test the independence of sets of variables, it is possible to write

$$\Lambda_1 = \prod_{k=1}^{m-1} \frac{|A_{kk}^*|}{|A_{kk}|\,|A_{k+1,k+1}^*|}$$

where

$$A_{kk}^* = \begin{bmatrix} A_{kk} & A_{k,k+1} & \cdots & A_{km} \\ A_{k+1,k} & A_{k+1,k+1} & \cdots & A_{k+1,m} \\ \vdots & \vdots & \ddots & \vdots \\ A_{mk} & A_{m,k+1} & \cdots & A_{mm} \end{bmatrix} \tag{71}$$

with $A_{11}^* = A$, and where, as in (5), $A_{kk}$ is the $k$-th diagonal block of $A$, of dimensions $p_k \times p_k$.

Each statistic

$$\Lambda_{1,k} = \frac{|A_{kk}^*|}{|A_{kk}|\,|A_{k+1,k+1}^*|} \qquad (k = 1, \ldots, m-1) \tag{72}$$

is indeed the LRT statistic to test the subhypothesis

$$H_{0k} : \bigwedge_{\ell=k+1}^{m} \Sigma_{k\ell} = 0$$

which is the null hypothesis of independence of $\underline{X}_k$, and the superset formed by joining $\underline{X}_{k+1}, \ldots, \underline{X}_m$, and where $\bigwedge_{\ell=k+1}^{m}$ represents a conjunction.

Note that one may then write the null hypothesis in (4) as

$$H_0 \equiv H_{0,m-1|\{m-2,\ldots,1\}} \circ \cdots \circ H_{03|\{2,1\}} \circ H_{02|1} \circ H_{01} \tag{73}$$

where $H_{0k|\{k-1,\ldots,1\}}$ represents $H_{0k}$, assuming that $H_{0,k-1}, \ldots, H_{01}$ hold.

But then, from Theorem 5 in Jensen (1988) it follows that not only, under $H_0$, $\Lambda_1$ is independent of $A_{11}, \ldots, A_{mm}$ but also each LRT statistic $\Lambda_{1,k}$ is, under $H_{0k}$, independent of $A_{kk}$ and $A_{k+1,k+1}^*$. Then, since $\Lambda_{1,k+1}$ is only built on $A_{k+1,k+1}^*$, this assures that $\Lambda_{1,k}$ and $\Lambda_{1,k+1}$ are independent, under $H_{0k}$, which makes all $m-1$ LRT statistics $\Lambda_{1,k}$ $(k = 1, \ldots, m-1)$ in (72) independent, under $H_0$ in (4) or (73).

As such,

$$E\left(\Lambda_1^h\right)=\prod_{k=1}^{m-1}E\left(\Lambda_{1,k}^h\right)$$

where $E\left(\Lambda_{1,k}^h\right)$ may be obtained in two different ways.

One possible way to obtain $E(\Lambda_{1,k}^h)$ is through a similar procedure to the one used to obtain $E(\Lambda_1^h)$ in Subsect. B.1 of this Appendix, from the fact that $A_{kk}^* \sim CW_{p_k+q_k}(n-1,\frac{1}{n}\Sigma_{kk}^*)$, for a similar partitioning and notation on $\Sigma$ to that used in (71) for the matrix $A$, and where, under $H_{0k}$, $|\Sigma_{kk}^*|=|\Sigma_{kk}||\Sigma_{k+1,k+1}^*|$, so that from (62)–(64) we have

$$\begin{aligned}
E\left(\Lambda_{1,k}^h\right) &= \frac{\Gamma_{p_k+q_k}(n-1+nh)}{\Gamma_{p_k+q_k}(n-1)}\,\frac{\Gamma_{p_k}(n-1)}{\Gamma_{p_k}(n-1+nh)}\,\frac{\Gamma_{q_k}(n-1)}{\Gamma_{q_k}(n-1+nh)}\\
&= \left\{\prod_{j=1}^{p_k+q_k}\frac{\Gamma(n+nh-j)}{\Gamma(n-j)}\right\}\left\{\prod_{j=1}^{p_k}\frac{\Gamma(n-j)}{\Gamma(n+nh-j)}\right\}\\
&\qquad\times\left\{\prod_{j=1}^{q_k}\frac{\Gamma(n-j)}{\Gamma(n+nh-j)}\right\}\\
&= \left\{\prod_{j=q_k+1}^{p_k+q_k}\frac{\Gamma(n+nh-j)}{\Gamma(n-j)}\right\}\left\{\prod_{j=1}^{p_k}\frac{\Gamma(n-j)}{\Gamma(n+nh-j)}\right\}\\
&= \prod_{j=1}^{p_k}\frac{\Gamma(n-j)\,\Gamma(n+nh-q_k-j)}{\Gamma(n-q_k-j)\,\Gamma(n+nh-j)}\,,
\end{aligned} \tag{74}$$

so that $E(\Lambda_1^h)$ comes immediately in the form given in (7).

The other way to obtain $E(\Lambda_{1,k}^h)$ is by writing

$$|A_{kk}^*|=|A_{kk}|\,|A_{k+1,k+1}^*-A_{k+1}^*A_{kk}^{-1}A_{k+1}^{*\prime}|$$

where

$$A_{k+1}^{*\prime}=\left[A_{k,k+1}\ \cdots\ A_{km}\right]$$

is the first row block of $A_{kk}^*$, without the block $A_{kk}$.

Then,

$$\Lambda_{1,k} = \frac{|A^*_{k+1,k+1} - A^*_{k+1}A^{-1}_{kk}A^{*\prime}_{k+1}|}{|A^*_{k+1,k+1}|}$$

where, under $H_{0k}$, $(A^*_{k+1,k+1} - A^*_{k+1}A^{-1}_{kk}A^{*\prime}_{k+1}) \sim CW_{q_k}(n-1-p_k, \Sigma^*_{k+1,k+1})$ and $A^*_{k+1}A^{-1}_{kk}A^{*\prime}_{k+1} \sim CW_{q_k}(p_k, \Sigma^*_{k+1,k+1})$ are two independent matrices, so that $A^*_{k+1,k+1} \sim CW_{q_k}(n-1, \Sigma^*_{k+1,k+1})$, and $E(\Lambda^h_{1,k})$ may be obtained from the results in Subsect. B.2 of this Appendix and from (66), with the due adaptations, as

$$E\left(\Lambda^h_{1,k}\right) = \prod_{j=1}^{q_k} \frac{\Gamma(n-p_k+nh-j)\,\Gamma(n-j)}{\Gamma(n-p_k-j)\,\Gamma(n+nh-j)},$$

which is equivalent to (74), since

$$\begin{aligned}
E\left(\Lambda^h_{1,k}\right) &= \prod_{j=1}^{q_k} \frac{\Gamma(n-p_k+nh-j)\,\Gamma(n-j)}{\Gamma(n-p_k-j)\,\Gamma(n+nh-j)} \\
&= \left\{\prod_{j=p_k+1}^{p_k+q_k} \frac{\Gamma(n+nh-j)}{\Gamma(n-j)}\right\} \left\{\prod_{j=1}^{q_k} \frac{\Gamma(n-j)}{\Gamma(n+nh-j)}\right\} \\
&= \left\{\prod_{j=1}^{p_k+q_k} \frac{\Gamma(n+nh-j)}{\Gamma(n-j)}\right\} \left\{\prod_{j=1}^{p_k} \frac{\Gamma(n-j)}{\Gamma(n+nh-j)}\right\} \\
&\quad \times \left\{\prod_{j=1}^{q_k} \frac{\Gamma(n-j)}{\Gamma(n+nh-j)}\right\},
\end{aligned}$$

which is the same as the second expression in (74).

## On the Moments of $\Lambda_4$

The null hypothesis in (32) may be written as

$$H_{02|1} \circ H_{01}$$

where

$$H_{01} : \Sigma = diag(\sigma_1^2, \ldots, \sigma_p^2)$$

is the null hypothesis of independence of the $p$ r.v.'s in $\underline{X}$ and

$$H_{02|1} : \sigma_1^2 = \cdots = \sigma_p^2 \quad \text{assuming } H_{01} : \Sigma = diag(\sigma_1^2, \ldots, \sigma_p^2)$$

is the null hypothesis of equality of the $p$ variances $\sigma_j^2$ $(j = 1, \ldots, p)$.

The LRT statistic to test $H_{01}$ is (see Sect. 2.1)

$$\Lambda_{4,1} = \left( \frac{|A|}{\prod_{j=1}^{p} a_{jj}} \right)^n$$

where $A$ is the MLE of $\Sigma$ and $a_{jj}$ its $j$-th diagonal element, and the LRT statistic to test $H_{02|1}$ is (see Sect. 4)

$$\Lambda_{4,2} = \left( p^p \frac{\prod_{j=1}^{p} a_{jj}}{\left(\sum_{j=1}^{p} a_{jj}\right)^p} \right)^n,$$

where, under $H_{01}$, the $p$ r.v.'s $a_{jj}$ are independent.

Since $\Lambda_4 = \Lambda_{4,1}\Lambda_{4,2}$, where, from Theorem 5 in Jensen (1988), $\Lambda_{4,1}$ is independent of $a_{11}, \ldots, a_{pp}$, so that, being $\Lambda_{4,2}$ built only on $a_{11}, \ldots, a_{pp}$, $\Lambda_{4,1}$ is independent of $\Lambda_{4,2}$, which entails that

$$E\left(\Lambda_4^h\right) = E\left(\Lambda_{4,1}^h\right) E\left(\Lambda_{4,2}^h\right)$$

where, from (7),

$$E\left(\Lambda_{4,1}^h\right) = \prod_{j=1}^{p-1} \frac{\Gamma(n-1)\,\Gamma(n-p+j-1+nh)}{\Gamma(n-p+j-1)\,\Gamma(n-1+nh)},$$

while from (38),

$$E\left(\Lambda_{4,2}^h\right) = \prod_{j=2}^{p} \frac{\Gamma(n-1+nh)\,\Gamma\left(n-1+\frac{j-1}{p}\right)}{\Gamma(n-1)\,\Gamma\left(n-1+\frac{j-1}{p}+nh\right)}.$$

The expression in (34) may be obtained by reversing the indexing in the expression of $E(\Lambda_{4,2}^h)$ and rearranging the indexes.

### On the Moments of $\Lambda_5$

Concerning the statistic $\Lambda_5$ in (51), one may write, for $A = A_1 + \cdots + A_q$ and $N = n_1 + \cdots + n_q$,

$$\Lambda_5 = \frac{N^{Np}}{\prod_{k=1}^{q} n_k^{n_k p}} \frac{\prod_{k=1}^{q} |A_k|^{n_k}}{|A|^N} = \prod_{k=1}^{q-1} \frac{(n_{k+1}^*)^{n_{k+1}^* p}}{(n_k^*)^{n_k^* p}\,(n_{k+1})^{n_{k+1} p}} \frac{|A_k^*|^{n_k^*}\,|A_{k+1}|^{n_{k+1}}}{|A_k^* + A_{k+1}|^{n_k^* + n_{k+1}}}$$
$$= \prod_{k=1}^{q-1} \Lambda_{5,k} \qquad (75)$$

for

$$A_k^* = \sum_{\ell=1}^{k} A_\ell \qquad \text{and} \qquad n_k^* = \sum_{\ell=1}^{k} n_\ell$$

with $A_q^* = A$ and $n_q^* = N$, and where

$$\Lambda_{5,k} = \frac{(n_{k+1}^*)^{n_{k+1}^* p}}{(n_k^*)^{n_k^* p}\,(n_{k+1})^{n_{k+1} p}} \frac{|A_k^*|^{n_k^*}\,|A_{k+1}|^{n_{k+1}}}{|A_k^* + A_{k+1}|^{n_k^* + n_{k+1}}}\,, \quad k = 1, \ldots, q-1\,, \qquad (76)$$

is the LRT statistic to test

$$H_{0k} : \Sigma_1 = \cdots = \Sigma_k = \Sigma_{k+1}$$
$$\text{assuming } \Sigma_1 = \cdots = \Sigma_k\,, \quad k = 1, \ldots, q-1\,.$$

In (76), under $H_0$ in (36),

$$A_k^* \sim CW_p(n_k^* - k, \Sigma) \qquad (77)$$

is independent of

$$A_{k+1} \sim CW_p(n_{k+1} - 1, \Sigma)\,. \qquad (78)$$

From Theorem 5 in Jensen (1988) it follows that $\Lambda_{5,k}$ is independent of $A_k^* + A_{k+1}$. Since $\Lambda_{5,k+1}$ is built only on $A_{k+1}^* = A_k^* + A_{k+1}$ and $A_{k+2}$, which are independent, given the independence of all $q$ matrices $A_k$ $(k = 1, \ldots, q)$, this makes $\Lambda_{5,k+1}$ independent of $\Lambda_{5,k}$, which in turn makes all $q-1$ statistics $\Lambda_{5,k}$ in (75) independent. As such,

$$E\left(\Lambda_5^h\right) = \prod_{k=1}^{q-1} E\left(\Lambda_{5,k}^h\right).$$

To simplify the notation, let us first obtain

$$E\left(\frac{|A|^r\,|B|^s}{|A+B|^{r+s}}\right)$$

where $A \sim CW_p(n_1, \Sigma)$ and $B \sim CW_p(n_2, \Sigma)$ are two independent matrices. From (59), the distributions of $A$ and $B$, and (61), one may write

$$\begin{aligned}
E&\left(\frac{|A|^r\,|B|^s}{|A+B|^{r+s}}\right)\\
&= \int_{\mathcal{A}>0}\int_{\mathcal{B}>0} \frac{|\mathcal{A}|^r\,|\mathcal{B}|^s}{|\mathcal{A}+\mathcal{B}|^{r+s}}\,\frac{e^{-tr(\Sigma^{-1}\mathcal{A})}\,|\mathcal{A}|^{n_1-p-1}}{|\Sigma|^{n_1}\,\Gamma_p(n_1)}\,\frac{e^{-tr(\Sigma^{-1}\mathcal{B})}\,|\mathcal{B}|^{n_2-p-1}}{|\Sigma|^{n_2}\,\Gamma_p(n_2)}\\
&\qquad d\mathcal{B}\,d\mathcal{A}\\
&= \frac{|\Sigma|^r\,\Gamma_p(n_1+r)}{\Gamma_p(n_1)}\,\frac{|\Sigma|^s\,\Gamma_p(n_2+s)}{\Gamma_p(n_2)}\\
&\qquad \int_{\mathcal{A}>0}\int_{\mathcal{B}>0}\frac{1}{|\mathcal{A}+\mathcal{B}|^{r+s}}\,\frac{e^{-tr(\Sigma^{-1}\mathcal{A})}\,|\mathcal{A}|^{n_1+r-p-1}}{|\Sigma|^{n_1+r}\,\Gamma_p(n_1+r)}\\
&\qquad\qquad \frac{e^{-tr(\Sigma^{-1}\mathcal{B})}\,|\mathcal{B}|^{n_2+s-p-1}}{|\Sigma|^{n_2+s}\,\Gamma_p(n_2+s)}\,d\mathcal{B}\,d\mathcal{A}\\
&= \frac{\Gamma_p(n_1+r)\,\Gamma_p(n_2+s)}{\Gamma_p(n_1)\,\Gamma_p(n_2)}\,|\Sigma|^{r+s}\,\underbrace{E\left(|A+B|^{-(r+s)}\right)}_{\text{for } A+B\sim CW_p(n_1+n_2+r+s,\,\Sigma)}\\
&= \frac{\Gamma_p(n_1+r)\,\Gamma_p(n_2+s)}{\Gamma_p(n_1)\,\Gamma_p(n_2)}\,|\Sigma|^{r+s}\,|\Sigma|^{-(r+s)}\,\frac{\Gamma_p(n_1+n_2+r+s-r-s)}{\Gamma_p(n_1+n_2+r+s)}\\
&= \frac{\Gamma_p(n_1+r)\,\Gamma_p(n_2+s)}{\Gamma_p(n_1)\,\Gamma_p(n_2)}\,\frac{\Gamma_p(n_1+n_2)}{\Gamma_p(n_1+n_2+r+s)}\,.
\end{aligned}$$

Given the distributions in (77) and (78), $E\left(\Lambda_{5,k}^h\right)$ may be written as

$$E\left(\Lambda_{5,k}^h\right) = \left(\frac{(n_{k+1}^*)^{n_{k+1}^* p}}{(n_k^*)^{n_k^* p}\,(n_{k+1})^{n_{k+1} p}}\right)^h E\left(\frac{|A_k^*|^{n_k^* h}\,|A_{k+1}|^{n_{k+1} h}}{|A_k^* + A_{k+1}|^{(n_k^* + n_{k+1})h}}\right)$$
$$= \frac{(n_{k+1}^*)^{n_{k+1}^* ph}}{(n_k^*)^{n_k^* ph}\,(n_{k+1})^{n_{k+1} ph}}\,\frac{\Gamma_p(n_k^* - k + n_k^* h)}{\Gamma_p(n_k^* - k)}\,\frac{\Gamma_p(n_{k+1} - 1 + n_{k+1} h)}{\Gamma_p(n_{k+1} - 1)}$$
$$\times \frac{\Gamma_p(n_k^* - k + n_{k+1} - 1)}{\Gamma_p(n_k^* - k + n_{k+1} - 1 + (n_k^* + n_{k+1})h)}.$$

Given that $n_k^* + n_{k+1} = n_{k+1}^*$ or $n_{k-1}^* + n_k = n_k^*$, $n_q^* = N = n_1 + \cdots + n_q$ and $n_1^* = n_1$, we may write

$$E\left(\Lambda_5^h\right) = \left\{\prod_{k=1}^{q-1} \frac{(n_{k+1}^*)^{n_{k+1}^* ph}}{(n_k^*)^{n_k^* ph}\,(n_{k+1})^{n_{k+1} ph}}\right\}\left\{\prod_{k=1}^{q-1} \frac{\Gamma_p(n_k^* - k + n_k^* h)}{\Gamma_p(n_k^* - k)}\right.$$
$$\left.\times \frac{\Gamma_p(n_{k+1} - 1 + n_{k+1} h)}{\Gamma_p(n_{k+1} - 1)}\,\frac{\Gamma_p(n_k^* - k + n_{k+1} - 1)}{\Gamma_p(n_k^* - k + n_{k+1} - 1 + (n_k^* + n_{k+1})h)}\right\}$$
$$= \frac{(n_q^*)^{n_q^* ph}}{\prod_{k=1}^q n_k^{n_k ph}}\left\{\prod_{k=1}^{q-1} \frac{\Gamma_p(n_k^* - k + n_k^* h)}{\Gamma_p(n_k^* - k)}\right.$$
$$\left.\times \frac{\Gamma_p(n_k^* - k + n_{k+1} - 1)}{\Gamma_p(n_k^* - k + n_{k+1} - 1 + (n_k^* + n_{k+1})h)}\right\}$$
$$\times \left\{\prod_{k=1}^{q-1} \frac{\Gamma_p(n_{k+1} - 1 + n_{k+1} h)}{\Gamma_p(n_{k+1} - 1)}\right\}$$
$$= \frac{N^{Nph}}{\prod_{k=1}^q n_k^{n_k ph}}\,\frac{\Gamma_p(n_1^* - 1 + n_1^* h)}{\Gamma_p(n_1^* - 1)}\,\frac{\Gamma_p(n_q^* - q)}{\Gamma_p(n_q^* - q + n_q^* h)}$$
$$\times \left\{\prod_{k=1}^{q-1} \frac{\Gamma_p(n_{k+1} - 1 + n_{k+1} h)}{\Gamma_p(n_{k+1} - 1)}\right\}$$
$$= \frac{N^{Nph}}{\prod_{k=1}^q n_k^{n_k ph}}\,\frac{\Gamma_p(N - q)}{\Gamma_p(N - q + n_q^* h)}\prod_{k=1}^q \frac{\Gamma_p(n_k - 1 + n_k h)}{\Gamma_p(n_k - 1)}$$
$$= \frac{N^{Nph}}{\prod_{k=1}^q n_k^{n_k ph}}\prod_{j=1}^p \left\{\frac{\Gamma(N - q - j + 1)}{\Gamma(N - q - j + 1 + n_q^* h)}\prod_{k=1}^q \frac{\Gamma(n_k - j + n_k h)}{\Gamma(n_k - j)}\right\}$$

which is (69).

# C Derivation of the Result in Expression (35)

The c.f. of $W_4 = -\log \Lambda_4$, using (11), may be written as

$$\begin{aligned}
\Phi_{W_4}(t) &= E\left(e^{\mathrm{i}t W_4}\right) = E\left(\Lambda_4^{-\mathrm{i}t}\right) \\
&= \prod_{j=1}^{p-1} \frac{\Gamma\left(n-1+\frac{j}{p}\right)}{\Gamma(n-j-1)} \frac{\Gamma(n-j-1-n\mathrm{i}t)}{\Gamma\left(n-1+\frac{j}{p}-n\mathrm{i}t\right)} \\
&= \prod_{j=1}^{p-1} \left\{ \frac{\Gamma(n-1)}{\Gamma(n-j-1)} \frac{\Gamma(n-j-1-n\mathrm{i}t)}{\Gamma(n-1-n\mathrm{i}t)} \frac{\Gamma\left(n-1+\frac{j}{p}\right)}{\Gamma(n-1)} \right. \\
&\qquad\qquad \left. \times \frac{\Gamma(n-1-n\mathrm{i}t)}{\Gamma\left(n-1+\frac{j}{p}-n\mathrm{i}t\right)} \right\} \\
&= \left\{ \prod_{j=1}^{p-1} \prod_{\ell=0}^{j-1} \frac{n-j-1+\ell}{n-j-1+\ell-n\mathrm{i}t} \right\} \\
&\qquad \times \left\{ \prod_{j=1}^{p-1} \frac{\Gamma\left(n-1+\frac{j}{p}\right)}{\Gamma(n-1)} \frac{\Gamma(n-1-n\mathrm{i}t)}{\Gamma\left(n-1+\frac{j}{p}-n\mathrm{i}t\right)} \right\} \\
&= \left\{ \prod_{j=1}^{p-1} \left( \frac{n-j-1}{n-j-1-n\mathrm{i}t} \right)^{p-j} \right\} \\
&\qquad \times \left\{ \prod_{j=1}^{p-1} \frac{\Gamma\left(n-1+\frac{j}{p}\right)\Gamma(n-1-n\mathrm{i}t)}{\Gamma(n-1)\,\Gamma\left(n-1+\frac{j}{p}-n\mathrm{i}t\right)} \right\} \\
&= \underbrace{\left\{ \prod_{j=1}^{p-1} \left( \frac{n-j-1}{n} \right)^{p-j} \left( \frac{n-j-1}{n} - \mathrm{i}t \right)^{-(p-j)} \right\}}_{\Phi_{1,W_4}(t)} \\
&\qquad \times \underbrace{\left\{ \prod_{j=1}^{p-1} \frac{\Gamma\left(n-1+\frac{j}{p}\right)\Gamma(n-1-n\mathrm{i}t)}{\Gamma(n-1)\,\Gamma\left(n-1+\frac{j}{p}-n\mathrm{i}t\right)} \right\}}_{\Phi_{2,W_4}(t)} .
\end{aligned}$$

## D Derivation of the Result in Expression (39)

The c.f. of $W_5 = -\log \Lambda_5$, using (11), may be written as

$$\Phi_{W_5}(t) = E\left(e^{-\mathrm{i}t W_5}\right) = E\left(\Lambda_5^{-\mathrm{i}t}\right) = \prod_{j=1}^{p}\prod_{k=1}^{q} \frac{\Gamma\left(n-1+\frac{k-j}{q}\right)\Gamma\left(n-j-n\mathrm{i}t\right)}{\Gamma\left(n-1+\frac{k-j}{q}-n\mathrm{i}t\right)\Gamma\left(n-j\right)}$$

$$= \prod_{j=1}^{p}\prod_{k=1}^{q}\left\{\frac{\Gamma\left(n-1+\left\lfloor\frac{k-j}{q}\right\rfloor\right)\Gamma\left(n-j-n\mathrm{i}t\right)}{\Gamma\left(n-j\right)\Gamma\left(n-1+\left\lfloor\frac{k-j}{q}\right\rfloor-n\mathrm{i}t\right)}\right.$$

$$\left.\times\frac{\Gamma\left(n-1+\frac{k-j}{q}\right)\Gamma\left(n-1+\left\lfloor\frac{k-j}{q}\right\rfloor-n\mathrm{i}t\right)}{\Gamma\left(n-1+\left\lfloor\frac{k-j}{q}\right\rfloor\right)\Gamma\left(n-1+\frac{k-j}{q}-n\mathrm{i}t\right)}\right\}$$

$$= \prod_{j=1}^{p}\prod_{k=1}^{q}\left\{\left\{\prod_{\ell=0}^{j-2+\left\lfloor\frac{k-j}{q}\right\rfloor}\frac{n-j+\ell}{n-j+\ell-n\mathrm{i}t}\right\}\right.$$

$$\left.\times\frac{\Gamma\left(n-1+\frac{k-j}{q}\right)\Gamma\left(n-1+\left\lfloor\frac{k-j}{q}\right\rfloor-n\mathrm{i}t\right)}{\Gamma\left(n-1+\left\lfloor\frac{k-j}{q}\right\rfloor\right)\Gamma\left(n-1+\frac{k-j}{q}-n\mathrm{i}t\right)}\right\}$$

$$= \prod_{j=1}^{p}\prod_{k=1}^{q}\left\{\left\{\prod_{\ell=0}^{j-2+\left\lfloor\frac{k-j}{q}\right\rfloor}\frac{n-j+\ell}{n}\left(\frac{n-j+\ell}{n}-\mathrm{i}t\right)^{-1}\right\}\right.$$

$$\left.\times\frac{\Gamma\left(n-1+\frac{k-j}{q}\right)\Gamma\left(n-1+\left\lfloor\frac{k-j}{q}\right\rfloor-n\mathrm{i}t\right)}{\Gamma\left(n-1+\left\lfloor\frac{k-j}{q}\right\rfloor\right)\Gamma\left(n-1+\frac{k-j}{q}-n\mathrm{i}t\right)}\right\}$$

$$= \underbrace{\left\{\prod_{j=1}^{p-1}\left(\frac{n-1-j}{n}\right)^{r_j}\left(\frac{n-1-j}{n}-\mathrm{i}t\right)^{-r_j}\right\}}_{\Phi_{1,W_5}(t)}$$

$$\times\underbrace{\left\{\prod_{j=1}^{p}\prod_{k=1}^{q}\frac{\Gamma\left(n-1+\frac{k-j}{q}\right)\Gamma\left(n-1+\left\lfloor\frac{k-j}{q}\right\rfloor-n\mathrm{i}t\right)}{\Gamma\left(n-1+\left\lfloor\frac{k-j}{q}\right\rfloor\right)\Gamma\left(n-1+\frac{k-j}{q}-n\mathrm{i}t\right)}\right\}}_{\Phi_{2,W_5}(t)}.$$

## E Sample Data and Weights $\pi_\nu$

$$\mathcal{A}_1 =$$

$$\begin{pmatrix} 125.565 & 2.188-23.064\,\mathrm{i} & 59.467+18.522\,\mathrm{i} & -12.051+7.813\,\mathrm{i} \\ 2.188+23.064\,\mathrm{i} & 100.543 & -26.184-16.905\,\mathrm{i} & -16.745+7.004\,\mathrm{i} \\ 59.467-18.522\,\mathrm{i} & -26.184+16.905\,\mathrm{i} & 77.689 & 8.963+1.653\,\mathrm{i} \\ -12.051-7.813\,\mathrm{i} & -16.745-7.004\,\mathrm{i} & 8.963-1.653\,\mathrm{i} & 19.570 \end{pmatrix}$$

$$\underline{\widetilde{\mathcal{X}}}_1 = \left[\,0.219+1.253\,\mathrm{i},\ -0.242-0.367\,\mathrm{i},\ -0.727-0.249\,\mathrm{i},\ 1.872+0.893\,\mathrm{i}\,\right]'$$

$$\mathcal{A}_2 =$$

$$\begin{pmatrix} 167.215 & 33.214+26.995\,\mathrm{i} & 92.885-8.800\,\mathrm{i} & 21.976-67.294\,\mathrm{i} \\ 33.214-26.995\,\mathrm{i} & 103.027 & 60.978-39.555\,\mathrm{i} & -7.911-7.932\,\mathrm{i} \\ 92.885+8.800\,\mathrm{i} & 60.978+39.555\,\mathrm{i} & 129.613 & 45.138-35.344\,\mathrm{i} \\ 21.976+67.294\,\mathrm{i} & -7.911+7.932\,\mathrm{i} & 45.138+35.344\,\mathrm{i} & 55.186 \end{pmatrix}$$

$$\underline{\widetilde{\mathcal{X}}}_2 = \left[\,-0.982+0.616\,\mathrm{i},\ -3.19+0.998\,\mathrm{i},\ -1.547+0.046\,\mathrm{i},\ 0.61+0.193\,\mathrm{i}\,\right]'$$

$$\mathcal{B} = \begin{pmatrix} 5.545 & 8.013+10.552\,\mathrm{i} & 2.391+2.630\,\mathrm{i} & 5.885-0.110\,\mathrm{i} \\ 8.013-10.552\,\mathrm{i} & 31.662 & 8.460-0.749\,\mathrm{i} & 8.295-11.359\,\mathrm{i} \\ 2.391-2.630\,\mathrm{i} & 8.46+0.749\,\mathrm{i} & 2.278 & 2.485-2.839\,\mathrm{i} \\ 5.885+0.110\,\mathrm{i} & 8.295+11.359\,\mathrm{i} & 2.485+2.839\,\mathrm{i} & 6.248 \end{pmatrix}$$

$$\begin{aligned}
\pi_0 = {} & 1.004283369747372270455188549688229025318002960305034664449022834133 1\\
& 204411422847863724310027002211047007008341262513688509992656805163 27\\
& 3916909722285312606991921609288075923547585040578\\
\pi_1 = {} & -0.03720093316004884944000763591825892338280635346280803719230967768\\
& 66123240681822646856268434506426552002722519584863828360836949545218\\
& 1639814985758843995994188346940588938471514184196431 3\\
\pi_2 = {} & 0.0344286826513966604634951473502539288050202631819641701492918261 12\\
& 801941581290894604243328994426265071826935764764751132143743907146 8\\
& 3120168070465953398448237127379752227578386602018344 5\\
\pi_3 = {} & -0.00312877952387464907775413891061179106271559814926749990028068611\\
& 742875246507096072629196654529906501159845177897106204927138810782 76\\
& 2160751878100985677188841086912764554903437873313478 51
\end{aligned}$$

# Testing the Hypothesis of a Double Exchangeable Covariance Matrix

**Carlos A. Coelho and Anuradha Roy**

**Abstract** In this paper the authors study the problem of testing the hypothesis of a doubly exchangeable covariance matrix for three-level multivariate observations, taken on $m$ variables over $u$ sites and over $v$ time/space points. Through the decomposition of the main hypothesis into a set of three sub-hypotheses, the likelihood ratio test statistic is defined, its exact moments are determined, and its exact distribution is studied. Because this distribution is very much intricate, a very precise near-exact distribution is developed. Numerical studies conducted to evaluate the closeness between this near-exact distribution and the exact distribution show the very good performance of this approximation even for very small sample sizes. A simulation study is also conducted and two real-data examples are presented.

**Keywords** Characteristic function · Composition of hypotheses · Distribution of likelihood ratio statistics · Mixtures · Near-exact distributions · Product distribution

---

Reprinted from Metrika, with permission from Springer Nature.

First published as Coelho, C.A., Roy, A. (2020). Testing the hypothesis of a doubly exchangeable covariance matrix. *Metrika*, **83**, 45–68. (https://doi.org/10.1007/s00184-019-00724-7)

**Supplementary Information** The online version contains supplementary material available at https://doi.org/10.1007/978-3-031-68008-3_13.

---

C. A. Coelho (✉)
Departamento de Matemática and Centro de Matemática e Aplicações, Faculdade de Ciências e Tecnologia, Universidade Nova de Lisboa, Caparica, Portugal
e-mail: cmac@fct.unl.pt

A. Roy
Department of Management Science and Statistics, The University of Texas at San Antonio, San Antonio, TX, USA

 
B. C. Arnold et al. (eds.), *Likelihood Ratio Tests and Related Topics in Multivariate Analysis*, Contributions to Statistics,
https://doi.org/10.1007/978-3-031-68008-3_13

**Mathematics Subject Classification** 62H15, 62H10, 62E15, 62E20, 62E10, 60E10

# 1 Introduction

Advances in computing power in the past few decades greatly encouraged the collection of multi-level multivariate data in all fields of science: biomedical, medical, environmental, social sciences and engineering. And, with these data sets complex multivariate testing problems occur frequently. It is common in clinical trial studies to collect measurements on more than one response variable at several locations taken repeatedly over time on one experimental unit to test the effectiveness of some medication, diet or treatment. These are called three-level multivariate data and the doubly exchangeable covariance structure (defined below) is a suitable variance-covariance matrix for this kind of data.

Good examples of such data may be found in osteopenia and osteoporosis studies, being estimated that one of every four post-menopausal women may suffer from osteoporosis. Although it is more common in white or Asian women older than 50 years, osteoporosis can occur in almost any person at any age, being estimated that in fact, more than 2 million American men have osteoporosis, the national cost for osteoporosis and related injuries being assessed at \$14 billion each year in the United States.

Let $\boldsymbol{y}$ be the $muv$-variate real-valued random vector of all measurements. We partition this vector $\boldsymbol{y}$ as follows:

$$\boldsymbol{y} = \begin{pmatrix} \boldsymbol{y}_1 \\ \vdots \\ \boldsymbol{y}_v \end{pmatrix}, \qquad \text{where} \quad \boldsymbol{y}_t = \begin{pmatrix} \boldsymbol{y}_{t1} \\ \vdots \\ \boldsymbol{y}_{tu} \end{pmatrix}, \qquad \text{with} \quad \boldsymbol{y}_{ts} = \begin{pmatrix} y_{ts1} \\ \vdots \\ y_{tsm} \end{pmatrix},$$

for $t = 1, \ldots, v$, $s = 1, \ldots, u$. The $m$-dimensional vector of measurements $\boldsymbol{y}_{ts}$ represents the replicate on the $s$th location and at the $t$th time point.

Let $\boldsymbol{\Theta} = \text{Cov}[\boldsymbol{y}]$ be the $(muv \times muv)$−dimensional partitioned covariance matrix. We say that the covariance matrix $\boldsymbol{\Theta}$ has a doubly exchangeable covariance structure (Roy and Leiva, 2007) if it can be written as

$$\begin{aligned} \boldsymbol{\Theta} &= \boldsymbol{I}_{uv} \otimes \boldsymbol{U}_0 + [\boldsymbol{I}_v \otimes (\boldsymbol{J}_u - \boldsymbol{I}_u))] \otimes \boldsymbol{U}_1 + [\boldsymbol{J}_{uv} - (\boldsymbol{I}_v \otimes \boldsymbol{J}_u)] \otimes \boldsymbol{W} \\ &= \boldsymbol{I}_{uv} \otimes (\boldsymbol{U}_0 - \boldsymbol{U}_1) + \boldsymbol{I}_v \otimes \boldsymbol{J}_u \otimes (\boldsymbol{U}_1 - \boldsymbol{W}) + \boldsymbol{J}_{uv} \otimes \boldsymbol{W} \qquad (1) \\ &= \boldsymbol{I}_v \otimes \boldsymbol{U} + (\boldsymbol{J}_v - \boldsymbol{I}_v) \otimes \boldsymbol{W}^*, \end{aligned}$$

where

$$\boldsymbol{U} = \boldsymbol{I}_u \otimes \boldsymbol{U}_0 + (\boldsymbol{J}_u - \boldsymbol{I}_u) \otimes \boldsymbol{U}_1$$

and

$$\boldsymbol{W}^* = \boldsymbol{J}_u \otimes \boldsymbol{W} ,$$

with $\boldsymbol{U}_0$ a positive definite symmetric $m \times m$ matrix, and $\boldsymbol{U}_1$ and $\boldsymbol{W}$ symmetric $m \times m$ matrices. The matrices $\boldsymbol{U}_0$, $\boldsymbol{U}_1$ and $\boldsymbol{W}$ are all unstructured. That is, the doubly exchangeable covariance structure can be written as a matrix having diagonal blocks of the same matrix $\boldsymbol{U}$ with component matrices $\boldsymbol{U}_0$ and $\boldsymbol{U}_1$ and otherwise being filled-up with matrices $\boldsymbol{W}$.

Thus, the vectors $\boldsymbol{y}_{11}, \ldots, \boldsymbol{y}_{1u}, \ldots, \boldsymbol{y}_{v1}, \ldots, \boldsymbol{y}_{vu}$ are doubly exchangeable if

$$\mathrm{Cov}\left[\boldsymbol{y}_{ts}; \boldsymbol{y}_{t^*s^*}\right] = \begin{cases} \boldsymbol{U}_0 & \text{if } t = t^* \text{ and } s = s^*, \\ \boldsymbol{U}_1 & \text{if } t = t^* \text{ and } s \neq s^*, \\ \boldsymbol{W} & \text{if } t \neq t^*, \end{cases}$$

with $\mathrm{Cov}\left[\boldsymbol{y}_{ts}; \boldsymbol{y}_{t^*s^*}\right] = \mathrm{Cov}\left[\boldsymbol{y}_{t^*s^*}; \boldsymbol{y}_{ts}\right]$.

The $m \times m$ diagonal blocks $\boldsymbol{U}_0$ in (1) represent the variance-covariance matrix of the $m$ response variables at any given location and at any given time point, whereas the $m \times m$ off-diagonal blocks $\boldsymbol{U}_1$ in (1) represent the covariance matrix of the $m$ response variables between any two different locations, at any given time point. We assume $\boldsymbol{U}_0$ is the same for all locations and time points, and $\boldsymbol{U}_1$ is the same for all time points. The $m \times m$ off-diagonal blocks $\boldsymbol{W}$ represent the covariance matrix of the $m$ response variables between any two different time points. It is assumed to be the same for any pair of different time points, irrespective of location.

A good and simple example of a three-level multivariate dataset on osteoporosis may be obtained from Johnson and Wichern (2007) who report data on a study where an investigator measures the mineral content of three bones, radius, humerus and ulna ($m = 3$) by photon absorptiometry to examine whether a particular dietary supplement increases bone mineral content and mass in older women. All three measurements are recorded on the dominant and non-dominant sides ($u = 2$) for each woman. These two-level multivariate measurements are then taken again one year after their first participation in the experimental program. Thus, this whole dataset has a three-level multivariate structure, with $m = 3$ variables, for $u = 2$ locations, over $v = 2$ time points and given what the variables involved represent, testing for a doubly exchangeable covariance structure may be an adequate goal. Another simple example is from a bone densitometry study where the interest may be to test the doubly exchangeable covariance structure for an osteopenia study on twelve patients. Bone mineral density were obtained from the femoral neck and trochanter ($m = 2$), for the right and left femur ($u = 2$), taken at two different times ($v = 2$), separated by about two years. This latter example is addressed in Sect. 6, while the example of the osteoporosis data from Johnson and Wichern (2007) is placed in the supplementary material.

Roy and Fonseca (2012) fitted a general linear model to three-level multivariate data, with a doubly exchangeable covariance structure for the error vector, and

Leiva and Roy (2011, 2012) used this doubly exchangeable covariance structure for classification of three-level multivariate data, while hypothesis testing on three-level multivariate data was first studied by Roy and Leiva (2008), where these two authors introduced parametrically parsimonious models for hypotheses testing of Kronecker product covariance structures.

However, none of these authors addressed directly the problem of testing the doubly exchangeable covariance structure.

The doubly exchangeable covariance structure is actually a very general and rich covariance structure which generalizes at the same time both the compound symmetry and the sphericity structures. As such it has many interesting covariance structures as particular cases. Some of these are depicted in Fig. 1. While some of these structures are rather well-known, as the above mentioned compound symmetry and sphericity structures, and also the block compound symmetry (Votaw, 1948; Szatrowski, 1976, 1982; Coelho and Roy, 2017) and the block-matrix sphericity structure (Moschopoulos, 1992; Cardeño and Nagar, 2001; Marques and Coelho, 2012), others are not even known in the literature as it is the case of the structure derived from the doubly exchangeable covariance structure when $\boldsymbol{U}_1 = \boldsymbol{0}$ but $\boldsymbol{W} \neq \boldsymbol{0}$, which we call multi-block-matrix sphericity (see Fig. 1). We may note how for either $u = 1$ or $v = 1$ we have the block compound symmetric structure for which a likelihood ratio test (l.r.t.) was developed in Coelho and Roy (2017), being thus the test in the present paper a generalization of the test in this reference.

Therefore, once the test for double exchangeability is developed, it may be used to test for (see Fig. 1)

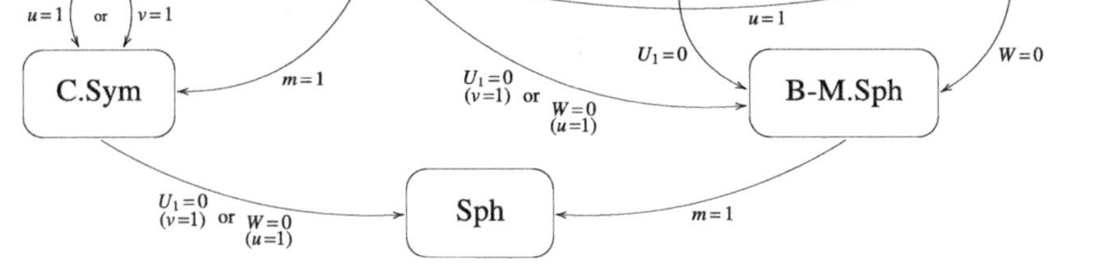


**Fig. 1** Web of relations of covariance structures which are particular cases of the doubly exchangeable structure. Covariance structures depicted and their acronyms: *D.E.* – Double exchangeable, *D.C.Sym* – Double compound symmetry, *B.C.Sym* – Block Compound Symmetry (or block exchangeability), *D-B.C.Sym* – Diagonal-block compound symmetry, *M.B-M.Sph* – Multi-Block-Matrix Sphericity, *C.Sym* – Compound symmetry, *B-M.Sph.* – Block-matrix sphericity, *Sph* – Sphericity

(i) block compound symmetry or block exchangeability, for $u=1$ or $v=1$,
(ii) double compound symmetry, for $m = 1$, or
(iii) compound symmetry, for $m = 1$ and either $u = 1$ or $v = 1$.

And, as such, in case we reject the doubly exchangeable structure for a given covariance matrix, we may then take our testing procedure one step further by testing for example for block compound symmetry or block exchangeability of submatrices, in trying to evaluate which may be the reason why the doubly exchangeable structure was rejected, and this can be done just by using one of the particular cases of this same test.

The use of the doubly exchangeable structure will also enable the use of a much smaller number of parameters to model the covariance structure, since for an unstructured covariance matrix of dimensions $muv \times muv$ there are $muv(muv+1)/2$ unknown parameters, whereas the doubly exchangeable covariance matrix has only $3m(m+1)/2$ unknown parameters, which does not depend on either $u$ or $v$. Perlman (1987) stresses the importance of carrying out tests for symmetry structures by saying that "if symmetries are known to be present, then sharper statistical inferences can be obtained".

These facts motivated the authors to develop a l.r.t. for the doubly exchangeable covariance structure for three-level multivariate data. The method used allows for an easy way to obtain the l.r.t. statistic as well as the characterization of its exact distribution, which is done in Sects. 2 and 3 of the paper. Given the extremely complicated structure of the exact distribution of the l.r.t. statistic, the development of sharp but manageable approximations stands up as a most desirable goal, in order to enable the practical application of the test. It happens that the approach undertaken also enables to obtain an extremely useful factorization of the characteristic function of the logarithm of the l.r.t. statistic, which in turn opens the way for the development of extremely sharp and manageable near-exact approximations for the distribution of the likelihood ratio statistic. This work is done in Sect. 4 of the paper. In Sect. 5 the authors carry out a number of numerical studies which show the extremely good performance of the near-exact distributions developed, even for very small samples and/or very large numbers of variables involved. In Sect. 6 two real-data examples are used to exemplify the practical application of the l.r.t. and of the near-exact approximations developed. Conclusions are drawn in Sect. 7.

## 2 Formulation of the Hypothesis and the Likelihood Ratio Test

Let $\boldsymbol{y} \sim N_{muv}(\boldsymbol{\mu}, \boldsymbol{\Sigma})$. We are interested in testing the hypothesis

$$H_0 : \boldsymbol{\Sigma} = \boldsymbol{\Theta}\,, \tag{2}$$

where $\boldsymbol{\Theta}$ is defined in (1).

In Lemma 3.1 in Roy and Fonseca (2012), it is shown that for $\boldsymbol{\Gamma}^* = \underset{v\times v}{\boldsymbol{C}'} \otimes \boldsymbol{I}_{mu}$ and $\boldsymbol{\Gamma}^\bullet = \boldsymbol{I}_v \otimes (\underset{u\times u}{\boldsymbol{C}^{*\prime}} \otimes \boldsymbol{I}_m)$, where $\boldsymbol{C}$ and $\boldsymbol{C}^*$ are orthogonal Helmert matrices whose first columns are proportional to $\mathbf{1}$'s,

$$\boldsymbol{\Gamma}^\bullet \boldsymbol{\Gamma}^* \boldsymbol{\Theta} \boldsymbol{\Gamma}^{*\prime} \boldsymbol{\Gamma}^{\bullet\prime} = \text{diag}(\boldsymbol{\Delta}_3, \boldsymbol{I}_{u-1} \otimes \boldsymbol{\Delta}_1, \boldsymbol{\Delta}_2, \boldsymbol{I}_{u-1} \otimes \boldsymbol{\Delta}_1, \boldsymbol{\Delta}_2, \ldots, \boldsymbol{I}_{u-1} \otimes \boldsymbol{\Delta}_1)$$

where

$$\begin{aligned}
\boldsymbol{\Delta}_1 &= \boldsymbol{U}_0 - \boldsymbol{U}_1, \\
\boldsymbol{\Delta}_2 &= \boldsymbol{U}_0 + (u-1)\,\boldsymbol{U}_1 - u\boldsymbol{W} = (\boldsymbol{U}_0 - \boldsymbol{U}_1) + u\,(\boldsymbol{U}_1 - \boldsymbol{W}), \\
\text{and} \quad \boldsymbol{\Delta}_3 &= \boldsymbol{U}_0 + (u-1)\,\boldsymbol{U}_1 + u\,(v-1)\,\boldsymbol{W} = (\boldsymbol{U}_0 - \boldsymbol{U}_1) + u\,(\boldsymbol{U}_1 - \boldsymbol{W}) + uv\boldsymbol{W}.
\end{aligned}$$

Since $\boldsymbol{\Gamma}^\bullet$ and $\boldsymbol{\Gamma}^*$ are not function of either $\boldsymbol{U}_0$, or $\boldsymbol{U}_1$ or $\boldsymbol{W}$, to test $H_0$ in (2) is equivalent to test

$$H_0 : \boldsymbol{\Sigma}^* = \boldsymbol{\Omega} \tag{3}$$

where

$$\boldsymbol{\Sigma}^* = \boldsymbol{\Gamma}^\bullet \boldsymbol{\Gamma}^* \boldsymbol{\Sigma} \boldsymbol{\Gamma}^{*\prime} \boldsymbol{\Gamma}^{\bullet\prime} \quad \text{and} \quad \boldsymbol{\Omega} = \boldsymbol{\Gamma}^\bullet \boldsymbol{\Gamma}^* \boldsymbol{\Theta} \boldsymbol{\Gamma}^{*\prime} \boldsymbol{\Gamma}^{\bullet\prime}.$$

We may split the null hypothesis in (3) as

$$H_0 \equiv \left(H_{0c|a} \,||\, H_{0b|a}\right) \circ H_{0a}\,, \tag{4}$$

where '$\circ$' means 'after' and '$||$' means 'parallel', meaning 'either after or before'.

In (4),

$$H_{0a} : \boldsymbol{\Sigma}^* = \text{block-diag}(\boldsymbol{\Sigma}^*_i,\ i = 1, \ldots, uv)\,, \tag{5}$$

is the hypothesis of independence of the $uv$ diagonal blocks $\boldsymbol{\Sigma}^*_i$ $(i = 1, \ldots, uv)$ of size $m\times m$ of $\boldsymbol{\Sigma}^*$;

$$\begin{array}{c} H_{0b|a} : \underbrace{\boldsymbol{\Sigma}^*_2 = \ldots = \boldsymbol{\Sigma}^*_u}_{u-1} = \underbrace{\boldsymbol{\Sigma}^*_{u+2} = \ldots = \boldsymbol{\Sigma}^*_{2u}}_{u-1} = \cdots = \underbrace{\boldsymbol{\Sigma}^*_{(v-1)u+2} = \ldots = \boldsymbol{\Sigma}^*_{vu}}_{u-1}, \\ \text{assuming } H_{0a}, \end{array} \tag{6}$$

is the hypothesis of equality of $v(u-1)$ covariance matrices of dimension $m\times m$, assuming $H_{0a}$, and

$$\begin{array}{c} H_{0c|a} : \boldsymbol{\Sigma}^*_{u+1} = \boldsymbol{\Sigma}^*_{2u+1} = \cdots = \boldsymbol{\Sigma}^*_{(v-1)u+1} \\ \text{assuming } H_{0a}, \end{array} \tag{7}$$

is the hypothesis of equality of the covariance matrices $\boldsymbol{\Sigma}^*_{u+1}, \boldsymbol{\Sigma}^*_{2u+1}, \ldots, \boldsymbol{\Sigma}^*_{(v-1)u+1}$, assuming $H_{0a}$.

The l.r.t. statistic to test $H_{0a}$ in (5) is (Anderson, 2003, Sect. 9.2)

$$\Lambda_a = \left(\frac{|\boldsymbol{A}|}{\prod_{j=1}^{uv}|\boldsymbol{A}_j|}\right)^{n/2}$$

where $\boldsymbol{A} = \boldsymbol{\Gamma}^{\bullet}\boldsymbol{\Gamma}^*\boldsymbol{A}^+\boldsymbol{\Gamma}^{*\prime}\boldsymbol{\Gamma}^{\bullet\prime}$ is the maximum likelihood estimator (m.l.e.) of $\boldsymbol{\Sigma}^*$, and $\boldsymbol{A}_j$ its $j$-th diagonal $m\times m$ block, being $\boldsymbol{A}^+$ the m.l.e. of $\boldsymbol{\Sigma}$.

The l.r.t. statistic to test $H_{0b|a}$ in (6) is (Anderson, 2003, Sect. 10.2)

$$\Lambda_b = \left((v(u-1))^{mv(u-1)}\,\frac{\prod_{\ell=1}^{v}\prod_{k=1}^{u-1}\left|\boldsymbol{A}_{(\ell-1)u+1+k}\right|}{|\boldsymbol{A}^*|^{v(u-1)}}\right)^{n/2}, \tag{8}$$

where

$$\boldsymbol{A}^* = \sum_{\ell=1}^{v}\sum_{k=1}^{u-1}\boldsymbol{A}_{(\ell-1)u+1+k}\,.$$

The l.r.t. statistic to test $H_{0c|a}$ in (7) is (Anderson, 2003, Sect. 10.2)

$$\Lambda_c = \left((v-1)^{m(v-1)}\,\frac{\prod_{k=1}^{v-1}|\boldsymbol{A}_{ku+1}|}{|\boldsymbol{A}^{**}|^{v-1}}\right)^{n/2} \tag{9}$$

where

$$\boldsymbol{A}^{**} = \sum_{k=1}^{v-1}\boldsymbol{A}_{ku+1}\,.$$

Then, through an extension of Lemma 10.3.1 in (Anderson, 2003, Sect. 10.3), the l.r.t. statistic to test $H_0$ in (3) will be

$$\begin{aligned}\Lambda &= \Lambda_a\Lambda_b\Lambda_c\\ &= \left((v(u-1))^{mv(u-1)}\,(v-1)^{m(v-1)}\,\frac{|\boldsymbol{A}|}{|\boldsymbol{A}_1||\boldsymbol{A}^*|^{v(u-1)}\,|\boldsymbol{A}^{**}|^{v-1}}\right)^{n/2},\end{aligned} \tag{10}$$

with

$$E\left(\Lambda^h\right) = E\left(\Lambda_a^h\right)E\left(\Lambda_b^h\right)E\left(\Lambda_c^h\right), \tag{11}$$

since on one hand, under $H_{0a}$, $\Lambda_a$ is independent of $\prod_{j=1}^{uv} |\boldsymbol{A}_j|$ (Marques and Coelho, 2012; Coelho and Marques, 2013), which makes $\Lambda_a$ independent of $\Lambda_b$ and $\Lambda_c$, while on the other hand, the $\boldsymbol{A}_j$ $(j = 1, \ldots, uv)$, under $H_{0a}$, are independent among themselves, which makes $\Lambda_b$ and $\Lambda_c$ independent because they are built on different $\boldsymbol{A}_j$'s.

In the following section we obtain the expressions for the moments of all three l.r.t. statistics $\Lambda_a$, $\Lambda_b$ and $\Lambda_c$, as well as their distributions.

## 3 On the Exact Distribution of the l.r.t. Statistic

Using the results in Coelho (2004), Coelho et al. (2010) and Marques et al. (2011) we may write the $h$-th moment of $\Lambda_a$ as

$$
\begin{aligned}
E\left(\Lambda_a^h\right) &= \prod_{k=1}^{uv-1} \prod_{j=1}^{m} \frac{\Gamma\left(\frac{n-j}{2}\right)\Gamma\left(\frac{n-(uv-k)m-j}{2}+\frac{n}{2}h\right)}{\Gamma\left(\frac{n-(uv-k)m-j}{2}\right)\Gamma\left(\frac{n-j}{2}+\frac{n}{2}h\right)} \\
&= \underbrace{\left\{\prod_{j=3}^{muv}\left(\frac{n-j}{n}\right)^{r_j}\left(\frac{n-j}{n}+h\right)^{-r_j}\right\}}_{\Phi_{a,1}(h)} \underbrace{\left(\frac{\Gamma\left(\frac{n-1}{2}\right)\Gamma\left(\frac{n-2}{2}+\frac{n}{2}h\right)}{\Gamma\left(\frac{n-1}{2}+\frac{n}{2}h\right)\Gamma\left(\frac{n-2}{2}\right)}\right)^{k^*}}_{\Phi_{a,2}(h)}
\end{aligned}
\tag{12}
$$

with

$$
k^* = \begin{cases} \left\lfloor \dfrac{uv}{2} \right\rfloor, & m \text{ odd} \\ 0, & m \text{ even}, \end{cases} \tag{13}
$$

where $\lfloor \cdot \rfloor$ represents the integer function, that is, the largest integer that does not exceed its argument, and

$$
r_j = \begin{cases} h_{j-2} + (-1)^j k^*, & j = 3, 4 \\ r_{j-2} + h_{j-2}, & j = 5, \ldots, muv\,, \end{cases}
$$

with

$$
h_j = \begin{cases} uv - 1, & j = 1, \ldots, m \\ -1, & j = m+1, \ldots, muv - 2\,. \end{cases}
$$

Now, using the results in Coelho and Marques (2012); Coelho et al. (2010); Marques et al. (2011) we obtain the expression for the $h$-th moment of $\Lambda_b$ as

$$E\left(\Lambda_b^h\right)=\prod_{j=1}^{m}\prod_{k=1}^{v(u-1)}\frac{\Gamma\left(\frac{n-1}{2}-\frac{j-1}{2v(u-1)}+\frac{k-1}{v(u-1)}\right)\Gamma\left(\frac{n-j}{2}+\frac{n}{2}h\right)}{\Gamma\left(\frac{n-1}{2}-\frac{j-1}{2v(u-1)}+\frac{k-1}{v(u-1)}+\frac{n}{2}h\right)\Gamma\left(\frac{n-j}{2}\right)}$$
$$=\underbrace{\left\{\prod_{j=2}^{m}\left(\frac{n-j}{n}\right)^{s_j}\left(\frac{n-j}{n}+h\right)^{-s_j}\right\}}_{\Phi_{b,1}(h)}$$
$$\underbrace{\times\left\{\prod_{j=1}^{\lfloor m/2\rfloor}\prod_{k=1}^{v(u-1)}\frac{\Gamma\left(n-1+\frac{k-2j}{v(u-1)}\right)\Gamma\left(n-1+\left\lfloor\frac{k-2j}{v(u-1)}\right\rfloor+nh\right)}{\Gamma\left(n-1+\frac{k-2j}{v(u-1)}+nh\right)\Gamma\left(n-1+\left\lfloor\frac{k-2j}{v(u-1)}\right\rfloor\right)}\right\}\times\left\{\prod_{k=1}^{v(u-1)}\frac{\Gamma\left(\frac{n-1}{2}+\frac{2k-m-1}{2v(u-1)}\right)\Gamma\left(\frac{n-m}{2}+\left\lfloor\frac{m-1}{2}+\frac{2k-m-1}{2v(u-1)}\right\rfloor+\frac{n}{2}h\right)}{\Gamma\left(\frac{n-1}{2}+\frac{2k-m-1}{2v(u-1)}+\frac{n}{2}h\right)\Gamma\left(\frac{n-m}{2}+\left\lfloor\frac{m-1}{2}+\frac{2k-m-1}{2v(u-1)}\right\rfloor\right)}\right\}^{\widetilde{m}}}_{\Phi_{b,2}(h)} \tag{14}$$

where $\widetilde{m}=m-2\lfloor m/2\rfloor$ and $s_j$ $(j=2,\ldots,m)$ are given in Appendix A in the supplementary material.

By looking at (8) and (9) we may see that the $h$-th moment of $\Lambda_c$ may be obtained, as follows, from the $h$-th moment of $\Lambda_b$ by first replacing $v$ by 1 and then replacing $u$ by $v$,

$$E\left(\Lambda_c^h\right)=\prod_{j=1}^{m}\prod_{k=1}^{v-1}\frac{\Gamma\left(\frac{n-1}{2}-\frac{j-1}{2(v-1)}+\frac{k-1}{v-1}\right)\Gamma\left(\frac{n-j}{2}+\frac{n}{2}h\right)}{\Gamma\left(\frac{n-1}{2}-\frac{j-1}{2(v-1)}+\frac{k-1}{v-1}+\frac{n}{2}h\right)\Gamma\left(\frac{n-j}{2}\right)}$$
$$=\underbrace{\left\{\prod_{j=2}^{m}\left(\frac{n-j}{n}\right)^{\delta_j}\left(\frac{n-j}{n}+h\right)^{-\delta_j}\right\}}_{\Phi_{c,1}(h)}$$
$$\underbrace{\times\left\{\prod_{j=1}^{\lfloor m/2\rfloor}\prod_{k=1}^{v-1}\frac{\Gamma\left(n-1+\frac{k-2j}{v-1}\right)\Gamma\left(n-1+\left\lfloor\frac{k-2j}{v-1}\right\rfloor+nh\right)}{\Gamma\left(n-1+\frac{k-2j}{v-1}+nh\right)\Gamma\left(n-1+\left\lfloor\frac{k-2j}{v-1}\right\rfloor\right)}\right\}\times\left\{\prod_{k=1}^{v-1}\frac{\Gamma\left(\frac{n-1}{2}+\frac{2k-m-1}{2(v-1)}\right)\Gamma\left(\frac{n-m}{2}+\left\lfloor\frac{m-1}{2}+\frac{2k-m-1}{2(v-1)}\right\rfloor+\frac{n}{2}h\right)}{\Gamma\left(\frac{n-1}{2}+\frac{2k-m-1}{2(v-1)}+\frac{n}{2}h\right)\Gamma\left(\frac{n-m}{2}+\left\lfloor\frac{m-1}{2}+\frac{2k-m-1}{2(v-1)}\right\rfloor\right)}\right\}^{\widetilde{m}}}_{\Phi_{c,2}(h)} \tag{15}$$

where $\widetilde{m} = m - 2\lfloor m/2 \rfloor$ and the shape parameters $\delta_j$ $(j = 2, \ldots, m)$ are given in Appendix A in the supplementary material.

Since the supports of $\Lambda_a$, $\Lambda_b$ and $\Lambda_c$ are delimited, their distributions are defined by their moments, and as such, from the first expression in (12) we may write

$$\Lambda_a \stackrel{\text{st}}{\sim} \prod_{j=1}^{m} \prod_{k=1}^{uv-1} (X_{jk})^{n/2}, \quad \text{where } X_{jk} \sim Beta\left(\frac{n-(uv-k)m-j}{2}, \frac{(uv-k)m}{2}\right), \tag{16}$$

where '$\stackrel{\text{st}}{\sim}$' means 'stochastically equivalent to' and $X_{jk}$ $(j = 1, \ldots, m;$ $k = 1, \ldots, uv - 1)$ are independent random variables, while from the first expression in (14) we may write

$$\Lambda_b \stackrel{\text{st}}{\sim} \prod_{j=1}^{m} \prod_{k=1}^{v(u-1)} \left(X^*_{jk}\right)^{n/2}, \quad \text{where} \quad X^*_{jk} \sim Beta\left(\frac{n-j}{2}, \frac{j-1}{2} + \frac{2k-j-1}{2v(u-1)}\right), \tag{17}$$

where $X^*_{jk}$ $(j = 1, \ldots, m; k = 1, \ldots, v(u - 1))$ are independent, and from the first expression in (15) we may write

$$\Lambda_c \stackrel{\text{st}}{\sim} \prod_{j=1}^{m} \prod_{k=1}^{v-1} \left(X^{**}_{jk}\right)^{n/2}, \quad \text{where} \quad X^{**}_{jk} \sim Beta\left(\frac{n-j}{2}, \frac{j-1}{2} + \frac{2k-j-1}{2(v-1)}\right), \tag{18}$$

where $X^{**}_{jk}$ $(j = 1, \ldots, m; k = 1, \ldots, v - 1)$ are independent, so that we may write for the overall l.r.t. statistic for $H_0$ in (4)

$$\Lambda \stackrel{\text{st}}{\sim} \prod_{j=1}^{m} \left\{ \left( \prod_{k=1}^{uv-1} X_{jk} \right)^{n/2} \times \left( \prod_{k=1}^{v(u-1)} X^*_{jk} \right)^{n/2} \times \left( \prod_{k=1}^{v-1} X^{**}_{jk} \right)^{n/2} \right\}, \tag{19}$$

where all random variables are independent.

On the other hand, based on the results in Appendix B in the supplementary material and from the second expressions in (12), (14) and (15) we may respectively write,

$$\Lambda_a \stackrel{\text{st}}{\sim} \left( \prod_{j=3}^{muv} e^{-Z_j} \right) \times \left( \prod_{j=1}^{k^*} (Y_j)^{n/2} \right) \tag{20}$$

where

$$Z_j \sim \Gamma\left(r_j, \frac{n-j}{n}\right) \quad \text{and} \quad Y_j \sim Beta\left(\frac{n-2}{2}, \frac{1}{2}\right) \tag{21}$$

are all independent random variables, while for $\Lambda_b$ we may write

$$\Lambda_b \overset{\text{st}}{\sim} \left(\prod_{j=2}^{m} e^{-Z_j^*}\right) \times \left(\prod_{j=1}^{\lfloor m/2 \rfloor} \prod_{k=1}^{v(u-1)} \left(Y_{1jk}^*\right)^n\right) \times \left(\prod_{k=1}^{v(u-1)} \left(Y_{2k}^*\right)^{n/2}\right)^{m \perp\!\!\perp 2} \tag{22}$$

where $m \perp\!\!\perp 2 = \text{mod}(p, 2)$,

$$Z_j^* \sim \Gamma\left(s_j, \frac{n-j}{n}\right), \quad Y_{1jk}^* \sim Beta\left(n-1+\left\lfloor \frac{k-2j}{v(u-1)} \right\rfloor, \frac{k-2j}{v(u-1)} - \left\lfloor \frac{k-2j}{v(u-1)} \right\rfloor\right), \tag{23}$$

and

$$Y_{2k}^* \sim Beta\left(\frac{n-m}{2} + \left\lfloor \frac{m-1}{2} + \frac{2k-m-1}{2v(u-1)} \right\rfloor, \right.$$
$$\left. \frac{m-1}{2} + \frac{2k-m-1}{2v(u-1)} - \left\lfloor \frac{m-1}{2} + \frac{2k-m-1}{2v(u-1)} \right\rfloor\right) \tag{24}$$

are all independent random variables, while for $\Lambda_c$ we may write

$$\Lambda_c \overset{\text{st}}{\sim} \left(\prod_{j=2}^{m} e^{-Z_j^{**}}\right) \times \left(\prod_{j=1}^{\lfloor m/2 \rfloor} \prod_{k=1}^{v-1} \left(Y_{1jk}^{**}\right)^n\right) \times \left(\prod_{k=1}^{v-1} \left(Y_{2k}^{**}\right)^{n/2}\right)^{m \perp\!\!\perp 2} \tag{25}$$

where

$$Z_j^{**} \sim \Gamma\left(\delta_j, \frac{n-j}{n}\right), \quad Y_{1jk}^{**} \sim Beta\left(n-1+\left\lfloor \frac{k-2j}{v-1} \right\rfloor, \frac{k-2j}{v-1} - \left\lfloor \frac{k-2j}{v-1} \right\rfloor\right), \tag{26}$$

and

$$Y_{2k}^{**} \sim Beta\left(\frac{n-m}{2} + \left\lfloor \frac{m-1}{2} + \frac{2k-m-1}{2(v-1)} \right\rfloor, \right.$$
$$\left. \frac{m-1}{2} + \frac{2k-m-1}{2(v-1)} - \left\lfloor \frac{m-1}{2} + \frac{2k-m-1}{2(v-1)} \right\rfloor\right) \tag{27}$$

are all independent random variables.

Thus, we have the following Theorem.

**Theorem 1** *The exact distribution of the overall l.r.t. statistic $\Lambda$ in (10), to test $H_0$ in (3) or (4) is, for general u, v and m, the same as that of*

$$\left(\prod_{j=2}^{muv} e^{-T_j}\right) \times \left(\prod_{j=1}^{k^*} Y_j\right)^{n/2} \times \left(\prod_{j=1}^{\lfloor m/2\rfloor}\prod_{k=1}^{v(u-1)} Y^*_{1jk}\right)^{n} \times \left(\prod_{j=1}^{\lfloor m/2\rfloor}\prod_{k=1}^{v-1} Y^{**}_{1jk}\right)^{n} \\ \times \left\{\left(\prod_{k=1}^{v(u-1)} Y^*_{2k}\right)^{n/2} \times \left(\prod_{k=1}^{v-1} Y^{**}_{2k}\right)^{n/2}\right\}^{\widetilde{m}} \tag{28}$$

*where $\widetilde{m} = m - 2\lfloor m/2\rfloor$ and, for $j = 2, \ldots, muv$,*

$$T_j \sim \Gamma\left(\mu_j, \frac{n-j}{n}\right),$$

*with*

$$\mu_j = \sum_{j=2}^{muv}\left(r_j^+ + s_j^+ + \delta_j^+\right) \tag{29}$$

*where*

$$r_j^+ = \begin{cases} 0, & j=2\\ r_j, & j=3,\ldots,muv \end{cases}$$

*and*

$$s_j^+ = \begin{cases} s_j, & j=2,\ldots,m\\ 0, & j=m+1,\ldots,muv\,, \end{cases} \qquad \delta_j^+ = \begin{cases} \delta_j, & j=2,\ldots,m\\ 0, & j=m+1,\ldots,muv\,, \end{cases}$$

*where the shape parameters $s_j$ and $\delta_j$ are defined in Appendix A in the supplementary material. The distributions of $Y_j$, $Y^*_{1jk}$, $Y^*_{2k}$, $Y^{**}_{1jk}$ and $Y^{**}_{2k}$ are defined in (21), (23), (24), (26) and (27).*

***Proof*** The proof of the above theorem is rather trivial, from the previously established results. We may only remark that the random variables $T_j$ are the sum of the random variables $Z_j$, $Z^*_j$ and $Z^{**}_j$, which are independent Gamma distributed random variables, all with the same rate parameters $\frac{n-j}{n}$, for a given $j$. As such, for a given $j$, their sum is a Gamma distributed random variable with that same rate parameter and a shape parameter which is the sum of the original shape parameters. □

The following Corollary refers to the particular cases for $v = 1$ and $v = 2$, which are particular cases of interest.

**Corollary 1** *For both $v = 1$ and $v = 2$, the hypothesis $H_{0c|a}$ vanishes and as such the distribution of $\Lambda$ is in this case the same as that of*

$$\left(\prod_{j=2}^{muv} e^{-T_j}\right) \times \left(\prod_{j=1}^{k^*} Y_j\right)^{n/2} \times \left(\prod_{j=1}^{\lfloor m/2 \rfloor} \prod_{k=1}^{v(u-1)} Y^*_{1jk}\right)^{n} \times \left\{\left(\prod_{k=1}^{v(u-1)} Y^*_{2k}\right)^{n/2}\right\}^{m \perp\!\!\!\perp 2} \tag{30}$$

*where, for $j = 2, \ldots, muv$,*

$$T_j \sim \Gamma\left(\mu^*_j, \frac{n-j}{n}\right),$$

*with*

$$\mu^*_j = \sum_{j=2}^{muv} \left(r^+_j + s^+_j\right) \tag{31}$$

*where*

$$r^+_j = \begin{cases} 0, & j = 2 \\ r_j, & j = 3, \ldots, muv, \end{cases} \qquad \text{and} \qquad s^+_j = \begin{cases} s_j, & j = 2, \ldots, m \\ 0, & j = m+1, \ldots, muv, \end{cases}$$

*where the shape parameters $s_j$ are defined in Appendix A in the supplementary material. The distributions of $Y_j$, $Y^*_{1jk}$ and $Y^*_{2k}$, are defined in (21), (23) and (24).*

The case for $v = 1$ is equivalent to the test for block compound symmetric covariance structure, addressed in Coelho and Roy (2017).

# 4 The Characteristic Function of $W = -\log \Lambda$ and the Development of Near-Exact Distributions

## *4.1 The Characteristic Function of $W = -\log \Lambda$*

In the previous section, the reason we obtain two equivalent representations for the exact distribution of $\Lambda$, one from (19) and the other one in Theorem 1, is because the first one of these representations neither yields a manageable cumulative distribution function nor it adequately leads to a sharp approximation to the exact distribution of $\Lambda$.

In order to obtain a very sharp and manageable approximation to the exact distribution of $\Lambda$, we will base our developments on the representations given by Theorem 1 and Corollary 1.

From Theorem 1 and expressions (12)–(15) we may write the characteristic function (c.f.) of $W = -\log \Lambda$ as

$$\begin{aligned}\Phi_W(t) &= E\left(e^{\mathrm{i}tW}\right) = E\left(\Lambda^{-\mathrm{i}t}\right)\\ &= \underbrace{\left\{\prod_{j=2}^{muv}\left(\frac{n-j}{n}\right)^{\mu_j}\left(\frac{n-j}{n}-\mathrm{i}t\right)^{-\mu_j}\right\}}_{\Phi_{W,1}(t)} \times \underbrace{\Phi_{a,2}(-\mathrm{i}t)\,\Phi_{b,2}(-\mathrm{i}t)\,\Phi_{c,2}(-\mathrm{i}t)}_{\Phi_{W,2}(t)}\end{aligned} \tag{32}$$

where $\mu_j$ is given by (29), $\Phi_{a,2}(\cdot)$, $\Phi_{b,2}(\cdot)$ and $\Phi_{c,2}(\cdot)$ are defined in (12)–(15), and $\Phi_{W,1}(t)$ is actually equal to $\Phi_{a,1}(-\mathrm{i}t)\Phi_{b,1}(-\mathrm{i}t)\Phi_{c,1}(-\mathrm{i}t)$.

For $v = 1$ and $v = 2$, according to Corollary 1 in the previous section $\Phi_W(t)$ reduces to

$$\Phi_W(t) = \underbrace{\left\{\prod_{j=2}^{muv}\left(\frac{n-j}{n}\right)^{\mu_j^*}\left(\frac{n-j}{n}-\mathrm{i}t\right)^{-\mu_j^*}\right\}}_{\Phi_{W,1}(t)} \times \underbrace{\Phi_{a,2}(-\mathrm{i}t)\,\Phi_{b,2}(-\mathrm{i}t)}_{\Phi_{W,2}(t)} \tag{33}$$

for $\mu_j^*$ given by (31) and where now $\Phi_{W,1}(t)$ is equal to $\Phi_{a,1}(-\mathrm{i}t)\Phi_{b,1}(-\mathrm{i}t)$.

Expressions (32) and (33), together with expressions (28) and (30), show that the exact distribution of $W = -\log \Lambda$ is the same as that of the sum of $muv - 1$ independent Gamma random variables with an independent random variable which itself is distributed as a sum of independent Logbeta random variables.

Then, in building the near-exact distributions in the next subsection we will keep $\Phi_{W,1}(t)$ untouched and will approximate $\Phi_{W,2}(t)$ asymptotically by the c.f. of a finite mixture of Gamma distributions.

## 4.2 Development of Near-Exact Distributions

Based on the result in Section 5 of Tricomi and Erdélyi (1951), we know that we may, for increasing values of $a$, asymptotically replace the distribution of any $Logbeta(a, b)$ distributed random variable by an infinite mixture of $\Gamma(b+\ell, a)$ distributions ($\ell = 0, 1, \ldots$). As such, we could replace $\Phi_{W,2}(t)$ in either (32) or (33) by the c.f. of the sum of infinite mixtures of Gamma distributions, which would be the same as the c.f. of an infinite mixture of sums of Gamma distributions. Although it happens that these Gamma distributions would have different rate parameters, these parameters would anyway be of comparable magnitude. As such, in building our near-exact distributions for $W = -\log \Lambda$ and $\Lambda$, while we will leave $\Phi_{W,1}(t)$ unchanged, we will replace $\Phi_{W,2}(t)$ in either (32) or (33), by

$$\Phi^*(t) = \sum_{\ell=0}^{m^*} \pi_\ell \, \nu^{r+\ell}(\nu - \mathrm{i}t)^{-(r+\ell)} \tag{34}$$

which is the c.f. of a finite mixture of $\Gamma(r+\ell, \nu)$ distributions, where for the general case in (32) and for $k^*$ in (13) and $v > 1$, we will take

$$\begin{aligned} r = \frac{k^*}{2} &+ \sum_{j=1}^{\lfloor m/2 \rfloor} \sum_{k=1}^{v(u-1)} \frac{k-2j}{v(u-1)} - \left\lfloor \frac{k-2j}{v(u-1)} \right\rfloor + \sum_{j=1}^{\lfloor m/2 \rfloor} \sum_{k=1}^{v-1} \frac{k-2j}{v(u-1)} - \left\lfloor \frac{k-2j}{v-1} \right\rfloor \\ &+ (m \perp\!\!\!\perp 2) \left( \sum_{k=1}^{v(u-1)} \frac{m-1}{2} + \frac{2k-m-1}{2v(u-1)} - \left\lfloor \frac{m-1}{2} + \frac{2k-m-1}{2v(u-1)} \right\rfloor \right. \\ &\qquad \left. + \sum_{k=1}^{v-1} \frac{m-1}{2} + \frac{2k-m-1}{2(v-1)} - \left\lfloor \frac{m-1}{2} + \frac{2k-m-1}{2(v-1)} \right\rfloor \right) \\ &= \begin{cases} \frac{m}{4}(uv-3), & \text{even } m \\ \frac{1}{2}\left\lfloor \frac{uv}{2} \right\rfloor + \frac{m+1}{4}(uv-3), & \text{odd } m\,, \end{cases} \end{aligned} \tag{35}$$

which is the sum of all the second parameters of the Logbeta distributions in $\Phi_{W,2}(t)$ in (32), while for the particular case in (33) we will take

$$\begin{aligned} r = \frac{k^*}{2} &+ \sum_{j=1}^{\lfloor m/2 \rfloor} \sum_{k=1}^{v(u-1)} \frac{k-2j}{v(u-1)} - \left\lfloor \frac{k-2j}{v(u-1)} \right\rfloor \\ &+ (m \perp\!\!\!\perp 2) \sum_{k=1}^{v(u-1)} \frac{m-1}{2} + \frac{2k-m-1}{2v(u-1)} - \left\lfloor \frac{m-1}{2} + \frac{2k-m-1}{2v(u-1)} \right\rfloor \\ &= \begin{cases} \frac{m}{4}\big(v(u-1)-1\big), & \text{even } m \\ \frac{1}{2}\left\lfloor \frac{uv}{2} \right\rfloor + \frac{m+1}{4}\big(v(u-1)-1\big), & \text{odd } m\,, \end{cases} \end{aligned} \tag{36}$$

which is the sum of all the second parameters of the Logbeta distributions in $\Phi_{W,2}(t)$ in (33).

The parameter $\nu$ in (34) is then taken as the rate parameter in

$$\Phi^{**}(t) = \theta \, \nu^{s_1}(\nu - \mathrm{i}t)^{-s_1} + (1-\theta)\nu^{s_2}(\nu - \mathrm{i}t)^{-s_2}$$

where $\theta$, $\nu$, $s_1$ and $s_2$ are determined in such a way that

$$\left.\frac{d\Phi_{W,2}(t)}{dt^h}\right|_{t=0} = \left.\frac{d\Phi^{**}(t)}{dt^h}\right|_{t=0} \quad \text{for} \quad h = 1, \ldots, 4,$$

while the weights $\pi_\ell$ ($\ell = 0, \ldots, m^* - 1$) in (34) will then be determined in such a way that

$$\left.\frac{d\Phi_{W,2}(t)}{dt^h}\right|_{t=0} = \left.\frac{d\Phi^{*}(t)}{dt^h}\right|_{t=0} \quad \text{for} \quad h = 1, \ldots, m^*,$$

with $\pi_{m^*} = 1 - \sum_{\ell=0}^{m^*-1} \pi_\ell$.

This procedure yields near-exact distributions for $W$ which will match the first $m^*$ exact moments of $W$ and which have c.f.

$$\Phi_{W,1}(t)\Phi^*(t),$$

with $\Phi_{W,1}(t)$ given by (32) or (33) and $\Phi^*(t)$ by (34), where $r$, given by (35) or (36) is always either an integer or a half-integer.

As such, the near-exact distributions developed yield, for $W$, distributions which, for non-integer $r$, are mixtures, with weights $\pi_\ell$ ($\ell = 0, \ldots, m^*$), of $m^* + 1$ Generalized Near-Integer Gamma (GNIG) distributions of depth $muv$ with integer shape parameters $\mu_j$ ($j = 2, \ldots, muv$) and real shape parameter $r$, in the general case, or shape parameters $\mu_j^*$ ($j = 2, \ldots, muv$) for the case of $v = 1$ or $v = 2$, and corresponding rate parameters $(n - j)/n$ ($j = 2, \ldots, muv$) and $\nu$, and which, for integer $r$, are similar mixtures but of Generalized Integer Gamma (GIG) distributions, with the same shape and rate parameters. See Coelho (1998, 2004) and Appendix C in the supplementary material for further details on the GIG and GNIG distributions and their probability density and cumulative distribution functions.

Using the notation in Appendix C in the supplementary material for the probability density and cumulative distribution functions of the GNIG distribution, the near-exact distributions obtained for $W$, for the general case of $v > 2$ and for the case of non-integer $r$, will have, for $w > 0$, probability density and cumulative distribution functions respectively of the form

$$f_W^*(w) = \sum_{\ell=0}^{m^*} \pi_\ell \, f^{GNIG}\Big(w \Big| \underline{s}; \underline{\lambda}; g\Big), \quad \text{and} \quad F_W^*(w) = \sum_{\ell=0}^{m^*} \pi_\ell \, F^{GNIG}\Big(w \Big| \underline{s}; \underline{\lambda}; g\Big),$$

where

$$\underline{s} = \{\mu_2, \ldots, \mu_{muv}, r + \ell\} \quad \text{and} \quad \underline{\lambda} = \left\{\frac{n-2}{n}, \ldots, \frac{n-muv}{n}, \nu\right\}$$

are respectively the sets of shape and rate parameters and $g = muv$ is the depth of the GNIG distributions, while, for $0<\lambda<1$ and $z = -\log\lambda$, the near-exact probability density and cumulative distribution functions of $\Lambda$ are respectively given by

$$f^*_\Lambda(\lambda)=\sum_{\ell=0}^{m^*}\pi_\ell f^{GNIG}\Big(z\Big|\underline{s};\underline{\lambda};g\Big)\frac{1}{\lambda}, \quad \text{and} \quad F^*_\Lambda(\lambda)=\sum_{\ell=0}^{m^*}\pi_\ell\Big(1-F^{GNIG}\Big(z\Big|\underline{s};\underline{\lambda};g\Big)\Big). \tag{37}$$

For the case $v = 1$ or $v = 2$ all we have to do is to replace the shape parameters $\mu_j$ given by (29) with the shape parameters $\mu_j^*$ given by (31) and use $r$ given by (36), instead of $r$ given by (35).

For integer $r$, all we have to do is to replace the GNIG probability density and cumulative distribution functions with their GIG counterparts (see Appendix C in the supplementary material).

## 5 Numerical Studies

In order to assess the performance of the near-exact distributions developed in the previous section we will use

$$\Delta = \frac{1}{2\pi}\int_{-\infty}^{+\infty}\left|\frac{\Phi_W(t)-\Phi_{W,1}(t)\Phi^*(t)}{t}\right| dt \tag{38}$$

with

$$\Delta \geq \max_{w>0}\left|F_W(w)-F^*_W(w)\right| = \max_{0<\lambda<1}\left|F_\Lambda(\lambda)-F^*_\Lambda(\lambda)\right| ,$$

as a measure of proximity between the exact and the near-exact distributions, where $\Phi_W(t)$ is the exact c.f. of $W$ in (32) or (33) and $F_W(\cdot)$ and $F^*_W(\cdot)$ represent respectively the exact and near-exact cumulative distribution functions of $W$, corresponding respectively to $\Phi_W(t)$ and $\Phi_{W,1}(t)\Phi^*(t)$, being $F_\Lambda(\cdot)$ and $F^*_\Lambda(\cdot)$ the corresponding exact and near-exact cumulative distribution functions of $\Lambda$.

In Tables 1, 2 and 3 we may analyze values of $\Delta$ for different combinations of values of $m$, $u$ and $v$ and different sample sizes. For each combination of values of $m$, $u$ and $v$, at least three different sample sizes $n$ exceeding the total number of variables $muv$ by 2, 30 and 100 are used. For larger combinations of values of $m$, $u$ and $v$, some larger values of $n$ are also used to illustrate the asymptotic behavior of the near-exact distributions in what concerns the sample size. For all near-exact distributions, values of $m^*$ equal to 4, 6 and 10 are used, that is, we use for each case near-exact distributions matching 4, 6 and 10 exact moments of $W$. Smaller values

**Table 1** Values of the measure $\Delta$ for the near-exact distributions for $m = 2$

| | $m^*$ | | | | $m^*$ | | |
|---|---|---|---|---|---|---|---|
| $n$ | 4 | 6 | 10 | $n$ | 4 | 6 | 10 |
| | $u = 2,\ v = 2$ | | | | $u = 2,\ v = 5$ | | |
| 10 | 4.23$E$−12 | 9.00$E$−15 | 2.64$E$−19 | 22 | 7.80$E$−17 | 8.63$E$−22 | 8.26$E$−30 |
| 38 | 1.68$E$−14 | 3.06$E$−18 | 9.23$E$−25 | 50 | 2.39$E$−17 | 2.04$E$−22 | 1.42$E$−31 |
| 108 | 9.17$E$−17 | 1.97$E$−21 | 8.66$E$−30 | 120 | 4.65$E$−19 | 9.74$E$−25 | 5.38$E$−35 |
| | $u = 5,\ v = 2$ | | | | $u = 2,\ v = 10$ | | |
| 22 | 1.80$E$−16 | 4.11$E$−21 | 1.39$E$−29 | 42 | 1.73$E$−19 | 1.88$E$−25 | 1.07$E$−36 |
| 50 | 6.28$E$−17 | 6.99$E$−22 | 6.21$E$−31 | 70 | 4.23$E$−19 | 4.80$E$−25 | 2.73$E$−36 |
| 120 | 1.30$E$−18 | 2.97$E$−24 | 1.10$E$−34 | 140 | 3.16$E$−20 | 1.19$E$−26 | 7.31$E$−39 |
| | $u = 5,\ v = 5$ | | | | $u = 5,\ v = 10$ | | |
| 52 | 1.78$E$−20 | 8.05$E$−27 | 5.49$E$−39 | 102 | 6.18$E$−24 | 1.72$E$−31 | 2.71$E$−46 |
| 80 | 6.71$E$−20 | 3.77$E$−26 | 3.56$E$−38 | 130 | 5.38$E$−23 | 2.67$E$−30 | 1.23$E$−44 |
| 150 | 7.77$E$−21 | 1.71$E$−27 | 2.37$E$−40 | 200 | 2.24$E$−23 | 7.15$E$−31 | 1.29$E$−45 |
| 250 | 8.33$E$−22 | 7.33$E$−29 | 1.58$E$−42 | 300 | 4.93$E$−24 | 8.25$E$−32 | 3.96$E$−47 |
| | $u = 10,\ v = 2$ | | | | $u = 10,\ v = 5$ | | |
| 42 | 1.10$E$−19 | 1.32$E$−25 | 8.00$E$−37 | 102 | 6.53$E$−24 | 1.80$E$−31 | 2.77$E$−46 |
| 70 | 2.73$E$−19 | 3.35$E$−25 | 1.79$E$−36 | 130 | 5.69$E$−23 | 2.80$E$−30 | 1.26$E$−44 |
| 140 | 2.06$E$−20 | 8.25$E$−27 | 4.20$E$−39 | 200 | 2.37$E$−23 | 7.50$E$−31 | 1.32$E$−45 |
| 240 | 1.83$E$−21 | 2.69$E$−28 | 1.74$E$−41 | 300 | 5.22$E$−24 | 8.66$E$−32 | 4.06$E$−47 |
| | $u = 10,\ v = 10$ | | | | | | |
| 202 | 3.92$E$−28 | 2.22$E$−36 | 1.02$E$−53 | | | | |
| 230 | 4.39$E$−27 | 5.35$E$−35 | 9.94$E$−52 | | | | |
| 300 | 4.53$E$−27 | 5.48$E$−35 | 8.66$E$−52 | | | | |
| 400 | 2.04$E$−27 | 1.82$E$−35 | 1.39$E$−52 | | | | |
| 500 | 8.98$E$−28 | 5.88$E$−36 | 2.23$E$−53 | | | | |
| 1000 | 4.50$E$−29 | 9.29$E$−38 | 2.99$E$−56 | | | | |

of $\Delta$ indicate a closer agreement with the exact distribution and as such, a better performance of the corresponding near-exact distribution.

We may see how the near-exact distributions developed provide very sharp approximations to the exact distribution even for very small samples, that is, for sample sizes hardly exceeding the total number of variables involved. Moreover, they also exhibit clear asymptotic behaviors not only for increasing sample sizes, but also for increasing values of $m$, $u$ and $v$. The asymptotic behavior in terms of sample size becoming apparent for larger sample sizes as the values of $m$, $u$ and $v$ get larger.

A Box asymptotic distribution (Box, 1949) is developed for $W = -\log \Lambda$ in Appendix D in the supplementary material. This distribution yields a mixture of two Gamma distributions as the asymptotic distribution for $W = -\log \Lambda$. As such, it should be compared with the near-exact distribution that matches only the

**Table 2** Values of the measure $\Delta$ for the near-exact distributions for $m = 5$

| | $m^*$ | | | | $m^*$ | | |
|---|---|---|---|---|---|---|---|
| $n$ | 4 | 6 | 10 | $n$ | 4 | 6 | 10 |
| | $u = 2,\ v = 2$ | | | | $u = 2,\ v = 5$ | | |
| 22 | $8.71E{-}16$ | $1.35E{-}19$ | $5.78E{-}27$ | 52 | $7.45E{-}21$ | $7.79E{-}27$ | $5.53E{-}38$ |
| 50 | $2.88E{-}16$ | $2.22E{-}20$ | $1.28E{-}27$ | 80 | $3.98E{-}20$ | $1.16E{-}25$ | $7.02E{-}37$ |
| 120 | $5.50E{-}18$ | $8.15E{-}23$ | $1.98E{-}31$ | 150 | $1.12E{-}20$ | $8.20E{-}27$ | $1.09E{-}38$ |
| | | | | 250 | $1.52E{-}21$ | $4.06E{-}28$ | $9.06E{-}41$ |
| | | | | 350 | $3.47E{-}22$ | $4.79E{-}29$ | $3.05E{-}42$ |
| | $u = 5,\ v = 2$ | | | | $u = 2,\ v = 10$ | | |
| 52 | $2.14E{-}19$ | $2.56E{-}25$ | $9.77E{-}38$ | 102 | $3.31E{-}22$ | $3.53E{-}29$ | $4.93E{-}43$ |
| 80 | $7.41E{-}19$ | $1.11E{-}24$ | $9.36E{-}37$ | 130 | $2.76E{-}21$ | $5.32E{-}28$ | $2.25E{-}41$ |
| 150 | $7.86E{-}20$ | $4.64E{-}26$ | $1.32E{-}38$ | 200 | $1.09E{-}21$ | $1.36E{-}28$ | $2.38E{-}42$ |
| 250 | $8.06E{-}21$ | $1.91E{-}27$ | $1.04E{-}40$ | 300 | $2.31E{-}22$ | $1.53E{-}29$ | $7.36E{-}44$ |
| 350 | $1.66E{-}21$ | $2.09E{-}28$ | $3.40E{-}42$ | 500 | $2.43E{-}23$ | $6.50E{-}31$ | $5.05E{-}46$ |
| | $u = 5,\ v = 5$ | | | | $u = 5,\ v = 10$ | | |
| 127 | $5.22E{-}23$ | $2.70E{-}30$ | $8.95E{-}45$ | 252 | $4.03E{-}26$ | $1.58E{-}34$ | $2.67E{-}51$ |
| 155 | $5.10E{-}22$ | $5.07E{-}29$ | $5.83E{-}43$ | 280 | $4.72E{-}25$ | $3.93E{-}33$ | $2.85E{-}49$ |
| 225 | $2.91E{-}22$ | $2.16E{-}29$ | $1.35E{-}43$ | 350 | $6.35E{-}25$ | $5.45E{-}33$ | $4.19E{-}49$ |
| 450 | $2.07E{-}23$ | $5.13E{-}31$ | $3.55E{-}46$ | 450 | $3.70E{-}25$ | $2.46E{-}33$ | $1.13E{-}49$ |
| 650 | $3.99E{-}24$ | $5.10E{-}32$ | $9.27E{-}48$ | 1000 | $1.70E{-}26$ | $3.15E{-}35$ | $1.11E{-}52$ |
| | $u = 10,\ v = 2$ | | | | $u = 10,\ v = 5$ | | |
| 102 | $3.18E{-}22$ | $3.16E{-}29$ | $4.45E{-}43$ | 252 | $4.41E{-}26$ | $1.76E{-}34$ | $3.14E{-}51$ |
| 130 | $2.70E{-}21$ | $4.83E{-}28$ | $2.05E{-}41$ | 280 | $5.16E{-}25$ | $4.38E{-}33$ | $3.34E{-}49$ |
| 200 | $1.09E{-}21$ | $1.26E{-}28$ | $2.21E{-}42$ | 350 | $6.94E{-}25$ | $6.09E{-}33$ | $4.92E{-}49$ |
| 500 | $2.49E{-}23$ | $6.13E{-}31$ | $4.78E{-}46$ | 450 | $4.05E{-}25$ | $2.75E{-}33$ | $1.32E{-}49$ |
| | | | | 1000 | $1.86E{-}26$ | $3.52E{-}35$ | $1.30E{-}52$ |
| | $u = 10,\ v = 10$ | | | | | | |
| 502 | $3.22E{-}29$ | $1.01E{-}38$ | $9.76E{-}58$ | | | | |
| 530 | $3.52E{-}28$ | $2.35E{-}37$ | $9.82E{-}56$ | | | | |
| 600 | $7.50E{-}28$ | $6.22E{-}37$ | $3.97E{-}55$ | | | | |
| 1000 | $3.27E{-}28$ | $1.76E{-}37$ | $4.67E{-}56$ | | | | |
| 5000 | $3.02E{-}31$ | $9.42E{-}42$ | $8.31E{-}63$ | | | | |

first exact moment, that is, the near-exact distribution with $m^* = 1$, since it is for this near-exact distribution that the part of the distribution that is approximated is asymptotically approximated with a mixture of two Gamma distributions.

In Tables D.1–D.3 in Appendix D in the supplementary material are displayed the values of the measure $\Delta$ in (38) for the Box asymptotic approximation and for the near-exact distributions that match the first $m^* = 1$ and $m^* = 2$ exact moments and we may see how the Box asymptotic distribution cannot match the quality of the

**Table 3** Values of the measure $\Delta$ for the near-exact distributions for $m = 10$

| | $m^*$ | | | | $m^*$ | | |
|---|---|---|---|---|---|---|---|
| $n$ | 4 | 6 | 10 | $n$ | 4 | 6 | 10 |
| | $u = 2,\ v = 2$ | | | | $u = 2,\ v = 5$ | | |
| 42 | $1.43E-17$ | $1.45E-22$ | $3.20E-32$ | 102 | $1.30E-23$ | $2.68E-31$ | $1.27E-46$ |
| 70 | $2.03E-17$ | $2.06E-22$ | $4.03E-32$ | 130 | $8.59E-22$ | $1.96E-28$ | $1.48E-41$ |
| 140 | $7.01E-19$ | $2.26E-24$ | $4.43E-35$ | 200 | $2.98E-22$ | $4.60E-29$ | $1.49E-42$ |
| | | | | 300 | $5.74E-23$ | $4.83E-30$ | $4.44E-44$ |
| | | | | 500 | $5.53E-24$ | $1.94E-31$ | $2.95E-46$ |
| | | | | 750 | $7.93E-25$ | $1.33E-32$ | $4.45E-48$ |
| | | | | 1000 | $1.95E-25$ | $1.90E-33$ | $2.13E-49$ |
| | $u = 5,\ v = 2$ | | | | $u = 2,\ v = 10$ | | |
| 102 | $2.49E-24$ | $3.43E-31$ | $2.19E-45$ | 202 | $2.19E-26$ | $5.75E-35$ | $3.99E-52$ |
| 130 | $1.57E-23$ | $5.03E-30$ | $1.02E-43$ | 230 | $2.57E-25$ | $1.40E-33$ | $4.03E-50$ |
| 200 | $3.21E-24$ | $1.21E-30$ | $1.09E-44$ | 300 | $2.86E-25$ | $1.46E-33$ | $3.73E-50$ |
| 300 | $2.45E-25$ | $1.30E-31$ | $3.41E-46$ | 500 | $6.25E-26$ | $1.61E-34$ | $1.04E-51$ |
| 500 | $1.37E-26$ | $5.31E-33$ | $2.34E-48$ | 1000 | $3.37E-27$ | $2.58E-36$ | $1.49E-54$ |
| | $u = 5,\ v = 5$ | | | | $u = 5,\ v = 10$ | | |
| 252 | $6.13E-26$ | $2.94E-34$ | $9.95E-51$ | 502 | $5.33E-30$ | $7.80E-40$ | $1.50E-59$ |
| 280 | $7.19E-25$ | $7.31E-33$ | $1.06E-48$ | 530 | $5.84E-29$ | $1.81E-38$ | $1.51E-57$ |
| 350 | $9.71E-25$ | $1.02E-32$ | $1.56E-48$ | 600 | $1.25E-28$ | $4.80E-38$ | $6.12E-57$ |
| 750 | $8.94E-26$ | $3.32E-34$ | $6.30E-51$ | 1000 | $5.49E-29$ | $1.37E-38$ | $7.25E-58$ |
| 1500 | $4.19E-27$ | $4.47E-36$ | $6.99E-54$ | 2000 | $3.55E-30$ | $2.85E-40$ | $1.55E-60$ |
| | $u = 10,\ v = 2$ | | | | $u = 10,\ v = 5$ | | |
| 202 | $1.20E-26$ | $3.09E-35$ | $1.91E-52$ | 502 | $4.91E-29$ | $1.64E-38$ | $2.66E-57$ |
| 230 | $1.37E-25$ | $7.41E-34$ | $1.90E-50$ | 530 | $5.38E-28$ | $3.81E-37$ | $2.67E-55$ |
| 300 | $1.47E-25$ | $7.51E-34$ | $1.72E-50$ | 600 | $1.15E-27$ | $1.01E-36$ | $1.08E-54$ |
| 500 | $3.03E-26$ | $7.93E-35$ | $4.68E-52$ | 1000 | $5.08E-28$ | $2.87E-37$ | $1.27E-55$ |
| 1000 | $1.57E-27$ | $1.24E-36$ | $6.54E-55$ | 2000 | $3.30E-29$ | $5.97E-39$ | $2.70E-58$ |
| | | | | 5000 | $4.77E-31$ | $1.55E-41$ | $2.26E-62$ |
| | $u = 10,\ v = 10$ | | | | | | |
| 1002 | $4.36E-33$ | $4.29E-44$ | $4.19E-66$ | | | | |
| 1030 | $4.03E-32$ | $8.04E-43$ | $3.14E-64$ | | | | |
| 1100 | $1.03E-31$ | $2.74E-42$ | $1.92E-63$ | | | | |
| 2000 | $6.59E-32$ | $1.25E-42$ | $4.37E-64$ | | | | |
| 5000 | $1.60E-33$ | $6.46E-45$ | $1.03E-67$ | | | | |

approximation provided even by the near-exact distribution that matches the single first exact moment.

The values used for $n$, $m$, $u$ and $v$ in these Tables were the same that were used in Tables 1, 2 and 3. We may notice that some of the values for the measure $\Delta$ for the Box asymptotic approximation even go above 1, particularly for cases where the number of variables involved is quite large, and simultaneously the sample size is rather small or not that large. Although this may seem to be something out of the norm, this happens because in these cases the distribution yielded by the Box approximation is not a legitimate distribution, with its 'p.d.f.' and 'c.d.f.' going below zero, so that the values reported for the measure $\Delta$ which keep being an upper bound between the exact and the asymptotic c.d.f.'s and are indeed correct.

We may also note how the asymptotic Box distribution quickly degrades its quality of approximation as the overall number of variables involved increases, in contrast to what happens with the near-exact distributions.

# 6 Two Real Data Examples

**Example 1 – Osteopenia data**
To illustrate our proposed testing method, we test the hypothesis (2) on a real data set. The original data consist of bone mineral density values obtained by a technique known as dual X-ray absorptiometry using a GE Lunar Prodigy machine. Measurements were obtained for the femoral neck and the trochanter ($m = 2$), for both the right and left femur ($u = 2$). These four measurements were observed a second time, approximately two years later ($v = 2$). The sample covariance matrix obtained from Roy and Leiva (2011) is matrix $\boldsymbol{A}^+$ in Appendix E.1 in the supplementary material, which we take as the m.l.e. of $\boldsymbol{\Sigma}$.

We see that the variance-covariance matrices ($\boldsymbol{U}_0$) of the two mineral contents for the femoral neck and trochanter appear very similar for the first as well as for the second year. Also, the covariance matrices ($\boldsymbol{U}_1$) of the left and right femurs seem to be fairly similar for both years. Finally, the covariance matrices ($\boldsymbol{W}$) of the two parts of the femur between the two years seem to be similar too. Thus, we will not be much surprised if the hypothesis that the population covariance matrix has a doubly exchangeable covariance structure is not rejected. In fact, as stated in Sect. 2, to test this hypothesis is equivalent to test the hypothesis in (3). We thus compute the m.l.e. of $\boldsymbol{\Sigma}^*$, which is the matrix $\boldsymbol{A}$ in Appendix E.1 in the supplementary material where the orthogonal matrices $\boldsymbol{\Gamma}^\bullet$ and $\boldsymbol{\Gamma}^*$ are

$$\boldsymbol{\Gamma}^\bullet = \boldsymbol{I}_v \otimes (\underset{u\times u}{\boldsymbol{C}^{*\prime}} \otimes \boldsymbol{I}_m) \qquad \text{and} \qquad \boldsymbol{\Gamma}^* = \underset{v\times v}{\boldsymbol{C}'} \otimes \boldsymbol{I}_{mu} = [\gamma_{ij}]$$

where, for $u = 2$,

$$\boldsymbol{C}^* = \begin{pmatrix} \frac{1}{\sqrt{2}} & \frac{1}{\sqrt{2}} \\ \frac{1}{\sqrt{2}} & -\frac{1}{\sqrt{2}} \end{pmatrix}, \tag{39}$$

and $\boldsymbol{C} = \boldsymbol{C}^*$, given that $v = u = 2$. In particular for $u = 2$ and $m = 2$ the elements $\gamma_{ij}$ of the symmetric matrix $\boldsymbol{\Gamma}^*$ are given by

$$\gamma_{ii} = (-1)^{I(i>mu)} 1/\sqrt{2} \quad \text{and} \quad \gamma_{ij} = I(j-i=mu) 1/\sqrt{2} \quad (i<j)\,,$$

for all $i, j = 1, \ldots, muv$, where $I(\cdot)$ represents the indicator function.

Now, from (10) we have

$$\Lambda = \left( 2^4 \frac{|\boldsymbol{A}|}{|\boldsymbol{A}_1|\,|\boldsymbol{A}_2 + \boldsymbol{A}_4|\,|\boldsymbol{A}_3|} \right)^{n/2},$$

where $\boldsymbol{A}_1, \ldots, \boldsymbol{A}_4$ denote the four diagonal blocks of dimension $2\times 2$ of $\boldsymbol{A}$ and where $n = 12$. The computed value of $\Lambda$ is $\lambda = 4.46637\times 10^{-12}$.

Using then the near-exact distribution for $\Lambda$ which matches $m^* = 4$ exact moments, with probability density and cumulative distribution functions given by (37), with $r = 2$, as given by (36), and the other shape parameters $\mu_j^*$ $(j = 2, \ldots, muv = 8)$ given by (31), with

$$\begin{aligned} r_j^+ &= \{0, 3, 3, 2, 2, 1, 1\} \\ s_j^+ &= \{1, 0, 0, 0, 0, 0, 0\} \\ \mu_j^* &= \{1, 3, 3, 2, 2, 1, 1\} = r_j^+ + s_j^+\,, \end{aligned}$$

we obtain a near-exact p-value of 0.29468. Thus, we should not reject the null hypothesis that the covariance structure is of the doubly exchangeable type.

In case we had used the common chi-square approximation for the distribution of l.r.t. statistics, we would have $-2\log\Lambda \overset{a}{\sim} \chi^2_{\{muv(muv+1)/2\}-\{3m(m+1)/2\}} \equiv \chi^2_{27}$, which would give a p-value of 0.00246. These results show how the chi-square approximation may be completely inadequate for small samples. Indeed, even for quite large samples the chi-square approximation may lead to completely inadequate p-values.

We should note that while the near-exact distribution that we used to compute the near-exact p-value yields a value of $7.73\times 10^{-12}$ for the measure $\Delta$ in (38), the chi-square approximation yields for this same measure the value of 0.70. These results clearly show that the common chi-square approximation, opposite to the near-exact approach, leads to too many rejections of the null hypothesis, or, equivalently, in general, to too low p-values, clearly inadequate for any practical purposes.

In Fig. 2 we may analyze the plots of the cumulative distribution function for the near-exact distribution for $-\log\Lambda$ and for the $\Gamma(27/2, 1)$ distribution which corresponds to the $\chi^2_{27}$ approximation for $-2\log\Lambda$, to see how much they differ.

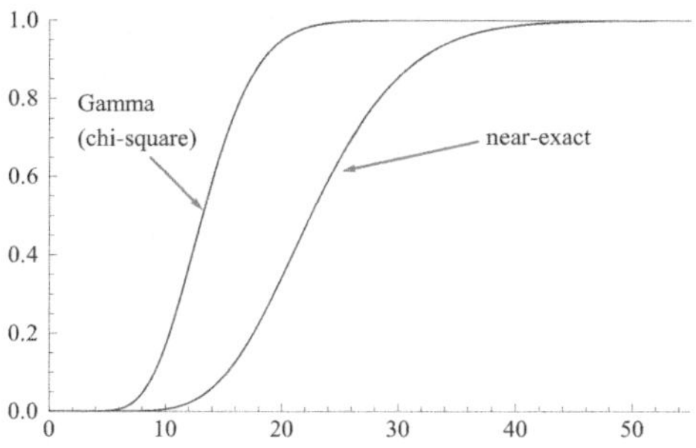


**Fig. 2** Plots of the cumulative distribution function for the near-exact distribution for $-\log \Lambda$ and for the $\Gamma(27/2, 1)$ distribution which corresponds to the $\chi^2_{27}$ approximation for $-2\log \Lambda$.

**Example 2 – Osteoporosis data**
See the supplementary material.

## 7 Discussion and Conclusions

We may see how the techniques used to handle the null hypothesis in (2), by first bringing it to the form in (3) and then using the decomposition in (4) enabled the development of very accurate near-exact distributions for the l.r.t. statistic.

From the results of the numerical studies carried out we see that the near-exact distributions developed show an interesting set of nice features. They not only have a good asymptotic behavior for increasing sample sizes, but also an extraordinary performance for very small sample sizes, as for example for sample sizes exceeding only by two the overall number of variables. Furthermore, opposite to common asymptotic distributions, these near-exact distributions also display a marked asymptotic behavior for increasing values of $m$, $u$ and $v$. All these features add up to make the developed near-exact approximations the best choice for practical applications of the test studied.

Moreover, given the discussion in Sections 9.11 and 10.11 of Anderson (2003), the results presented concerning the exact distribution of the l.r.t. statistic as well the near-exact distributions developed may be made extensive to cases where the vector $\boldsymbol{y}$ has an elliptically contoured distribution.

As future research of interest, stemming out of the work developed in the present paper the authors point out the tests to the hypotheses

$$H_0: \boldsymbol{W} = \boldsymbol{0}, \text{ assuming } \boldsymbol{\Sigma} = \boldsymbol{\Theta} \text{ (that is, } \boldsymbol{\Sigma} \text{ is doubly-exchangeable)}$$
vs
$$H_1: \boldsymbol{\Sigma} = \boldsymbol{\Theta}\,,$$

which is the test between the doubly exchangeable covariance structure, in $H_1$, and the diagonal-block compound symmetry or diagonal-block exchangeable covariance structure, in $H_0$, as well as the test to the hypotheses

$$\begin{aligned} & H_0 : \boldsymbol{U}_1 = \mathbf{0}, \text{ assuming } \boldsymbol{\Sigma} = \boldsymbol{\Theta} \\ \text{vs} & \\ & H_1 : \boldsymbol{\Sigma} = \boldsymbol{\Theta}\,, \end{aligned}$$

which is the test between the doubly exchangeable covariance structure, in $H_1$, and the multi-block-matrix sphericity covariance structure, in $H_0$.

Also the tests to the hypotheses

$$\begin{aligned} & H_0 : \boldsymbol{W} = \mathbf{0}, \boldsymbol{U}_1 = \mathbf{0}, \text{ assuming } \boldsymbol{\Sigma} = \boldsymbol{\Theta} \\ \text{vs} & \\ & H_1 : \boldsymbol{\Sigma} = \boldsymbol{\Theta}, \text{ with } \boldsymbol{U}_1 = \mathbf{0} \end{aligned}$$

which is the test between the multi-block-matrix sphericity covariance structure, in $H_1$, and the block-matrix sphericity covariance structure, in $H_0$, and yet the test to the hypotheses

$$\begin{aligned} & H_0 : \boldsymbol{W} = \mathbf{0}, \boldsymbol{U}_1 = \mathbf{0}, \text{ assuming } \boldsymbol{\Sigma} = \boldsymbol{\Theta} \\ \text{vs} & \\ & H_1 : \boldsymbol{\Sigma} = \boldsymbol{\Theta}, \text{ with } \boldsymbol{W} = \mathbf{0} \end{aligned}$$

which is the test between the multi-block compound symmetry covariance structure, in $H_1$, and the block-matrix sphericity covariance structure, in $H_0$, are tests whose development is of interest.

Since all these tests lay upon the assumption of a doubly exchangeable structure, these are a few more reasons why developing a test for the doubly exchangeable structure was an imperative goal. And once these four tests are developed we will be able to test between any two of the covariance structures in Fig. 1.

**Acknowledgments** This research was partially supported by FCT—Fundação para a Ciência e a Tecnologia (Portuguese Foundation for Science and Technology), projects UID/MAT/00297/2013 and UID/MAT/00297/2019, through Centro de Matemática e Aplicações (CMA/FCT/UNL). The second author thanks the support for the summer research grant from the College of Business at the University of Texas at San Antonio. The authors also want to thank the Editor-in-Chief and the two anonymous reviewers for their careful reading, valuable comments and suggestions that led to a quite improved version of the manuscript.

## References

Anderson, T.W. (2003). *An Introduction to Multivariate Statistical Analysis*, 3rd edn. Wiley, New Jersey

Box, G.E.P. (1949). A General Distribution Theory for a Class of Likelihood Criteria. *Biometrika* **36**, 317–346

Cardeño, L., Nagar, D.K. (2001). Testing block sphericity of a covariance matrix. *Divulg. Math.* **9**, 25–34

Coelho, C.A. (1998). The Generalized Integer Gamma distribution – a basis for distributions in Multivariate Stastistics. *J. Multivariate Anal.* **64**, 86–102

Coelho, C.A. (2004). The Generalized Near-Integer Gamma distribution: a basis for 'near-exact' approximations to the distributions of statistics which are the product of an odd number of independent Beta random variables. *J. Multivariate Anal.* **89**, 191–218

Coelho, C.A., Arnold, B.C., Marques, F.J. (2010). Near-exact distributions for certain likelihood ratio test statistics. *J. Stat. Theory Pract.* **4**, 711–725

Coelho, C.A., Marques, F.J. (2012). Near-exact distributions for the likelihood ratio test statistic to test equality of several variance-covariance matrices in elliptically contoured distributions. *Comput. Statist.* **27**, 627–659

Coelho, C.A., Marques, F.J. (2013). The multi-sample block-scalar sphericity test — exact and near-exact distributions for its likelihood ratio test statistic. *Comm. Statist. Theory Methods* **42**, 1153–1175

Coelho, C.A., Roy, A. (2017). Testing the hypothesis of a block compound symmetric covariance matrix for elliptically contoured distributions. *TEST* **26**, 308–330

Johnson, R.A., Wichern, D.W. (2007). *Applied Multivariate Statistical Analysis*, 6th edn., Prentice Hall, New Jersey

Leiva, R., Roy, A. (2011). Linear discrimination for multi-level multivariate data with separable means and jointly equicorrelated covariance structure. *J. Statist. Plann. Inference* **141**, 1910–1924

Leiva, R., Roy, A. (2012). Linear discrimination for three-level multivariate data with separable additive mean vector and doubly exchangeable covariance structure. *Comput. Statist. Data Anal.* **56**, 1644–1661

Marques, F.J., Coelho, C.A. (2012). Near-exact distributions for the likelihood ratio test statistic of the multi-sample block-matrix sphericity test. *Appl. Math. Comput. 219*, 2861–2874

Marques, F.J., Coelho, C.A., Arnold, B.C. (2011). A general near-exact distribution theory for the most common likelihood ratio test statistics used in Multivariate Analysis. *TEST* **20**, 180–203

Moschopoulos, P.-G. (1992). The Hypothesis of multisample block sphericity. *Sankhyā* **54**, 260–270

Muirhead, R.J.(2005). *Aspects of Multivariate Statistical Theory*, Wiley, New Jersey.

Perlman, M.D. (1987). Group symmetry covariance models. *Statist. Sci.* **2**, 421–425

Roy, A., Fonseca, M. (2012). Linear models with doubly exchangeable distributed errors. *Comm. Statist. Theory Methods* **41**, 2545–2569

Roy, A., Leiva, R. (2007). Discrimination with Jointly Equicorrelated Multi-level Multivariate Data. *Adv. Data Anal. Classif.* **1**, 175–199

Roy, A., Leiva, R. (2008). Likelihood Ratio Tests for Triply Multivariate Data with Structured Correlation on Spatial Repeated Measurements. *Statist. Probab. Lett.* **78**, 1971–1980

Roy, A., Leiva, R. (2011). Estimating and Testing a Structured Covariance Matrix for Three-Level Multivariate Data. *Comm. Statist. Theory Methods* **40**, 1945–1963

Starnes, J.W., Neidre, D.B., Nyman, J.S., Roy, A., Nelson, M.J., Gutierrez, G., Wang, X. (2013). Synergistic Effect of Exercise and Statins on Femoral Strength in Rats. *Exp. Gerontol.* **48**, 751–755

Szatrowski, T.H. (1976). Estimation and testing for block compound symmetry and other patterned covariance matrices with linear and nonlinear structure. Tech Rep No. 107, Stanford University.

Szatrowski, T.H. (1982). Testing and estimation in the block compound symmetry problem. J. Edu. Statist. **7**, 3–18

Tricomi, F.G., Erdélyi, A. (1951). The asymptotic expansion of a ratio of Gamma functions. *Pacific. J. Math.* **1**, 133–142

Votaw, D. (1948). Testing Compound Symmetry in a Normal Multivariate Distribution. *Ann. Math. Statist.* **19**, 447–473

Wilks, S.S. (1946). Sample criteria for testing equality of means, equality of variances, and equality of covariances in a Normal multivariate distribution. *Ann. Math. Statist.* **17**, 257–281

# On the Distribution of Linear Combinations of Chi-Square Random Variables

**Carlos A. Coelho**

**Abstract** The distribution of linear combinations of independent chi-square random variables is intimately related with the distribution of quadratic forms in normal random variables. As such, this distribution has been studied by many authors. However, there is still some room left for improvement, since while some simpler approximations do not yield sufficiently good results, other approximations which show a better performance are sometimes too complicated to be implemented in practical terms. In this paper the exact distribution of linear combinations of independent chi-square random variables is obtained, for some particular cases, in closed finite highly manageable forms, while for more general cases near-exact approximations are obtained, which are able to yield very manageable and well-performing approximations. Numerical studies compare the performance of these near-exact distributions with other existing approximations and distributions and show how sharp are the approximations provided by these near-exact distributions. A useful subproduct that is obtained is closed form expressions for the distribution of quadratic forms and for some instances of ratios of quadratic forms, useful in ANOVA and other linear or mixed-linear models where heterocedasticity is present or assumed. Solutions for the problem of the distribution of the statistic associated with the Behrens-Fisher problem are then in turn obtained as a much useful subproduct of the distribution of ratios of quadratic forms. Modules programmed in Mathe-

Reprinted with permission from Springer Nature. First published as Coelho, C. A. (2020). On the Distribution of Linear Combinations of Chi-Square Random Variables, in 'Computational and Methodological Statistics and Biostatistics', Bekker, A., Chen, D.-G., Ferreira, J.T., (eds.), Springer Series in Emerging Topics in Statistics and Biostatistics, pp. 211–252. (https://doi.org/10.1007/978-3-030-42196-0_9)

C. A. Coelho (✉)
Mathematics Department (DM) and Centro de Matemática e Aplicações (CMA), Faculdade de Ciências e Tecnologia, NOVA University of Lisbon, Caparica, Portugal
e-mail: cmac@fct.unl.pt



B. C. Arnold et al. (eds.), *Likelihood Ratio Tests and Related Topics in Multivariate Analysis*, Contributions to Statistics,
https://doi.org/10.1007/978-3-031-68008-3_14

matica®, Maxima and R for the implementation of the distributions developed are made available at the site https://sites.google.com/site/lincombchisquares.

## 1 Introduction

The distribution of linear combinations of independent chi-square r.v.'s (random variables) has been addressed by a large number of authors and it has been obtained under a number of different forms.

This distribution is of interest not only given its relation with the distribution of quadratic forms in normal r.v.'s but also because it appears associated with the distribution of a number of statistics. Indeed in many cases it appears associated with the distribution of quadratic forms in normal r.v.'s, as it is for example the case of the works by Robbins (1948), Robbins and Pitman (1949), Box (1954), Ruben (1960, Sect. 7), Imhof (1961), Ruben (1962), Baldesari (1967), Kotz et al. (1967a,b), Johnson and Kotz (1968), Shah (1970), Solomon and Stephens (1977), Bock (1984), Bock and Solomon (1987), Morin-Wahhab (1988), Mathai and Provost (1992), Baksalary et al. (1994), Provost and Rudiuk (1996), and Lu (2006), to mention only a few, namely some of the former and some recent authors. In what concerns its relation with the distribution of statistics of interest, Chernoff and Lehmann (1954) show that the distribution of the common chi-square goodness-of-fit statistic, when maximum likelihood estimates based on the original data are used, is better approximated by the distribution of a linear combination of independent chi-square r.v.'s, all of them but one with one degree-of-freedom, and Durbin et al. (1975) show that the distribution of the Cramer-von Mises goodness-of-fit statistic may be written as that of an infinite linear combination of independent chi-square r.v.'s also all with one degree-of-freedom (d.f.). Ruben (1960, Sect. 7) refers cases where statistics of interest may have the distribution of linear combinations of independent chi-square r.v.'s with an even number of d.f.'s, and Alvo et al. (1982) show that the asymptotic distribution of the Kendall tau is the same as the distribution of a linear combination of two independent chi-square r.v.'s.

Robbins (1948) obtains the distribution of a linear combination of independent chi-square r.v.'s, all with one d.f., under the form of an infinite power series with a rate parameter that is equal to the geometric mean of the coefficients associated with the chi-square r.v.'s. Embedded in this series are some multinomial coefficients, which as the order of the terms increase get more and more complicated to be computed, but the author develops a recurrence relation which allows for a simpler computation. However, a power series will always need many more terms to converge than a series involving exponentials, as the ones that involve chi-square or Gamma densities. As such, the author also obtains the same distribution in the form of an infinite mixture of chi-square distributions. But one other problem that may arise is when some of the coefficients in the linear combination are equal, leading to chi-square r.v.'s with more than just one d.f., in which situation the author calls for the use of limits on the expressions used to define the weights in the mixture distribution. However, this may be a not completely straightforward and easy job, mainly when several of the coefficients in the linear combination are equal to other coefficients, leading to the existence of several chi-square r.v.'s with more than one d.f.. Robbins and Pitman (1949) obtain a somewhat similar representation where the geometric mean of the coefficients is now replaced just by their minimum and where the weights in the mixture are to be obtained through the solution of a product identity, which solution may be not so simple for terms of larger order in the mixture. Ruben (1962) suggests the use of a similar mixture of Gamma distributions but with an arbitrary scale parameter, without giving explicit formulation for the weights.

Box (1954, Sect. 2, Thm. 2.4) obtains an interesting expression for the distribution of the linear combination of independent chi-square r.v.'s for the case when all these have an even number of d.f.'s. This distribution has close relations with the GIG (Generalized Integer Gamma) distribution (Coelho, 1998) proposed in Sect. 2 for the same situation, but opposite to the GIG distribution, which has all the involved coefficients specified by their algebraic expressions, Box leaves some of the coefficients in the distribution to be obtained from derivatives of a function which involves the coefficients in the linear combination and the d.f.'s of the chi-square r.v.'s. It happens that for larger values of the d.f.'s these derivatives may be not completely so easy to obtain, and a general expression for them is not easy to spot, although they are indeed intimately related with the $c_{j,k}$ coefficients in the probability density function (p.d.f.) and in the cumulative distribution function (c.d.f.) of the GIG distribution (see e.g. Appendix 1).

Imhof (1961) refers in Sect. 2 of his paper this same case and also presents a finite closed form solution but he remarks not being able to establish a complete identity between his result and the one presented by Box (1954), although there is a clear similitude between the two solutions. While Box used the partial fraction technique, Imhof bases his result on the inversion of the characteristic function (c.f.). However, once again, the weights in the final mixture are to be obtained from successive derivatives of a function that involves the coefficients in the linear combination and the d.f.'s of the chi-square r.v.'s and which present the same problem as the ones in the mixture obtained by Box (1954), since, as remarked by the author "with increasing $v_k$'s [the number of d.f.'s], the labour of computing the corresponding derivatives ... rapidly becomes considerable. The numerical method of the next section is then often easier to apply". Indeed then in Sect. 3 Imhof resorts to the numerical inversion of the c.f., based on Gil-Pelaez (1951) inversion formula. This is indeed a framework which we do not intend to engage into the present paper but which was pursued further by several authors, among which we may point out Davies (1980) and Lu (2006) who also use Gil-Pelaez (1951) inversion formula.

Kotz et al. (1967a), address, as other authors do, the case where all chi-square r.v.'s in the linear combination have just one only d.f., this because of the relation of this distribution with that of quadratic forms, and obtain in Sect. 3 expansions of the distribution of linear combinations of independent chi-square r.v.'s in terms of power series and series based on Laguerre polynomials which end up being infinite mixtures of chi-square distributions. These latter ones are indeed quite close to the series representation later obtained by Moschopoulos (1985) for the distribution of a linear combination of Gamma r.v.'s. Actually we will use Moschopoulos (1985) mixture distribution as a benchmark for comparison with the near-exact approximations developed in Sect. 4 of our paper, given its accuracy, the good convergence properties of the series and also its ease of implementation, given the fact that the weights in the mixture and all other coefficients involved are explicitly given by algebraic expressions. While Kotz et al. (1967a) address the central case, Kotz et al. (1967b) address the non-central case, being the case that we will restrict our attention in the present paper to the case of linear combinations of central chi-square r.v.'s, which will allow for the obtention of simpler expressions, with closed forms for some particular cases.

Shah (1970) centers his attention on the distribution of matrix quadratic forms and obtains their distributions in the form of infinite series of zonal polynomials. Solomon and Stephens (1977) approximate the distribution of quadratic forms and that of a linear combination of independent chi-squares by a Pearson curve that matches the first four moments and by the distribution of a single chi-square multiplied by a constant, matching the first three exact moments of the quadratic form or linear combination of chi-squares, but these may indeed give only quite rough approximations. Bock (1984) expresses the distribution of quadratic forms in terms of the distributions of linear combinations of other quadratic forms and linear combinations of independent chi-square r.v.'s, but in Sect. 3 she deals with the distribution of a linear combination of only two independent chi-square r.v.'s, for which she proposes an expression involving the Kummer confluent hypergeometric function, and finds a finite sum representation when at least one of the two chi-square r.v.'s has an even number of d.f.'s, but this representation involves Dawson's integral. In Sect. 4 she addresses the distribution of a linear combination of chi-squares where all but one of the chi-squares have just 1 d.f..

Bock and Solomon (1987) obtain the distribution of a linear combination of only two independent chi-square r.v.'s in several forms, most of them involving modified Bessel functions and the distribution function of a linear combination of two independent chi-square r.v.'s, both with 1 d.f., which has then to be dealt with. Mathai and Provost (1992) provide several forms for the distribution of quadratic forms and as such also for the distribution of linear combinations of independent chi-square r.v.'s, in Chap. 4 of their book, in terms of power series and Laguerre polynomial series expansions, as well as in terms of chi-square distributions. For this latter case, although the authors refer that their result is similar to the one obtained by Ruben (1962), all coefficients in the expansion and the weights in the final mixture representation are given explicit expressions. It happens that since a chi-square r.v. multiplied by any constant is indeed a Gamma r.v., there is an intimate relation between the result obtained by Mathai and Provost and the result obtained by Moschopoulos (1985) for the linear combination of Gamma r.v.'s, which given our approach followed in Sect. 4, gives us, as already remarked, a more adequate and express benchmark for our work.

Provost and Rudiuk (1996) address the general case of the distribution of the difference of two linear combinations of independent chi-square r.v.'s, and express it in terms of a series of Whittaker functions. This distribution may indeed appear as a particular case of the distribution of linear combinations of independent chi-square r.v.'s when the coefficients in the linear combination have different signs, and it is addressed in Sect. 3 of the present paper, where a finite closed form is given for this distribution for the case where all chi-square r.v.'s involved have an even number of d.f.'s.

In fact, besides the literature that addresses the distribution of linear combinations of independent chi-square r.v.'s in the domain of the distribution of quadratic forms, there is also a host of other authors who along the years have addressed the distribution of linear combinations of independent chi-square r.v.'s just by itself, as it is the case for example of Fleiss (1971) who obtained the distribution of a linear combination of chi-square r.v.'s in the form of a number of nested integrals. However, this form is not too adequate for implementation, since all the integrals, in equal number to the chi-square r.v.'s in the linear combination, have to be numerically solved. Davis (1977) uses a differential equation approach that gives a power series distribution which matches that of Robbins (1948) when all coefficients in the linear combination are positive. Moschopoulos and Canada (1984) obtained an infinite Gamma series, with intimate relations with the more embracing result in Moschopoulos (1985) on the distribution of a linear combination of independent Gamma r.v.'s. To address the non-central case, Castaño-Martínez and López-Blázquez (2005) use series expansions involving Laguerre polynomials, while more recently Ha and Provost (2013) use a linear combination of Laguerre polynomials together with a moment matching technique.

Gabler and Wolff (1987) approximate the distribution of a linear combination of chi-squares by a combination of a distribution function of a unique scaled chi-square and that of a finite mixture of scaled Gamma distributions. Other simple approximations for the distribution of a linear combination of independent chi-square r.v.'s by a single chi-square distribution, eventually with a non-integer number of d.f.'s, that is, actually by a Gamma distribution, are the ones by Satterthwaite (1941) and Welch (1947).

Other references to works on the distribution of linear combinations of independent chi-square r.v.'s, previous to 1994, may be found in Sect. 18.8 of Johnson et al. (1994).

Since indeed all linear combinations of independent chi-square r.v.'s are either sums or linear combinations of independent Gamma r.v.'s, also all the vast literature on the distribution of linear combinations or sums of independent Gamma r.v.'s is of interest. To mention only a few of those we feel as more important references, Mathai (1982), who obtains this distribution as a finite sum but involving zonal polynomials or hypergeometric functions, while Moschopoulos (1985), as already mentioned, obtains it as an infinite mixture of Gamma distributions with rate parameters equal to the sum of the original rate parameters plus the index of the mixture, running from zero to infinity, and with a scale parameter that is the minimum of the scale parameters of the original Gamma r.v.'s in the linear combination. Also, as already mentioned, due to its good convergence properties (that is, the need for much lesser terms to obtain a good approximation to the exact distribution, than other series representations) and ease of implementation, since all parameters, coefficients and weights in the mixture are fully specified, this distribution will be used as a benchmark in Sect. 4 to be compared with the performance of the near-exact approximations there developed. Provost (1988) obtains the distribution of a linear combination of independent Gamma r.v.'s in terms of integrals and Whittaker functions, while the same author in a 1989 paper (Provost, 1989) obtains the distribution of the sum of independent Gamma r.v.'s as an infinite series, with some relations with the series obtained by Moschopoulos (1985), but with the common scale parameter of the Gamma distributions in the mixture as the geometric or harmonic means of the original scale parameters, instead of its minimum (see Sect. 4 of the present paper).

It is the aim of the present contribution to lend some new results and to sharpen some existing ones giving them a more workable form, which will surely be useful in applications and that may also be used to establish new results concerning the distribution of ratios of linear combinations of chi-squares and the distribution of the Behrens-Fisher statistic. Finite closed forms are provided for both the probability density and cumulative distribution functions of some cases of linear combinations of chi-square r.v.'s, being provided sharp approximations for the other cases. As a by-product we also obtain a form to build near-exact distributions for sums or linear combinations of independent Gamma r.v.'s.

Intentionally, only the central case is addressed, in order to be able to provide simpler results, including some simple finite closed form results, and to keep the contribution within an acceptable length.

In precise terms, we will address the problem of obtaining explicit manageable expressions for both the p.d.f. and the c.d.f. of the distribution of a r.v.

$$Z = \sum_{j=1}^{p} w_j Y_j \tag{1}$$

where $Y_j \sim \chi^2_{k_j}$ $(j = 1, \ldots, p)$, form a set of $p$ independent r.v.'s and where $w_j$ $(j = 1, \ldots, p)$ are real weights. We will consider a number of cases of interest where punctually some restrictions may be placed on the weights $w_j$ or on the d.f.'s $k_j$ in order to be able to obtain simpler results.

In Sect. 2 we will consider the case where all $w_j$ are positive and all or all but one of the $k_j$ are even. In Sect. 3 we will consider the case where all $k_j$ are even and the $w_j$ are real, and in Sect. 4 we will address the case where all $w_j$ are positive and the d.f.'s $k_j$ are left to be either even or odd. Then in Sect. 5 the distribution of ratios of linear combinations of chi-squares is addressed, which is intimately related with the distribution of statistics used in ANOVA and other models when heterocedasticity is present and in Sect. 6 is addressed the distribution of the statistic associated with the Behrens-Fisher problem, obtained as a by-product of the results in Sect. 5. In Sect. 7 are presented some examples of situations where the distributions developed in Sects. 2–5 arise and in Sect. 8 some conclusions are drawn.

## 2 The Case of Positive $w_j$ with at Most One Odd $k_j$

Let us first address the case where all $w_j$ are positive and all $Y_j$ have an even number of d.f.'s, and let us denote by $X \sim \Gamma(r, \lambda)$ the fact that the r.v. $X$ has a Gamma distribution with shape parameter $r$ and rate parameter $\lambda$, that is, a r.v. with p.d.f.

$$f_X(x) = \frac{\lambda^r}{\Gamma(r)} e^{-\lambda x} x^{r-1} \quad (r, \lambda > 0; x > 0) .$$

Then we have, for $j = 1, \dots, p$,

$$Y_j \sim \Gamma\left(\frac{k_j}{2}, \frac{1}{2}\right) \implies Z_j = w_j Y_j \sim \Gamma\left(\frac{k_j}{2}, \frac{1}{2w_j}\right) \tag{2}$$

where all $k_j/2$ are integers. In this case $Z$ in (1) is thus a sum of $p$ independent Gamma r.v.'s, all with integer shape parameters, and as such its distribution is a GIG distribution, with shape parameters $k_1/2, \dots, k_p/2$, rate parameters $1/(2w_1), \dots, 1/(2w_p)$ and depth $p$, this in case all $1/(2w_j)$ are different (see Appendix 1 and Coelho (1998) for details on the GIG distribution, its p.d.f. and c.d.f.).

In case some of the $w_j$'s are equal, without any loss of generality, since the order of the r.v.'s $Y_j$ in the sum in (1) is irrelevant, let us assume these are the ones indexed $1, \dots, g(\le p)$ and let us call $w$ the common value of these $g$ $w_j$'s. Then all we have to do is to first add these $g$ Gamma r.v.'s to obtain a $\Gamma\left(\sum_{j=1}^{g} \frac{k_j}{2}, \frac{1}{2w}\right)$, then yielding for $Z$ a GIG distribution of depth $p - g + 1$, with rate parameters $\sum_{j=1}^{g} \frac{k_j}{2}, \frac{k_{g+1}}{2}, \dots, \frac{k_p}{2}$ and shape parameters $\frac{1}{2w}, \frac{1}{2w_{g+1}}, \dots, \frac{1}{2w_p}$. Of course, rate parameters may also be equal by 'sectors'. Then the above reasoning should be adequately adapted.

Assuming the case where all $w_j$ are different and using the notation in Appendix 1 for the GIG p.d.f and c.d.f., the p.d.f. and c.d.f. of $Z$ are written as

$$f_Z(z) = f^{GIG}\Big(z \,\Big|\, \{k_j/2\}_{j=1:p}; \{1/(2w_j)\}_{j=1:p}; p\Big), \quad (z > 0)$$

and

$$F_Z(z) = F^{GIG}\Big(z \,\Big|\, \{k_j/2\}_{j=1:p}; \{1/(2w_j)\}_{j=1:p}; p\Big), \quad (z > 0) .$$

In case among the $p$ r.v.'s $Y_j$ only one of them has an odd number of d.f.'s, then only one of the $Z_j$ r.v.'s in (2) has a Gamma distribution with a non-integer shape parameter, and the distribution of $Z$ will be what we call a GNIG (Generalized Near-Integer Gamma) distribution (see Appendix 1 and Coelho (2004) for details on this distribution, its p.d.f. and c.d.f.). In case all $w_j$ are different, without any loss of generality, since as already stated, the ordering of the r.v.'s $Y_j$ in (1) is irrelevant, let the r.v. $Y_j$ with an odd number of d.f.'s be indexed as $Y_p$. Then $Z$ will have a GNIG distribution of depth $p$, with p.d.f. and c.d.f. which, using the notation in Appendix 1, may be respectively written as

$$f^{GNIG}\left(z \,\middle|\, \frac{k_1}{2}, \dots, \frac{k_{p-1}}{2}; \frac{k_p}{2}; \frac{1}{2w_1}, \dots, \frac{1}{2w_{p-1}}; \frac{1}{2w_p}; p\right)$$

and

$$F^{GNIG}\left(z \,\middle|\, \frac{k_1}{2}, \dots, \frac{k_{p-1}}{2}; \frac{k_p}{2}; \frac{1}{2w_1}, \dots, \frac{1}{2w_{p-1}}; \frac{1}{2w_p}; p\right).$$

In case some of the $w_j$ are equal, say $g$ of them, then the depth of the GNIG distribution will be equal to $p - g + 1$, being the case that the adjustment will be done in a similar way to that described above for the case of all even $k_j$.

The Mathematica® modules to implement the GIG and the GNIG distributions are available at https://sites.google.com/site/nearexactdistributions, with the modules for the GIG distribution also available in Sect. 2.7 of Coelho and Mexia (2010) and in Appendix 4.A of Coelho and Arnold (2019). The MAXIMA and R modules for the GIG distribution are also available at this book supplementary material site, at https://sites.google.com/site/meijerfoxfiniteforms.

## 3 The Case Where All $k_j$ Are Even But the $w_j$ May Be Both Positive and Negative

Let us suppose that all chi-square r.v.'s $Y_j$ in (1) have an even number of d.f.'s but some of the $w_j$ are positive and some are negative. Without any loss of generality, since, as already mentioned, the order of the $Y_j$ r.v.'s in (1) is irrelevant, let us consider that the $w_j$ that are positive are the ones for $j = 1, \dots, p_1 (< p)$, the others being negative. Let also $p_2 = p - p_1$. Then let us consider the r.v.'s

$$Z_1^* = \sum_{j=1}^{p_1} w_j Y_j \quad \text{and} \quad Z_2^* = \sum_{j=p_1+1}^{p} -w_j Y_j\,.$$

Since all $k_j$ $(j = 1, \dots, p)$ are even, both $Z_1^*$ and $Z_2^*$ have GIG distributions, which assuming that all $w_j$ are different will have respectively depths $p_1$ and $p_2$. Then we have $Z = Z_1^* - Z_2^*$, where $Z_1^*$ and $Z_2^*$ are two independent GIG r.v.'s. Thus, the r.v. $Z$ will have in this case what Coelho and Mexia (2010) call a DGIG (Difference of two GIG) distribution, with p.d.f. (see Appendix 2 for details on the DGIG distribution and the expressions for its p.d.f. and c.d.f.)

$$f_Z(z) = f^{DGIG}\left(z \middle| \left\{\frac{k_j}{2}\right\}_{j=1:p_1}; \left\{\frac{k_j}{2}\right\}_{j=p_1+1:p}; \left\{\frac{1}{2w_j}\right\}_{j=1:p_1}; \left\{\frac{1}{2w_j}\right\}_{j=p_1+1:p}; p_1, p_2\right),$$

and c.d.f.

$$F_Z(z) = F^{DGIG}\left(z \middle| \left\{\frac{k_j}{2}\right\}_{j=1:p_1}; \left\{\frac{k_j}{2}\right\}_{j=p_1+1:p}; \left\{\frac{1}{2w_j}\right\}_{j=1:p_1}; \left\{\frac{1}{2w_j}\right\}_{j=p_1+1:p}; p_1, p_2\right),$$

with $z \in \mathbb{R}$.

A caution note is here in place: as mentioned in Coelho and Mexia (2010), it is in this reference where the correct expressions for the p.d.f. and c.d.f. of the DGIG distribution are to be found, since the ones in Coelho and Mexia (2007) suffer from some typos in their writings.

In case some of the $w_j$ for $j = 1, \ldots, p_1$, or some of the $w_j$ for $j = p_1 + 1, \ldots, p$ are equal, then adjustments similar to the ones indicated in Sect. 2 should be made.

Mathematica® modules to implement the DGIG distribution are available in Sect. 2.7 of Coelho and Mexia (2010). MAXIMA and R modules are also available from the author or at https://sites.google.com/site/lincombchisquares.

## 4 The Case Where All $w_j$ Are Positive and the $k_j$ Are Both Even and Odd or All Odd

Let us take

$$\ell_j = \left\lfloor \frac{k_j}{2} \right\rfloor \quad \text{and} \quad \ell_j^* = \frac{k_j}{2} - \ell_j \quad (j = 1, \ldots, p) \tag{3}$$

respectively as the integer and non-integer parts of $k_j/2$ $(j = 1, \ldots, p)$, and let

$$Z_j^* \sim \Gamma\left(\ell_j, \frac{1}{2w_j}\right) \quad \text{and} \quad Z_j^{**} \sim \Gamma\left(\ell_j^*, \frac{1}{2w_j}\right), \quad j = 1, \ldots, p$$

be, for a given $j$, two independent r.v.'s, where $Z_j^{**}$ vanishes if $\ell_j^* = 0$ and $Z_j^*$ vanishes if $k_j = 1$. Then we have

$$Z_j = Z_j^* + Z_j^{**}.$$

Furthermore, let

$$Z^* = \sum_{j=1}^{p} Z_j^* \quad \text{and} \quad Z^{**} = \sum_{j=1}^{p} Z_j^{**} . \tag{4}$$

Then we have

$$Z = Z^* + Z^{**} ,$$

where $Z^*$ and $Z^{**}$ are two independent r.v.'s, and where, in case all $w_j$ $(j = 1, \ldots, p)$ are different, $Z^*$ has a GIG distribution of depth $p$, with shape parameters $\ell_j$ and rate parameters $1/(2w_j)$, or a smaller depth, in case some of the $w_j$ are equal, in which case a similar approach to the one described in Sect. 2 has to be taken.

In order to obtain a near-exact distribution for $Z$ we will leave $Z^*$ untouched and we will approximate the distribution of $Z^{**}$ by a finite mixture of say $m^* + 1$ Gamma distributions, all with the same rate parameter $\lambda$, which may be obtained in a number of different ways, which will be discussed next. The shape parameters in this mixture will be equal to the sum of the shape parameters of the $p$ r.v.'s $Z_j^{**}$, to which we add $0, 1, \ldots, m^*$.

We will be able to explain better and more precisely which will be the shape parameters of the distributions in this mixture and how the weights in this mixture will be determined if we consider the characteristic function (c.f.) of $Z^{**}$, which is

$$\Phi_{Z^{**}}(t) = \prod_{j=1}^{p} \left(\frac{1}{2w_j}\right)^{\ell_j^*} \left(\frac{1}{2w_j} - \mathrm{i}t\right)^{-\ell_j^*} , \tag{5}$$

where it is clear that if some $\ell_j^*$ is zero, then the corresponding r.v. vanishes, being its contribution for the product in (5) equal to 1.

Since the c.f. in (5) is the c.f. of a sum of independent $\Gamma(\ell_j^*, 1/(2w_j))$ r.v.'s and since from the results in Moschopoulos (1985) it is possible to show that this distribution is also the distribution of an infinite mixture of Gamma distributions with shape parameters $r + k$ $(k = 0, 1, \ldots)$, where

$$r = \sum_{j=1}^{p} \ell_j^* , \tag{6}$$

we will approximate the c.f. in (5) by the c.f.

$$\Phi^*(t) = \sum_{k=0}^{m^*} \pi_k \lambda^{r+k} (\lambda - \mathrm{i}t)^{-(r+k)} , \tag{7}$$

which is the c.f. of a mixture with $m^* + 1$ components which are $\Gamma(r + k, \lambda)$ distributions, for $k = 0, 1, \ldots, m^*$, where $r$ is given by (6) and where, as already mentioned, $\lambda$ may be determined in a few different ways, which will be addressed and discussed next. The weights $\pi_k$ $(k = 0, \dots, m^*)$ in (6) will be determined in such a way that the first $m^*$ non-central moments of the distributions corresponding to $\Phi_{Z^{**}}(t)$ and $\Phi^*(t)$ are the same, that is, $\pi_k$ for $k = 0, \dots, m^* - 1$ will be determined as the solutions of the system of $m^*$ equations

$$\left.\frac{\partial^h}{\partial t^h}\Phi_{Z^{**}}(t)\right|_{t=0} = \left.\frac{\partial^h}{\partial t^h}\Phi^*(t)\right|_{t=0}, \quad h = 1, \dots, m^*, \tag{8}$$

taking then $\pi_{m^*} = 1 - \sum_{k=0}^{m^*-1} \pi_k$.

This approach will give as near-exact c.f. for $Z$ the c.f.

$$\begin{aligned}\Phi_Z^*(t) &= \Phi_{Z^*}(t)\,\Phi^*(t)\\ &= \underbrace{\left\{\prod_{j=1}^{p}\left(\frac{1}{2w_j}\right)^{\ell_j}\left(\frac{1}{2w_j} - \mathrm{i}t\right)^{-\ell_j}\right\}}_{\Phi_{Z^*}(t)} \underbrace{\sum_{k=0}^{m^*} \pi_k \lambda^{r+k} (\lambda - \mathrm{i}t)^{-(r+k)}}_{\Phi^*(t)}\\ &= \sum_{k=0}^{m^*} \pi_k \left\{\prod_{j=1}^{p}\left(\frac{1}{2w_j}\right)^{\ell_j}\left(\frac{1}{2w_j} - \mathrm{i}t\right)^{-\ell_j}\right\} \lambda^{r+k} (\lambda - \mathrm{i}t)^{-(r+k)}\end{aligned} \tag{9}$$

which yields as near-exact distributions for $Z$ mixtures with $m^* + 1$ components, each of which is either a GIG or a GNIG distribution of depth $p + 1$, according to the case of $r$ being integer or not (we may note that it may happen quite often $r$ to be an integer since $\ell_j^*$ is either zero, if $k_j$ is even, or 1/2, if $k_j$ is odd), with shape parameters $\ell_1, \dots, \ell_p, r + k$ $(k = 0, \dots, m^*)$ and rate parameters $1/(2w_1), \dots, 1/(2w_p), \lambda$, and weights $\pi_k$ $(k = 0, \dots, m^* + 1)$. That is, from (9), the near-exact p.d.f.'s and c.d.f.'s of $Z$ will be respectively written as

$$f_Z^*(z) = \sum_{k=0}^{m^*} \pi_k\, f^{\,GNIG}\left(z \,\middle|\, \ell_1, \dots, \ell_p; r + k; \frac{1}{2w_1}, \dots, \frac{1}{2w_p}; \lambda; p + 1\right)$$

and

$$F_Z^*(z) = \sum_{k=0}^{m^*} \pi_k\, F^{\,GNIG}\left(z \,\middle|\, \ell_1, \dots, \ell_p; r + k; \frac{1}{2w_1}, \dots, \frac{1}{2w_p}; \lambda; p + 1\right),$$

using the notation in Appendix 1 for the GNIG p.d.f. and c.d.f., and where the GNIG p.d.f. and c.d.f. are to be replaced by the GIG p.d.f. and c.d.f. in case $r$ in (6) is an integer.

In implementing this near-exact approximation, the c.f. of $Z^*$, $\Phi_{Z^*}(t)$, which will be kept unchanged, is a 'major' part of the whole c.f. of $Z$, $\Phi_Z(t)$, in case all

$k_j$ are larger than 1, while the c.f. of $Z^{**}$, $\Phi_{Z^{**}}(t)$, is a 'smaller' part, since in this case we have

$$\int_{-\infty}^{+\infty} |\Phi_Z(t) - \Phi_{Z^*}(t)|\, dt < \int_{-\infty}^{+\infty} |\Phi_Z(t) - \Phi_{Z^{**}}(t)|\, dt\,.$$

The distribution of $Z^{**}$ is that of a sum of independent Gamma r.v.'s with possibly different rate parameters and non-integer, that is, half-integer, shape parameters. Such a distribution was obtained by several authors as an infinite mixture of Gamma distributions, as it is the case of Moschopoulos (1985), who obtained it in the form of an infinite mixture of Gamma distributions all with the same rate parameter, which is the maximum of the rate parameters of the Gamma distributions involved in the sum, and shape parameters $r + k$ ($k = 0, 1, \ldots$) for $r$ in (6).

As such, the approach followed in approximating $\Phi_{Z^{**}}(t)$ by $\Phi^*(t)$ follows similar lines, to which was added what may be seen as a somewhat heuristic approach, by turning the infinite mixture into a finite mixture, which is then rendered extremely close to the exact distribution through the way its weights are determined.

Concerning the determination of the rate parameter $\lambda$ in (7), several strategies may be taken, among which we will address the following ones:

(i) $\max_{j\in\{1,\ldots,p\}} \frac{1}{2w_j}$

(ii) $\min_{j\in\{1,\ldots,p\}} \frac{1}{2w_j}$

(iii) arithmetic mean of the $\frac{1}{2w_j}$ $(j = 1, \ldots, p)$

(iv) harmonic mean of the $\frac{1}{2w_j}$ $(j = 1, \ldots, p)$

(v) geometric mean of the $\frac{1}{2w_j}$ $(j = 1, \ldots, p)$

(vi) the rate parameter $\lambda$ in $\widetilde{\Phi}(t) = \lambda^r(\lambda - \mathrm{i}t)^{-r}$, where

$$\left.\frac{\partial^h}{\partial t^h}\Phi_{Z^{**}}(t)\right|_{t=0} = \left.\frac{\partial^h}{\partial t^h}\widetilde{\Phi}(t)\right|_{t=0}, \quad h = 1, 2 \tag{10}$$

(vii) the rate parameter $\lambda$ in $\widetilde{\Phi}(t) = \theta\lambda^{r_1}(\lambda - \mathrm{i}t)^{-r_1} + (1-\theta)\lambda^{r_2}(\lambda - \mathrm{i}t)^{-r_2}$, where

$$\left.\frac{\partial^h}{\partial t^h}\Phi_{Z^{**}}(t)\right|_{t=0} = \left.\frac{\partial^h}{\partial t^h}\widetilde{\Phi}(t)\right|_{t=0}, \quad h = 1, \ldots, 4\,.$$

These choices for $\lambda$ will be denoted respectively by (i) 'Max', (ii) 'Min', (iii) 'A', (iv) 'H', (v) 'G', (vi) '2m' and (vii) '4m' in the Tables that appear in Sect. 4.1.

The choice in (i) is based on the paper by Moschopoulos (1985), where we should note that the author uses in the notation of the p.d.f. of the Gamma distribution a scale parameter instead of a rate parameter, so that his $\beta_j$ are the reciprocals of our rate parameters $1/(2w_j)$. The choice in (ii) is made to show that in some cases this choice works better than the choice in (i). The choice in (iii), which as we will see shortly ahead works quite often better than the two previous ones, is inspired in the well-known fact that if all $w_j$ are equal, say to a common value $w$, then the distribution of $Z^{**}$ (and actually the distribution of $Z^*$ and of $Z$) will be just a Gamma distribution with a rate parameter equal to $1/(2w)$. So, we may think about taking the arithmetic average of the rate parameters $1/(2w_j)$ in case they are different. However, the choice in (iv) may be a better one, given the fact that the c.f. of $Z^{**}$, as shown in (5) is indeed a product, and it is in this product that we are kind of 'replacing' the set of $p$ rate parameters $1/(2w_j)$ by just a single parameter, making then most sense to take the harmonic mean of the original parameters, instead of the arithmetic mean. The choice in (v) is made to see how the choice of the geometric mean compares with the harmonic mean and also say to 'complete' the set of the most common means. The choice in (vi) is a simplification of the choice in (vii), which is known to the author to provide generally very good results.

As we will see, options (iv), (v), (vi) and (vii) give the best results, in many situations just with slight differences among them. In these cases, options (iv) or (v) will be the preferred ones given their computational simplicity, compared with options (vi) and (vii). However, also option (i) may give very good results in some particular situations.

We may note that in case all $k_j$ are equal to 1, then all $\ell_j$ will be equal to zero and in this case $Z^*$ vanishes and the procedure developed will then provide a quite sharp asymptotic distribution for $Z$, rather than a near-exact one.

Modules programmed in Mathematica® to implement these near-exact distributions are available at https://sites.google.com/site/lincombchisquares.

## 4.1 On the Behavior of the Near-Exact Distributions for the Several Different Choices of λ in (10)

To assess the quality of the approximation provided by different near-exact distributions we use the measure

$$\Delta = \frac{1}{2\pi} \int_{-\infty}^{+\infty} \left| \frac{\Phi_Z(t) - \Phi_Z^*(t)}{t} \right| dt \tag{11}$$

where

$$\Phi_Z(t) = \prod_{j=1}^{p} \left(\frac{1}{2w_j}\right)^{k_j/2} \left(\frac{1}{2w_j} - \mathrm{i}t\right)^{-k_j/2} = \Phi_{Z^*}(t)\,\Phi_{Z^{**}}(t)$$

is the exact c.f. of $Z$ and $\Phi_Z^*(t)$ is the near-exact c.f. of $Z$ in (9).

The measure $\Delta$, as defined in (11), is a sharp upper-bound on the difference between the exact and the near-exact c.d.f.'s, since we may write

$$\Delta \geq \max_{z>0} \left| F_Z(z) - F_Z^*(z) \right| , \tag{12}$$

where $F_Z(z)$ and $F_Z^*(z)$ represent respectively the exact and the near-exact c.d.f.'s of $Z$. Actually, both in (11) as well as in (12) one may use respectively $\Phi_Z^*(t)$ and $F_Z^*(z)$ as the c.f. and the c.d.f. for any approximate distribution of $Z$ as for example an asymptotic or a truncated distribution. This is exactly what we will do next when comparing the performance of the near-exact distributions with truncations of the infinite mixture obtained by Moschopoulos (1985).

To illustrate the behavior of the seven possible different choices for $\lambda$ suggested in (10) we have chosen to use ten different scenarios, the first eight of which all deal with a set of $p = 5$ independent chi-square r.v.'s $Y_j$. In the first four scenarios the chi-square r.v's have $k_j = \{2, 7, 16, 17, 19\}$ d.f.'s, and then we study four different choices for the coefficients $w_j$ $(j = 1, \ldots, 5)$. These four scenarios illustrate the following situations:

I scenario with values of $w_j = \{2.3, 3.4, 5.6, 7.8, 8.9\}$, that is, *larger* than 1 and which *increase* with increasing values for the d.f.'s

II scenario with values of $w_j = \{0.11, 0.13, 0.18, 0.29, 0.43\}$ *smaller* than 1 and which *increase* with increasing values for the d.f.'s

III scenario with values of $w_j = \{8.9, 7.8, 5.6, 3.4, 2.3\}$ *larger* than 1 and which *decrease* with increasing values for the d.f.'s

IV scenario with values of $w_j = \{0.43, 0.29, 0.18, 0.13, 0.11\}$ *smaller* than 1 and which *decrease* with increasing values for the d.f.'s.

For scenarios V–VIII we use respectively exactly the same $w_j$ that are used for scenarios I–IV, with the difference that now the $p = 5$ independent chi-square r.v.'s $Y_j$ have $k_j = \{12, 17, 26, 27, 29\}$ d.f.'s, that is, each one of them has ten more d.f.'s than in scenarios I–IV.

Then for scenarios IX and X we use four chi-square r.v.'s, respectively with $k_j = \{3, 1, 2, 2\}$ and $k_j = \{9, 3, 6, 6\}$ d.f.'s, in both cases with $w_j = \{2/5, 1/2, 1, 9\}$, in order to not only consider situations where the $w_j$ take a wide range of values but also where they take values both smaller and larger than 1.

For each of the ten above scenarios the measure $\Delta$ in (11) was computed for the Moschopoulos (1985) distribution with truncations at 100, 150 and 200 terms and for the near-exact distributions with the seven different choices for $\lambda$ in (10) (denoted by the descriptors listed right after (10)), for values of $m^* = 10$, $m^* = 15$ and $m^* = 20$, that is, for near-exact distributions that match $m^* = 10, 15$ or 20 of the exact moments of $Z$. The values for the measure $\Delta$ in (11) are shown in Tables 1, 2 and 3 and we will analyze the results for each scenario individually.

Our intent is to compare the near-exact distributions that match 10, 15 and 20 exact moments respectively with the Moschopoulos (1985) series distribution truncated to 100, 150 and 200 terms.

**Table 1** Values of the measure $\Delta$ in (11) for scenarios I–IV, for truncations of the Moschopoulos (1985) series distribution with 100, 150 and 200 terms (respectively the columns marked with $m^* = 10$, $m^* = 15$ and $m^* = 20$), denoted by 'Mosc', and for the near-exact distributions with $\lambda$ defined by the seven choices in (10), denoted by the descriptors indicated right after (10) in Sect. 4, for linear combinations of $p = 5$ independent chi-square r.v.'s with $k_j = \{2, 7, 16, 17, 19\}$ d.f.'s and with coefficients $w_j = \{2.3, 3.4, 5.6, 7.8, 8.9\}$ for Scenario I, $w_j = \{0.11, 0.13, 0.18, 0.29, 0.43\}$ for Scenario II, $w_j = \{8.9, 7.8, 5.6, 3.4, 2.3\}$ for Scenario III and $w_j = \{0.43, 0.29, 0.18, 0.13, 0.11\}$ for Scenario IV

| | Scenario I | | | Scenario II | | |
|---|---|---|---|---|---|---|
| | $m^* = 10$ | $m^* = 15$ | $m^* = 20$ | $m^* = 10$ | $m^* = 15$ | $m^* = 20$ |
| Mosc | $2.61\times10^{-1}$ | $4.43\times10^{-5}$ | $1.02\times10^{-9}$ | $1.19\times10^{-2}$ | $3.71\times10^{-7}$ | $2.81\times10^{-12}$ |
| Max | $2.04\times10^{-6}$ | $4.39\times10^{-7}$ | $3.58\times10^{-7}$ | $5.52\times10^{-6}$ | $2.60\times10^{-6}$ | $4.35\times10^{-6}$ |
| Min | $7.67\times10^{-9}$ | $4.65\times10^{-11}$ | $4.77\times10^{-13}$ | $1.05\times10^{-7}$ | $1.60\times10^{-9}$ | $3.93\times10^{-11}$ |
| A | $1.04\times10^{-8}$ | $1.34\times10^{-10}$ | $4.30\times10^{-12}$ | $1.62\times10^{-7}$ | $1.05\times10^{-8}$ | $1.71\times10^{-9}$ |
| H | $1.25\times10^{-10}$ | $1.92\times10^{-13}$ | $6.31\times10^{-16}$ | $6.60\times10^{-9}$ | $8.11\times10^{-11}$ | $2.16\times10^{-12}$ |
| G | $1.36\times10^{-9}$ | $6.45\times10^{-12}$ | $7.06\times10^{-14}$ | $4.22\times10^{-8}$ | $1.34\times10^{-9}$ | $9.94\times10^{-11}$ |
| 2m | $4.62\times10^{-10}$ | $1.11\times10^{-12}$ | $4.85\times10^{-15}$ | $3.15\times10^{-9}$ | $1.63\times10^{-11}$ | $1.48\times10^{-13}$ |
| 4m | $2.48\times10^{-11}$ | $2.37\times10^{-14}$ | $4.05\times10^{-17}$ | $1.31\times10^{-10}$ | $4.49\times10^{-13}$ | $1.28\times10^{-15}$ |

| | Scenario III | | | Scenario IV | | |
|---|---|---|---|---|---|---|
| | $m^* = 10$ | $m^* = 15$ | $m^* = 20$ | $m^* = 10$ | $m^* = 15$ | $m^* = 20$ |
| Mosc | $9.57\times10^{-7}$ | $4.96\times10^{-13}$ | $1.76\times10^{-19}$ | $8.97\times10^{-10}$ | $3.44\times10^{-16}$ | $1.32\times10^{-22}$ |
| Max | $8.33\times10^{-6}$ | $4.44\times10^{-6}$ | $6.73\times10^{-6}$ | $1.13\times10^{-6}$ | $1.80\times10^{-7}$ | $6.87\times10^{-8}$ |
| Min | $2.63\times10^{-6}$ | $8.38\times10^{-8}$ | $3.92\times10^{-9}$ | $1.84\times10^{-5}$ | $8.80\times10^{-7}$ | $5.57\times10^{-8}$ |
| A | $7.19\times10^{-9}$ | $8.63\times10^{-11}$ | $2.00\times10^{-12}$ | $1.06\times10^{-9}$ | $4.90\times10^{-12}$ | $4.13\times10^{-14}$ |
| H | $5.19\times10^{-9}$ | $2.55\times10^{-11}$ | $2.15\times10^{-13}$ | $3.34\times10^{-9}$ | $9.63\times10^{-12}$ | $4.60\times10^{-14}$ |
| G | $3.17\times10^{-10}$ | $2.81\times10^{-12}$ | $3.95\times10^{-15}$ | $1.70\times10^{-10}$ | $2.91\times10^{-13}$ | $3.88\times10^{-16}$ |
| 2m | $8.12\times10^{-9}$ | $4.58\times10^{-11}$ | $4.42\times10^{-13}$ | $8.79\times10^{-10}$ | $1.62\times10^{-12}$ | $5.07\times10^{-15}$ |
| 4m | $2.17\times10^{-9}$ | $1.58\times10^{-11}$ | $2.00\times10^{-13}$ | $1.17\times10^{-9}$ | $5.59\times10^{-12}$ | $4.93\times10^{-14}$ |

From Table 1 we may see that for Scenario I the Moschopoulos (1985) series truncated at 100 terms does have a very poor performance, with a quite large value of $\Delta$, and that the Moschopoulos series distribution truncated at 150 terms is overpowered by all the near-exact distributions that match 15 exact moments, while the truncated series with 200 terms only beats the near-exact distribution that uses $\lambda$ as the maximum of the rate parameters $1/(2w_j)$, with the near-exact distribution that uses $\lambda$ defined as in (vii) in (10) showing the best performance among all near-exact distributions, and the one that uses $\lambda$ as the harmonic mean of the $1/(2w_j)$ coming second.

For Scenario II the distributions have a somewhat similar behavior as it happens for Scenario I, with all near-exact distributions that match 10 exact moments outperforming Moschopoulos series truncated at 100 terms and all but the one

**Table 2** Values of the measure $\Delta$ in (11) for scenarios V–VIII, for truncations of the Moschopoulos (1985) series distribution with 100, 150 and 200 terms (respectively the columns marked with $m^* = 10$, $m^* = 15$ and $m^* = 20$), denoted by 'Mosc', and for the near-exact distributions with $\lambda$ defined by the seven choices in (10), denoted by the descriptors indicated right after (10) in Sect. 4, for linear combinations of $p = 5$ independent chi-square r.v.'s with $k_j = \{12, 17, 26, 27, 29\}$ d.f.'s and with coefficients $w_j = \{2.3, 3.4, 5.6, 7.8, 8.9\}$ for Scenario V, $w_j = \{0.11, 0.13, 0.18, 0.29, 0.43\}$ for Scenario VI, $w_j = \{8.9, 7.8, 5.6, 3.4, 2.3\}$ for Scenario VII and $w_j = \{0.43, 0.29, 0.18, 0.13, 0.11\}$ for Scenario VIII

| | Scenario V | | | Scenario VI | | |
|---|---|---|---|---|---|---|
| | $m^* = 10$ | $m^* = 15$ | $m^* = 20$ | $m^* = 10$ | $m^* = 15$ | $m^* = 20$ |
| Mosc | $1.39\times10^{1}$ | $1.35\times10^{-1}$ | $6.90\times10^{-5}$ | $1.93\times10^{0}$ | $1.35\times10^{-3}$ | $8.96\times10^{-8}$ |
| Max | $8.92\times10^{-8}$ | $3.42\times10^{-9}$ | $4.23\times10^{-10}$ | $2.41\times10^{-7}$ | $2.09\times10^{-8}$ | $5.87\times10^{-9}$ |
| Min | $8.56\times10^{-10}$ | $2.39\times10^{-12}$ | $1.22\times10^{-14}$ | $1.35\times10^{-8}$ | $1.03\times10^{-10}$ | $1.39\times10^{-12}$ |
| A | $6.35\times10^{-10}$ | $2.26\times10^{-12}$ | $2.08\times10^{-14}$ | $9.50\times10^{-9}$ | $1.62\times10^{-10}$ | $7.34\times10^{-12}$ |
| H | $9.35\times10^{-12}$ | $4.81\times10^{-15}$ | $5.67\times10^{-18}$ | $4.74\times10^{-10}$ | $1.89\times10^{-12}$ | $1.76\times10^{-14}$ |
| G | $9.21\times10^{-11}$ | $1.34\times10^{-13}$ | $4.72\times10^{-16}$ | $2.71\times10^{-9}$ | $2.50\times10^{-11}$ | $5.78\times10^{-13}$ |
| 2m | $4.32\times10^{-11}$ | $4.27\times10^{-14}$ | $8.33\times10^{-17}$ | $3.08\times10^{-10}$ | $6.78\times10^{-13}$ | $2.89\times10^{-15}$ |
| 4m | $2.00\times10^{-12}$ | $7.10\times10^{-16}$ | $4.87\times10^{-19}$ | $1.02\times10^{-11}$ | $1.42\times10^{-14}$ | $1.62\times10^{-17}$ |

| | Scenario VII | | | Scenario VIII | | |
|---|---|---|---|---|---|---|
| | $m^* = 10$ | $m^* = 15$ | $m^* = 20$ | $m^* = 10$ | $m^* = 15$ | $m^* = 20$ |
| Mosc | $2.72\times10^{-1}$ | $2.89\times10^{-5}$ | $3.63\times10^{-10}$ | $1.27\times10^{-3}$ | $1.02\times10^{-8}$ | $2.72\times10^{-14}$ |
| Max | $4.86\times10^{-8}$ | $2.35\times10^{-9}$ | $3.56\times10^{-10}$ | $5.26\times10^{-9}$ | $9.16\times10^{-11}$ | $4.65\times10^{-12}$ |
| Min | $1.03\times10^{-7}$ | $1.24\times10^{-9}$ | $2.55\times10^{-11}$ | $8.01\times10^{-7}$ | $1.63\times10^{-8}$ | $5.05\times10^{-10}$ |
| A | $9.20\times10^{-11}$ | $2.09\times10^{-13}$ | $1.11\times10^{-15}$ | $9.90\times10^{-12}$ | $8.62\times10^{-15}$ | $1.72\times10^{-17}$ |
| H | $8.72\times10^{-11}$ | $1.04\times10^{-13}$ | $2.60\times10^{-16}$ | $4.25\times10^{-11}$ | $2.99\times10^{-14}$ | $4.30\times10^{-17}$ |
| G | $3.40\times10^{-12}$ | $8.99\times10^{-15}$ | $3.22\times10^{-18}$ | $1.62\times10^{-12}$ | $6.35\times10^{-16}$ | $2.09\times10^{-19}$ |
| 2m | $1.43\times10^{-10}$ | $2.03\times10^{-13}$ | $5.92\times10^{-16}$ | $9.88\times10^{-12}$ | $4.15\times10^{-15}$ | $3.66\times10^{-18}$ |
| 4m | $3.06\times10^{-11}$ | $4.48\times10^{-14}$ | $1.38\times10^{-16}$ | $1.08\times10^{-11}$ | $9.73\times10^{-15}$ | $2.02\times10^{-17}$ |

with $\lambda$ defined as the maximum of the $1/(2w_j)$ showing better performances than the Moschopoulos series truncated at 150 terms. However, only the near-exact distributions that use $\lambda$ defined as in (vii), (vi) and (iv) in (10) are able to outperform the Moschopoulos series truncated at 200 terms. In all cases it is the near-exact distribution with $\lambda$ given by (vii) in (10) that has the best performance, followed by the one with $\lambda$ given by (vi) in (10) and by the one that uses $\lambda$ as the harmonic mean of the $1/(2w_j)$, with the one that uses $\lambda$ as the geometric mean of the $1/(2w_j)$ coming in fourth place for the near-exact distributions that match 10 or 15 exact moments, but being beaten by the one that uses $\lambda$ as the minimum of the $1/(2w_j)$ when we consider the near-exact distributions that match 20 exact moments.

In Scenario III while all near-exact distributions that match 10 exact moments, except the ones that use $\lambda$ as the maximum or the minimum of the $1/(2w_j)$, beat

**Table 3** Values of the measure $\Delta$ in (11) for scenarios IX–X, for truncations of the Moschopoulos (1985) series distribution with 100, 150 and 200 terms (respectively the columns marked with $m^* = 10$, $m^* = 15$ and $m^* = 20$), denoted by 'Mosc', and for the near-exact distributions with $\lambda$ defined by the seven choices in (10), denoted by the descriptors indicated right after (10) in Sect. 4, for linear combinations of $p = 4$ independent chi-square r.v.'s respectively with $k_j = \{3, 1, 2, 2\}$ and $k_j = \{9, 3, 6, 6\}$ d.f.'s and with coefficients $w_j = \{2/5, 1/2, 1, 9\}$ for both scenarios

| | Scenario IX | | | | Scenario X | | |
|---|---|---|---|---|---|---|---|
| | $m^* = 10$ | $m^* = 15$ | $m^* = 20$ | | $m^* = 10$ | $m^* = 15$ | $m^* = 20$ |
| Mosc | $3.92\times10^{-1}$ | $4.04\times10^{-2}$ | $4.16\times10^{-3}$ | Mosc | $6.58\times10^{0}$ | $1.35\times10^{0}$ | $2.31\times10^{-1}$ |
| Max | $3.76\times10^{-12}$ | $1.35\times10^{-15}$ | $6.43\times10^{-19}$ | Max | $1.94\times10^{-16}$ | $1.36\times10^{-20}$ | $2.04\times10^{-24}$ |
| Min | $4.92\times10^{-2}$ | $2.92\times10^{-2}$ | $1.85\times10^{-2}$ | Min | $4.91\times10^{-3}$ | $2.08\times10^{-3}$ | $1.02\times10^{-3}$ |
| A | $2.39\times10^{-8}$ | $1.61\times10^{-10}$ | $1.38\times10^{-12}$ | A | $9.33\times10^{-12}$ | $1.98\times10^{-14}$ | $7.30\times10^{-17}$ |
| H | $2.00\times10^{-3}$ | $5.45\times10^{-4}$ | $1.66\times10^{-4}$ | H | $2.60\times10^{-5}$ | $4.00\times10^{-6}$ | $8.01\times10^{-7}$ |
| G | $1.18\times10^{-5}$ | $6.60\times10^{-7}$ | $4.45\times10^{-8}$ | G | $1.99\times10^{-8}$ | $4.64\times10^{-10}$ | $1.63\times10^{-11}$ |
| 2m | $1.72\times10^{-15}$ | $1.95\times10^{-20}$ | $2.58\times10^{-25}$ | 2m | $1.36\times10^{-19}$ | $3.48\times10^{-25}$ | $1.56\times10^{-30}$ |
| 4m | $1.24\times10^{-15}$ | $1.58\times10^{-20}$ | $1.94\times10^{-25}$ | 4m | $9.89\times10^{-20}$ | $2.70\times10^{-25}$ | $1.11\times10^{-30}$ |

the Moschopoulos series truncated at 100 terms, which already exhibits a much good performance, none of the near-exact distributions that match 15 exact moments is able to beat the Moschopoulos series truncated at 150 terms, neither none of the near-exact distributions that match 20 exact moments is able to outperform the Moschopoulos series truncated at 200 terms, although the near-exact distributions that match 20 exact moments and use $\lambda$ given by (v), (vii), (iv) and (vi) in (10) beat the Moschopoulos series distribution truncated at 150 terms, with performances in that order, with the near-exact distribution with $\lambda$ given by the geometric mean of the rate parameters $1/(2w_j)$ displaying the best behavior.

In Scenario IV, among the near-exact distributions that match 10 exact moments, only the one that uses $\lambda$ as the geometric mean of the $1/(2w_j)$ is able to outperform the Moschopoulos series truncated at 100 terms. Indeed for this scenario the Moschopoulos series distribution displays an extraordinary good performance, with none of the near-exact distributions that match 15 or even 20 exact moments being able to beat the Moschopoulos series distribution truncated to 150 terms. Anyway, among the near-exact distributions it is the one that uses $\lambda$ as the geometric mean of the rate parameters $1/(2w_j)$, that shows the best performance, followed by the one that uses $\lambda$ given by (vi) in (10), leaving the one that uses $\lambda$ given by (vii) in (10) with the third place among those that match 10 or 15 exact moments or the fourth place among those that match 20 exact moments, where the third place is taken by the near-exact distribution that uses $\lambda$ given by the harmonic mean of the $1/(2w_j)$, which takes fourth place among the near-exact distributions that match 10 or 15 exact moments.

In the overall, for Scenarios I–IV, we may see that the Moschopoulos distribution shows very good performances for cases where the products $w_j k_j$ $(j = 1, \ldots, p)$ show a more balanced sequence of values, preferably with smaller values, that is, for Scenarios III and IV, where the $w_j$ are inversely proportional to the $k_j$, while the near-exact distributions show clear advantages in the opposite situations, that is, in situations where the values of $w_j k_j$ ($j = 1, \ldots, p$) are more unbalanced, which are the situations in Scenarios I and II, where the $w_j$ have values proportional to the $k_j$.

Of course that since the numerical computation of the exact moments of $Z$ is rather simple and the numerical solution of the system of equations in (8) is quite simple and quick, since it is a linear system of equations, we may always easily extend the value of $m^*$ in (7) and (8), that is, the number of exact moments that the near-exact distributions match, in order to obtain a better fitting distribution. In this way we may always be able to beat the performance of any Moschopoulos series distribution truncation. For example if in Scenario IV we take the near-exact distribution with $\lambda$ defined as the geometric mean of the $1/(2w_j)$ that matches 35 exact moments, we would obtain for this distribution a value of the measure $\Delta$ equal to $2,51 \times 10^{-23}$, this way beating the performance of the Moschopoulos series distribution truncated to 200 terms.

For Scenarios V–VIII, where the only change from Scenarios I–IV is in the d.f.'s of the chi-square r.v.'s involved, which now have ten more d.f.'s each, we may see, from Table 2, that the performance of the truncations of the Moschopoulos series distribution has worsened quite much, with the truncations with 100 terms even displaying for Scenarios V and VI values of the measure $\Delta$ which do not make sense, since these values are supposed to lie between zero and 1. Going in the opposite direction, all near-exact distributions show sharp improvements in the $\Delta$ values, when compared with the corresponding distributions in Scenarios I–IV.

We may see that now for Scenario V all near-exact distributions that match 10 exact moments even show a better performance than the Moschopoulos series truncated at 200 terms, a situation that is almost repeated for Scenario VI, with the only exception being the near-exact distribution with $\lambda$ given by the maximum of the $1/(2w_j)$ and that matches 10 exact moments, which shows a less good performance than the Moschopoulos series truncated at 200 terms, but still showing a better performance than the Moschopoulos series truncated at 150 terms.

For Scenario VII only the near-exact distributions with $\lambda$ given either by the maximum or the minimum of the $1/(2w_j)$, which match 10 or 15 exact moments are not able to beat the Moschopoulos series truncated at 200 terms, but with all near-exact distributions that match 10, 15 or 20 exact moments outperforming respectively the Moschopoulos series expansions with 100, 150 or 200 terms.

Even for Scenario VIII, which among the four Scenarios V–VIII is the most favorable one for the Moschopoulos series expansion, all near-exact distributions that match 10 exact moments are able to outperform the Moschopoulos series expansions with 100 terms, and among those that match 15 exact moments, only the near-exact distribution with $\lambda$ defined as the minimum of the $1/(2w_j)$ is not able to outperform the Moschopoulos series truncated at 150 terms. Also, among the near-exact distributions that match 20 exact moments, only those with $\lambda$ defined by either the maximum or the minimum of the $1/(2w_j)$ are not able to outperform the Moschopoulos series distribution truncated at 200 terms.

For Scenarios V–VIII the behavior of the near-exact distributions for the different choices of $\lambda$ in (10) is similar in their interrelations to the ones described for Scenarios I–IV. For Scenarios I and V, the near-exact distributions ranking 1–4 in their performance are, in this order, the ones with $\lambda$ given by (vii), (iv), (vi) and (v) in (10), while for Scenarios II and VI the near-exact distributions ranking 1–3 are the ones with $\lambda$ given by (vii), (vi) and (iv) in (10), with a swap between the ones with $\lambda$ given by (ii) and (v) for fourth and fifth place. Also for Scenarios III and VII the near-exact distributions ranking 1–5 in their performance are the same, being, in this order, the ones with $\lambda$ given by (v), (vii), (iv), (vi) and (iii) in (10). Only for Scenarios IV and VIII there is a swap between the near-exact distributions with $\lambda$ given by (iii) and (iv) in (10), for third and fifth performance places, but with the ones with $\lambda$ given by (v), (vi) and (vii) in (10) keeping the first, second and fourth performance places. For these two scenarios there is a somewhat strange inversion in the performance of the near-exact distributions with $\lambda$ given by (vii) and (vi) in (10), with this last one showing a better performance.

From the values in Table 3 we may see how truncations of the Moschopoulos distribution, actually as most other approximations and series distributions, have severe trouble handling the situations in scenarios IX and X and also how when the numbers of d.f.'s increase the situation becomes even worse.

By contrast, the near-exact distributions, namely those with $\lambda$ defined as the maximum or the average of the $1/(2w_j)$ or those which use $\lambda$ defined by (vi) or (vii) in (10) give extremely sharp approximations, and, as expected, even increase their performance when the $k_j$, that is, the number of d.f.'s, increase.

For these situations depicted in scenarios IX and X we would certainly choose to use $\lambda$ defined as the maximum of the $1/(2w_j)$ or given by (vi) in (10).

## 4.2 A Note on Computation Times and the Use of the GNIG p.d.f. and c.d.f.

We may note that although the general expressions for the p.d.f. and the c.d.f. of the GNIG distribution involve the Kummer confluent hypergeometric function

$$
{}_1F_1(a, b; z) = \sum_{i=0}^{\infty} \frac{\Gamma(a+i)}{\Gamma(a)} \frac{\Gamma(b)}{\Gamma(b+i)} \frac{1}{z^i},
$$

which is nowadays possible to compute precisely and efficiently with most symbolic softwares as it is for example the case of Mathematica® and also MAXIMA, in our case it is possible to use an even more efficient and convenient representation for these functions since their arguments $a$ and $b$ will always be half-integers. As such we may use the fact that for any odd integer $k$ we may write

$$
{}_1F_1\left(\frac{k}{2},\frac{k}{2}+1,z\right)=\frac{k\,e^z}{2^{(k+1)/2}\,(-z)^{k/2}}\left(\left\{\prod_{j=1}^{\frac{k-1}{2}}(2j-1)\right\}e^{-z}\sqrt{\pi}\,\mathrm{Erf}(\sqrt{-z})\right.
$$
$$
\left.-\sum_{j=1}^{\frac{k-1}{2}}(-z)^{\frac{2j-1}{2}}\,2^j\left\{\prod_{\ell=0}^{\frac{k-3}{2}-j}2\left(\frac{k-1}{2}-\ell\right)-1\right\}\right) \tag{13}
$$

where

$$
\mathrm{Erf}(z)=\frac{2}{\sqrt{\pi}}\int_0^z e^{-t^2}\,dt
$$

is the error function. This expression is computationally very convenient since most softwares can nowadays compute this error function very efficiently and then we may use the fact that

$$
{}_1F_1(a,a,-z)=e^{-z}
$$

together with expression 13.4.2 in Slater (1972), which, for $j=1,2,\ldots$, in our case may be written as

$$
\begin{aligned}
{}_1F_1\left(\tfrac{k}{2},\tfrac{k}{2}+j+1,z\right)=-\tfrac{1}{z}\left(\tfrac{k}{2j}+1\right)\Big\{&\left(\tfrac{k}{2}+j-1\right){}_1F_1\left(\tfrac{k}{2},\tfrac{k}{2}+j-1,z\right)\\
&+\left(1-\tfrac{k}{2}-j-z\right){}_1F_1\left(\tfrac{k}{2},\tfrac{k}{2}+j,z\right)\Big\},
\end{aligned}
$$

to compute iteratively all Kummer confluent hypergeometric functions that appear in the expressions for the p.d.f. and the c.d.f. of the GNIG distributions used.

## 5 The Distribution of Some Instances of Ratios of Two Independent Linear Combinations of Independent Chi-Squares

The distribution of ratios of two independent linear combinations of independent chi-squares appears mostly as the distribution of ratios of independent quadratic forms in multivariate normal variates, each one of which has the distribution of a linear combination of independent chi-squares. In turn these ratios have since long been associated with tests in Analysis of Variance models where different variances are assumed for the populations associated with the different levels of the factor or factors (Box, 1954; Welch, 1947; Satterthwaite, 1941). Based on the results obtained in Sects. 2 and 4 it is possible to obtain the exact distribution of a ratio of two independent linear combinations of independent chi-squares in one situation and an asymptotic distribution for another situation.

First of all we need to recognize that actually the GIG distribution may be seen as a mixture of Gamma distributions, all with integer shape parameters, since we may write the p.d.f. of the GIG distribution in (28) as

$$\begin{aligned} f_Z(z) &= K \sum_{j=1}^{p} \sum_{k=1}^{r_j} c_{j,k}\, e^{-\lambda_j z}\, z^{k-1} \\ &= \sum_{j=1}^{p} \sum_{k=1}^{r_j} \underbrace{\frac{K c_{j,k} \Gamma(k)}{\lambda_j^k}}_{p_{j,k}} \underbrace{\frac{\lambda_j^k}{\Gamma(k)} e^{-\lambda_j z} z^{k-1}}_{\text{p.d.f. of a } \Gamma(k,\lambda_j) \text{ dist.}} = \sum_{j=1}^{p} \sum_{k=1}^{r_j} p_{j,k} \frac{\lambda_j^k}{\Gamma(k)} e^{-\lambda_j z} z^{k-1} \end{aligned} \qquad (14)$$

which is the p.d.f. of a mixture of $\sum_{j=1}^{p} r_j$ distributions $\Gamma(k, \lambda_j)$, with weights $p_{j,k} = K c_{j,k} \Gamma(k)/\lambda_j^k$ $(j = 1, \ldots, p; k = 1, \ldots, r_j)$.

Second of all, we may note that if $X \sim \Gamma(r_1, \lambda_1)$ and $Y \sim \Gamma(r_2, \lambda_2)$ are two independent r.v.'s, then the r.v. $W = X/Y$ has p.d.f.

$$f_W(w) = \frac{k^{r_1}}{B(r_1, r_2)} (1 + kw)^{-r_1-r_2}\, w^{r_1-1} \quad (w > 0) \qquad (15)$$

and c.d.f.

$$\begin{aligned} F_W(w) &= \frac{k^{r_1}}{B(r_1, r_2)} \frac{w^{r_1}}{r_1}\, {}_2F_1(r_1 + r_2, r_1; r_1 + 1; -kw) \\ &= BR\left(r_1, r_2; \frac{kw}{kw+1}\right) \quad (w > 0)\,, \end{aligned} \qquad (16)$$

where $k = \lambda_1/\lambda_2$, $B(\cdot\,,\,\cdot)$ is the common Beta function, ${}_2F_1(\cdot\,,\,\cdot;\,\cdot)$ is the Gauss hypergeometric function and $BR(\cdot\,,\,\cdot\,; z)$ is the regularized incomplete Beta function, with

$$BR(r_1, r_2; z) = \frac{1}{B(r_1, r_2)} \int_0^z x^{r_1-1}(1-x)^{r_2-1}\, dx\,.$$

Then, we have two important cases where it is possible to obtain closed form expressions for the p.d.f. and c.d.f. of the ratio of two independent linear combinations of independent chi-squares.

Let us suppose that

$$Z_1 = \sum_{j=1}^{p_1} w_j Y_j \quad \text{and} \quad Z_2 = \sum_{\ell=1}^{p_2} w_\ell^* Y_\ell^* \qquad (17)$$

are two independent linear combinations of independent chi-squares, where

$$Y_j \sim \chi^2_{k_j} \quad \text{and} \quad Y^*_\ell \sim \chi^2_{k^*_\ell} \quad j = 1, \ldots, p_1;\ \ell = 1, \ldots, p_2$$

where all of $k_j$ $(j = 1, \ldots, p_1)$ and $k^*_\ell$ $(\ell = 1, \ldots, p_2)$ are even integers.

Then the exact distribution of both $Z_1$ and $Z_2$ are GIG distributions, which in case all $w_j$ $(j = 1, \ldots, p_1)$ and also all $w^*_\ell$ $(\ell = 1, \ldots, p_2)$ are different, will have respectively depths $p_1$ and $p_2$. But then, from the mixture representation of the GIG distribution we may see that in this case the exact distribution of

$$V = Z_1/Z_2$$

will be a double mixture of ratios of independent Gamma distributed r.v.'s, which, from (15), will have, for $v > 0$, p.d.f.

$$f_V(v) = \sum_{j=1}^{p_1} \sum_{\ell=1}^{p_2} \sum_{h=1}^{k_j/2} \sum_{i=1}^{k^*_\ell/2} p_{j,h}\, p^*_{\ell,i} \frac{\left(\frac{w^*_\ell}{w_j}\right)^h}{B(h,i)} \left(1 + \frac{w^*_\ell}{w_j} v\right)^{-h-i} v^{\,h-1} \tag{18}$$

and from (16), c.d.f.

$$F_V(v) = \sum_{j=1}^{p_1} \sum_{\ell=1}^{p_2} \sum_{h=1}^{k_j/2} \sum_{i=1}^{k^*_\ell/2} p_{j,h}\, p^*_{\ell,i}\ BR\left(h, i, \frac{\frac{w^*_\ell}{w_j} v}{\frac{w^*_\ell}{w_j} v + 1}\right) \tag{19}$$

where, from (14) and from (29),

$$p_{j,h} = \left\{ \prod_{j=1}^{p_1} \left(\frac{1}{2w_j}\right)^{k_j/2} \right\} \frac{c_{j,h}\, \Gamma(h)}{\left(\frac{1}{2w_j}\right)^h}, \quad j = 1, \ldots, p_1;\ h = 1, \ldots, k_j/2$$

and

$$p^*_{\ell,i} = \left\{ \prod_{\ell=1}^{p_2} \left(\frac{1}{2w^*_\ell}\right)^{k^*_\ell/2} \right\} \frac{c_{\ell,i}\, \Gamma(i)}{\left(\frac{1}{2w^*_\ell}\right)^i}, \quad \ell = 1, \ldots, p_2;\ i = 1, \ldots, k^*_\ell/2, \tag{20}$$

for $c_{j,h}$ and $c_{\ell,i}$ defined according to (30)-(32) in Appendix 1.

One other situation of interest, as we will see in Sect. 6, is when $Z_1$ in (17) is a linear combination of independent chi-squares, all with one d.f.. In this case, if we take the approach described in Sect. 4, all the parameters $\ell_j$ in (3) will be equal to zero and the r.v. $Z^*$ in (4) vanishes, while $\Phi_{Z^{**}}(\,\cdot\,)$, the c.f. to be asymptotically approximated, will be the whole c.f. of $Z_1$. Thus, in this case we will obtain for $Z_1$ not a near-exact distribution but simply an asymptotic distribution, as noted in Sect. 4. This asymptotic distribution, which will correspond to the c.f. $\Phi^*(\cdot)$ in (7) will then be a mixture of $m^* + 1$ independent $\Gamma(r + k, \lambda)$ distributions ($k = 0, \ldots, m^*$), where $r = p_1/2$ and $\lambda$ is to be computed according to one of the strategies in (10).

Then, if the exact distribution of $Z_2$ remains being the one in (17), with all $k_\ell^*$ even, based on this asymptotic distribution for $Z_1$, the asymptotic distribution of $V = Z_1/Z_2$ will have p.d.f.

$$f_V(v) = \sum_{j=0}^{m^*}\sum_{\ell=1}^{p_2}\sum_{i=1}^{k_\ell^*/2} \pi_j\, p_{\ell,i}^*\, \frac{(2\lambda w_\ell^*)^{r+j}}{B(r+j,i)} \left(1 + 2\lambda w_\ell^* v\right)^{-r-j-i} v^{r+j-1} \tag{21}$$

and c.d.f.

$$F_V(v) = \sum_{j=0}^{m^*}\sum_{\ell=1}^{p_2}\sum_{i=1}^{k_\ell^*/2} \pi_j\, p_{\ell,i}^*\, BR\left(r + j, i; \frac{2\lambda w_\ell^* v}{1 + 2\lambda w_\ell^* v}\right), \tag{22}$$

for $v > 0$.

We may note that the part of the incomplete Beta function in both regularized incomplete Beta functions in (19) and (22), given the fact that the second argument is an integer, may be written as a finite sum, that is, in both cases we may write, for any $a > 0$ and any $z > 0$,

$$BR(a, i; z) = \frac{1}{B(a,i)} \sum_{j=0}^{i-1} \frac{(-1)^j}{a+j} \binom{i-1}{j} z^{a+j}.$$

## 6 The Distribution of the Behrens-Fisher Statistic

In this section we will address the distribution of the statistic used to test the equality of the expected values of two Normal r.v.'s, when their variances are assumed to be different. This testing problem is commonly known as the Behrens-Fisher problem (Behrens, 1929; Fisher, 1939), and we will call "Behrens-Fisher statistic" the associated test statistic.

We will consider this statistic to be the statistic $V = Q_B/(Q_W/(n_1 + n_2 - 2))$ where $n_1$ and $n_2$ are the sample sizes respectively for the samples of the first and second populations, that is, of $X_1 \sim N(\mu_1, \sigma_1^2)$ and $X_2 \sim N(\mu_2, \sigma_2^2)$, and where $Q_B$ and $Q_W$ are respectively the between and within sums of squares in Sect. 7 of Box (1954), for the case $k = 2$.

This statistic $V$ may be seen as the square of the $T$ statistic for the Behrens-Fisher problem. From Sect. 7 of Box (1954), $Q_B$ is distributed as $w_1\chi_1^2$, with

$$w_1 = (n_2\sigma_1^2 + n_1\sigma_2^2)/(n_1 + n_2), \tag{23}$$

which is the non-null eigenvalue of the matrix $VM$ on page 298 of Box (1954), where

$$V = \begin{bmatrix} \sigma_1^2/n_1 & 0 \\ 0 & \sigma_2^2/n_2 \end{bmatrix} \quad \text{and} \quad M = \begin{bmatrix} n_1 - \frac{n_1^2}{n_1+n_2} & -\frac{n_1 n_2}{n_1+n_2} \\ -\frac{n_1 n_2}{n_1+n_2} & n_2 - \frac{n_2^2}{n_1+n_2} \end{bmatrix},$$

that is, $Q_B \sim \Gamma(1/2, \lambda)$, with $\lambda = 1/(2w_1)$, while $Q_W$ is distributed as $\sigma_1^2 Y_1 + \sigma_2^2 Y_2$ where $Y_1$ and $Y_2$ are two independent r.v.'s, with $Y_1 \sim \chi^2_{n_1-1}$ and $Y_2 \sim \chi^2_{n_2-1}$.

The population variances, $\sigma_1^2$ and $\sigma_2^2$, in case they are not known, will appear as nuisance parameters in the distribution of $V$. In this case they may be replaced by their estimates (preferably the maximum likelihood or the common centered ones).

If one uses the statistic $V$ to carry out the test, one will basically be testing the null hypothesis $H_0 : \mu_1 = \mu_2$ versus the alternative hypothesis $H_1 : \mu \neq \mu_2$, rejecting $H_0$ if the computed value of $V$ exceeds the $\alpha$-quantile of $V$. However, one may take the square root of the computed value of $V$ affected by the sign of $\overline{X}_1 - \overline{X}_2$ (where $\overline{X}_1$ and $\overline{X}_2$ represent the sample means of the samples from $X_1$ and $X_2$) to carry out a one-sided test, where then the quantile to be used in order to carry out an $\alpha$ level test will be the square root of the 1-$\alpha$ quantile of $V$ affected by the same sign. Alternatively, if one wants to take a p-value approach, this means that, taking $F_V(\cdot)$ as the c.d.f. of $V$ and $v$ as the computed value of $V$, while one would use $1 - F_V(v)$ as the p-value for the two-sided test, one would have to use $(1 - F_V(v))/2$ as the p-value for the one-sided test.

## 6.1 The Case of Both $n_1$ and $n_2$ Odd

In case $n_1$ and $n_2$ are both odd, the exact p.d.f. and c.d.f. of $V$ may be obtained respectively either from (18) and (19) or from (21) and (22), since in this case the exact distribution of $Q_W$ is a GIG distribution of depth 2, with $r_1 = (n_1 - 1)/2$, $r_2 = (n_2 - 1)/2$, $\lambda_1 = (n_1 + n_2 - 2)/(2\sigma_1^2)$ and $\lambda_2 = (n_1 + n_2 - 2)/(2\sigma_2^2)$. Let us then take (21) and (22) as the basis for the expressions of the p.d.f. and c.d.f. of $V$. Then, the exact p.d.f. of $V = Q_B/(Q_W/(n_1 + n_2 - 2))$ is given by (21) with $m^* = 0$, $\pi_0 = 1$, $p_2 = 2$, $k_1^* = n_1 - 1$, $k_2^* = n_2 - 1$, $r = 1/2$, $\lambda = 1/(2w_1)$, $w_1^* = \sigma_1^2/(n_1 + n_2 - 2)$ and $w_2^* = \sigma_2^2/(n_1 + n_2 - 2)$, that is, it will be given by

$$f_V(v) = \sum_{\ell=1}^{2} \sum_{i=1}^{(n_\ell-1)/2} p_{\ell,i}^* \frac{\sigma_\ell^2 \frac{n_1+n_2}{n_2\sigma_1^2+n_1\sigma_2^2}}{B\left(\frac{1}{2}, i\right)} \left(1 + \sigma_\ell^2 \frac{(n_1+n_2)\,v}{n_2\sigma_1^2 + n_1\sigma_2^2}\right)^{-\frac{1}{2}-i} v^{-1/2}$$

and the exact c.d.f. by

$$F_V(v) = \sum_{\ell=1}^{2} \sum_{i=1}^{(n_\ell-1)/2} p^*_{\ell,i} \; BR\left(\frac{1}{2}, i; \frac{\sigma_\ell^2\, v\,(n_1+n_2)}{n_2\sigma_1^2 + n_1\sigma_2^2 + \sigma_\ell^2 v(n_1+n_2)}\right),$$

where the $p^*_{\ell,i}$ ($\ell = 1, 2; i = 1, \ldots, (n_\ell - 1)/2$) are given by (20).

## 6.2 The Case of Either $n_1$ or $n_2$ Odd and the Other One Even

In this case the distribution of $Q_W$ is a GNIG distribution of depth 2 with parameters, which if, without any loss of generality, we assume $n_1$ to be odd and $n_2$ even, will be $r_1 = (n_1 - 1)/2$, $r_2 = (n_2 - 1)/2$, $\lambda_1 = (n_1 + n_2 - 2)/(2\sigma_1^2)$ and $\lambda_2 = (n_1 + n_2 - 2)/(2\sigma_2^2)$.

It happens that in this case, where the GIG part of the GNIG distribution has only depth 1, that is, is formed only by a single integer Gamma r.v., the coefficients $c_{j,k}$, which would only be $c_{1,k}$, are all null except $c_{1,r_1}$, a fact that is easily verified from (30)–(32) in Appendix 1. Furthermore, in this case we have $c_{1,r_1} = 1/(r_1 - 1)!$, which simplifies the expression for the p.d.f. of $Q_W$ to (see (33) in Appendix 1)

$$f_{Q_W}(q) = K e^{-\lambda_1 q} \frac{1}{\Gamma(r_1 + r_2)} q^{r_1+r_2-1} \, {}_1F_1(r_2, r_2 + r_1, -(\lambda_2 - \lambda_1)q)\,, \tag{24}$$

for $K = \lambda_1^{r_1}\lambda_2^{r_2}$, with $r_1$, $r_2$, $\lambda_1$ and $\lambda_2$ as defined above.

Using then the series expansion of the Kummer confluent hypergeometric function, we are able to write the distribution of $Q_W$ as an infinite mixture of Gamma distributions, since we may write

$$\begin{aligned} f_{Q_W}(q) &= \sum_{i=0}^{\infty} K \frac{\Gamma(r_2+i)}{\Gamma(r_2)\,\Gamma(r_2+r_1+i)} \frac{(\lambda_1-\lambda_2)^i}{i!} q^{r_1+r_2+i-1} e^{-\lambda_1 q} \\ &= \sum_{i=0}^{\infty} \underbrace{K \frac{\Gamma(r_2+i)}{\Gamma(r_2)\,\lambda_1^{r_1+r_2+i}} \frac{(\lambda_1-\lambda_2)^i}{i!}}_{p_i} \underbrace{\frac{\lambda_1^{r_1+r_2+i}}{\Gamma(r_1+r_2+i)} e^{-\lambda_1 q} q^{r_1+r_2+i-1}}_{\text{p.d.f. of } \Gamma(r_1+r_2+i,\lambda_1)}, \end{aligned}$$

which is the p.d.f. of an infinite mixture of $\Gamma(r_1 + r_2 + i, \lambda_1)$ distributions, with weights $p_i$ ($i = 0, 1, \ldots$).

Thus, since $Q_B \sim \Gamma(1/2, \lambda)$ with $\lambda = 1/(2w_1)$, for $w_1$ given by (23), the exact distribution of $V$ is that of an infinite mixture of ratios of $\Gamma(1/2, \lambda)$ and $\Gamma(r_1+r_2+i, \lambda_1)$ distributions, with weights $p_i$ $(i = 0, 1, \dots)$, with p.d.f.

$$
\begin{aligned}
f_V(v) &= \sum_{i=0}^{\infty} K \frac{\Gamma(r_2+i)}{\Gamma(r_2)} \frac{\lambda^{1/2}}{B\left(\frac{1}{2}, r_1+r_2+i\right)} \frac{(\lambda_1-\lambda_2)^i}{i!} \left(\frac{1}{\lambda_1+\lambda v}\right)^{r_1+r_2+1/2+i} v^{-1/2} \\
&= K \left(\frac{\lambda}{v}\right)^{1/2} \left(\frac{1}{\lambda_1+\lambda v}\right)^{r_1+r_2+1/2} \frac{1}{B\left(\frac{1}{2}, r_1+r_2\right)} \\
&\quad \times \sum_{i=0}^{\infty} \frac{\Gamma(r_2+i)}{\Gamma(r_2)} \frac{\Gamma(r_1+r_2)}{\Gamma(r_1+r_2+i)} \frac{\Gamma(r_1+r_2+1/2+i)}{\Gamma(r_1+r_2+1/2)} \left(\frac{\lambda_1-\lambda_2}{\lambda_1+\lambda v}\right)^i \frac{1}{i!} \\
&= K \left(\frac{\lambda}{v}\right)^{1/2} \left(\frac{1}{\lambda_1+\lambda v}\right)^{r_1+r_2+1/2} \frac{1}{B\left(\frac{1}{2}, r_1+r_2\right)} \\
&\quad \times {}_2F_1\left(r_2, r_1+r_2+\frac{1}{2}; r_1+r_2; \frac{\lambda_1-\lambda_2}{\lambda_1+\lambda v}\right)
\end{aligned}
$$

and c.d.f.

$$
F_V(v) = K \sum_{i=0}^{\infty} \frac{\Gamma(r_2+i)}{\Gamma(r_2)\,\lambda_1^{r_1+r_2+i}} \frac{(\lambda_1-\lambda_2)^i}{i!} BR\left(\frac{1}{2}, r_1+r_2+i; \frac{\lambda v}{\lambda_1+\lambda v}\right)
$$

where the infinite summation for the c.d.f. may usually be cut short to just a few dozen terms and is nowadays easy to compute in a precise manner using any available symbolic software.

## 6.3 The Case of Both $n_1$ and $n_2$ Even

In this case there are indeed two strategies we may follow. We may use the near-exact approach in Sect. 4 to obtain a near-exact distribution for $Q_W$, which would then be a mixture of $m^* + 1$ GIG distributions of depth 3, with shape parameters $(n_1 - 2)/2$, $(n_2 - 2)/2$ and $1 + k$ $(k = 0, \dots, m^*)$ and rate parameters $\lambda_1 = (n_1+n_2-2)/(2\sigma_1^2)$, $\lambda_2 = (n_1+n_2-2)/(2\sigma_2^2)$ and $\lambda$ given by one of the strategies in (10). Since each of these GIG distributions may then be seen as a finite mixture of Gamma distributions, a near-exact type of distribution for $V$ would then be obtained as a double finite mixture of ratios of Gamma r.v.'s. But, we may take a different approach, in order to obtain the exact p.d.f. and c.d.f. of $V$ in this case. Indeed, if we pay heed to the expression for the p.d.f. of $Q_W$ in (24), we may see that it remains valid even if $r_1$ is not an integer. As such even when both $n_1$ and $n_2$ are even, the p.d.f. of $Q_W$ remains given by (24). As such, all the results about the distribution of

$V$ obtained in the previous subsection remain valid for the case where $n_1$ and $n_2$ are both even.

# 7 Some Examples for the Distributions Mentioned in Sects. 2–5

In this section we will use a number of times the result in the following Theorem on the distribution of quadratic forms on Normal r.v.'s from Box (1954).

**Theorem (Box, 1954, Thm. 2.1)** *If* $\underline{X} \sim N_p(\underline{0}, \Sigma)$ *and* $Q = \underline{X}'M\underline{X}$, *where* $M$ *is a* $p{\times}p$ *symmetric matrix with* $rank(M) = r \leq p$, *then*

$$Q \stackrel{d}{\equiv} \sum_{j=1}^{r} \mu_j Y_j$$

*where* $Y_j \sim \chi^2_1$ *are independent r.v.'s and* $\mu_j (j = 1, \ldots, r)$ *are the non-null eigenvalues of* $\Sigma M$.

## 7.1 An Example of a Linear Combination of Chi-Squares with a GIG Distribution

Let us take Example 1 from Ha and Provost (2013). In this example the authors consider an approximation to the distribution of a quadratic form $Q = \underline{X}'M\underline{X}$ where $\underline{X} \sim N_6(\underline{0}, I_6)$ and $M$ is a $6{\times}6$ positive-definite matrix with eigenvalues $\mu_1^* = \mu_2^* = 1$, $\mu_3^* = \mu_4^* = 2.5$ and $\mu_5^* = \mu_6^* = 9$. Since in this case we have only three different eigenvalues, each one with a multiplicity of 2, the distribution of the quadratic form $Q$ is that of

$$\sum_{j=1}^{3} \mu_j Y_j$$

where $\mu_1 = 1$, $\mu_2 = 2.5$ and $\mu_3 = 9$ and $Y_j \sim \chi^2_2$ $(j = 1, 2, 3)$.

As such, the exact distribution of $Q$ is obtained right away from the result in Sect. 2 as a GIG distribution of depth 3, with $p = 3$, $r_1 = r_2 = r_3 = 1$, $\lambda_1 = 1/2$, $\lambda_2 = 1/(2{\times}\frac{5}{2}) = 1/5$ and $\lambda_3 = 1/(2{\times}9) = 1/18$. The coefficients $c_{j,k}$ are in this case very easy to compute, even by hand, and they are

$$c_{1,1} = \frac{15}{2}, \qquad c_{2,1} = -\frac{300}{13}, \qquad c_{3,1} = \frac{405}{26},$$

so that the exact p.d.f. of $Q$ is

$$f_Q(z) = \left(\frac{1}{2}\times\frac{1}{2}\times\frac{1}{18}\right)\left[\frac{15}{2}\,e^{-z/2} - \frac{300}{13}\,e^{-z/5} + \frac{405}{26}\,e^{-z/18}\right]$$

and its c.d.f.

$$F_Q(z) = 1 - \left(\frac{1}{2}\times\frac{1}{2}\times\frac{1}{18}\right)\left[\frac{15}{2}\times 2\times e^{-z/2} - \frac{300}{13}\times 5\times e^{-z/5} + \frac{405}{26}\times 18\times e^{-z/18}\right].$$

Quantiles for any GIG distribution may be readily and very quickly obtained using the Mathematica® module at https://sites.google.com/site/lincombchisquares. This module uses some faster modules for the implementation of the p.d.f. and c.d.f. of the GIG distribution which are available in Appendix 4.A of Chapter 4 of Coelho and Arnold (2019). The exact quantiles computed from this distribution confirm the exact quantiles in Table 1 of Ha and Provost (2013).

## *7.2 An Example of a Linear Combination of Chi-Squares with a GNIG Distribution*

Let us consider a simple case where $\underline{X} \sim N_7(\underline{0}, \Sigma)$ with

$$\Sigma = \frac{1}{12}\begin{bmatrix} 44 & -16 & -16 & 0 & 0 & 0 & 0 \\ -16 & 44 & -16 & 0 & 0 & 0 & 0 \\ -16 & -16 & 44 & 0 & 0 & 0 & 0 \\ 0 & 0 & 0 & 45 & -9 & -3 & -21 \\ 0 & 0 & 0 & -9 & 45 & -21 & -3 \\ 0 & 0 & 0 & -3 & -21 & 45 & -9 \\ 0 & 0 & 0 & -21 & -3 & -9 & 45 \end{bmatrix}$$

and where we are interested in the distribution of the quadratic form $Q = \underline{X}'M\underline{X}$, with

$$M = \frac{1}{12}\begin{bmatrix} 8 & 2 & 2 & 0 & 0 & 0 & 0 \\ 2 & 8 & 2 & 0 & 0 & 0 & 0 \\ 2 & 2 & 8 & 0 & 0 & 0 & 0 \\ 0 & 0 & 0 & 18 & -9 & -3 & 6 \\ 0 & 0 & 0 & -9 & 18 & 6 & -3 \\ 0 & 0 & 0 & -3 & 6 & 18 & -9 \\ 0 & 0 & 0 & 6 & -3 & -9 & 18 \end{bmatrix}.$$

Since the eigenvalues of $\Sigma M$ are $\mu_1^* = \mu_2^* = 1$, $\mu_3^* = \mu_4^* = \mu_5^* = 5/2$ and $\mu_6^* = \mu_7^* = 9$, the exact distribution of $Q$ is the same as that of

$$\sum_{j=1}^{3} \mu_j Y_j$$

where $\mu_1 = 1$, $\mu_2 = 9$, $\mu_3 = 5/2$ and $Y_1 \sim \chi^2_2$, $Y_2 \sim \chi^2_2$ and $Y_3 \sim \chi^2_3$. As such, from Sect. 2, the exact distribution of $Q$ is a GNIG distribution of depth 3, with, using the notation in Appendix 1, $r_1 = r_2 = 1$ and $r_3 = 3/2$ and $\lambda_1 = 1/2$, $\lambda_2 = 1/18$ and $\lambda_3 = 1/5$, which makes the $c_{j,k}$ $(j = 1, 2; k = 1)$ parameters to be very easy to compute, with

$$c_{1,1} = -\frac{9}{4}, \qquad c_{2,1} = \frac{9}{4}.$$

Thus, the quadratic form $Q$ has p.d.f.

$$f_Q(z) = \frac{1}{2\times 18}\left(\frac{1}{5}\right)^{3/2} \frac{z^{3/2}}{\Gamma(5/2)} \left[ -e^{-\frac{1}{2}z} \frac{9}{4}\, {}_1F_1\left(\frac{3}{2}, \frac{5}{2}, \frac{3}{10}z\right) + e^{-\frac{1}{18}z} \frac{9}{4}\, {}_1F_1\left(\frac{3}{2}, \frac{5}{2}, -\frac{13}{90}z\right)\right],$$

which using (13), that for our case may assume the form,

$${}_1F_1\left(\tfrac{3}{2}, \tfrac{5}{2}, z\right) = \tfrac{3}{4}\left((-z)^{-3/2}\sqrt{\pi}\,\mathrm{Erf}(\sqrt{-z}) + 2\,e^{z}\,\tfrac{1}{z}\right), \tag{25}$$

may be written as

$$f_Q(z) = \frac{1}{80}\left(\frac{1}{5}\right)^{1/2} \left[ -e^{-\frac{1}{2}z}\left(-\frac{3}{10}\right)^{-3/2} \mathrm{Erf}\left(\sqrt{-\frac{3}{10}z}\right) + e^{-\frac{1}{18}z}\left(\frac{13}{90}\right)^{-3/2} \mathrm{Erf}\left(\sqrt{\frac{13}{90}z}\right) - \frac{800}{39\sqrt{\pi}}\, e^{-\frac{1}{5}z}\, z^{1/2}\right]$$

and c.d.f.

$$F_Q(z) = \frac{\left(\frac{1}{5}\right)^{3/2} z^{3/2}}{\Gamma(5/2)} \left\{ {}_1F_1\left(\frac{3}{2}, \frac{5}{2}, -\frac{1}{5}z\right) - \frac{1}{2}\frac{1}{18}\frac{9}{4}\left[ -2\,e^{-\frac{1}{2}z}\, {}_1F_1\left(\frac{3}{2}, \frac{5}{2}, \frac{3}{10}z\right) + 18\,e^{-\frac{1}{18}z}\, {}_1F_1\left(\frac{3}{2}, \frac{5}{2}, -\frac{13}{90}z\right)\right]\right\},$$

which, using (25), may be written as

$$F_Q(z) = \mathrm{Erf}\left(\sqrt{\frac{z}{5}}\right) - \frac{1}{8}\left[-e^{-\frac{1}{2}z}\left(-\frac{3}{2}\right)^{-3/2}\mathrm{Erf}\left(\sqrt{-\frac{3z}{10}}\right) + e^{-\frac{1}{18}z}\left(\frac{13}{18}\right)^{-3/2}\mathrm{Erf}\left(\sqrt{\frac{13z}{90}}\right)\right] + \frac{60\sqrt{5}}{234\sqrt{\pi}}e^{-\frac{1}{5}z}z^{1/2}.$$

## 7.3 An Example of a Linear Combination of Chi-Squares with a DGIG Distribution

To build a simple example where a quadratic form in Normal variables has a DGIG distribution we consider $\underline{X} \sim N_6(\underline{0}, \Sigma)$ and the quadratic form $Q = \underline{X}'M\underline{X}$ where

$$\Sigma = \frac{1}{2}\begin{bmatrix} 7 & -1 & -3 & -1 & 0 & 0 \\ -1 & 7 & -1 & -3 & 0 & 0 \\ -3 & -1 & 7 & -1 & 0 & 0 \\ -1 & -3 & -1 & 7 & 0 & 0 \\ 0 & 0 & 0 & 0 & 4 & -2 \\ 0 & 0 & 0 & 0 & -2 & 4 \end{bmatrix} \quad \text{and} \quad M = \frac{1}{4}\begin{bmatrix} -1 & 4 & -3 & 4 & 0 & 0 \\ 4 & -1 & 4 & -3 & 0 & 0 \\ -3 & 4 & -1 & 4 & 0 & 0 \\ 4 & -3 & 4 & -1 & 0 & 0 \\ 0 & 0 & 0 & 0 & -4 & 8 \\ 0 & 0 & 0 & 0 & 8 & -4 \end{bmatrix}.$$

Then the eigenvalues of $\Sigma M$ are $\mu_1^* = \mu_2^* = 1$, $\mu_3^* = \mu_4^* = 5/2$ and $\mu_5^* = \mu_6^* = -9$. As such, in this case, from Theorem 2.1 in Box (1954), restated at the beginning of the present section, the quadratic form $Q$ has the same distribution as $Z_1 - Z_2$, where $Z_1$ and $Z_2$ are independent, with

$$Z_1 = Y_1 + \frac{5}{2}Y_2 \quad \text{and} \quad Z_2 = 9Y_3$$

where $Y_1$, $Y_2$ and $Y_3$ are all chi-squares with 2 d.f.'s, so that the exact distribution of $Q$ is a DGIG distribution with $p_1 = 2$ and $p_2 = 1$, and with $r_1 = r_2 = 1$, $s_1 = 1$, $\lambda_1 = 1/2$, $\lambda_2 = 1/5$ and $\nu_1 = 1/18$. It is then easy to compute

$$c_{1,1} = -\frac{10}{3}, \quad c_{2,1} = \frac{10}{3}, \quad d_{1,1} = 1$$

and obtain the p.d.f. of $Q$ given by

$$f_Q(z) = \begin{cases} \frac{1}{180}\left(-\frac{10}{3}\frac{1}{\frac{1}{2}+\frac{1}{18}}e^{-\frac{1}{2}z} + \frac{10}{3}\frac{1}{\frac{1}{5}+\frac{1}{18}}e^{-\frac{1}{5}z}\right) \\ \frac{1}{180}\left(-\frac{10}{3}\frac{1}{\frac{1}{2}+\frac{1}{18}}e^{\frac{1}{18}z} + \frac{10}{3}\frac{1}{\frac{1}{5}+\frac{1}{18}}e^{\frac{1}{18}z}\right) \end{cases} = \begin{cases} \frac{1}{180}\left(-6\,e^{-\frac{1}{2}z} + \frac{300}{23}\,e^{-\frac{1}{5}z}\right), & z \geq 0 \\ \frac{9}{230}\,e^{\frac{1}{18}z}, & z \leq 0 \end{cases}$$

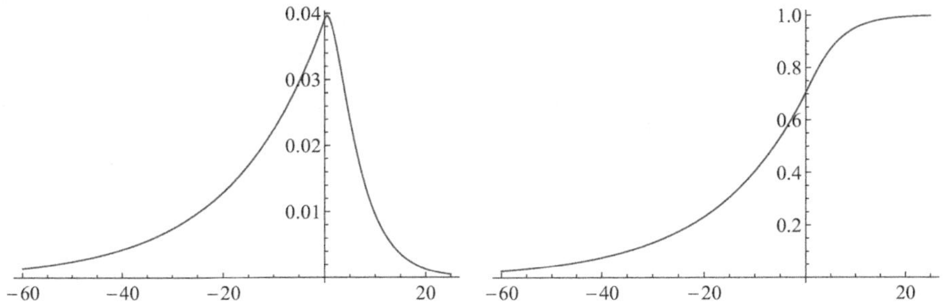


**Fig. 1** Plots of the p.d.f. and c.d.f. of the DGIG distribution of $Q$.

and c.d.f.

$$F_Q(z) = \begin{cases} 1 - \frac{1}{180}\left(-\frac{10}{3}\frac{1}{\frac{1}{2}+\frac{1}{18}}\frac{e^{-\frac{1}{2}z}}{1/2} + \frac{10}{3}\frac{1}{\frac{1}{5}+\frac{1}{18}}\frac{e^{-\frac{1}{5}z}}{1/5}\right), & z \geq 0 \\ \frac{18}{180}\, e^{\frac{1}{18}z}\left(-\frac{10}{3}\frac{1}{\frac{1}{2}+\frac{1}{18}} + \frac{10}{3}\frac{1}{\frac{1}{5}+\frac{1}{18}}\right), & z \leq 0 \end{cases}$$
$$= \begin{cases} 1 - \frac{1}{180}\left(-12\, e^{-\frac{1}{2}z} + \frac{1500}{23}\, e^{-\frac{1}{5}z}\right), & z \geq 0 \\ \frac{81}{115}\, e^{\frac{1}{18}z}, & z \leq 0\,. \end{cases}$$

The plots of the above p.d.f. and c.d.f. are displayed in Fig. 1.

## 7.4 *An Example of a Linear Combination of Chi-Squares with a Near-Exact Distribution Which Is a Mixture of GIG Distributions*

Let now $\underline{X} \sim N_8(\underline{0}, \Sigma)$ and $Q = \underline{X}' M \underline{X}$ where

$$\Sigma = \frac{1}{10}\begin{bmatrix} 11 & -5 & -5 & 9 & 0 & 0 & 0 & 0 \\ -5 & 11 & 9 & -5 & 0 & 0 & 0 & 0 \\ -5 & 9 & 11 & -5 & 0 & 0 & 0 & 0 \\ 9 & -5 & -5 & 11 & 0 & 0 & 0 & 0 \\ 0 & 0 & 0 & 0 & 16 & 9 & -4 & -1 \\ 0 & 0 & 0 & 0 & 9 & 16 & -1 & -4 \\ 0 & 0 & 0 & 0 & -4 & -1 & 16 & 9 \\ 0 & 0 & 0 & 0 & -1 & -4 & 9 & 16 \end{bmatrix},$$

$$M = \frac{1}{24}\begin{bmatrix} 8 & -2 & -2 & 0 & 0 & 0 & 0 & 0 \\ -2 & 8 & 0 & -2 & 0 & 0 & 0 & 0 \\ -2 & 0 & 8 & -2 & 0 & 0 & 0 & 0 \\ 0 & -2 & -2 & 8 & 0 & 0 & 0 & 0 \\ 0 & 0 & 0 & 0 & 5 & 2 & -2 & -3 \\ 0 & 0 & 0 & 0 & 2 & 5 & -3 & -2 \\ 0 & 0 & 0 & 0 & -2 & -3 & 5 & 2 \\ 0 & 0 & 0 & 0 & -3 & -2 & 2 & 5 \end{bmatrix}.$$

Then, the eigenvalues of $\Sigma M$ are $\mu_1^* = \mu_2^* = \mu_3^* = 2/5$, $\mu_4^* = 1/2$, $\mu_5^* = \mu_6^* = 1$, $\mu_7^* = \mu_8^* = 9$, so that the distribution of $Q$ is that of

$$\sum_{j=1}^{4} \mu_j Y_j$$

where $\mu_1 = 2/5$, $\mu_2 = 1/2$, $\mu_3 = 1$, $\mu_4 = 9$ and

$$Y_1 \sim \chi_3^2, \quad Y_2 \sim \chi_1^2, \quad Y_3 \sim \chi_2^2, \quad Y_4 \sim \chi_2^2$$

are four independent r.v.'s.

As such, we will have, from Sect. 4, the distribution of $Q$ as that of

$$Z^* + Z^{**}$$

where

$$Z^* = Y_1^* + \mu_3 Y_3^* + \mu_4 Y_4^* \quad \text{and} \quad Z^{**} = \mu_1 Y_1^{**} + \mu_2 Y_2^{**}$$

where $Y_1^*$, $Y_3^*$ and $Y_4^*$ are all $\chi_2^2$ r.v.'s and $Y_1^{**}$ and $Y_2^{**}$ are both $\chi_1^2$ r.v.'s, and where the r.v.'s $Y_2^*$, $Y_3^{**}$ and $Y_4^{**}$ vanished.

As such, the exact distribution of $Z^*$ is a GIG distribution of depth 3 with shape parameters $\ell_1 = \ell_2 = \ell_3 = 1$ and rate parameters $\lambda_1 = 5/4$, $\lambda_2 = 1/2$ and $\lambda_3 = 1/18$, and we will approximate the distribution of $Z^{**}$ by a finite mixture of $m^* + 1$ $\Gamma(r + k, \lambda)$ distributions, for $k = 0, \ldots, m^*$, $r = \sum_{j=1}^{2} \ell_j^* = \frac{1}{2} + \frac{1}{2} = 1$ and $\lambda$ given by one of the choices in (10). Since the case we are dealing with is exactly the one in scenario IX in Sect. 4.1, we know that good choices for $\lambda$ are the maximum of $\lambda_1$, $\lambda_2$ and $\lambda_3$, that is, $\lambda = 5/4$, or the value obtained from (vi) in (10), which rounded to 6 decimal places is 1.097561.

Hence, the near-exact distribution obtained for $Q$ is a mixture of $m^* + 1$ GIG distributions, which, if our choice is to use the value 1.097561 for $\lambda$, will have depth 4, with shape parameters $r_1 = r_2 = r_3 = 1$ and $r_4 = 1 + k$ $(k = 0, \ldots, m^*)$ and rate parameters $\lambda_1 = 5/4$, $\lambda_2 = 1/2$, $\lambda_3 = 1/18$ and $\lambda_4 = 1.097561$, yielding for $Z$ a p.d.f. which may be written as

$$f_Q(z) = \sum_{k=0}^{m^*} \pi_k \, f^{GIG}\left(z \,\{1, 1, 1, 1+k\}; \{5/4, 1/2, 1/18, 1.097561\}; 4\right).$$

If our choice is to use $\lambda = 5/4$, the mixture components will be GIG distributions of depth 3, with shape parameters $r_1 = 2 + k$, $r_2 = r_3 = 1$ and rate parameters $\lambda_1 = 5/4$, $\lambda_2 = 1/2$, $\lambda_3 = 1/18$, yielding for $Z$ a p.d.f. which may be written as

$$f_Q(z) = \sum_{k=0}^{m^*} \pi_k \, f^{GIG}\left(z \left\{2+k, 1, 1\right\}; \{5/4, 1/2, 1/18\}; 3\right).$$

The weights $\pi_k$ $(k = 0, \ldots, m^* - 1)$ are then computed through the solution of the system of equations in (8), taking then $\pi_{m^*} = 1 - \sum_{k=0}^{m^*-1} \pi_k$.

## 7.5 An Example for the Distribution of Ratios of Linear Combinations of Chi-Squares

We take here the example in Sect. 7 of Box (1954) of a one-way ANOVA model "in which the observations are normally distributed but the variances differ from group to group" and we will show how using the results obtained we will be able to obtain near-exact distributions for the test statistic (which is the ratio of the between and the within sums of squares, each of which is both a quadratic form in Normal r.v.'s and a linear combination of independent chi-squares), and how these near-exact distributions will exactly reproduce the exact probabilities presented by Box (1954) in his paper.

Box (1954) states that the between groups sum of squares, $Q_B$, has, under the null hypothesis of equality of the $k$ group means, the same distribution as that of

$$\sum_{\ell=1}^{k-1} \mu_\ell Y_\ell$$

where $Y_\ell \sim \chi^2_1$ are $k - 1$ independent r.v.'s and $\mu_\ell$ are the non-null eigenvalues of the matrix $\Sigma M$, where

$$\Sigma = diag\left(\sigma^2_\ell / n_\ell, \ell = 1, \ldots, k\right)$$

is the variance-covariance matrix of the observations, where $\sigma_\ell^2$ is the variance in the $\ell$-th group and $n_\ell$ the sample size in that group ($\ell = 1, \ldots, k$), and

$$M = diag\,(n_\ell, \ell = 1, \ldots, k) - \frac{1}{N}\,[n_1, n_2, \ldots, n_k]\,[n_1, n_2, \ldots, n_k]'$$

where $N = \sum_{\ell=1}^{k} n_\ell$, so that

$$\Sigma M = diag(\sigma_\ell^2) - \frac{1}{N} M^*$$

where $M^*$ has its $\ell j$-th element equal to $\sigma_\ell^2 n_j$ ($\ell = 1, \ldots, k;\ j = 1, \ldots, k$).

Box (1954) also states that the within groups sum of squares, $Q_W$ is distributed as

$$\sum_{\ell=1}^{k} \sigma_\ell^2 Y_\ell^*$$

where $Y_\ell^*$ are $k$ independent $\chi^2_{n_\ell - 1}$ r.v.'s.

We will use example (4) in Table 4 of Box (1954), where $k = 3$, $\sigma_1^2 = 1$, $\sigma_2^2 = 2$, $\sigma_3^2 = 3$ and $n_1 = 7$, $n_2 = 5$, $n_3 = 3$, to illustrate how using a near-exact type of distribution we will be able to obtain the exact probabilities reported in that Table.

The exact distribution of $Q_W$ is in this case a GIG distribution of depth 3, with shape parameters $r_1 = (n_1 - 1)/2 = 3$, $r_2 = (n_2 - 1)/2 = 2$, $r_3 = (n_3 - 1)/2 = 1$ and rate parameters $\lambda_1 = 1/(2\sigma_1^2) = 1/2$, $\lambda_2 = 1/(2\sigma_2^2) = 1/4$ and $\lambda_3 = 1/(2\sigma_3^2) = 1/6$, from which it is quite simple to compute

$$c_{1,1} = -3888\,,\ c_{1,2} = -528\,,\ c_{1,3} = -24\,,\ c_{2,1} = 0\,,\ c_{2,2} = -768\,,\ c_{3,1} = 3888\,, \tag{26}$$

which may also be easily computed using module `Makec`, available in the references and web-pages referenced at the end of Sect. 2.

For $Q_B$, since the two chi-square r.v.'s involved have only 1 d.f., we obtain from Sect. 4, for $m^* = 4$ (an arbitrary choice, which we are sure to be good enough to assure the computation of tail probabilities with great accuracy), an asymptotic distribution as a mixture of five $\Gamma(1 + k, \lambda)$ distributions, for $k = 0, \ldots, 4$ and $\lambda = 15/64$, taken as the harmonic mean of $\lambda_1 = 1/(2\mu_1) = \frac{15}{2}/(32 + \sqrt{79})$ and $\lambda_2 = 1/(2\mu_1) = \frac{15}{2}/(32 - \sqrt{79})$, where $\mu_1 = \frac{1}{15}(32 + \sqrt{79})$ and $\mu_2 = \frac{1}{15}(32-\sqrt{79})$ are the eigenvalues of the matrix $\Sigma M$. The weights $\pi_k$ ($k = 0, \ldots, 4$), rounded to 4 decimal places are

$$\begin{aligned} \pi_0 = 1.0408\,,\ \ \pi_1 = -0.0861\,,\ \ \pi_2 = 0.0520\,,\ \ \pi_3 = -0.0089 \\ \text{and}\quad \pi_4 = 1 - (\pi_0 + \pi_1 + \pi_2 + \pi_3) = 0.0022\,. \end{aligned} \tag{27}$$

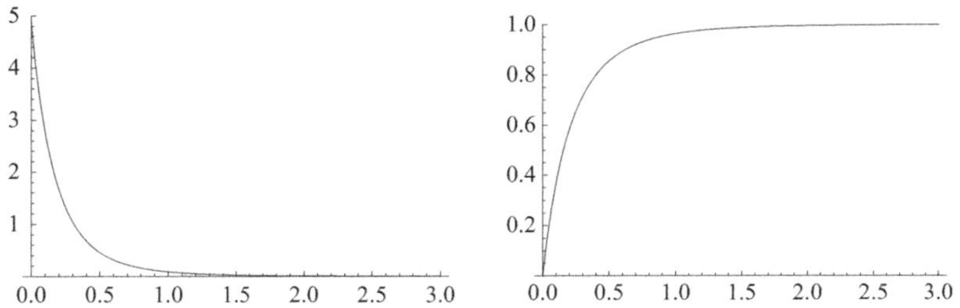


**Fig. 2** Plots of the asymptotic p.d.f. and c.d.f. of $Q_B/Q_W$ for example (4) in Table 4 of Box (1954).

The asymptotic p.d.f. and c.d.f. of $Q_B/Q_W$ are thus given respectively by (21) and (22) with $\lambda = 15/64$, $m^* = 4$, $\pi_j$ $(j = 0, \ldots, 4)$ given by (27), $k_1^* = 6$, $k_2^* = 4$, $k_3^* = 2$, $w_1^* = 1$, $w_2^* = 2$ and $w_3^* = 3$, with $p_{\ell,i}^*$ given by (20) with the $c_{\ell,i}$ given by (26), for $\ell = 1, 2, 3$ and $i = 1, \ldots, k_\ell^*/2$.

The plots of this p.d.f. and c.d.f. are displayed in Fig. 2.

Then, since d.f.$(Q_B) = k - 1 = 2$ and d.f.$(Q_W) = N - k = 12$ and the 0.95 quantile of an $F_{2,12}$ distribution, rounded to 6 decimal places is equal to 3.885294, the probability that the ratio of quadratic forms $V = Q_B/Q_W$ exceeds the 0.95 quantile of the $F_{2,12}$ distribution is (rounded to 6 decimal places)

$$1 - F_V\left(\frac{1}{6}\times 3.885296\right) = 0.092514$$

which exactly matches the exact probability reported by Box (1954) of 0.0925, thus showing the usefulness of this approach in obtaining distributions which have quite simple forms and that lie very close to the exact distribution. We should note that we need to multiply the quantile of the $F_{2,12}$ distribution by 1/6 because the relation that may be established with the $F$ distribution is indeed that of the distribution of the statistic $\frac{Q_B}{d.f.(Q_B)} \Big/ \frac{Q_W}{d.f.(Q_W)}$, where as stated above, $d.f.(Q_B) = 2$ and $d.f.(Q_W) = 12$ so that we have to multiply $Q_B/Q_W$ by $12/2 = 6$, or divide the quantile of the $F_{2,12}$ distribution by 6 in order that the relation may be established.

In Fig. 3 is shown the Mathematica® implementation of this c.d.f., together with the result obtained.

Using a similar procedure, all exact probabilities in Table 4 of Box (1954) may be easily computed and confirmed, with the only remark that the exact probability for example (5) in that Table should rather be 3.25 and not 4.03.

```
k={6,4,2};
w={1,2,3};
c=Makec[k/2,1/(2*w),3];
pi={1.0408,-.0861,.0520,-.0089,.0022};
lambda=15/64;
r=1;
CDFV[z_]:=Sum[Sum[Sum[pi[[j+1]]*
  (Product[(1/(2*w[[ell]]))^(k[[ell]]/2),{ell,1,3}]*
              c[[ell]][[i]]*Gamma[i]/(1/(2*w[[ell]]))^i)*
  BetaRegularized[2*lambda*w[[ell]]*z/(2*lambda*w[[ell]]*z+1),
              r+j,i],{i,1,k[[ell]]/2}],{ell,1,3}],{j,0,4}]

1 - CDFV[1/6*3.885294]
0.0925144
```

**Fig. 3** Mathematica® commands used to implement the c.d.f. of the ratio of quadratic forms $Q_B/Q_W$ and to compute the probability reported in Table 4 of Box (1954).

## 8 Conclusions

Closed form expressions, with all parameters in the distributions with express algebraic expressions, and as such easy to implement with any symbolic software, are provided for the distribution of the linear combination of independent chi-square r.v.'s for the cases where all the chi-square r.v.'s involved have an even number of d.f.'s, both in the case that all coefficients in the linear combination are positive as well as in the case that some are positive and some negative.

Also for the case where all but one of the chi-square r.v.'s have an even number of d.f.'s and all the coefficients in the linear combination are positive, closed form representations are given for both the p.d.f. and the c.d.f. of the distribution of linear combinations of independent chi-square r.v.'s.

For the cases where the d.f.'s of the chi-square r.v.'s in the linear combination may be either even or odd, sharp near-exact approximations are provided, based on mixtures of sums of independent Gamma r.v.'s, all or all but one of them, with integer shape parameters, and thus with closed form expressions for the distribution of these sums. In these cases one may have a choice to make in terms of the common rate parameter to be used for those Gamma r.v.'s. Numerical studies show that in general the best choices are: (i) for cases where the values $w_j k_j$ (where the $w_j$ are the coefficients in the linear combination of the chi-square r.v.'s and the $k_j$ their d.f.'s – see (1)) display an unbalanced set of values, the harmonic mean of the $1/(2w_j)$ or the definition of $\lambda$ as in (vii) in (10), while (ii) for cases where the $w_j k_j$ display a quite balanced set of values, the geometric mean of the $1/(2w_j)$ is the best choice and (iii) for cases where the $w_j$ exhibit values both above and below 1, the choices of $\lambda$ as the maximum of the $1/(2w_j)$ or given by (vi) or (vii) in (10). In any case the choice of defining $\lambda$ as in (vii) of (10) always leads to quite good behaviors, which in cases it is not the best choice, will anyway give very good results.

The exact and near-exact distributions obtained may then be used to obtain exact or asymptotic distributions for ratios of linear combinations of chi-squares and thus also for ratios of quadratic forms, some of which are important statistics in many ANOVA and other statistical models under heterocedasticity. Moreover, these distributions also allowed us to obtain the exact distribution of the Behrens-Fisher statistic.

**Acknowledgments** The author wants to thank the National Research Foundation from South-Africa for its support through Grant number 71199 and also two anonymous referees whose comments and remarks much contributed to a better paper.

## Appendix 1: Notation and Expressions for the p.d.f. and c.d.f. of the GIG and GNIG Distributions

Let $Z_j \sim \Gamma(r_j, \lambda_j)$ $(j = 1, \ldots, p)$ be a set of $p$ independent r.v.'s and consider the r.v.

$$Z = \sum_{j=1}^{p} Z_j .$$

In case all the $r_j \in \mathbb{N}$, the distribution of $Z$ is what we call a GIG (Generalized Integer Gamma) distribution (Coelho, 1998). If all the $\lambda_j$ are different, $Z$ has a GIG distribution of depth $p$, with shape parameters $r_j$ and rate parameters $\lambda_j$, with p.d.f.

$$f_Z(z) = f^{\,GIG}\Big(z \,\Big|\, \{r_j\}_{j=1:p};\, \{\lambda_j\}_{j=1:p};\, p\Big) = K \sum_{j=1}^{p} P_j(z)\, e^{-\lambda_j z}\,, \qquad (z > 0) \tag{28}$$

and c.d.f.

$$F_Z(z) = F^{GIG}\Big(z \,\Big|\, \{r_j\}_{j=1:p};\, \{\lambda_j\}_{j=1:p};\, p\Big) = 1 - K \sum_{j=1}^{p} P_j^*(z)\, e^{-\lambda_j z}\,, \qquad (z > 0)$$

where

$$K = \prod_{j=1}^{p} \lambda_j^{r_j}\,, \qquad P_j(z) = \sum_{k=1}^{r_j} c_{j,k}\, z^{k-1} \tag{29}$$

and

$$P_j^*(z) = \sum_{k=1}^{r_j} c_{j,k}(k-1)! \sum_{i=0}^{k-1} \frac{z^i}{i!\,\lambda_j^{k-i}},$$

with

$$c_{j,r_j} = \frac{1}{(r_j-1)!} \prod_{i=1_{i\neq j}}^{p} (\lambda_i - \lambda_j)^{-r_i}, \quad j = 1, \ldots, p, \tag{30}$$

and, for $k = 1, \ldots, r_j - 1$ and $j = 1, \ldots, p$,

$$c_{j,r_j-k} = \frac{1}{k} \sum_{i=1}^{k} \frac{(r_j-k+i-1)!}{(r_j-k-1)!} R(i, j, p)\, c_{j,r_j-(k-i)}, \tag{31}$$

where

$$R(i, j, p) = \sum_{\substack{k=1\\k\neq j}}^{p} r_k(\lambda_j - \lambda_k)^{-i} \quad (i = 1, \ldots, r_j - 1). \tag{32}$$

In case some of the $\lambda_j$ assume the same value as other $\lambda_j$'s, the distribution of $Z$ still is a GIG distribution, but in this case with a reduced depth. In this more general case, let $\{\lambda_\ell;\, \ell = 1, \ldots, g(\leq p)\}$ be the set of different $\lambda_j$'s and let $\{r_\ell;\, \ell = 1, \ldots, g(\leq p)\}$ be the set of the corresponding shape parameters, with $r_\ell$ being the sum of all $r_j$ $(j \in \{1, \ldots, p\})$ which correspond to the $\lambda_j$ assuming the value $\lambda_\ell$. In this case $Z$ will have a GIG distribution of depth $g$, with shape parameters $r_\ell$ and rate parameters $\lambda_\ell$ $(\ell = 1, \ldots, g)$.

If all r.v.'s have Gamma distributions with different rate parameters $\lambda_j$ and $Z_p$ has a Gamma distribution with a non-integer shape parameter $r_p$, then we will say that the r.v. $Z$ has a GNIG (Generalized Near-Integer Gamma) distribution of depth $p$. The probability density and cumulative distribution functions of $Z$ are, for $z > 0$, respectively given by (Coelho, 2004)

$$f^{GNIG}(z \,|\, r_1, \ldots, r_{p-1}; r_p;\, \lambda_1, \ldots, \lambda_{p-1}; \lambda_p;\, p)$$

$$= K \sum_{j=1}^{p-1} e^{-\lambda_j z} \sum_{k=1}^{r_j} \left\{ c_{j,k} \frac{\Gamma(k)}{\Gamma(k+r_p)} z^{k+r_p-1} {}_1F_1(r_p, k+r_p, -(\lambda_p - \lambda_j)z) \right\}, \tag{33}$$

and

$$F^{GNIG}(z\,|\,r_1,\ldots,r_{p-1};r_p;\;\lambda_1,\ldots,\lambda_{p-1};\lambda_p;\,p)=\frac{\lambda_p^{r_p}\,z^{r_p}}{\Gamma(r_p+1)}\,{}_1F_1(r_p,r_p+1,-\lambda_p z)$$

$$-K\sum_{j=1}^{p-1}e^{-\lambda_j z}\sum_{k=1}^{r_j}c^*_{j,k}\sum_{i=0}^{k-1}\frac{z^{r_p+i}\lambda_j^i}{\Gamma(r_p+1+i)}\,{}_1F_1(r_p,r_p+1+i,-(\lambda_p-\lambda_j)z)\,,$$

with $K=\prod_{j=1}^{p}\lambda_j^{r_j}$ and $c^*_{j,k}=\frac{c_{j,k}\Gamma(k)}{\lambda_j^k}$ and where ${}_1F_1(a,b,z)$ represents the Kummer confluent hypergeometric function.

## Appendix 2: Notation and Expressions for the p.d.f. and c.d.f. of the DGIG Distribution

Let,

$$Z_1^*\sim GIG(r_j,\lambda_j;\;j=1,\ldots,p_1)\,,\quad\text{and}\quad Z_2^*\sim GIG(s_\ell,\nu_\ell;\;\ell=1,\ldots,p_2)\,,$$

be two independent r.v.'s and let

$$Z\;=\;Z_1^*-Z_2^*\,.$$

Then $Z$ has what we call a DGIG (Difference of two GIG) distribution, whose p.d.f. and c.d.f are given by Coelho and Mexia (2010, Sect. 2.5, Thm. 2.1).

Taking

$$K_1=\prod_{j=1}^{p_1}\lambda_j^{r_j}\quad\text{and}\quad K_2=\prod_{\ell=1}^{p_2}\nu_\ell^{s_\ell}\,,$$

and taking $c_{j,k}$ defined in a similar manner to $c_{j,k}$ in (30)–(32), with $p$ replaced by $p_1$ and $d_{\ell,h}$ defined in a corresponding manner, with $p$ replaced by $p_2$, $\lambda_j$ ($j=1,\ldots,p$) replaced by $\nu_\ell$ ($\ell=1,\ldots,p_2$) and $r_j$ ($j=1,\ldots,p$) replaced by $s_\ell$ ($\ell=1,\ldots,p_2$), the p.d.f. of $Z$ is given by

$$f_Z(z)\;=\;\begin{cases}K_1K_2\displaystyle\sum_{j=1}^{p_1}P^{**}_{1j}(z)\,e^{-\lambda_j z} & z\geq 0\\ K_1K_2\displaystyle\sum_{\ell=1}^{p_2}P^{**}_{2\ell}(z)\,e^{\nu_\ell z} & z\leq 0\end{cases}$$

where[1]

$$P^{**}_{1j}(z) = \sum_{k=1}^{r_j} c_{j,k} \sum_{\ell=1}^{p_2}\sum_{h=1}^{s_\ell} d_{\ell,h} \sum_{i=0}^{k-1} \binom{k-1}{i} z^{k-1-i} \frac{(h+i-1)!}{(\lambda_j+\nu_\ell)^{h+i}}$$

and

$$P^{**}_{2\ell}(z) = \sum_{h=1}^{s_\ell} d_{\ell,h} \sum_{j=1}^{p_1}\sum_{k=1}^{r_j} c_{j,k} \sum_{i=0}^{h-1} \binom{h-1}{i} (-z)^{h-1-i} \frac{(k+i-1)!}{(\lambda_j+\nu_\ell)^{k+i}}$$

and the c.d.f. by

$$F_Z(z) = \begin{cases} 1 - K_1K_2 \sum_{j=1}^{p_1} P^{***}_{1j}(z)\, e^{-\lambda_j z} & z \geq 0 \\ K_1K_2 \sum_{\ell=1}^{p_2} P^{***}_{2\ell}(z)\, e^{\nu_\ell z} & z \leq 0 \end{cases}$$

with

$$P^{***}_{1j}(z) = \sum_{k=1}^{r_j} c_{j,k} \sum_{\ell=1}^{p_2}\sum_{h=1}^{s_\ell} d_{\ell,h} \sum_{i=0}^{k-1} \binom{k-1}{i} \frac{(h+i-1)!}{(\lambda_j+\nu_\ell)^{h+i}} \sum_{t=0}^{k-1-i} \frac{(k-1-i)!}{t!} \frac{z^t}{\lambda_j^{k-i-t}}$$

and

$$P^{***}_{2\ell}(z) = \sum_{h=1}^{s_\ell} d_{\ell,h} \sum_{j=1}^{p_1}\sum_{k=1}^{r_j} c_{j,k} \sum_{i=0}^{h-1} \binom{h-1}{i} \frac{(k+i-1)!}{(\lambda_j+\nu_\ell)^{k+i}} \sum_{t=0}^{h-1-i} \frac{(h-1-i)!}{t!} \frac{(-z)^t}{\nu_\ell^{h-i-t}}\,.$$

As a short notation for the p.d.f. and c.d.f. of the r.v. $Z$ we will use

$$f_Z(z) = f^{\,DGIG}\Big(z \,\Big|\, \{r_j\}_{j=1:p_1}; \{s_\ell\}_{\ell=1:p_2}; \{\lambda_j\}_{j=1:p_1}; \{\nu_\ell\}_{\ell=1:p_2}; p_1, p_2\Big), \quad (z \in \mathbb{R})$$

and

$$F_Z(z) = F^{DGIG}\Big(z \,\Big|\, \{r_j\}_{j=1:p_1}; \{s_\ell\}_{\ell=1:p_2}; \{\lambda_j\}_{j=1:p_1}; \{\nu_\ell\}_{\ell=1:p_2}; p_1, p_2\Big), \quad (z \in \mathbb{R}).$$

[1] in the original chapter there were some small typos in $P^{**}_{1j}(z)$, $P^{**}_{2\ell}(z)$, $P^{***}_{1j}(z)$ and $P^{***}_{2\ell}(z)$, that were corrected here, where $r_{1j}$ should be replace by just $r_j$, and $r_{2\ell}$ by $s_\ell$

## References


Alvo, M., Cabilio, P., Feigin, P. D. (1982). Asymptotic theory for measures of concordance with special reference to average Kendall Tau. *Ann. Statist.*, 10, 1269–1276.

Baksalary, J. K., Hauke, J., Styan, G. P. H. (1994). On some distributional properties of quadratic forms in normal variables and on some associated matrix partial orderings. In *Multivariate Analysis and its Applications* (T. W. Anderson, K. T. Fang & I. Olkin, eds.), *IMS Lecture Notes – Monograph Series*, Institute of Mathematical Statistics, Hayward, California, vol. 24, 111–121.

Baldesari, B. (1967). The Distribution of a Quadratic Form of Normal Random Variables. *Ann. Math. Statist.*, 38, 1700–1704.

Behrens, W. H. V. (1929). Ein Beitrang zur Fehlerberchnung bei wenigen Beobachtungen. *Landwirtsch Jahrbucher*, 68, 807–837.

Bock, M. E. (1984). Distribution results for positive definite quadratic forms with repeated roots. Tech. Rep. No. 347, 1984, Department of Statistics, Stanford University, CA.

Bock, M. E., Solomon, H. (1987). Distributions of quadratic forms. Tech. Rep. No. 398, 1987, Department of Statistics, Stanford University, CA.

Box, G. E. P. (1954). Some theorems on quadratic forms applied in the study of Analysis of Variance problems, I. Effect of inequality of variance in the one-way classification. *Ann. Math. Stat.*, 25, 290–302.

Castaño-Martínez, A., López-Blázquez, F. (2005). Distribution of a sum of weighted noncentral chi-square variables. *Test*, 14, 397–415.

Chernoff, H., Lehmann, E. L. (1954). The use of maximum likelihood estimates in $\chi^2$ tests for goodness of fit. *Ann. Math. Statist.*, 25, 579–586.

Coelho, C.A. (1998). The Generalized Integer Gamma Distribution – a Basis for Distributions in Multivariate Statistics. *J. Multivariate Anal.*, 64, 86–102.

Coelho, C.A. (2004). The generalized near-integer Gamma distribution: a basis for 'near-exact' approximations to the distribution of statistics which are the product of an odd number of independent Beta random variables. *J. Multivariate Anal.*, 89, 191–218.

Coelho, C. A., Arnold, B. C. (2019). *Finite Forms Representations for Meijer G and Fox H Functions – Applied to Multivariate Likelihood Ratio Tests Using Mathematica®*, MAXIMA *and* R. Lecture Notes in Statistics, Springer.

Coelho, C.A., Mexia, J.T. (2007). On the Distribution of the Product and Ratio of Independent Generalized Gamma-Ratio Random Variables, *Sankhyā*, 69, 221–255.

Coelho, C.A., Mexia, J.T. (2010). *Product and Ratio of Generalized Gamma-Ratio Random Variables: Exact and Near-exact Distributions - Applications*. Lambert Academic Publishing AG & Co. KG, Saarbrücken, Germany, 145+v pp. (ISBN: 978-3-8383-5846-8).

Davies, R. B. (1980). Algorithm AS 155. The Distribution of a Linear Combination of $\chi^2$ Random Variables. *J. Roy. Statist. Soc. Ser. C*, 29, 323–333.

Davis, A. W. (1977). A differential equation approach to linear combinations of independent chi-squares. *J. Amer. Statist. Assoc.*, 72, 212–214.

Durbin, J., Knott, M., Taylor, C. C. (1975). Components of Cramer-von Mises statistics. II. *J. Roy. Statist. Soc. Ser. B*, 37, 216–237.

Fisher, R. A. (1939). The comparison of samples with possibly unequal variances. *Ann. Eugenics*, 9, 174–180.

Fleiss, J. L. (1971). On the distribution of a linear combination of independent chi-squares. *J. Amer. Statist. Assoc.*, 66, 142–144.

Gabler, S., Wolff, C. (1987). A quick and easy approximation to the distribution of a sum of weighted chi-square variables. *Statist. Papers*, 28, 317–325.

Gil-Pelaez, J. (1951). Note on the Inversion Theorem. *Biometrika*, 38, 481–482.

Ha, H.-T., Provost, S. B. (2013). An accurate approximation to the distribution of a linear combination of non-central chi-square random variables. *REVSTAT – Stat. Journal*, 11, 231–254.

Imhof, J. P. (1961). Computing the distribution of quadratic forms in normal variables. *Biometrika*, 48, 419–426.

Johnson, N. L., Kotz, S. (1968). Tables of distributions of positive definite quadratic forms in central normal variables. *Sankhyā*, Ser. B, 30, 303–314.

Johnson, L., Kotz, S., Balakrishnan, N. (1994). *Continuous Univariate Distributions*, Vol. 1, 2nd Ed. Wiley series in Probability and Statistics, J. Wiley & Sons, New York.

Kotz, S., Johnson, N. L., Boyd, D. W. (1967a). Series representations of distributions of quadratic forms in normal variables. I. Central case. *Ann. Math. Statist.*, 38, 823–837.

Kotz, S., Johnson, N. L., Boyd, D. W. (1967b). Series representations of distributions of quadratic forms in normal variables. II. Non-central case. *Ann. Math. Statist.*, 38, 838–848.

Lu, Z.-H. (2006). The numerical evaluation of the probability density function of a quadratic form in normal variables. *Comput. Statist. Data Anal.*, 51, 1986–1996.

Mathai, A. M. (1982). Storage capacity of a dam with gamma type inputs. *Ann. Inst. Statist. Math.*, 34, 591–597.

Mathai, A. M., Provost, S. B. (1992). *Quadratic Forms in Random Variables*. Marcel Dekker, Inc., New York.

Morin-Wahhab, D. (1988). On some functions of quadratic forms and related topics. Ph. D. Thesis, Department of Mathematics and Statistics, McGill University, Montréal, Québec, Canada.

Moschopoulos, P. G. (1985). The distribution of the sum of independent Gamma random variables. *Ann. Inst. Statist. Math.*, 37, 541–544.

Moschopoulos, P. G., Canada, W. B. (1984). The distribution function of a linear combination of chi-squares. *Comput. Math. Appl.*, 10, 383–386.

Provost, S. B. (1988). The exact density of a general linear combination of gamma variates. *Metron*, 46, 61–69.

Provost, S. B. (1989). On sums of independent gamma random variables. *Statistics*, 20, 583–591.

Provost, S. B., Rudiuk, E. M. (1996). The exact distribution of indefinite quadratic forms in noncentral normal vectors. *Ann. Inst. Statist. Math.*, 48, 381–394.

Robbins, H. (1948). The distribution of a definite quadratic form. *Ann. Math. Statist.*, 19, 266–270.

Robbins, H., Pitman, E. J. G. (1949). Application of the method of mixtures to quadratic forms in normal variates. *Ann. Math. Statist.*, 20, 552–560.

Ruben, H. (1960). Probability content of regions under spherical normal distributions, I. *Ann. Math. Statist.*, 31, 598–618.

Ruben, H. (1962). Probability content of regions under spherical normal distributions, IV: The distribution of homogeneous and non-homogeneous quadratic functions of normal variables. *Ann. Math. Statist.*, 33, 542–570.

Satterthwaite, F. E. (1941). Synthesis of variance. *Psychometrika*, 6, 309–316.

Shah, B. K. (1970). Distribution theory of a positive definite quadratic form with matrix argument. *Ann. Math. Statist.*, 41, 692–697.

Slater, L. J. (1972). Confluent Hypergeometric Functions, in Abramowitz, M. and Stegun, I. A. (Eds.) Handbook of Mathematical Functions with Formulas, Graphs, and Mathematical Tables, National Bureau of Standards Applied Mathematics Series, # 55. U.S. Government Printing Office, Washington, D.C. 10th Printing, 1972.

Solomon, H., Stephens, M. A. (1977). Distribution of a sum of weighted chi-square variables. *J. Amer. Statist. Assoc.*, 72, 881–885.

Welch, B. L. (1947). The generalization of 'Student's' problem when several different population variances are involved. *Biometrika*, 34, 28–35.

# On the Distribution of the Product of Independent Beta Random Variables — Applications

**Carlos A. Coelho and Rui P. Alberto**

**Abstract** A first approach, based on recently obtained asymptotic expansions of ratios of gamma functions, enables the obtention of the distribution of the product of independent and identically distributed random variables in a much manageable form. However, for the general case, this approach leads to a form which although being very manageable and in line with some previous results, suffers from serious problems of precision and convergence, which have been completely overlooked by other authors and which in most cases prevent its practical use. Nevertheless, it is based on these first results that the authors, using the concept of near-exact distribution, are able to obtain highly manageable but extremely accurate approximations for all cases of the distribution of the product of independent Beta random variables. These near-exact approximations, given their high manageability, accuracy and proximity to the exact distribution, may in practice be used instead of the exact distribution.

Reprinted with permission from Springer Nature. First published as Coelho, C.A., Alberto, R.P.: On the Distribution of the Product of Independent Beta Random Variables – Applications, in 'Advances in Statistics – Theory and Applications – Honoring the Contributions of Barry C. Arnold in Statistical Science', Ghosh, I., Balakrishnan, N., Ng, H.K.T. (eds.), Springer Series in Emerging Topics in Statistics and Biostatistics, 85–117 (2021). (https://doi.org/10.1007/978-3-030-62900-7_5)

**Supplementary Information** The online version contains supplementary material available at https://doi.org/10.1007/978-3-031-68008-3_15.

C. A. Coelho (✉)
Departamento de Matemática and Centro de Matemática e Aplicações, FCT NOVA, NOVA University of Lisbon, Lisbon, Portugal
e-mail: cmac@fct.unl.pt

R. P. Alberto
Centro de Matemática e Aplicações, FCT NOVA, NOVA University of Lisbon, Lisbon, Portugal
e-mail: rui.pesado@sapo.pt

 
B. C. Arnold et al. (eds.), *Likelihood Ratio Tests and Related Topics in Multivariate Analysis*, Contributions to Statistics,
https://doi.org/10.1007/978-3-031-68008-3_15

# 1 Introduction

The distribution of the product of independent Beta random variables (r.v.'s) is a distribution which plays a key role in Statistics. There are many likelihood ratio test (l.r.t.) statistics, namely in Multivariate Analysis, whose distribution has been shown to be that of the product of a number of independent Beta r.v.'s, as for example the l.r.t. statistics to test, under multivariate normality, or rather, under multivariate elliptically contoured or left orthogonal-invariant distributions, the independence of several sets of variables, the equality of several mean vectors, the equality of several variance-covariance matrices or sphericity of the covariance matrix (Anderson, 2003; Kshirsagar, 1972; Marques et al., 2011; Muirhead, 2005), the l.r.t. statistics to test similar hypotheses under the complex multivariate normal setting (Coelho et al., 2015; Gupta, 1971; Khatri, 1965a; Pillai and Jouris, 1971), the l.r.t. statistics to test multi-sample sphericity or block-scalar or block-matrix sphericity (Coelho and Marques, 2011a; Marques and Coelho, 2011a,b), the l.r.t. statistic to test reality of a covariance matrix in a complex Normal distribution (Khatri, 1965b) or the l.r.t. statistics to test circularity of the covariance matrix and stationarity (Olkin and Press, 1961), or even the l.r.t. statistic to test the equality of several exponential distributions, in the case of equal sample sizes (Coelho and Marques, 2011b).

Although along the years many authors have worked on obtaining the distribution of the product of independent Beta r.v.'s in a manageable form or otherwise obtaining manageable approximations for its distribution (Bharbava and Khatri, 1981; Carter and Springer, 1977; Mathai, 1984; Nagar et al., 1985; Nagarsenker and Das, 1975; Nagarsenker and Suniaga, 1983; Nandi, 1980; Pederzoli, 1985; Springer, 1979; Springer and Thompson, 1966, 1970; Tang and Gupta, 1984, 1986; Tretter and Walster, 1975; Tukey and Wilks, 1946; Walster and Tretter, 1980), obtaining an explicit, accurate, and highly manageable expression for both the probability density function (p.d.f.) and the cumulative distribution function (c.d.f.) of this distribution has been a hard task and we are absolutely sure that there is still much room left for improvement.

Our aim is exactly to obtain explicit highly manageable expressions for both the p.d.f. and c.d.f. of extremely well-fitting approximations for this distribution, based on highly accurate and manageable mixture expansions.

One popular form to represent the exact p.d.f. and c.d.f. of the distribution of the product of independent Beta r.v.'s has been the use of either the Meijer G or Fox's H functions (Carter and Springer, 1977; Springer, 1979). However, although this is a very handy way to represent these distributions, these representations are not very adequate in practical terms since both Fox's H function and Meijer G function implementations have their drawbacks not only in terms of precision, but mainly in terms of computation time in all the commonly available software.

Other authors like Bharbava and Khatri (1981) and Pederzoli (1985) used multiple series representations where each summation corresponds to a different Beta r.v. in the product, yielding a too much complicated structure for the distribution, mainly when the number of Beta r.v.'s involved is rather large.

Yet other authors like Nagarsenker and Das (1975), Nandi (1980), Nagarsenker and Suniaga (1983) and Tang and Gupta (1984) express the distribution of the product of independent Beta r.v.'s as infinite mixtures of Beta distributions. However these representations have the drawback of needing, in the general case, a large number of terms to attain the desired accuracy and not allowing for the development of further better performing approximations, while are also not able to take advantage from those cases where some of the Beta r.v.'s have the same distribution. Furthermore, the distribution in Nandi (1980) and Tang and Gupta (1984) depends on the particular ordering of the Beta r.v.'s considered, while the series in Nagarsenker and Das (1975) and Nagarsenker and Suniaga (1983) have highly complicated coefficients.

The papers by Tang and Gupta (1986), Mathai (1984), and Nagar et al. (1985) present results which match the results of what is our first approach to the problem; however without any reference to accuracy and convergence issues which indeed occur with the series obtained. Moreover, opposite to the results obtained in the present paper, in Nagar et al. (1985) the weights in the infinite mixture have a very complicated formulation, while in Tang and Gupta (1986) there is a parameter left with a not well-defined value.

Yet, our approach enables the use of more adequate expansions for the cases where some of the Beta r.v.'s involved in the product have the same distribution. In a further step we will determine the weights in our mixture distributions by matching some of the first exact moments, what will enable us to obtain distributions which will converge on the whole support of the r.v., overcoming the problems mentioned in Lemma 3.1 and Remark 3.1 in Tang and Gupta (1986). We may note that this moment matching approach is supported by the well known fact that the product of independent Beta r.v.'s, having a bounded support, has its distribution determined by its moments.

In this paper the authors show how, based on asymptotic expansions of the ratio of two gamma functions, presented by Burić and Elezović (2011) but for which much simpler proofs are shown in Appendix A, it is possible to obtain a single mixture of exponentiated gamma distributions as an asymptotic approximation for the exact distribution of the product of any number of independent Beta r.v.'s, with easily computable coefficients. However, in practical terms this approach leads to almost impossible to handle difficulties, arising from the fact that the resulting series distributions are either very slowly convergent or even divergent.

To overcome these difficulties the authors, recover the concept of near-exact distribution (see Sect. 4), and based on this approach they develop near-exact approximations which lie very close to the exact distribution of the product of independent Beta r.v.'s, although remaining highly manageable, while displaying much better performances than previous approaches proposed by different authors.

As already mentioned, many likelihood ratio test statistics have the same distribution as that of a product of independent Beta r.v.'s. In these cases, the first parameter in the distribution of these Beta r.v.'s is directly related with the sample sizes, while the second parameter is commonly directly related with the number of variables. The near-exact distributions developed by the authors, when applied to these settings, show very good performances even for situations in which the sample sizes are very small, that is, barely exceeding the number of variables, even when the number of variables involved is large.

The setup: let

$$X_j \sim Beta(a_j, b_j) \quad \text{and} \quad Y_j = -\log X_j\,, \quad j = 1, \dots, p\,, \tag{1}$$

be a set of $p$ independent random variables and let

$$Z = \prod_{j=1}^{p} X_j\,, \quad W = -\log Z = -\sum_{j=1}^{p} \log X_j = \sum_{j=1}^{p} Y_j\,. \tag{2}$$

We will say that the random variable $Y = -\log X$ has a $Logbeta(a, b)$ distribution and we are interested in the distribution of $Z$, or, somehow equivalently, in the distribution of $W$.

At first sight we may think that a good idea to work around the difficulties of the representation of the distribution of $Z$ might be to try to express the distribution of each $X_j$ or $Y_j$ as a mixture.

In order to obtain an asymptotic expansion for the distribution of a Logbeta r.v., we might think about using expression following expression (18) together with expression (19) in Tricomi and Erdélyi (1951) or expression (1) in Fields (1966) which may be written, for $|arg(z+\alpha)| < \pi$, as $z \to \infty$, as

$$\frac{\Gamma(z+\alpha)}{\Gamma(z+\beta)} \approx \sum_{k=0}^{\infty} \frac{1}{k!} \frac{\Gamma(1+\alpha-\beta)}{\Gamma(1+\alpha-\beta-k)} B_k^{(1+\alpha-\beta)}(\alpha)\, z^{\alpha-\beta-k}\,, \tag{3}$$

where $B_j^{(n)}(\cdot)$ is the generalized Bernoulli polynomial of degree $j$ and order $n$.

In fact, an application of expression (3), with $z = a - \mathrm{i}t$, $\alpha = 0$ and $\beta = b$, yields

$$\frac{\Gamma(a-\mathrm{i}t)}{\Gamma(a+b-\mathrm{i}t)} \approx \sum_{k=0}^{\infty} p_k(b)\,(a-\mathrm{i}t)^{-b-k} \quad (\text{as } a \to \infty) \tag{4}$$

where

$$p_k(b) = \frac{1}{k!} \frac{\Gamma(1-b)}{\Gamma(1-b-k)} B_k^{(1-b)}(0)$$

with $B_k^{(1-b)}(0)$ being the generalized Bernoulli number of degree $k$ and order $1-b$.

The result in (4) enables us to represent asymptotically any $Logbeta(a,b)$ distribution as an infinite mixture of $\Gamma(b+k,a)$ distributions ($k=0,1,\dots$) — see Appendix B in the supplementary material for the notation used for the gamma distribution. Indeed, from (4) we may then write the characteristic function (c.f.) of $Y_j$ as

$$\begin{aligned}\Phi_{Y_j}(t) &= E\left(e^{\mathrm{i}tY_j}\right) = E\left(X_j^{-\mathrm{i}t}\right) = \frac{\Gamma(a_j+b_j)}{\Gamma(a_j)}\,\frac{\Gamma(a_j-\mathrm{i}t)}{\Gamma(a_j+b_j-\mathrm{i}t)}\\ &\approx \sum_{k=0}^{\infty}\underbrace{\frac{\Gamma(a_j+b_j)}{\Gamma(a_j)}\,\frac{p_k(b_j)}{a_j^{b_j+k}}}_{p_k^*(a_j,b_j)}\,a_j^{b_j+k}\,(a_j-\mathrm{i}t)^{-(b_j+k)}\,, \quad (\text{as } a_j\to\infty)\end{aligned}$$

which is the c.f. of an infinite mixture of $\Gamma(b_j+k,a_j)$ distributions, with weights $p_k^*(a_j,b_j)$ ($k=0,1,\dots$).

Then in order to obtain an asymptotic representation for the distribution of $W$ we might think about convoluting $p$ independent Logbeta distributions in the form of mixtures of Gamma distributions, to obtain then, by exponentiation, the distribution of the corresponding product of independent Beta r.v.'s.

However, the problem is that in this case the asymptotic distribution of $W$ would be an infinite mixture of sums of independent gamma random variables, with possibly different rate parameters $a_j$. Then, the distribution of each of these sums would have itself to be expressed in the form of a mixture, rendering the final expression for the whole distribution not manageable at all.

To overcome these difficulties, our aim is to approximate asymptotically the distribution of $W$ by a single infinite mixture of gamma distributions. This will be achieved with the approach followed in the next section.

## 2 Some Results Concerning Ratios of Gamma Functions

Let, for the $a_j$ in (1),

$$a = \min(a_1, a_2, \dots, a_p)\,, \tag{5}$$

and then, for

$$\alpha_j = a_j - a \quad \text{and} \quad \beta_j = a_j + b_j - a = \alpha_j + b_j\,, \quad (j = 1,\dots,p)\,, \tag{6}$$

using (3) with $z = a - it$, write (as $a \to \infty$)

$$\frac{\Gamma(a_j - \mathrm{i}t)}{\Gamma(a_j + b_j - \mathrm{i}t)} = \frac{\Gamma(a - \mathrm{i}t + \alpha_j)}{\Gamma(a - \mathrm{i}t + \beta_j)}$$
$$\approx \sum_{k=0}^{\infty} \frac{1}{k!} \frac{\Gamma(1 - b_j)}{\Gamma(1 - b_j - k)} B_k^{(1-b_j)}(\alpha_j)\,(a - \mathrm{i}t)^{-b_j-k}$$
$$= \Gamma(1-b_j)\,(a-\mathrm{i}t)^{-b_j} \sum_{k=0}^{\infty} \frac{1}{k!\,\Gamma(1-b_j-k)} B_k^{(1-b_j)}(\alpha_j)\,(a-\mathrm{i}t)^{-k}\,.$$

Then, for $W$ in (2), we may write, as $a \to \infty$,

$$\Phi_W(t) = \prod_{j=1}^{p} \frac{\Gamma(a_j + b_j)}{\Gamma(a_j)} \frac{\Gamma(a_j - \mathrm{i}t)}{\Gamma(a_j + b_j - \mathrm{i}t)} \tag{7}$$

$$\approx C\,(a - \mathrm{i}t)^{-b} \prod_{j=1}^{p} \sum_{k=0}^{\infty} \underbrace{\frac{1}{k!\,\Gamma(1 - b_j - k)} B_k^{(1-b_j)}(\alpha_j)}_{q_{jk}}\,(a - \mathrm{i}t)^{-k}\,, \tag{8}$$

where

$$C = \prod_{j=1}^{p} \frac{\Gamma(1 - b_j)\,\Gamma(a_j + b_j)}{\Gamma(a_j)} \quad \text{and} \quad b = \sum_{j=1}^{p} b_j\,. \tag{9}$$

In (8) we have

$$\prod_{j=1}^{p} \sum_{k=0}^{\infty} q_{jk}\,(a - \mathrm{i}t)^{-k} = \sum_{k=0}^{\infty} w_k\,(a - \mathrm{i}t)^{-k}\,,$$

where

$$w_k = \sum_{r_1+r_2+\cdots+r_p=k} q_{1r_1} q_{2r_2} \ldots q_{pr_p}\,,$$

with each sequence $\{r_1, r_2, \ldots, r_p\}$ being a weak composition of the integer $k$ into $p$ parts, and being the sum extended to the complete set of the above compositions, which cardinality is $\binom{k+p-1}{p-1} = \binom{k+p-1}{k}$ (see for example Heuback and Mansour (2009)).

We may thus write

$$\Phi_W(t) \approx C \sum_{k=0}^{\infty} \frac{w_k}{a^{b+k}}\, a^{b+k}\,(a-\mathrm{i}t)^{-(b+k)} \quad (\text{as } a \to \infty) \tag{10}$$

which shows that for large enough $a$ and for $b$ in (9) the c.f. of $W$ in (2) should be well approximated by the c.f. of an infinite mixture of $\Gamma(b+k,a)$ $(k=0,1,\dots)$ distributions, with weights $Cw_k/a^{b+k}$.

However, in the more general case we may have each pair $(a_j, b_j)$ in (7) repeated say $m_j$ times. Although in this case we could still use the same approach as above, it happens that for each set of random variables with the same set of parameters $(a_j, b_j)$ the computation of the weights in the mixture will be much more efficient if carried out through a slightly different way. In this case, instead of (3) we should use the more general expression

$$\left(\frac{\Gamma(z+\alpha)}{\Gamma(z+\beta)}\right)^m \approx \sum_{k=0}^{\infty} \nu_{k,m}(\alpha,\beta)\, z^{m(\alpha-\beta)-k}\,, \quad (\text{as } z\to\infty,\ |arg(z+\alpha)|<\pi)\,, \tag{11}$$

with

$$\nu_{k,m}(\alpha,\beta) = \frac{1}{k}\sum_{j=1}^{k} j\,\delta_{j,m}(\alpha,\beta)\,\nu_{k-j,m}(\alpha,\beta)\,, \quad (k=1,2,\dots), \quad \nu_{0,m}(\alpha,\beta)=1\,, \tag{12}$$

where

$$\delta_{j,m}(\alpha,\beta) = (-1)^j m\, \frac{B_{j+1}(\beta)-B_{j+1}(\alpha)}{j(j+1)}\,, \tag{13}$$

and where $B_j(a)$ represents the value of the Bernoulli polynomial of degree $j$ in $a$.

Expression (11) is expression (2.1) in Burić and Elezović (2011) and it may be quite easily obtained from Barnes expression (37). See Appendix A for a simpler and straightforward demonstration. The use of (11) is shown in the next section.

# 3 Main Results Concerning the Exact Distribution of the Product of Independent Beta Random Variables

## 3.1 The General Case

In this "general case" we will allow some of the Beta r.v.'s to be identically distributed. Let then

$$X_\ell \sim Beta(a_\ell, b_\ell) \qquad \ell = 1, \dots, p$$

be a set of $p$ independent random variables. Let then $q$ be the number of distinct pairs of parameters $(a_\ell, b_\ell)$ among the $p$ random variables $X_\ell$ and let $(a_j, b_j)$ $(j = 1, \dots, q)$ represent the set of distinct pairs $(a_\ell, b_\ell)$ $(\ell = 1, \dots, p)$. Let further $m_j$ be the number of times the pair of parameters $(a_j, b_j)$ appears among the $p$ random variables $X_\ell$, thus with

$$p = \sum_{j=1}^{q} m_j\,,$$

and of course with $q = p$ if all $m_j = 1$.

Since from (11), taking $a$ as defined in (5), $z = a - \mathrm{i}t$, and $\alpha_j$ and $\beta_j$ as in (6), we may write

$$\left(\frac{\Gamma(a_j - \mathrm{i}t)}{\Gamma(a_j + b_j - \mathrm{i}t)}\right)^{m_j} = \left(\frac{\Gamma(a - \mathrm{i}t + \alpha_j)}{\Gamma(a - \mathrm{i}t + \beta_j)}\right)^{m_j}$$
$$\approx \sum_{k=0}^{\infty} \nu_{k,m_j}(\alpha_j, \beta_j)\,(a - \mathrm{i}t)^{-m_j b_j - k}, \tag{14}$$

we may thus write, for $W$ in (2),

$$\Phi_W(t) = \prod_{j=1}^{q} \left(\frac{\Gamma(a_j + b_j)}{\Gamma(a_j)}\right)^{m_j} \left(\frac{\Gamma(a_j - \mathrm{i}t)}{\Gamma(a_j + b_j - \mathrm{i}t)}\right)^{m_j} \tag{15}$$

$$\approx C\,(a - \mathrm{i}t)^{-b} \prod_{j=1}^{q} \sum_{k=0}^{\infty} \nu_{k,m_j}(\alpha_j, \beta_j)\,(a - \mathrm{i}t)^{-k} \quad (\text{as } a \to \infty) \tag{16}$$

where $\nu_{k,m_j}(\alpha_j, \beta_j)$ $(k = 0, 1, \dots;\ j = 1, \dots, q)$ are given by (12)–(13) and

$$C = \prod_{j=1}^{q} \left(\frac{\Gamma(a_j + b_j)}{\Gamma(a_j)}\right)^{m_j} \quad \text{and} \quad b = \sum_{j=1}^{q} m_j b_j\,. \tag{17}$$

In (16), we may set

$$\prod_{j=1}^{q}\sum_{k=0}^{\infty} \nu_{k,m_j}(\alpha_j,\beta_j)\,(a-\mathrm{i}t)^{-k} = \sum_{k=0}^{\infty}\omega_k\,(a-\mathrm{i}t)^{-k}$$

with

$$\omega_k = \sum_{r_1+r_2+\dots+r_q=k} \nu_{r_1,m_1}(\alpha_1,\beta_1)\,\nu_{r_2,m_2}(\alpha_2,\beta_2)\,\dots\,\nu_{r_q,m_q}(\alpha_q,\beta_q)\,, \tag{18}$$

where, as in the previous section, each sequence $\{r_1, r_2, \dots, r_q\}$ is a weak composition of the integer $k$ into $q$ parts, and the sum is extended to the complete set of these compositions, whose cardinality is $\binom{k+q-1}{q-1} = \binom{k+q-1}{k}$.

From (16) and (18) we may write

$$\Phi_W(t) \approx C\sum_{k=0}^{\infty}\frac{\omega_k}{a^{b+k}}\,a^{b+k}\,(a-\mathrm{i}t)^{-(b+k)} \quad (\text{as } a\to\infty) \tag{19}$$

for $a$ as in (5), $C$ and $b$ given by (17) and $\omega_k$ given by (18).

Expression (19) shows that for sufficiently large $a$ the exact distribution of $W$ is well approximated by an infinite mixture of $\Gamma(b+k,a)$ distributions, with weights $C\omega_k/a^{b+k}$ $(k=0,1,\dots)$, and as such, the exact distribution of $Z=e^{-W}$ will be also well approximated by the corresponding mixture, that is, with the same weights, of exponentiated $\Gamma(b+k,a)$ distributions.

From (19) we may thus write

$$f_W(w) \approx C\sum_{k=0}^{\infty}\frac{\omega_k}{\Gamma(b+k)}\,e^{-aw}\,w^{b+k-1}\,, \qquad (w>0)\,,$$

and

$$F_W(w) \approx C\sum_{k=0}^{\infty}\omega_k\,\frac{\Gamma^*(b+k,aw)}{\Gamma(b+k)}\,, \qquad (w>0)\,,$$

where

$$\Gamma^*(b+k,aw) = \int_0^{aw} e^{-aw}\,w^{b+k-1}\,dw$$

is a version of the incomplete gamma function, and therefore also

$$f_Z(z) \approx C\sum_{k=0}^{\infty}\frac{\omega_k}{\Gamma(b+k)}\,z^{a-1}\,(-\log z)^{b+k-1}\,, \qquad (0<z<1)\,,$$

and

$$F_Z(z) \approx 1 - C \sum_{k=0}^{\infty} \omega_k \, \frac{\Gamma^*(b+k, -a \log z)}{\Gamma(b+k)}, \qquad (0 < z < 1).$$

However, as we will see in Sect. 6, with the help of some numerical studies, the practical implementation of these representations faces some problems which are not easy to solve and which are related with the facts that

i) while for sets of Beta r.v.'s with values of $a_j$ which show a moderately large variability we may need an unsoundly large number of terms in the series to get a good approximation,
ii) on the other hand, the use of a larger number of terms may start to give worse approximations, namely for sets of Beta r.v.'s with values of $a_j$ which show a small variability, while the use of a moderately large number of terms may still not give the desired precision.

We should remark that these problems are not brought to our attention by other authors who, although using different approaches, obtain similar results, as it is the case of Tang and Gupta (1986), Mathai (1984) and Nagar et al. (1985).

We should also note here that the approach followed, in case all $m_j$ in (14) and (16) are equal to 1, yields the approach of Moschopoulos (1986). Although the use of the $m_j$ with values different from 1 may seem a minor detail, that is indeed not so, since this approach, where general $m_j$ are used, may allow for much better approximations than the ones that would be obtained if such approach is not taken. This is true even when we will be interested in developing near-exact distributions, rather than just common asymptotic distributions, as it is shown later in Sect. 7.1, where the use of such expressions enables the obtention of even better near-exact distributions than the ones previously obtained.

If some restrictions are met by the parameters of the Beta distributed r.v.'s in (1) and (2), the distribution of $Z$ and also that of $W$ in (2) may then have a finite closed form representation, and a number of l.r.t. statistics fall in this case. For details on this topic we refer the reader to Coelho and Arnold (2019).

## 3.2 The Particular Case of All Different Sets of Pairs of Parameters

In case each Beta r.v. has a different set of parameters, then all the $m_j$ defined in the previous subsection will be equal to 1 and we will have $q = p$. In this case $W$ will have an approximate c.f. still given by (16) and (19) in the previous subsection, for all $m_j = 1$, or, if we prefer, given by (8) or (10) in Sect. 2.

### 3.3 The Particular Case of All Equal Sets of Pairs of Parameters

In case all $p$ random variables $X_j$ have the same distribution, although remaining independent, then we may either take $q = p$, with all $m_j = 1$ and then apply the results in Sect. 3.1, or take $q = 1$, with $m_1 = p$ and then apply the results in Sect. 3.1. From the derivations in Sect. 2 we are able to see that the second choice is far more efficient in terms of the computation of the weights $\omega_k$.

However, using expression (45) in Appendix A, we may obtain a much faster converging series or mixture distribution. This expression is expression (7.1) in Burić and Elezović (2011), for which we provide a much simpler and straightforward demonstration path in Appendix A.

In this case, if we take $a_j = a$ and $b_j = b$ ($j = 1, \ldots, p$), and if we also take in (45), $z = a - \mathrm{i}t$, $\alpha = 0$ and $\beta = b$, we may write

$$
\begin{aligned}
\Phi_W(t) &= \left(\frac{\Gamma(a+b)}{\Gamma(a)}\,\frac{\Gamma(a-\mathrm{i}t)}{\Gamma(a+b-\mathrm{i}t)}\right)^p \\
&\approx C\sum_{k=0}^{\infty} \nu_{2k,p}(-k, b-k)\left(a - \mathrm{i}t + \frac{b-1}{2}\right)^{-pb-2k} \\
&= C\sum_{k=0}^{\infty} \frac{\nu_{2k,p}(-k, b-k)}{\left(a+\frac{b-1}{2}\right)^{pb+2k}}\left(a+\frac{b-1}{2}\right)^{pb+2k}\left(a+\frac{b-1}{2}-\mathrm{i}t\right)^{-(pb+2k)}
\end{aligned}
$$

with $\nu_{2k,p}(-k, b-k)$ given by (12)–(13) and

$$
C = \left(\frac{\Gamma(a+b)}{\Gamma(a)}\right)^p,
$$

which shows that in this case the distribution of $W$ is asymptotically approximated, for increasing $a+\frac{b-1}{2}$, by an infinite mixture of $\Gamma\left(pb+2k, a+\frac{b-1}{2}\right)$ distributions with weights $C\nu_{2k,p}(-k, b-k)\Big/\left(a+\frac{b-1}{2}\right)^{pb+2k}$ $(k = 0, 1, \ldots)$.

### 3.4 The Distribution of the Product of Powers of Independent Beta Random Variables

So far we have only dealt with the distribution of the product of independent Beta r.v.'s as expressed in (2), that is, with the Beta r.v.'s raised to the power 1. Of course, the results obtained are easily extended to the case where all the Beta r.v.'s are raised to some common positive power. In this case one only has to consider the distribution of the product of the random variable $W = -\log Z$ considered so far, multiplied by that power or the distribution of the exponential of this random variable, if the distribution of $Z$ is desired, being both very simple to obtain by simple transformation.

However, there may be cases where we are interested in the distribution of the random variable

$$Z = \prod_{j=1}^{p} X_j^{c_j}$$

where $X_j$ are as in (2) and the $c_j$ are positive reals. In this case the problem becomes much harder to tackle.

As in Sect. 3.1, let us suppose that among the $p$ random variables $X_\ell$ ($\ell = 1, \ldots, p$) there are $q \leq p$ of them with different first and second parameter or raised to a different power, that is, let us suppose that there are $q \leq p$ different triplets $(a_j, b_j, c_j)$ ($j = 1, \ldots, q$).

Let us further suppose that the triplet $(a_j, b_j, c_j)$ appears $m_j$ times ($j = 1, \ldots, q$), once again with $p = \sum_{j=1}^{q} m_j$. Then, for $a = \min(a_1, \ldots, a_q)$ and for $\alpha_j$ and $\beta_j$ given by (6), we may write the c.f. of $W = -\log Z$ as

$$\begin{aligned}
\Phi_W(t) &= \prod_{j=1}^{q} \left(\frac{\Gamma(a_j + b_j)}{\Gamma(a_j)}\right)^{m_j} \left(\frac{\Gamma(a_j - c_j\mathrm{i}t)}{\Gamma(a_j + b_j - c_j\mathrm{i}t)}\right)^{m_j} \\
&\approx C \prod_{j=1}^{q} \sum_{k=0}^{\infty} \nu_{k,m_j}(\alpha_j, \beta_j)\, (a - c_j\mathrm{i}t)^{-(m_j b_j + k)} \\
&= C \prod_{j=1}^{q} \sum_{k=0}^{\infty} \nu_{k,m_j}(\alpha_j, \beta_j) \left(\frac{a}{c_j} - \mathrm{i}t\right)^{-(m_j b_j + k)} (c_j)^{-(m_j b_j + k)}\,.
\end{aligned}$$

The problem now is hard to tackle because the rate parameters $a/c_j$ are now function of $j$, that is, their value varies now with $j$, and there is no simple way to work around this fact.

Of course we may take each set of Beta r.v.'s with parameters $(a_j, b_j, c_j)$ individually and take the approach first outlined in Sect. 2, but this, as stated in Sect. 2, would lead us to an almost intractable infinite mixture of infinite mixtures, with the further concomitant even arduous problems related with the convergence of truncations of these distributions, if we try to use them in applications.

Although in this case the Fox $H$ function gives a nice way to express both the exact p.d.f. and c.d.f. of $Z$, there seems to be no really satisfactory formulation for the distribution of $Z$ for practical uses, in this case. However, as we will see in Sects. 5 and 7, the near-exact approach is able to handle this case with no big problems and extremely satisfactory results. See Sect. 5.2 for the theoretical developments and Sect. 7.3 for an application and some numerical results.

# 4 The Practical Implementation of the Exact Distribution in Sect. 3 — Another Look at the Exact Distribution on the Way to Near-Exact Distributions

Since in any of the cases addressed in Sect. 3 the exact distribution always takes the form of an infinite mixture, in practice when using these distributions we will have to truncate them.

For a truncation corresponding to a c.f. of the form

$$C\sum_{k=0}^{m}\frac{\omega_k}{a^{b+k}}\,a^{b+k}\,(a-\mathrm{i}t)^{-(b+k)} \tag{20}$$

an upper-bound for the truncation error is

$$1-C\sum_{k=0}^{m}\frac{\omega_k}{a^{b+k}}\,.$$

But, as we may see from the results in Sect. 6, we may do much better by using a truncation with completion of the weights in such a way that they add up to 1, that is, by using in (20) $\omega_k$ given by (18) and (12)–(13) for $k = 0, \ldots, m-1$, and

$$\omega_m = \frac{a^{b+m}}{C} - \sum_{k=0}^{m-1}\omega_k\,a^{m-k}\,. \tag{21}$$

Although for quite small values of $k$ the computation of the weights $\omega_k$ does not pose any serious problem, that is not anymore the case for large values of $k$, where the weights $\omega_k$, given the summation in (18), become quite heavy to compute.

Since we have indeed the exact c.f. of $W$ at hand in the form in (15), we may think about determining the weights $\omega_k$ in (20), for $k = 0, \ldots, m-1$, in such a manner that the first $m$ derivatives of the c.f. in (20), at $t = 0$, match the corresponding derivatives of $\Phi_W(t)$ in (15), setting then $\omega_m$ to the value given by (21).

This approach will indeed give much better results than the computation of the weights $\omega_k$ through their original expression. See the results in Sect. 6.

Yet, in order to better analyze the behavior of these approximations we will consider in Sect. 6 truncations where the first $\omega_k$ are computed from (18) and (12)–(13) and the remaining are computed by matching some of the exact moments, that is, we will use a c.f.

$$C\sum_{k=0}^{m^*}\frac{\omega_k}{a^{b+k}}\,a^{b+k}\,(a-\mathrm{i}t)^{-(b+k)}+\sum_{k=m^*+1}^{m^{**}}\pi_{k-m^*}\,a^{b+k}\,(a-\mathrm{i}t)^{-(b+k)}\,, \tag{22}$$

where $\omega_k$ ($k = 0, \ldots, m^*$) are computed from (18) and (12)–(13) and $\pi_{k-m^*}$ ($k = m^* + 1, \ldots, m^{**}$) are computed by matching the first $m^{**} - m^*$ derivatives at $t = 0$ of the c.f. in (22) and $\Phi_W(t)$.

One other thing we may notice is that, either computing the weights through their original expression or by matching derivatives, that is, by matching exact moments, for a given number of terms used in (20), the approximations are always much better when the $b_j$'s fall between zero and one. The ascertainment of this fact may lead us to a different approach in which we would "extract" the integer part of the $b_j$'s and somehow set it apart, whenever any of the $b_j$'s exceeds the value of one. This is indeed the approach pursued in the next section.

We should remark that although the notation used in this section is the one used in Sect. 3.1, in order to encompass the case treated in Sect. 3.3 one only has to consider $\omega_k = \nu_{2k,p}(-k, b - k)$.

# 5 Near-Exact Distributions for the Product of Independent Beta Random Variables

## *5.1 Near-Exact Distributions for Z in (2)*

In trying to "extract" the integer part of the $b_j$'s and set it apart, whenever any of the $b_j$'s exceeds the value of one, this will indeed lead us to the use of what has been called 'near-exact distributions'.

In simple terms, near-exact distributions are distributions which keep intact a good part of the exact distribution and which approximate asymptotically the remaining part.

In order to try to keep intact a good part of the exact c.f. when at least one of the $b_j$'s is greater than 1, we may work through the exact c.f. of $W$ as it is done below, using the fact that for any real or complex $a$ and positive integer $n$ we may write

$$\frac{\Gamma(a+n)}{\Gamma(a)} = \prod_{\ell=0}^{n-1} (a+\ell)\,. \tag{23}$$

In fact, using (23), taking $b_j^* = \lfloor b_j \rfloor$ and $b_j^{**} = b_j - b_j^*$, we may write, for the general case of the distribution of $W$ in Sect. 3.1,

$$\Phi_W(t) = \prod_{j=1}^{q} \left(\frac{\Gamma(a_j+b_j)}{\Gamma(a_j)}\right)^{m_j} \left(\frac{\Gamma(a_j - \mathrm{i}t)}{\Gamma(a_j+b_j-\mathrm{i}t)}\right)^{m_j}$$

$$= \prod_{j=1}^{q} \left(\frac{\Gamma(a_j+b_j)}{\Gamma(a_j+b_j^*)}\,\frac{\Gamma(a_j+b_j^*)}{\Gamma(a_j)}\right)^{m_j}$$

$$\times\left(\frac{\Gamma(a_j-\mathrm{i}t)}{\Gamma(a_j+b_j^*-\mathrm{i}t)}\,\frac{\Gamma(a_j+b_j^*-\mathrm{i}t)}{\Gamma(a_j+b_j-\mathrm{i}t)}\right)^{m_j}$$

$$=\prod_{j=1}^{q}\left(\frac{\Gamma(a_j+b_j)}{\Gamma(a_j+b_j^*)}\,\frac{\Gamma(a_j+b_j^*-\mathrm{i}t)}{\Gamma(a_j+b_j-\mathrm{i}t)}\right)^{m_j}\prod_{\ell=0}^{b_j^*-1}(a_j+\ell)^{m_j}(a_j+\ell-\mathrm{i}t)^{-m_j}$$

$$=\underbrace{\left\{\prod_{j=1}^{q}\prod_{\ell=0}^{b_j^*-1}(a_j+\ell)^{m_j}(a_j+\ell-\mathrm{i}t)^{-m_j}\right\}}_{\Phi_{1,W}(t)}$$

$$\times\underbrace{\left\{\prod_{j=1}^{q}\left(\frac{\Gamma(a_j+b_j)}{\Gamma(a_j+b_j^*)}\,\frac{\Gamma(a_j+b_j^*-\mathrm{i}t)}{\Gamma(a_j+b_j-\mathrm{i}t)}\right)^{m_j}\right\}}_{\Phi_{2,W}(t)} \tag{24}$$

where $\Phi_{1,W}(t)$ is the c.f. of a sum of $\sum_{j=1}^{q} b_j^*$ independent Gamma random variables, with integer shape parameters $m_j$ and rate parameters $a_j+\ell$ ($\ell=0,\ldots,b_j^*-1;\ j=1,\ldots,q$), which is a GIG distribution (Coelho, 1998) of depth $\sum_{j=1}^{q} b_j^*$, with rate parameters $a_j+\ell$ and shape parameters $m_j$ ($j=1,\ldots,q;\ \ell=0,\ldots,b_j^*-1$) and $\Phi_{2,W}(t)$ is the c.f. of a sum of $p=\sum_{j=1}^{q} m_j$ independent Logbeta r.v.'s, $m_j$ of which have the set of parameters $(a_j+b_j^*, b_j^{**})$ ($j=1,\ldots,q$).

We may note that if in (24) some $b_j$ is smaller than 1, then the corresponding $b_j^*$ equals zero and the corresponding term in $\Phi_{1,W}(t)$ simply vanishes, i.e., equals 1.

Since, opposite to $\Phi_{2,W}(t)$, $\Phi_{1,W}(t)$ corresponds to a very manageable distribution, in building the near-exact distribution for $W$ we will leave $\Phi_{1,W}(t)$ unchanged and we will approximate $\Phi_{2,W}(t)$ in a similar manner to the one used in Sect. 3 to approximate $\Phi_W(t)$.

We will take

$$a=\min(a_1+b_1^*,\ldots,a_q+b_q^*),\quad \alpha_j=a_j+b_j-a\,,\quad \text{and}\quad \beta_j=b_j^{**}=b_j-b_j^*\,, \tag{25}$$

and we will write

$$\Phi_{2,W}(t)\approx C\sum_{k=0}^{\infty}\frac{\omega_k}{a^{b+k}}\,a^{b+k}\,(a-\mathrm{i}t)^{-(b+k)}\,,$$

where

$$C=\prod_{j=1}^{q}\left(\frac{\Gamma(a_j+b_j)}{\Gamma(a_j+b_j^*)}\right)^{m_j} \quad \text{and} \quad b=\sum_{j=1}^{q} m_j(b_j-b_j^*)\,, \tag{26}$$

and where $\omega_k$ is computed using (18), now with $\alpha_j$ and $\beta_j$ given by (25). We will then use as near-exact c.f. for $W$ the c.f.

$$\Phi_W^*(t)=\Phi_{1,W}(t)\; C\sum_{k=0}^{m}\pi_k\, a^{b+k}\,(a-\mathrm{i}t)^{-(b+k)} \tag{27}$$

where the weights $\pi_k$ will be determined in such a way that

$$\left.\frac{d^h}{dt^h}\,C\sum_{k=0}^{m}\pi_k\, a^{b+k}\,(a-\mathrm{i}t)^{-(b+k)}\right|_{t=0}=\left.\frac{d^h}{dt^h}\Phi_{2,W}(t)\right|_{t=0}, \qquad h=1,\dots,m$$

or, equivalently, in such a way that

$$\left.\frac{d^h}{dt^h}\Phi_W^*(t)\right|_{t=0}=\left.\frac{d^h}{dt^h}\Phi_W(t)\right|_{t=0}, \qquad h=1,\dots,m\,.$$

The reason to compute the weights $\pi_k$ in this way is twofold: computed this way, instead of using expressions (18) and (12)–(13), the weights $\pi_k$ (i) will yield better approximations and (ii) will be easier to compute for moderately large values of $k$.

The distribution corresponding to the c.f. $\Phi_W^*(t)$ in (27) is a finite mixture of GNIG (Generalized Near-Integer Gamma) distributions (see Coelho (2004) for the definition of this distribution and Appendix B of the supplementary material for a definition of its p.d.f. and c.d.f.) which in case all $a_j+\ell$ ($j=1,\dots,q$; $\ell=0,\dots,b_j^*-1$) are different will have depth $1+\sum_{j=1}^{q} b_j^*$ with shape parameters

$$\underbrace{m_1,\dots,m_1}_{b_1^*\text{ times}},\dots,\underbrace{m_j,\dots,m_j}_{b_j^*\text{ times}},\dots,\underbrace{m_q,\dots,m_q}_{b_q^*\text{ times}},b+k$$

and rate parameters $\{a_j+0,\dots,a_j+b_j^*-1(j=1,\dots,q),a\}$.

For the general case, where some of the $a_j+\ell$ may have the same value, let

$$\left\{a_\nu^*;\ \nu=1,\dots,q^*\le\textstyle\sum_{j=1}^{q} b_j^*\right\}=\{\{a_j+\ell;\ \ell=0,\dots,b_j^*-1;\ j=1,\dots,q\}\}$$

represent the set of all different rate parameters $a_j+\ell$, that is, the set of all different values of $a_j+\ell$ ($\ell=0,\dots,b_j^*-1$; $j=1,\dots,q$) and let

$$\left\{m_\nu^*;\ \nu=1,\dots,q^*\le\sum_{j=1}^q b_j^*\right\}$$
$$=\{\{m_{j\ell}=m_j;\ \ell=0,\dots,b_j^*-1;\ j=1,\dots,q\}\}_{\{a_j+\ell\}}$$

be the set of the corresponding shape parameters, where the shape parameter $m_\nu^*$ is the shape parameter corresponding to $a_\nu^*$, that is, it is the sum of all $m_{j\ell}=m_j$ corresponding to the $\nu$-th distinct value $a_j+\ell$. Then the p.d.f. and c.d.f. corresponding to the c.f. in (27), using the notation in Appendix B of the supplementary material for the GNIG distribution would be respectively

$$f_W^*(w)=\sum_{k=0}^m \pi_k\, f^{GNIG}\left(w\,|\,\{m_\nu^*\}_{\nu=1:q^*}, b+k;\ \{a_\nu^*\}_{\nu=1:q^*}, a;\ q^*+1\right)$$

and

$$F_W^*(w)=\sum_{k=0}^m \pi_k\, F^{GNIG}\left(w\,|\,\{m_\nu^*\}_{\nu=1:q^*}, b+k;\ \{a_\nu^*\}_{\nu=1:q^*}, a;\ q^*+1\right)$$

and the corresponding p.d.f. and c.d.f. for $Z=e^{-W}$ would be

$$f_Z^*(z)=\sum_{k=0}^m \pi_k\, f^{GNIG}\left(-\log z\,|\,\{m_\nu^*\}_{\nu=1:q^*}, b+k;\ \{a_\nu^*\}_{\nu=1:q^*}, a;\ q^*+1\right)\frac{1}{z}$$

and[1]

$$F_Z^*(z)=1-\sum_{k=0}^m \pi_k\, F^{GNIG}\left(-\log z\,|\,\{m_\nu^*\}_{\nu=1:q^*}, b+k;\ \{a_\nu^*\}_{\nu=1:q^*}, a;\ q^*+1\right).$$

We may note that in case $b$ in (26) is an integer the GNIG distributions will indeed be GIG (Generalized Integer Gamma) distributions.

Built in this way, the near-exact distribution which c.f. is in (27) will provide extremely good approximations to the exact distribution, while remaining quite manageable. Some of the several vast advantages of these near-exact distributions are that:

[1] in the original publication there are small typos in these last four expressions, where the sets of shape parameters $\{m_\nu^*\}$ appear represented as $\{a_\nu^*\}$

(i) they provide the exact distribution in cases where all $b_j$ are integer (in the form of a GIG distribution for $W = -\log Z$ or an EGIG distribution (Arnold et al., 2013) for $Z$),
(ii) the approximations provided will be even better for cases where the $b_j$ are larger,
(iii) they have a much stable performance for a wide range of values of the parameters $a_j$, $b_j$ and $m_j$,
(iv) they will perform even slightly better in the situations where the other approximations show severe difficulties.

Actually, since the weights $\pi_k$ in (27) are not going to be computed from (18) and (12)–(13), we may also think about a couple of alternative ways to compute the parameter $a$ in (27). These may be to take $a$ in (27) as one of the following choices:

$$\begin{array}{c}
\text{i) } a = \min_{j=1,\dots,q}(a_j + b_j^*) \quad \text{ii) } a = \frac{1}{q}\sum_{j=1}^{q}(a_j + b_j^*) \\
\text{iii) } a = \frac{\sum_{j=1}^{q}(b_j - b_j^*)(a_j + b_j^*)}{\sum_{j=1}^{q}(b_j - b_j^*)} \\
\text{iv) } a = \frac{\sum_{j=1}^{q} m_j(a_j + b_j^*)}{\sum_{j=1}^{q} m_j} \qquad \text{v) } a = \frac{\sum_{j=1}^{q} m_j(b_j - b_j^*)(a_j + b_j^*)}{\sum_{j=1}^{q} m_j(b_j - b_j^*)} \\
\text{vi) the rate parameter } a \text{ in } \Phi^{**}(t) = a^s(a - \mathrm{i}t)^{-s}, \text{ where} \\
\left.\frac{\partial^h}{\partial t^h}\Phi^{**}(t)\right|_{t=0} = \left.\frac{\partial^h}{\partial t^h}\Phi_{2,W}(t)\right|_{t=0}, \quad h = 1, 2 \\
\text{vii) the rate parameter } a \text{ in } \Phi^{***}(t) = \pi a^{s_1}(a - \mathrm{i}t)^{-s_1} + (1-\pi)a^{s_2}(a - \mathrm{i}t)^{-s_2}, \\
\text{where} \\
\left.\frac{\partial^h}{\partial t^h}\Phi^{***}(t)\right|_{t=0} = \left.\frac{\partial^h}{\partial t^h}\Phi_{2,W}(t)\right|_{t=0}, \quad h = 1, \dots, 4.
\end{array} \tag{28}$$

Details on the performance of such near-exact approximations may be analyzed in the next section and also in Sect. 7.

Mathematica modules to compute these as well as all the other approximations developed in this manuscript are available, on demand, from the authors.

These different strategies for the choice of $a$ will be used only for approximations which correspond to the situations studied in Sects. 3.1 and 3.2, since for the situation addressed in Sect. 3.3 there is no doubt in the choice of $a$, having only to consider that in this case the parameter $a$ in (27) has to be taken as $a + \frac{b-1}{2}$, where $b$ is the common value of the second parameter for all the Beta r.v.'s involved.

In cases where all the $b_j$ are smaller than 1, the near-exact distribution will then provide an asymptotic distribution of the same kind of the ones in Sect. 3. Anyway, for these cases the distributions in Sect. 3, which are indeed asymptotic distributions, work very well, as we may see from the results in Sect. 6. Furthermore, these are not the most common cases that occur, mainly when these distributions refer to likelihood ratio test statistics, and, as we may also see from the results in Sect. 6, by determining some of the weights through equating some of the first exact moments we may obtain approximations which lie very close to the exact distribution.

Actually, in these latter cases where all $b_j$ are smaller than 1, as it may be seen from the results analyzed in Sect. 6 and reported in Appendix C in the supplementary material, a good balance between a number of weights determined through (18) and (12)–(13) and a number of weights determined by equating some of the first exact moments leads usually to the best results and although this approach might also be applied to the general near-exact distributions, for reasons of simplicity and extent of the manuscript we decided to use in the near-exact distributions only weights determined by equating exact moments. The adaptation of the approach followed to this other one of determining some of the weights through their formulation in (18) and (12)–(13) is quite easy and straightforward, although the computation of the weights through (18) and (12)–(13), namely for higher orders, may require more computation than their determination by equating exact moments.

## 5.2 *Near-Exact Distributions for the Product of Independent Beta Random Variables Raised to Different Powers*

However, it is interesting to note that this case of different powers $c_j$ poses indeed no big problems when we adopt the near-exact approach. It indeed poses no problem at all in what concerns the handling of the part of the c.f. $\Phi_W(t)$ which will be left unchanged, that is, $\Phi_{1,W}(t)$, anyway posing a similar problem to the one described above for $\Phi_{2,W}(t)$, the part of the c.f. $\Phi_W(t)$ to be asymptotically approximated, since in the present case, following the same lines as in (24) we have

$$
\begin{aligned}
\Phi_W(t) &= \left\{\prod_{j=1}^{q}\prod_{\ell=0}^{b_j^*-1}(a_j+\ell)^{m_j}\,(a_j+\ell-c_j\mathrm{i}t)^{-m_j}\right\} \\
&\qquad\times\left\{\prod_{j=1}^{q}\left(\frac{\Gamma(a_j+b_j)}{\Gamma(a_j+b_j^*)}\,\frac{\Gamma(a_j+b_j^*-c_j\mathrm{i}t)}{\Gamma(a_j+b_j-c_j\mathrm{i}t)}\right)^{m_j}\right\} \\
&= \underbrace{\left\{\prod_{j=1}^{q}\prod_{\ell=0}^{b_j^*-1}\left(\frac{a_j+\ell}{c_j}\right)^{m_j}\left(\frac{a_j+\ell}{c_j}-\mathrm{i}t\right)^{-m_j}\right\}}_{\Phi_{1,W}(t)} \\
&\qquad\times\underbrace{\left\{\prod_{j=1}^{q}\left(\frac{\Gamma(a_j+b_j)}{\Gamma(a_j+b_j^*)}\,\frac{\Gamma(a_j+b_j^*-c_j\mathrm{i}t)}{\Gamma(a_j+b_j-c_j\mathrm{i}t)}\right)^{m_j}\right\}}_{\Phi_{2,W}(t)}
\end{aligned}
$$

where now $\Phi_{1,W}(t)$ is the c.f. of a sum of $\sum_{j=1}^{q} b_j^*$ independent Gamma random variables, with integer shape parameters $m_j$ and rate parameters $(a_j+\ell)/c_j$ ($\ell = 0,\ldots,b_j^*-1$; $j = 1,\ldots,q$), which is a GIG distribution of depth $\sum_{j=1}^{q} b_j^*$, with rate parameters $(a_j + \ell)/c_j$ and shape parameters $m_j$ $(j = 1, \ldots, q; \ell = 0, \ldots, b_j^* - 1)$ and $\Phi_{2,W}(t)$ is the c.f. of a sum of $p = \sum_{j=1}^{q} m_j$ independent Logbeta r.v.'s, $m_j$ of which have parameters $(a_j + b_j^*, b_j^{**})$ and are multiplied by $c_j$ $(j = 1, \ldots, q)$.

Then, based on a heuristic approach, in building the near-exact distribution for $W$ we will leave $\Phi_{1,W}(t)$ unchanged and we will now approximate $\Phi_{2,W}(t)$ in exactly the same way we did in the previous subsection, now taking $a$ in (27) as one of the choices in (28), with $a_j + b_j^*$ replaced by $(a_j + b_j^*)/c_j$.

As it happened in the previous subsection, these different strategies for the choice of $a$ will be used only for approximations which correspond to the situations studied in Sects. 3.1 and 3.2, since for the situation addressed in Sect. 3.3, once again there is no doubt in the choice of $a$, having now to consider that in this case not only the first and second parameters of all the Beta r.v.'s have to be equal but also all the powers to which they are raised have also to be the same. Then we may either take the parameter $a$ in (27) as $\left(a + \frac{b-1}{2}\right)/c$, where $a$ and $b$ are respectively the common values of the first and second parameters for all the Beta r.v.'s involved and $c$ the common power, or just take the transformation procedure outlined in Sect. 3.4.

From the results in the next section as well as in Sect. 7.3 we may see that this approach really works well in practice.

## 6 Some Evidence Better Analyzed from Numerical Studies

While we clearly expect the "completed" truncation to perform much better than the simple truncation, and the version with the weights computed from the moments to perform even much better, we also expect the near-exact distributions to display an even much better performance than any of these. The results coming out of a few numerical studies may indeed help us in better figuring out the fine behavior of each of these approximations and also to better see that the near-exact approximations are indeed the only ones with a remarkable outstanding performance in all situations, that is, for any combination of values of the parameters involved. These results may be analyzed in detail in the supplementary material.

In order to assess the proximity between the exact distributions of $W$ and $Z$ and the approximations suggested in this manuscript we use the measure

$$\Delta = \frac{1}{2\pi} \int_{-\infty}^{+\infty} \left| \frac{\Phi_W(t) - \Phi^*(t)}{t} \right| dt \tag{29}$$

with

$$\max_{w \in S_W} \left| F_W(w) - F_W^*(w) \right| = \max_{z \in S_Z} \left| F_Z(z) - F_Z^*(z) \right| \leq \Delta \,,$$

where $\Phi_W(t)$ represents the exact c.f. of $W$ given by (7) or (16) and $\Phi^*(t)$ the approximate c.f. corresponding to the approximation being used, and where $F_W(w)$ and $F_W^*(w)$ represent the c.d.f.'s corresponding to $\Phi_W(t)$ and $\Phi^*(t)$, and $F_Z(z)$ and $F_Z^*(z)$ the corresponding c.d.f.'s of $Z$. Therefore smaller values of $\Delta$ correspond to better approximations.

The measure in (29) may be seen as related with the Berry-Esseen bound (Berry, 1941; Esseen, 1945; Hsu, 1945; Hwang, 1998) and has been used in several studies as a measure of proximity between distributions (Coelho et al., 2010, 2015; Marques and Coelho, 2011a; Marques et al., 2011).

From the results in Tables 1-23 in Appendix C in the supplementary material, we may see how the approximations based on the c.f. in (20) only give somewhat satisfactory results for the scenarios with $b_j$'s smaller than 1, with completely nonsense results for the scenarios with larger spans of the $a_j$, with values of $\Delta$ larger than 1, indicating that indeed the supposedly approximating c.f.'s are not indeed true c.f.'s, that is, they do not even correspond to true distributions. Anyway the behavior of the approximation based on (20) that uses the completion of weights is in all cases much better than the one that does not use the weight completion, and the approximation based on (20) with the weights computed by matching exact moments has always a much better performance than the approximations that compute the weights through (18) and (12)–(13), with quite satisfactory results for all scenarios. However, all these approaches suffer from the problem of starting to yield worse approximations when the number of terms $m$ goes above some threshold for which it is not easy to obtain an "a priori" guess. This problem is more pronounced for the approximation based on (20) with weights given by (18) and (12)–(13) without weight completion and much less pronounced for the approximation that uses weights obtained by equating moments, and it also starts at much larger values of $m$ for this latter approximation.

The results show that in general there is no advantage of having some of the first weights computed through their exact expressions, that is, they show that it is indeed more advantageous to have all weights computed by matching exact moments.

However for the scenarios with $b_j$'s smaller than 1 some combinations of moderate values of $m^*$ with somehow larger values of $m^{**} - m^*$ in (22) may give somewhat better results than those obtained when all the weights are obtained by equating the first exact moments.

This shows that this approach might be useful to be implemented for the near-exact approximations, where the remaining part of the c.f. left to be asymptotically approximated corresponds indeed to a product of independent Beta r.v.'s, all with $b_j$ parameters smaller than 1. However, for reasons related with the already large length of this manuscript, such approaches were not explored, although they may be easily implemented from the approaches developed.

For the same number of exact moments matched, and for the same number of terms in the mixtures, the near-exact distributions yield approximations which exhibit values of $\Delta$ which are several millions of times smaller than those for the corresponding approximations based on simple truncations of the c.f., that is the ones based on the c.f.'s in (20) and (22). Moreover, the near-exact distributions show a much stable behavior across all scenarios, actually tending to show slightly better performances in those scenarios where the other approaches tend to behave worse, which are the scenarios where the $a_j$ show larger spans, or scenarios where the $b_j$ have larger values, or yet scenarios where there are more Beta r.v.'s involved, showing this way their great ability and adequacy to handle more complicated situations.

Although for situations where we consider the distribution of the product of different powers of independent Beta r.v.'s, the near-exact approximations exhibit rather larger values of the measure $\Delta$ than what they do for situations where no powers are considered, they still have very good performances, with the near-exact approximations with $a$ in (27) given by the minimum of the $(a_j - b_j^*)/c_j$ being the one with a less good performance and the near-exact approximations with $a$ in (27) given by vi) or vii) in (28) being the ones with the best performance. This also shows that there are good reasons to consider the different choices for $a$ in (28), although introduced then in a somewhat heuristic way. As we will also see in Sect. 7.3, these near-exact approximations also perform very well when applied to the distribution of likelihood ratio test statistics.

## 7 Applications

In this section we have chosen the likelihood ratio test statistics to test (i) the independence of several sets of variables, (ii) sphericity and (iii) equality of several covariance matrices, under multivariate normal or elliptically contoured distributions to illustrate the performance of the near-exact distributions suggested in Sect. 5 and the different cases addressed in Sects. 3 and 5. In this section we will indeed only use near-exact distributions, given the evidence that these distributions give much better approximations, while remaining highly manageable.

For each type of near-exact distribution studied, 4, 6, 10 or 15 exact moments are matched, in order to make it easier to compare the results obtained with the ones published elsewhere for other near-exact distributions which equated exactly the same numbers of exact moments. As it may be seen from the numerical results, some of the near-exact distributions suggested in this paper, for the general case of the product of independent Beta r.v.'s, yield even slightly better results than the best performing near-exact distributions earlier expressly developed for each of the statistics used as examples.

### 7.1 The Distribution of the Likelihood Ratio Test Statistic to Test Independence of Sets of Variables

According to Marques et al. (2011), the distribution of $\Lambda_1$, the likelihood ratio test statistic to test the independence of $m$ groups of variables, the $k$-th of them having $p_k$ variables, based on a sample of size $n$ from a multivariate normal or elliptically contoured distribution is the same as the distribution of

$$\prod_{k=1}^{m-1}\prod_{j=1}^{p_k}\left(Y_{jk}\right)^{n/2} \qquad \text{where} \qquad Y_{jk} \sim Beta\left(\frac{n-q_k-j}{2}, \frac{q_k}{2}\right) \tag{30}$$

or the distribution of

$$\left\{\prod_{j=3}^{p} e^{-Z_j}\right\}\left\{\prod_{j=1}^{k^*}\left(Y_j^*\right)^{n/2}\right\} \text{ where } Z_j \sim \Gamma\left(r_j, \frac{n-j}{n}\right)$$
$$\text{and } Y_j^* \sim Beta\left(\frac{n-2}{2}, \frac{1}{2}\right) \tag{31}$$

with

$$p = \sum_{k=1}^{m} p_k\,, \qquad q_k = p_{k+1} + \cdots + p_m \quad \text{and} \quad k^* = \left\lfloor \frac{\ell}{2} \right\rfloor$$

where $\ell$ is the number of sets of variables with an odd number of variables and where $r_j$ $(j = 3, \ldots, p)$ are given by (A.2)-(A.3) in Marques et al. (2011).

We may note that the non-integer part of the second parameters of the Beta r.v.'s in (30) is always either zero or $1/2$, thus rendering the near-exact approach for the distribution in (31) quite simple and much adequate, while in case $k^* = 0$, that is, when at most one of the sets has an odd number of variables, it is clear from (31) above, that we have in this case the exact distribution of $\Lambda_1$ as an EGIG distribution.

The way the distribution of $\Lambda_1$ is presented in (31), although requiring more elaborate work on the c.f. of $-\log \Lambda_1$, it will allow for much more precise approximations than the ones obtained so far. The c.f. corresponding to $\prod_{j=3}^{p} e^{Z_j}$ in (31) plays the role of $\Phi^*_{1,W}(t)$ in (24), while the c.f. corresponding to $\prod_{j=1}^{k^*}\left(Y_j^*\right)^{n/2}$ in (31) plays the role of $\Phi_{2,W}(t)$ in (24). In this case, in order to approximate $\Phi_{2,W}(t)$ we will use the approach outlined in Sect. 5, corresponding to the case studied in Sect. 3.3.

In Table 28 in Appendix D of the supplementary material we have values of the measure $\Delta$ in (29) for near-exact distributions for $\Lambda_1$ for different numbers of sets of variables, different numbers of variables in each set and different sample sizes. In this table, NE-1, NE-2, NE-6 and NE-7 refer to the near-exact distributions based on the expression of the distribution of $\Lambda_1$ in the form in (30) and corresponding different choices for $a$ in (28), while NE-II refers to the near-exact distribution obtained using the form of the distribution in (31) and the approach described in Sect. 3.3, which uses a mixture of gamma distributions, all with rate parameters equal to $\frac{n-2}{2} - \frac{1}{4}$ and shape parameters $\frac{k^*}{2} + 2k$, for $k = 0, \ldots, m$. We should note that given the fact that for the representation of the distribution of $\Lambda_1$ in (30) all $b_j^* = \lfloor b_j \rfloor$ are equal to either zero or 1/2 and all $m_j$ are equal to 1, then the near-exact distributions NE-2 through NE-5, corresponding to the choices for $a$ in ii) through v) in (28) yield exactly the same near-exact distributions.

We may note that the form of the distribution in (31) allows for much better approximations. Actually, the results obtained for this form of the distribution by using the approach described in Sects. 5 and 3.3 allows for much better approximations than any of the ones obtained before for this statistic. However, it is interesting to note that the bus approach of building near-exact approximations based directly on the form of the distribution in (30) gives very sharp near-exact approximations which are only barely worse than the ones previously obtained for this statistic when using a dedicated approach (see Coelho et al. (2010) and Marques et al. (2011)).

Some more detailed comments on the performance of the several near-exact distributions are:

- as expected, all the near-exact distributions show a clear asymptotic behavior both for increasing values of $m$, the number of exact moments matched, and $n$, the sample size,
- NE-II is by far the best performing near-exact distribution, with a performance even much better than the near-exact distributions in Coelho et al. (2010) and Marques et al. (2011).
- NE-1 through NE-7, which use what we may call a "bus" approach, do not give much worse results than the near-exact distributions in Coelho et al. (2010) and Marques et al. (2011), specifically developed for $\Lambda_1$,
- for increasing values of $p = \sum_{k=1}^{\nu} p_k$, keeping $n - p$ fixed, even for very small values of $n - p$ and reduced values of $m$ (the number of exact moments matched), the performance of the near-exact approximations NE-1 through NE-7 does not worsen while the performance of NE-II even improves, what may be seen as a much desirable feature,
- NE-1 is the worse performing near-exact approximation, what amply justifies the choices for the rate parameter $a$ in ii)–vii) of (28) in Sect. 5, which then seemed to be a somewhat heuristic approach,
- the near-exact approximation NE-6, with the rate parameter $a$ based on only two moments, performs always better than NE-7, with a rate parameter $a$ based on four moments, and this latter one better than NE-2,

## 7.2 The Distribution of the Likelihood Ratio Test Statistic to Test Sphericity

Once again according to Marques et al. (2011), the distribution of $\Lambda_2$, the likelihood ratio test statistic to test sphericity in a $p$-multivariate normal or elliptically contoured distribution, based on a sample of size $n$ has the same distribution as

$$\prod_{j=2}^{p} (Y_j)^{n/2} \qquad \text{where} \qquad Y_j \sim Beta\left(\frac{n-j}{2}, \frac{j-1}{p} + \frac{j-1}{2}\right) \tag{32}$$

or

$$\left\{\prod_{j=2}^{p} e^{-Z_j}\right\}\left\{\prod_{j=2}^{p-k^*} \left(Y_j^*\right)^{n/2}\right\}\left\{\prod_{j=p-k^*+1}^{p} \left(Y_j^{**}\right)^{n/2}\right\}, \tag{33}$$

where $k^* = \lfloor p/2 \rfloor$, and

$$Z_j \sim \Gamma\left(r_j, \frac{n-j}{n}\right), \quad Y_j^* \sim Beta\left(\frac{n-1}{2}, \frac{j-1}{p}\right) \quad \text{and} \quad Y_j^{**} \sim Beta\left(\frac{n}{2}, \frac{j-1}{p} - \frac{1}{2}\right)$$

with

$$r_j = \left\lfloor \frac{p-j+2}{2} \right\rfloor, \qquad j = 2, \ldots, p\,.$$

Since all Beta r.v.'s in either (32) or (33) are all different, with different first and second parameters, in this case all the $m_j$ will be equal to 1 and the near-exact distributions corresponding to the choices of $a$ in ii) and iv) of (28), on one side, and the ones corresponding to iii) and v) in (28) will yield the same result.

As such, in Tables 29-31, in Subsect. D.2 of Appendix D in the supplementary material, we refer to the near-exact distributions for the distribution of $\Lambda_2$ as depicted in (32) by NE-1 through NE-7, not using NE-4 neither NE-5, and by NE2-1 through NE2-7 for the representation of the distribution in (33), not using NE2-4 neither NE2-5. For this statistic,

- as it also happens with the statistic $\Lambda_1$ in the previous subsection, and as it was expected, all near-exact approximations show a marked asymptotic behavior for increasing values of $n$, the sample size, and $m$, the number of exact moments matched,
- an important characteristic of all the near-exact approximations for $\Lambda_2$ is that they are also asymptotic in $p$, that is, for all values of $m$, even the smaller ones, they perform even better for larger values of $p$, when keeping $n - p$ constant, even for very small values of $n - p$.

- for the first approach, the NE-2 near-exact distribution presents the worse performance, while NE-1 is comparable to NE-3, with NE-6 and NE-7, which are comparable, exhibiting the best performance,
- for the second approach, NE2-6 and NE2-7 are comparable, showing a better performance than NE2-2 and NE2-3, which are comparable,
- also for the second approach, and for the larger values of $m$, NE2-1 generally performs better than NE2-6 and NE2-7 for smaller values of $n$, and the other way around for larger values of $n$,
- generally, the near-exact distributions based on the second approach show a better performance than the ones based on the first approach, although the gain may be rather slim and tends to reduce for larger values of $m$,

As the near-exact distributions based on the first approach, based on (32), need less work on the exact c.f. and the gains with the second approach, based on (33), tend to be rather slim, the "bus" near-exact distributions based on the first approach may represent a good choice.

## 7.3 *The Distribution of the Likelihood Ratio Test Statistic to Test Equality of Covariance Matrices*

Once again according to Marques et al. (2011), the distribution of $\Lambda_3$, the likelihood ratio test statistic to test the equality of $q$ covariance matrices, based on $q$ independent samples, all with size $n$, from as many $p$-multivariate normal or elliptically contoured distributions, has the same distribution as

$$\prod_{j=1}^{p}\prod_{k=1_{\text{except } j=k=1}}^{q} \left(Y_{jk}\right)^{n/2} \quad \text{where } Y_{jk} \sim Beta\left(\frac{n-j}{2}, \frac{j(q-1)+2k-1-q}{2q}\right), \tag{34}$$

or

$$\left\{\prod_{j=2}^{p} e^{-Z_j}\right\}\left\{\prod_{j=1}^{\lfloor p/2\rfloor}\prod_{k=1}^{q}\left(Y_{jk}^*\right)^{n}\right\}\left\{\prod_{k=1}^{q}\left(Y_k^*\right)^{n/2}\right\}^{p \perp\!\!\!\perp 2} \tag{35}$$

where $p \perp\!\!\!\perp 2$ is the remainder of the integer division of $p$ by 2, or equivalently, the indicator function of $p$ being odd,

$$Z_j \sim \Gamma\left(r_j, \frac{n-j}{n}\right), \quad Y_{jk}^* \sim Beta\left(a_j + b_{jk}^*, b_{jk} - b_{jk}^*\right)$$
$$\text{and } Y_j^{**} \sim Beta\left(a_p + b_{pk}^*, b_{pk} - b_{pk}^*\right)$$

with

$$a_j = n - 2j, \quad b_{jk} = 2j - 1 + \frac{k - 2j}{q}, \quad b_{jk}^* = \lfloor b_{jk} \rfloor,$$

$$a_p = \frac{n - p}{2}, \quad b_{pk} = \frac{pq - q - p + 2k - 1}{2q}, \quad b_{pk}^* = \lfloor b_{pk} \rfloor,$$

and $r_j$ given by (A.12)-(A.16) in Marques et al. (2011).

Also in the distribution of $\Lambda_3$ all Beta r.v.'s in either (34) or (35) are all different. As such, in Tables 32-34 in Appendix D of the supplementary material, similar to what was done for $\Lambda_2$, we do not use the near-exact distributions NE-4 neither NE-5, NE2-4, NE2-5. In these Tables NE-1 through NE-7 refer to the form of the distribution of $\Lambda_3$ in (34) and NE2-1 through NE2-7 refer to the form of the distribution of $\Lambda_3$ in (35).

We should also note that for odd $p$, when using the representation of the distribution of $\Lambda_3$ in (35), we will have two products of Beta r.v.'s where in each product the Beta r.v.'s involved are raised to a different power. As such, when using this representation of the distribution for odd $p$, we will have to use the approach outlined in Sect. 5.2. As overall comments we may see that

- once again, and as expected, all near-exact approximations show a marked asymptotic behavior for increasing values of $n$, the sample size, and $m$, the number of exact moments matched,
- NE-6, NE-7, NE2-6 and NE2-7, which are the best performing approximations, even seem to improve quite a bit with the increase in $q$, what is a token in favor of these near-exact approximations, or, if we want, in favor of the choice of the computation of the parameter $a$ as in vi) and vii) of (28), and, as in Sect. 7.1, it shows that what might have seemed then as mostly heuristic choices for the computation of the parameter $a$, are indeed well justified, moreover since this behavior even seems to be more accentuated for larger values of $m$ and $n$,
- generally the second approach, based on (35) yields somewhat better results, although the difference to the first approach is rather slim mainly for NE-6 and NE-7, what shows that these two near-exact approximations, based on a simpler and more crude approach may indeed constitute good alternatives to NE2-6 and NE2-7,
- very good approximations are obtained, namely when using NE2-6 and NE2-7, but also NE-6 and NE-7, even for small values of $n - p$, which is a much desirable feature.

# 8 Conclusions

Although the asymptotic distributions which are finite mixtures of beta or exponentiated gamma distributions, are simpler and as such easier to compute, in most situations they do not provide the necessary precision, yielding in many situations completely inadequate approximations. However, opposite to these, the near-exact approximations provide high quality approximations for all situations, even though remaining much manageable. Mainly in situations where the parameters $b_j$ have large values, the $a_j$ present large spans and/or the number of Beta r.v's involved in the product is moderately large or large, the near-exact distributions, which are finite mixtures of GNIG or GIG distributions, provide the sensible approach and the adequate answer to the problem of approximating the distribution of the product of independent Beta r.v.'s with high quality but manageable approximations. These approximations also show very stable performances for all kinds of situations. That is, they show values of the measure $\Delta$ which seem to be not much affected by changes in the number of Beta r.v.'s in the product, or the span of the values of the $a_j$, actually giving even better approximations for situations where the number of Beta r.v.'s involved in the product increase or the values of the parameters $b_j$ increase, and this in clear contrast with all other asymptotic approximations.

In what concerns the l.r.t. statistics used in Multivariate Analysis addressed in Sect. 7, all near-exact distributions show very good performances even for very small sample sizes, and an asymptotic behavior not only for increasing sample sizes but also for increasing number of variables involved. Among all near-exact distributions, the ones with the parameter $a$ defined by vi) or vii) in (28) are usually the best performing ones. The fact that NE-II was by far the best performing near-exact approximation for $\Lambda_1$ and that the near-exact distributions corresponding to the "second approaches" for both $\Lambda_2$ and $\Lambda_3$ yield in general the best approximations, indicates that specific further work on the exact c.f.'s in order to develop near-exact distributions which may lead to the ability of keeping as much as possible of the original c.f. under its exact form, usually pays off. However, for both $\Lambda_1$ and $\Lambda_2$ this gain is rather slim, while for $\Lambda_3$ this gain is extremely slim for smaller values of $p$ (the number of variables involved) and larger for larger values of $p$, indicating that the application of the near-exact approach and near-exact distributions to a more crude process obtained by working directly on the original product of independent Beta r.v.'s also leads to very accurate approximations, with c.d.f.'s which may easily lie apart from the original by much less than a hundredth of a millionth part, which would never be possible to attain with any other approach.

In fact, the results in this chapter have an even much wider scope than the one analyzed so far. Actually, the distribution of $W$ in (15) is not only the distribution of the negative logarithm of a large number of l.r.t. statistics but it is also the distribution of a linear combination of independent gamma random variables.

Indeed, although in the setting that expression (15) is taken in Sect. 3, all $m_j$ were there taken as positive integers, $m$ in (11) does not have to be an integer. In fact, if in (15) we take $b_j = 1$ and $m_j$ as positive reals ($j = 1, \ldots, q$), the c.f. of $W$ may be written as

$$\Phi_W(t) = \prod_{j=1}^{q} (a_j)^{m_j} (a_j - \mathrm{i}t)^{-m_j} \tag{36}$$

which is the c.f. of a sum of independent $\Gamma(m_j, a_j)$ random variables, which is also the c.f. of a linear combination of independent gamma distributed random variables.

If in (36) we take $m_j = k_j/2$ for some positive integers $m_j$ $(j = 1, \ldots, q)$, we may write

$$\Phi_W(t) = \prod_{j=1}^{q} (a_j)^{k_j/2} (a_j - \mathrm{i}t)^{-k_j/2} = \prod_{j=1}^{q} \left(\frac{1}{2}\right)^{k_j/2} \left(\frac{1}{2} - \frac{1}{2a_j}\mathrm{i}t\right)^{-k_j/2},$$

which is the c.f. of a linear combination of independent $\chi^2_{k_j}$ random variables with coefficients $\frac{1}{2a_j}$ $(j = 1, \ldots, q)$. This distribution is of key importance since it is intimately related with the distribution of quadratic forms, more precisely, quadratic forms in normal variables (Baksalary et al., 1994; Imhof, 1961; Kotz et al., 1967a,b; Shah, 1963) which arise in many estimation and testing problems related with Gaussian processes and normal models, or which arise as limiting distributions in non-normal processes (Jensen and Solomon, 1972; Khatri, 1980).

**Acknowledgments** This work was partially supported by Fundação para a Ciência e a Tecnologia (Portuguese Foundation for Science and Technology) through projects UID/MAT/00297/2013 and UIDB/00297/2020 (Centro de Matemática e Aplicações, CMA-FCT/UNL).

## Appendix A: Simple Proofs for Expressions (11) and (45)

The asymptotic expansions in (39) and (45) were first proved by Burić and Elezović (2011). We propose here simpler and straightforward demonstrations.

Barnes (1899) established an asymptotic expansion for the logarithm of the gamma function in the form

$$\log \Gamma(x+h) \approx \log \sqrt{2\pi} + \left(x + h - \frac{1}{2}\right) \log x - x - \sum_{r=1}^{\infty} (-1)^r \frac{B_{r+1}(h)}{r(r+1)x^r}, \quad (x \to \infty) \tag{37}$$

for any $x, h \in \mathbb{C}$ and where $B_r(\cdot)$ is the Bernoulli polynomial of degree r.

Assuming $z, \alpha, \beta \in \mathbb{C}$ and $m \in \mathbb{R}$, the application of (37) leads to

$$\log\left[\frac{\Gamma(z+\alpha)}{\Gamma(z+\beta)}\right]^m \approx m\,(\alpha-\beta)\log z + \sum_{r=1}^{\infty}\delta_{r,m}(\alpha,\beta)\, z^{-r}, \qquad (z\to\infty)$$

where

$$\delta_{r,m}(\alpha,\beta) = (-1)^r m\,\frac{B_{r+1}(\beta)-B_{r+1}(\alpha)}{r(r+1)}\,. \qquad (38)$$

Therefore, we may write

$$\left[\frac{\Gamma(z+\alpha)}{\Gamma(z+\beta)}\right]^m \approx z^{m\,(\alpha-\beta)}\, e^{\sum_{r=1}^{\infty}\delta_{r,m}(\alpha,\beta)\, z^{-r}}, \qquad (z\to\infty)\,,$$

from which, expanding the exponential function according to expressions (2.7) and (2.8) in Moschopoulos (1985) we obtain the asymptotic expansion

$$\left[\frac{\Gamma(z+\alpha)}{\Gamma(z+\beta)}\right]^m \approx \sum_{k=0}^{\infty}\nu_{k,m}(\alpha,\beta)\, z^{m\,(\alpha-\beta)-k}, \qquad (z\to\infty)\,, \qquad (39)$$

for the power of a ratio of two gamma functions, with $\nu_{k,m}(\alpha,\beta)$ given by (12).

However, it is indeed possible to improve the series expansion in (39), achieving a faster convergence series after a convenient parameter manipulation, in case we are willing to use a power basis which is also function of $\alpha$ and $\beta$.

From the property of the Bernoulli polynomials stated in expression 23.1.8 in Abramowitz and Stegun (1972), we note that, when $n$ is even,

$$B_n(1-x) = B_n(x)\,. \qquad (40)$$

Consider now $c \in \mathbb{C}$ and let $z^* = z + c$, $\alpha^* = \alpha - c$ and $\beta^* = \beta - c$. Then, from (39) we may write

$$\left[\frac{\Gamma(z^*+\alpha^*)}{\Gamma(z^*+\beta^*)}\right]^m \approx \sum_{k=0}^{\infty}\nu_{k,m}(\alpha^*,\beta^*)\,(z^*)^{m\,(\alpha-\beta)-k}, \qquad (z\to\infty)\,. \qquad (41)$$

Now, the proper choice of $c$ will allow us to reduce the number of terms in the above series. For this purpose we will force $\nu_{1,m}(\alpha^*,\beta^*) = 0$. Since $\nu_{1,m}(\alpha^*,\beta^*) = \delta_{1,m}(\alpha^*,\beta^*)$, it is enough to determine $c$ such that

$$B_2(\beta-c) = B_2(\alpha-c)\,. \qquad (42)$$

But then, according to (40), we have

$$B_2(\beta - c) = B_2(1 - (\beta - c)),$$

which, together with (42), entails

$$c = (\alpha + \beta - 1)/2. \tag{43}$$

Moreover, this choice of $c$ implies that, for every odd $j$

$$\delta_{j,m}(\alpha - k, \beta - k) = 0. \tag{44}$$

Thus, from (43) and (44), we may prove that, for every odd $k$

$$\nu_{k,m}(\alpha - c, \beta - c) = 0,$$

and that therefore all the odd terms of the series represented in (41) vanish.

The proof is done by induction. As induction basis, the statement $\nu_{1,m}(\alpha - c, \beta - c) = 0$ holds, according to the choice of $c$ in (43). Assuming as induction hypothesis that, for any given odd $k$, the statement

$$\nu_{1,m}(\alpha - c, \beta - c) = \nu_{3,m}(\alpha - c, \beta - c) = \ldots = \nu_{k-2,m}(\alpha - c, \beta - c) \\ = \nu_{k,m}(\alpha - c, \beta - c) = 0$$

is true, we have, by (38),

$$\nu_{k+2,m}(\alpha - c, \beta - c) = \frac{1}{k+2} \sum_{j=1}^{k+2} j\, \delta_{j,m}(\alpha - c, \beta - c)\, \nu_{k+2-j,m}(\alpha - c, \beta - c).$$

Now all the terms in the summation are zero since, when $j$ is odd the factor $\delta_{j,m}(\alpha - c, \beta - c)$ is zero, according to (44). Otherwise, when $j$ is even, the index $k + 2 - j$ is odd and, by the induction hypothesis, the correspondent coefficient $\nu_{k+2-j,m}(\alpha - c, \beta - c)$ is also zero.

Finally, expression (41) may be written as

$$\left[\frac{\Gamma(z+\alpha)}{\Gamma(z+\beta)}\right]^m \approx \sum_{k=0}^{\infty} \nu_{2k,m}(\alpha - c, \beta - c)\,(z + \tfrac{\alpha+\beta-1}{2})^{m\,(\alpha-\beta)-2k}, \quad (z \to \infty), \tag{45}$$

where each coefficient $\nu_{2k,m}(\alpha - c, \beta - c)$ is given by (12), with $c = (\alpha + \beta - 1)/2$.

Setting $m = 1$ in (39) we obtain a power expansion which is equivalent to the asymptotic series expansion for the ratio of two gamma functions proposed by Tricomi and Erdélyi (1951), here in the form given by Fields (1966), while by setting

$m = 1$ in (45) we obtain the equivalent to the asymptotic series expansion for the ratio of two gamma functions proposed by Fields (1966).

## References


Abramowitz, M., Stegun, I. (eds.) (1972). *Handbook of Mathematical Functions: with Formulas, Graphs, and Mathematical Tables*. New York: Dover.

Anderson, T. W. (2003). *An Introduction to Multivariate Statistical Analysis* (3rd edn.). New York: J. Wiley & Sons.

Arnold, B.C., Coelho, C.A., & Marques, F.J. (2013). The distribution of the product of powers of independent uniform random variables – a simple but useful tool to address and better understand the structure of some distributions. *Journal of Multivariate Analysis*, 113, 19–36.

Baksalary, J. K., Hauke, J., & Styan, G. P. H. (1994). On some distributional properties of quadratic forms in normal variables and on some associated matrix partial orderings. In *Multivariate Analysis and Its Applications, IMS Lecture Notes Monogr. Ser.* (Vol. 24, pp. 111–121). Hayward, CA: Inst. Math. Statist.

Barnes, E. W. (1899). The theory of the gamma function. *Messenger of Mathematics*, 29, 64–128.

Berry, A. C. (1941). The accuracy of the Gaussian approximation to the sum of independent variates. *Transactions of the American Mathematical Society*, 49, 122–136.

Bhargava, R. P., & Khatri, C. G. (1981). The distribution of product of independent Beta r.v.'s with applications to multivariate analysis. *Annals of the Institute of Statistical Mathematics, Series A*, 33, 287–296.

Burić, T., & Elezović, N. (2011). Bernoulli polynomials and asymptotic expansions of the quotient of gamma functions. *Journal of Computational and Applied Mathematics*, 235, 3315–3331.

Carter, B. D., & Springer, M. D. (1977). The distribution of products, quotients and powers of independent H-function variates. *SIAM Journal on Applied Mathematics*, 33, 542–558.

Coelho, C. A. (1998). The generalized integer gamma distribution – a basis for distributions in multivariate statistics. *Journal of Multivariate Analysis*, 64, 86–102.

Coelho, C. A. (2004). The generalized near-integer gamma distribution: a basis for 'near-exact' approximations to the distribution of statistics which are the product of an odd number of independent Beta r.v.'s. *Journal of Multivariate Analysis*, 89, 191–218.

Coelho, C. A., & Arnold, B. C. (2019). *Finite Form Representations for Meijer G and Fox H Functions – Applied to Multivariate Likelihood Ratio Tests Using Mathematica®, MAXIMA and R*. Springer International Publishing.

Coelho, C. A., & Marques, F. J. (2011a). The multi-sample block-scalar sphericity test: exact and near-exact distributions for its likelihood ratio test statistic. *Communications in Statistics: Theory and Methods*, 42, 1153–1175.

Coelho, C. A., & Marques, F. J. (2011b). On the exact, asymptotic and near-exact distributions for the likelihood ratio statistics to test equality of several exponential distributions. *AIP Conference Proceedings*, 1389, 1471–1474.

Coelho, C. A., Marques, F. J., & Arnold, B. C. (2010). Near-exact distributions for certain likelihood ratio test statistics. *Journal of Statistical Theory and Practice*, 4, 711–725.

Coelho, C. A., Marques, F. J., & Arnold, B. C. (2015). The exact and near-exact distributions of the main likelihood ratio test statistics used in the complex multivariate normal setting. *Test*, 24, 386–416.

Esseen, C.-G. (1945). Fourier Analysis of distribution functions - a mathematical study of the Laplace-Gaussian law. *Acta Mathematica*, 77, 1–125.

Fields, J.L. (1966). A note on the asymptotic expansion of a ratio of gamma functions. *Proceedings of the Edinburgh Mathematical Society*, 15, 43–45.

Gupta, A. K. (1971). Distribution of Wilks' likelihood-ratio criterion in the complex case. *Annals of the Institute of Statistical Mathematics*, 23, 77–87.

Heuback, S., & Mansour, T. (2009). *Combinatorics of Compositions and Words*. Boca Raton: Chapman and Hall/CRC Press.

Hsu, P. L. (1945). The approximate distributions of the mean and variance of a sample of independent variables. *Annals of Mathematical Statistics*, 15, 1–29.

Hwang, H.-K. (1998). On convergence rates in the central limit theorems for combinatorial structures. *European Journal of Combinatorics*, 19, 329–343.

Imhof, J. P. (1961). Computing the distribution of quadratic forms in normal variables. *Biometrika*, 48, 419–426.

Jensen, D. R., & Solomon, H. (1972). A Gaussian approximation to the distribution of a definite quadratic form. *Journal of the American Statistical Association*, 67, 898–902.

Khatri, C. G. (1965a). Classical Statistical Analysis Based on a Certain Multivariate Complex Gaussian Distribution. *Annals of Mathematical Statistics*, 36, 98–114.

Khatri, C. G. (1965b). A test for reality of a covariance matrix in a certain complex Gaussian distribution. *Annals of Mathematical Statistics*, 36, 115–119.

Khatri, C. G. (1980). Quadratic forms in normal variables. In P.R. Krishnaiah (Ed.), *Handbook of Statistics, Volume 1: Analysis of Variance* (pp. 443–469). Amsterdam: North-Holland.

Kotz, S., Johnson, N. L., & Boyd, D. W. (1967a). Series representations of distributions of quadratic forms in Normal variables I - central case. *Annals of Mathematical Statistics*, 38, 823–837 (1967a)

Kotz, S., Johnson, N. L., & Boyd, D. W. (1967b). Series representations of distributions of quadratic forms in Normal variables II - non-central case. *Annals of Mathematical Statistics*, 38, 838–848.

Kshirsagar, A. M. (1972). *Multivariate Analysis*. New York: Marcel Dekker.

Marques, F. J., & Coelho, C. A. (2011a). The block sphericity test – exact and near-exact distributions for the likelihood ratio statistic. *Mathematical Methods in the Applied Sciences*, 35, 373–383.

Marques, F. J., & Coelho, C. A. (2011b). The multi-sample block-matrix sphericity test. *AIP Conference Proceedings*, 1389, 1479–1482.

Marques, F. J., Coelho, C. A., & Arnold, B. C. (2011). A general near-exact distribution theory for the most common likelihood ratio test statistics used in Multivariate Analysis. *TEST*, 20, 180–203.

Mathai, A. M. (1984). Extensions of Wilks' integral equations and distributions of test statistics. *Annals of the Institute of Statistical Mathematics*, 36 (Part A), 271–288.

Moschopoulos, P. G. (1985). The distribution of the sum of independent gamma random variables. *Annals of the Institute of Statistical Mathematics*, 37 (Part A), 541–544.

Moschopoulos, P. G. (1986). New Representations for the Distribution Function of a Class of Likelihood Ratio Criteria. *Journal of Statistical Research*, 20, 13–20.

Muirhead, R. J. (2005). *Aspects of Multivariate Statistical Theory* (2nd edn.). New York: J. Wiley & Sons.

Nagar, D. K., Jain, S. K., & Gupta, A. K. (1985). Distribution of LRC for testing sphericity of a complex multivariate Gaussian model. *International Journal of Mathematics and Mathematical Sciences*, 8, 555–562.

Nagarsenker, B. N., & Das, M. M. (1975). Exact distribution of sphericity criterion in the complex case and its percentage points. *Communications in Statistics*, 4, 363–374.

Nagarsenker, B. N., & Suniaga, J. (1983). Distributions of a class of statistics useful in multivariate analysis. *Journal of the American Statistical Association*, 78, 472–475.

Nandi, S. B. (1980). The exact distributions of certain likelihood ratio criteria. *Sankhya, Series B*, 42, 187–196.

Olkin, I., & Press, S. J. (1961). Testing and estimation for a circular stationary model. *Annals of Mathematical Statistics*, 40, 1358–1373.

Pederzoli, G. (1985). Some properties of beta functions and the distribution for the product of independent beta random variables. *Trabajos de Estadistica y de Investigacion Operativa*, 36, 122–128.

Pillai, K. C. S., & Jouris, G. M. (1971). Distribution of Wilks' likelihood-ratio criterion in the complex case. *Annals of Mathematical Statistics*, 42, 517–525.

Shah, B. K. (1963). Distribution of definite and indefinite quadratic forms from a non-central normal distribution. *Annals of Mathematical Statistics*, 34, 186–190.

Springer, M. D. (1979). *The Algebra of Random Variables*. New York: J. Wiley & Sons.

Springer, M. D., & Thompson, W. E. (1966). The distribution of products of independent random variables. *SIAM Journal of Applied Mathematics*, 14, 511–526.

Springer, M. D., & Thompson, W. E. (1970). The distribution of products of beta, gamma and Gaussian random variables. *SIAM Journal of Applied Mathematics*, 18, 721–737.

Tang, J., & Gupta, A. K. (1984). On the distribution of the product of independent Beta random variables. *Statistics and Probability Letters*, 2, 165–168.

Tang, J., & Gupta, A. K. (1986). Exact distribution of certain general test statistics in multivariate analysis. *Australian Journal of Statistics*, 28, 107–114.

Tretter, M. J., & Walster, G. W. (1975). Central and noncentral distribution of Wilks' statistics in MANOVA as mixtures of incomplete beta functions. *Annals of Statistics*, 3, 467–472.

Tricomi, F. G., & Erdélyi, A. (1951). The asymptotic expansion of a ratio of gamma functions. *Pacific Journal of Mathematics*, 1, 133–142.

Tukey, J. W., & Wilks, S. S. (1946). Approximation of the distribution of the product of beta variables by a single beta variable. *Annals of Mathematical Statistics*, 17, 318–324.

Walster, G. W., & Tretter, M. J. (1980). Exact noncentral distributions of Wilks Lambda and Wilks-Lawley U criteria as mixtures of incomplete Beta functions: for three tests. *Annals of Statistics*, 8, 1388–1390.

# Testing Equality of Mean Vectors with Block-Circular and Block Compound-Symmetric Covariance Matrices

**Carlos A. Coelho**

**Abstract** While the likelihood ratio test (LRT) for the equality of mean vectors when no particular structure is assumed for the covariance matrices is a well-known and well-studied test, the same is not true when some structure, namely a block structure, is assumed for the covariance matrices. In the present work, the author obtains the expressions for the LRT statistics to test the equality of mean vectors when the covariance matrices are assumed to be block-circular or block compound-symmetric and it is shown that actually in most cases the distributions of these statistics have closed finite form representations. For the other cases, families of near-exact distributions are developed and their performance is then numerically assessed. It is shown that these families of near-exact distributions lie very close to the exact distribution, even for very small samples and that they have an asymptotic behavior not only for increasing sample sizes but also for increasing numbers of populations involved and increasing numbers of sets of variables and variables in each set.

## 1 Introduction

The circular or circulant covariance structure is the covariance structure for any finite stretch of a stationary time series (Brillinger, 2001, Sect. 3.7; Kedem, 1994,

Reprinted with permission from Springer Nature. First published as Coelho, C.A.: Testing equality of mean vectors with block-circular and block compound-symmetric covariance matrices, in 'Multivariate, Multilinear and Mixed Linear Models', Filipiak, K., Markiewicz, A., von Rosen, D. (eds.), Springer series in Contributions to Statistics, 157–201 (2021). (https://doi.org/10.1007/978-3-030-75494-5_7)

C. A. Coelho (✉)
Mathematics Department (DM) and Centro de Matemática e Aplicações (CMA), Faculdade de Ciências e Tecnologia, NOVA University of Lisbon, Caparica, Portugal
e-mail: cmac@fct.unl.pt

 
B. C. Arnold et al. (eds.), *Likelihood Ratio Tests and Related Topics in Multivariate Analysis*, Contributions to Statistics,
https://doi.org/10.1007/978-3-031-68008-3_16

Sect. 3.2; Olkin and Press, 1969; Pollock, 2002). But this structure is also commonly assumed in a number of other situations of interest in many areas as for example in cyclic designs (John, 1987), in serially correlated time series (Anderson, 1972; Olkin and Press, 1969), and in a wealth of other applications (Khattree, 1996), and the block version is of interest or appears as a good or a much plausible model in applications in geology and seismology when for example several, say $r$, variables are measured at $m$ different time points or vertices of a regular polygon. Suppose that in a seismological study we have $q$ different points of interest, in each of which a total of $mr$ continuous variables are measured. Suppose that these $mr$ variables are indeed a set of $r$ variables which are measured on a set of $m$ points situated at the vertices of a regular polygon whose center is placed on each one of the $q$ points of interest. Suppose also that these polygons are exactly the same for each of these $q$ points of interest and suppose that we take independent samples on each of these $q$ points of interest. Then it would make sense to assume that the $r$ variables measured on each of the $mq$ points that surround the $q$ points of interest have the same variance-covariance matrix in each of these $mq$ points, while for each set of two given different points of the $m$ points surrounding each of the $q$ points of interest, the covariance matrices of the $r$ variables will be only a function of the distance between these two points and will remain the same for the set of $q$ points. Then, if we number consecutively the set of $m$ points around any given point of interest, the variance-covariance matrix for the $m$ sets of $r$ variables measured around each one of the $q$ points of interest will have a block-circular structure. For example, for $m = 5$ and $m = 6$, we would have for the $k$-th point of interest ($k = 1, \dots, q$) a structure as the one shown in Fig. 1, where

$$\boldsymbol{\Sigma}_0^* = \mathrm{Var}(\boldsymbol{x}_{1k}) = \cdots = \mathrm{Var}(\boldsymbol{x}_{mk}) \tag{1}$$

and, for $i = 1, \dots, m$ and $\ell = 1, \dots, \lfloor m/2 \rfloor$,

$$\boldsymbol{\Sigma}_\ell^* = \mathrm{Cov}(\boldsymbol{x}_{ik}, \boldsymbol{x}_{\mathrm{mod}^*(i+\ell,m),k}) = \mathrm{Cov}(\boldsymbol{x}_{ik}, \boldsymbol{x}_{\mathrm{mod}^*(i+m-\ell,m),k}), \tag{2}$$

where $\boldsymbol{x}_{1k}, \dots, \boldsymbol{x}_{mk}$ are each an $r \times 1$ random vector with the $r$ random variables measured on each of the $m$ measuring points defined around the $k$-th point of interest ($k = 1, \dots, q$) and

$$\mathrm{mod}^*(a, b) = \begin{cases} \mathrm{mod}(a, b) & \text{for } \mathrm{mod}(a, b) \neq 0 \\ b & \text{for } \mathrm{mod}(a, b) = 0\,. \end{cases} \tag{3}$$

Let then

$$\boldsymbol{x}_k = [\boldsymbol{x}'_{1k}, \dots, \boldsymbol{x}'_{mk}]'$$

be the random vector of dimension $mr$ for local $k$ ($k = 1, \dots, q$).

$$\Sigma = \begin{bmatrix} \Sigma_0^* & \Sigma_1^* & \Sigma_2^* & \Sigma_2^* & \Sigma_1^* \\ \Sigma_1^* & \Sigma_0^* & \Sigma_1^* & \Sigma_2^* & \Sigma_2^* \\ \Sigma_2^* & \Sigma_1^* & \Sigma_0^* & \Sigma_1^* & \Sigma_2^* \\ \Sigma_2^* & \Sigma_2^* & \Sigma_1^* & \Sigma_0^* & \Sigma_1^* \\ \Sigma_1^* & \Sigma_2^* & \Sigma_2^* & \Sigma_1^* & \Sigma_0^* \end{bmatrix} \qquad \Sigma = \begin{bmatrix} \Sigma_0^* & \Sigma_1^* & \Sigma_2^* & \Sigma_3^* & \Sigma_2^* & \Sigma_1^* \\ \Sigma_1^* & \Sigma_0^* & \Sigma_1^* & \Sigma_2^* & \Sigma_3^* & \Sigma_2^* \\ \Sigma_2^* & \Sigma_1^* & \Sigma_0^* & \Sigma_1^* & \Sigma_2^* & \Sigma_3^* \\ \Sigma_3^* & \Sigma_2^* & \Sigma_1^* & \Sigma_0^* & \Sigma_1^* & \Sigma_2^* \\ \Sigma_2^* & \Sigma_3^* & \Sigma_2^* & \Sigma_1^* & \Sigma_0^* & \Sigma_1^* \\ \Sigma_1^* & \Sigma_2^* & \Sigma_3^* & \Sigma_2^* & \Sigma_1^* & \Sigma_0^* \end{bmatrix}$$

**Fig. 1** Illustration of a set of $m = 5$ and a set of $m = 6$ points placed at the vertices of a regular polygon centered at the point of interest and the corresponding variance-covariance matrices for the $r \cdot m$ variables measured at the given point of interest (○ – point of interest, • – one of the $m$ measurement points)

Then we have

$$\mathrm{Var}(\boldsymbol{x}_k) = \boldsymbol{\Sigma}\,,$$

and if we take

$$\mathrm{E}(\boldsymbol{x}_k) = \boldsymbol{\mu}_k\,,$$

we will be interested in testing the null hypothesis of equality of the $q$ expected value vectors $\boldsymbol{\mu}_k$, that is, the null hypothesis

$$H_0 : \boldsymbol{\mu}_1 = \cdots = \boldsymbol{\mu}_q\,, \tag{4}$$

assuming that $\boldsymbol{\Sigma}$ has a block-circular structure as that in Fig. 1, that is, as given by (1) and (2).

In Sect. 2, the estimation process and the likelihood ratio test (LRT) statistic are developed and then in Sect. 3, the distribution of the LRT statistic is characterized. It is shown that in most cases the distribution of the LRT statistic has a finite closed form representation through the EGIG (Exponentiated Generalized Integer Gamma) distribution and a family of near-exact distributions which are asymptotic not only for increasing sample sizes but also for increasing values of $r$, $m$ and $q$ is developed for the other cases. In Sect. 4 are shown the asymptotic properties of these near-exact distributions and that they indeed lie extremely close to the exact distribution, even for very small sample sizes. Then in Sect. 5, the test for equality of mean vectors, with block compound-symmetric covariance matrices is addressed and in Sect. 6 some conclusions are drawn and some concluding remarks are made.

## 2 The Likelihood Ratio Test and Its Statistic

There are a number of different ways in which the LRT statistic to test the null hypothesis in (4) may be obtained. One of them is, of course, the common direct way, starting by obtaining the maximum likelihood estimators (MLEs) of $\boldsymbol{\Sigma}$ under the null and the alternative hypotheses and obtaining then the ratio of the maximum of the likelihoods under these two hypotheses. However, given the structure of the covariance matrix $\boldsymbol{\Sigma}$, this is, in this case, a quite long and involved process and as such we will undertake a different approach.

Let us suppose that $\boldsymbol{x}_k \sim N_p(\boldsymbol{\mu}_k, \boldsymbol{\Sigma})$ $(k = 1, \ldots, q)$, where $p = r \cdot m$. Then the one that may be the simplest way to obtain the LRT statistic to test $H_0$ in (4) is to start by considering the LRT statistic used to test $H_0$ in (4) when no structure is assumed for $\boldsymbol{\Sigma}$. This statistic is (see eg. Kshirsagar (1972, Sect. 9.1))

$$\Lambda = \frac{|\boldsymbol{A}|}{|\boldsymbol{A} + \boldsymbol{B}|} \tag{5}$$

where

$$\boldsymbol{A} = \sum_{k=1}^{q} (n_k - 1)\boldsymbol{S}_k \quad \text{and} \quad \boldsymbol{B} = \sum_{k=1}^{q} n_k \, (\overline{\boldsymbol{x}}_k - \overline{\boldsymbol{x}}) \, (\overline{\boldsymbol{x}}_k - \overline{\boldsymbol{x}})' \tag{6}$$

are, respectively, the 'within' and 'between' sum of squares and sum of products matrices, where $n_k$ is the size of the sample from $\boldsymbol{x}_k$ $(k = 1, \ldots, q)$ and $\boldsymbol{S}_k$ and $\overline{\boldsymbol{x}}_k$ are, respectively, the sample covariance matrix and mean vector of the sample from $\boldsymbol{x}_k$, and where we assume the $q$ samples, one from each $\boldsymbol{x}_k$ $(k = 1, \ldots, q)$, to be independent, and where

$$\overline{\boldsymbol{x}} = \frac{1}{n} \sum_{k=1}^{q} n_k \, \overline{\boldsymbol{x}}_k$$

is the overall sample mean vector, for

$$n = \sum_{k=1}^{q} n_k \,.$$

While $\boldsymbol{A}$ is the MLE of $\boldsymbol{\Sigma}$ under the alternative hypothesis, $\boldsymbol{A} + \boldsymbol{B}$ is the MLE of $\boldsymbol{\Sigma}$ under $H_0$ in (4). But, as it is clear from their definitions in (6), these are the unstructured MLEs of $\boldsymbol{\Sigma}$, that is, they do not account for the block-circular structure of $\boldsymbol{\Sigma}$.

One can use the LRT statistic in (5) to test $H_0$ in (4); however, failure in accounting for the structure of $\boldsymbol{\Sigma}$ will lead to a loss in power.

But then, one may easily argue that in our case the LRT statistic to test $H_0$ in (4), accounting for the block-circulant structure of $\boldsymbol{\Sigma}$, is

$$\Lambda = \frac{|\boldsymbol{A}^*|}{|\boldsymbol{A}^* + \boldsymbol{B}^*|} \tag{7}$$

where $\boldsymbol{A}^*$ and $\boldsymbol{B}^*$ are now block-circulant matrices, still with $\boldsymbol{A}^*$ and $\boldsymbol{A}^*+\boldsymbol{B}^*$ being the MLEs of $\boldsymbol{\Sigma}$, respectively under the alternative and the null hypotheses, but now accounting for the block-circular structure of $\boldsymbol{\Sigma}$.

But so, the question now is: "how to find a simple way to obtain $\boldsymbol{A}^*$ and $\boldsymbol{B}^*$?".

Let $\boldsymbol{\Sigma}$ be a block-circular matrix with $m\times m$ blocks of dimensions $r\times r$. Then we may write

$$\boldsymbol{\Sigma} = \sum_{i=0}^{m-1} \Big( \underset{(m\times m)}{\boldsymbol{W}_i} \otimes \underset{(r\times r)}{\boldsymbol{\Sigma}_i^*} \Big), \tag{8}$$

where $\boldsymbol{\Sigma}_0^*$ is an $r\times r$ positive-definite matrix,

$$\boldsymbol{\Sigma}_i^* = \boldsymbol{\Sigma}_{m-i}^*, \quad i = 1, \ldots, m-1, \tag{9}$$

are symmetric matrices, and

$$\boldsymbol{W}_{i\,(m\times m)} = \begin{bmatrix} 0 & \boldsymbol{I}_{m-i} \\ \boldsymbol{I}_i & 0 \end{bmatrix}, \quad i = 0, \ldots, m-1 \qquad (\text{with } \boldsymbol{W}_0 = \boldsymbol{I}_m),$$

where $\boldsymbol{I}_m$ stands for an identity matrix of order $m$ (see Olkin (1973, Sect. 4) and see the representations of $\boldsymbol{\Sigma}$ in Fig. 1).

But then, by the invariance property of MLEs, since the matrices $\boldsymbol{\Sigma}_i^*$ ($i = 0, \ldots, m-1$) in (8) are well identified, we may write

$$\widehat{\boldsymbol{\Sigma}} = \sum_{i=0}^{m-1} \Big( \underset{(m\times m)}{\boldsymbol{W}_i} \otimes \underset{(r\times r)}{\widehat{\boldsymbol{\Sigma}}_i^*} \Big), \tag{10}$$

so that we will take

$$\boldsymbol{A}^* = \sum_{i=0}^{m-1} (\boldsymbol{W}_i \otimes \boldsymbol{A}_i^*) \quad \text{and} \quad \boldsymbol{B}^* = \sum_{i=0}^{m-1} (\boldsymbol{W}_i \otimes \boldsymbol{B}_i^*) \tag{11}$$

where the question now is: how to find a simple way to obtain $\boldsymbol{A}_i^*$ and $\boldsymbol{B}_i^*$ $(i = 0, \dots, m-1)$?

From an extension of the results of Lv and Huang (2007) and also from the results in Lv and Huang (2013), as well as from the block spectral decomposition of the matrix $\boldsymbol{\Sigma}$ (see Appendix A), we know that the inverse of a block-circulant matrix is a block-circulant matrix, and as such both $\boldsymbol{\Sigma}$ and $\boldsymbol{\Sigma}^{-1}$ may be written in the form in (8). A fact that may also be derived from the simple fact that the space of block-circular matrices is a quadratic space (Seely, 1971). We may then use the results in Szatrowski (1978, 1980) together with the facts that (see (8)–(10) and Fig. 1)

- there are always $m$ blocks $\boldsymbol{\Sigma}_0^*$
- for odd $m$ there are $2m$ of all other blocks
- for even $m$
  - there are also $m$ blocks $\boldsymbol{\Sigma}_{m/2}^*$
  - and there are $2m$ of all other blocks,

to take

$$\boldsymbol{A}_0^* = \frac{1}{m}\sum_{j=1}^{m} \widetilde{\boldsymbol{A}}_{jj}\,, \quad \boldsymbol{B}_0^* = \frac{1}{m}\sum_{j=1}^{m} \widetilde{\boldsymbol{B}}_{jj}\,, \tag{12}$$

and for $i = 1, \dots, m-1$, for odd $m$, or for $i = 1, \dots, m/2-1, m/2+1, \dots, m$, for even $m$,

$$\boldsymbol{A}_i^* = \frac{1}{2m}\sum_{j=1}^{m} \widetilde{\boldsymbol{A}}_{j,\mathrm{mod}^*(j+i,m)} + \widetilde{\boldsymbol{A}}_{\mathrm{mod}^*(j+i,m),j}\,, \tag{13}$$

and

$$\boldsymbol{B}_i^* = \frac{1}{2m}\sum_{j=1}^{m} \widetilde{\boldsymbol{B}}_{j,\mathrm{mod}^*(j+i,m)} + \widetilde{\boldsymbol{B}}_{\mathrm{mod}^*(j+i,m),j}\,,$$

with $\boldsymbol{A}_i^* = \boldsymbol{A}_{m-i}^*$ and $\boldsymbol{B}_i^* = \boldsymbol{B}_{m-i}^*$, for $i = 1, \dots, m-1$, and for even $m$

$$\boldsymbol{A}_{m/2}^* = \frac{1}{m}\sum_{j=1}^{m} \widetilde{\boldsymbol{A}}_{j,\mathrm{mod}^*(j+m/2,m)}\,, \quad \boldsymbol{B}_{m/2}^* = \frac{1}{m}\sum_{j=1}^{m} \widetilde{\boldsymbol{B}}_{j,\mathrm{mod}^*(j+m/2,m)}\,, \tag{14}$$

where $\mathrm{mod}^*(\cdot\,,\,\cdot)$ is given by (3) and

$$\widetilde{\boldsymbol{A}}_{jk} \quad \text{and} \quad \widetilde{\boldsymbol{B}}_{jk}\,, \quad j,k \in \{1,\dots,m\}$$

denote the $r \times r$ blocks of $\boldsymbol{A}$ and $\boldsymbol{B}$ starting at row $r(j-1)+1$ and column $r(k-1)+1$.

## 3 The Distribution of the Likelihood Ratio Statistic

Although (7) provides a good way to compute the LRT statistic, we will need to resort to an equivalent formulation in order to be able to derive the distribution of $\Lambda$.

Let us consider the matrices

$$\boldsymbol{A}^{**} = (\boldsymbol{\Gamma}_m \otimes \boldsymbol{I}_r)\, \boldsymbol{A}\, (\boldsymbol{\Gamma}_m \otimes \boldsymbol{I}_r) \quad \text{and} \quad \boldsymbol{B}^{**} = (\boldsymbol{\Gamma}_m \otimes \boldsymbol{I}_r)\, \boldsymbol{B}\, (\boldsymbol{\Gamma}_m \otimes \boldsymbol{I}_r) \tag{15}$$

where $\Gamma_m$ is an orthogonal symmetrical matrix with elements

$$\gamma_{jk} = \frac{1}{\sqrt{m}}\left[\sin\left(\tfrac{2\pi}{m}(j-1)(k-1)\right) + \cos\left(\tfrac{2\pi}{m}(j-1)(k-1)\right)\right], \qquad j,k \in \{1,\ldots,m\}, \tag{16}$$

and let $\boldsymbol{A}_j^{**}$ and $\boldsymbol{B}_j^{**}$ $(j = 1,\ldots,m)$ denote the $j$-th diagonal block of dimensions $r{\times}r$ respectively of $\boldsymbol{A}^{**}$ and $\boldsymbol{B}^{**}$. The matrix $\boldsymbol{\Gamma}_m$ is the matrix used by Olkin and Press (1969) to diagonalize circular symmetric matrices and it also diagonalizes all matrices $\boldsymbol{W}_i + \boldsymbol{W}_{m-i}$ $(i = 0,\ldots,m)$ (see Olkin (1973, Sect. 4)).

Then, going through some algebra (see Appendix B), it is possible to show that, for $m^+ = \lfloor p/2 \rfloor$,

$$|\boldsymbol{A}^*| = |\boldsymbol{A}_1^{**}| \left(|\boldsymbol{A}_{1+m^+}^{**}|\right)^{\mathrm{mod}(m+1,2)} \prod_{j=2}^{m-m^+} \left|\frac{\boldsymbol{A}_j^{**} + \boldsymbol{A}_{m-j+2}^{**}}{2}\right|^2 \tag{17}$$

and

$$|\boldsymbol{A}^* + \boldsymbol{B}^*| = |\boldsymbol{A}_1^{**} + \boldsymbol{B}_1^{**}| \left(|\boldsymbol{A}_{1+m^+}^{**} + \boldsymbol{B}_{1+m^+}^{**}|\right)^{\mathrm{mod}(m+1,2)} \times \prod_{j=2}^{m-m^+} \left|\frac{\boldsymbol{A}_j^{**} + \boldsymbol{B}_j^{**} + \boldsymbol{A}_{m-j+2}^{**} + \boldsymbol{B}_{m-j+2}^{**}}{2}\right|^2, \tag{18}$$

so that we have for $\Lambda$ in (7),

$$\Lambda = \underbrace{\frac{|\boldsymbol{A}_1^{**}|}{|\boldsymbol{A}_1^{**}+\boldsymbol{B}_1^{**}|}}_{\Lambda_1} \underbrace{\left(\frac{|\boldsymbol{A}_{1+m^+}^{**}|}{|\boldsymbol{A}_{1+m^+}^{**}+\boldsymbol{B}_{1+m^+}^{**}|}\right)^{\mathrm{mod}(m+1,2)}}_{\Lambda_2} \times \underbrace{\prod_{j=2}^{m-m^+}\left(\frac{\left|\boldsymbol{A}_j^{**}+\boldsymbol{A}_{m-j+2}^{**}\right|}{\left|\boldsymbol{A}_j^{**}+\boldsymbol{B}_j^{**}+\boldsymbol{A}_{m-j+2}^{**}+\boldsymbol{B}_{m-j+2}^{**}\right|}\right)^2}_{\Lambda_3} . \tag{19}$$

Then we may take into account that, on one hand, in (19), the $\boldsymbol{A}_j^{**}$ and $\boldsymbol{B}_j^{**}$ are independent, given the independence between $\boldsymbol{A}$ and $\boldsymbol{B}$, and that on the other hand, the $\boldsymbol{A}_j^{**}$ $(j=1,\ldots,m)$ form a set of $m$ independent matrices and also the $\boldsymbol{B}_j^{**}$ $(j=1,\ldots,m)$ form another set of independent matrices. This is so given that we know that for $n=\sum_{k=1}^q n_k$,

$$\boldsymbol{A} \sim W_p(n-q,\boldsymbol{\Sigma}) \implies \boldsymbol{A}^{**} \sim W_p\big(n-q,(\boldsymbol{\Gamma}_m\otimes\boldsymbol{I}_r)\,\boldsymbol{\Sigma}\,(\boldsymbol{\Gamma}_m\otimes\boldsymbol{I}_r)\big)$$

and

$$\boldsymbol{B} \sim W_p(q-1,\boldsymbol{\Sigma}) \implies \boldsymbol{B}^{**} \sim W_p\big(q-1,(\boldsymbol{\Gamma}_m\otimes\boldsymbol{I}_r)\,\boldsymbol{\Sigma}\,(\boldsymbol{\Gamma}_m\otimes\boldsymbol{I}_r)\big)$$

with (see Olkin (1973, Sect. 4, expressions (4.6) and (4.7)))

$$(\boldsymbol{\Gamma}_m\otimes\boldsymbol{I}_r)\,\boldsymbol{\Sigma}\,(\boldsymbol{\Gamma}_m\otimes\boldsymbol{I}_r)=\mathrm{BDiag}\,(\boldsymbol{\Psi}_1,\boldsymbol{\Psi}_2,\ldots,\boldsymbol{\Psi}_m)\,,$$

where, "BDiag$(\cdots)$" stands for a block-diagonal matrix with the diagonal blocks shown inside the parenthesis and where $\boldsymbol{\Psi}_j$ $(j=1,\ldots,m)$ are $r\times r$ symmetric matrices (see Olkin (1973, Sect. 4)), so that, for $j=1,\ldots,m$,

$$\boldsymbol{A}_j^{**} \sim W_r(n-q,\boldsymbol{\Psi}_j) \quad \text{and} \quad \boldsymbol{B}_j^{**} \sim W_r(q-1,\boldsymbol{\Psi}_j)\,, \tag{20}$$

with $\boldsymbol{A}_j^{**}$ independent of $\boldsymbol{A}_{j'}^{**}$, for $j\neq j'$, and $\boldsymbol{B}_j^{**}$ independent of $\boldsymbol{B}_{j'}^{**}$.

Furthermore, since for $j=2,\ldots,m$, $\boldsymbol{\Psi}_j=\boldsymbol{\Psi}_{m-j+2}$ (see Olkin (1973, Sect. 4)), we have, for $j=2,\ldots,m$,

$$\boldsymbol{A}_j^{**}+\boldsymbol{A}_{m-j+2}^{**} \sim W_r(2(n-q),\boldsymbol{\Psi}_j)$$

and

$$\boldsymbol{B}_j^{**}+\boldsymbol{B}_{m-j+2}^{**} \sim W_r(2(q-1),\boldsymbol{\Psi}_j)\,.$$

As a consequence, the statistics $\Lambda_1$, $\Lambda_2$ and $\Lambda_3$ in (19) are independent, with (see eg. Coelho and Arnold (2019, Chap. 5, App. 1))

$$\Lambda_1 \stackrel{d}{\equiv} \Lambda_2 \stackrel{st}{\sim} \prod_{j=1}^{r} Y_j$$

where, " $\stackrel{st}{\sim}$ " is to be read as "is stochastically equivalent to" or "has the same distribution as" and where for $n > q + r - 1$,

$$Y_j \sim Beta\left(\frac{n-q+1-j}{2}, \frac{q-1}{2}\right), \quad j = 1, \ldots, r$$

form a set of $r$ independent RVs, and where $r$ and $q - 1$ are interchangeable (see Coelho and Arnold (2019, Chap. 5, App. 1)), and

$$\Lambda_3 = \prod_{j=2}^{m-m^+} \Lambda_j^* \quad \text{with} \quad \Lambda_j^* = \left(\frac{\left|\boldsymbol{A}_j^{**} + \boldsymbol{A}_{m-j+2}^{**}\right|}{\left|\boldsymbol{A}_j^{**} + \boldsymbol{B}_j^{**} + \boldsymbol{A}_{m-j+2}^{**} + \boldsymbol{B}_{m-j+2}^{**}\right|}\right)^2$$

where the $m - m^+ - 1$ statistics $\Lambda_j^*$ $(j = 2, \ldots, m - m^+)$ are also independent. Furthermore, from the distributions of $A_j^{**}$ and $B_j^{**}$ in (20), and given that $\boldsymbol{\Psi}_j = \boldsymbol{\Psi}_{m-j+2}$ for $j = 2, \ldots, m$, we have

$$\boldsymbol{A}_j^{**} + \boldsymbol{A}_{m-j+2}^{**} \sim W_r(2(n-q), \boldsymbol{\Psi}_j), \quad j = 2, \ldots, m - m^+$$

and

$$\boldsymbol{B}_j^{**} + \boldsymbol{B}_{m-j+2}^{**} \sim W_r(2(q-1), \boldsymbol{\Psi}_j), \quad j = 2, \ldots, m - m^+$$

so that (see Coelho and Arnold (2019, Sect. 5, App. 1))

$$\Lambda_j^* \stackrel{st}{\sim} \prod_{k=1}^{r} (Y_k^*)^2$$

where

$$Y_k^* \sim Beta\left(n - q + \frac{1-k}{2}, q - 1\right), \quad k = 1, \ldots, r$$

also form a set of $r$ independent RVs.

But then we can write

$$\Lambda \stackrel{st}{\sim} \left\{\prod_{j=1}^{r} Y_j\right\}^{1+\text{mod}(m+1,2)} \left\{\prod_{j=2}^{m-m^+} \prod_{k=1}^{r} (Y_k^*)^2\right\}$$

where $m^+ = \lfloor m/2 \rfloor$,

$$Y_j \sim Beta\left(\frac{n-q+1-j}{2}, \frac{q-1}{2}\right), \quad j = 1, \ldots, r$$

and

$$Y_k^* \sim Beta\left(n-q+\frac{1-k}{2}, q-1\right), \quad k = 1, \ldots, r,$$

form a set of $r(m-m^+) + 1 + \text{mod}(m+1, 2)$ independent RVs.

Hence, for any $q, m, r$ and $n$, the probability density function (p.d.f.) of $\Lambda$ may be written in terms of the Fox $H$ function (Fox, 1961; Mathai, 1993, Sect. 3.11; Mathai and Haubold, 2008, Sect. 1.9; Mathai and Saxena, 1978, Sect. 1.1; Prudnikov et al., 1990, Sect. 8.3), using the notation in Coelho and Arnold (2019, Sect. 2.1) as

$$f_\Lambda(z) = \left\{\prod_{j=1}^{r} \frac{\Gamma\left(\frac{n-j}{2}\right)}{\Gamma\left(\frac{n-q+1-j}{2}\right)}\right\}^{1+\lfloor\lfloor m+1 \rfloor\rfloor} \left\{\prod_{j=2}^{m-m^+} \prod_{k=1}^{r} \frac{\Gamma\left(n-1+\frac{1-k}{2}\right)}{\Gamma\left(n-q+\frac{1-k}{2}\right)}\right\}$$
$$\times H^{p^*,0}_{p^*,p^*}\left(\left.\begin{array}{l} \left\{\left\{\left(\frac{n-j-2}{2}, 1\right)\right\}_{j=1:r(1+\lfloor\lfloor m+1 \rfloor\rfloor)}, \left\{\left(n-2+\frac{1-k}{2}, 2\right)\right\}_{\substack{j=2:m-m^+ \\ k=1:r}}\right\} \\ \left\{\left\{\left(\frac{n-q-1-j}{2}, 1\right)\right\}_{j=1:r(1+\lfloor\lfloor m+1 \rfloor\rfloor)}, \left\{\left(n-q-1+\frac{1-k}{2}, 2\right)\right\}_{\substack{j=2:m-m^+ \\ k=1:r}}\right\} \end{array}\right| z\right)$$

for $\lfloor\lfloor m+1 \rfloor\rfloor = \text{mod}(m+1, 2)$ and $p^* = r(m-m^+) + \text{mod}(m+1, 2)$, and its cumulative distribution function (c.d.f.) as

$$F_\Lambda(z) = \left\{\prod_{j=1}^{r} \frac{\Gamma\left(\frac{n-j}{2}\right)}{\Gamma\left(\frac{n-q+1-j}{2}\right)}\right\}^{1+\lfloor\lfloor m+1 \rfloor\rfloor} \left\{\prod_{j=2}^{m-m^+} \prod_{k=1}^{r} \frac{\Gamma\left(n-1+\frac{1-k}{2}\right)}{\Gamma\left(n-q+\frac{1-k}{2}\right)}\right\}$$
$$\times H^{p^*,1}_{p^*+1,p^*+1}\left(\left.\begin{array}{l} \left\{(1,1), \left\{\left(\frac{n-j}{2}, 1\right)\right\}_{j=1:r(1+\lfloor\lfloor m+1 \rfloor\rfloor)}, \left\{\left(n-1+\frac{1-k}{2}, 2\right)\right\}_{\substack{j=2:m-m^+ \\ k=1:r}}\right\} \\ \left\{\left\{\left(\frac{n-q+1-j}{2}, 1\right)\right\}_{j=1:r(1+\lfloor\lfloor m+1 \rfloor\rfloor)}, \left\{\left(n-q+\frac{1-k}{2}, 2\right)\right\}_{\substack{j=2:m-m^+ \\ k=1:r}}, (0,1)\right\} \end{array}\right| z\right)$$

where

$$H^{m,n}_{p,q}\left(\begin{matrix}\{(a_k,\alpha_k)\}_{k=1:p}\\ \{(b_j,\beta_j)\}_{j=1:q}\end{matrix}\middle|\, z\right)=\frac{1}{2\pi \mathrm{i}}\oint_L \frac{\left\{\prod_{j=1}^{m}\Gamma(b_j+\beta_j s)\right\}\left\{\prod_{k=1}^{n}\Gamma(1-a_k-\alpha_k s)\right\}}{\left\{\prod_{j=m+1}^{q}\Gamma(1-b_j-\beta_j s)\right\}\left\{\prod_{k=n+1}^{p}\Gamma(a_k+\alpha_k s)\right\}}\, z^s\, ds$$

represents the Fox $H$ function with arguments $m, n, p, q \in \mathbb{N}_0$ $(m \le q; n \le p)$, $z \neq 0$, and any real or complex $a_k$ and $b_j$ and positive reals $\alpha_k$ and $\beta_j$ $(k = 1, \dots, p;\ j = 1, \dots, q)$, such that for $\nu \in \mathbb{N}_0$, $a_k - (b_j + \nu)\alpha_k/\beta_j \notin \mathbb{N}$ for $k = 1, \dots, n$ and $j = 1, \dots, m$.

Although the Fox $H$ function provides a very handy representation for both the p.d.f. and the c.d.f. of $\Lambda$, it does not provide a useful means of computation of these functions since even for small values of the parameters $q$, $m$, $r$ and $n$, any computation will take too much long time if at all possible. As such we will have to resort to a different approach.

### 3.1 The Exact Distribution of $\Lambda$ for Odd $q$ or Even $r$

We start by noticing that we can write the $h$-th moment of $\Lambda_3$ as

$$\begin{aligned}
\mathrm{E}(\Lambda_3^h) &= \prod_{j=2}^{m-m^+}\prod_{k=1}^{r}\mathrm{E}\left((Y_k^*)^{2h}\right)\\
&= \prod_{j=2}^{m-m^+}\prod_{k=1}^{r}\frac{\Gamma\left(n-\frac12-\frac k2\right)\Gamma\left(n-q+\frac{1-k}{2}+2h\right)}{\Gamma\left(n-q+\frac{1-k}{2}\right)\Gamma\left(n-\frac12-\frac k2+2h\right)}\\
&= \left\{\prod_{k=1}^{r}\frac{\Gamma\left(n-\frac12-\frac k2\right)\Gamma\left(n-q+\frac{1-k}{2}+2h\right)}{\Gamma\left(n-q+\frac{1-k}{2}\right)\Gamma\left(n-\frac12-\frac k2+2h\right)}\right\}^{m-m^+-1},
\end{aligned}$$

which, using the relation

$$\frac{\Gamma(a+n)}{\Gamma(a)}=\prod_{\ell=0}^{n-1}(a+\ell)\,, \tag{21}$$

valid for any real or complex $a$ and positive integer $n$, may be rewritten as

$$
\begin{aligned}
\mathrm{E}(\Lambda_3^h) &= \left\{\prod_{k=1}^{r} \frac{\Gamma\left(n-\frac{1}{2}-\frac{k}{2}\right)\,\Gamma\left(n-q+\frac{1-k}{2}+2h\right)}{\Gamma\left(n-q+\frac{1-k}{2}\right)\,\Gamma\left(n-\frac{1}{2}-\frac{k}{2}+2h\right)}\right\}^{m-m^+-1} \\
&= \left\{\prod_{k=1}^{r}\prod_{\ell=0}^{q-2}\left(n-q+\frac{1-k}{2}+\ell\right)\left(n-q+\frac{1-k}{2}+\ell+2h\right)^{-1}\right\}^{m-m^+-1} \\
&= \left\{\prod_{k=1}^{r}\prod_{\ell=0}^{q-2}\left(\frac{n-q+\ell}{2}+\frac{1-k}{4}\right)\left(\frac{n-q+\ell}{2}+\frac{1-k}{4}+h\right)^{-1}\right\}^{m-m^+-1} \\
&= \prod_{j=1}^{r+2q-4}\left(\frac{2n-2q-r+j}{4}\right)^{r_j(m-m^+-1)}\left(\frac{2n-2q-r+j}{4}+h\right)^{-r_j(m-m^+-1)}
\end{aligned}
\tag{22}
$$

where

$$
r_j \equiv r(j; r, 2(q-1), 2)
$$

with

$$
r(j; a, b, h^*) = \begin{cases} h_j\,, & j = 1, \ldots, h^* \\ h_j + r_{j-h^*}\,, & j = h^*+1, \ldots, a+b-h^* \end{cases} \tag{23}
$$

where

$$
h_j \equiv h(j; a, b, h^*) = \begin{cases} +1\,, & j = 1, \ldots, \min(a,b) \\ \phantom{+}0\,, & j = 1+\min(a,b), \ldots, \max(a,b) \\ -1\,, & j = 1+\max(a,b), \ldots, a+b-h^*\,. \end{cases} \tag{24}
$$

Then, given that all Gamma functions in (22) remain valid when we replace $h$ by any complex value, we may write the characteristic function (c.f.) of $W_3 = -\log \Lambda_3$ as

$$
\begin{aligned}
\Phi_{W_3}(t) &= \mathrm{E}\left(e^{\mathrm{i}tW_3}\right) = \mathrm{E}\left(\Lambda_3^{-\mathrm{i}t}\right) \\
&= \prod_{j=1}^{r+2q-4}\left(\frac{2n-2q-r+j}{4}\right)^{r_j(m-m^+-1)}\left(\frac{2n-2q-r+j}{4}-\mathrm{i}t\right)^{-r_j(m-m^+-1)}
\end{aligned}
$$

which shows that the exact distribution of $W_3 = -\log \Lambda_3$ is a GIG (Generalized Integer Gamma) distribution of depth $2q+r-4$, with shape parameters $r_j$ and rate parameters $\frac{2n-2q-r+j}{4}$ $(j = 1, \ldots, 2q+r-4)$ (see Appendix C and Coelho (1998) for the definition of the GIG distribution, its p.d.f. and c.d.f.).

The same result might have been obtained using a modified version of Theorem 3.2 in Coelho and Arnold (2019) to encompass the distribution of $\Lambda_3$ in the realm of the EGIG distribution, which may be stated as follows.

**Theorem 1** *For positive integers $n_\nu$, $k_\nu$ and $m_\nu$ $(\nu = 1, \ldots, m^*)$ and for real*

$$a_{\nu\ell} = a_\nu + \ell + \frac{1-\ell}{h^* k_\nu}$$

*with $h^* \in \mathbb{N}$ and with $a_\nu > \frac{n_\nu - 1}{h^* k_\nu}$ $(\nu = 1, \ldots, m^*)$, let*

$$Z = \prod_{\nu=1}^{m^*} \prod_{\ell=1}^{n_\nu} \prod_{j=1}^{k_\nu} Y_{\nu\ell j}$$

*where*

$$Y_{\nu\ell j} \sim Beta\left(a_{\nu\ell}+1-\ell-\frac{j}{k_\nu}, \frac{m_\nu}{k_\nu}\right), \quad \nu=1,\ldots,m^*;\ \ell=1,\ldots,n_\nu;\ j=1,\ldots,k_\nu,$$

*are independent r.v.'s. Then,*

$$Z \stackrel{d}{\equiv} \prod_{\nu=1}^{m^*} \prod_{\ell=1}^{n_\nu} \prod_{j=1+k_\nu(\ell-1)}^{k_\nu \ell} Y^*_{\nu\ell j} \stackrel{d}{\equiv} \prod_{\nu=1}^{m^*} \prod_{\ell=1}^{n_\nu} (Y^{**}_{\nu\ell})^{k_\nu} \stackrel{d}{\equiv} \prod_{\nu=1}^{m^*} \prod_{j=1}^{n_\nu + h^*(m_\nu - 1)} e^{-W_{\nu j}} \tag{25}$$

*where*

$$Y^*_{\nu\ell j} \sim Beta\left(a_{\nu\ell} - \frac{j}{k_\nu}, \frac{m_\nu}{k_\nu}\right), \qquad Y^{**}_{\nu\ell} \sim Beta\left((a_{\nu\ell} - \ell)k_\nu, m_\nu\right)$$

*and*

$$W_{\nu j} \sim \Gamma\left(r_{\nu j}, a_\nu + \frac{j - n_\nu}{h^* k_\nu}\right)$$

*for*

$$r_{\nu j} \equiv r(j; n_\nu, h^* m_\nu, h^*)$$

*where $r(j; a, b, h^*)$ is defined in (23)–(24) and where $j = 1, \ldots, n_\nu + h^*(m_\nu - 1)$ and $\nu = 1, \ldots, m^*$.*

*If $a_\nu = a$, $n_\nu = n$, $k_\nu = k$ and $m_\nu = m$ for $\nu = 1, \ldots, m^*$, then*

$$Z \equiv \prod_{j=1}^{n+h^*(m-1)} e^{-W_j} \tag{26}$$

*with*

$$W_j \sim \Gamma\left(m^* r_j, a + \frac{j-n}{h^* k}\right)$$

*where, for* $j = 1, \ldots, n + h^*(m-1)$,

$$r_j \equiv r(j; n, h^* m, h^*) \tag{27}$$

*for the definition of* $r(j; a, b, h^*)$ *in (23)–(24).*

The proof of this Theorem follows similar lines to that of Theorem 3.1 in Coelho and Arnold (2019) and is shown in Appendix D. Its most significant contribution resides in the fact that the last equivalence in (25) shows that the distribution of $W = -\log Z$ is a GIG distribution, or, equivalently, that the distribution of $Z$ is an EGIG distribution.

Using the above Theorem with $h^* = 2$, $a_\nu = (n-q)/2$, $k_\nu = 2$, $n_\nu = r$, $m_\nu = q - 1$ and $m^* = m - m^+ - 1$, we obtain immediately the distribution of $W_3 = -\log \Lambda_3$ as a GIG distribution, or that of $\Lambda_3$.

We will now show that the exact distribution of $W_1 = -\log \Lambda_1$ and $W_2 = -\log \Lambda_2$ for odd $q$ or even $r$ are also GIG distributions and that, as such, the exact distribution of $W = -\log \Lambda$ is in this case a GIG distribution, and consequently, that of $\Lambda$ an EGIG distribution.

The statistics $\Lambda_1$ and $\Lambda_2$ have the same distribution, and for example, for $\Lambda_1$, we may write its $h$-th moment as

$$\mathrm{E}\left(\Lambda_1^h\right) = \prod_{j=1}^{r} \mathrm{E}(Y_j^h) = \prod_{j=1}^{r} \frac{\Gamma\left(\frac{n-j}{2}\right) \Gamma\left(\frac{n-q+1-j}{2} + h\right)}{\Gamma\left(\frac{n-q+1-j}{2}\right) \Gamma\left(\frac{n-j}{2} + h\right)}$$

which, for odd $q$, using (21), may then be written as

$$\begin{aligned}
\mathrm{E}\left(\Lambda_1^h\right) &= \prod_{j=1}^{r} \prod_{\ell=0}^{\frac{q-1}{2}-1} \left(\frac{n-q+1-j}{2} + \ell\right)\left(\frac{n-q+1-j}{2} + \ell + h\right)^{-1} \\
&= \prod_{j=1}^{r+q-3} \left(\frac{n-r-q+j}{2}\right)^{s_j} \left(\frac{n-r-q+j}{2} + h\right)^{-s_j}
\end{aligned}$$

where, for $j = 1, \ldots, r + q - 3$,

$$s_j \equiv r(j; r, q-1, 2) \tag{28}$$

for the definition of $r(j; a, b, h^*)$ in (23)–(24).

As such, we may write the c.f. of $W_1 = -\log \Lambda_1$ as

$$\Phi_{W_1}(t) = \mathrm{E}\left(\Lambda_1^{-\mathrm{i}t}\right) = \prod_{j=1}^{r+q-3} \left(\frac{n-r-q+j}{2}\right)^{s_j} \left(\frac{n-r-q+j}{2} - \mathrm{i}t\right)^{-s_j} \tag{29}$$

which is the c.f. of a GIG distribution of depth $r+q-3$, with shape parameters $s_j$ and rate parameters $\frac{n-r-q+j}{2}$ $(j = 1, \ldots, r+q-3)$.

And it is not hard to show that, since $q-1$ and $r$ are interchangeable in the distributions of $\Lambda_1$ and $\Lambda_2$ (see Coelho and Arnold (2019, Chap. 5, App. 1)), a similar result may be obtained when $r$ is even.

Indeed, a similar result for the distribution of $\Lambda_1$ and $\Lambda_2$ may be obtained by using Theorem 3.2 in Coelho and Arnold (2019) with $m^* = 1$, and then, for even $r$

$$a_1 = \frac{n-q+1}{2}, \quad n_1 = r/2, \quad k_1 = 2, \quad m_1 = q-1$$

or, for odd $q$,

$$a_1 = \frac{n-r}{2}, \quad n_1 = (q-1)/2, \quad k_1 = 2, \quad m_1 = r.$$

But then we may write the c.f. of $W = -\log \Lambda$, for odd $q$ or even $r$, as

$$\begin{aligned}
\Phi_W(t) &= \Phi_{W_1}(t) \left\{\Phi_{W_2}(t)\right\}^{\mathrm{mod}(m+1,2)} \Phi_{W_3}(t) \\
&= \left\{ \prod_{j=1}^{r+q-3} \left(\frac{n-r-q+j}{2}\right)^{s_j} \left(\frac{n-r-q+j}{2} - \mathrm{i}t\right)^{-s_j} \right\}^{1+\mathrm{mod}(m+1,2)} \\
&\quad \times \prod_{k=1}^{r+2q-4} \left(\frac{2n-2q-r+k}{4}\right)^{r_k(m-m^*-1)} \left(\frac{2n-2q-r+k}{4} - \mathrm{i}t\right)^{-r_k(m-m^*-1)}
\end{aligned}$$

which is the c.f. of a GIG distribution, where however some of the rate parameters may be "repeated", so that in order to devise the correct representation for this GIG distribution (where equal rate parameters "are not allowed"), we have to write the c.f. of $W$ as

$$\Phi_W(t) = \prod_{j=1}^{r+q-3} \left(\frac{n-r-q+j}{2}\right)^{s_j^*} \left(\frac{n-r-q+j}{2} - \mathrm{i}t\right)^{-s_j^*}$$
$$\times \prod_{\substack{k=1+\mathrm{mod}(r,2)\\ \text{step } 2}}^{r+2q-5} \left(\frac{2n-2q-r+k}{4}\right)^{r_k^*} \left(\frac{2n-2q-r+k}{4} - \mathrm{i}t\right)^{-r_k^*}$$
$$\times \left(\frac{n-2}{2}\right)^{r_{r+2q-4}^*} \left(\frac{n-2}{2} - \mathrm{i}t\right)^{-r_{r+2q-4}^*} \tag{30}$$

where

$$s_j^* = \begin{cases} s_j(1+\mathrm{mod}(m+1,2)), & j = 1, \dots, \lfloor r/2 \rfloor \\ s_j(1+\mathrm{mod}(m+1,2)) + r^*_{2(j-\lfloor r/2 \rfloor)-\mathrm{mod}(r,2)}, & j = \lfloor r/2 \rfloor + 1, \dots, r+q-3 \end{cases}$$

and

$$r_k^* = r_k(m - m^* - 1), \quad k = 1, \dots, r + 2q - 4 .$$

The c.f. in (30) shows that the exact distribution of $W$ is for odd $q$ or even $r$ a GIG distribution of depth $r + 2q + \left\lfloor \frac{r+1}{2} \right\rfloor - 3$, with rate parameters

$$\left\{ \left\{ \frac{n-r-q+j}{2} \right\}_{j=1:r+q-3}, \left\{ \frac{2n-2q-r+k}{4} \right\}_{\substack{k=1+\mathrm{mod}(r,2):r+2q-5\\ \text{step } 2}}, \frac{n-2}{2} \right\}$$

and corresponding shape parameters

$$\left\{ \left\{ s_j^* \right\}_{j=1:r+q-3}, \left\{ r_k^* \right\}_{\substack{k=1+\mathrm{mod}(r,2):r+2q-5\\ \text{step } 2}}, r^*_{r+2q-4} \right\} .$$

Equivalently, the exact distribution of $\Lambda$ is, in these cases, an EGIG distribution (Arnold et al., 2013) with the above depth and the above shape and rate parameters. For the definition and expressions of the p.d.f. and c.d.f. of the GIG and EGIG distributions, see Appendix C.

## 3.2 *Near-Exact Distributions for $\Lambda$ for the Case when $q$ Is Even and $r$ Is Odd*

For even $q$ and odd $r$, the expressions for the exact p.d.f. and c.d.f. of $\Lambda$ become non-manageable and the best solution is to resort to the development and use of very sharp near-exact distributions.

These near-exact distributions will be asymptotic not only for increasing sample sizes but also for increasing values of $r$, $m$ and $q$.

For even $q$ and odd $r$, the distribution of $\Lambda_3$ remains the same and in what concerns that of $\Lambda_1$ and $\Lambda_2$, we may write, for example, the $h$-th moment of $\Lambda_1$ as

$$\mathrm{E}\left(\Lambda_1^h\right)=\left\{\prod_{j=1}^{r+q-4}\left(\frac{n-r-q+j}{2}\right)^{s_j}\left(\frac{n-r-q+j}{2}+h\right)^{-s_j}\right\}\frac{\Gamma\left(\frac{n-1}{2}\right)}{\Gamma\left(\frac{n-2}{2}\right)}\frac{\Gamma\left(\frac{n-2}{2}+h\right)}{\Gamma\left(\frac{n-1}{2}+h\right)} \tag{31}$$

where

$$s_j=\begin{cases} h_j^*, & j=1,2\\ h_j^*+s_{j-2}, & j=3,\dots,r+q-4\end{cases} \tag{32}$$

with

$$h_j^*=(\#\text{ of elements in }\{r,q-1\}\geq j)-1\,,\quad j=1,\dots,r+q-4\,. \tag{33}$$

Therefore, for $W_1=-\log\Lambda_1$, $W_2=-\log\Lambda_2$ and $W_3=-\log\Lambda_3$, we may write the exact c.f. of $W=-\log\Lambda$ as

$$\begin{aligned}\Phi_W(t)&=\Phi_{W_1}(t)\left(\Phi_{W_2}(t)\right)^{\mathrm{mod}(m+1,2)}\Phi_{W_3}(t)\\
&=\underbrace{\begin{aligned}&\prod_{j=1}^{r+q-4}\left(\frac{n-r-q+j}{2}\right)^{s_j^*}\left(\frac{n-r-q+j}{2}-\mathrm{i}t\right)^{-s_j^*}\\
&\times\prod_{\substack{k=2\\ \text{step }2}}^{r+2q-5}\left(\frac{2n-2q-r+k}{4}\right)^{r_k^*}\left(\frac{2n-2q-r+k}{4}-\mathrm{i}t\right)^{-r_k^*}\\
&\times\left(\frac{n-3}{2}\right)^{r_{r+2q-6}^*}\left(\frac{n-3}{2}-\mathrm{i}t\right)^{-r_{r+2q-6}^*}\\
&\times\left(\frac{n-2}{2}\right)^{r_{r+2q-4}^*}\left(\frac{n-2}{2}-\mathrm{i}t\right)^{-r_{r+2q-4}^*}\end{aligned}}_{\Phi_{W,1}(t)}\\
&\qquad\times\underbrace{\left(\frac{\Gamma\left(\frac{n-1}{2}\right)}{\Gamma\left(\frac{n-2}{2}\right)}\frac{\Gamma\left(\frac{n-2}{2}-\mathrm{i}t\right)}{\Gamma\left(\frac{n-1}{2}-\mathrm{i}t\right)}\right)^{1+\mathrm{mod}(m+1,2)}}_{\Phi_{W,2}(t)}\,.\end{aligned}$$

Then, in order to build the near-exact distributions for $W$ and $\Lambda$, we will

- keep $\Phi_{W,1}(t)$ unchanged, since it is the c.f. of a GIG distribution, and
- asymptotically approximate $\Phi_{W,2}(t)$, which is the c.f. of the sum of $1 + \text{mod}(m+1,2)$ $Logbeta\left(\frac{n-2}{2},\frac{1}{2}\right)$ distributions, with a very sharp asymptotic approximation.

In fact, from the two first expressions in Sect. 5 of Tricomi and Erdélyi (1951) and also from expressions (11) and (14) in the same reference, we may write

$$\frac{\Gamma(a-\mathrm{i}t)}{\Gamma(a+b-\mathrm{i}t)} \simeq \sum_{\ell=0}^{\infty} p_\ell(b)(a-\mathrm{i}t)^{-b-\ell}$$

where $p_0(b)=1$ and for $\ell=1,2,\ldots,$

$$p_\ell(b)=\frac{1}{\ell}\sum_{m=0}^{\ell-1}\left(\frac{\Gamma(1-b-m)}{\Gamma(-b-\ell)(\ell-m+1)!}+(-1)^{\ell+m}b^{\ell-m+1}\right)p_m(b)\,,$$

so that, since the $h$-th moment of a RV $X$ with a $Beta(a,b)$ distribution is given by

$$\mathrm{E}(X^h)=\frac{\Gamma(a+b)}{\Gamma(a)}\,\frac{\Gamma(a+h)}{\Gamma(a+b+h)}\,,$$

expression that remains valid for any $h\in\mathbb{C}$, we may write the c.f. of a RV $Y=-\log X$, which will be a RV with a $Logbeta(a,b)$ distribution, as

$$\begin{aligned}\Phi_Y(t)&=\mathrm{E}\left(e^{\mathrm{i}tY}\right)=\mathrm{E}\left(e^{-\mathrm{i}t\log X}\right)=\mathrm{E}\left(X^{-\mathrm{i}t}\right)\\&=\frac{\Gamma(a+b)}{\Gamma(a)}\,\frac{\Gamma(a-\mathrm{i}t)}{\Gamma(a+b-\mathrm{i}t)}\simeq\frac{\Gamma(a+b)}{\Gamma(a)}\sum_{\ell=0}^{\infty}p_\ell(b)\,(a-\mathrm{i}t)^{-(b+\ell)}\\&=\sum_{\ell=0}^{\infty}\underbrace{\frac{\Gamma(a+b)}{\Gamma(a)}\,\frac{p_\ell(b)}{a^{b+\ell}}}_{p^*_\ell(a,b)}\,a^{b+\ell}(a-\mathrm{i}t)^{-(b+\ell)}\,,\end{aligned}$$

which is the c.f. of an infinite mixture of $\Gamma(b+\ell,a)$ $(\ell=0,1,\ldots)$ distributions, with weights $p^*_\ell(\alpha,\beta)$.

As such, since, as noted above, $\Phi_{W,2}(t)$ is the c.f. of a sum of $1{+}\text{mod}(m{+}1,2)$ $Logbeta\left(\frac{n-2}{2},\frac{1}{2}\right)$ distributions, using the above result, we may approximate $\Phi_{W,2}(t)$ by the c.f. of the sum of $1{+}\text{mod}(m{+}1,2)$ infinite mixtures of $\Gamma\left(\frac{1}{2}+\ell,\frac{n-2}{2}\right)$ distributions, which is an infinite mixture of $\Gamma\left(\frac{1+\text{mod}(m+1,2)}{2}+\ell,\frac{n-2}{2}\right)$ distributions.

This way, using a somewhat heuristic approach, we will approximate $\Phi_{W,2}(t)$ by the c.f. of a finite mixture of $\Gamma\left(\frac{1+\text{mod}(m+1,2)}{2}+\ell,\frac{n-2}{2}\right)$ distributions, or, more precisely, we will, for a given positive integer $\nu$, approximate $\Phi_{W,2}(t)$ by

$$\Phi_2^*(t) = \sum_{\ell=0}^{\nu} \pi_\ell \left(\frac{n-2}{2}\right)^{\frac{1+\text{mod}(m+1,2)}{2}+\ell} \left(\frac{n-2}{2} - \text{i}t\right)^{-\left(\frac{1+\text{mod}(m+1,2)}{2}+\ell\right)}$$

where the weights $\pi_\ell$, $\ell = 0, \ldots, \nu - 1$ will be determined in such a way that the first $\nu$ derivatives of $\Phi_{W,2}(t)$ and $\Phi_2^*(t)$ relative to $t$, at $t = 0$, are equal, that is, in such a way that

$$\left.\frac{\partial^h}{\partial t^h}\Phi_{W,2}(t)\right|_{t=0} = \left.\frac{\partial^h}{\partial t^h}\Phi_2^*(t)\right|_{t=0}, \qquad h = 1, \ldots, \nu,$$

and with $\pi_\nu = 1 - \sum_{\ell=0}^{\nu-1} \pi_\ell$.

By doing this, one obtains as near-exact c.f. for $W$

$$\begin{aligned}
\Phi_W^*(t) &= \Phi_{W,1}(t)\, \Phi_2^*(t) \\
&= \sum_{\ell=0}^{\nu} \pi_\ell \left(\frac{n-2}{2}\right)^{\frac{1+\text{mod}(m+1,2)}{2}+\ell} \left(\frac{n-2}{2} - \text{i}t\right)^{-\left(\frac{1+\text{mod}(m+1,2)}{2}+\ell\right)} \times \Phi_{W,1}(t) \\
&= \sum_{\ell=0}^{\nu} \pi_\ell \left(\frac{n-2}{2}\right)^{r^*_{r+2q-4}+\frac{1+\text{mod}(m+1,2)}{2}+\ell} \left(\frac{n-2}{2} - \text{i}t\right)^{-\left(r^*_{r+2q-4}+\frac{1+\text{mod}(m+1,2)}{2}+\ell\right)} \\
&\quad \times \prod_{j=1}^{r+q-4} \left(\frac{n-r-q+j}{2}\right)^{s_j^*} \left(\frac{n-r-q+j}{2} - \text{i}t\right)^{-s_j^*} \\
&\quad \times \prod_{\substack{k=2 \\ \text{step } 2}}^{r+2q-5} \left(\frac{2n-2q-r+k}{4}\right)^{r_k^*} \left(\frac{2n-2q-r+k}{4} - \text{i}t\right)^{-r_k^*} \\
&\quad \times \left(\frac{n-3}{2}\right)^{r^*_{r+2q-6}} \left(\frac{n-3}{2} - \text{i}t\right)^{-r^*_{r+2q-6}}
\end{aligned}$$

which is the c.f. of a mixture with $\nu + 1$ components, each of which is a GNIG distribution in case $m$ is odd, or a GIG distribution in case $m$ is even, in either case with depth $r + 2q - 4 + (r - 1)/2$.

The p.d.f.'s and c.d.f.'s of these near-exact distributions for $W = -\log \Lambda$ are, for odd $m$, respectively given by (see Appendix C for the definition of the GNIG distribution and its p.d.f. and c.d.f.)

$$f_W(w) = \sum_{\ell=0}^{\nu} \pi_\ell \, f^{GNIG}\Big(w \,\Big|\, \{u_j\}_{j=1:g}, r^*_{r+2q-4} + \frac{1+\text{mod}(m+1,2)}{2} + \ell; \; \{\lambda_j\}_{j=1:g}, \frac{n-2}{2}; g\Big)$$

and

$$F_W(w) = \sum_{\ell=0}^{\nu} \pi_\ell \, F^{GNIG}\Big(w \,\Big|\, \{u_j\}_{j=1:g}, r^*_{r+2q-4} + \frac{1+\text{mod}(m+1,2)}{2} + \ell; \; \{\lambda_j\}_{j=1:g}, \frac{n-2}{2}; g\Big)$$

for $w > 0$, and for $\Lambda$ by

$$f_\Lambda(\lambda) = \sum_{\ell=0}^{\nu} \pi_\ell \, f^{GNIG}\Big(-\log \lambda \,\Big|\, \{u_j\}_{j=1:g}, r^*_{r+2q-4} + \frac{1+\text{mod}(m+1,2)}{2} + \ell; \; \{\lambda_j\}_{j=1:g}, \frac{n-2}{2}; g\Big) \frac{1}{\lambda}$$

and[1]

$$F_\Lambda(\lambda) = (\nu+1) - \sum_{\ell=0}^{\nu} \pi_\ell \, F^{GNIG}\Big(-\log \lambda \,\Big|\, \{u_j\}_{j=1:g}, r^*_{r+2q-4} + \frac{1+\text{mod}(m+1,2)}{2} + \ell; \; \{\lambda_j\}_{j=1:g}, \frac{n-2}{2}; g\Big)$$

for $0 < \lambda < 1$, where $g = r + 2q - 4 + (r-1)/2$,

$$\{u_j\}_{j=1:g} = \left\{ \left\{s^*_j\right\}_{j=1:r+q-4}, \left\{r^*_k\right\}_{\substack{k=2:r+2q-5 \\ \text{step } 2}}, r^*_{r+2q-6}, r^*_{r+2q-4} \right\},$$

and

$$\{\lambda_j\}_{j=1:g} = \left\{ \left\{\frac{n-r-q+j}{2}\right\}_{j=1:r+q-4}, \left\{\frac{2n-2q-r+k}{4}\right\}_{\substack{k=2:r+2q-5 \\ \text{step } 2}}, \frac{n-3}{2}, \frac{n-2}{2} \right\},$$

and where the p.d.f. and c.d.f. of the GNIG distribution are to be replaced by the p.d.f. and c.d.f. of the GIG distribution for the case of even $m$.

[1] see author and editors note at the end of the Chapter

We may note that these near-exact distributions match the $\nu$ first exact moments of $W$.

## 4 Numerical Studies for the Near-Exact Distributions

Even though we do not have the expression for the exact c.d.f. of $\Lambda$ or $W$, we can measure the distance between the near-exact c.d.f.'s and the exact c.d.f., through the use of the measure

$$\Delta = \frac{1}{2\pi}\int_{-\infty}^{+\infty}\left|\frac{\Phi_W(t)-\Phi_W^*(t)}{t}\right| dt$$

with

$$\Delta \geq \sup_{w>0}\left|F_W(w)-F_W^*(w)\right| = \sup_{0<\ell<1}\left|F_\Lambda(\ell)-F_\Lambda^*(\ell)\right|,$$

where $\Phi_W(t)$ and $\Phi_W^*(t)$ represent, respectively, the exact and the near-exact c.f.'s of $W$, and where $F_W(\,\cdot\,)$ and $F_\Lambda(\,\cdot\,)$ are, respectively, the exact c.d.f. of $W = -\log\Lambda$ and $\Lambda$, and $F_W^*(\,\cdot\,)$ is the near-exact c.d.f. which corresponds to the c.f. $\Phi_W^*(\,\cdot\,)$, with $F_\Lambda^*(\ell) = 1 - F_W^*(-\log\ell)$.

In Tables 1, 2 and 3 are shown values of the measure $\Delta$ for various values of $n$, the sample size, $r$, the number of variables in each subset, $q$, the number of populations in the study, and $m$, the number of subsets of variables. In each of these tables, two of these parameters are kept constant, while the sample size and the other parameters evolve in value. This is done in order to allow the reader to evaluate the asymptotic properties of the near-exact distributions developed.

In each table are considered near-exact distributions that match the $\nu = 2, 6$ and 10 first exact moments of $W = -\log\Lambda$, and while from each of the three tables, the asymptotic behavior of the near-exact distributions for increasing sample sizes is clear, from Table 1 we may ascertain the asymptotic behavior of the near-exact distributions developed for increasing values of $r$, the number of variables in each subset, from Table 2 the asymptotic behavior of these distributions for increasing values of $q$, the number of populations involved in the test, and from Table 3 the asymptotic behavior for increasing values of $m$, the number of subsets of variables.

## 5 The Case of Block Compound-Symmetric Matrices

Using a similar approach, we may address rather simply the case of block compound-symmetric covariance matrices.

**Table 1** Values of the measure $\Delta$ for increasing values of $r$ and increasing sample sizes, for the near-exact distributions that match $\nu$ exact moments of $W$

| $n$ | $r$ | $q$ | $m$ | $\nu$ (# of exact moments of $W$ matched) | | |
|---|---|---|---|---|---|---|
| | | | | 2 | 6 | 10 |
| 7 | 3 | 4 | 4 | $2.83\times10^{-7}$ | $2.30\times10^{-12}$ | $7.40\times10^{-16}$ |
| 57 | | | | $1.76\times10^{-9}$ | $2.46\times10^{-17}$ | $9.95\times10^{-24}$ |
| 107 | | | | $2.62\times10^{-10}$ | $2.96\times10^{-19}$ | $1.11\times10^{-26}$ |
| 11 | 7 | 4 | 4 | $9.33\times10^{-9}$ | $1.47\times10^{-16}$ | $1.26\times10^{-20}$ |
| 61 | | | | $3.78\times10^{-10}$ | $8.18\times10^{-19}$ | $6.23\times10^{-26}$ |
| 111 | | | | $6.44\times10^{-11}$ | $1.35\times10^{-20}$ | $1.12\times10^{-28}$ |
| 19 | 15 | 4 | 4 | $2.19\times10^{-10}$ | $2.72\times10^{-19}$ | $2.28\times10^{-26}$ |
| 69 | | | | $7.20\times10^{-11}$ | $1.87\times10^{-20}$ | $1.93\times10^{-28}$ |
| 119 | | | | $1.55\times10^{-11}$ | $5.26\times10^{-22}$ | $7.73\times10^{-31}$ |
| 29 | 25 | 4 | 4 | $1.39\times10^{-11}$ | $6.46\times10^{-22}$ | $6.08\times10^{-31}$ |
| 79 | | | | $1.89\times10^{-11}$ | $8.61\times10^{-22}$ | $1.66\times10^{-30}$ |
| 129 | | | | $5.15\times10^{-12}$ | $4.16\times10^{-23}$ | $1.51\times10^{-32}$ |

**Table 2** Values of the measure $\Delta$ for increasing values of $q$ and increasing sample sizes, for the near-exact distributions that match $\nu$ exact moments of $W$

| $n$ | $r$ | $q$ | $m$ | $\nu$ (# of exact moments of $W$ matched) | | |
|---|---|---|---|---|---|---|
| | | | | 2 | 6 | 10 |
| 19 | 15 | 4 | 4 | $2.19\times10^{-10}$ | $2.72\times10^{-19}$ | $2.28\times10^{-26}$ |
| 69 | | | | $7.20\times10^{-11}$ | $1.87\times10^{-20}$ | $1.93\times10^{-28}$ |
| 119 | | | | $1.55\times10^{-11}$ | $5.26\times10^{-22}$ | $7.73\times10^{-31}$ |
| 21 | 15 | 6 | 4 | $8.08\times10^{-11}$ | $2.53\times10^{-20}$ | $2.59\times10^{-28}$ |
| 71 | | | | $2.94\times10^{-11}$ | $2.37\times10^{-21}$ | $7.82\times10^{-30}$ |
| 121 | | | | $6.68\times10^{-12}$ | $7.59\times10^{-23}$ | $3.85\times10^{-32}$ |
| 25 | 15 | 10 | 4 | $1.94\times10^{-11}$ | $9.49\times10^{-22}$ | $6.77\times10^{-32}$ |
| 75 | | | | $9.51\times10^{-12}$ | $1.73\times10^{-22}$ | $1.33\times10^{-31}$ |
| 125 | | | | $2.39\times10^{-12}$ | $7.03\times10^{-24}$ | $9.41\times10^{-34}$ |

We say that the matrix $\boldsymbol{\Sigma}$ has a block compound-symmetric (BCS) structure, with $m$ blocks $\boldsymbol{\Sigma}_0$ of dimensions $r\times r$ on the diagonal and off-diagonal blocks $\boldsymbol{\Sigma}_1$, also of dimensions $r\times r$, if

$$\begin{aligned}\boldsymbol{\Sigma} &= \boldsymbol{I}_m \otimes (\boldsymbol{\Sigma}_0 - \boldsymbol{\Sigma}_1) + \boldsymbol{J}_m \otimes \boldsymbol{\Sigma}_1 \\ &= \boldsymbol{I}_m \otimes \boldsymbol{\Sigma}_0 + (\boldsymbol{J}_m - \boldsymbol{I}_m) \otimes \boldsymbol{\Sigma}_1\end{aligned}$$

**Table 3** Values of the measure $\Delta$ for increasing values of $m$ and increasing sample sizes, for the near-exact distributions that match $\nu$ exact moments of $W$

| $n$ | $r$ | $q$ | $m$ | $\nu$ (# of exact moments of $W$ matched) 2 | 6 | 10 |
|---|---|---|---|---|---|---|
| 19 | 15 | 4 | 4 | $2.19\times10^{-10}$ | $2.72\times10^{-19}$ | $2.28\times10^{-26}$ |
| 69 | | | | $7.20\times10^{-11}$ | $1.87\times10^{-20}$ | $1.93\times10^{-28}$ |
| 119 | | | | $1.55\times10^{-11}$ | $5.26\times10^{-22}$ | $7.73\times10^{-31}$ |
| 19 | 15 | 4 | 6 | $1.81\times10^{-10}$ | $1.55\times10^{-19}$ | $8.14\times10^{-27}$ |
| 69 | | | | $4.06\times10^{-11}$ | $4.97\times10^{-21}$ | $2.46\times10^{-29}$ |
| 119 | | | | $8.59\times10^{-12}$ | $1.36\times10^{-22}$ | $9.46\times10^{-32}$ |
| 19 | 15 | 4 | 10 | $1.33\times10^{-10}$ | $6.64\times10^{-20}$ | $1.81\times10^{-27}$ |
| 69 | | | | $1.94\times10^{-11}$ | $9.01\times10^{-22}$ | $1.71\times10^{-30}$ |
| 119 | | | | $4.06\times10^{-12}$ | $2.39\times10^{-23}$ | $6.35\times10^{-33}$ |

where $\boldsymbol{I}_m$ is an identity matrix of order $m$ and $\boldsymbol{J}_m$ an $m\times m$ matrix of 1's, and where $\boldsymbol{\Sigma}_0$ is a positive-definite matrix and $\boldsymbol{\Sigma}_1$ a symmetric matrix such that $\boldsymbol{\Sigma}_0 - \boldsymbol{\Sigma}_1$ and $\boldsymbol{\Sigma}_0 + (m-1)\boldsymbol{\Sigma}_1$ are both positive-definite matrices.

We are now interested in testing a similar null hypothesis to that in (4), but now assuming that $\boldsymbol{\Sigma}$ has a BCS structure.

From Appendix B, we may easily see that if all $\boldsymbol{A}_i^*$ are equal for $i = 1, \ldots, m-1$, that is, if $\boldsymbol{A}^*$ is BCS, we will have $\widehat{\boldsymbol{\Psi}}_2 = \cdots = \widehat{\boldsymbol{\Psi}}_m$, so that if $\boldsymbol{\Sigma}$ has a BCS structure, then

$$(\boldsymbol{\Gamma}_m \otimes \boldsymbol{I}_r)\boldsymbol{\Sigma}(\boldsymbol{\Gamma}_m \otimes \boldsymbol{I}_r) = \boldsymbol{\Psi} = \text{BDiag}(\boldsymbol{\Psi}_1, \underbrace{\boldsymbol{\Psi}_2, \ldots, \boldsymbol{\Psi}_2}_{m-1}) \tag{34}$$

where $\boldsymbol{\Psi}_1$ and $\boldsymbol{\Psi}_2$ are positive-definite matrices, actually with $\boldsymbol{\Psi}_1 = \boldsymbol{\Sigma}_0+(m-1)\boldsymbol{\Sigma}_1$ and $\boldsymbol{\Psi}_2 = \boldsymbol{\Sigma}_0 - \boldsymbol{\Sigma}_1$, and where $\boldsymbol{\Gamma}_m$ is the $m\times m$ orthogonal symmetric matrix that was used in Sect. 3, with elements $\gamma_{jk}$ $(j, k \in \{1, \ldots, m\})$ given by (16) and $\boldsymbol{I}_r$ is an identity matrix of order $r$.

Using a similar procedure to the one used in Appendix A, it is easy to show that also $\boldsymbol{\Sigma}^{-1}$ is a BCS matrix. Thus, we may use the results in Szatrowski (1978, 1980) in estimating $\boldsymbol{\Sigma}$ and take the MLE of $\boldsymbol{\Sigma}$ given by

$$\widehat{\boldsymbol{\Sigma}} = \boldsymbol{I}_m \otimes \widehat{\boldsymbol{\Sigma}}_0 + (\boldsymbol{J}_m - \boldsymbol{I}_m) \otimes \widehat{\boldsymbol{\Sigma}}_1$$

where under $H_1$

$$\widehat{\boldsymbol{\Sigma}}_{0|H_1} = \frac{1}{m}\sum_{j=1}^{m} \widetilde{\boldsymbol{A}}_{jj} \quad \text{and} \quad \widehat{\boldsymbol{\Sigma}}_{1|H_1} = \frac{1}{m(m-1)}\sum_{j=1}^{m}\sum_{k=j+1}^{m} (\widetilde{\boldsymbol{A}}_{jk} + \widetilde{\boldsymbol{A}}_{kj})$$

and under $H_0$,[2]

$$\widehat{\boldsymbol{\Sigma}}_{0|H_1} = \frac{1}{m}\sum_{j=1}^{m}(\widetilde{\boldsymbol{A}}_{jj} + \widetilde{\boldsymbol{B}}_{jj})$$

and

$$\widehat{\boldsymbol{\Sigma}}_{1|H_1} = \frac{1}{m(m-1)}\sum_{j=1}^{m}\sum_{k=j+1}^{m}(\widetilde{\boldsymbol{A}}_{jk} + \widetilde{\boldsymbol{A}}_{kj} + \widetilde{\boldsymbol{B}}_{jk} + \widetilde{\boldsymbol{B}}_{kj}),$$

where, as before, $\widetilde{\boldsymbol{A}}_{jk}$ and $\widetilde{\boldsymbol{B}}_{jk}$ $(j, k \in \{1, \ldots, m\})$ are the $r\times r$ blocks of $\boldsymbol{A}$ and $\boldsymbol{B}$ in (6), starting at row $r(j-1)+1$ and column $r(k-1)+1$. That is, the MLEs of $\boldsymbol{\Sigma}_0$, respectively under $H_1$ and $H_0$, are the averages of the diagonal blocks of the matrices $\boldsymbol{A}$ and $\boldsymbol{A}+\boldsymbol{B}$, while the MLEs of $\boldsymbol{\Sigma}_1$ are the averages of the off-diagonal blocks of the same matrices.

Then, the LRT statistic to test $H_0$ in (4), assuming the common variance-covariance matrix to be BCS, will be

$$\Lambda = \frac{|\boldsymbol{A}^*|}{|\boldsymbol{A}^* + \boldsymbol{B}^*|}$$

where

$$\boldsymbol{A}^* = \boldsymbol{I}_m \otimes \widehat{\boldsymbol{\Sigma}}_{0|H_1} + (\boldsymbol{J}_m - \boldsymbol{I}_m) \otimes \widehat{\boldsymbol{\Sigma}}_{1|H_1}$$

and

$$\boldsymbol{A}^* + \boldsymbol{B}^* = \boldsymbol{I}_m \otimes \widehat{\boldsymbol{\Sigma}}_{0|H_0} + (\boldsymbol{J}_m - \boldsymbol{I}_m) \otimes \widehat{\boldsymbol{\Sigma}}_{1|H_0}.$$

Let us consider the matrices $\boldsymbol{A}^{**}$ and $\boldsymbol{B}^{**}$ in (15). Then, is possible to show that (see Appendix E)

$$|\boldsymbol{A}^*| = |\boldsymbol{A}_1^{**}| \left| \frac{\sum_{j=2}^{m} \boldsymbol{A}_j^{**}}{m-1} \right|^{m-1} \tag{35}$$

and, in a similar manner, that

$$|\boldsymbol{A}^* + \boldsymbol{B}^*| = |\boldsymbol{A}_1^{**} + \boldsymbol{B}_1^{**}| \left| \frac{\sum_{j=2}^{m}(\boldsymbol{A}_j^{**} + \boldsymbol{B}_j^{**})}{m-1} \right|^{m-1},$$

[2] see author and editors note at the end of the Chapter

where $\boldsymbol{A}_j^{**}$ and $\boldsymbol{B}_j^{**}$ $(j = 1, \dots, m)$ denote, as before, the $j$-th diagonal block respectively of $\boldsymbol{A}^{**}$ and $\boldsymbol{B}^{**}$ of order $r \times r$.

Thus, we may then write

$$\Lambda = \underbrace{\frac{|\boldsymbol{A}_1^{**}|}{|\boldsymbol{A}_1^{**} + \boldsymbol{B}_1^{**}|}}_{\Lambda_1} \underbrace{\left( \frac{\left|\sum_{j=2}^{m} \boldsymbol{A}_j^{**}\right|}{\left|\sum_{j=2}^{m} (\boldsymbol{A}_j^{**} + \boldsymbol{B}_j^{**})\right|} \right)^{m-1}}_{\Lambda_2}, \tag{36}$$

where, we now have

$$\boldsymbol{A} \sim W_p(n-q, \boldsymbol{\Sigma}) \Longrightarrow \boldsymbol{A}^{**} \sim W_p(n-q, \boldsymbol{\Psi})$$

and

$$\boldsymbol{B} \sim W_p(q-1, \boldsymbol{\Sigma}) \Longrightarrow \boldsymbol{B}^{**} \sim W_p(q-1, \boldsymbol{\Psi})$$

for $\boldsymbol{\Psi}$ in (34). But so, as before, the diagonal blocks $\boldsymbol{A}_j^{**}$ and $\boldsymbol{B}_l^{**}$ are all independent among themselves, now with

$$\begin{array}{ll} \boldsymbol{A}_1^{**} \sim W_r(n-q, \boldsymbol{\Psi}_1), & \boldsymbol{B}_1^{**} \sim W_r(q-1, \boldsymbol{\Psi}_1) \\ \boldsymbol{A}_j^{**} \sim W_r(n-q, \boldsymbol{\Psi}_2), & \boldsymbol{B}_j^{**} \sim W_r(q-1, \boldsymbol{\Psi}_2) \quad j = 2, \dots, m, \end{array}$$

and as such with

$$\boldsymbol{A}_1^{**} + \boldsymbol{B}_1^{**} \sim W_r(n-1, \boldsymbol{\Psi}_1)$$

and

$$\boldsymbol{A}_j^{**} + \boldsymbol{B}_j^{**} \sim W_r(n-1, \boldsymbol{\Psi}_2), \quad j = 2, \dots, m,$$

and also

$$\sum_{j=2}^{m} \boldsymbol{A}_j^{**} \sim W_r\left((m-1)(n-q), \boldsymbol{\Psi}_2\right)$$

and

$$\sum_{j=2}^{m} (\boldsymbol{A}_j^{**} + \boldsymbol{B}_j^{**}) \sim W_r\left((m-1)(n-1), \boldsymbol{\Psi}_2\right).$$

Hence, we have that

$$\Lambda_1 = \frac{|\boldsymbol{A}_1^{**}|}{|\boldsymbol{A}_1^{**} + \boldsymbol{B}_1^{**}|} \sim \prod_{j=1}^{r} Y_j \tag{37}$$

where

$$Y_j \sim Beta\left(\frac{n-q+1-j}{2}, \frac{q-1}{2}\right), \quad j = 1, \ldots, r,$$

are a set of independent RVs, where $r$ and $q - 1$ are interchangeable (see Coelho and Arnold (2019, Chap. 5, App. 1)), and

$$\Lambda_2 = \left(\frac{\left|\sum_{j=2}^{m} \boldsymbol{A}_j^{**}\right|}{\left|\sum_{j=2}^{m}(\boldsymbol{A}_j^{**} + \boldsymbol{B}_j^{**})\right|}\right)^{m-1} \sim \prod_{k=1}^{r} \left(Y_k^*\right)^{m-1} \tag{38}$$

where

$$Y_k^* \sim Beta\left(\frac{(m-1)(n-q)}{2} + \frac{1-j}{2}, \frac{(m-1)(q-1)}{2}\right), \quad k = 1, \ldots, r,$$

also form a set of independent RVs, and where $r$ and $(m - 1)(q - 1)$ are interchangeable, so that we may also write

$$\Lambda_2 \sim \prod_{k=1}^{(m-1)(q-1)} \left(Y_k^{**}\right)^{m-1}$$

where

$$Y_k^{**} \sim Beta\left(\frac{(m-1)(n-1)+1-k-r}{2}, \frac{r}{2}\right), \quad k = 1, \ldots, (m-1)(q-1),$$

form a set of independent RVs, and where $\Lambda_1$ and $\Lambda_2$ are two independent statistics.

## 5.1 The Exact Distribution of $\Lambda$ in (36) for Odd $q$ or Even $r$

The distribution of the statistic $\Lambda_1$ in (36) and (37) is exactly the same as that of the statistic $\Lambda_1$ in (19), and as such the c.f. of $W_1 = -\log \Lambda_1$ may be written, for odd $q$ or even $r$ as in (29), with $s_j$ given by (28), which shows that its distribution is a GIG distribution of depth $r+q-3$ with rate parameters $\frac{n-q}{2} + \frac{j-r}{2}$ $(j = 1, \ldots, r+q-3)$ and corresponding shape parameters $s_j$, given by (28), a result that, also as already stated in Sect. 3.1 may also be obtained from Theorem 3.2 in Coelho and Arnold (2019).

In what concerns the distribution of $\Lambda_2$ in (36) and (38), for odd $q$ or odd $m$, its distribution may be obtained from Theorem 1 in Sect. 3 with

$$m^*=1,\ a_1=\frac{n-q}{2},\ n_1=r,\ k_1=m-1,\ m_1=\frac{(m-1)(q-1)}{2},\ h=2$$

and for even $r$ with

$$m^*=1,\ a_1=\frac{n-1}{2}-\frac{r}{2(m-1)},\ n_1=(m-1)(q-1),\ k_1=m-1,\ m_1=\frac{r}{2},\ h=2$$

as an EGIG distribution of depth $r+(m-1)(q-1)-2$, with rate parameters

$$\frac{n-q}{2}+\frac{k-r}{2(m-1)},\qquad k=1,\ldots,r+(m-1)(q-1)-2$$

and shape parameters

$$r_k=\begin{cases} h_k\,, & k=1,2\\ h_k+r_{k-2}\,, & k=3,\ldots,r+(m-1)(q-1)-2 \end{cases} \tag{39}$$

with

$$h_k=(\#\text{ of elements in }\{r,(m-1)(q-1)\}\geq k)-1 \tag{40}$$

for $k=1,\ldots,r+(m-1)(q-1)-2$.

As such, for even $r$ or odd $q$ the c.f. of $W=-\log\Lambda$, for $\Lambda$ in (36), may be written as

$$\begin{aligned}\Phi_W(t)=&\prod_{j=1}^{r+q-3}\left(\frac{n-r-q+j}{2}\right)^{s_j}\left(\frac{n-r-q+j}{2}+h\right)^{-s_j}\\ &\times\prod_{k=1}^{r+(m-1)(q-1)-2}\left(\frac{n-q}{2}+\frac{k-r}{2(m-1)}\right)^{r_k}\left(\frac{n-q}{2}+\frac{k-r}{2(m-1)}-\mathrm{i}t\right)^{-r_k}\end{aligned} \tag{41}$$

which is the c.f. of a GIG distribution. That is, for even $r$ or odd $q$ the exact distribution of $W=-\log\Lambda$ is a GIG distribution with rate parameters $\frac{n-q}{2}+\frac{j-r}{2}$ ($j=1,\ldots,r+q-3$) and $\frac{n-q}{2}+\frac{k-r}{2(m-1)}$ ($k=1,\ldots,(m-1)(q-1)-2$) and corresponding rate parameters $s_j$ ($j=1,\ldots,r+q-3$) and $r_k$ ($k=1,\ldots,r+(m-1)(q-1)-2$). However, we should be aware that the depth of this GIG distribution is indeed smaller than $r+q-3+r+(m-1)(q-1)-2=2r+mq-m-4$ since some of the rate parameters $\frac{n-q}{2}+\frac{k-r}{2(m-1)}$ match some, or even all, of the rate parameters $\frac{n-q}{2}+\frac{j-r}{2}$ in the first product. More precisely, the rate parameters in the first product in (41) will be matched by rate parameters in the second product for $j = j_m = 1 + \left\lfloor \frac{1+(m-2)r}{m-1} \right\rfloor$ through $j = j_M = q + r - 3$, and as such, the rate parameters in the second product that match rate parameters in the first product will be those with

$$k = (m-1)\left(1 + \left\lfloor \frac{1+(m-2)r}{m-1} \right\rfloor\right) - r(m-2)$$

through

$$k = (m-1)(q+r-3) - r(m-2)$$

with a step of $m-1$. As such the depth of this GIG distribution that yields the exact distribution of $W = -\log \Lambda$ is indeed equal to $r+(m-1)(q-1)-2+\left\lfloor \frac{1+(m-2)r}{m-1} \right\rfloor$. In fact, using a notation where rate parameters do not appear repeated we may write $\Phi_W(t)$ as

$$\begin{aligned}\Phi_W(t) = & \prod_{j=1}^{\left\lfloor \frac{1+(m-2)r}{m-1} \right\rfloor} \left(\frac{n-r-q+j}{2}\right)^{s_j} \left(\frac{n-r-q+j}{2} + h\right)^{-s_j} \\ & \times \prod_{k=1}^{r+(m-1)(q-1)-2} \left(\frac{n-q}{2} + \frac{k-r}{2(m-1)}\right)^{r_k^*} \left(\frac{n-q}{2} + \frac{k-r}{2(m-1)} - \mathrm{i}t\right)^{-r_k^*}\end{aligned}$$

where, if we denote by $K^*$ the set

$$\begin{aligned}\Bigg\{k : k = (m-1)\left(1 + \left\lfloor \frac{1+(m-2)r}{m-1} \right\rfloor\right) - r(m-2), \ldots, \\ (m-1)(q+r-3) - r(m-2), \text{ step } m-1\Bigg\},\end{aligned}$$

then we have

$$r_k^* = \begin{cases} r_k\,, & \text{for } k \in \{1, \ldots, r+(m-1)(q-1)-2\} \backslash K^* \\ r_k + s_{(k+r(m-2))/(m-1)}\,, & \text{for } k \in K^* \end{cases} \tag{42}$$

for $r_k$ ($k = 1, \ldots, r+(m-1)(q-1)-2$) given by (39)–(40) and $s_j$ ($j = 1, \ldots, r+q-3$) given by (28).

Therefore, the exact distribution of $\Lambda$ is, in this case, an EGIG distribution of the same depth with the rate and shape parameters above, being thus very easy to handle with any of the available symbolic software.

## 5.2 Near-Exact Distributions for $\Lambda$ in (36) for Odd $r$ and Even $q$

In case $r$ is odd and $q$ is even, the exact distribution of $\Lambda$ in (36) is not manageable and a good choice is to resort to the use of near-exact distributions as we did for the distribution of $\Lambda$ in (19) for the same case.

We should note that even when $r$ is odd and $q$ is even, if $m$ is odd, the exact distribution of $W_2 = -\log\,\Lambda_2$ for $\Lambda_2$ in (36) and (38) is still the GIG distribution in the previous subsection, while the c.f. of $W_1 = -\log\,\Lambda_1$ is given by (31), with $h$ replace by $-it$. Therefore, for $r$ odd, $q$ even, and $m$ odd, we may write the c.f. of $W = -\log\,\Lambda$ as

$$\Phi_W(t) = \prod_{j=1}^{r+q-4}\left(\frac{n-r-q+j}{2}\right)^{s_j^*}\left(\frac{n-r-q+j}{2}-\mathrm{i}t\right)^{-s_j^*}$$
$$\times \prod_{k=1}^{r+(m-1)(q-1)-2}\left(\frac{n-q}{2}+\frac{k-r}{2(m-1)}\right)^{r_k}\left(\frac{n-q}{2}+\frac{k-r}{2(m-1)}-\mathrm{i}t\right)^{-r_k}$$
$$\times \frac{\Gamma\left(\frac{n-1}{2}\right)\Gamma\left(\frac{n-2}{2}-\mathrm{i}t\right)}{\Gamma\left(\frac{n-2}{2}\right)\Gamma\left(\frac{n-1}{2}-\mathrm{i}t\right)},$$

where now the $s_j$ are given by (32)–(33).

However, in the expression above, there is repetition of some of the rate parameters in both products. If one wants to avoid this repetition we write $\Phi_W(t)$ as

$$\Phi_W(t) = \underbrace{\prod_{j=1}^{\left\lfloor\frac{1+(m-2)r}{m-1}\right\rfloor}\left(\frac{n-r-q+j}{2}\right)^{s_j}\left(\frac{n-r-q+j}{2}+h\right)^{-s_j} \times \prod_{k=1}^{r+(m-1)(q-1)-2}\left(\frac{n-q}{2}+\frac{k-r}{2(m-1)}\right)^{r_k^*}\left(\frac{n-q}{2}+\frac{k-r}{2(m-1)}-\mathrm{i}t\right)^{-r_k^*}}_{\Phi_{W,1}(t)}$$
$$\times \underbrace{\frac{\Gamma\left(\frac{n-1}{2}\right)\Gamma\left(\frac{n-2}{2}-\mathrm{i}t\right)}{\Gamma\left(\frac{n-2}{2}\right)\Gamma\left(\frac{n-1}{2}-\mathrm{i}t\right)}}_{\Phi_{W,2}(t)},$$

where the $r_k^*$ have a similar definition to the one in (42) but now for $K^*$ defined as being the set

$$\left\{k : k = (m-1)\left(1+\left\lfloor\frac{1+(m-2)r}{m-1}\right\rfloor\right) - r(m-2), \ldots, \right.$$
$$\left. (m-1)(q+r-4) - r(m-2), \text{ step } m-1\right\}.$$

In order to build the near-exact distributions we will use a procedure in all similar to the one adopted in Sect. 3.2, keeping $\Phi_{W,1}(t)$ untouched and asymptotically approximating $\Phi_{W,2}(t)$ by the c.f.

$$\Phi_2^*(t) = \sum_{\ell=0}^{\nu} \pi_\ell \left(\frac{n-2}{2}\right)^{\frac{1}{2}+\ell} \left(\frac{n-2}{2} - \mathrm{i}t\right)^{-\left(\frac{1}{2}+\ell\right)}$$

where, similar to what was done in Sect. 3.2, the weights $\pi_\ell$, $\ell = 0, \ldots, \nu-1$, will be determined in such a way that the first $\nu$ derivatives of $\Phi_{W,2}(t)$ and $\Phi_2^*(t)$ relative to $t$, at $t = 0$, are equal, that is, in such a way that

$$\left.\frac{\partial^h}{\partial t^h}\Phi_{W,2}(t)\right|_{t=0} = \left.\frac{\partial^h}{\partial t^h}\Phi_2^*(t)\right|_{t=0}, \qquad h = 1, \ldots, \nu,$$

and with $\pi_\nu = 1 - \sum_{\ell=0}^{\nu-1} \pi_\ell$.

This procedure will yield near-exact c.f.'s for $W$ of the form

$$\begin{aligned}
\Phi_W^*(t) &= \Phi_{W,1}(t)\,\Phi_2^*(t)\\
&= \sum_{\ell=0}^{\nu} \pi_\ell \left(\frac{n-2}{2}\right)^{\frac{1}{2}+\ell} \left(\frac{n-2}{2} - \mathrm{i}t\right)^{-\left(\frac{1}{2}+\ell\right)}\\
&\quad\times \prod_{j=1}^{\left\lfloor\frac{1+(m-2)r}{m-1}\right\rfloor} \left(\frac{n-r-q+j}{2}\right)^{s_j} \left(\frac{n-r-q+j}{2} + h\right)^{-s_j}\\
&\quad\times \prod_{k=1}^{r+(m-1)(q-1)-2} \left(\frac{n-q}{2} + \frac{k-r}{2(m-1)}\right)^{r_k^*} \left(\frac{n-q}{2} + \frac{k-r}{2(m-1)} - \mathrm{i}t\right)^{-r_k^*}
\end{aligned}$$

to which correspond a mixture of $\nu+1$ GNIG distributions of depth $r+(m-1)(q-1)-1+\left\lfloor\frac{1+(m-2)r}{m-1}\right\rfloor$, with weights $\pi_\ell$ ($\ell = 0, \ldots, \nu$), rate parameters $\frac{n-r-q+j}{2}$ $\left(j = 1, \ldots, \left\lfloor\frac{1+(m-2)r}{m-1}\right\rfloor\right)$, $\frac{n-q}{2} + \frac{k-r}{2(m-1)}$ ($k = 1, \ldots, r+(m-1)(q-1)-2$) and $\frac{n-2}{2}$, to which correspond the shape parameters $s_j$ $\left(j = 1, \ldots, \left\lfloor\frac{1+(m-2)r}{m-1}\right\rfloor\right)$, $r_k^*$ ($k = 1, \ldots, r+(m-1)(q-1)-2$) and $\frac{1}{2}+\ell$ ($\ell = 0, \ldots, \nu$).

We have to note that for $q = 2$ the length of the GNIG distributions will be one less, because then the rate parameter $\frac{n-2}{2}$ will be matched by the $r$-th rate parameter in the second product, to which will then correspond a shape parameter of $(m-1)/2$ if $m \leq r$ or of $(r+1)/2$ if $m > r$.

Similar to what happens with the near-exact distributions developed in Sect. 3.2, also these near-exact distributions are asymptotic for increasing sample sizes as well as for increasing values of $q$, $r$ and $m$, yielding approximations of very good quality, with performances of a comparable caliber to the ones developed in Sect. 3.2 for the test with block-circular matrices.

In case $r$ is odd and both $m$ and $q$ are even, we will use the duplication formula for the Gamma function, which is the multiplication formula in (56) for $n = 2$, and (21) to write the c.f. of $W_2 = -\log \Lambda_2$ as

$$
\begin{aligned}
\Phi_{W_2}(t) &= \prod_{k=1}^{(m-1)(q-1)} \frac{\Gamma\left(\frac{(m-1)(n-1)+1-k}{2}\right)\Gamma\left(\frac{(m-1)(n-1)+1-k-r}{2}-(m-1)\mathrm{i}t\right)}{\Gamma\left(\frac{(m-1)(n-1)+1-k-r}{2}\right)\Gamma\left(\frac{(m-1)(n-1)+1-k}{2}-(m-1)\mathrm{i}t\right)} \\
&= \frac{\Gamma\left(\frac{(m-1)(n-1)}{2}\right)}{\Gamma\left(\frac{(m-1)(n-1)-r}{2}\right)} \frac{\Gamma\left(\frac{(m-1)(n-1)-r}{2}-(m-1)\mathrm{i}t\right)}{\Gamma\left(\frac{(m-1)(n-1)}{2}-(m-1)\mathrm{i}t\right)} \\
&\quad \times \prod_{k=2}^{(m-1)(q-1)} \frac{\Gamma\left(\frac{(m-1)(n-1)+1-k}{2}\right)\Gamma\left(\frac{(m-1)(n-1)+1-k-r}{2}-(m-1)\mathrm{i}t\right)}{\Gamma\left(\frac{(m-1)(n-1)+1-k-r}{2}\right)\Gamma\left(\frac{(m-1)(n-1)+1-k}{2}-(m-1)\mathrm{i}t\right)} \\
&= \frac{\Gamma\left(\frac{(m-1)(n-1)-r+1}{2}\right)}{\Gamma\left(\frac{(m-1)(n-1)-r}{2}\right)} \frac{\Gamma\left(\frac{(m-1)(n-1)-r}{2}-(m-1)\mathrm{i}t\right)}{\Gamma\left(\frac{(m-1)(n-1)-r+1}{2}-(m-1)\mathrm{i}t\right)} \\
&\quad \times \frac{\Gamma\left(\frac{(m-1)(n-1)}{2}\right)}{\Gamma\left(\frac{(m-1)(n-1)-r+1}{2}\right)} \frac{\Gamma\left(\frac{(m-1)(n-1)-r+1}{2}-(m-1)\mathrm{i}t\right)}{\Gamma\left(\frac{(m-1)(n-1)}{2}-(m-1)\mathrm{i}t\right)} \\
&\quad \times \prod_{\substack{k=2 \\ \text{step } 2}}^{(m-1)(q-1)} \frac{\Gamma\left((m-1)(n-1)-k\right)}{\Gamma\left((m-1)(n-1)-k-r\right)} \\
&\qquad\qquad \times \frac{\Gamma\left((m-1)(n-1)-k-r-2(m-1)\mathrm{i}t\right)}{\Gamma\left((m-1)(n-1)-k-2(m-1)\mathrm{i}t\right)} \\
&= \frac{\Gamma\left(\frac{(m-1)(n-1)-r+1}{2}\right)}{\Gamma\left(\frac{(m-1)(n-1)-r}{2}\right)} \frac{\Gamma\left(\frac{(m-1)(n-1)-r}{2}-(m-1)\mathrm{i}t\right)}{\Gamma\left(\frac{(m-1)(n-1)-r+1}{2}-(m-1)\mathrm{i}t\right)}
\end{aligned}
$$

$$\times \prod_{\ell=0}^{\frac{r-1}{2}-1} \left(\frac{(m-1)(n-1)-r+1}{2}+\ell\right) \times \left(\frac{(m-1)(n-1)-r+1}{2}+\ell-(m-1)\mathrm{i}t\right)^{-1}$$

$$\times \prod_{\substack{k=2\\ \text{step } 2}}^{(m-1)(q-1)} \prod_{\ell=0}^{r-1} \left(\frac{(m-1)(n-1)-k-r}{2}+\ell\right) \times \left(\frac{(m-1)(n-1)-k-r}{2}+\ell-(m-1)\mathrm{i}t\right)^{-1}$$

$$= \frac{\Gamma\left(\frac{(m-1)(n-1)-r+1}{2}\right)}{\Gamma\left(\frac{(m-1)(n-1)-r}{2}\right)} \frac{\Gamma\left(\frac{(m-1)(n-1)-r}{2}-(m-1)\mathrm{i}t\right)}{\Gamma\left(\frac{(m-1)(n-1)-r+1}{2}-(m-1)\mathrm{i}t\right)}$$

$$\times \prod_{\ell=0}^{\frac{r-1}{2}-1} \left(\frac{n-1}{2}+\frac{1-r}{2(m-1)}+\frac{\ell}{m-1}\right)\left(\frac{n-1}{2}+\frac{1-r}{2(m-1)}+\frac{\ell}{m-1}-\mathrm{i}t\right)^{-1}$$

$$\times \prod_{\substack{k=2\\ \text{step } 2}}^{(m-1)(q-1)} \prod_{\ell=0}^{r-1} \left(\frac{n-1}{2}-\frac{k+r}{2}+\frac{\ell}{m-1}\right)\left(\frac{n-1}{2}-\frac{k+r}{2}+\frac{\ell}{m-1}-\mathrm{i}t\right)^{-1}$$

$$= \frac{\Gamma\left(\frac{(m-1)(n-1)-r+1}{2}\right)}{\Gamma\left(\frac{(m-1)(n-1)-r}{2}\right)} \frac{\Gamma\left(\frac{(m-1)(n-1)-r}{2}-(m-1)\mathrm{i}t\right)}{\Gamma\left(\frac{(m-1)(n-1)-r+1}{2}-(m-1)\mathrm{i}t\right)}$$

$$\times \prod_{\ell=0}^{\frac{r-1}{2}-1} \left(\frac{n-1}{2}+\frac{1-r}{2(m-1)}+\frac{\ell}{m-1}\right)\left(\frac{n-1}{2}+\frac{1-r}{2(m-1)}+\frac{\ell}{m-1}-\mathrm{i}t\right)^{-1}$$

$$\times \prod_{j=1}^{r+(m-1)(q-1)-3} \left(\frac{n-1}{2}-\frac{j+2}{2(m-1)}\right)^{r_j} \left(\frac{n-1}{2}-\frac{j+2}{2(m-1)}-\mathrm{i}t\right)^{-r_j}$$

$$= \frac{\Gamma\left(\frac{(m-1)(n-1)-r+1}{2}\right)}{\Gamma\left(\frac{(m-1)(n-1)-r}{2}\right)} \frac{\Gamma\left(\frac{(m-1)(n-1)-r}{2}-(m-1)\mathrm{i}t\right)}{\Gamma\left(\frac{(m-1)(n-1)-r+1}{2}-(m-1)\mathrm{i}t\right)}$$

$$\times \prod_{j=1}^{r+(m-1)(q-1)-3} \left(\frac{n-1}{2}-\frac{j+2}{2(m-1)}\right)^{r_j^*} \left(\frac{n-1}{2}-\frac{j+2}{2(m-1)}-\mathrm{i}t\right)^{-r_j^*} \tag{43}$$

where, for $j = 1, \ldots, r + (m-1)(q-1) - 3$,[3]

$$r_j \equiv (j; r, (m-1)(q-1) - 1, 2)$$

for the definition of $r(j; a, b, h^*)$ in (23)–(24), and

$$r_j^* = \begin{cases} 1, & j = 0 \\ r_j, & j = 1, \ldots, r-4, \text{ step } 2 \\ r_j + 1, & j = 2, \ldots, r-3, \text{ step } 2 \\ r_j, & j = r-2, \ldots, r + (m-1)(q-1) - 3 . \end{cases}$$

From (31) and (43), we may then write the exact c.f. of $W = -\log \Lambda$ as

$$\begin{aligned} \Phi_W(t) = & \prod_{j=1}^{r+q-4} \left( \frac{n-j-1}{2} - 1 \right)^{s_{r+q-3-j}} \left( \frac{n-j-1}{2} - 1 - \mathrm{i}t \right)^{-s_{r+q-3-j}} \\ & \times \prod_{j=1}^{r+(m-1)(q-1)-3} \left( \frac{n-1}{2} - \frac{j+2}{2(m-1)} \right)^{r_j^*} \left( \frac{n-1}{2} - \frac{j+2}{2(m-1)} - \mathrm{i}t \right)^{-r_j^*} \\ & \times \frac{\Gamma\left(\frac{n-1}{2}\right) \Gamma\left(\frac{n-2}{2} - \mathrm{i}t\right)}{\Gamma\left(\frac{n-2}{2}\right) \Gamma\left(\frac{n-1}{2} - \mathrm{i}t\right)} \\ & \times \frac{\Gamma\left(\frac{(m-1)(n-1)-r+1}{2}\right)}{\Gamma\left(\frac{(m-1)(n-1)-r}{2}\right)} \frac{\Gamma\left(\frac{(m-1)(n-1)-r}{2} - (m-1)\mathrm{i}t\right)}{\Gamma\left(\frac{(m-1)(n-1)-r+1}{2} - (m-1)\mathrm{i}t\right)} , \end{aligned}$$

which, in order to avoid the repetition of rate parameters between the first and the second product, may be rewritten as

$$\begin{aligned} \Phi_W(t) = & \underbrace{\prod_{j=q-2+\left\lfloor \frac{r-1}{m-1} \right\rfloor}^{r+q-4} \left( \frac{n-j-1}{2} - 1 \right)^{s_{r+q-3-j}} \left( \frac{n-j-1}{2} - 1 - \mathrm{i}t \right)^{-s_{r+q-3-j}} \times \prod_{j=1}^{r+(m-1)(q-1)-3} \left( \frac{n-1}{2} - \frac{j+2}{2(m-1)} \right)^{r_j^{**}} \left( \frac{n-1}{2} - \frac{j+2}{2(m-1)} - \mathrm{i}t \right)^{-r_j^{**}}}_{\Phi_{W_1}(t)} \end{aligned}$$

[3] see author and editors note at the end of the Chapter

$$\underbrace{\times \frac{\Gamma\left(\frac{(m-1)(n-1)-r+1}{2}\right)}{\Gamma\left(\frac{(m-1)(n-1)-r}{2}\right)} \frac{\Gamma\left(\frac{(m-1)(n-1)-r}{2}-(m-1)\mathrm{i}t\right)}{\Gamma\left(\frac{(m-1)(n-1)-r+1}{2}-(m-1)\mathrm{i}t\right)} \times \frac{\Gamma\left(\frac{n-1}{2}\right)}{\Gamma\left(\frac{n-2}{2}\right)} \frac{\Gamma\left(\frac{n-2}{2}-\mathrm{i}t\right)}{\Gamma\left(\frac{n-1}{2}-\mathrm{i}t\right)}}_{\Phi_{W_2}(t)},$$

where the $s_j$ are given by (32)–(33) and

$$r_j^{**} = \begin{cases} r_j^* + s_{r+q-1-\frac{k+1}{m-1}}, & \text{for } k = 3m-5, \ldots, r+(m-1)(q-1)-3, \\ & \qquad \text{step } m-1 \\ r_j^*, & \text{for other values of } k \in \{0, \ldots, r+(m-1)(q-1)-3\}. \end{cases}$$

Then in order to build the near-exact distributions we will use a procedure in all similar to the ones adopted before, keeping $\Phi_{W,1}(t)$ unchanged and approximating asymptotically $\Phi_{W,2}(t)$, now by the c.f.

$$\Phi_2^*(t) = \sum_{\ell=0}^{\nu} \pi_\ell\, (\lambda)^{1+\ell}\, (\lambda - \mathrm{i}t)^{-(1+\ell)}$$

where now $\lambda$ will be taken as the harmonic mean of $\frac{n-2}{2}$ and $\frac{n-1}{2} - \frac{r}{2(m-1)}$ and where, also similar to what was done before, the weights $\pi_\ell$, $\ell = 0, \ldots, \nu-1$ will be determined in such a way that the first $\nu$ derivatives of $\Phi_{W,2}(t)$ and $\Phi_2^*(t)$ relative to $t$, at $t=0$, are equal, that is, in such a way that

$$\left.\frac{\partial^h}{\partial t^h}\Phi_{W,2}(t)\right|_{t=0} = \left.\frac{\partial^h}{\partial t^h}\Phi_2^*(t)\right|_{t=0}, \qquad h = 1, \ldots, \nu,$$

taking then $\pi_\nu = 1 - \sum_{\ell=0}^{\nu-1} \pi_\ell$.

This procedure generates near-exact c.f.'s for $W$ of the form

$$\begin{aligned} \Phi_W^*(t) &= \Phi_{W,1}(t)\, \Phi_2^*(t) \\ &= \sum_{\ell=0}^{\nu} \pi_\ell\, (\lambda)^{1+\ell}\, (\lambda - \mathrm{i}t)^{-(1+\ell)} \end{aligned}$$

$$\times \prod_{j=q-2+\left\lfloor\frac{r-1}{m-1}\right\rfloor}^{r+q-4} \left(\frac{n-j-1}{2}-1\right)^{s_{r+q-3-j}} \left(\frac{n-j-1}{2}-1-\mathrm{i}t\right)^{-s_{r+q-3-j}}$$

$$\times \prod_{j=1}^{r+(m-1)(q-1)-3} \left(\frac{n-1}{2}-\frac{j+2}{2(m-1)}\right)^{r_j^{**}} \left(\frac{n-1}{2}-\frac{j+2}{2(m-1)}-\mathrm{i}t\right)^{-r_j^{**}}$$

to which corresponds a mixture of $\nu+1$ GIG distributions of depth $2r+(m-1)(q-1)-4-\left\lfloor\frac{r-1}{m-1}\right\rfloor$, with weights $\pi_\ell$ ($\ell=0,\ldots,\nu$), rate parameters $\frac{n-j-1}{2}-1$ $\left(j=q-2+\left\lfloor\frac{r-1}{m-1}\right\rfloor,\ldots,r+q-4\right)$, $\frac{n-1}{2}-\frac{j+2}{2(m-1)}$ ($j=1,\ldots,r+(m-1)(q-1)-3$), and $\lambda$, to which correspond the shape parameters $s_{r+q-3-j}$ $\left(j=q-2+\left\lfloor\frac{r-1}{m-1}\right\rfloor,\ldots,\ r+q-4\right)$, $r_j^{**}$ ($j=1,\ldots,r+(m-1)(q-1)-3$), and $1+\ell$ ($\ell=0,\ldots,\nu$).

Once again, similar to what happens with other near-exact distributions developed, also these are asymptotic for increasing sample sizes as well as for increasing values of $q$, $r$ and $m$, yielding once again approximations with comparable performances to the other near-exact distributions developed, only with a slightly less marked asymptotic behavior. Slightly better results may be generally obtained if one chooses to use $\lambda=\frac{\gamma_1}{\gamma_2-\gamma_1^2}$, where $\gamma_1$ and $\gamma_2$ represent respectively the first and second moments of the distribution that corresponds to $\Phi_{W_2}(t)$.

## 6 Conclusions

The procedures adopted allowed not only for a simple way to obtain the MLEs of the block covariance matrices involved in the tests as well as for a quite simple way to derive the LRT statistics and their distributions. It was then possible to show that the exact distribution of these statistics, no matter how complicated they may seem to be at first glance, may have indeed finite form representations in most cases, more precisely, in cases where $r$, the number of variables in each subgroup of variables, is even or when $q$, the number of populations involved in the test, is odd. In order to make this task a bit simpler, Theorem 1 in Sect. 3, was developed, which may be seen as an extended version of Theorem 3.2 in Coelho and Arnold (2019). For the remaining cases, that is, for cases where $r$ is odd and simultaneously $q$ is even, near-exact distributions were developed for the LRT statistics. These distributions are asymptotic not only for increasing sample sizes but also for increasing numbers of variables in each subgroup, that is, for increasing values of $r$, as well as for increasing values of $q$ and $m$, which are the number of populations involved in the tests and the number of subgroups of variables. These near-exact distributions besides being asymptotic for increasing values of $n$, $r$, $q$ and $m$, also show very good performances for very small samples, being thus a good choice for the situations where the exact distribution of the LRT statistics are not manageable.

We should also note that for $m = 2$ and $m = 3$ the test in Sect. 5, for equality of mean vectors, with block compound-symmetric matrices, is equivalent to the test for equality of mean vectors with block-circular matrices since in these cases the block-circular matrices are indeed also block compound-symmetric matrices.

## Appendix A: The Block Spectral Decomposition of a Block-Circular Matrix and Its Inverse

It is known that the matrix $\boldsymbol{\Sigma}$ is a block-circular matrix with $m \times m$ blocks of dimensions $r \times r$ if and only if

$$(\boldsymbol{\Gamma}_m \otimes \boldsymbol{I}_r)\Sigma(\boldsymbol{\Gamma}_m \otimes \boldsymbol{I}_r) = \boldsymbol{\Psi} = \text{BDiag}(\boldsymbol{\Psi}_1, \boldsymbol{\Psi}_2, \dots, \boldsymbol{\Psi}_m)$$

where $\boldsymbol{I}_r$ is an identity matrix of order $r$, $\boldsymbol{\Gamma}_m$ is an $m \times m$ orthogonal symmetric matrix with elements $\gamma_{jk}$ $(j, k \in \{1, \dots, m\})$ given by (16) and $\boldsymbol{\Psi}_1, \dots, \boldsymbol{\Psi}_m$ are positive-definite matrices with

$$\boldsymbol{\Psi}_j = \boldsymbol{\Psi}_{m-j+2}, \quad j = 2, \dots, m.$$

But then, given that $\boldsymbol{\Gamma}_m$ is orthogonal and symmetric, we may write

$$\boldsymbol{\Sigma} = (\boldsymbol{\Gamma}_m \otimes \boldsymbol{I}_r)\boldsymbol{\Psi}(\boldsymbol{\Gamma}_m \otimes \boldsymbol{I}_r),$$

and then also we have

$$\boldsymbol{\Sigma}^{-1} = (\boldsymbol{\Gamma}_m \otimes \boldsymbol{I}_r)\boldsymbol{\Psi}^{-1}(\boldsymbol{\Gamma}_m \otimes \boldsymbol{I}_r)$$

where

$$\boldsymbol{\Psi}^{-1} = \text{BDiag}(\boldsymbol{\Psi}_1^{-1}, \boldsymbol{\Psi}_2^{-1}, \dots, \boldsymbol{\Psi}_m^{-1}),$$

and as such $\boldsymbol{\Sigma}^{-1}$ is also a block-circular matrix.

A similar procedure may be used to show that the inverse of a block compound-symmetric matrix is also a block compound-symmetric matrix.

We may note that actually both results may also be directly obtained from the fact that the spaces of both the block-circular and block compound-symmetric structures are quadratic spaces (see Seely (1971)).

## Appendix B: Proof of Expression (17)

We may note that from (11), and from the fact that $\boldsymbol{A}_i^* = \boldsymbol{A}_{m-i}^*$ for $i = 1, \ldots, m-1$, we may write

$$\boldsymbol{A}^* = \left(\boldsymbol{W}_0 \otimes \boldsymbol{A}_0^*\right) + \sum_{i=1}^{\frac{m-1}{2}} (\boldsymbol{W}_i + \boldsymbol{W}_{m-i}) \otimes \boldsymbol{A}_i^* \tag{44}$$

for odd $m$, or as

$$\boldsymbol{A}^* = \left(\boldsymbol{W}_0 \otimes \boldsymbol{A}_0^*\right) + \left(\boldsymbol{W}_{m/2} \otimes \boldsymbol{A}_{m/2}^*\right) + \sum_{i=1}^{\frac{m}{2}-1} (\boldsymbol{W}_i + \boldsymbol{W}_{m-i}) \otimes \boldsymbol{A}_i^* \tag{45}$$

for even $m$.

We may also note that since $\boldsymbol{W}_0 = \boldsymbol{I}_m$, we have

$$\boldsymbol{\Gamma}_m \boldsymbol{W}_0 \boldsymbol{\Gamma}_m = \boldsymbol{\Gamma}_m \boldsymbol{\Gamma}_m = \boldsymbol{I}_m\,, \tag{46}$$

where $\boldsymbol{\Gamma}_m$ is the matrix in (15), with elements given by (16), and given that for $i = 1, \ldots, m-1$ we have

$$\boldsymbol{W}_{m-i} = \boldsymbol{W}_i'\,,$$

so that

$$\boldsymbol{W}_i + \boldsymbol{W}_{m-i} = \boldsymbol{W}_i + \boldsymbol{W}_i' \quad (i = 1, \ldots, m-1)$$

are symmetric matrices, and moreover, for $i = 1, \ldots, m-1$,

$$\boldsymbol{W}_i = \boldsymbol{W}_1^i$$

so that all matrices $\boldsymbol{W}_i + \boldsymbol{W}_{m-i}$ are commutative, and as such may be all diagonalized by the same orthogonal matrix (see Olkin (1973, Sect. 2)). This matrix is exactly the matrix $\boldsymbol{\Gamma}_m$, which, given its structure, yields

$$\boldsymbol{\Gamma}_m (\boldsymbol{W}_i + \boldsymbol{W}_{m-i}) \boldsymbol{\Gamma}_m = \boldsymbol{\Delta}_i\,, \quad i = 1, \ldots, m-1 \tag{47}$$

where

$$\boldsymbol{\Delta}_i = \text{diag}(\lambda_{i\ell}^{**}) \quad i = 1, \ldots, m-1;\ \ell = 1, \ldots, m \tag{48}$$

with

$$\lambda^{**}_{i\ell} = 2\cos\left(2\pi i(\ell-1)/m\right), \quad i=1,\dots,m-1;\, \ell=1,\dots,m\,. \tag{49}$$

But then, given the definition of $\boldsymbol{A}^*$ in (44) and (45), the matrix $\boldsymbol{\Gamma}_m \otimes \boldsymbol{I}_r$ is the matrix that block-diagonalizes $\boldsymbol{A}^*$, with

$$(\boldsymbol{\Gamma}_m \otimes \boldsymbol{I}_r)\boldsymbol{A}^*(\boldsymbol{\Gamma}_m \otimes \boldsymbol{I}_r) = \widehat{\boldsymbol{\Psi}} = \text{BDiag}\left(\widehat{\boldsymbol{\Psi}}_1, \widehat{\boldsymbol{\Psi}}_2, \dots, \widehat{\boldsymbol{\Psi}}_m\right) \tag{50}$$

where

$$\widehat{\boldsymbol{\Psi}}_j = \widehat{\boldsymbol{\Psi}}_{m-j+2}\,, \quad j=2,\dots,m$$

and where, for $j=1,\dots,m$,

$$\widehat{\boldsymbol{\Psi}}_j = \sum_{i=0}^{m-1} \lambda_{ji}\boldsymbol{A}^*_i \tag{51}$$

with

$$\lambda_{ji} = \cos\left(2\pi(j-1)i/m\right), \quad i=0,\dots,m-1;\, j=1,\dots,m\,. \tag{52}$$

But then from (50) we have

$$|\boldsymbol{A}^*| = \left|\widehat{\boldsymbol{\Psi}}\right| = \prod_{j=1}^{m}\left|\widehat{\boldsymbol{\Psi}}_j\right|.$$

On the other hand, from (12), (13) and (14), using the definition of BTr (block trace operator) in Filipiak et al. (2018), we may write

$$\boldsymbol{A}^*_i = \frac{1}{2m}\,\text{BTr}\left(\left(\left(\boldsymbol{W}_i + \boldsymbol{W}_{m-i}\right)\otimes \boldsymbol{I}_r\right)A\right), \quad i=0,\dots,m-1 \tag{53}$$

where we take $\boldsymbol{W}_0 = \boldsymbol{W}_m = \boldsymbol{I}_m$. Then, if we take

$$\boldsymbol{A}^{**} = (\boldsymbol{\Gamma}_m \otimes \boldsymbol{I}_r)\boldsymbol{A}(\boldsymbol{\Gamma}_m \otimes \boldsymbol{I}_r)\,,$$

since the matrix $(\boldsymbol{\Gamma}_m \otimes \boldsymbol{I}_r)$ is orthogonal and symmetric, we may write

$$\boldsymbol{A} = (\boldsymbol{\Gamma}_m \otimes \boldsymbol{I}_r)\boldsymbol{A}^{**}(\boldsymbol{\Gamma}_m \otimes \boldsymbol{I}_r)\,,$$

so that, using the properties of BTr (see Filipiak et al. (2018)), and of the Kronecker product, we have from (53), for $i=0,\dots,m-1$,

$$\begin{aligned}A_i^* &= \frac{1}{2m}\,\mathrm{BTr}\left(((W_i + W_{m-i}) \otimes I_r)\,(\Gamma_m \otimes I_r) A^{**} (\Gamma_m \otimes I_r)\right)\\ &= \frac{1}{2m}\,\mathrm{BTr}\left((\Gamma_m \otimes I_r)\,((W_i + W_{m-i}) \otimes I_r)\,(\Gamma_m \otimes I_r) A^{**}\right)\\ &= \frac{1}{2m}\,\mathrm{BTr}\left(((\Gamma_m (W_i + W_{m-i}) \Gamma_m) \otimes I_r)\, A^{**}\right)\end{aligned}$$

where, from (46) and (47)–(49) we have

$$\Gamma_m (W_i + W_{m-i}) \Gamma_m = \Delta_i^*\,, \quad i = 0, \ldots, m-1$$

with

$$\Delta_i^* = \mathrm{diag}\left(\lambda_{i\ell}^*\right)\,, \quad i = 0, \ldots, m-1;\, \ell = 1, \ldots, m$$

where

$$\lambda_{i\ell}^* = 2\cos\left(2\pi i(\ell-1)/m\right)\,, \quad \ell = 1, \ldots, m;\, i = 0, \ldots, m-1\,. \tag{54}$$

Therefore, we may write for $i = 0, \ldots, m-1$,

$$A_i^* = \frac{1}{2m} \sum_{\ell=1}^{m} \lambda_{i\ell}^* A_\ell^{**}\,,$$

where $A_\ell^{**}$ denotes the $\ell$-th diagonal block of dimensions $r \times r$ of $A^{**}$.
But then, from (51) we may write, for $j = 1, \ldots, m$,

$$\widehat{\Psi}_j = \frac{1}{2m} \sum_{i=0}^{m-1} \lambda_{ji} \sum_{\ell=1}^{m} \lambda_{i\ell}^* A_\ell^{**} = \frac{1}{2m} \sum_{\ell=1}^{m} \left( \sum_{i=0}^{m-1} \lambda_{ji} \lambda_{i\ell}^* \right) A_\ell^{**}$$

where, from (52) and (54) we have, for $j = 1$,

$$\sum_{i=0}^{m-1} \lambda_{ji} \lambda_{i\ell}^* = \begin{cases} 2m\,, & \ell = 1 \\ 0\,, & \ell = 2, \ldots, m \end{cases}$$

for $j = 2, \ldots, m$ for odd $m$, or for $j = 2, \ldots, m/2, 2 + m/2, \ldots, m$ for even $m$,

$$\sum_{i=0}^{m-1} \lambda_{ji} \lambda_{i\ell}^* = \begin{cases} m\,, & \ell = j \text{ or } \ell = m - j + 2 \\ 0\,, & \text{other values of } \ell \end{cases}$$

and for $j = m/2$ for even $m$

$$\sum_{i=0}^{m-1} \lambda_{ji}\lambda^*_{i\ell} = \begin{cases} 2m\,, & \ell = m/2 \\ 0\,, & \text{other values of } \ell \end{cases}$$

so that we have

$$\widehat{\boldsymbol{\Psi}}_1 = \boldsymbol{A}_1^{**}$$

and

$$\widehat{\boldsymbol{\Psi}}_j = \frac{1}{2}\left(\boldsymbol{A}_j^{**} + \boldsymbol{A}_{m-j+2}^{**}\right)$$

for $j = 2, \ldots, m$ for odd $m$, or for $j = 2, \ldots, m/2, 2 + m/2, \ldots, m$ for even $m$, and

$$\widehat{\boldsymbol{\Psi}}_{1+m/2} = \boldsymbol{A}_{1+m/2}^{**}$$

for even $m$.

We may thus write

$$|\boldsymbol{A}^*| = |\boldsymbol{A}_1^{**}| \left(|\boldsymbol{A}_{1+m^+}^{**}|\right)^{\text{mod}(m+1,2)} \prod_{j=2}^{m-m^+} \left|\frac{\boldsymbol{A}_j^{**} + \boldsymbol{A}_{m-j+2}^{**}}{2}\right|^2,$$

which is (17).

The equality in expression (18) may be shown in a similar manner.

## Appendix C: Definition, Notation and Expressions for the p.d.f. and c.d.f. of the GIG and GNIG Distributions

We say that the RV $X$ has a Gamma distribution with shape parameter $r$ ($> 0$) and rate parameter $\lambda$ ($> 0$), and we will denote this fact by $X \sim \Gamma(r, \lambda)$, if the p.d.f. of $X$ is

$$f_X(x) = \frac{\lambda^r}{\Gamma(r)}\, e^{-\lambda x}\, x^{r-1} \quad (x > 0)\,.$$

Let $Z_j \sim \Gamma(r_j, \lambda_j)$ ($j = 1, \ldots, p$) be a set of $p$ independent RVs and consider the RV

$$Z = \sum_{j=1}^{p} Z_j\,.$$

In case all the $r_j \in \mathbb{N}$, the distribution of $Z$ is what we call a GIG (Generalized Integer Gamma) distribution (Coelho, 1998). If all the $\lambda_j$ are different, $Z$ has a GIG distribution of depth $p$, with shape parameters $r_j$ and rate parameters $\lambda_j$, with p.d.f.

$$f_Z(z) = f^{GIG}\Big(z \,\Big|\, \{r_j\}_{j=1:p}; \{\lambda_j\}_{j=1:p}; p\Big) = K \sum_{j=1}^{p} P_j(z)\, e^{-\lambda_j z}, \qquad (z > 0)$$

and c.d.f.

$$F_Z(z) = F^{GIG}\Big(z \,\Big|\, \{r_j\}_{j=1:p}; \{\lambda_j\}_{j=1:p}; p\Big) = 1 - K \sum_{j=1}^{p} P_j^*(z)\, e^{-\lambda_j z}, \qquad (z > 0)$$

where

$$K = \prod_{j=1}^{p} \lambda_j^{r_j}, \qquad P_j(z) = \sum_{k=1}^{r_j} c_{j,k}\, z^{k-1}$$

and

$$P_j^*(z) = \sum_{k=1}^{r_j} c_{j,k}(k-1)! \sum_{i=0}^{k-1} \frac{z^i}{i!\,\lambda_j^{k-i}},$$

with

$$c_{j,r_j} = \frac{1}{(r_j - 1)!} \prod_{\substack{i=1\\ i\neq j}}^{p} (\lambda_i - \lambda_j)^{-r_i}, \qquad j = 1, \ldots, p,$$

and, for $k = 1, \ldots, r_j - 1$ and $j = 1, \ldots, p$,

$$c_{j,r_j-k} = \frac{1}{k} \sum_{i=1}^{k} \frac{(r_j - k + i - 1)!}{(r_j - k - 1)!} R(i, j, p) c_{j,r_j-(k-i)},$$

where

$$R(i, j, p) = \sum_{\substack{k=1\\ k\neq j}}^{p} r_k (\lambda_j - \lambda_k)^{-i} \quad (i = 1, \ldots, r_j - 1).$$

In case some of the $\lambda_j$ assume the same value as other $\lambda_j$'s, the distribution of $Z$ still is a GIG distribution, but in this case with a reduced depth. In this more general case, let $\{\lambda_\ell; \ell = 1, \ldots, g(\leq p)\}$ be the set of different $\lambda_j$'s and let $\{r_\ell; \ell = 1, \ldots, g(\leq p)\}$ be the set of the corresponding shape parameters, with $r_\ell$ being the sum of all $r_j$ $(j \in \{1, \ldots, p\})$ which correspond to the $\lambda_j$ assuming the value $\lambda_\ell$. In this case $Z$ will have a GIG distribution of depth $g$, with shape parameters $r_\ell$ and rate parameters $\lambda_\ell$ $(\ell = 1, \ldots, g)$.

If all RVs have Gamma distributions with different rate parameters $\lambda_j$ and $Z_p$ has a Gamma distribution with a non-integer shape parameter $r_p$, then we will say that the RV $Z$ has a GNIG (Generalized Near-Integer Gamma) distribution of depth $p$. The probability density and cumulative distribution functions of $Z$ are, for $z > 0$, respectively given by (Coelho, 2004)

$$
\begin{aligned}
&f^{GNIG}(z \mid r_1, \ldots, r_{p-1}; r_p; \lambda_1, \ldots, \lambda_{p-1}; \lambda_p; p)\\
&\quad = K \sum_{j=1}^{p-1} e^{-\lambda_j z} \sum_{k=1}^{r_j} \left\{ c_{j,k} \frac{\Gamma(k)}{\Gamma(k+r_p)} z^{k+r_p-1} \, {}_1F_1(r_p, k+r_p, -(\lambda_p-\lambda_j)z) \right\},
\end{aligned}
$$

and

$$
\begin{aligned}
F^{GNIG}(z \mid r_1, \ldots, r_{p-1}; r_p; \lambda_1, \ldots, \lambda_{p-1}; \lambda_p; p) &= \frac{\lambda_p^{r_p} z^{r_p}}{\Gamma(r_p+1)} \, {}_1F_1(r_p, r_p+1, -\lambda_p z)\\
-K \sum_{j=1}^{p-1} e^{-\lambda_j z} \sum_{k=1}^{r_j} c^*_{j,k} \sum_{i=0}^{k-1} &\frac{z^{r_p+i} \lambda_j^i}{\Gamma(r_p+1+i)} \, {}_1F_1(r_p, r_p+1+i, -(\lambda_p-\lambda_j)z),
\end{aligned}
$$

with $K = \prod_{j=1}^{p} \lambda_j^{r_j}$ and $c^*_{j,k} = \frac{c_{j,k}\Gamma(k)}{\lambda_j^k}$ and where ${}_1F_1(a, b, z)$ represents the Kummer confluent hypergeometric function.

Let $Z$ have a GIG distribution of depth $p$ and let us consider the RV $Z = e^{-W}$. Then the RV $Z$ has what Arnold et al. (2013) call an Exponentiated Generalized Integer Gamma (EGIG) distribution of depth $p$, with p.d.f.

$$
\begin{aligned}
f_Z(z) &= f^{EGIG}\Big(z \,\Big|\, \{r_j\}_{j=1:p}; \{\lambda_j\}_{j=1:p}; p\Big)\\
&= f^{GIG}\Big(-\log z \,\Big|\, \{r_j\}_{j=1:p}; \{\lambda_j\}_{j=1:p}; p\Big) \frac{1}{z} \qquad (0 < z < 1)
\end{aligned}
$$

and c.d.f.

$$
\begin{aligned}
F_Z(z) &= F^{EGIG}\Big(z \,\Big|\, \{r_j\}_{j=1:p}; \{\lambda_j\}_{j=1:p}; p\Big)\\
&= 1 - F^{GIG}\Big(-\log z \,\Big|\, \{r_j\}_{j=1:p}; \{\lambda_j\}_{j=1:p}; p\Big) \qquad (0 < z < 1).
\end{aligned}
$$

## Appendix D: Proof of Theorem 1 in Sect. 3

Let $W = -\log Z$. Then we may write the c.f. of $Z$ as

$$\begin{aligned}\Phi_W(t) &= E\left(e^{Wit}\right) = E\left(e^{-it\log Z}\right) = E\left(Z^{-it}\right) = \prod_{\nu=1}^{m^*}\prod_{\ell=1}^{n_\nu}\prod_{j=1}^{k_\nu} E\left(Y_{\nu\ell j}^{-it}\right)\\ &= \prod_{\nu=1}^{m^*}\prod_{\ell=1}^{n_\nu}\prod_{j=1}^{k_\nu} \frac{\Gamma\left(a_\nu+\ell+\frac{1-\ell}{h^*k_\nu}+1-\ell+\frac{m_\nu-j}{k_\nu}\right)}{\Gamma\left(a_\nu+\ell+\frac{1-\ell}{h^*k_\nu}+1-\ell-\frac{j}{k_\nu}\right)}\\ &\qquad\times\frac{\Gamma\left(a_\nu+\ell+\frac{1-\ell}{h^*k_\nu}+1-\ell-\frac{j}{k_\nu}-\text{i}t\right)}{\Gamma\left(a_\nu+\ell+\frac{1-\ell}{h^*k_\nu}+1-\ell+\frac{m_\nu-j}{k_\nu}-\text{i}t\right)}\end{aligned} \tag{55}$$

which, just by reindexing in $j$ may be written as

$$\begin{aligned}\Phi_W(t) &= \prod_{\nu=1}^{m^*}\prod_{\ell=1}^{n_\nu}\prod_{j=1+k_\nu(\ell-1)}^{k_\nu\ell} \frac{\Gamma\left(a_\nu+\frac{1-\ell}{h^*k_\nu}+1+\frac{m_\nu-j+k_\nu(\ell-1)}{k_\nu}\right)}{\Gamma\left(a_\nu+\frac{1-\ell}{h^*k_\nu}+1-\frac{j+k_\nu(\ell-1)}{k_\nu}\right)}\\ &\qquad\times\frac{\Gamma\left(a_\nu+\frac{1-\ell}{h^*k_\nu}+1-\frac{j+k_\nu(\ell-1)}{k_\nu}-\text{i}t\right)}{\Gamma\left(a_\nu+\frac{1-\ell}{h^*k_\nu}+1+\frac{m_\nu-j+k_\nu(\ell-1)}{k_\nu}-\text{i}t\right)}\\ &= \prod_{\nu=1}^{m^*}\prod_{\ell=1}^{n_\nu}\prod_{j=1+k_\nu(\ell-1)}^{k_\nu\ell} \frac{\Gamma\left(a_\nu+\frac{1-\ell}{h^*k_\nu}+\frac{m_\nu-j}{k_\nu}\right)\,\Gamma\left(a_\nu+\frac{1-\ell}{h^*k_\nu}-\frac{j}{k_\nu}-\text{i}t\right)}{\Gamma\left(a_\nu+\frac{1-\ell}{h^*k_\nu}-\frac{j}{k_\nu}\right)\,\Gamma\left(a_\nu+\frac{1-\ell}{h^*k_\nu}+\frac{m_\nu-j}{k_\nu}-\text{i}t\right)}\\ &= \prod_{\nu=1}^{m^*}\prod_{\ell=1}^{n_\nu}\prod_{j=1+k_\nu(\ell-1)}^{k_\nu\ell} \frac{\Gamma\left(a_{\nu\ell}+\frac{m_\nu-j}{k_\nu}\right)\,\Gamma\left(a_{\nu\ell}-\frac{j}{k_\nu}-\text{i}t\right)}{\Gamma\left(a_{\nu\ell}-\frac{j}{k_\nu}\right)\,\Gamma\left(a_{\nu\ell}+\frac{m_\nu-j}{k_\nu}-\text{i}t\right)},\end{aligned}$$

which shows the first equivalence in (25). Then, from (55), rearranging the Gamma arguments and using the multiplication formula for the Gamma function in the second equality in expression (3.5) of Coelho and Arnold (2019),

$$\prod_{j=1}^{n}\Gamma\left(z+\frac{j-1}{n}\right) = \prod_{j=1}^{n}\Gamma\left(z+\frac{n-j}{n}\right) = \Gamma(nz)\,(2\pi)^{\frac{n-1}{2}}\,n^{\frac{1}{2}-nz}\,, \tag{56}$$

we may write the c.f. of $W$ as

$$\Phi_W(t) = \prod_{\nu=1}^{m^*}\prod_{\ell=1}^{n_\nu}\prod_{j=1}^{k_\nu} \frac{\Gamma\left(a_\nu+\frac{1-\ell}{h^*k_\nu}+\frac{k_\nu-j}{k_\nu}+\frac{m_\nu}{k_\nu}\right)\,\Gamma\left(a_\nu+\frac{1-\ell}{h^*k_\nu}+\frac{k_\nu-j}{k_\nu}-\text{i}t\right)}{\Gamma\left(a_\nu+\frac{1-\ell}{h^*k_\nu}+\frac{k_\nu-j}{k_\nu}\right)\,\Gamma\left(a_\nu+\frac{1-\ell}{h^*k_\nu}+\frac{k_\nu-j}{k_\nu}+\frac{m_\nu}{k_\nu}-\text{i}t\right)}$$

$$= \prod_{\nu=1}^{m^*}\prod_{\ell=1}^{n_\nu} \frac{\Gamma\left(a_\nu\, k_\nu + \frac{1-\ell}{h^*} + m_\nu\right)\, \Gamma\left(a_\nu\, k_\nu + \frac{1-\ell}{h^*} - k_\nu\, \mathrm{i}t\right)}{\Gamma\left(a_\nu\, k_\nu + \frac{1-\ell}{h^*}\right)\, \Gamma\left(a_\nu\, k_\nu + \frac{1-\ell}{h^*} + m_\nu - k_\nu\, \mathrm{i}t\right)}$$

$$= \prod_{\nu=1}^{m^*}\prod_{\ell=1}^{n_\nu} \frac{\Gamma((a_{\nu\ell}-\ell)k_\nu + m_\nu)\, \Gamma((a_{\nu\ell}-\ell)k_\nu - k_\nu\, \mathrm{i}t)}{\Gamma\,((a_{\nu\ell}-\ell)k_\nu)\, \Gamma((a_{\nu\ell}-\ell)k_\nu + m_\nu - k_\nu\, \mathrm{i}t)}, \qquad (57)$$

which shows the distribution in the second product in (25).

Then, from (57), using (21), we may write

$$\Phi_W(t) = \prod_{\nu=1}^{m^*}\prod_{\ell=1}^{n_\nu}\prod_{j=0}^{m_\nu-1} ((a_{\nu\ell}-\ell)k_\nu + j)\,((a_{\nu\ell}-\ell)k_\nu + j - k_\nu\, \mathrm{i}t)^{-1}$$

$$= \prod_{\nu=1}^{m^*}\prod_{\ell=1}^{n_\nu}\prod_{j=0}^{m_\nu-1} \left(a_{\nu\ell} - \ell + \frac{j}{k_\nu}\right)\left(a_{\nu\ell} - \ell + \frac{j}{k_\nu} - \mathrm{i}t\right)^{-1}$$

$$= \prod_{\nu=1}^{m^*}\prod_{\ell=1}^{n_\nu}\prod_{j=0}^{m_\nu-1} \left(a_\nu + \frac{1-\ell}{h^* k_\nu} + \frac{j}{k_\nu}\right)\left(a_\nu + \frac{1-\ell}{h^* k_\nu} + \frac{j}{k_\nu} - \mathrm{i}t\right)^{-1}$$

$$= \prod_{\nu=1}^{m^*}\prod_{j=1}^{n_\nu+h^*(m_\nu-1)} \left(a_\nu + \frac{j-n_\nu}{h^* k_\nu}\right)^{r_{\nu j}}\left(a_\nu + \frac{j-n_\nu}{h^* k_\nu} - \mathrm{i}t\right)^{-r_{\nu j}}, \qquad (58)$$

where, for $j = 1, \ldots, n_\nu + h^*(m_\nu - 1)$ and $\nu = 1, \ldots, m^*$,

$$r_{\nu j} \equiv r(j; n_\nu, h^* m_\nu, h^*)$$

for the definition of $r(j; a, b, h^*)$ in (23)–(24), and where the last equality is obtained by identifying the different rate parameters in the expression above and their multiplicities, in much the same way that Wald and Brookner (1941) identify the different poles and their multiplicities in the Meijer $G$ function representation of p.d.f.'s of distributions of this type.

Expression (58) shows that, in case all $a_\nu$ are different, the distribution of $W = -\log Z$ is a GIG distribution, and consequently that of $Z$ an EGIG distribution of depth $\sum_{\nu=1}^{m^*} n_\nu + h^*(m_\nu - 1)$, with shape parameters $r_{\nu j}$ and rate parameters $a_\nu + \frac{j-n_\nu}{h^* k_\nu}$, for $\nu = 1, \ldots, m^*$ and $j = 1, \ldots, n_\nu + h^*(m_\nu - 1)$.

In case $a_\nu$, $n_\nu$, $k_\nu$ and $m_\nu$ are not function of $\nu$, say with $a_\nu = a$, $n_\nu = n$, $k_\nu = k$ and $m_\nu = m$, we then have for $Z$ the EGIG distribution depicted in (26), which is an EGIG distribution of depth $n + h^*(m-1)$, with shape parameters $m^* r_j$ and rate parameters $a + \frac{j-n}{h^* k}$, for $j = 1, \ldots, n + h^*(m-1)$, where the $r_j$ are given by (27).

## Appendix E: Proof of Expression (35)

Actually expression (35) is only expression (17) for the case where all $\boldsymbol{A}_i^*$ are equal for $i = 1, \ldots, m-1$, and as such the proof of (17) is indeed enough to prove also (35).

However, there is a different and simpler way to prove (35), if we consider

$$\mathrm{BTr}(\boldsymbol{A}) = \sum_{j=1}^{m} \boldsymbol{A}_{jj} \quad \text{and} \quad \mathrm{BSum}(\boldsymbol{A}) = \sum_{j=1}^{m}\sum_{k=1}^{m} \boldsymbol{A}_{jk} = \mathrm{BTr}\left((\boldsymbol{J}_m \otimes \boldsymbol{I}_r)\boldsymbol{A}\right),$$

where $\boldsymbol{A}_{jk}$ denotes the $r \times r$ block of $\boldsymbol{A}$ starting at row $r(j-1)+1$ and column $r(k-1)+1$, and as such $\boldsymbol{A}_{jj}$ denotes the $j$-th diagonal block of $\boldsymbol{A}$ of dimensions $r \times r$.

Then, in a proof due to Daniel Klein, with the collaboration of Augustyn Markiewicz and Ivan Žežula, we may write

$$\boldsymbol{A}_0^* = \widehat{\boldsymbol{\Sigma}}_{0|H_1} = \frac{1}{m}\sum_{j=1}^{m} \boldsymbol{A}_{jj} = \frac{1}{m}\mathrm{BTr}(\boldsymbol{A})$$

and

$$\begin{aligned}
\boldsymbol{A}_1^* = \widehat{\boldsymbol{\Sigma}}_{1|H_1} &= \frac{1}{m(m-1)}\sum_{j=1}^{m}\sum_{k=j+1}^{m}\left(\boldsymbol{A}_{jk} + \boldsymbol{A}_{kj}\right) \\
&= \frac{1}{m(m-1)}\left(\mathrm{BSum}(\boldsymbol{A}) - \mathrm{BTr}(\boldsymbol{A})\right) \\
&= \frac{1}{m(m-1)}\left(\mathrm{BTr}\left((\boldsymbol{J}_m \otimes \boldsymbol{I}_r)\boldsymbol{A}\right) - \mathrm{BTr}(\boldsymbol{A})\right) \\
&= \frac{1}{m(m-1)}\mathrm{BTr}\left(\left((\boldsymbol{J}_m - \boldsymbol{I}_m) \otimes \boldsymbol{I}_r\right)\boldsymbol{A}\right)
\end{aligned}$$

so that, using

$$\boldsymbol{Q}_m = \boldsymbol{I}_m - \boldsymbol{P}_m \qquad \text{with} \qquad \boldsymbol{P}_m = \frac{1}{m}\boldsymbol{J}_m,$$

we may write

$$\begin{aligned}
\boldsymbol{A}_0^* - \boldsymbol{A}_1^* &= \frac{1}{m}\mathrm{BTr}(\boldsymbol{A}) - \frac{1}{m(m-1)}\mathrm{BTr}\left(\left(\left(\boldsymbol{J}_m - \boldsymbol{I}_m\right)\otimes \boldsymbol{I}_r\right)\boldsymbol{A}\right)\\
&= \frac{1}{m}\mathrm{BTr}\left(\left(\left(\boldsymbol{I}_m - \frac{1}{m-1}\boldsymbol{J}_m + \frac{1}{m-1}\boldsymbol{I}_m\right)\otimes \boldsymbol{I}_r\right)\boldsymbol{A}\right)\\
&= \frac{1}{m}\mathrm{BTr}\left(\left(\left(\frac{m}{m-1}\boldsymbol{I}_m - \frac{1}{m-1}\boldsymbol{J}_m\right)\otimes \boldsymbol{I}_r\right)\boldsymbol{A}\right)\\
&= \frac{1}{m-1}\mathrm{BTr}\left((\boldsymbol{Q}_m \otimes \boldsymbol{I}_r)\boldsymbol{A}\right)
\end{aligned}$$

and

$$\begin{aligned}
\boldsymbol{A}_0^* + (m-1)\boldsymbol{A}_1^* &= \frac{1}{m}\mathrm{BTr}(\boldsymbol{A}) + \frac{1}{m}\mathrm{BTr}\left(\left(\left(\boldsymbol{J}_m - \boldsymbol{I}_m\right)\otimes \boldsymbol{I}_r\right)\boldsymbol{A}\right)\\
&= \frac{1}{m}\mathrm{BTr}\left((\boldsymbol{J}_m \otimes \boldsymbol{I}_r)\boldsymbol{A}\right) = \mathrm{BTr}\left((\boldsymbol{P}_m \otimes \boldsymbol{I}_r)\boldsymbol{A}\right).
\end{aligned}$$

But then we may write

$$\begin{aligned}
\boldsymbol{A}^* &= \boldsymbol{I}_m \otimes (\boldsymbol{A}_0^* - \boldsymbol{A}_1^*) + \boldsymbol{J}_m \otimes \boldsymbol{A}_1^*\\
&= \boldsymbol{Q}_m \otimes (\boldsymbol{A}_0^* - \boldsymbol{A}_1^*) + \boldsymbol{P}_m \otimes (\boldsymbol{A}_0^* - \boldsymbol{A}_1^*) + m\boldsymbol{P}_m \otimes \boldsymbol{A}_1^*\\
&= \boldsymbol{Q}_m \otimes (\boldsymbol{A}_0^* - \boldsymbol{A}_1^*) + \boldsymbol{P}_m \otimes \left(\boldsymbol{A}_0^* + (m-1)\boldsymbol{A}_1^*\right),
\end{aligned}$$

where

$$\boldsymbol{Q}_m\boldsymbol{P}_m = \left(\boldsymbol{I}_m - \frac{1}{m}\boldsymbol{J}_m\right)\frac{1}{m}\boldsymbol{J}_m = \frac{1}{m}\boldsymbol{J}_m - \frac{1}{m^2}\underbrace{\boldsymbol{J}_m\boldsymbol{J}_m}_{=m\boldsymbol{J}_m} = \frac{1}{m}\boldsymbol{J}_m - \frac{1}{m}\boldsymbol{J}_m = 0\,,$$

with

$$\mathrm{rank}(\boldsymbol{Q}_m) = m-1 \quad \text{and} \quad \mathrm{rank}(\boldsymbol{P}_m) = 1\,,$$

so that

$$\begin{aligned}
|\boldsymbol{A}^*| &= \left|\boldsymbol{A}_0^* + (m-1)\boldsymbol{A}_1^*\right| \left|\boldsymbol{A}_0^* - \boldsymbol{A}_1^*\right|^{m-1}\\
&= \left|\mathrm{BTr}\left((\boldsymbol{P}_m \otimes \boldsymbol{I}_r)\boldsymbol{A}\right)\right| \left|\frac{1}{m-1}\mathrm{BTr}\left((\boldsymbol{Q}_m \otimes \boldsymbol{I}_r)\boldsymbol{A}\right)\right|^{m-1}.
\end{aligned}$$

Then, taking

$$\boldsymbol{A}^{**} = (\boldsymbol{\Gamma}_m \otimes \boldsymbol{I}_r)\,\boldsymbol{A}\,(\boldsymbol{\Gamma}_m \otimes \boldsymbol{I}_r)$$

with $\boldsymbol{\Gamma}_m$ orthogonal and symmetric, we may write

$$\boldsymbol{A} = (\boldsymbol{\Gamma}_m \otimes \boldsymbol{I}_r)\, \boldsymbol{A}^{**}\, (\boldsymbol{\Gamma}_m \otimes \boldsymbol{I}_r)\,,$$

while, on the other hand, using the properties of BTr (see Filipiak et al. (2018)), we may write

$$\begin{aligned}
\mathrm{BTr}\,((\boldsymbol{P}_m \otimes \boldsymbol{I}_r)\boldsymbol{A}) &= \mathrm{BTr}\left((\boldsymbol{P}_m \otimes \boldsymbol{I}_r)\,(\boldsymbol{\Gamma}_m \otimes \boldsymbol{I}_r)\,\boldsymbol{A}^{**}\,(\boldsymbol{\Gamma}_m \otimes \boldsymbol{I}_r)\right)\\
&= \mathrm{BTr}\left((\boldsymbol{\Gamma}_m \otimes \boldsymbol{I}_r)\,(\boldsymbol{P}_m \otimes \boldsymbol{I}_r)\,(\boldsymbol{\Gamma}_m \otimes \boldsymbol{I}_r)\,\boldsymbol{A}^{**}\right)\\
&= \mathrm{BTr}\left((\boldsymbol{Q}_m^* \otimes \boldsymbol{I}_r)\boldsymbol{A}^{**}\right) = \boldsymbol{A}_1^{**}
\end{aligned}$$

where

$$\boldsymbol{Q}_m^* = \boldsymbol{\Gamma}_m \boldsymbol{P}_m \boldsymbol{\Gamma}_m = \left[\begin{array}{c|c} 1 & \boldsymbol{0}'_{m-1} \\ \hline \boldsymbol{0}_{m-1} & \boldsymbol{0}_{(m-1)\times(m-1)} \end{array}\right]$$

and

$$\begin{aligned}
\tfrac{1}{m-1}\mathrm{BTr}\left((\boldsymbol{Q}_m \otimes \boldsymbol{I}_r)\boldsymbol{A}\right) &= \tfrac{1}{m-1}\mathrm{BTr}\left((\boldsymbol{Q}_m \otimes \boldsymbol{I}_r)\,(\boldsymbol{\Gamma}_m \otimes \boldsymbol{I}_r)\,\boldsymbol{A}^{**}\,(\boldsymbol{\Gamma}_m \otimes \boldsymbol{I}_r)\right)\\
&= \tfrac{1}{m-1}\mathrm{BTr}\left((\boldsymbol{\Gamma}_m \otimes \boldsymbol{I}_r)\,(\boldsymbol{Q}_m \otimes \boldsymbol{I}_r)\,(\boldsymbol{\Gamma}_m \otimes \boldsymbol{I}_r)\,\boldsymbol{A}^{**}\right)\\
&= \tfrac{1}{m-1}\mathrm{BTr}\left((\boldsymbol{P}_m^* \otimes \boldsymbol{I}_r)\boldsymbol{A}^{**}\right) = \tfrac{1}{m-1}\sum_{j=2}^{m}\boldsymbol{A}_j^{**}
\end{aligned}$$

where

$$\boldsymbol{P}_m^* = \boldsymbol{\Gamma}_m \boldsymbol{Q}_m \boldsymbol{\Gamma}_m = \left[\begin{array}{c|c} 0 & \boldsymbol{0}'_{m-1} \\ \hline \boldsymbol{0}_{m-1} & \boldsymbol{I}_{m-1} \end{array}\right]$$

so that we may indeed write

$$|\boldsymbol{A}^*| = \left|\boldsymbol{A}_1^{**}\right| \left|\frac{1}{m-1}\sum_{j=2}^{m}\boldsymbol{A}_j^{**}\right|^{m-1}.$$

**Acknowledgments** The author wants to thank the support provided by the Banach Center of the Institute of Mathematics of the Polish Academy of Sciences and by CMA/UNL (Centro de Matemática e Aplicações, NOVA University of Lisbon) through the project UID/MAT/00297/2019.

## References


Anderson, T.W.: The Statistical Analysis of Time Series. Wiley, New York (1972)

Arnold, B.C., Coelho, C.A., Marques, F.J.: The distribution of the product of powers of independent uniform random variables — a simple but useful tool to address and better understand the structure of some distributions. J. Multivar. Anal. **113**, 19–36 (2013)

Brillinger, D.R.: Time Series – Data Analysis and Theory. SIAM (2001)

Coelho, C.A.: The Generalized Integer Gamma Distribution — a Basis for Distributions in Multivariate Statistics. J. Multivar. Anal. **64**, 86–102 (1998)

Coelho, C.A.: The generalized near-integer Gamma distribution: a basis for 'near-exact' approximations to the distribution of statistics which are the product of an odd number of independent Beta random variables. J. Multivar. Anal. **89**, 191–218 (2004)

Coelho, C.A., Arnold, B.C.: Finite Forms Representations for Meijer G and Fox H Functions — Applied to Multivariate Likelihood Ratio Tests Using Mathematica®, MAXIMA and R. Lecture Notes in Statistics. Springer (2019)

Filipiak, K., Klein, D., Vojtková, E.: The properties of partial trace and block trace operators of partitioned matrices. Electron. J. Linear Algebra **33**, 3–15 (2018)

Fox, C.: The *G* and *H* functions as symmetrical kernels. Trans. Am. Math. Soc. **98**, 395–429 (1961)

John, J.A.: Cyclic designs. Chapman and Hall, London (1987)

Khattree, R.: Multivariate statistical inference involving circulant matrices: a review. In: Gupta, A.K., Girko, V.L. (eds.) Multidimensional Statistical Analysis and Theory of Random Matrices, pp. 101–110. VSP, The Netherlands (1996)

Kedem, B.: Time Series Analysis by Higher Order Crossings. IEEE Press, Inc., New York (1994)

Kshirsagar, A.M.: Multivariate Analysis. Marcel Dekker, New York (1972)

Lv, X.-G., Huang, T.-Z.: A note on inversion of Toeplitz matrices. Appl. Math. Lett. **20**, 1189–1193 (2007)

Lv, X.-G., Huang, T.-Z.: The inverses of block Toeplitz matrices. J. Math. Article ID 207176 (2013)

Mathai, A.M.: A Handbook of Generalized Special Functions for Statistical and Physical Sciences. Oxford University Press, New York (1993)

Mathai, A.M., Haubold, H.J.: Special Functions for Applied Scientists. Springer, New York (2008)

Mathai, A.M., Saxena, R.K.: The H-function with Applications in Statistics and Other Disciplines. J. Wiley & Sons, New York (1978)

Olkin, I.: Testing and estimation for structures which are circularly symmetric in blocks. In: Kabe, D.G., Gupta, R.P. (eds.) Multivariate Statistical Inference, pp. 183–195. North Holland, New York (1973)

Olkin, I., Press, S.J.: Testing and estimation for a circular stationary model. Ann. Math. Stat. **40**, 1358–1373 (1969)

Pollock, D.S.G.: Circulant matrices and time-series analysis. Int. J. Math. Educ. Sci. Technol. **33**, 213–230 (2002)

Prudnikov, A.P., Brychkov, Y.A., Marichev, O.J.: Integrals and Series, Vol. 3: More Special Functions. Gordon and Breach, Newark (1990)

Seely, J.: Quadratic subspaces and completeness. Ann. Math. Stat. **42**, 710–721 (1971)

Szatrowski, T.D.: Explicit solutions, one iteration convergence and averaging in the multivariate Normal estimation problem for patterned means and covariances. Ann. Inst. Stat. Math. **30**, 81–88 (1978)

Szatrowski, T.D.: Necessary and sufficient conditions for explicit solutions in the multivariate Normal estimation problem for patterned means and covariances. Ann. Stat. **8**, 802–810 (1980)

Tricomi, F.G., Erdélyi, A.: The asymptotic expansion of a ratio of gamma functions. Pac. J. Math. **1**, 133–142 (1951)

Wald, A., Brookner, R.J.: On the distribution of Wilks' statistic for testing the independence of several groups of variates. Ann. Math. Stat. **12**, 137–152 (1941)

***Author and Editors notes:***

[1] there is a typo in the expression for $F_\Lambda(\lambda)$, where $(\nu+1)$ should be replaced just by 1, that is, this expression should rather read

$$F_\Lambda(\lambda) = 1-\sum_{\ell=0}^{\nu} \pi_\ell \, F^{GNIG}\left(-\log \lambda \,\middle|\, \{u_j\}_{j=1:g}, r^*_{r+2q-4}+\frac{1+\bmod(m+1,2)}{2}+\ell;\ \{\lambda_j\}_{j=1:g}, \frac{n-2}{2}; g\right)$$

[2] there is a small typo in the expressions for the estimators of $\Sigma$, where $\widehat{\boldsymbol{\Sigma}}_{0|H1}$ and $\widehat{\boldsymbol{\Sigma}}_{1|H1}$ should rather be $\widehat{\boldsymbol{\Sigma}}_{0|H0}$ and $\widehat{\boldsymbol{\Sigma}}_{1|H0}$.

[3] there is a typo in the expression for the $r_j$, where an $r$ is missing in the expression, after the equivalent sign, that is, the expressions should rather read as

$$r_j \equiv r(j; r, (m-1)(q-1)-1, 2)$$

# The Likelihood Ratio Test of Equality of Mean Vectors with a Doubly Exchangeable Covariance Matrix

**Carlos A. Coelho and Jolanta Pielaszkiewicz**

**Abstract** The authors derive the LRT statistic for the test of equality of mean vectors when the covariance matrix has what is called a double exchangeable structure. A second expression for this statistic, based on determinants of Wishart matrices with a block-diagonal parameter matrix, allowed for the expression of the distribution of this statistic as that of a product of independent Beta random variables. Moreover, the split of the LRT statistic into three independent components, induced by this second representation, will then allow for the expression of the exact distribution of the LRT statistic in a very manageable finite closed form for most cases and the obtention of very sharp near-exact distributions for the other cases. Numerical studies show that, as expected, due to the way they are built, these near-exact distributions are indeed asymptotic not only for increasing sample sizes but also for increasing values of all other parameters in the distribution, besides lying very close to the exact distribution even for extremely small samples.

**Keywords** Asymptoticity for all parameters · Exact distribution · Near-exact distributions · Product of Betas · Quadratic space · Small samples

---

Reprinted with permission from Springer Nature. First published as Coelho, C.A., Pielaszkiewicz, J.: The Likelihood Ratio Test of Equality of Mean Vectors with a Doubly Exchangeable Covariance Matrix, in 'Methodology and Applications of Statistics — A Volume in Honor of C. R. Rao on the Occasion of his 100th Birthday', Arnold, B.C., Balakrishnan, N., Coelho, C.A. (eds.), Springer series in Contributions to Statistics, 151–191 (2021). (https://doi.org/10.1007/978-3-030-83670-2_8)

---

C. A. Coelho (✉)
Mathematics Department, NOVA FCT and CMA-FCT/UNL, Caparica, Portugal
e-mail: cmac@fct.unl.pt

J. Pielaszkiewicz
Department of Computer and Information Science, Linköping University, Linköping, Sweden
e-mail: jolanta.pielaszkiewicz@liu.se



B. C. Arnold et al. (eds.), *Likelihood Ratio Tests and Related Topics in Multivariate Analysis*, Contributions to Statistics,
https://doi.org/10.1007/978-3-031-68008-3_17

# 1 Introduction

Let us suppose we have a multivariate random vector $\underline{X}$ which is split into $m$ subvectors $\underline{X}_i$ $(i = 1, \dots, m)$, all of the same length $vr$, each of which is, in turn, split into $v$ subvectors $\underline{X}_{ij}$ $(j = 1, \dots, v)$, each of them of length $r$, that is, we have

$$\underline{X} = [\, \underbrace{\underline{X}_{11}, \underline{X}_{12}, \dots, \underline{X}_{1v}}_{\underline{X}_1}, \underbrace{\underline{X}_{21}, \underline{X}_{22}, \dots, \underline{X}_{2v}}_{\underline{X}_2}, \dots, \underbrace{\underline{X}_{m1}, \underline{X}_{m2}, \dots, \underline{X}_{mv}}_{\underline{X}_m} \,]'$$

where each $\underline{X}_{ij}$ $(i = 1, \dots, m;\ j = 1, \dots, v)$ is of dimension $r$.

The $\underline{X}_i$ may correspond for example to different times of measurement $(i = 1, \dots, m)$, with the $\underline{X}_{ij}$ $(j = 1, \dots, v)$, for a given $i$, corresponding to the set of $r$ variables measured at different $v$ locations at that given time $i$.

Then it may make sense to assume a covariance structure where we have, for $i = 1, \dots, m$,

$$Var(\underline{X}_i) = I_v \otimes U + (I_v - J_v) \otimes V = \begin{bmatrix} U & V & \dots & V \\ V & U & \dots & V \\ \vdots & \vdots & \ddots & \vdots \\ V & V & \dots & U \end{bmatrix}$$

and for $i \neq i'$ $(i, i' \in \{1, \dots, m\})$

$$Cov(\underline{X}_i, \underline{X}_{i'}) = J_v \otimes M = \begin{bmatrix} M & M & \dots & M \\ M & M & \dots & M \\ \vdots & \vdots & \ddots & \vdots \\ M & M & \dots & M \end{bmatrix}$$

where $I_p$ represents an identity matrix of order $p$ and $J_p$ a matrix of 1's of dimensions $p \times p$, so that for any $i = 1, \dots, m$ and $j = 1, \dots, v$

$$Var(\underline{X}_{ij}) = U$$

and for $j \neq j'$ $(j, j' \in \{1, \dots, v\})$ and $i \neq i'$ $(i, i' \in \{1, \dots, m\})$

$$Cov(\underline{X}_{ij}, \underline{X}_{ij'}) = V \quad \text{and} \quad Cov(\underline{X}_{ij}, \underline{X}_{i'j'}) = Cov(\underline{X}_{ij}, \underline{X}_{i'j}) = M\,.$$

In this case the covariance matrix for our random vector $\underline{X}$ will have what we call a double exchangeable structure, with

$$\Sigma = I_{vm} \otimes U + \{I_m \otimes (J_v - I_v)\} \otimes V + \{J_{vm} - (I_m \otimes J_v)\} \otimes M\,, \tag{1}$$

for any positive-definite $r \times r$ matrices $U$ and symmetric $r \times r$ matrices $V$ and $M$, such that $U - V$, $U - V + v(V - M)$ and $U - V + v(V - M) + vmM$ are positive-definite.

An account of the properties of this covariance structure was given by several authors and its applicability in spatial/time analyses established (Roy and Fonseca, 2012; Pavlenko and Roy, 2017; Coelho and Roy, 2020). It also has as particular cases several covariance structures of interest as for example the well-known compound-symmetric structure, for $r = 1$ and $m = 1$ or $v = 1$, the block-compound symmetric structure, for $m = 1$ or $v = 1$, and the double complete symmetry (Coelho and Singull, 2020) for $r = 1$.

For example for $m = 4$ and $v = 3$, $\Sigma$ would have the following structure

$$\Sigma = \begin{bmatrix} U & V & V & M & M & M & M & M & M & M & M & M \\ V & U & V & M & M & M & M & M & M & M & M & M \\ V & V & U & M & M & M & M & M & M & M & M & M \\ M & M & M & U & V & V & M & M & M & M & M & M \\ M & M & M & V & U & V & M & M & M & M & M & M \\ M & M & M & V & V & U & M & M & M & M & M & M \\ M & M & M & M & M & M & U & V & V & M & M & M \\ M & M & M & M & M & M & V & U & V & M & M & M \\ M & M & M & M & M & M & V & V & U & M & M & M \\ M & M & M & M & M & M & M & M & M & U & V & V \\ M & M & M & M & M & M & M & M & M & V & U & V \\ M & M & M & M & M & M & M & M & M & V & V & U \end{bmatrix} .$$

Let us suppose that we have $q$ independent replicates of $\underline{X}$, which we will denote by $\underline{X}^{(k)}$, for $k = 1, \ldots, q$, with

$$E\left(\underline{X}^{(k)}\right) = \underline{\mu}^{(k)} \quad \text{and} \quad Var(\underline{X}^{(k)}) = \Sigma \, ,$$

where $\Sigma$ has the above double exchangeable structure, and let us suppose that we are then interested in testing the null hypothesis

$$H_0 : \underline{\mu}^{(1)} = \cdots = \underline{\mu}^{(q)} \, , \tag{2}$$

based on $q$ independent samples, one from each $\underline{X}^{(k)}$, accounting for the structure of $\Sigma$ in (1).

In the next section we obtain the LRT statistic for this test, deriving the MLE's for $\Sigma$, under the null and the alternative hypotheses, using a strategy based on the fact that both $\Sigma$ and $\Sigma^{-1}$ have a similar structure. Then in Sect. 3 we first rewrite the expression for this LRT statistic, expressing it in terms of determinants of diagonal blocks of Wishart matrices with a block-diagonal parameter matrix. This will enable us to obtain the distribution of the LRT statistic in terms of the distribution of a product of independent Beta r.v.'s. This form of the distribution will in turn enable us to obtain the exact distribution of this LRT statistic in a closed finite and very manageable form for all cases where $q$ is odd or $r$ is even and also to develop very sharp near-exact distributions for all other cases. Finally in Sect. 4 conclusions are drawn and a discussion is undertaken upon the methods and techniques used.

## 2 The LRT Statistic

Let $\underline{X}^{(k)} \sim N_{vmr}(\underline{\mu}^{(k)}, \Sigma)$ $(k = 1, \ldots, q)$, where $\Sigma$ has the double exchangeable structure in (1) and let us suppose that we are interested in testing the null hypothesis in (2).

Let

$$A = \sum_{k=1}^{q} (n_k - 1) S_k \quad \text{and} \quad B = \sum_{k=1}^{q} n_k (\overline{\underline{X}}^{(k)} - \overline{\underline{X}})(\overline{\underline{X}}^{(k)} - \overline{\underline{X}})' \tag{3}$$

where $n_k$ is the size of the sample from $\underline{X}^{(k)}$ and $S_k$ and $\overline{\underline{X}}^{(k)}$ are respectively the sample covariance matrix and mean vector of the $k$-th sample and where

$$\overline{\underline{X}} = \frac{1}{n} \sum_{k=1}^{q} n_k \overline{\underline{X}}^{(k)},$$

for $n = \sum_{k=1}^{q} n_k$. Then, see for example Kshirsagar (1972, Sect. 9.1), the $(2/n)$-th power of the LRT statistic to test $H_0$ in (2), without accounting for the structure in $\Sigma$, is

$$\Lambda = \frac{|A|}{|A + B|},$$

where $A$ and $A + B$ are the unstructured MLEs of $\Sigma$, respectively under the alternative and under the null hypotheses.

So, the question now is: which will be the ($(2/n)$-th power of the) LRT statistic to test $H_0$ in (2), accounting for the structure of $\Sigma$ in (1)?

It is not too hard to show that the answer is: it is a statistic of the form

$$\Lambda = \frac{|A^*|}{|A^* + B^*|}, \tag{4}$$

where $A^*$ and $A^* + B^*$ will be the MLEs of $\Sigma$, respectively under the alternative and under the null hypotheses, both of them accounting for the double exchangeable structure of $\Sigma$.

So, the next question is: how will we be able to obtain $A^*$ and $B^*$?

We may note that, not only the space of doubly exchangeable covariance matrices is itself a quadratic space (Seely, 1971), as well as if we let $\Gamma_p$ denote an orthogonal Helmert matrix whose first row is proportional to a vector of 1's, and consider the orthogonal matrix

$$\Gamma = \Gamma_m \otimes \Gamma_v \otimes I_r \,, \tag{5}$$

then we have, for $\Sigma$ in (1) (Roy and Fonseca, 2012),

$$\Gamma \Sigma \Gamma' = \Psi \,,$$

with

$$\Psi = bdiag(\Psi_1, \underbrace{\Psi_3, \ldots, \Psi_3}_{v-1}, \underbrace{\Psi_2, \underbrace{\Psi_3, \ldots, \Psi_3}_{v-1}, \ldots, \Psi_2, \underbrace{\Psi_3, \ldots, \Psi_3}_{v-1}}_{m-1}) \,, \tag{6}$$

where $bdiag$ denotes a block-diagonal matrix and where $\Psi_1$, $\Psi_2$ and $\Psi_3$ are all $r \times r$ positive-definite matrices, with

$$\begin{aligned} \Psi_1 &= U + (v-1)V + v(m-1)M \\ \Psi_2 &= U + (v-1)V - vM \\ \Psi_3 &= U - V \,. \end{aligned} \tag{7}$$

But then, since we will have

$$\Psi^{-1} = bdiag(\Psi_1^{-1}, \underbrace{\Psi_3^{-1}, \ldots, \Psi_3^{-1}}_{v-1}, \underbrace{\Psi_2^{-1}, \underbrace{\Psi_3^{-1}, \ldots, \Psi_3^{-1}}_{v-1}, \ldots, \Psi_2^{-1}, \underbrace{\Psi_3^{-1}, \ldots, \Psi_3^{-1}}_{v-1}}_{m-1}),$$

we may write

$$\Sigma^{-1} = \Gamma \Psi^{-1} \Gamma' \,,$$

which shows that both $\Sigma$ and $\Sigma^{-1}$ have the double exchangeable structure, what allows us to use the results in Szatrowski (1978, 1980) and Seely (1971) in order to obtain the MLEs of $\Sigma$ under $H_0$, as well as under the alternative hypothesis, built from the MLEs of the corresponding blocks of dimensions $r \times r$, which will then be obtained by averaging the corresponding blocks of the unstructured MLE.

This way, under the alternative hypothesis $H_1 : \exists_{k,k' \in \{1,\dots,q\}} : \underline{\mu}^{(k)} \neq \underline{\mu}^{(k')}$, the MLE of $\Sigma$ is

$$A^* = I_{vm} \otimes \widehat{U}_{|H_1} + [I_m \otimes (J_v - I_v)] \otimes \widehat{V}_{|H_1} + [J_{vm} - (I_m \otimes J_v)] \otimes \widehat{M}_{|H_1}\,, \tag{8}$$

where

$$\widehat{U}_{|H_1} = \frac{1}{mv} \sum_{j=1}^{mv} A_{jj}\,, \tag{9}$$

$$\widehat{V}_{|H_1} = \frac{1}{v(v-1)m} \sum_{j=0}^{m-1} \sum_{k=1}^{v-1} \sum_{\ell=1}^{v-k} A_{jv+k,jv+k+\ell} + A'_{jv+k,jv+k+\ell}\,, \tag{10}$$

and

$$\widehat{M}_{|H_1} = \frac{1}{v^2 m(m-1)} \sum_{j=0}^{m-2} \sum_{k=1}^{v} \sum_{\ell=0}^{vm-(j+1)v-1} A_{jv+k,(j+1)v+1+\ell} + A'_{jv+k,(j+1)v+1+\ell}\,, \tag{11}$$

with $A_{jk}$ representing the $r \times r$ block of $A$ starting at row $r(j-1)+1$ and column $r(k-1)+1$.

Under the null hypothesis in (2) the MLE of $\Sigma$ is

$$A^* + B^* = I_{vm} \otimes \widehat{U}_{|H_0} + [I_m \otimes (J_v - I_v)] \otimes \widehat{V}_{|H_0} + [J_{vm} - (I_m \otimes J_v)] \otimes \widehat{M}_{|H_0}\,,$$

where

$$\widehat{U}_{|H_0} = \frac{1}{mv} \sum_{j=1}^{mv} A_{jj} + B_{jj}\,,$$

$$\widehat{V}_{|H_0} = \frac{1}{v(v-1)m} \sum_{j=0}^{m-1} \sum_{k=1}^{v-1} \sum_{\ell=1}^{v-k} \Big( A_{jv+k,jv+k+\ell} + A'_{jv+k,jv+k+\ell} + B_{jv+k,jv+k+\ell} + B'_{jv+k,jv+k+\ell} \Big)$$

and

$$\widehat{M}_{|H_0} = \frac{1}{v^2 m(m-1)} \sum_{j=0}^{m-2} \sum_{k=1}^{v} \sum_{\ell=0}^{vm-(j+1)v-1} \big( A_{jv+k,(j+1)v+1+\ell} + A'_{jv+k,(j+1)v+1+\ell} + B_{jv+k,(j+1)v+1+\ell} + B'_{jv+k,(j+1)v+1+\ell} \big),$$

where $A_{jk}$ and $B_{jk}$ represent respectively the $r\times r$ blocks of $A$ and $B$ starting at row $r(j-1)+1$ and column $r(k-1)+1$.

Then, for a sample of size $n=\sum_{k=1}^{q} n_k$, the $(2/n)$-th power of the LRT statistic to test $H_0$ in (2) is the statistic $\Lambda$ in (4).

However, although that expression may be used to obtain the computed value of the statistic, it is not much useful in helping us derive the distribution of $\Lambda$. In order to do this the best way is to take the approach in the next section.

## 3 The Distribution of the l.r.t. Statistic

The results on the distribution of the statistic $\Lambda$ in (4), which will be used to obtain its distribution in a finite closed form for most cases and near-exact approximations for the remaining cases, are summarized in the following Theorem, whose Corollary shows that the results in Coelho (2017) are indeed a particular case of the present results.

**Theorem 1** *The distribution of* $\Lambda$ *in (4) is the same as that of*

$$\left\{\prod_{j=1}^{r} Y_j\right\}\times\left\{\prod_{j=1}^{r}\left(Y_j^*\right)^{m-1}\right\}\times\left\{\prod_{j=1}^{r}\left(Y_j^{**}\right)^{(v-1)m}\right\}, \tag{12}$$

*where, for* $j=1,\dots,r$, *and for* $n>q+r-1$,

$$\begin{aligned} Y_j &\sim Beta\left(\frac{n-q+1-j}{2},\frac{q-1}{2}\right)\\ Y_j^* &\sim Beta\left(\frac{(m-1)(n-q)}{2}+\frac{1-j}{2},\frac{(m-1)(q-1)}{2}\right)\\ Y_j^{**} &\sim Beta\left(\frac{(v-1)m(n-q)}{2}+\frac{1-j}{2},\frac{(v-1)m(q-1)}{2}\right)\end{aligned} \tag{13}$$

*form three sets of independent r.v.'s, also all independent among themselves.*

***Proof*** Let us consider the matrices

$$A^{**}=\Gamma A\Gamma' \quad\text{and}\quad B^{**}=\Gamma B\Gamma', \tag{14}$$

for $\Gamma$ in (5). Then, we should remark that, given the fact that the matrices $A$ and $B$ in (3) are independent, with

$$A\sim W_{vmr}(n-q,\Sigma) \quad\text{and}\quad B\sim W_{vmr}(q-1,\Sigma),$$

then, $A^{**}$ and $B^{**}$ will be two independent matrices, with

$$A^{**} \sim W_{vmr}(n-q,\Psi) \quad \text{and} \quad B^{**} \sim W_{vmr}(q-1,\Psi)\,, \tag{15}$$

for $\Psi$ in (6).

Then, it is possible to show that (see Appendix A)

$$|A^*| = \underbrace{|\,A_1^{**}\,|}_{A_1^{***}}\Bigg(\underbrace{\left|\frac{1}{m-1}\sum_{j=1}^{m-1}A_{jv+1}^{**}\right|}_{A_2^{***}}\Bigg)^{m-1} \times\Bigg(\underbrace{\left|\frac{1}{(v-1)m}\sum_{j=1}^{m}\sum_{k=1}^{v-1}A_{(j-1)v+1+k}^{**}\right|}_{A_3^{***}}\Bigg)^{(v-1)m}, \tag{16}$$

and

$$|A^*+B^*| = \underbrace{|\,A_1^{**}+B_1^{**}\,|}_{A_1^{***}+B_1^{***}}\Bigg(\underbrace{\left|\frac{1}{m-1}\sum_{j=1}^{m-1}A_{jv+1}^{**}+B_{jv+1}^{**}\right|}_{A_2^{***}+B_2^{***}}\Bigg)^{m-1} \times\Bigg(\underbrace{\left|\frac{1}{(v-1)m}\sum_{j=1}^{m}\sum_{k=1}^{v-1}A_{(j-1)v+1+k}^{**}+B_{(j-1)v+1+k}^{**}\right|}_{A_3^{***}+B_3^{***}}\Bigg)^{(v-1)m}, \tag{17}$$

where $A_j^{**}$ and $B_j^{**}$ represent the $j$-th diagonal block of dimensions $r \times r$ respectively of $A^{**}$ and $B^{**}$.

But then, from (4), (16) and (17) we may write

$$\Lambda = \underbrace{\frac{|A_1^{***}|}{|A_1^{***}+B_1^{***}|}}_{\Lambda_1}\underbrace{\left(\frac{|A_2^{***}|}{|A_2^{***}+B_2^{***}|}\right)^{m-1}}_{\Lambda_2}\underbrace{\left(\frac{|A_3^{***}|}{|A_3^{***}+B_3^{***}|}\right)^{(v-1)m}}_{\Lambda_3} \tag{18}$$

where, under $H_0$ in (2), given the distributions of $A^{**}$ and $B^{**}$ in (15) and the fact that $\Psi$ is a block-diagonal matrix, $A_1^{***}$, $A_2^{***}$, $A_3^{***}$, $B_1^{***}$, $B_2^{***}$ and $B_3^{***}$ are all independent, with

$$\begin{aligned} A_1^{***} &\sim W_r(n-q,\Psi_1) \\ A_2^{***} &\sim W_r\big((m-1)(n-q),\Psi_2\big) \\ A_3^{***} &\sim W_r\big((r-1)m(n-q),\Psi_3\big) \end{aligned}$$

and

$$\begin{aligned} B_1^{***} &\sim W_r(q-1, \Psi_1) \\ B_2^{***} &\sim W_r\big((m-1)(q-1), \Psi_2\big) \\ B_3^{***} &\sim W_r\big((r-1)m(q-1), \Psi_3\big) \end{aligned}$$

for $\Psi_1$, $\Psi_2$ and $\Psi_3$ in (6), so that, under $H_0$ in (2), we have (Coelho and Arnold, 2019, Chap. 5, App. A)

$$\Lambda_1 \stackrel{\text{st}}{\sim} \prod_{j=1}^{r} Y_j, \quad \Lambda_2 \stackrel{\text{st}}{\sim} \prod_{j=1}^{r} \left(Y_j^*\right)^{m-1} \quad \text{and} \quad \Lambda_3 \stackrel{\text{st}}{\sim} \prod_{j=1}^{r} \left(Y_j^{**}\right)^{(v-1)m},$$

where, for $j = 1, \ldots, r$, and for $n > q + r - 1$, the r.v.'s $Y_j$, $Y_j^*$ and $Y_j^{**}$ form three sets of independent r.v.'s, which, given the independence of $\Lambda_1$, $\Lambda_2$ and $\Lambda_3$, consequence of the block-diagonal structure of the matrix $\Psi$, are also independent among themselves, with the distributions in (13). □

For $r = 1$ and $v = 1$ or for $r = 1$ and $m = 1$, the matrix $\Sigma$ is a compound symmetric matrix, and as such, for these values of the parameters it is possible to establish the following Corollary, which shows that the results in Coelho (2017) are a particular case of the results in the present paper.

**Corollary 1** *For $r = 1$, $v = 1$ and $m = p$, or for $r = 1$, $m = 1$ and $v = p$, we may write, from (16)–(18), the statistic $\Lambda$ as*

$$\Lambda = \frac{a_{11}^{***}}{c_{11}^{***}} \frac{(a^{***})^{p-1}}{(c^{***})^{p-1}}$$

*for*

$$c_{11}^{***} = a_{11}^{***} + b_{11}^{***} \quad \text{and} \quad c^{***} = a^{***} + b^{***},$$

*where $a_{11}^{***}$ and $b_{11}^{***}$ stand for the only diagonal element in $A_1^{***}$ and $B_1^{***}$ respectively, which are now $1 \times 1$ matrices, and where $a^{***}$ and $b^{***}$ represent respectively either the sum or the average of the $A_j^{**}$ and the $B_j^{**}$ for $j = 2, \ldots, p$.*

*Furthermore, for $r = 1$, $v = 1$ and $m = p$ we have, from (12) and (13), the distribution of $\Lambda$ as that of*

$$Y_1 \left(Y_1^*\right)^{p-1},$$

*where*

$$Y_1 \sim Beta\left(\frac{n-q}{2}, \frac{q-1}{2}\right) \quad \text{and} \quad Y_1^* \sim Beta\left(\frac{(p-1)(n-q)}{2}, \frac{(p-1)(q-1)}{2}\right),$$

*are two independent r.v.'s, and for* $r = 1$, $m = 1$ *and* $\upsilon = p$, *as that of*

$$Y_1 \left(Y_1^{**}\right)^{p-1} ,$$

*where* $Y_1^{**}$ *has exactly the same distribution as that of* $Y_1^*$ *above, being also independent of* $Y_1$.

*These results establish those in Coelho (2017) as a particular case of the present ones.*

In addressing the distribution of the l.r.t. statistic in (18) we will have to consider four different scenarios, split into two main scenarios, the second of which, itself split into three sub-scenarios.

The two main scenarios to consider are:

(i) the case of odd $q$ or even $r$, where we will be able to obtain the exact distribution of $\Lambda$ in a closed manageable form,

(ii) the case of even $q$ and odd $r$, where we will have to approximate the distribution of $\Lambda$ using near-exact distributions, which will be shown to lay very close to the exact distribution, and where we will have to consider three sub-scenarios:

  (ii.a) where $m$ and $\upsilon$ are odd, in which case we may still obtain the distribution of $\Lambda_1$ and $\Lambda_2$ in a closed finite manageable form,

  (ii.b) where $m$ is odd and $\upsilon$ is even, in which case we may obtain only the distribution of $\Lambda_2$ in a closed finite manageable form, and

  (ii.c) where $m$ is even, in which case only the distribution of $\Lambda_3$ may be obtained in a closed finite manageable form.

This enchaining of scenarios is addressed in the following subsections.

## 3.1 *The Case of Odd q and Even r*

When either $q$ is odd or $r$ is even we are actually able to obtain the exact distribution of $\Lambda$ in a closed finite and very manageable form.

This is so because in this case we may use Theorem 3.2 in Coelho and Arnold (2019) to obtain, for odd $q$ or even $r$ the exact distribution of $\Lambda_1$, $\Lambda_2$ and $\Lambda_3$ in a closed finite and very manageable form, which moreover, given the independence of the three statistics, will then open the way for obtaining the exact distribution of $\Lambda$ in a closed finite and very manageable form.

We have thus the following Theorem, which gives the exact distribution of $\Lambda$ for odd $q$ or even $r$ in a closed finite and very manageable form.

**Theorem 2** *For odd $q$ or even $r$ the exact distribution of $\Lambda$ is an EGIG distribution (Arnold et al., 2013) of depth at most $3r + vm(q-1) - 6$, with rate parameters*

$$\lambda_{jk} = \frac{n-1}{2} - \frac{j+1}{2c_k}, \quad j = 1, \ldots, r + c_k(q-1) - 2;\ k = 1, \ldots, 3, \tag{19}$$

*for*

$$c_1 = 1, \quad c_2 = m - 1 \quad \text{and} \quad c_3 = (v-1)m, \tag{20}$$

*and shape parameters*

$$r_{jk} = \begin{cases} h_{jk} & j = 1, 2 \\ h_{jk} + r_{j-2,k} & j = 3, \ldots, r + c_k(q-1) - 2 \end{cases} \tag{21}$$

*where*

$$\begin{aligned} h_{jk} &= (\#\ \textit{of elements in}\ \{r, c_k(q-1)\} \geq j) - 1 \\ &= \begin{cases} 1 & j = 1, \ldots, \min(r, c_k(q-1)) \\ 0 & j = 1 + \min(r, c_k(q-1)), \ldots, \max(r, c_k(q-1)) \\ -1 & j = 1 + \max(r, c_k(q-1)), \ldots, r + c_k(q-1) - 2, \end{cases} \end{aligned} \tag{22}$$

*with p.d.f. and c.d.f. respectively given by*

$$f_\Lambda(z) = f^{EGIG}\left(z \,\middle|\, \overset{\approx}{\{r_{jk}\}}, \widetilde{\{\lambda_{jk}\}}; \#\widetilde{\{\lambda_{jk}\}}\right) \tag{23}$$

*and*

$$F_\Lambda(z) = F^{EGIG}\left(z \,\middle|\, \overset{\approx}{\{r_{jk}\}}, \widetilde{\{\lambda_{jk}\}}; \#\widetilde{\{\lambda_{jk}\}}\right), \tag{24}$$

*for $0 < z < 1$, and where $f^{EGIG}$ and $F^{EGIG}$ denote respectively the p.d.f. and the c.d.f. of the EGIG distribution, with the full notation being explained in Appendix B, and with $\widetilde{\{\lambda_{jk}\}}$ denoting the set of unique $\lambda_{jk}$ in (19) (given the fact that for different $k$ some of the $\lambda_{jk}$ may be equal) and with $\overset{\approx}{\{r_{jk}\}}$ denoting the set of corresponding shape parameters, where the original shape parameters $r_{jk}$ that correspond to equal $\lambda'_{jk}s$ are added together (this has to be done because the definition of the EGIG p.d.f. and c.d.f. requests a set of unique rate parameters). The depth of this distribution, which is the number of different Gamma distributions involved and the product of whose exponentials form the EGIG distribution, is then equal to $\#\widetilde{\{\lambda_{jk}\}}$ or $\#\ \overset{\approx}{\{r_{jk}\}}$, where # denotes the cardinality of the set.*

***Proof*** Let $W_k = -\log \Lambda_k$, for $k = 1, \ldots, 3$. Then, for even $r$, the distribution of $\Lambda_k$ and $W_k$ $(k = 1, \ldots, 3)$ are given, for $c_k$ in (20), by Theorem 3.2 in Coelho and

Arnold (2019) for

$$m^* = 1\,,\;\; k_1 = 2\,,\;\; n_1 = r/2\,,\;\; m_1 = c_k(q-1) \;\text{ and }\; a_1 = \frac{c_k(n-q)}{2} + \frac{1}{2}\,,$$

or, for odd $q$, for

$$m^* = 1\,,\;\; k_1 = 2\,,\;\; n_1 = c_k(q-1)/2\,,\;\; m_1 = r \;\text{ and }\; a_1 = \frac{c_k(n-1)-r}{2} + \frac{1}{2}\,, \tag{25}$$

yielding, in either case,

$$W_k \stackrel{d}{\equiv} \sum_{j=1}^{r+c_k(q-1)-2} Z_{jk} \quad \text{or} \quad \Lambda_k \stackrel{d}{\equiv} \prod_{j=1}^{r+c_k(q-1)-2} e^{-Z_{jk}} \tag{26}$$

where "$\stackrel{d}{\equiv}$" stands for "is equivalent in distribution to" and

$$Z_{jk} \sim \Gamma(r_{jk}, \lambda_{jk})\,, \quad j = 1, \ldots, r + c_k(q-1) - 2 \tag{27}$$

are a set of independent r.v.'s, with $r_{jk}$ given by (21)–(22) and

$$\lambda_{jk} = \frac{n-q}{2} + \frac{j-r}{2c_k}\,, \quad j = 1, \ldots, r + c_k(q-1) - 2;\, k = 1, \ldots, 3\,,$$

thus yielding for $W_k$ a GIG distribution (Coelho, 1998) and for $\Lambda_k$ an EGIG distribution (Arnold et al., 2013) of depth $r + mv(q-1) - 2$ (where $mv = c_1 + c_2 + c_3$), with rate parameters $\lambda_{jk}$ and shape parameters $r_{jk}$ ($j = 1, \ldots, r + c_k(q-1) - 2$; $k = 1, 2, 3$).

Given the fact that the shape parameters $r_{jk}$ are symmetric for $j = 1, \ldots, r + c_k(q-1)-2$, for any $k = 1, \ldots, 3$, we may reverse the index $j$ in the expression for $\lambda_{jk}$, writing these parameters as in (19). This will render simpler expressions and a simpler approach later in Sect. 3.2.

The exact characteristic function (c.f.) of $W_k$ ($k = 1, \ldots, 3$) is thus,

$$\Phi_{W_k}(t) = \prod_{j=1}^{r+c_k(q-1)-2} \lambda_{jk}^{r_{jk}} \left(\lambda_{jk} - \mathrm{i}t\right)^{-r_{jk}}\,, \tag{28}$$

for $r_{jk}$ given by (21)–(22), $\lambda_{jk}$ given by (19) and $c_k$ given by (20), and therefore, given the independence of $\Lambda_1$, $\Lambda_2$ and $\Lambda_3$, and as such also that of $W_1$, $W_2$ and $W_3$, the exact c.f. of $W = -\log \Lambda = W_1 + W_2 + W_3$ is

$$\Phi_W(t) = \Phi_{W_1}(t)\, \Phi_{W_2}(t)\, \Phi_{W_3}(t)$$

and the exact distribution of $W$ is, therefore, a GIG distribution of depth at most $3r + (c_1 + c_2 + c_3)(q - 1) - 6$, with rate parameters $\lambda_{jk}$ and shape parameters $r_{jk}$ $(k = 1, \ldots, 3;\ j = 1, \ldots, r + c_k(q - 1) - 2)$ and that of $\Lambda$ an EGIG distribution of the same depth and with the same rate and shape parameters. We say that the depth of this GIG distribution, that is, the number of different rate parameters it has, is at most $3r + (c_1 + c_2 + c_3)(q - 1) - 6$, since for some combinations of values of the parameters $m$, $q$, $v$ and $r$, some of the rate parameters $\lambda_{jk}$ may be equal for different $k$'s. In this case the depth of the GIG and of the EGIG distributions will be smaller than $3r + (c_1 + c_2 + c_3)(q - 1) - 6$ since we will have to "collapse", or more precisely, to add together the Gamma r.v.'s with the same rate parameters into a single Gamma r.v. with that same rate parameter and a shape parameter that is the sum of the shape parameters for the original Gamma r.v.'s with that same rate parameter. We could arrange for a rather intricate play of indexes to index all these different rate parameters and also to indicate which would collapse for different possible relations among the parameters $m$, $q$, $v$ and $r$, but since we will anyway need a software to compute the values of the p.d.f. and the c.d.f. of $\Lambda$ or $W$, this "selection" of the different rate parameters may then be easily done by this software using a function similar to the function `Tally` in Mathematica®.

Using the notation in Appendix B for the GIG p.d.f. and c.d.f., and using for the sets of rate and shape parameters the notation used and explained in the statement of the Theorem, we will write, for $w > 0$, the exact p.d.f. of $W$ as

$$f_W(w) = f^{GIG}\left(w \,\middle|\, \overset{\approx}{\widetilde{\{r_{jk}\}}}, \overset{\approx}{\widetilde{\{\lambda_{jk}\}}}; \#\widetilde{\{\lambda_{jk}\}}\right)$$

and its c.d.f. as

$$F_W(w) = F^{GIG}\left(w \,\middle|\, \overset{\approx}{\widetilde{\{r_{jk}\}}}, \overset{\approx}{\widetilde{\{\lambda_{jk}\}}}; \#\widetilde{\{\lambda_{jk}\}}\right).$$

The corresponding exact p.d.f. and c.d.f. of $\Lambda = e^{-W}$ are then easily obtained by simple transformation, and may be expressed in terms of the p.d.f. and c.d.f. of the EGIG distribution, as in (23) and (24). □

The following Corollary shows that the exact distribution in Theorem 2 fully matches the exact distribution obtained for odd $q$ in Coelho (2017) for the case when $\Sigma$ is a compound symmetric matrix, that is, when $r = 1$ and either $v = 1$ or $m = 1$. Although in face of Corollary 1 such result would be fully expected, it may be not immediately discernible from the result in Theorem 2, besides the fact that there is a small typo in expression (13) in Coelho (2017) which matters to correct.

**Corollary 2** *For $r = 1$, $v = 1$ and $m = p$, or for $r = 1$, $m = 1$ and $v = p$, and odd $q$, the exact distribution of $\Lambda$ is an EGIG distribution of depth $(q-1)(p-1)/2$, with rate parameters*

$$\lambda_j = \frac{n-q}{2} + \frac{j-1}{p-1}, \quad j = 1, \ldots, (q-1)(p-1)/2 \tag{29}$$

*and shape parameters*

$$r_j = \begin{cases} 1, & j = 1, \ldots, (q-1)(p-1)/2 \\ & j \neq (\ell-1)(p-1)+1 \textit{ for } \ell = 1, \ldots, (q-1)/2 \\ 2, & j = (\ell-1)(p-1)+1 \textit{ for } \ell = 1, \ldots, (q-1)/2 \end{cases} \tag{30}$$

*with p.d.f. and c.d.f.*

$$f_\Lambda(z) = f^{EGIG}\left(z \,\middle|\, \{r_j\}_{j=1:g}, \left\{\frac{n-q}{2} - \frac{j-1}{p-1}\right\}_{j=1:g}; g\right) \tag{31}$$

*and*

$$F_\Lambda(z) = F^{EGIG}\left(z \,\middle|\, \{r_j\}_{j=1:g}, \left\{\frac{n-q}{2} - \frac{j-1}{p-1}\right\}_{j=1:g}; g\right) \tag{32}$$

*for $g = (q-1)(p-1)/2$ and $r_j$ given by (30).*

***Proof*** Being a direct Corollary of Theorem 2, no formal proof is indeed necessary. All one has to do is to set $r = 1$, $v = 1$ and $m = p$ in Theorem 2, and notice that in this case we have, from (20),

$$c_1 = 1\,, \quad c_2 = p-1 \quad \text{and} \quad c_3 = 0\,,$$

so that we have, from (19),

$$\lambda_{j1} = \frac{n-q}{2} + \frac{j-1}{2}, \quad j = 1, \ldots, q-2$$

and

$$\lambda_{j2} = \frac{n-q}{2} + \frac{j-1}{2(p-1)}, \quad j = 1, \ldots, (p-1)(q-1)-1\,,$$

while the $\lambda_{j3}$ vanish. Furthermore, from (22) we have

$$h_{j1} = \begin{cases} 1, & j = 1 \\ 0, & j = 2, \ldots, q-2 \end{cases}$$

so that, from (21) we obtain

$$r_{j1} = \begin{cases} 1\,, & \text{for odd } j,\ \text{i.e. } j = 1, \ldots, q-2,\ \text{step } 2 \\ 0\,, & \text{for even } j,\ \text{i.e. } j = 2, \ldots, q-3,\ \text{step } 2 \end{cases}$$

and

$$h_{j2} = \begin{cases} 1\,, & j = 1 \\ 0\,, & j = 2, \ldots, (p-1)(q-1)-1 \end{cases}$$

which gives

$$r_{j2} = \begin{cases} 1\,, & \text{for odd } j,\ \text{i.e. } j = 1, \ldots, (p-1)(q-1)-1,\ \text{step } 2 \\ 0\,, & \text{for even } j,\ \text{i.e. } j = 2, \ldots, (p-1)(q-1)-2,\ \text{step } 2\,. \end{cases}$$

We may thus write the c.f. of $W_1$ as

$$\Phi_{W_1}(t) = \prod_{j=1}^{q-2} \lambda_{j1}^{r_{j1}} \left(\lambda_{j1} - \mathrm{i}t\right)^{-r_{j1}} = \prod_{j=0}^{\frac{q-1}{2}} \left(\frac{n-q}{2} + j\right)^1 \left(\frac{n-q}{2} + j - \mathrm{i}t\right)^{-1} \tag{33}$$

and the c.f. of $W_2$ as

$$\begin{aligned} \Phi_{W_2}(t) &= \prod_{j=1}^{(p-1)(q-1)-2} \lambda_{j2}^{r_{j2}} \left(\lambda_{j2} - \mathrm{i}t\right)^{-r_{j2}} \\ &= \prod_{j=1}^{(p-1)(q-1)/2} \left(\frac{n-q}{2} + \frac{j-1}{p-1}\right)^1 \left(\frac{n-q}{2} + \frac{j-1}{p-1} - \mathrm{i}t\right)^{-1}, \end{aligned}$$

while $\Lambda_3$ and $W_3$ vanish. This gives for $W$ a c.f. which is

$$\Phi_W(t) = \Phi_{W_1}(t)\,\Phi_{W_2}(t) = \prod_{j=1}^{(p-1)(q-1)/2} \lambda_j^{r_j} \left(\lambda_j - \mathrm{i}t\right)^{-r_j} \tag{34}$$

for $\lambda_j$ in (29) and $r_j$ in (30), which is exactly the c.f. on page 24 of Coelho (2017), thus confirming that result as a particular case of the result in Theorem 5. In this reference there is a small typo on the second row of expression (13), where a $j$ appears where there should be an $\ell$.

For $r = 1$, $m = 1$ and $v = p$, in Theorem 2,we have

$$c_1 = 1\,,\quad c_2 = 0 \quad \text{and} \quad c_3 = p - 1$$

so that we will have again the c.f. of $W_1$ given by (33) and $W_3$ with a similar distribution to that of $W_2$ for the case $r = 1$, $v = 1$ and $m = p$, while $\Lambda_2$ and $W_2$ vanish, this way giving again the c.f. of $W$ as that in (34), confirming, also for this case, the result in Coelho (2017) as a particular case of the result in Theorem 2. □

## 3.2 *The Case of Even q and Odd r*

When $q$ is even and $r$ is odd there are indeed a number of sub-cases to consider. That is what we do in the next subsections. On our study of these cases we will need to obtain the c.f.'s of $W_1$, $W_2$ and $W_3$, for situations where the distributions of these statistics will not have a closed finite form representation, in a form suitable for the development of near-exact approximations.

In order to fulfill this aim we may use the results in Appendix A2 of Marques et al. (2011) concerning the statistic $\Lambda_2$ in that paper.

Using these results, we may write the c.f. of $W_1$, when $r$ is odd and $q$ is even, the c.f. of $W_2$ when $r$ is odd, $q$ is even and $m$ is also even or the c.f. of $W_3$ when $r$ is odd, $q$ is even, $m$ is odd and $v$ is even, as

$$\Phi_{W_k}(t) = \underbrace{\left\{ \prod_{j=1}^{r+c_k(q-1)-2} \lambda_{jk}^{r_{jk}} \left(\lambda_{jk} - \mathrm{i}t\right)^{-r_{jk}} \right\}}_{\Phi_{W_k,1}(t)} \times \underbrace{\frac{\Gamma\left(\frac{c_k(n-1)}{2}\right)\Gamma\left(\frac{c_k(n-1)-1}{2} - c_k\mathrm{i}t\right)}{\Gamma\left(\frac{c_k(n-1)-1}{2}\right)\Gamma\left(\frac{c_k(n-1)}{2} - c_k\mathrm{i}t\right)}}_{\Phi_{W_k,2}(t)} \tag{35}$$

now with

$$r_{jk} = \begin{cases} h_{jk} + (-1)^j\,, & j = 1, 2 \\ h_{jk} + r_{j-2,k}\,, & j = 3, \ldots, r + c_k(q-1) - 2 \end{cases} \tag{36}$$

for $h_{jk}$ still given by (22), and thus yielding $r_{1k} = 0$ and $r_{2k} = 2$ $(k = 1, \ldots, 3)$, and $\lambda_{jk}$ and $c_k$ respectively given by (19) and (20).

In all other cases the c.f.'s of $W_1$, $W_2$ and $W_3$ will be given by (28), with $\lambda_{jk}$ and $r_{jk}$ respectively given by (19) and (21)–(22).

### 3.2.1 The Case of Even $q$, Odd $r$, Odd $m$, and Odd $v$

We should note that in this case we still have the exact distribution of $\Lambda_2$ and $\Lambda_3$, and as such also that of $W_2$ and $W_3$, given by Theorem 3.2 in Coelho and Arnold (2019), for $m^* = 1$ and $k_1, n_1, m_1$ and $a_1$ given by (25).

Then we have the following Theorem, which gives near-exact distributions for $\Lambda$ when $q$ is even, $r$ is odd and $m$ and $v$ are also both odd.

**Theorem 3** *For even $q$ and odd $r$, $m$ and $v$, near-exact distributions for $\Lambda$ will have p.d.f.'s and c.d.f.'s respectively of the form*

$$f_\Lambda(z) = \sum_{\ell=0}^{v} \pi_\ell\, f^{GNIG}\left(-\log z \,\Big|\, \overset{\approx}{\{r_{jk}\}}, \frac{1}{2}+2\ell; \widetilde{\{\lambda_{jk}\}}, \frac{n-2}{2}-\frac{1}{4}; \#\widetilde{\{\lambda_{jk}\}}+1\right)\frac{1}{z} \tag{37}$$

*and*

$$F_\Lambda(z) = \sum_{\ell=0}^{v} \pi_\ell\left(1 - F^{GNIG}\left(-\log z \,\Big|\, \overset{\approx}{\{r_{jk}\}}, \frac{1}{2}+2\ell; \widetilde{\{\lambda_{jk}\}}, \frac{n-2}{2}-\frac{1}{4}; \#\widetilde{\{\lambda_{jk}\}}+1\right)\right), \tag{38}$$

*for $0 < z < 1$, where $f^{GNIG}$ and $F^{GNIG}$ stand respectively for the p.d.f. and c.d.f. of the GNIG distribution (Coelho, 2004), using the notation in Appendix B, and where $\lambda_{jk}$ are given by (19) and $r_{jk}$ are given by (36) and (22) for $k = 1$ and by (21)–(22) for $k = 2, 3$, and where the notations $\widetilde{\{\lambda_{jk}\}}$ and $\overset{\approx}{\{r_{jk}\}}$ are the same as the ones used in Theorem 2.*

***Proof*** Given the remark right before the statement of this Theorem, on the distributions of $W_2$ and $W_3$, in order to build a near-exact distribution for $\Lambda$ we will use, for even $q$, odd $r$, and odd $m$ and $v$, the c.f. of $W_1$ given by (35) and those of $W_2$ and $W_3$ given by (28) and write

$$\Phi_W(t) = \underbrace{\Phi_{W_{1,1}}(t)\,\Phi_{W_2}(t)\,\Phi_{W_3}(t)}_{\Phi_1(t)}\ \underbrace{\Phi_{W_{1,2}}(t)}_{\Phi_2(t)}, \tag{39}$$

where we will keep untouched $\Phi_1(t)$ and will approximate asymptotically $\Phi_2(t)$, in such a way that the c.f. that will work as its asymptotic replacement may then be "joined" back together with $\Phi_1(t)$ to produce the c.f. of a manageable distribution. Since $\Phi_2(t)$ is the c.f. of a single *Logbeta* $\left(\frac{n-2}{2}, \frac{1}{2}\right)$ r.v., we will use the approach in Sect. 3.3 in Coelho and Alberto (2021) and we will replace $\Phi_2(t)$ in (39) by

$$\widetilde{\Phi}_2(t) = \sum_{\ell=0}^{\nu} \pi_\ell \left(\frac{n-2}{2} - \frac{1}{4}\right)^{\frac{1}{2}+2\ell} \left(\frac{n-2}{2} - \frac{1}{4} - \mathrm{i}t\right)^{-\left(\frac{1}{2}+2\ell\right)} \quad (40)$$

which is the c.f. of a finite mixture with $\nu + 1$ components which are Gamma distributions, all with the same rate parameter $\frac{n-2}{2} - \frac{1}{4}$ and shape parameters $\frac{1}{2} + 2\ell$ ($\ell = 0, \ldots, \nu$), whose weights $\pi_\ell$ for $\ell = 0, \ldots, \nu - 1$ will be obtained through the numerical solution of the linear system of $\nu$ equations

$$\frac{\partial^\ell}{\partial t^\ell} \Phi_2(t)\bigg|_{t=0} = \frac{\partial^\ell}{\partial t^\ell} \widetilde{\Phi}_2(t)\bigg|_{t=0}, \quad \ell = 1, \ldots, \nu,$$

with $\pi_\nu = 1 - \sum_{\ell=0}^{\nu-1} \pi_\ell$.

This approach will lead to near-exact distributions that match the first exact moments of $W = -\log \Lambda$ and yields for this r.v. the near-exact c.f.

$$\widetilde{\Phi}_W(t) = \Phi_1(t)\,\widetilde{\Phi}_2(t) = \sum_{\ell=0}^{\nu} \pi_\ell\, \Phi_1(t) \left(\frac{n-2}{2} - \frac{1}{4}\right)^{\frac{1}{2}+2\ell} \left(\frac{n-2}{2} - \frac{1}{4} - \mathrm{i}t\right)^{-\left(\frac{1}{2}+2\ell\right)}$$

with $\Phi_1(t)$ given by (39) and $\widetilde{\Phi}_2(t)$ by (40), which gives as near-exact distributions for $W$ a mixture of $\nu + 1$ GNIG (Generalized Near-Integer Gamma) distributions (Coelho, 2004) with integer shape parameters $r_{jk}$ ($k = 1, \ldots, 3$; $j = 1, \ldots, r + c_k(q-1) - 2$), given by (21) and (22) for $k = 2, 3$ and by (36) and (22) for $k = 1$, and non-integer shape parameter $\frac{1}{2} + 2\ell$ ($\ell = 0, \ldots, \nu$) and corresponding rate parameters $\lambda_{jk}$ ($k = 1, \ldots, 3$; $j = 1, \ldots, r + c_k(q-1) - 2$), given by (19), and $\frac{n-2}{2} - \frac{1}{4}$.

This yields near-exact distributions for $W$ with p.d.f.'s given by

$$f_W(w) = \sum_{\ell=0}^{\nu} \pi_\ell\, f^{GNIG}\left(w \,\middle|\, \widetilde{\{r_{jk}\}}, \frac{1}{2} + 2\ell; \widetilde{\{\lambda_{jk}\}}, \frac{n-2}{2} - \frac{1}{4}; \#\widetilde{\{\lambda_{jk}\}} + 1\right)$$

and c.d.f.'s given by

$$F_W(w) = \sum_{\ell=0}^{\nu} \pi_\ell\, F^{GNIG}\left(w \,\middle|\, \widetilde{\{r_{jk}\}}, \frac{1}{2} + 2\ell; \widetilde{\{\lambda_{jk}\}}, \frac{n-2}{2} - \frac{1}{4}; \#\widetilde{\{\lambda_{jk}\}} + 1\right)$$

for $w > 0$, and where we use for the GNIG p.d.f. and c.d.f. the notation in Appendix B.

The corresponding near-exact distributions for $\Lambda$ may then be easily obtained through the transformation $\Lambda = e^{-W}$ and these have p.d.f.'s and c.d.f.'s given by (37) and (38). □

**Table 1** Values of $\Delta$ for near-exact distributions of $\Lambda$ or $W$ for the case of even $q$, odd $r$, odd $m$, and odd $v$, for increasing values of $q$ and $r$ and sample sizes of $r+q+\{0, 100, 200, 500\}$

| | | | | | $\nu$ (number of exact moments matched) | | | |
|---|---|---|---|---|---|---|---|---|
| $n$ | $r$ | $q$ | $m$ | $v$ | 2 | 4 | 6 | 10 |
| 7 | 3 | 4 | 5 | 7 | $5.58\times10^{-9}$ | $6.28\times10^{-12}$ | $3.25\times10^{-14}$ | $9.35\times10^{-18}$ |
| 107 | | | | | $9.83\times10^{-17}$ | $1.79\times10^{-24}$ | $3.03\times10^{-31}$ | $5.52\times10^{-43}$ |
| 207 | | | | | $1.76\times10^{-18}$ | $2.20\times10^{-27}$ | $2.55\times10^{-35}$ | $2.23\times10^{-49}$ |
| 507 | | | | | $7.81\times10^{-21}$ | $2.64\times10^{-31}$ | $8.31\times10^{-41}$ | $5.35\times10^{-58}$ |
| 13 | 9 | 4 | 5 | 7 | $9.98\times10^{-12}$ | $3.20\times10^{-16}$ | $7.32\times10^{-20}$ | $1.03\times10^{-25}$ |
| 113 | | | | | $1.37\times10^{-17}$ | $6.87\times10^{-26}$ | $3.26\times10^{-33}$ | $4.97\times10^{-46}$ |
| 213 | | | | | $2.88\times10^{-19}$ | $1.11\times10^{-28}$ | $4.03\times10^{-37}$ | $3.66\times10^{-52}$ |
| 513 | | | | | $1.42\times10^{-21}$ | $1.58\times10^{-32}$ | $1.68\times10^{-42}$ | $1.29\times10^{-60}$ |
| 23 | 19 | 4 | 5 | 7 | $5.11\times10^{-14}$ | $6.12\times10^{-20}$ | $6.52\times10^{-25}$ | $3.67\times10^{-33}$ |
| 123 | | | | | $2.63\times10^{-18}$ | $4.45\times10^{-27}$ | $7.15\times10^{-35}$ | $1.27\times10^{-48}$ |
| 223 | | | | | $7.10\times10^{-20}$ | $1.08\times10^{-29}$ | $1.57\times10^{-38}$ | $2.29\times10^{-54}$ |
| 523 | | | | | $4.13\times10^{-22}$ | $2.04\times10^{-33}$ | $9.57\times10^{-44}$ | $1.48\times10^{-62}$ |
| 29 | 19 | 10 | 5 | 7 | $1.81\times10^{-15}$ | $2.32\times10^{-22}$ | $2.71\times10^{-28}$ | $2.04\times10^{-38}$ |
| 129 | | | | | $3.53\times10^{-19}$ | $1.57\times10^{-28}$ | $6.65\times10^{-37}$ | $8.32\times10^{-52}$ |
| 229 | | | | | $1.12\times10^{-20}$ | $4.99\times10^{-31}$ | $2.13\times10^{-40}$ | $2.72\times10^{-57}$ |
| 529 | | | | | $7.32\times10^{-23}$ | $1.14\times10^{-34}$ | $1.71\times10^{-45}$ | $2.68\times10^{-65}$ |
| 39 | 19 | 20 | 5 | 7 | $5.93\times10^{-17}$ | $7.93\times10^{-25}$ | $9.87\times10^{-32}$ | $9.41\times10^{-44}$ |
| 139 | | | | | $6.51\times10^{-20}$ | $9.38\times10^{-30}$ | $1.29\times10^{-38}$ | $1.71\times10^{-54}$ |
| 239 | | | | | $2.64\times10^{-21}$ | $4.50\times10^{-32}$ | $7.35\times10^{-42}$ | $1.38\times10^{-59}$ |
| 539 | | | | | $2.07\times10^{-23}$ | $1.40\times10^{-35}$ | $9.02\times10^{-47}$ | $2.65\times10^{-67}$ |

These near-exact distributions will be asymptotic not only for increasing sample sizes but also for increasing values of all other parameters in the distribution of $\Lambda$ and $W$, that is, for $r$, $q$, $m$ and $v$, as it is shown by the values in Tables 1 and 2, where are displayed values of the measure[1]

$$\Delta = \int_{-\infty}^{+\infty} \left| \frac{\Phi_W(t) - \widetilde{\Phi}_W(t)}{t} \right| dt \,. \tag{41}$$

This measure gives a very sharp upper-bound on the absolute value of the difference between the exact and the near-exact c.d.f.'s of $W$ or $\Lambda$, with

$$\Delta \geq \max_{w>0} \left| F_W(w) - \widetilde{F}_W(w) \right| = \max_{0<z<1} \left| F_\Lambda(z) - \widetilde{F}_\Lambda(z) \right| ,$$

[1] see author and editors note at the end of the Chapter

**Table 2** Values of $\Delta$ for near-exact distributions of $\Lambda$ or $W$ for the case of even $q$, odd $r$, odd $m$, and odd $v$, for increasing values of $m$ and $v$ and sample sizes of $r + q + \{0, 100, 200, 500\}$

| | | | | | $\nu$ (number of exact moments matched) | | | |
|---|---|---|---|---|---|---|---|---|
| $n$ | $r$ | $q$ | $m$ | $v$ | 2 | 4 | 6 | 10 |
| 39 | 19 | 20 | 5 | 7 | $5.93\times10^{-17}$ | $7.93\times10^{-25}$ | $9.87\times10^{-32}$ | $9.41\times10^{-44}$ |
| 139 | | | | | $6.51\times10^{-20}$ | $9.38\times10^{-30}$ | $1.29\times10^{-38}$ | $1.71\times10^{-54}$ |
| 239 | | | | | $2.64\times10^{-21}$ | $4.50\times10^{-32}$ | $7.35\times10^{-42}$ | $1.38\times10^{-59}$ |
| 539 | | | | | $2.07\times10^{-23}$ | $1.40\times10^{-35}$ | $9.02\times10^{-47}$ | $2.65\times10^{-67}$ |
| 39 | 19 | 20 | 15 | 7 | $1.45\times10^{-17}$ | $7.50\times10^{-26}$ | $3.59\times10^{-33}$ | $4.99\times10^{-46}$ |
| 139 | | | | | $1.27\times10^{-20}$ | $6.17\times10^{-31}$ | $2.86\times10^{-40}$ | $4.31\times10^{-57}$ |
| 239 | | | | | $5.13\times10^{-22}$ | $2.93\times10^{-33}$ | $1.61\times10^{-43}$ | $3.40\times10^{-62}$ |
| 539 | | | | | $4.01\times10^{-24}$ | $9.02\times10^{-37}$ | $1.95\times10^{-48}$ | $6.43\times10^{-70}$ |
| 39 | 19 | 20 | 25 | 7 | $7.11\times10^{-18}$ | $2.29\times10^{-26}$ | $6.79\times10^{-34}$ | $3.62\times10^{-47}$ |
| 139 | | | | | $5.93\times10^{-21}$ | $1.73\times10^{-31}$ | $4.82\times10^{-41}$ | $2.63\times10^{-58}$ |
| 239 | | | | | $2.39\times10^{-22}$ | $8.19\times10^{-34}$ | $2.70\times10^{-44}$ | $2.06\times10^{-63}$ |
| 539 | | | | | $1.86\times10^{-24}$ | $2.52\times10^{-37}$ | $3.27\times10^{-49}$ | $3.89\times10^{-71}$ |
| 39 | 19 | 20 | 25 | 17 | $1.98\times10^{-18}$ | $2.70\times10^{-27}$ | $3.41\times10^{-35}$ | $3.28\times10^{-49}$ |
| 139 | | | | | $1.57\times10^{-21}$ | $1.89\times10^{-32}$ | $2.17\times10^{-42}$ | $2.02\times10^{-60}$ |
| 239 | | | | | $6.32\times10^{-23}$ | $8.94\times10^{-35}$ | $1.21\times10^{-45}$ | $1.57\times10^{-65}$ |
| 539 | | | | | $4.93\times10^{-25}$ | $2.74\times10^{-38}$ | $1.47\times10^{-50}$ | $2.96\times10^{-73}$ |
| 39 | 19 | 20 | 25 | 27 | $1.00\times10^{-18}$ | $8.69\times10^{-28}$ | $6.97\times10^{-36}$ | $2.70\times10^{-50}$ |
| 139 | | | | | $7.85\times10^{-22}$ | $5.95\times10^{-33}$ | $4.31\times10^{-43}$ | $1.59\times10^{-61}$ |
| 239 | | | | | $3.16\times10^{-23}$ | $2.81\times10^{-35}$ | $2.40\times10^{-46}$ | $1.24\times10^{-66}$ |
| 539 | | | | | $2.46\times10^{-25}$ | $8.63\times10^{-39}$ | $2.90\times10^{-51}$ | $2.33\times10^{-74}$ |

where $F_W(\cdot)$ and $\widetilde{F}_W(\cdot)$ represent, respectively, the exact and the near-exact c.d.f.'s of $W$, and $F_\Lambda(\cdot)$ and $\widetilde{F}_\Lambda(\cdot)$ represent, respectively, the exact and near-exact c.d.f.'s of $\Lambda$.

As we may see from the values of the measure $\Delta$ in Tables 1 and 2, the near-exact distributions exhibit very low values of $\Delta$, showing their great closeness to the exact distribution, even for extremely small sample sizes, as it is the case of the sample sizes of $n = q + r$, which are the smallest possible ones, and even for the smaller values of $\nu$, the number of exact moments of $W$ matched by the near-exact distributions. The values in these Tables also show the asymptotic behavior of these near-exact distributions for increasing values of all other parameters, $q$, $r$, $m$ and $v$, with a more marked asymptotic trend for larger values of $\nu$, that is, with a more accentuated decrease in the values of $\Delta$ for the near-exact distributions that match more of the exact moments of $W$.

Although it is possible to obtain different forms of analytical upper-bounds on $\max_{\{w>0\}}|F_W(w) - \widetilde{F}_W(w)|$ or $\max_{\{0<z<1\}}|F_\Lambda(z) - \widetilde{F}_\Lambda(z)|$, the ones obtained so far are not sharp enough. Given the fact that the measure $\Delta$ in (41) gives a really sharp upper-bound on these quantities, the analysis of the values of this measure in Tables 1 and 2, for different values of the parameters in the distribution, gives a much better idea of the asymptotic characteristics of the near-exact distributions obtained and their closeness to the exact distribution.

For $r = 1$, $m = 1$ and $v = p$, the near-exact distributions in Theorem 3 may be used as near-exact distributions for the statistic studied in Coelho (2017), for even $q$ and odd $p$, exhibiting some advantages, in terms of closeness to the exact distribution, over the near-exact distributions developed in Sect. 5.2 of that reference.

### 3.2.2 The Case of Even $q$, Odd $r$, Odd $m$ and Even $v$

In this case we still have the exact distribution of $\Lambda_2$ and $W_2$ given by Theorem 3.2 in Coelho and Arnold (2019), for $m^* = 1$ and $k_1, n_1, m_1$ and $a_1$ given by (25), while for $W_1$ and $W_3$ we will have to take their c.f.'s given by (35), with $c_k$ given by (20) for $k = 1$ and $k = 3$.

We then have the following Theorem, which gives near-exact distributions for $\Lambda$ when $q$ is even, $r$ and $m$ are odd and $v$ is even.

**Theorem 4** *For even $q$ and odd $r$ and $m$ and even $v$, near-exact distributions for $\Lambda$ will have p.d.f.'s and c.d.f.'s respectively of the form*

$$f_\Lambda(z) = \sum_{\ell=0}^{v} \pi_\ell \, f^{EGIG}\left(z \,\middle|\, \overset{\approx}{\{r_{jk}\}}, 1+\ell; \widetilde{\{\lambda_{jk}\}}, \lambda; \#\widetilde{\{\lambda_{jk}\}}+1\right) \tag{42}$$

*and*

$$F_\Lambda(z) = \sum_{\ell=0}^{v} \pi_\ell \, F^{EGIG}\left(z \,\middle|\, \overset{\approx}{\{r_{jk}\}}, 1+\ell; \widetilde{\{\lambda_{jk}\}}, \lambda; \#\widetilde{\{\lambda_{jk}\}}+1\right), \tag{43}$$

*for $0 < z < 1$, where $f^{EGIG}$ and $F^{EGIG}$ stand, respectively, for the p.d.f. and c.d.f. of the EGIG distribution, using the notation in Appendix B, and where $\lambda_{jk}$ are given by (19) and $r_{jk}$ are given by (36) and (22) for $k = 1, 3$ and by (21)–(22) for $k = 2$, and $\lambda$ is obtained from the numerical solution of the system of equations (46), and where the notations $\widetilde{\{\lambda_{jk}\}}$ and $\overset{\approx}{\{r_{jk}\}}$ are the same as the ones used in Theorem 2.*

***Proof*** Given the remark at the beginning of the present subsection, in order to build near-exact distributions for $\Lambda$ and $W = -\log \Lambda$, for the case of even $q$, odd $r$ and $m$ and even $v$, we will take $\Phi_{W_2}(t)$ given by (28) and $\Phi_{W_1}(t)$ and $\Phi_{W_3}(t)$ given by (35), and write

$$\Phi_W(t) = \underbrace{\Phi_{W_1,1}(t)\,\Phi_{W_2}(t)\,\Phi_{W_3,1}(t)}_{\Phi_1(t)}\;\underbrace{\Phi_{W_1,2}(t)\,\Phi_{W_3,2}(t)}_{\Phi_2(t)}, \tag{44}$$

where, as we did in the previous subsection, we will keep $\Phi_1(t)$ unchanged and will approximate asymptotically $\Phi_2(t)$. We may note that in this case $\Phi_2(t)$ is the c.f. of the sum of two independent Logbeta r.v.'s, one with parameters $\frac{n-2}{2}$ and $\frac{1}{2}$ and the other with parameters $\frac{(v-1)m(n-1)-1}{2}$ and $\frac{1}{2}$, this last one multiplied by $(v-1)m$. As such, we will replace $\Phi_2(t)$ by

$$\widetilde{\Phi}_2(t)=\sum_{\ell=0}^{\nu}\pi_\ell\,\lambda^{1+\ell}(\lambda-\mathrm{i}t)^{-(1+\ell)} \tag{45}$$

where the parameter $\lambda$ is obtained as the rate parameter in a mixture of two Gamma distributions.

This approximation is based on a somewhat heuristic approach, which was already used in similar situations where it proved to be extremely effective (Coelho et al., 2010, 2015; Coelho and Marques, 2013; Coelho and Roy, 2020) and which is based on the fact that from the results in Sect. 5 of Tricomi and Erdélyi (1951) we may asymptotically approximate a $Logbeta(a, b)$ distribution by a mixture of $\Gamma(b+k, a)$ $(k=0,1,\dots)$ distributions. The parameter $\lambda$ in (45) is itself obtained through the numerical solution of the system of four equations

$$\left.\frac{\partial^\ell}{\partial t^\ell}\,\Phi_2(t)\right|_{t=0}=\left.\frac{\partial^\ell}{\partial t^\ell}\,\Phi^*(t)\right|_{t=0},\qquad \ell=1,\dots,4\,, \tag{46}$$

where

$$\Phi^*(t)=\theta\,\lambda^{s_1}(\lambda-\mathrm{i}t)^{-s_1}+(1-\theta)\lambda^{s_2}(\lambda-\mathrm{i}t)^{-s_2}$$

is the c.f. of a mixture of two Gamma distributions, both with the same rate parameter $\lambda$. The weights $\pi_\ell$ $(\ell=0,\dots,\nu)$ are obtained in a similar manner to the one used in Sect. 3.2.1, where now $\widetilde{\Phi}_2(t)$ is given by (45) and where the 1 in $1+\ell$ comes exactly from the sum of the second parameters of the Logbeta r.v.'s in $\Phi_2(t)$.

This way we obtain, for $\Phi_1(t)$ in (44), as near-exact c.f. for $W$

$$\widetilde{\Phi}_W(t)=\Phi_1(t)\,\widetilde{\Phi}_2(t)=\sum_{\ell=0}^{\nu}\pi_\ell\,\Phi_1(t)\,\lambda^{1+\ell}(\lambda-\mathrm{i}t)^{-(1+\ell)} \tag{47}$$

which is the c.f. of a mixture of GIG distributions with integer shape parameters $r_{jk}$ $(k=1,\dots,3;\ j=1,\dots,r+c_k(q-1)-2)$, given by (21) and (22) for $k=2$ and by (36) and (22) for $k=1,3$, and $1+\ell$ $(\ell=0,\dots,\nu)$ and rate parameters $\lambda_{jk}$, given by (19), and $\lambda$.

The corresponding near-exact p.d.f.'s and c.d.f.'s for $W$ are given by (using the notation in Appendix B)

$$f_W(w)=\sum_{\ell=0}^{\nu}\pi_\ell\,f^{GIG}\left(w\,\middle|\,\overset{\approx}{\{r_{jk}\}},\,1+\ell;\,\widetilde{\{\lambda_{jk}\}},\,\lambda;\,\#\widetilde{\{\lambda_{jk}\}}+1\right)$$

and c.d.f.'s given by

$$F_W(w) = \sum_{\ell=0}^{\nu} \pi_\ell \, F^{GIG}\left(w \,\middle|\, \overset{\approx}{\{r_{jk}\}}, 1+\ell; \widetilde{\{\lambda_{jk}\}}, \lambda; \#\widetilde{\{\lambda_{jk}\}}+1\right)$$

for $w > 0$.

The corresponding near-exact distributions for $\Lambda$, may then, once again, be easily obtained through the transformation $\Lambda = e^{-W}$ and will have p.d.f.'s and c.d.f.'s given by (42) and (43) respectively. □

Given the fact that these near-exact distributions are based on the approximation of the distribution of a sum of two independent Logbeta r.v.'s and not on the approximation of the distribution of a single Logbeta r.v., as it happened in the previous subsection, they will yield larger values of the measure $\Delta$ in (41) than those near-exact distributions, anyway still showing very low values of this measure, mainly when four or more of the exact moments of $W$ are matched by these distributions, thus showing also for this case an extreme closeness to the exact distribution. Values of the measure $\Delta$ for these near-exact distributions are shown in Tables 3 and 4.

**Table 3** Values of $\Delta$ for near-exact distributions of $\Lambda$ or $W$ for the case of even $q$, odd $r$, odd $m$ and even $v$, for increasing values of $q$ and $r$ and sample sizes of $r+q+\{0, 100, 200, 500\}$

| | | | | | $\nu$ (number of exact moments matched) | | | |
|---|---|---|---|---|---|---|---|---|
| $n$ | $r$ | $q$ | $m$ | $v$ | 2 | 4 | 6 | 10 |
| 7 | 3 | 4 | 5 | 8 | $2.63\times10^{-8}$ | $9.29\times10^{-13}$ | $1.24\times10^{-15}$ | $2.49\times10^{-21}$ |
| 107 | | | | | $1.24\times10^{-11}$ | $5.32\times10^{-19}$ | $6.49\times10^{-23}$ | $4.87\times10^{-32}$ |
| 207 | | | | | $1.70\times10^{-12}$ | $5.96\times10^{-21}$ | $5.94\times10^{-25}$ | $3.04\times10^{-35}$ |
| 507 | | | | | $1.15\times10^{-13}$ | $3.10\times10^{-23}$ | $1.07\times10^{-27}$ | $1.44\times10^{-39}$ |
| 13 | 9 | 4 | 5 | 8 | $8.85\times10^{-10}$ | $5.43\times10^{-15}$ | $1.44\times10^{-18}$ | $2.58\times10^{-25}$ |
| 113 | | | | | $2.01\times10^{-12}$ | $2.41\times10^{-20}$ | $9.55\times10^{-25}$ | $6.70\times10^{-35}$ |
| 213 | | | | | $3.00\times10^{-13}$ | $3.13\times10^{-22}$ | $1.06\times10^{-26}$ | $5.66\times10^{-38}$ |
| 513 | | | | | $2.14\times10^{-14}$ | $1.93\times10^{-24}$ | $2.16\times10^{-29}$ | $3.26\times10^{-42}$ |
| 23 | 19 | 4 | 5 | 8 | $4.17\times10^{-11}$ | $2.39\times10^{-17}$ | $1.61\times10^{-21}$ | $2.71\times10^{-30}$ |
| 123 | | | | | $5.03\times10^{-13}$ | $2.12\times10^{-21}$ | $3.76\times10^{-26}$ | $4.20\times10^{-37}$ |
| 223 | | | | | $8.48\times10^{-14}$ | $3.39\times10^{-23}$ | $5.59\times10^{-28}$ | $5.60\times10^{-40}$ |
| 523 | | | | | $6.58\times10^{-15}$ | $2.81\times10^{-25}$ | $1.38\times10^{-30}$ | $4.39\times10^{-44}$ |
| 29 | 19 | 10 | 5 | 8 | $3.30\times10^{-12}$ | $2.74\times10^{-19}$ | $3.93\times10^{-24}$ | $3.68\times10^{-34}$ |
| 129 | | | | | $7.80\times10^{-14}$ | $8.84\times10^{-23}$ | $4.85\times10^{-28}$ | $4.54\times10^{-40}$ |
| 229 | | | | | $1.45\times10^{-14}$ | $1.65\times10^{-24}$ | $9.07\times10^{-30}$ | $8.65\times10^{-43}$ |
| 529 | | | | | $1.20\times10^{-15}$ | $1.69\times10^{-26}$ | $2.64\times10^{-32}$ | $8.75\times10^{-47}$ |
| 39 | 19 | 20 | 5 | 8 | $2.95\times10^{-13}$ | $3.61\times10^{-21}$ | $1.32\times10^{-26}$ | $6.54\times10^{-38}$ |
| 139 | | | | | $1.81\times10^{-14}$ | $6.88\times10^{-24}$ | $1.60\times10^{-29}$ | $2.13\times10^{-42}$ |
| 239 | | | | | $3.89\times10^{-15}$ | $1.63\times10^{-25}$ | $4.22\times10^{-31}$ | $6.97\times10^{-45}$ |
| 539 | | | | | $3.61\times10^{-16}$ | $2.36\times10^{-27}$ | $1.59\times10^{-33}$ | $1.06\times10^{-48}$ |

**Table 4** Values of $\Delta$ for near-exact distributions of $\Lambda$ or $W$ for the case of even $q$, odd $r$, odd $m$ and even $v$, for increasing values of $m$ and $v$ and sample sizes of $r+q+\{0, 100, 200, 500\}$

| | | | | | $\nu$ (number of exact moments matched) | | | |
|---|---|---|---|---|---|---|---|---|
| $n$ | $r$ | $q$ | $m$ | $v$ | 2 | 4 | 6 | 10 |
| 39 | 19 | 20 | 5 | 8 | $2.95\times10^{-13}$ | $3.61\times10^{-21}$ | $1.32\times10^{-26}$ | $6.54\times10^{-38}$ |
| 139 | | | | | $1.81\times10^{-14}$ | $6.88\times10^{-24}$ | $1.60\times10^{-29}$ | $2.13\times10^{-42}$ |
| 239 | | | | | $3.89\times10^{-15}$ | $1.63\times10^{-25}$ | $4.22\times10^{-31}$ | $6.97\times10^{-45}$ |
| 539 | | | | | $3.61\times10^{-16}$ | $2.36\times10^{-27}$ | $1.59\times10^{-33}$ | $1.06\times10^{-48}$ |
| 39 | 19 | 20 | 15 | 8 | $7.13\times10^{-14}$ | $1.68\times10^{-22}$ | $4.51\times10^{-28}$ | $3.22\times10^{-40}$ |
| 139 | | | | | $3.60\times10^{-15}$ | $7.41\times10^{-25}$ | $3.46\times10^{-31}$ | $5.17\times10^{-45}$ |
| 239 | | | | | $7.68\times10^{-16}$ | $8.14\times10^{-26}$ | $9.01\times10^{-33}$ | $1.65\times10^{-47}$ |
| 539 | | | | | $7.10\times10^{-17}$ | $1.91\times10^{-27}$ | $3.35\times10^{-35}$ | $2.45\times10^{-51}$ |
| 39 | 19 | 20 | 25 | 8 | $3.49\times10^{-14}$ | $4.09\times10^{-23}$ | $8.41\times10^{-29}$ | $2.29\times10^{-41}$ |
| 139 | | | | | $1.68\times10^{-15}$ | $2.76\times10^{-25}$ | $5.81\times10^{-32}$ | $3.13\times10^{-46}$ |
| 239 | | | | | $3.59\times10^{-16}$ | $2.80\times10^{-26}$ | $1.51\times10^{-33}$ | $9.93\times10^{-49}$ |
| 539 | | | | | $3.31\times10^{-17}$ | $6.33\times10^{-28}$ | $5.60\times10^{-36}$ | $1.47\times10^{-52}$ |
| 39 | 19 | 20 | 25 | 18 | $1.08\times10^{-14}$ | $4.53\times10^{-24}$ | $5.40\times10^{-30}$ | $3.08\times10^{-43}$ |
| 139 | | | | | $5.01\times10^{-16}$ | $4.45\times10^{-26}$ | $3.41\times10^{-33}$ | $3.63\times10^{-48}$ |
| 239 | | | | | $1.07\times10^{-16}$ | $4.31\times10^{-27}$ | $8.82\times10^{-35}$ | $1.15\times10^{-50}$ |
| 539 | | | | | $9.86\times10^{-18}$ | $9.51\times10^{-29}$ | $3.27\times10^{-37}$ | $1.69\times10^{-54}$ |
| 39 | 19 | 20 | 25 | 28 | $5.64\times10^{-15}$ | $1.42\times10^{-24}$ | $1.18\times10^{-30}$ | $2.83\times10^{-44}$ |
| 139 | | | | | $2.58\times10^{-16}$ | $1.54\times10^{-26}$ | $7.26\times10^{-34}$ | $3.20\times10^{-49}$ |
| 239 | | | | | $5.51\times10^{-17}$ | $1.48\times10^{-27}$ | $1.88\times10^{-35}$ | $1.01\times10^{-51}$ |
| 539 | | | | | $5.09\times10^{-18}$ | $3.25\times10^{-29}$ | $6.96\times10^{-38}$ | $1.48\times10^{-55}$ |

From these values we can see that once again these near-exact distributions display a very good performance even for extremely small sample sizes and an asymptotic behavior not only for increasing sample sizes but also for increasing values of all the other parameters in the distributions of $\Lambda$ and $W$, once again with this asymptotic character being more accentuated for larger values of $\nu$, the number of exact moments of $W$ matched by these near-exact distributions.

For $r = 1$, $m = 1$ and $v = p$, the near-exact distributions in Theorem 4 may be used as near-exact distributions for the statistic studied in Coelho (2017), yielding equivalent near-exact distributions to the ones obtained in Sect. 5.1 of that reference.

### 3.2.3 The Case of Even $q$, Odd $r$ and Even $m$

In this case we have the exact distribution of $\Lambda_3$ and $W_3$ given by Theorem 3.2 in Coelho and Arnold (2019), for $m^* = 1$ and $k_1$, $n_1$, $m_1$ and $a_1$ given by (25), while for $W_1$ and $W_2$ we will have to take their c.f.'s given by (35), with $c_k$ given by (20) for $k = 1$ and $k = 2$. Hence we have the following Theorem.

**Theorem 5** *For even $q$, odd $r$ and even $m$ we have near-exact distributions for $\Lambda$ with p.d.f.'s and c.d.f.'s similar to the ones in (42) and (43), now with $r_{jk}$ given by (28) for $k = 3$ and by (36) for $k = 1, 2$.*

***Proof*** For even $q$, odd $r$ and even $m$, in order to build near-exact distributions for $\Lambda$ and $W = -\log \Lambda$ we will follow an approach in all similar to the one taken in the previous subsection, taking into account the remark at the beginning of this subsection and thus taking $\Phi_{W_3}(t)$ given by (28) and $\Phi_{W_1}(t)$ and $\Phi_{W_2}(t)$ given by (35) and writing

$$\Phi_W(t) = \underbrace{\Phi_{W_{1,1}}(t)\, \Phi_{W_{2,1}}(t)\, \Phi_{W_3}(t)}_{\Phi_1(t)}\ \underbrace{\Phi_{W_{1,2}}(t)\, \Phi_{W_{2,2}}(t)}_{\Phi_2(t)}, \tag{48}$$

where, as we did in the previous subsections, we will once again keep $\Phi_1(t)$ unchanged and will approximate asymptotically $\Phi_2(t)$. As we said, we will take an approach in all similar to the one taken in the previous subsection. In the present case $\Phi_2(t)$ is the c.f. of the sum of two independent Logbeta r.v.'s, one with parameters $\frac{n-2}{2}$ and $\frac{1}{2}$ and the other with parameters $\frac{(m-1)(n-1)-1}{2}$ and $\frac{1}{2}$, this last one multiplied by $(m-1)$.

So, following the same approach as we did in the previous subsection, we will replace $\Phi_2(t)$ by $\widetilde{\Phi}_2(t)$, with a similar structure to the one in (45) and with $\lambda$ obtained in a similar manner to the one described in Sect. 3.2.2, and where the 1 in $1 + \ell$ comes once again from the sum of the second parameters of the Logbeta r.v.'s in $\Phi_2(t)$.

The weights $\pi_\ell$ ($\ell = 0, \ldots, \nu$) are obtained in a similar manner to the one used in Sects. 3.2.1 and 3.2.2, and in this way we obtain, for $\Phi_1(t)$ in (48), as near-exact c.f. for $W$ a c.f. with a similar expression to the one in (47), which is, as in the previous subsection, the c.f. of a mixture of GIG distributions with integer shape parameters $r_{jk}$ ($k = 1, \ldots, 3$; $j = 1, \ldots, r + c_k(q-1) - 2$), but now given by (21) and (22) for $k = 3$ and by (36) and (22) for $k = 1, 2$, and $1 + \ell$ ($\ell = 0, \ldots, \nu$), and rate parameters $\lambda_{jk}$, given by (19), and $\lambda$, which now, although being computed in a somewhat similar manner, will have different values than the ones obtained in the previous subsection.

The corresponding near-exact p.d.f.'s and c.d.f.'s for $W$ and $\Lambda$ will have similar expressions to the ones in the previous subsection, now with the differences in the parameters pointed out above. □

In Tables 5 and 6 we may analyze the values of the measure $\Delta$ in (41) for these near-exact distributions. We may see how generally they have quite similar values to the ones in Tables 3 and 4 for the near-exact distributions in the previous subsection, which was indeed expected, given the similitude in structure and building approach between the near-exact distributions obtained in the present subsection and the ones obtained in the previous subsection. Once again, these near-exact distributions show their asymptotic character not only for increasing sample sizes but also for increasing values of all other parameters, that is, for increasing values of the overall

**Table 5** Values of $\Delta$ for near-exact distributions of $\Lambda$ or $W$ for the case of even $q$, odd $r$ and even $m$, for increasing values of $q$ and $r$ and sample sizes of $r+q+\{0, 100, 200, 500\}$

| | | | | | $\nu$ (number of exact moments matched) | | | |
|---|---|---|---|---|---|---|---|---|
| $n$ | $r$ | $q$ | $m$ | $\upsilon$ | 2 | 4 | 6 | 10 |
| 7 | 3 | 4 | 6 | 8 | $2.03\times10^{-8}$ | $1.09\times10^{-12}$ | $2.32\times10^{-15}$ | $1.02\times10^{-20}$ |
| 107 | | | | | $7.57\times10^{-12}$ | $4.51\times10^{-18}$ | $3.82\times10^{-23}$ | $2.00\times10^{-32}$ |
| 207 | | | | | $1.03\times10^{-12}$ | $1.59\times10^{-19}$ | $3.59\times10^{-25}$ | $1.33\times10^{-35}$ |
| 507 | | | | | $6.96\times10^{-14}$ | $1.75\times10^{-21}$ | $6.54\times10^{-28}$ | $6.63\times10^{-40}$ |
| 13 | 9 | 4 | 6 | 8 | $6.47\times10^{-10}$ | $7.60\times10^{-15}$ | $5.35\times10^{-19}$ | $1.74\times10^{-25}$ |
| 113 | | | | | $1.23\times10^{-12}$ | $2.20\times10^{-19}$ | $5.63\times10^{-25}$ | $2.73\times10^{-35}$ |
| 213 | | | | | $1.82\times10^{-13}$ | $8.90\times10^{-21}$ | $6.39\times10^{-27}$ | $2.46\times10^{-38}$ |
| 513 | | | | | $1.29\times10^{-14}$ | $1.07\times10^{-22}$ | $1.32\times10^{-29}$ | $1.48\times10^{-42}$ |
| 23 | 19 | 4 | 6 | 8 | $2.95\times10^{-11}$ | $4.97\times10^{-17}$ | $9.33\times10^{-22}$ | $4.47\times10^{-32}$ |
| 123 | | | | | $3.08\times10^{-13}$ | $2.18\times10^{-20}$ | $2.23\times10^{-26}$ | $1.74\times10^{-37}$ |
| 223 | | | | | $5.15\times10^{-14}$ | $1.09\times10^{-21}$ | $3.37\times10^{-28}$ | $2.44\times10^{-40}$ |
| 523 | | | | | $3.98\times10^{-15}$ | $1.50\times10^{-23}$ | $8.45\times10^{-31}$ | $1.99\times10^{-44}$ |
| 29 | 19 | 10 | 6 | 8 | $2.25\times10^{-12}$ | $6.58\times10^{-19}$ | $2.31\times10^{-24}$ | $9.77\times10^{-35}$ |
| 129 | | | | | $4.77\times10^{-14}$ | $9.75\times10^{-22}$ | $2.88\times10^{-28}$ | $1.88\times10^{-40}$ |
| 229 | | | | | $8.79\times10^{-15}$ | $5.71\times10^{-23}$ | $5.47\times10^{-30}$ | $3.77\times10^{-43}$ |
| 529 | | | | | $7.27\times10^{-16}$ | $8.84\times10^{-25}$ | $1.61\times10^{-32}$ | $3.97\times10^{-47}$ |
| 39 | 19 | 20 | 6 | 8 | $1.95\times10^{-13}$ | $1.10\times10^{-20}$ | $7.92\times10^{-27}$ | $2.34\times10^{-38}$ |
| 139 | | | | | $1.11\times10^{-14}$ | $8.51\times10^{-23}$ | $9.50\times10^{-30}$ | $8.89\times10^{-43}$ |
| 239 | | | | | $2.36\times10^{-15}$ | $6.38\times10^{-24}$ | $2.54\times10^{-31}$ | $3.04\times10^{-45}$ |
| 539 | | | | | $2.18\times10^{-16}$ | $1.19\times10^{-25}$ | $9.67\times10^{-34}$ | $4.79\times10^{-49}$ |

number of variables involved in the test, at the same time that they keep showing very good performances even for very small sample sizes.

For $r = 1$, $\upsilon = 1$ and $m = p$, the near-exact distributions in Theorem 5 will yield for the statistic studied in Coelho (2017), similar near-exact distributions to the ones obtained for this statistic from the results in Theorem 4, and thus also similar to the ones developed in Sect. 5.1 of that reference, for even $q$ and even $p$.

## 4 Conclusions and Discussion

The problem addressed seemed to pose a couple of interesting challenges. The first of these challenges was the question of the approach that had to be followed in order to be able to express the distribution of the LRT statistic in the form of that of a product of independent Beta distributed r.v.'s. Then, there was the question of the possible existence of several scenarios to be considered, in terms of different combinations of values of the five parameters in the distribution of the LRT statistic, and if for any of them it would be possible to obtain the distribution of the LRT

**Table 6** Values of $\Delta$ for near-exact distributions of $\Lambda$ or $W$ for the case of even $q$, odd $r$ and even $m$, for increasing values of $m$ and $\upsilon$ and sample sizes of $r+q+\{0, 100, 200, 500\}$

| | | | | | $\nu$ (number of exact moments matched) | | | |
|---|---|---|---|---|---|---|---|---|
| $n$ | $r$ | $q$ | $m$ | $\upsilon$ | 2 | 4 | 6 | 10 |
| 39 | 19 | 20 | 6 | 8 | $1.95\times10^{-13}$ | $1.10\times10^{-20}$ | $7.92\times10^{-27}$ | $2.34\times10^{-38}$ |
| 139 | | | | | $1.11\times10^{-14}$ | $8.51\times10^{-23}$ | $9.50\times10^{-30}$ | $8.89\times10^{-43}$ |
| 239 | | | | | $2.36\times10^{-15}$ | $6.38\times10^{-24}$ | $2.54\times10^{-31}$ | $3.04\times10^{-45}$ |
| 539 | | | | | $2.18\times10^{-16}$ | $1.19\times10^{-25}$ | $9.67\times10^{-34}$ | $4.79\times10^{-49}$ |
| 39 | 19 | 20 | 16 | 8 | $6.23\times10^{-14}$ | $5.43\times10^{-22}$ | $3.72\times10^{-28}$ | $2.25\times10^{-40}$ |
| 139 | | | | | $3.08\times10^{-15}$ | $2.29\times10^{-24}$ | $2.89\times10^{-31}$ | $3.88\times10^{-45}$ |
| 239 | | | | | $6.56\times10^{-16}$ | $1.56\times10^{-25}$ | $7.57\times10^{-33}$ | $1.26\times10^{-47}$ |
| 539 | | | | | $6.05\times10^{-17}$ | $2.67\times10^{-27}$ | $2.83\times10^{-35}$ | $1.90\times10^{-51}$ |
| 39 | 19 | 20 | 26 | 8 | $3.21\times10^{-14}$ | $1.16\times10^{-22}$ | $7.46\times10^{-29}$ | $1.84\times10^{-41}$ |
| 139 | | | | | $1.53\times10^{-15}$ | $2.90\times10^{-25}$ | $5.21\times10^{-32}$ | $2.63\times10^{-46}$ |
| 239 | | | | | $3.27\times10^{-16}$ | $1.62\times10^{-26}$ | $1.36\times10^{-33}$ | $8.45\times10^{-49}$ |
| 539 | | | | | $3.01\times10^{-17}$ | $2.19\times10^{-28}$ | $5.04\times10^{-36}$ | $1.26\times10^{-52}$ |
| 39 | 19 | 20 | 26 | 18 | $9.91\times10^{-15}$ | $1.63\times10^{-23}$ | $4.78\times10^{-30}$ | $2.45\times10^{-43}$ |
| 139 | | | | | $4.55\times10^{-16}$ | $3.83\times10^{-26}$ | $3.06\times10^{-33}$ | $3.07\times10^{-48}$ |
| 239 | | | | | $9.69\times10^{-17}$ | $2.14\times10^{-27}$ | $7.95\times10^{-35}$ | $9.81\times10^{-51}$ |
| 539 | | | | | $8.94\times10^{-18}$ | $2.89\times10^{-29}$ | $2.96\times10^{-37}$ | $1.46\times10^{-54}$ |
| 39 | 19 | 20 | 26 | 28 | $5.17\times10^{-15}$ | $5.51\times10^{-24}$ | $1.05\times10^{-30}$ | $2.25\times10^{-44}$ |
| 139 | | | | | $2.35\times10^{-16}$ | $1.27\times10^{-26}$ | $6.53\times10^{-34}$ | $2.71\times10^{-49}$ |
| 239 | | | | | $4.99\times10^{-17}$ | $7.09\times10^{-28}$ | $1.70\times10^{-35}$ | $8.65\times10^{-52}$ |
| 539 | | | | | $4.61\times10^{-18}$ | $9.58\times10^{-30}$ | $6.30\times10^{-38}$ | $1.29\times10^{-55}$ |

statistic in a closed finite and manageable form and yet if for the other cases, using the so-called near-exact distributions, it would be possible to build approximations that would be asymptotic for all five parameters involved in the distribution of $\Lambda$, while performing very well even for small samples.

The way around that was taken in trying to express the distribution of $\Lambda$ in terms of that of a product of independent Beta r.v.'s, not taking an approach directly based on the expression for $\Lambda$ in (4) but rather using a different approach by rewriting the determinants that appear in the numerator and denominator in that expression in terms of the diagonal blocks of other matrices which have a Wishart distribution with a block-diagonal parameter matrix, proved itself very effective. Indeed somewhat similar approaches had already proven effective in somewhat similar situations, but in our case the proof of the relation in (16) or (17) was quite a challenge.

Then, once the distribution of $\Lambda$ was obtained in the form of that of a product of independent Beta distributed r.v.'s, Theorem 3.2 in Coelho and Arnold (2019) proved itself extremely useful in being able to very quickly and quite easily obtain the exact distribution of $\Lambda_1$, $\Lambda_2$ and $\Lambda_3$ and as such also of $\Lambda$ in a closed finite and manageable form for the cases where $q$ is odd or $r$ is even.

It remained then the problem of building near-exact distributions for the remaining cases. A detailed check of these cases showed that there were indeed three sub-cases to be taken into account and which should be considered individually since this would allow for a more dedicated approach to each sub-case. This allowed us to see that in each of these three sub-cases there was always one or two of the partial statistics $\Lambda_1$, $\Lambda_2$ or $\Lambda_3$, for which we could obtain the exact distribution in a closed finite form, while for the other or others it was possible to obtain a decomposition of the c.f. of its or their logarithm which might lead to very sharp near-exact distributions. Typically these distributions, when properly built, are supposed to be asymptotic for increasing values of all the parameters involved in the exact distribution, but anyway this was a challenge for a distribution that has five parameters, four of which related with the number of variables in the analysis. Indeed, the approach followed proved effective in obtaining near-exact distributions that not only perform very well for very small samples, but which besides that are asymptotic not only for increasing sample sizes but also for increasing values of all other parameters, which are related with the overall number of variables involved in the testing procedure.

Given the ease of computation for the form of the exact distribution for the cases where $q$ is odd or $r$ is even, which are indeed the majority of the cases, and given the quite easy computability of the near-exact distributions obtained, the authors feel that the results in the present work may not only be useful for practical applications but also they may encourage the pursual of the obtention of similar results for situations that at first may seem very complicated or very challenging to approach, while the techniques used showed to be quite effective in addressing the problem and as such seem to be possible to be applied in a wide range of situations.

The results in Sect. 11 of Chap. 8 of Anderson (2003), namely the one in Theorem 8.11.3, together with the properties of the double exchangeable covariance matrix, namely the one that makes it part of a quadratic space, allow us to foresee that the present test statistic should have the same null distribution for elliptically contoured and left-spherical distributions.

Taking into account the particular cases of the double exchangeable covariance structure enumerated in the Introduction, for $r = 1$ and $m = 1$ or $v = 1$ we have the test of equality of mean vectors with compound symmetric matrices, and for just $m = 1$ or $v = 1$, the test of equality of mean vectors for block compound symmetric covariance matrices, while for $r = 1$ we have the same test for double complete symmetric covariance matrices.

**Acknowledgments** This work is funded by Portuguese National Funds through FCT - Fundação para a Ciência e a Tecnologia, I.P., under the scope of the project UIDB/00297/ 2020 (CMA-FCT/UNL – Center for Mathematics and Applications)

## Appendix A: Proof of Expression (16)

For ease of exposition let us split the proof of (16) in the proof of three consecutive results. First let us prove the following result.
*Result 1*: The MLE of $\Sigma$ under $H_1$, which is $A^*$ in (8), may be equivalently written as

$$A^* = P_{vm} \otimes \widehat{\Psi}_{1|H_1} + Q_{vm} \otimes \widehat{\Psi}_{2|H_1} + R_{vm} \otimes \widehat{\Psi}_{3|H_1}, \tag{A.1}$$

where, from (7),

$$\begin{aligned} \widehat{\Psi}_{1|H_1} &= \widehat{U}_{|H_1} + (v-1)\widehat{V}_{|H_1} + v(m-1)\widehat{M}_{|H_1} \\ \widehat{\Psi}_{2|H_1} &= \widehat{U}_{|H_1} + (v-1)\widehat{V}_{H_1} - v\widehat{M}_{|H_1} \\ \widehat{\Psi}_{3|H_1} &= \widehat{U}_{|H_1} - \widehat{V}_{|H_1}\,, \end{aligned} \tag{A.2}$$

and

$$\begin{aligned} P_{vm} &= \frac{1}{mv} J_{vm} \\ Q_{vm} &= \frac{1}{v}(I_m \otimes J_v) - P_{vm} \\ R_{vm} &= I_{vm} - Q_{vm} - P_{vm}\,. \end{aligned} \tag{A.3}$$

***Proof*** The proof is easier if we start from (A.1) and try to end with (8), by using (A.2) and (A.3) to write

$$\begin{aligned}
&P_{vm} \otimes \widehat{\Psi}_{1|H_1} + Q_{vm} \otimes \widehat{\Psi}_{2|H_1} + R_{vm} \otimes \widehat{\Psi}_{3|H_1} \\
&\quad = \frac{1}{mv} J_{vm} \otimes \left(\widehat{U}_{|H_1} + (v-1)\widehat{V}_{|H_1} + v(m-1)\widehat{M}_{|H_1}\right) \\
&\qquad + \left(\frac{1}{v}(I_m \otimes J_v) - \frac{1}{mv} J_{vm}\right) \otimes \left(\widehat{U}_{|H_1} + (v-1)\widehat{V}_{|H_1} - v\widehat{M}_{|H_1}\right) \\
&\qquad + \left(I_{vm} - \tfrac{1}{v}(I_m \otimes J_v) + P_{vm} - P_{vm}\right) \otimes \left(\widehat{U}_{|H_1} - \widehat{V}_{|H_1}\right) \\
&\quad = \left(\frac{1}{mv} J_{vm} + \frac{1}{v}(I_m \otimes J_v) - \frac{1}{mv} J_{vm} + I_{vm} - \frac{1}{v}(I_m \otimes J_v)\right) \otimes \widehat{U}_{|H_1} \\
&\qquad + \left(\frac{v-1}{mv} J_{vm} + \frac{v-1}{v}(I_m \otimes J_v) - \frac{v-1}{mv} J_{vm} - I_{vm} + \frac{1}{v}(I_m \otimes J_v)\right) \otimes \widehat{V}_{|H_1} \\
&\qquad + \left(\frac{m-1}{m} J_{vm} - (I_m \otimes J_v) + \frac{1}{m} J_{vm}\right) \otimes \widehat{M}_{|H_1} \\
&\quad = I_{vm} \otimes \widehat{U}_{|H_1} + (I_m \otimes (J_v - I_v)) \otimes \widehat{V}_{|H_1} + (J_{vm} - (I_m \otimes J_m)) \otimes \widehat{M}_{|H_1}
\end{aligned}$$

which is (8). □

We now prove that $P_{vm}$, $Q_{vm}$ and $R_{vm}$ in (A.3) are indeed mutually orthogonal projectors.

*Result 2*: $P_{vm}$, $Q_{vm}$ and $R_{vm}$ in (A.3) are mutually orthogonal projectors with $\text{rank}(P_{vm}) = 1$, $\text{rank}(Q_{vm}) = m - 1$, $\text{rank}(R_{vm}) = m(v - 1)$.

***Proof*** $P_{vm}$, $Q_{vm}$ and $R_{vm}$ in (A.3) are idempotent, since

$$P_{vm}P_{vm} = \frac{1}{mv}J_{vm}\frac{1}{mv}J_{vm} = \frac{1}{(mv)^2}mvJ_{vm} = \frac{1}{mv}J_{vm} = P_{vm},$$

$$\begin{aligned} Q_{vm}Q_{vm} &= \left(\frac{1}{v}(I_m \otimes J_v) - P_{vm}\right)\left(\frac{1}{v}(I_m \otimes J_v) - P_{vm}\right) \\ &= \frac{1}{v^2}(I_m \otimes J_v)(I_m \otimes J_v) - \frac{1}{v}(I_m \otimes J_v)P_{vm} - \frac{1}{v}P_{vm}(I_m \otimes J_v) + P_{vm} \\ &= \frac{1}{v}(I_m \otimes J_v) - \frac{2}{vm}J_{vm} + \frac{1}{vm}J_{vm} = \frac{1}{v}(I_m \otimes J_v) - P_{vm} = Q_{vm}, \end{aligned}$$

$$\begin{aligned} R_{vm}R_{vm} &= (I_{vm} - Q_{vm} - P_{vm})(I_{vm} - Q_{vm} - P_{vm}) = I_{vm} - Q_{vm} - P_{vm} \\ &\quad - Q_{vm} + Q_{vm}Q_{vm} + Q_{vm}P_{vm} - P_{vm} + P_{vm}Q_{vm} + P_{vm}P_{vm} \\ &= I_{vm} - Q_{vm} - P_{vm} - Q_{vm} + Q_{vm} - P_{vm} + P_{vm} \\ &= I_{vm} - Q_{vm} - P_{vm} = R_{vm}, \end{aligned}$$

and they are mutually orthogonal since

$$\begin{aligned} P_{vm}Q_{vm} &= P_{vm}\left(\tfrac{1}{v}(I_m \otimes J_v) - P_{vm}\right) = \tfrac{1}{v}P_{vm}(I_m \otimes J_v) - P_{vm} \\ &= \frac{1}{vm}J_{vm}\left(\frac{1}{v}(I_m \otimes J_v) - I_{vm}\right) = 0_{vm}, \\ R_{vm}P_{vm} &= (I_{vm} - Q_{vm} - P_{vm})P_{vm} = P_{vm} - P_{vm} = 0_{vm}, \\ R_{vm}Q_{vm} &= (I_{vm} - Q_{vm} - P_{vm})Q_{vm} = Q_{vm} - Q_{vm} = 0_{vm}. \end{aligned}$$

Since $P_{vm}$, $Q_{vm}$ and $R_{vm}$ are idempotent, we have

$$\begin{aligned} \text{rank}(P_{vm}) &= \text{tr}(P_{vm}) = \frac{1}{vm}\text{tr}(J_{vm}) = 1 \\ \text{rank}(Q_{vm}) &= \text{tr}(Q_{vm}) = \frac{1}{v}\text{tr}(I_m \otimes J_v) - \text{tr}(P_{vm}) = \frac{mv}{v} - 1 = m - 1 \\ \text{rank}(R_{vm}) &= \text{tr}(R_{vm}) = \text{tr}(I_{vm}) - \text{tr}(Q_{vm}) - \text{tr}(P_{vm}) = vm - m = (v - 1)m. \end{aligned}$$

□

Setting together the previous two results with the next one, we will finally be able to prove (16).

*Result 3*: For the matrix $A^{**}$ in (14), we may write

$$\begin{aligned}
\widehat{\Psi}_{1|H_1} &= A_1^{**}\,,\\
\widehat{\Psi}_{2|H_1} &= \frac{1}{m-1}\sum_{j=1}^{m-1} A_{jv+1}^{**}\,,\\
\widehat{\Psi}_{3|H_1} &= \frac{1}{m(v-1)}\sum_{j=1}^{m}\sum_{k=1}^{v-1} A_{(j-1)v+k+1}^{**}\,.
\end{aligned}$$

***Proof*** Let $\mathrm{BTr}_r(A)$ denote the sum of all $r\times r$ diagonal blocks of $A$. Then, for $U_{|H_1}$, $V_{|H_1}$ and $W_{|H_1}$ in (9)–(11) we may write

$$\begin{aligned}
\widehat{U}_{|H_1} &= \frac{1}{mv}\mathrm{BTr}_r(A),\\
\widehat{V}_{|H_1} &= \frac{1}{mv(v-1)}\left(\mathrm{BTr}_r((I_m\otimes J_v\otimes I_r)A)-\mathrm{BTr}_r(A)\right),\\
\widehat{M}_{|H_1} &= \frac{1}{mv^2(m-1)}\left(\mathrm{BTr}_r((J_{vm}\otimes I_r)A)-\mathrm{BTr}_r((I_m\otimes J_v\otimes I_r)A)\right).
\end{aligned}$$

Then, since from (14) we may write

$$A=\Gamma' A^{**}\Gamma\,, \tag{A.4}$$

from (A.2), and from the definition of the matrix $\Gamma$ in (5), we have

$$\begin{aligned}
\widehat{\Psi}_{1|H_1} &= \widehat{U}_{|H_1}+(v-1)\widehat{V}_{|H_1}+v(m-1)\widehat{M}_{|H_1}\\
&= \frac{1}{mv}\mathrm{BTr}_r(A)+\frac{1}{mv}\left(\mathrm{BTr}_r((I_m\otimes J_v\otimes I_r)A)-\mathrm{BTr}_r(A)\right)\\
&\quad+\frac{1}{mv}\left(\mathrm{BTr}_r((J_{vm}\otimes I_r)A)-\mathrm{BTr}_r((I_m\otimes J_v\otimes I_r)A)\right)\\
&= \frac{1}{mv}\mathrm{BTr}_r((J_{vm}\otimes I_r)A)=\mathrm{BTr}_r((P_{vm}\otimes I_r)A)\\
&= \mathrm{BTr}_r((P_{vm}\otimes I_r)\Gamma' A^{**}\Gamma)=\mathrm{BTr}_r(\Gamma(P_{vm}\otimes I_r)\Gamma' A^{**})\\
&= \mathrm{BTr}_r((\underbrace{((\Gamma_m\otimes\Gamma_v)P_{vm}(\Gamma_m\otimes\Gamma_v)')}_{=diag(\underbrace{1,0,\ldots,0}_{vm})}\otimes I_r)A^{**})=A_1^{**}\,,
\end{aligned}$$

where $A_1^{**}$ represents the first diagonal block of $A^{**}$ of dimensions $r\times r$, since, given the definition of a Helmert matrix, $\Gamma_m \otimes \Gamma_v$ is a $vm\times vm$ matrix whose first row is equal to $\frac{1}{\sqrt{vm}}\underline{1}'_{vm}$, where $\underline{1}_{vm}$ denotes a vector of 1's of dimension $vm$, and as such $(\Gamma_m \otimes \Gamma_v)P_{vm}$ is a $vm\times vm$ matrix whose first row is equal to $\frac{1}{\sqrt{vm}}\underline{1}'_{vm}$ and all other rows are null, so that

$$(\Gamma_m \otimes \Gamma_v)P_{vm}(\Gamma_m \otimes \Gamma_v)' = diag(\underbrace{1, 0, \ldots, 0}_{vm}), \tag{A.5}$$

what implies

$$((\Gamma_m \otimes \Gamma_v)P_{vm}(\Gamma_m \otimes \Gamma_v)') \otimes I_r = bdiag(I_r, 0_{v(m-1)\times v(m-1)}).$$

Using (A.4), and once again the definition of the matrix $\Gamma$ in (5), we also have

$$\begin{aligned}
\widehat{\Psi}_{2|H_1} &= \widehat{U}_{|H_1} + (v-1)\widehat{V}_{|H_1} - v\widehat{M}_{|H_1} \\
&= \frac{1}{mv}\mathrm{BTr}_r(A) + \frac{1}{mv}(\mathrm{BTr}_r((I_m \otimes J_v \otimes I_r)A) - \mathrm{BTr}_r(A)) \\
&\quad - \frac{1}{mv(m-1)}(\mathrm{BTr}_r((J_{vm} \otimes I_r)A) - \mathrm{BTr}_r((I_m \otimes J_v \otimes I_r)A)) \\
&= \frac{1}{mv}\left(1 + \frac{1}{m-1}\right)\mathrm{BTr}_r((I_m \otimes J_v \otimes I_r)A) \\
&\quad - \frac{1}{mv(m-1)}\mathrm{BTr}_r((J_{vm} \otimes I_r)A) \\
&= \frac{1}{m-1}\mathrm{BTr}_r\left(\left(\left(\frac{1}{v}(I_m \otimes J_v) - \frac{1}{mv}J_{vm}\right) \otimes I_r\right) A\right) \\
&= \frac{1}{m-1}\mathrm{BTr}_r\left((Q_{vm} \otimes I_r)\, A\right) = \frac{1}{m-1}\mathrm{BTr}_r((Q_{vm} \otimes I_r)\Gamma' A^{**}\Gamma) \\
&= \frac{1}{m-1}\mathrm{BTr}_r(\Gamma(Q_{vm} \otimes I_r)\Gamma' A^{**}) \\
&= \frac{1}{m-1}\mathrm{BTr}_r((((\Gamma_m \otimes \Gamma_v)Q_{vm}(\Gamma_m \otimes \Gamma_v)) \otimes I_r)A^{**}) \\
&= \frac{1}{m-1}\sum_{j=1}^{m-1} A^{**}_{jv+1},
\end{aligned}$$

where $A_j^{**}$ represents the $j$-th diagonal block of $A^{**}$ of dimensions $r\times r$, since

$$(\Gamma_m\otimes\Gamma_v)Q_{vm}(\Gamma_m\otimes\Gamma_v)' = diag(\underbrace{0,\ldots,0}_{v},\underbrace{\underbrace{1,0,\ldots,0}_{v},\ldots,\underbrace{1,0,\ldots,0}_{v}}_{m-1 \text{ times}}) \qquad (A.6)$$

and as such,

$$\begin{aligned}((\Gamma_m\otimes\Gamma_v)Q_{vm}(\Gamma_m\otimes\Gamma_v)')\otimes I_r \\ = bdiag(I_{vr\times vr},\underbrace{\underbrace{I_r,0_{(v-1)r\times(v-1)r}},\ldots,\underbrace{I_r,0_{(v-1)r\times(v-1)r}}}_{m-1}).\end{aligned}$$

This is so because, given the definition of a Helmert matrix, for a $v\times v$ Helmert matrix $\Gamma_v$, we have that $\Gamma_v J_v$ is a matrix whose first row is equal to $\frac{1}{\sqrt{v}}\underline{1}'_v$, and as such we have

$$\Gamma_v J_v \Gamma_v' = diag(\underbrace{v,0,\ldots,0}_{v}),$$

so that we have

$$\begin{aligned}&\frac{1}{v}\left(I_m\otimes\Gamma_v J_v\Gamma_v'\right)\\ &\quad= bdiag(\underbrace{diag(\underbrace{1,0,\ldots,0}_{v}),diag(\underbrace{1,0,\ldots,0}_{v}),\ldots,diag(\underbrace{1,0,\ldots,0}_{v})}_{m}))\\ &\quad= diag(\underbrace{\underbrace{1,0,\ldots,0}_{v},\underbrace{1,0,\ldots,0}_{v},\ldots,\underbrace{1,0,\ldots,0}_{v}}_{m})\end{aligned}$$

and as such we may write, using the relation in (A.5),

$$\begin{aligned}&(\Gamma_m\otimes\Gamma_v)Q_{vm}(\Gamma_m\otimes\Gamma_v)'\\ &\quad=(\Gamma_m\otimes\Gamma_v)\left(\frac{1}{v}(I_m\otimes J_v)-p_{vm}\right)(\Gamma_m\otimes\Gamma_v)'\\ &\quad=\left(\frac{1}{v}(\Gamma_m\otimes\Gamma_v J_v)-(\Gamma_m\otimes\Gamma_v)P_{vm}\right)(\Gamma_m\otimes\Gamma_v)'\\ &\quad=\frac{1}{v}(\Gamma_m\otimes\Gamma_v J_v)(\Gamma_m\otimes\Gamma_v)'-(\Gamma_m\otimes\Gamma_v)P_{vm}(\Gamma_m\otimes\Gamma_v)'\\ &\quad=\frac{1}{v}(\Gamma_m\Gamma_m'\otimes\Gamma_v J_v\Gamma_v')-(\Gamma_m\otimes\Gamma_v)P_{vm}(\Gamma_m\otimes\Gamma_v)'\end{aligned}$$

$$
\begin{aligned}
&= \frac{1}{v}(I_m \otimes \Gamma_v J_v \Gamma_v')) - diag(\underbrace{1, 0, \ldots, 0}_{vm}) \\
&= diag(\underbrace{\underbrace{1, 0, \ldots, 0}_{v}, \underbrace{1, 0, \ldots, 0}_{v}, \ldots, \underbrace{1, 0, \ldots, 0}_{v}}_{m}) - diag(\underbrace{1, 0, \ldots, 0}_{vm}) \\
&= diag(\underbrace{0, \ldots, 0}_{v}, \underbrace{\underbrace{1, 0, \ldots, 0}_{v}, \ldots, \underbrace{1, 0, \ldots, 0}_{v}}_{m-1 \text{ times}}) .
\end{aligned}
$$

And finally, using once again (A.4), and the definition of $R_{vm}$ in (A.3), we have

$$
\begin{aligned}
\widehat{\Psi}_{3|H_1} &= \widehat{U}_{|H_1} - \widehat{V}_{|H_1} \\
&= \frac{1}{mv}\mathrm{BTr}_r(A) - \frac{1}{mv(v-1)}\left(\mathrm{BTr}_r((I_m \otimes J_v \otimes I_r)A) - \mathrm{BTr}_r(A)\right) \\
&= \frac{1}{mv}\left(1 + \frac{1}{v-1}\right)\mathrm{BTr}_r(A) - \frac{1}{mv(v-1)}\mathrm{BTr}_r((I_m \otimes J_v \otimes I_r)A) \\
&= \frac{1}{m(v-1)}\left(\mathrm{BTr}_r(A) - \frac{1}{v}\mathrm{BTr}_r((I_m \otimes J_v \otimes I_r)A)\right) \\
&= \frac{1}{m(v-1)}\mathrm{BTr}_r\left((I_{mvr} - \frac{1}{v}(I_m \otimes J_v \otimes I_r))A\right) \\
&= \frac{1}{m(v-1)}\mathrm{BTr}_r\left(((I_{mv} - \frac{1}{v}(I_m \otimes J_v)) \otimes I_r)A\right) \\
&= \frac{1}{m(v-1)}\mathrm{BTr}_r\left((R_{vm} \otimes I_r)A\right) \\
&= \frac{1}{m(v-1)}\mathrm{BTr}_r\left((R_{vm} \otimes I_r)\Gamma' A^{**} \Gamma\right) \\
&= \frac{1}{m(v-1)}\mathrm{BTr}_r\left(\Gamma(R_{vm} \otimes I_r)\Gamma' A^{**}\right) \\
&= \frac{1}{m(v-1)}\mathrm{BTr}_r((((\Gamma_m \otimes \Gamma_v)R_{vm}(\Gamma_m \otimes \Gamma_v)') \otimes I_r)A^{**}) \\
&= \frac{1}{m(v-1)}\sum_{j=1}^{m}\sum_{k=1}^{v-1} A^{**}_{(j-1)v+k+1} ,
\end{aligned}
$$

where, once again, $A_j^{**}$ represents the $j$-th diagonal block of $A^{**}$ of dimensions $r\times r$.

This is so given that from the definition of $R_{vm}$ in (A.3) and from (A.5) and (A.6) it is clear that we have

$$
\begin{aligned}
&(\Gamma_m \otimes \Gamma_v)R_{vm}(\Gamma_m \otimes \Gamma_v)' \\
&\quad = (\Gamma_m \otimes \Gamma_v)(\Gamma_m \otimes \Gamma_v)' - (\Gamma_m \otimes \Gamma_v)Q_{vm}(\Gamma_m \otimes \Gamma_v)' \\
&\qquad\qquad\qquad\qquad\qquad\qquad - (\Gamma_m \otimes \Gamma_v)P_{vm}(\Gamma_m \otimes \Gamma_v)' \\
&\quad = I_{vm} - diag(\underbrace{0,\ldots,0}_{v},\underbrace{\underbrace{1,0,\ldots,0}_{v},\ldots,\underbrace{1,0,\ldots,0}_{v}}_{m-1 \text{ times}}) - diag(\underbrace{1,0,\ldots,0}_{vm}) \\
&\quad = diag(\underbrace{0,\underbrace{1,\ldots,1}_{v-1},0,\underbrace{1,\ldots,1}_{v-1},\ldots,0,\underbrace{1,\ldots,1}_{v-1}}_{m \text{ times}}),
\end{aligned}
$$

so that

$$
\begin{aligned}
&((\Gamma_m \otimes \Gamma_v)R_{vm}(\Gamma_m \otimes \Gamma_v)') \otimes I_r \\
&\qquad = bdiag(\underbrace{0_{r\times r}, I_{(v-1)r}, 0_{r\times r}, I_{(v-1)r}, \ldots, 0_{r\times r}, I_{(v-1)r}}_{m \text{ times}}). \quad \square
\end{aligned}
$$

Finally, expression (16) holds since from Result 1 and the mutual orthogonality of $P_{vm}$, $Q_{vm}$ and $R_{vm}$ proven in Result 2 and yet the rank results therein, we may write

$$
|A^*| = |\widehat{\Psi}_{1|H_1}|\,|\widehat{\Psi}_{2|H_1}|^{m-1}\,|\widehat{\Psi}_{3|H_1}|^{m(v-1)},
$$

which given the expressions obtained in Result 3 for $\widehat{\Psi}_{1|H_1}$, $\widehat{\Psi}_{2|H_1}$ and $\widehat{\Psi}_{3|H_1}$ yields (16).

Expression (17) would be proven in a completely similar way.

## Appendix B: The GIG, EGIG and GNIG Distributions

In this appendix we define the GIG (Generalized Integer Gamma), EGIG (Exponentiated Generalized Integer Gamma) and GNIG (Generalized Near-Integer Gamma) distributions and establish the notation used concerning their p.d.f.'s and c.d.f.'s.

We say that the r.v. (random variable) $X$ has a Gamma distribution with shape parameter $r$ ($> 0$) and rate parameter $\lambda$ ($> 0$), and we will denote this fact by $X \sim \Gamma(r, \lambda)$, if the p.d.f. of $X$ is

$$f_X(x) = \frac{\lambda^r}{\Gamma(r)}\, e^{-\lambda x}\, x^{r-1} \qquad (x > 0)\,.$$

Let $X_j \sim \Gamma(r_j, \lambda_j)$ $(j = 1, \dots, p)$ be a set of $p$ independent r.v.'s and consider the r.v.

$$W = \sum_{j=1}^{p} X_j\,.$$

In case all the $r_j \in \mathbb{N}$, the distribution of $W$ is what we call a GIG distribution (Coelho, 1998). If all the $\lambda_j$ are different, $W$ has a GIG distribution of depth $p$, with shape parameters $r_j$ and rate parameters $\lambda_j$, with p.d.f.

$$f_W(w) = f^{GIG}\Big(w \,\big|\, \{r_j\}_{j=1:p};\, \{\lambda_j\}_{j=1:p};\, p\Big) = K \sum_{j=1}^{p} P_j(w)\, e^{-\lambda_j w}\,,$$

and c.d.f.

$$F_W(w) = F^{GIG}\Big(w \,\big|\, \{r_j\}_{j=1:p};\, \{\lambda_j\}_{j=1:p};\, p\Big) = 1 - K \sum_{j=1}^{p} P_j^*(w)\, e^{-\lambda_j w}\,,$$

for $w > 0$, where

$$K = \prod_{j=1}^{p} \lambda_j^{r_j}\,, \qquad P_j(w) = \sum_{k=1}^{r_j} c_{j,k}\, w^{k-1} \tag{B.1}$$

and

$$P_j^*(w) = \sum_{k=1}^{r_j} c_{j,k}(k-1)! \sum_{i=0}^{k-1} \frac{w^i}{i!\, \lambda_j^{k-i}}\,, \tag{B.2}$$

with

$$c_{j,r_j} = \frac{1}{(r_j - 1)!} \prod_{\substack{i=1 \\ i \neq j}}^{p} (\lambda_i - \lambda_j)^{-r_i}\,, \qquad j = 1, \dots, p\,, \tag{B.3}$$

and, for $k = 1, \ldots, r_j - 1$ and $j = 1, \ldots, p$,

$$c_{j,r_j-k} = \frac{1}{k}\sum_{i=1}^{k} \frac{(r_j-k+i-1)!}{(r_j-k-1)!} R(i,j,p) c_{j,r_j-(k-i)}\,, \tag{B.4}$$

where

$$R(i,j,p) = \sum_{\substack{k=1\\k\neq j}}^{p} r_k(\lambda_j-\lambda_k)^{-i} \quad (i=1,\ldots,r_j-1)\,. \tag{B.5}$$

In case some of the $\lambda_j$ assume the same value as other $\lambda_j$'s, the distribution of $W$ still is a GIG distribution, but in this case with a reduced depth. In this more general case, let $\widetilde{\{\lambda_\ell\}}_{\ell=1:g(\le p)}$ be the set of different $\lambda_j$'s and let $\overset{\approx}{\{r_\ell\}}_{\ell=1:g(\le p)}$ be the set of the corresponding shape parameters, with $r_\ell$ ($\ell = 1, \ldots, g$) being the sum of all $r_j$ ($j \in \{1, \ldots, p\}$) which correspond to the $\lambda_j$ assuming the value $\lambda_\ell$. In this case $W$ will have a GIG distribution of depth $g$, with shape parameters $r_\ell$ and rate parameters $\lambda_\ell$ ($\ell = 1, \ldots, g$).

The r.v. $Z = e^{-W}$ has then what Arnold et al. (2013) call an Exponentiated Generalized Integer Gamma (EGIG) distribution of depth $g$, with p.d.f.

$$\begin{aligned} f_Z(z) &= f^{EGIG}\left(z \;\middle|\; \overset{\approx}{\{r_\ell\}}_{\ell=1:g};\widetilde{\{\lambda_\ell\}}_{\ell=1:g};g\right)\\ &= f^{GIG}\left(-\log z \;\middle|\; \overset{\approx}{\{r_\ell\}}_{j=1:g};\widetilde{\{\lambda_\ell\}}_{\ell=1:g};g\right)\frac{1}{z}\\ &= K^*\sum_{\ell=1}^{g} P_\ell(-\log z)\, z^{\lambda_\ell-1} \qquad (0<z<1) \end{aligned} \tag{B.6}$$

and c.d.f.

$$\begin{aligned} F_Z(z) &= F^{EGIG}\left(z \;\middle|\; \overset{\approx}{\{r_\ell\}}_{\ell=1:g};\widetilde{\{\lambda_\ell\}}_{\ell=1:g};g\right)\\ &= 1-F^{GIG}\left(-\log z \;\middle|\; \overset{\approx}{\{r_\ell\}}_{\ell=1:g};\widetilde{\{\lambda_\ell\}}_{\ell=1:g};g\right)\\ &= K^*\sum_{\ell=1}^{g} P^*_\ell(-\log z)\, z^{\lambda_\ell-1} \qquad (0<z<1)\,, \end{aligned} \tag{B.7}$$

for

$$K^* = \prod_{\ell=1}^{g} \lambda_\ell^{r_\ell} \tag{B.8}$$

and $P_\ell(\cdot)$ and $P^*_\ell(\cdot)$ given by (B.1) and (B.2), with $j$ replaced by $\ell$.

Then let $W$ be a r.v. with a GIG distribution of depth $g$, with rate parameters $\widetilde{\{\lambda_\ell\}}_{\ell=1:g}$ and shape parameters $\overset{\approx}{\{r_\ell\}}_{\ell=1:g}$ ($\in \mathbb{N}$) and let $W^* \sim \Gamma(r, \lambda)$, with $r \in \mathbb{R}^+ \backslash \mathbb{N}$. Let further $W$ and $W^*$ be two independent r.v.'s. Then the r.v.

$$Y = W + W^*$$

has a Generalized Near-Integer Gamma (GNIG) distribution (Coelho, 2004) of depth $g+1$, with rate parameters $\widetilde{\{\lambda_\ell\}}_{\ell=1:g}$ and $\lambda$ and corresponding shape parameters $\overset{\approx}{\{r_\ell\}}_{\ell=1:g}$ ($\in \mathbb{N}$) and $r$, with p.d.f.

$$\begin{aligned} f_Y(y) &= f^{GNIG}\left(y \,|\, \overset{\approx}{\{r_\ell\}}_{\ell=1:g}, r; \widetilde{\{\lambda_\ell\}}_{\ell=1:g}, \lambda; g+1\right) \\ &= K^* \lambda^r \sum_{\ell=1}^{g} e^{-\lambda_\ell y} \sum_{k=1}^{r_\ell} \left\{ c_{\ell,k} \frac{\Gamma(k)}{\Gamma(k+r)} y^{k+r-1} {}_1F_1\big(r, k+r, -(\lambda-\lambda_\ell) y\big) \right\} \end{aligned}$$

and c.d.f.

$$\begin{aligned} F_Y(y) &= F^{GNIG}\left(y \,|\, \overset{\approx}{\{r_\ell\}}_{\ell=1:g}, r; \widetilde{\{\lambda_\ell\}}_{\ell=1:g}, \lambda; g+1\right) \\ &= \frac{\lambda^r\, y^r}{\Gamma(r+1)} {}_1F_1\big(r, r+1, -\lambda y\big) \\ &\quad - K^{**} \sum_{\ell=1}^{g} e^{-\lambda_\ell y} \sum_{k=1}^{r_\ell} c^*_{\ell,k} \sum_{i=0}^{k-1} \left\{ \frac{y^{r+i}\, \lambda_\ell^i}{\Gamma(r^*+i)} {}_1F_1\big(r, r^*+i, -(\lambda-\lambda_\ell) y\big) \right\} \end{aligned}$$

where $y > 0$, $K^{**} = K^* \lambda^r$, for $K^*$ in (B.8), $r^* = r + 1$, and

$$c^*_{\ell,k} = \frac{c_{\ell,k}}{\lambda_\ell^k} \Gamma(k)$$

for $c_{\ell,k}$ given by (B.3)–(B.5) with $j$ replaced by $\ell$, and

$${}_1F_1(a, b, y) = \sum_{i=0}^{\infty} \frac{\Gamma(a+i)}{\Gamma(b+i)} \frac{\Gamma(b)}{\Gamma(a)} \frac{z^i}{i!}$$

is the Kummer confluent hypergeometric function.

## References

Anderson, T.W.: An Introduction to Multivariate Statistical Analysis, 3rd edn. Wiley, Hoboken, New Jersey (2003)

Arnold, B.C., Coelho, C.A., Marques, F.J.: The distribution of the product of powers of independent uniform random variables – a simple but useful tool to address and better understand the structure of some distributions. J. Multivariate Anal. **113**, 19–36 (2013)

Coelho, C.A.: The Generalized Integer Gamma distribution – a basis for distributions in multivariate statistics. J. Multivariate Anal. **64**, 86–102 (1998)

Coelho, C.A.: The generalized near-integer gamma distribution: a basis for 'near-exact' approximations to the distribution of statistics which are the product of an odd number of independent Beta random variables. J. Multivariate Anal. **89**, 191–218 (2004)

Coelho, C.A.: The Likelihood Ratio Test for Equality of Mean Vectors with Compound Symmetric Covariance Matrices. In: Gervasi, O., Murgante, B., Misra, S., Borruso, G., Torre, C.M., Rocha, A.M.A.C., Taniar, D., Apduhan, B.O., Stankova, E., Cuzzocrea, A. (eds.) Computational Science and Its Applications, Part V, pp. 20–32 (2017)

Coelho, C.A., Alberto, R.P.: On the Distribution of the Product of Independent Beta Random Variables – Applications. In: Ghosh, I., Balakrishnan, N., Ng, H.K. (eds.) Advances in Statistics - Theory and Applications: Honoring the Contributions of Barry C. Arnold in Statistical Science, (in print) Springer (2021)

Coelho, C.A., Arnold, B.C.: Finite Form Representations for Meijer G and Fox H Functions – Applied to Multivariate Likelihood Ratio Tests Using Mathematica®, MAXIMA and R. Lecture Notes in Statistics. Springer, Cham (2019)

Coelho, C.A., Marques, F.J.: The multi-sample block-scalar sphericity test: exact and near-exact distributions for its likelihood ratio test statistic. Commun. Statist. Theory Methods **42**, 1153–1175 (2013)

Coelho, C.A., Roy, A.: Testing the hypothesis of a doubly exchangeable covariance matrix. Metrika **83**, 45–68 (2020)

Coelho, C.A., Singull, M.: Testing for double complete symmetry. In: Holgersson, T., Singull, M. (eds.) Recent Developments in Multivariate and Random Matrix Analysis, pp. 17–39 (2020)

Coelho, C.A., Arnold, B.C., Marques, F.J.: Near-exact distributions for certain likelihood ratio test statistics. J. Stat. Theory Pract. **4**, 711–725 (2010)

Coelho, C.A., Marques, F.J., Arnold, B.C.: The exact and near-exact distributions of the main likelihood ratio test statistics used in the complex multivariate normal setting. Test **24**, 386–416 (2015)

Kshirsagar, A.M.: Multivariate Analysis. Marcel Dekker, New York (1972)

Marques, F.J., Coelho, C.A., Arnold, B.C.: A general near-exact distribution theory for the most common likelihood ratio test statistics used in Multivariate Analysis. TEST **20**, 180–203 (2011)

Pavlenko, T., Roy, A.: Supervised classifiers for high-dimensional higher-order data with locally doubly exchangeable covariance structure. Comm. Statist. Theory Methods **46**, 11612–11634 (2017)

Roy A., Fonseca, M.: Linear models with doubly exchangeable distributed errors. Comm. Statist. Theory Methods **41**, 2545–2569 (2012)

Seely, J.: Quadratic subspaces and completeness. Ann. Math. Statist. **42**, 710–721 (1971)

Szatrowski, T.D.: Explicit solutions, one iteration convergence and averaging in the multivariate Normal estimation problem for patterned means and covariances. Ann. Inst. Statist. Math. **30**, 81–88 (1978)

Szatrowski, T.D.: Necessary and sufficient conditions for explicit solutions in the multivariate Normal estimation problem for patterned means and covariances. Ann. Statist. **8**, 802–810 (1980)

Tricomi, F.G., Erdélyi, A.: The asymptotic expansion of a ratio of Gamma functions. Pacific J. Math. **1**, 133–142 (1951)

***Author and Editors note:***

[1] the expression for $\Delta$ in (41) should rather be

$$\Delta = \frac{1}{2\pi} \int_{-\infty}^{+\infty} \left| \frac{\Phi_W(t) - \widetilde{\Phi}_W(t)}{t} \right| dt \,, \tag{41}$$

that is, the ratio $\frac{1}{2\pi}$ is missing. However all the values in Tables 1-6 are correct since they were computed with the right expression.